Theory of Simple Liquids
with Applications to Soft Matter

Fourth Edition

Theory of Simple Liquids

with Applications to Soft Matter

Fourth Edition

Jean-Pierre Hansen

Université Pierre et Marie Curie, Paris, France
& University of Cambridge, UK

Ian R. McDonald

University of Cambridge, UK

AMSTERDAM • BOSTON • HEIDELBERG • LONDON
NEW YORK • OXFORD • PARIS • SAN DIEGO
SAN FRANCISCO • SINGAPORE • SYDNEY • TOKYO

Academic Press is an imprint of Elsevier

Academic Press is an imprint of Elsevier
The Boulevard, Langford Lane, Kidlington, Oxford OX5 1GB, UK
Radarweg 29, PO Box 211, 1000 AE Amsterdam, The Netherlands
225 Wyman Street, Waltham, MA 02451, USA
525 B Street, Suite 1800, San Diego, CA 92101-4495, USA

Fourth edition

British Library Cataloguing in Publication Data
A catalogue record for this book is available from the British Library

Library of Congress Cataloging-in-Publication Data
A catalog record for this book is available from the Library of Congress

For information on all Academic Press publications
visit our web site at store.elsevier.com

ISBN: 978-0-12-387032-2

Working together
to grow libraries in
developing countries

www.elsevier.com • www.bookaid.org

Contents

5. Perturbation Theory   149

6. Inhomogeneous Fluids   203

7. Time-dependent Correlation and Response Functions   265

Preface to the Fourth Edition

The remarkable way in which the theory of simple liquids has developed since the first edition of our book appeared can be traced in the prefaces of succeeding editions that are reprinted in the pages which follow. Nonetheless, new developments have continued to emerge, while at the same time there has been a growing interest among liquid-state theorists in more complex, mesoscopic systems. In preparing a further edition we have therefore undertaken a dual task. On the one hand we have significantly expanded a number of existing sections and have added new but related sections on binary systems, the asymptotic decay of the pair correlation function, singular perturbations, surface tension, wetting phenomena, fluid flow at the interface with a solid, the thermodynamics of supercooled liquids, and fluids adsorbed in porous media. At the same time we have taken account of recent shifts in emphasis by inclusion of a long, new chapter entitled "Applications to Soft Matter," which is designed to illustrate the growing convergence of two, related fields of work that initially had to a large extent evolved independently of each other. The new chapter is not a systematic introduction to the topic of soft matter, though we do provide an extensive list of references. We have a more limited objective in mind: that of providing examples to show how methods used previously for simple atomic and molecular systems can be successfully adapted, often with surprising ease, to studies of complex fluids. The examples chosen are taken from work on polymer solutions and polymer melts, colloidal dispersions and colloidal liquid crystals, clustering and gelation, together with two sections on the dynamics of colloidal systems. The selection of topics is again a product of our own interests and experience but it does, we believe, reflect fairly what has been achieved while providing hints of what further possibilities exist.

As has been the case for earlier editions we have benefited greatly from the advice and encouragement of colleagues and from the practical assistance that many of them have provided. We are therefore glad to take this opportunity to record our debt to Andy Archer, Jean-Louis Barrat, Dor Ben-Amotz, Ludovic Berthier, Emanuela Bianchi, Peter Bolhuis, Philip Camp, David Chandler, Jure Dobnikar, Marjolein Dijkstra, Jeppe Dyre, Bob Evans, Daan Frenkel, George Jackson, David Heyes, John Molina, Anna Oleksy, Alberto Parola, Roberto Piazza, Benjamin Rotenberg, Francesco Sciortino and Tullio Scopigno. Special mention must be made of Mathieu Salanne, who supplied us with a very large body of unpublished, molecular dynamics results for molten salts, and of Eduardo Waisman, whose solution of the Percus-Yevick equation first appeared in the second edition but can be found for the third time among the appendices.

We also thank those publishers of journals from which we have reproduced figures; detailed acknowledgements are contained in the figure captions.

J.P. HANSEN
I.R. MCDONALD
March 2013

Preface to the Third Edition

At the time when the second edition of this book was published the study of the liquid state was a rapidly expanding field of research. In the twenty years since then, the subject has matured both theoretically and experimentally to a point where a real understanding exists of the behavior of "simple" liquids at the microscopic level. Although there has been a shift towards more complex systems, there remains in our view a place for a book that deals with the principles of liquid-state theory, covering both statics and dynamics. Thus, in preparing a third edition, we have resisted the temptation to broaden too far the scope of the book, and the focus remains firmly on simple systems, though many of the methods we describe continue to find as wider range of application, Nonetheless, some reorganization of the book has been required in order to give proper weight to more recent developments. The most obvious change is in the space devoted to the theory of inhomogeneous fluids, an area in which considerable progress has been made since 1986. Other major additions are sections on the properties of supercooled liquids, which include a discussion of the mode coupling theory of the kinetic glass transition, on theories of condensation and freezing and on the electric double layer. To make way for this and other new material, some sections of the second edition have either been reduced in length or omitted altogether. In particular, we no longer see the need to include a complete chapter on molecular simulation, the publication of several excellent texts on the subject having filled what was previously a serious gap in the literature. Our aim has been to emphasise what seems to be work of lasting interest. Such judgements are inevitably somewhat arbitrary and, as before, the choice of topics is colored by our own experience and tastes. We make no attempt to provide an exhaustive list of references, limiting ourselves to what we consider to be the fundamental papers in different areas, along with selected applications.

We are grateful to a number of colleagues who have helped us in different ways: Dor Ben-Amotz, Teresa Head-Gordon David Heyes, David Grier, Bill Jorgensen, Gerhard Kahl, Peter Monson, Anna Oleksy, Albert Reiner, Phil Salmon, Ilja Siepmann, Alan Soper, George Stell and Jens-Boie Suck. Bob Evans made many helpful suggestions concerning the much revised chapter on ionic liquids, George Jackson acted as our guide to the literature on the theory of associating liquids, Alberto Parola provided a valuable set of notes on hierarchical reference theory, and Jean-Jacques Weis undertook on our behalf new Monte Calculations of the dielectric constant of dipolar hard spheres. Our task could not have been completed without the support, encouragement and advice of these and other colleagues, to all of whom we give our thanks. Finally, we thank the respective publishers to reproduce figures from Journal of Chemical Physics, Journal of Non-Crystalline Solids, Physical Review and Physical Review Letters.

J.P. HANSEN
I.R. MCDONALD

November 2005

Preface to the Second Edition

The first edition of this book was written in the wake of an unprecedented advance in our understanding of the microscopic structure and dynamics of simple liquids. The rapid progress made in a number of different experimental and theoretical areas had led to a rather clear and complete picture of the properties of simple, atomic liquids. In the ten years that have passed since then, interest in the liquid state has remained very active, and the methods described in our book have been successfully generalized and applied to a variety of more complicated systems. Important development have therefore been seen in the theory of ionic, molecular and polar liquids, of liquid metals, and of the liquid surface, while the quantitative reliability of theories of atomic fluids has also improved.

In an attempt to give a balanced account of the basic theory and of the advances of the past decade, this new edition has been rearranged and considerably expanded relative to the earlier one. Every chapter has been completely rewritten, and three new chapters have been added, devoted to ionic, metallic and molecular liquids, together with substantial new sections on the theory of inhomogeneous fluids. The material contained in Chapter 10 of the first edition, which dealt with phase transitions, has been omitted, since it proved impossible to do justice to such a large field in the limited space available. Although many excellent review articles and monographs have appeared in recent years, a comprehensive and up-to-date treatment of the theory of "simple" liquids appears to be lacking, and we would hope that the new edition of our book will fill this gap. The choice of material again reflects our own tastes, but we have aimed at presenting the main ideas of modern liquid-state theory in a way that is both pedagogical and, so far as possible, self-contained. The book should be accessible to graduate students and research workers, both experimentalists and theorists, who have a good background in elementary statistical mechanics. We are well aware, however, that certain sections, notably in Chapters 4, 6, 9, and 12 require more concentration from the reader than others. Although the book is not intended to be exhaustive, we give many references to material that is not covered in depth in the text. Even at this level, it is impossible to include all the relevant work. Omissions may reflect our ignorance or a lack of good judgment, but we consider that our goal will have been achieved if the book serves as an introduction and guide to a continuously growing field.

While preparing the new edition, we have benefited from the advice, criticism and help of many colleagues. We give our sincere thanks to all. There are, alas, too many names to list individually, but we wish to acknowledge our particular debt to Marc Baus, David Chandler, Giovanni Ciccotti, Bob Evans, Paul

Madden and Dominic Tildesley, who have read large parts of the manuscript; to Susan O'Gorman, for her help with Chapter 4; and to Eduardo Waisman, who wrote the first (and almost final) version of Appendix B. We are also grateful to those colleagues who have supplied references, preprints and material for figures and tables, and to authors and publishers for permission to reproduce diagrams from published papers. The last stages of the work were carried out at the Institut Laue-Langevin in Grenoble, and we thank Philippe Nozières for the invitations that made our visits possible. Finally, we are greatly indebted to Martine Hansen, Christiane Lanceron, Rehda Mazighi and Susan O'Gorman for their help and patience in the preparation of the manuscript and figures.

J.P. HANSEN
I.R. MCDONALD

May 1986

Preface to the First Edition

The past ten years or so have seen a remarkable growth in our understanding of the statistical mechanics of simple liquids. Many of these advances have not yet been treated fully in any book and the present work is aimed at filling this gap at a level similar to that of Egelstaff's "The Liquid State," though with a greater emphasis on theoretical developments. We discuss both static and dynamic properties, but no attempt is made at completeness and the choice of topics naturally reflects our own interests. The emphasis throughout is placed on theories that have been brought to a stage at which numerical comparison with experiment can be made. We have attempted to make the book as self-contained as possible, assuming only a knowledge of statistical mechanics at a final-year undergraduate level. We have also included a large number of references to work which lack of space has prevented us discussing in detail. Our hope is that the book will prove useful to all those interested in the physics and chemistry of liquids.

Our thanks go to many friends for their help and encouragement. We wish, in particular, to express our gratitude to Loup Verlet for allowing us to make unlimited use of his unpublished lecture notes on the theory of liquids. He, together with Dominique Levesque, Konrad Singer, and George Stell, have read several parts of the manuscript and made suggestions for its improvement. We are also greatly indebted to Jean-Jacques Weis for his help with the section on molecular liquids. The work was completed during a summer spent as visitors to the Chemistry Division of the National Research Council of Canada; it is a pleasure to have this opportunity to thank Mike Klein for his hospitality at that time and for making the visit possible. Thanks go finally to Susan O'Gorman for her help with mathematical problems and for checking the references; to John Copley, Jan Sengers and Sidney Yip for sending us useful material; and to Mrs K.L. Hales for so patiently typing the many drafts.

A number of figures and tables have been reproduced, with permission, from Physical Review, Journal of Chemical Physics, Molecular Physics and Physica; detailed acknowledgements are made at appropriate points in the text.

J.P. HANSEN
I.R. McDONALD

June 1976

Chapter 1

Introduction

1.1 THE LIQUID STATE

The liquid state of matter is intuitively perceived as one that is intermediate in nature between a gas and a solid. Given that point of view, a natural starting point for discussion of the properties of a given substance is the relationship between pressure P, number density ρ and temperature T in its different phases, summarised in the equation of state $f(P, \rho, T) = 0$. The phase diagram in the density-temperature plane typical of a simple, one-component system is sketched in Figure 1.1. The region of existence of the liquid phase is bounded above by the critical point (subscript c) and below by the triple point (subscript t). Above the critical point there is only a single fluid phase, so a continuous path exists from liquid to fluid to vapour. This is not true of the transition from liquid to solid because the solid-fluid coexistence line (the melting curve) does not end at a critical point. In many respects the properties of the dense, supercritical fluid are not very different from those of the liquid and much of the theory we develop in later chapters applies equally well to the two cases.

We shall be concerned in this book almost exclusively with classical liquids, that is to say with liquids that can to a good approximation be treated theoretically by the methods of classical statistical mechanics. A simple test of the classical hypothesis is provided by the value of the de Broglie thermal wavelength Λ, defined for a particle of mass m as

$$\Lambda = \left(\frac{2\pi \beta \hbar^2}{m} \right)^{1/2} \tag{1.1.1}$$

with $\beta = 1/k_B T$, where k_B is the Boltzmann constant. To justify a classical treatment of static properties Λ must be much smaller than a, where $a \approx \rho^{-1/3}$ is the mean nearest-neighbour separation. Some results for a variety of atomic and simple molecular liquids are shown in Table 1.1; hydrogen and neon apart, quantum effects should be small for all the systems listed. In the case of time-dependent processes it is necessary in addition that the time scale involved be much longer than $\beta \hbar$, which at room temperature, for example, means for times

Theory of Simple Liquids, Fourth Edition. http://dx.doi.org/10.1016/B978-0-12-387032-2.00001-5

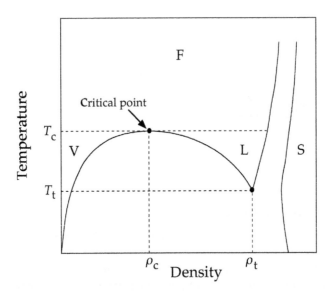

FIGURE 1.1 Schematic phase diagram of a typical monatomic substance, showing the boundaries between solid (S), liquid (L) and vapour (V) or fluid (F) phases.

TABLE 1.1 Test of the classical hypothesis.

Liquid	T_t (K)	Λ (Å)	Λ/a
H_2	14.1	3.3	0.97
Ne	24.5	0.78	0.26
CH_4	91	0.46	0.12
N_2	63	0.42	0.11
Li	454	0.31	0.11
Ar	84	0.30	0.083
HCl	159	0.23	0.063
Na	371	0.19	0.054
Kr	116	0.18	0.046
CCl_4	250	0.09	0.017

$t \gg 10^{-14}$ s. This second condition is somewhat more restrictive than the first, but where translational motion is concerned the problem is again severe only in extreme cases such as hydrogen.

Use of the classical approximation leads to an important simplification insofar as the contributions to thermodynamic properties arising from thermal motion can be separated from those due to interactions between particles.

The separation of kinetic and potential terms suggests a simple means of characterising the liquid state. Let V_N be the total potential energy of a system, where N is the number of particles, and let K_N be the total kinetic energy. Then in the liquid state we find that $K_N/|V_N| \approx 1$, whereas $K_N/|V_N| \gg 1$ corresponds to the dilute gas and $K_N/|V_N| \ll 1$ to the low-temperature solid. Alternatively, if we characterise a given system by a length σ and an energy ϵ, corresponding roughly to the range and strength of the intermolecular forces, we find that in the liquid region of the phase diagram the reduced number density $\rho^* = N\sigma^3/V$, where V is the volume, and reduced temperature $T^* = k_B T/\epsilon$ are both of order unity. Liquids and dense fluids are also distinguished from dilute gases by the greater importance of collisional processes and short-range, positional correlations, and from crystalline solids by the absence of the long-range order associated with a periodic lattice; their structure is in many cases dominated by the 'excluded volume' effect associated with the packing together of particles with hard cores.

Selected properties of a simple monatomic liquid (argon), a simple molecular liquid (nitrogen) and a simple liquid metal (sodium) are listed in Table 1.2. Not unexpectedly, the properties of the liquid metal are in certain respects very different from those of the other systems, notably in the values of the thermal conductivity, isothermal compressibility, surface tension, heat of vaporisation and the ratio of critical to triple-point temperatures; the source of

TABLE 1.2 Selected properties of typical simple liquids.

Property	Ar	Na	N_2
T_t/K	84	371	63
T_b/K ($P = 1$ atm)	87	1155	77
T_c/K	151	2600	126
T_c/T_t	1.8	7.0	2.0
ρ_t/nm^{-3}	21	24	19
c_P/c_V	2.2	1.1	1.6
$L_{vap}/kJ\ mol^{-1}$	6.5	99	5.6
$\chi_T/10^{-12}\ cm^2\ dyn^{-1}$	200	19	180
$c/m\ s^{-1}$	863	2250	995
$\gamma/dyn\ cm^{-1}$	13	191	12
$D/10^{-5}\ cm^2\ s^{-1}$	1.6	4.3	1.0
$\eta/mg\ cm^{-1}\ s^{-1}$	2.8	7.0	3.8
$\lambda/mW\ cm^{-1}\ K^{-1}$	1.3	8800	1.6
$(k_B T/2\pi D\eta)/\text{Å}$	4.1	2.7	3.6

χ_T = isothermal compressibility, c = speed of sound, γ = surface tension, D = self-diffusion coefficient, η = shear viscosity and λ = thermal conductivity, all at $T = T_t$; L_{vap} = heat of vaporisation at $T = T_b$.

these differences will become clear in Chapter 10. The quantity $k_B T/2\pi D\eta$ in the table provides a Stokes-law estimate of the particle diameter.

1.2   INTERMOLECULAR FORCES AND MODEL POTENTIALS

The most important feature of the pair potential between atoms or molecules is the harsh repulsion that appears at short range and has its origin in the overlap of the outer electron shells. The effect of these strongly repulsive forces is to create the short-range order characteristic of the liquid state. The attractive forces, which act at long range, vary much more smoothly with the distance between particles and play only a minor role in determining the structure of the liquid. They provide, instead, an essentially uniform, attractive background that gives rise to the cohesive energy required to stabilise the liquid. This separation of the effects of repulsive and attractive forces is a very old-established concept. It lies at the heart of the ideas of van der Waals, which in turn form the basis of the very successful perturbation theories of the liquid state discussed in Chapter 5.

The simplest model of a fluid is a system of hard spheres, for which the pair potential $v(r)$ at a separation r is

$$v(r) = \infty, \quad r < d$$
$$= 0, \quad r > d \tag{1.2.1}$$

where d is the hard-sphere diameter. This simple potential is ideally suited to the study of phenomena in which the hard core of the potential is the dominant factor. Much of our understanding of the properties of the hard-sphere model comes from computer simulations. Such calculations have revealed very clearly that the structure of a hard-sphere fluid does not differ in any significant way from that corresponding to more complicated interatomic potentials, at least under conditions close to crystallisation. The model also has some relevance to real, physical systems. For example, the osmotic equation of state of a suspension of micron-sized silica spheres in an organic solvent matches almost exactly that of a hard-sphere fluid.[1] However, although simulations show that the hard-sphere fluid undergoes a freezing transition at $\rho^*(=\rho d^3) \approx 0.945$, the absence of attractive forces means that there is only one fluid phase. A model that can describe a true liquid is obtained by supplementing the hard-sphere potential with a square-well attraction, as illustrated in the left-hand panel of Figure 1.2. This introduces two additional parameters, ϵ and γ; ϵ is the depth of the well and $(\gamma - 1)d$ is the width, where γ typically has a value of about 1.5. An alternative to the square-well potential with features that are of particular interest theoretically is the hard-core Yukawa potential, given by

$$v(r) = \infty, \qquad\qquad\qquad r < d$$
$$= -\frac{\epsilon d}{r} \exp[-\lambda(r/d - 1)], \quad r > d \tag{1.2.2}$$

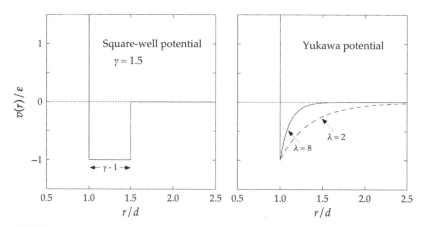

FIGURE 1.2 Simple potential models for monatomic systems. See text for details.

where the parameter λ measures the inverse range of the attractive tail in the potential. The two examples plotted in the right-hand panel of the figure are drawn for values of λ appropriate either to the interaction between rare-gas atoms ($\lambda = 2$) or to the short-range, attractive forces[2] characteristic of certain colloidal systems ($\lambda = 8$). The limit in which the range of the attraction tends to zero whilst the well depth goes to infinity corresponds to a 'sticky sphere' model, an early version of which was introduced by Baxter.[3] Models of this type have proved useful in studies of the clustering of colloidal particles and the formation of gels.

A more realistic potential for neutral atoms can be constructed by a detailed quantum-mechanical calculation. At large separations the dominant contribution to the potential comes from the multipolar dispersion interactions between the instantaneous electric moments on one atom, created by spontaneous fluctuations in the electronic charge distribution, and moments induced in the other. All terms in the multipole series represent attractive contributions to the potential. The leading term, varying as r^{-6}, describes the dipole-dipole interaction. Higher-order terms represent dipole-quadrupole (r^{-8}), quadrupole-quadrupole (r^{-10}) interactions, and so on, but these are generally small in comparison with the term in r^{-6}.

A rigorous calculation of the short-range interaction presents greater difficulty, but over relatively small ranges of r it can be adequately represented by an exponential function of the form $\exp(-r/r_0)$, where r_0 is a range parameter. This approximation must be supplemented by requiring that $v(r) \to \infty$ for r less than some arbitrarily chosen, small value. In practice, largely for reasons of mathematical convenience, it is more usual to represent the short-range repulsion by an inverse-power law, i.e. r^{-n}, where for closed-shell atoms n lies in the range from about 9 to 15. The behaviour of $v(r)$ in the limiting cases $r \to \infty$

and $r \to 0$ may therefore be incorporated in a potential function of the form

$$v(r) = 4\epsilon \left[(\sigma/r)^{12} - (\sigma/r)^6 \right] \qquad (1.2.3)$$

which is the famous 12–6 potential of Lennard-Jones. Equation (1.2.3) involves two parameters: the collision diameter σ, which is the separation of the particles where $v(r) = 0$; and ϵ, the depth of the potential well at the minimum in $v(r)$. The Lennard-Jones potential provides a fair description of the interaction between pairs of rare-gas atoms and of quasi-spherical molecules such as methane. Computer simulations[4] have shown that the triple point of the Lennard-Jones fluid is at $\rho^* \approx 0.85$, $T^* \approx 0.68$.

Experimental information on the pair interaction can be extracted from a study of any phenomenon that involves collisions between particles. The most direct method involves the measurement of atom-atom scattering cross-sections as a function of incident energy and scattering angle; inversion of the data allows, in principle, a determination of the pair potential at all separations. A simpler procedure is to assume a specific form for the potential and determine the parameters by fitting to the results of gas phase measurements of quantities such as the second virial coefficient (see Chapter 3) or shear viscosity.[5] In this way, for example, the parameters ϵ and σ in the Lennard-Jones potential have been determined for a large number of gases.

The theoretical and experimental methods we have mentioned all relate to the properties of an isolated pair of molecules. Use of the resulting pair potentials in calculations for the liquid state involves the neglect of many-body forces, an approximation that is difficult to justify. In the rare-gas liquids the three-body, triple-dipole dispersion term is the most important many-body interaction; the net effect of triple-dipole forces is repulsive, amounting in the case of liquid argon to a small percentage of the total potential energy due to pair interactions. Moreover, careful measurements, particularly those of second virial coefficients at low temperatures, have shown that the true pair potential for rare-gas atoms[6] is not of the Lennard-Jones form, but has a deeper bowl and a weaker tail, as illustrated by the curves plotted in Figure 1.3. Apparently the success of the Lennard-Jones potential in accounting for many of the macroscopic properties of argon-like liquids is the consequence of a fortuitous cancellation of errors. A number of more accurate pair potentials have been developed for the rare gases, but their use in the calculation of properties the liquid or solid requires the explicit incorporation of three-body interactions.

Although the true pair potential for rare-gas atoms is not the same as the effective pair potential used in liquid state theory, the difference is a relatively minor, quantitative one. The situation in the case of liquid metals is different because the form of the effective ion-ion interaction is strongly influenced by the presence of a degenerate gas of conduction electrons that does not exist before the liquid is formed. The calculation of the ion-ion interaction is a complicated problem, as we shall see in Chapter 10. The ion-electron interaction is first

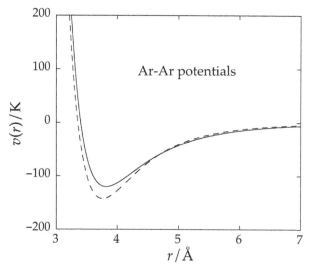

FIGURE 1.3 Pair potentials for argon in temperature units. Full curve: the Lennard-Jones potential with parameter values $\epsilon/k_B = 120$ K, $\sigma = 3.4$ Å, which is a good effective potential for the liquid; dashes: a potential based on gas phase data.[7]

described in terms of a 'pseudopotential' that incorporates both the coulombic attraction and the repulsion due to the Pauli exclusion principle. Account must then be taken of the way in which the pseudopotential is modified by interaction between the conduction electrons. The end result is a potential which represents the interaction between screened, electrically neutral 'pseudoatoms'. Irrespective of the detailed assumptions made, the main features of the potential are always the same: a soft repulsion, a deep attractive well and a long-range oscillatory tail. The potential, and in particular the depth of the well, are strongly density dependent but only weakly dependent on temperature. Figure 1.4 shows an effective potential for liquid potassium. The differences compared with the potentials for argon are clear, both at long range and in the core region.

For molten salts and other ionic liquids in which there is no shielding of the electrostatic forces of the type found in liquid metals, the coulombic interaction provides the dominant contribution to the interionic potential. There must, in addition, be a short-range repulsion between ions of opposite charge, without which the system would collapse, but the detailed way in which the repulsive forces are treated is of minor importance. Polarisation of the ions by the internal electric field also plays a role, but such effects are essentially many body in nature and cannot be adequately represented by an additional term in the pair potential.

Description of the interaction between two molecules poses greater problems than that between spherical particles because the pair potential is a function of both the separation of the molecules and their mutual orientation.

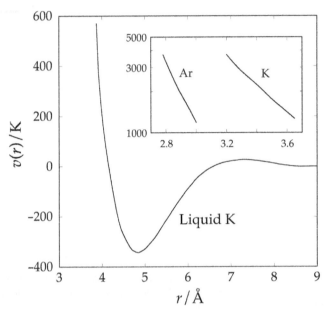

FIGURE 1.4 Main figure: effective ion-ion potential (in temperature units) for liquid potassium at high density.[8] Inset: comparison on a logarithmic scale of potentials for argon and potassium in the core region.

The model potentials discussed in this book mostly fall into one of two classes. The first consists of idealised models of polar liquids in which a point dipole-dipole interaction is superimposed on a spherically symmetric potential. In this case the pair potential for particles labelled 1 and 2 has the general form

$$v(1,2) = v_0(R) - \mu_1 \cdot \mathbf{T(R)} \cdot \mu_2 \qquad (1.2.4)$$

where $\mathbf{R}$ is the vector separation of the molecular centres, $v_0(R)$ is the spherically symmetric term, μ_i is the dipole moment vector of particle i and $\mathbf{T(R)}$ is the dipole-dipole interaction tensor:

$$\mathbf{T(R)} = 3\mathbf{RR}/R^5 - \mathbf{I}/R^3 \qquad (1.2.5)$$

where $\mathbf{I}$ is the unit tensor.

Two examples of (1.2.4) that are of particular interest are those of dipolar hard spheres, where $v_0(R)$ is the hard-sphere potential, and the Stockmayer potential, where $v_0(R)$ takes the Lennard-Jones form. Both these models, together with extensions that include, for example, dipole-quadrupole and quadrupole-quadrupole terms, have received much attention from theoreticians. Their main limitation as models of real molecules is the fact that they ignore the anisotropy of the short-range forces. One way to take account of such effects is through the use of potentials of the second main type with which we shall be concerned. These are models in which the molecule is represented by a set of discrete *interaction sites* that are commonly, but not invariably, located at the sites of the atomic nuclei. The total potential energy of two interaction-site molecules is then obtained as the sum of spherically symmetric, interaction-site potentials. Let $\mathbf{r}_{i\alpha}$ be the coordinates of site α in molecule i and let $\mathbf{r}_{j\beta}$ be the coordinates of site β in molecule j. Then the total intermolecular potential energy is

$$v(1,2) = \frac{1}{2} \sum_{\alpha} \sum_{\beta} v_{\alpha\beta}(|\mathbf{r}_{2\beta} - \mathbf{r}_{1\alpha}|) \qquad (1.2.6)$$

where $v_{\alpha\beta}(r)$ is a site-site potential and the sums on α and β run over all interaction sites in the respective molecules. Electrostatic interactions are easily allowed for by inclusion of coulombic terms in the site-site potentials.

Let us take as an example of the interaction-site approach the simple case of a homonuclear diatomic, such as that pictured in Figure 1.5. A crude interaction-site model would be that of a 'hard dumb-bell', consisting of two overlapping hard spheres of diameter d with their centres separated by a distance $L < 2d$. This should be adequate to describe the main structural features of a liquid such as nitrogen. An obvious improvement would be to replace the hard spheres by two Lennard-Jones interaction sites, with potential parameters chosen to fit, say, the experimentally determined equation of state. Some homonuclear diatomics also have a large quadrupole moment, which can play a significant role in determining the short-range angular correlations in the liquid. The model could in that case be further refined by placing point charges q at the Lennard-Jones sites, together with a compensating charge $-2q$ at the mid-point of the internuclear bond; such a charge distribution has zero dipole moment but a non-vanishing quadrupole moment proportional to qL^2. Models of this general type have proved remarkably successful in describing the properties of a wide variety of molecular liquids, both simple and complicated.

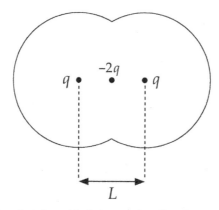

FIGURE 1.5　An interaction-site model of a homonuclear diatomic.

1.3   EXPERIMENTAL METHODS

The experimental methods available for studying the properties of simple liquids fall into one of two broad categories, depending on whether they are concerned with measurements on the macroscopic or microscopic scale. In general, values obtained theoretically for microscopic properties are more sensitive to the approximations made and the assumed form of the interparticle potentials, but macroscopic properties can usually be measured with considerably greater accuracy. The two classes of experiment are therefore complementary, each providing information that is useful in the development of a statistical mechanical theory of the liquid state.

The classic macroscopic measurements are those of thermodynamic properties, particularly of the equation of state. Integration of accurate P-ρ-T data yields information on other thermodynamic quantities, which can be supplemented by calorimetric measurements. For most liquids the pressure is known as a function of temperature and density only in the vicinity of the liquid-vapour equilibrium line, but for certain systems of particular theoretical interest experiments have been carried out at much higher pressures; the low compressibility of a liquid near its triple point means that highly specialised techniques are required.

The second main class of macroscopic measurements are those relating to transport coefficients. A variety of experimental methods are used. The shear viscosity, for example, can be determined from the observed damping of torsional oscillations or from capillary flow experiments, whilst the thermal conductivity can be obtained from a steady-state measurement of the transfer of heat between a central filament and a surrounding cylinder or between parallel plates. A direct method of determining the coefficient of self-diffusion involves the use of radioactive tracers, which places it in the category of microscopic measurements; in favourable cases the diffusion coefficient can be measured by nuclear magnetic resonance (NMR). NMR and other spectroscopic methods (infrared and Raman) are also useful in the study of reorientational motion in molecular liquids, whilst dielectric response measurements provide information on the slow, structural relaxation in supercooled liquids near the glass transition.

Much the most important class of microscopic measurements, at least from the theoretical point of view, are the radiation scattering experiments. Elastic scattering of neutrons or X-rays, in which the scattering cross-section is measured as a function of momentum transfer between the radiation and the sample, is the source of our experimental knowledge of the static structure of a fluid. In the case of inelastic scattering the cross-section is measured as a function of both momentum and energy transfer. It is thereby possible to extract information on wavenumber and frequency-dependent fluctuations in liquids at wavelengths comparable with the spacing between particles. This provides a very powerful method of studying microscopic time-dependent processes in liquids. Inelastic light scattering experiments provide similar information, but the accessible range of momentum transfer limits the method to the study of fluctuations of wavelength of order 10^{-5} cm, which lie in the hydrodynamic regime. Such experiments are, however, of considerable value in the study of colloidal dispersions or of critical phenomena.

Finally, there are a range of techniques of a quasi-experimental character, referred to collectively as computer simulation, the importance of which in the development of liquid state theory can hardly be overstated. Simulation provides what are essentially exact results for a given potential model; its usefulness rests ultimately on the fact that a sample containing a few hundred or few thousand particles is in many cases sufficiently large to simulate the behaviour of a macroscopic system. There are two classic approaches: the *Monte Carlo* method and the method of *molecular dynamics*. There are many variants of each, but in broad terms a Monte Carlo calculation is designed to generate particle configurations corresponding to a target, equilibrium distribution, most commonly the Boltzmann distribution, whilst molecular dynamics involves the solution of the classical equations of motion of the particles. Molecular dynamics therefore has the advantage of allowing the study of time-dependent processes, but for the calculation of static properties a Monte Carlo method may be more efficient. Chapter 2 contains a discussion of the principles underlying the two types of calculation and some details of their implementation.

REFERENCES

[1] Vrij, A., Jansen, J.W., Dhont, J.K.G., Pathmamanoharan, C., Kops-Werkhoven, M.M. and Fijnaut, H.M., *Faraday Disc.* **76**, 19 (1983).

[2] See, e.g., Meijer, E.J. and Frenkel D., *Phys. Rev. Lett.* **67**, 1110 (1991). The interactions in a charge-stabilised colloidal suspension can be modelled by a Yukawa potential with a positive tail.

[3] Baxter, R.J., *J. Chem. Phys.* **49**, 2770 (1968).

[4] Hansen, J.P. and Verlet, L., *Phys. Rev.* **184**, 151 (1969).

[5] Maitland, G.C., Rigby, M., Smith, E.B. and Wakeham, W.A., 'Intermolecular Forces'. Clarendon Press, Oxford, 1981.

[6] For a history of the efforts to construct an accurate pair potential for argon, see Rowlinson, J.S., 'Cohesion'. Cambridge University Press, Cambridge, 2002, Section 5.2.

[7] Model MS of Ref. 5, pp. 497–8.

[8] Dagens, L., Rasolt, M. and Taylor, R., *Phys. Rev. B* **11**, 2726 (1975).

Chapter 2

Statistical Mechanics

The greater part of this chapter is devoted to a summary of the principles of classical statistical mechanics, a discussion of the link between statistical mechanics and thermodynamics, and the definition of certain equilibrium and time-dependent distribution functions of fundamental importance in the theory of liquids. It also establishes much of the notation used in later parts of the book. The emphasis is on atomic systems; some of the complications that arise in the study of molecular liquids are discussed in Chapter 11. The last two sections deal with computer simulation, an approach that can be described as "numerical" statistical mechanics and which has played a major role in improving our understanding of the liquid state.

2.1 TIME EVOLUTION AND KINETIC EQUATIONS

Consider an isolated, macroscopic system consisting of N identical, spherical particles of mass m enclosed in a volume V. An example would be a one-component, monatomic gas or liquid. In classical mechanics the dynamical state of the system at any instant is completely specified by the $3N$ coordinates $\mathbf{r}^N \equiv \mathbf{r}_1, \ldots, \mathbf{r}_N$ and $3N$ momenta $\mathbf{p}^N \equiv \mathbf{p}_1, \ldots, \mathbf{p}_N$ of the particles. The values of these $6N$ variables define a *phase point* in a $6N$-dimensional *phase space*. Let $\mathcal{H}$ be the hamiltonian of the system, which we write in general form as

$$\mathcal{H}(\mathbf{r}^N, \mathbf{p}^N) = K_N(\mathbf{p}^N) + V_N(\mathbf{r}^N) + \Phi_N(\mathbf{r}^N) \qquad (2.1.1)$$

where

$$K_N = \sum_{i=1}^{N} \frac{|\mathbf{p}_i|^2}{2m} \qquad (2.1.2)$$

is the kinetic energy, V_N is the interatomic potential energy and Φ_N is the potential energy arising from the interaction of the particles with some spatially varying, external field. If there is no external field, the system will be both spatially uniform and isotropic. The motion of the phase point along its *phase*

Theory of Simple Liquids, Fourth Edition. http://dx.doi.org/10.1016/B978-0-12-387032-2.00002-7

trajectory is determined by Hamilton's equations:

$$\dot{\mathbf{r}}_i = \frac{\partial \mathcal{H}}{\partial \mathbf{p}_i}, \quad \dot{\mathbf{p}}_i = -\frac{\partial \mathcal{H}}{\partial \mathbf{r}_i} \tag{2.1.3}$$

These equations are to be solved subject to $6N$ initial conditions on the coordinates and momenta. Since the trajectory of a phase point is wholly determined by the values of $\mathbf{r}^N$, $\mathbf{p}^N$ at any given time, it follows that two different trajectories cannot pass through the same point in phase space.

The aim of equilibrium statistical mechanics is to calculate observable properties of a system of interest either as averages over a phase trajectory (the method of Boltzmann), or as averages over an ensemble of systems, each of which is a replica of the system of interest (the method of Gibbs). The main features of the two methods are reviewed in later sections of this chapter. Here it is sufficient to recall that in Gibbs's formulation of statistical mechanics the distribution of phase points of systems of the ensemble is described by a *phase space probability density* $f^{[N]}(\mathbf{r}^N, \mathbf{p}^N; t)$. The quantity $f^{[N]} \, \mathrm{d}\mathbf{r}^N \, \mathrm{d}\mathbf{p}^N$ is the probability that at time t the physical system is in a microscopic state represented by a phase point lying in the infinitesimal, $6N$-dimensional phase space element $\mathrm{d}\mathbf{r}^N \, \mathrm{d}\mathbf{p}^N$. This definition implies that the integral of $f^{[N]}$ over phase space is

$$\iint f^{[N]}(\mathbf{r}^N, \mathbf{p}^N; t) \, \mathrm{d}\mathbf{r}^N \mathrm{d}\mathbf{p}^N = 1 \tag{2.1.4}$$

for all t. Given a complete knowledge of the probability density it would be possible to calculate the average value of any function of the coordinates and momenta.

The time evolution of the probability density at a fixed point in phase space is governed by the Liouville equation, which is a $6N$-dimensional analogue of the equation of continuity of an incompressible fluid; it describes the fact that phase points of the ensemble are neither created nor destroyed as time evolves. The Liouville equation may be written either as

$$\frac{\partial f^{[N]}}{\partial t} + \sum_{i=1}^{N} \left(\frac{\partial f^{[N]}}{\partial \mathbf{r}_i} \cdot \dot{\mathbf{r}}_i + \frac{\partial f^{[N]}}{\partial \mathbf{p}_i} \cdot \dot{\mathbf{p}}_i \right) = 0 \tag{2.1.5}$$

or, more compactly, as

$$\frac{\partial f^{[N]}}{\partial t} = \{\mathcal{H}, f^{[N]}\} \tag{2.1.6}$$

where $\{A, B\}$ denotes the Poisson bracket:

$$\{A, B\} \equiv \sum_{i=1}^{N} \left(\frac{\partial A}{\partial \mathbf{r}_i} \cdot \frac{\partial B}{\partial \mathbf{p}_i} - \frac{\partial A}{\partial \mathbf{p}_i} \cdot \frac{\partial B}{\partial \mathbf{r}_i} \right) \tag{2.1.7}$$

Alternatively, by introducing the Liouville operator $\mathcal{L}$, defined as

$$\mathcal{L} \equiv i\{\mathcal{H}, \ \} \tag{2.1.8}$$

the Liouville equation becomes

$$\frac{\partial f^{[N]}}{\partial t} = -i\mathcal{L}f^{[N]} \tag{2.1.9}$$

the formal solution to which is

$$f^{[N]}(t) = \exp(-i\mathcal{L}t)f^{[N]}(0) \tag{2.1.10}$$

The Liouville equation can be expressed even more concisely in the form

$$\frac{df^{[N]}}{dt} = 0 \tag{2.1.11}$$

where d/dt denotes the total derivative with respect to time. This result is called the Liouville theorem; it shows that the probability density, as seen by an observer moving with a phase point along its phase space trajectory, is independent of time. To see its further significance, consider the phase points that at time $t = t_0$, say, are contained in the region of phase space labelled $\mathcal{D}_0$ in Figure 2.1 and which at time t_1 are contained in the region $\mathcal{D}_1$. The region will have changed in shape but no phase points will have entered or left, since that would require phase space trajectories to have crossed. The Liouville theorem therefore implies that the volumes (in $6N$ dimensions) of $\mathcal{D}_0$ and $\mathcal{D}_1$ must be the same. Volume in phase space is said to be 'conserved', which is equivalent to saying that the jacobian corresponding to the coordinate transformation $\mathbf{r}^N(t_0)\mathbf{p}^N(t_0) \to \mathbf{r}^N(t_1)\mathbf{p}^N(t_1)$ is equal to unity; this is a direct consequence of Hamilton's equations and is easily proved explicitly.[1]

The time dependence of any function of the phase space variables, $B(\mathbf{r}^N, \mathbf{p}^N)$ say, may be represented in a manner similar to (2.1.10). Although B is not an explicit function of t, it will in general change with time as the system

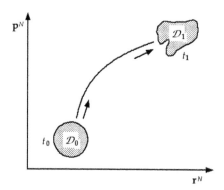

FIGURE 2.1 Conservation of volume in phase space. The phase points contained in the region $\mathcal{D}_0$ at a time $t = t_0$ move along their phase space trajectories in the manner prescribed by Hamilton's equations to occupy the region $\mathcal{D}_1$ at $t = t_1$. The Liouville theorem shows that the two regions have the same volume.

moves along its phase space trajectory. The time derivative of B is therefore given by

$$\frac{\mathrm{d}B}{\mathrm{d}t} = \sum_{i=1}^{N} \left(\frac{\partial B}{\partial \mathbf{r}_i} \cdot \dot{\mathbf{r}}_i + \frac{\partial B}{\partial \mathbf{p}_i} \cdot \dot{\mathbf{p}}_i \right) \qquad (2.1.12)$$

or, from Hamilton's equations:

$$\frac{\mathrm{d}B}{\mathrm{d}t} = \sum_{i=1}^{N} \left(\frac{\partial B}{\partial \mathbf{r}_i} \cdot \frac{\partial \mathcal{H}}{\partial \mathbf{p}_i} - \frac{\partial B}{\partial \mathbf{p}_i} \cdot \frac{\partial \mathcal{H}}{\partial \mathbf{r}_i} \right) = i\mathcal{L}B \qquad (2.1.13)$$

which has as its solution

$$B(t) = \exp(i\mathcal{L}t)B(0) \qquad (2.1.14)$$

Note the change of sign in the propagator compared with (2.1.10).

The description of the system that the full phase space probability density provides is for many purposes unnecessarily detailed. Normally we are interested only in the behaviour of a subset of particles of size n, say, and the redundant information can be eliminated by integrating $f^{[N]}$ over the coordinates and momenta of the other $(N - n)$ particles. We therefore define a *reduced phase space distribution function* $f^{(n)}(\mathbf{r}^n, \mathbf{p}^n; t)$ by

$$f^{(n)}(\mathbf{r}^n, \mathbf{p}^n; t) = \frac{N!}{(N-n)!} \int\int f^{[N]}(\mathbf{r}^N, \mathbf{p}^N; t) \mathrm{d}\mathbf{r}^{(N-n)} \mathrm{d}\mathbf{p}^{(N-n)} \qquad (2.1.15)$$

where $\mathbf{r}^n \equiv \mathbf{r}_1, \ldots, \mathbf{r}_n$ and $\mathbf{r}^{(N-n)} \equiv \mathbf{r}_{n+1}, \ldots, \mathbf{r}_N$, etc. The quantity $f^{(n)} \mathrm{d}\mathbf{r}^n \mathrm{d}\mathbf{p}^n$ determines the probability of finding a subset of n particles in the reduced phase space element $\mathrm{d}\mathbf{r}^n \mathrm{d}\mathbf{p}^n$ at time t irrespective of the coordinates and momenta of the remaining particles; the combinatorial factor $N!/(N - n)!$ is the number of ways of choosing a subset of size n.

To find an equation of motion for $f^{(n)}$ we consider the special case when the total force acting on particle i is the sum of an external force $\mathbf{X}_i$, arising from an external potential $\phi(\mathbf{r}_i)$, and of pair forces $\mathbf{F}_{ij}$ due to other particles j, with $\mathbf{F}_{ii} = 0$. The second of Hamilton's equations (2.1.3) then takes the form

$$\dot{\mathbf{p}}_i = \mathbf{X}_i + \sum_{j=1}^{N} \mathbf{F}_{ij} \qquad (2.1.16)$$

and the Liouville equation becomes

$$\left(\frac{\partial}{\partial t} + \sum_{i=1}^{N} \frac{\mathbf{p}_i}{m} \cdot \frac{\partial}{\partial \mathbf{r}_i} + \sum_{i=1}^{N} \mathbf{X}_i \cdot \frac{\partial}{\partial \mathbf{p}_i} \right) f^{[N]} = - \sum_{i=1}^{N} \sum_{j=1}^{N} \mathbf{F}_{ij} \cdot \frac{\partial f^{[N]}}{\partial \mathbf{p}_i} \qquad (2.1.17)$$

We now multiply through by $N!/(N - n)!$ and integrate over the $3(N - n)$ coordinates $\mathbf{r}_{n+1}, \ldots, \mathbf{r}_N$ and $3(N-n)$ momenta $\mathbf{p}_{n+1}, \ldots, \mathbf{p}_N$. The probability density $f^{[N]}$ is zero when $\mathbf{r}_i$ lies outside the volume occupied by the system and must vanish as $\mathbf{p}_i \rightarrow \infty$ to ensure convergence of the integrals over momenta in (2.1.4). Thus $f^{[N]}$ vanishes at the limits of integration and the derivative of $f^{[N]}$ with respect to any component of position or momentum will contribute nothing to the result when integrated with respect to that component. On integration, therefore, all terms disappear for which $i > n$ in (2.1.17). What remains, given the definition of $f^{(n)}$ in (2.1.15), is

$$\left(\frac{\partial}{\partial t} + \sum_{i=1}^{n} \frac{\mathbf{p}_i}{m} \cdot \frac{\partial}{\partial \mathbf{r}_i} + \sum_{i=1}^{n} \mathbf{X}_i \cdot \frac{\partial}{\partial \mathbf{p}_i}\right) f^{(n)}$$

$$= -\sum_{i=1}^{n}\sum_{j=1}^{n} \mathbf{F}_{ij} \cdot \frac{\partial f^{(n)}}{\partial \mathbf{p}_i}$$

$$-\frac{N!}{(N-n)!}\sum_{i=1}^{n}\sum_{j=n+1}^{N} \iint \mathbf{F}_{ij} \cdot \frac{\partial f^{[N]}}{\partial \mathbf{p}_i} \, d\mathbf{r}^{(N-n)} \, d\mathbf{p}^{(N-n)} \quad (2.1.18)$$

Because the particles are identical, $f^{[N]}$ is symmetric with respect to interchange of particle labels and the sum of terms for $j = n + 1$ to N on the right-hand side of (2.1.18) may be replaced by $(N - n)$ times the value of any one term. This simplification makes it possible to rewrite (2.1.18) in a manner which relates the behaviour of $f^{(n)}$ to that of $f^{(n+1)}$:

$$\left(\frac{\partial}{\partial t} + \sum_{i=1}^{n} \frac{\mathbf{p}_i}{m} \cdot \frac{\partial}{\partial \mathbf{r}_i} + \sum_{i=1}^{n} \left(\mathbf{X}_i + \sum_{j=1}^{n} \mathbf{F}_{ij}\right) \cdot \frac{\partial}{\partial \mathbf{p}_i}\right) f^{(n)}$$

$$= -\sum_{i=1}^{n} \iint \mathbf{F}_{i,n+1} \cdot \frac{\partial f^{(n+1)}}{\partial \mathbf{p}_i} \, d\mathbf{r}_{n+1} \, d\mathbf{p}_{n+1} \quad (2.1.19)$$

The system of coupled equations represented by (2.1.19) was first obtained by Yvon and subsequently rederived by others. It is known as the Bogoliubov–Born–Green–Kirkwood–Yvon or BBGKY hierarchy. The equations are exact, though limited in their applicability to systems for which the particle interactions are pairwise additive. They are not immediately useful, however, because they merely express one unknown function, $f^{(n)}$, in terms of another, $f^{(n+1)}$. Some approximate 'closure relation' is therefore needed.

In practice the most important member of the BBGKY hierarchy is that corresponding to $n = 1$:

$$\left(\frac{\partial}{\partial t} + \frac{\mathbf{p}_1}{m} \cdot \frac{\partial}{\partial \mathbf{r}_1} + \mathbf{X}_1 \cdot \frac{\partial}{\partial \mathbf{p}_1}\right) f^{(1)}(\mathbf{r}_1, \mathbf{p}_1; t)$$

$$= -\iint \mathbf{F}_{12} \cdot \frac{\partial}{\partial \mathbf{p}_1} f^{(2)}(\mathbf{r}_1, \mathbf{p}_1, \mathbf{r}_2, \mathbf{p}_2; t) d\mathbf{r}_2 \, d\mathbf{p}_2 \quad (2.1.20)$$

Much effort has been devoted to finding approximate solutions to (2.1.20) on the basis of expressions that relate the two-particle distribution function $f^{(2)}$ to the single-particle function $f^{(1)}$. From the resulting *kinetic equations* it is possible to calculate the hydrodynamic transport coefficients, but the approximations made are rarely appropriate to liquids because correlations between particles are mostly treated in a very crude way.[2] The simplest possible approximation is to ignore pair correlations altogether by writing

$$f^{(2)}(\mathbf{r}, \mathbf{p}, \mathbf{r}', \mathbf{p}'; t) \approx f^{(1)}(\mathbf{r}, \mathbf{p}; t) f^{(1)}(\mathbf{r}', \mathbf{p}'; t) \tag{2.1.21}$$

This leads to the Vlasov equation:

$$\left(\frac{\partial}{\partial t} + \frac{\mathbf{p}}{m} \cdot \frac{\partial}{\partial \mathbf{r}} + [\mathbf{X}(\mathbf{r}, t) + \bar{\mathbf{F}}(\mathbf{r}, t)] \cdot \frac{\partial}{\partial \mathbf{p}} \right) f^{(1)}(\mathbf{r}, \mathbf{p}; t) = 0 \tag{2.1.22}$$

where the quantity

$$\bar{\mathbf{F}}(\mathbf{r}, t) = \iint \mathbf{F}(\mathbf{r}, \mathbf{r}'; t) f^{(1)}(\mathbf{r}', \mathbf{p}'; t) \mathrm{d}\mathbf{r}' \mathrm{d}\mathbf{p}' \tag{2.1.23}$$

is the average force exerted by other particles, situated at points $\mathbf{r}'$, on a particle that at time t is at a point $\mathbf{r}$; this is an approximation of classic, mean field type. Though obviously not suitable for liquids, the Vlasov equation is widely used in plasma physics, where the long-range character of the Coulomb potential justifies a mean field treatment of the interactions.

Equation (2.1.20) may be rewritten schematically in the form

$$\left(\frac{\partial}{\partial t} + \frac{\mathbf{p}_1}{m} \cdot \frac{\partial}{\partial \mathbf{r}_1} + \mathbf{X}_1 \cdot \frac{\partial}{\partial \mathbf{p}_1} \right) f^{(1)} = \left(\frac{\partial f^{(1)}}{\partial t} \right)_{\mathrm{coll}} \tag{2.1.24}$$

where the term $(\partial f^{(1)}/\partial t)_{\mathrm{coll}}$ is the rate of change of $f^{(1)}$ due to collisions between particles. The collision term is given rigorously by the right-hand side of (2.1.20) but in the Vlasov equation it is eliminated by replacing the true external force $\mathbf{X}(\mathbf{r}, t)$ by an effective force – the quantity inside square brackets in (2.1.22) – which depends in part on $f^{(1)}$ itself. For this reason the Vlasov equation is called a 'collisionless' approximation.

In the most famous of all kinetic equations, derived by Boltzmann in 1872, the collision term is evaluated with the help of two assumptions, which in general are justified only at low densities: that two-body collisions alone are involved and that successive collisions are uncorrelated.[2] The second of these assumptions, that of 'molecular chaos', corresponds formally to supposing that the factorisation represented by (2.1.21) applies prior to any collision, though not subsequently. In simple terms it means that when two particles collide, no memory is retained of any previous encounters between them, an assumption that breaks down when recollisions are frequent events. A binary collision at a point $\mathbf{r}$ is characterised by the momenta $\mathbf{p}_1, \mathbf{p}_2$ of the two particles before collision and their momenta $\mathbf{p}'_1, \mathbf{p}'_2$ afterwards; the post-collisional momenta are related to their pre-collisional values by the laws of classical mechanics. With Boltzmann's approximations the collision term in (2.1.24) becomes

$$
\left(\frac{\partial f^{(1)}}{\partial t} \right)_{\text{coll}} = \frac{1}{m} \iint \sigma(\Omega, \Delta p)[f^{(1)}(\mathbf{r}, \mathbf{p}'_1; t) f^{(1)}(\mathbf{r}, \mathbf{p}'_2; t)
$$
$$
- f^{(1)}(\mathbf{r}, \mathbf{p}_1; t) f^{(1)}(\mathbf{r}, \mathbf{p}_2; t)] \mathrm{d}\Omega \, \mathrm{d}\mathbf{p}_2 \qquad (2.1.25)
$$

where $\Delta p \equiv |\mathbf{p}_2 - \mathbf{p}_1|$ and $\sigma(\Omega, \Delta p)$ is the differential cross-section for scattering into a solid angle $\mathrm{d}\Omega$. As Boltzmann showed, this form of the collision term is able to account for the fact that many-particle systems evolve irreversibly towards an equilibrium state. That irreversibility is described by Boltzmann's H-theorem; its source is the assumption of molecular chaos.

Solution of the Boltzmann equation leads to explicit expressions for the hydrodynamic transport coefficients in terms of certain 'collision integrals'.[3] The differential scattering cross-section and hence the collision integrals themselves can be evaluated numerically for a given choice of two-body interaction, though for hard spheres they have a simple, analytical form. The results, however, are applicable only to dilute gases. In the case of hard spheres the Boltzmann equation was later modified semi-empirically by Enskog in a manner that extends its range of applicability to considerably higher densities. Enskog's theory retains the two key assumptions involved in the derivation of the Boltzmann equation, but it also corrects in two ways for the finite size of the colliding particles. First, allowance is made for the modification of the collision rate by the hard-sphere interaction. Because the same interaction is also responsible for the increase in pressure over its ideal gas value, the enhancement of the collision rate relative to its low-density limit can be calculated if the hard-sphere equation of state is known. Secondly, 'collisional transfer' is incorporated into the theory by rewriting (2.1.25) in a form in which the distribution functions for the two colliding particles are evaluated not at the same point, $\mathbf{r}$, but at points separated by a distance equal to the hard-sphere diameter. This is an important modification of the theory, since at high densities interactions rather than particle displacements provide the dominant mechanism for the transport of energy and momentum.

The phase space probability density of a system in thermodynamic equilibrium is a function of the time-varying coordinates and momenta, but is independent of t at each point in phase space. We shall use the symbol $f_0^{[N]}(\mathbf{r}^N, \mathbf{p}^N)$ to denote the equilibrium probability density; it follows from (2.1.6) that a sufficient condition for a probability density to be descriptive of a system in equilibrium is that it should be some function of the hamiltonian. Integration of $f_0^{[N]}$ over a subset of coordinates and momenta in the manner of (2.1.15) yields a set of equilibrium phase space distribution functions

$f_0^{(n)}(\mathbf{r}^n, \mathbf{p}^n)$. The case when $n = 1$ corresponds to the equilibrium single-particle distribution function; if there is no external field the distribution is independent of $\mathbf{r}$ and has the familiar maxwellian form, i.e.

$$f_0^{(1)}(\mathbf{r}, \mathbf{p}) = \frac{\rho \exp(-\beta|\mathbf{p}|^2/2m)}{(2\pi m k_B T)^{3/2}}$$

$$\equiv \rho f_M(\mathbf{p}) \tag{2.1.26}$$

where $f_M(\mathbf{p})$ is the Maxwell distribution of momenta, normalised such that

$$\int f_M(\mathbf{p})d\mathbf{p} = 1 \tag{2.1.27}$$

The corresponding distribution of particle velocities, $\mathbf{u}$, is

$$\phi_M(\mathbf{u}) = \left(\frac{m}{2\pi k_B T}\right)^{3/2} \exp\left(-\frac{1}{2}m\beta|\mathbf{u}|^2\right) \tag{2.1.28}$$

2.2 TIME AVERAGES AND ENSEMBLE AVERAGES

Certain thermodynamic properties of a physical system may be written as averages of functions of the coordinates and momenta of the constituent particles. These are the so-called 'mechanical' properties, which include internal energy and pressure; 'thermal' properties such as entropy are not expressible in this way. In a state of thermal equilibrium such averages must be independent of time. To avoid undue complication we again suppose that the system of interest consists of N identical, spherical particles. If the system is isolated from its surroundings, its total energy is constant, i.e. the hamiltonian is a constant of the motion.

As before, let $B(\mathbf{r}^N, \mathbf{p}^N)$ be some function of the $6N$ phase space variables and let $\langle B \rangle$ be its average value, where the angular brackets represent an averaging process of a nature as yet unspecified. Given the coordinates and momenta of the particles at some instant, their values at any later (or earlier) time can in principle be obtained as the solution to Newton's equations of motion, i.e. to a set of $3N$ coupled, second-order, differential equations that, in the absence of an external field, have the form

$$m\ddot{\mathbf{r}}_i = \mathbf{F}_i = -\nabla_i V_N(\mathbf{r}^N) \tag{2.2.1}$$

where $\mathbf{F}_i$ is the total force on particle i. It is therefore natural to view $\langle B \rangle$ as a time average over the dynamical history of the system, i.e.

$$\langle B \rangle_t = \lim_{\tau \to \infty} \frac{1}{\tau} \int_0^\tau B\left[\mathbf{r}^N(t), \mathbf{p}^N(t)\right] dt \tag{2.2.2}$$

A simple example of the use of (2.2.2) arises in the calculation of the thermodynamic temperature of the system from the time average of the total kinetic energy. If

$$T(t) = \frac{2}{3Nk_B} K_N(t) = \frac{1}{3Nk_Bm} \sum_{i=1}^{N} |\mathbf{p}_i(t)|^2 \qquad (2.2.3)$$

then

$$T \equiv \langle T \rangle_t = \lim_{\tau \to \infty} \frac{1}{\tau} \int_0^\tau T(t) \, dt \qquad (2.2.4)$$

As a more interesting example we can use (2.2.2) and (2.2.4) to show that the equation of state is related to the time average of the *virial function* of Clausius. The virial function is defined as

$$\mathcal{V}(\mathbf{r}^N) = \sum_{i=1}^{N} \mathbf{r}_i \cdot \mathbf{F}_i \qquad (2.2.5)$$

From previous formulae, together with an integration by parts, we find that

$$\langle \mathcal{V} \rangle_t = \lim_{\tau \to \infty} \frac{1}{\tau} \int_0^\tau \sum_{i=1}^{N} \mathbf{r}_i(t) \cdot \mathbf{F}_i(t) dt = \lim_{\tau \to \infty} \frac{1}{\tau} \int_0^\tau \sum_{i=1}^{N} \mathbf{r}_i(t) \cdot m\ddot{\mathbf{r}}_i(t) \, dt$$

$$= -\lim_{\tau \to \infty} \frac{1}{\tau} \int_0^\tau \sum_{i=1}^{N} m|\dot{\mathbf{r}}_i(t)|^2 \, dt = -3Nk_B T \qquad (2.2.6)$$

or

$$\langle \mathcal{V} \rangle_t = -2 \langle K_N \rangle_t \qquad (2.2.7)$$

which is the virial theorem of classical mechanics. The total virial function may be separated into two parts: one, $\mathcal{V}_{int}$, comes from the forces between particles; the other, $\mathcal{V}_{ext}$, arises from the forces exerted by the walls and is related in a simple way to the pressure, P. The force exerted by a surface element dS located at $\mathbf{r}$ is $-P\mathbf{n} \, dS$, where $\mathbf{n}$ is a unit vector directed outwards, and its contribution to the average virial is $-P\mathbf{r} \cdot \mathbf{n} \, dS$. On integrating over the surface we find that

$$\langle \mathcal{V}_{ext} \rangle = -P \int \mathbf{r} \cdot \mathbf{n} \, dS = -P \int \nabla \cdot \mathbf{r} \, dV = -3PV \qquad (2.2.8)$$

Equation (2.2.7) may therefore be rearranged to give the *virial equation*:

$$PV = Nk_B T + \frac{1}{3} \langle \mathcal{V}_{int} \rangle_t = Nk_B T - \frac{1}{3} \left\langle \sum_{i=1}^{N} \mathbf{r}_i(t) \cdot \nabla_i V_N \left[\mathbf{r}^N(t) \right] \right\rangle_t \qquad (2.2.9)$$

or

$$\frac{\beta P}{\rho} = 1 - \frac{\beta}{3N} \left\langle \sum_{i=1}^{N} \mathbf{r}_i(t) \cdot \nabla_i V_N \left[\mathbf{r}^N(t) \right] \right\rangle_t \qquad (2.2.10)$$

In the absence of interactions between particles, i.e. when $V_N = 0$, the virial equation reduces to the equation of state of an ideal gas, $PV = Nk_BT$.

The alternative to the time-averaging procedure described by (2.2.2) is to average over a suitably constructed *ensemble*. A statistical mechanical ensemble is an arbitrarily large collection of imaginary systems, each of which is a replica of the physical system of interest and characterised by the same macroscopic parameters. The systems of the ensemble differ from each other in the assignment of coordinates and momenta of the particles and the dynamics of the ensemble as a whole is represented by the motion of a cloud of phase points distributed in phase space according to the probability density $f^{[N]}(\mathbf{r}^N, \mathbf{p}^N; t)$ introduced in Section 2.1. The equilibrium ensemble average of the function $B(\mathbf{r}^N, \mathbf{p}^N)$ is therefore given by

$$\langle B \rangle_e = \iint B(\mathbf{r}^N, \mathbf{p}^N) f_0^{[N]}(\mathbf{r}^N, \mathbf{p}^N) \, d\mathbf{r}^N \, d\mathbf{p}^N \tag{2.2.11}$$

where $f_0^{[N]}$ is the equilibrium probability density. For example, the thermodynamic internal energy is the ensemble average of the hamiltonian:

$$U \equiv \langle \mathcal{H} \rangle_e = \iint \mathcal{H} f_0^{[N]} \, d\mathbf{r}^N \, d\mathbf{p}^N \tag{2.2.12}$$

The explicit form of the equilibrium probability density depends on the macroscopic parameters that describe the ensemble. The simplest case is when the systems of the ensemble are assumed to have the same number of particles, the same volume and the same total energy, E say. An ensemble constructed in this way is called a *microcanonical* ensemble and describes a system that exchanges neither heat nor matter with its surroundings. The microcanonical equilibrium probability density is

$$f_0^{[N]}(\mathbf{r}^N, \mathbf{p}^N) = C\delta(\mathcal{H} - E) \tag{2.2.13}$$

where $\delta(\cdots)$ is the Dirac δ-function and C is a normalisation constant. The systems of a microcanonical ensemble are therefore uniformly distributed over the region of phase space corresponding to a total energy E; from (2.2.13) we see that the internal energy is equal to the value of the parameter E. The constraint of constant total energy is reminiscent of the condition of constant total energy under which time averages are taken. Indeed, time averages and ensemble averages are identical if the system is *ergodic*, by which is meant that after a suitable lapse of time the phase trajectory of the system will have passed an equal number of times through every phase space element in the region defined by (2.2.13). In practice, however, it is almost always easier to calculate ensemble averages in one of the ensembles described in the next two sections.

2.3 CANONICAL AND ISOTHERMAL–ISOBARIC ENSEMBLES

A *canonical* ensemble is a collection of systems characterised by the same values of N, V and T. It therefore represents a system immersed in a heat bath of fixed temperature. The equilibrium probability density for a system of identical, spherical particles is now

$$f_0^{[N]}(\mathbf{r}^N, \mathbf{p}^N) = \frac{1}{h^{3N} N!} \frac{\exp(-\beta \mathcal{H})}{Q_N} \tag{2.3.1}$$

where h is Planck's constant and the normalisation constant Q_N is the canonical *partition function*, given by

$$Q_N = \frac{1}{h^{3N} N!} \iint \exp(-\beta \mathcal{H}) \, d\mathbf{r}^N \, d\mathbf{p}^N \tag{2.3.2}$$

Inclusion of the factor $1/h^{3N}$ in these definitions ensures that both $f_0^{[N]} \, d\mathbf{r}^N \, d\mathbf{p}^N$ and Q_N are dimensionless and consistent in form with the corresponding quantities of quantum statistical mechanics, while division by $N!$ ensures that microscopic states are correctly counted.

The *thermodynamic potential* appropriate to a situation in which N, V and T are chosen as independent thermodynamic variables is the Helmholtz free energy, F, defined as

$$F = U - TS \tag{2.3.3}$$

where S is the entropy. Use of the term 'potential' refers to the fact that equilibrium at constant values of N, V and T is reached when F is a minimum with respect to variations in any internal constraint. The link between statistical mechanics and thermodynamics is established via a relation between the thermodynamic potential and the partition function:

$$F = -k_B T \ln Q_N \tag{2.3.4}$$

Let us assume that there is no external field and hence that the system of interest is homogeneous. Then the change in internal energy arising from infinitesimal changes in N, V and S is

$$dU = T \, dS - P \, dV + \mu \, dN \tag{2.3.5}$$

where μ is the chemical potential. Since N, V and S are all extensive variables it follows that

$$U = TS - PV + \mu N \tag{2.3.6}$$

Combination of (2.3.5) with the differential form of (2.3.3) shows that the change in free energy in an infinitesimal process is

$$dF = -S \, dT - P \, dV + \mu \, dN \tag{2.3.7}$$

Thus N, V and T are the natural variables of F; if F is a known function of those variables, all other thermodynamic functions can be obtained by differentiation:

$$S = -\left(\frac{\partial F}{\partial T}\right)_{V,N}, \quad P = -\left(\frac{\partial F}{\partial V}\right)_{T,N}, \quad \mu = \left(\frac{\partial F}{\partial N}\right)_{T,V} \quad (2.3.8)$$

and

$$U = F + TS = \left(\frac{\partial(F/T)}{\partial(1/T)}\right)_{V,N} \quad (2.3.9)$$

To each such thermodynamic relation there corresponds an equivalent relation in terms of the partition function. For example, it follows from (2.2.12) and (2.3.1) that

$$U = \frac{1}{h^{3N}N!Q_N}\iint \mathcal{H}\exp(-\beta\mathcal{H})\,\mathrm{d}\mathbf{r}^N\mathrm{d}\mathbf{p}^N = -\left(\frac{\partial\ln Q_N}{\partial\beta}\right)_V \quad (2.3.10)$$

This result, together with the fundamental relation (2.3.4), is equivalent to the thermodynamic formula (2.3.9). Similarly, the expression for the pressure given by (2.3.8) can be rewritten as

$$P = k_{\mathrm{B}}T\left(\frac{\partial\ln Q_N}{\partial V}\right)_{T,N} \quad (2.3.11)$$

and shown to be equivalent to the virial equation (2.2.10).[4]

If the hamiltonian is separated into kinetic and potential energy terms in the manner of (2.1.1), the integrations over momenta in the definition (2.3.2) of Q_N can be carried out analytically, yielding a factor $(2\pi m k_{\mathrm{B}}T)^{1/2}$ for each of the $3N$ degrees of freedom. This allows the partition function to be rewritten as

$$Q_N = \frac{1}{N!}\frac{Z_N}{\varLambda^{3N}} \quad (2.3.12)$$

where $\varLambda$ is the de Broglie thermal wavelength defined by (1.1.1) and

$$Z_N = \int\exp(-\beta V_N)\,\mathrm{d}\mathbf{r}^N \quad (2.3.13)$$

is the *configuration integral*. If $V_N = 0$:

$$Z_N = \int\cdots\int\mathrm{d}\mathbf{r}_1\cdots\mathbf{r}_N = V^N \quad (2.3.14)$$

Hence the partition function of a uniform, ideal gas is

$$Q_N^{\mathrm{id}} = \frac{1}{N!}\frac{V^N}{\varLambda^{3N}} = \frac{q^N}{N!} \quad (2.3.15)$$

where $q = V/\Lambda^3$ is the single-particle translational partition function, familiar from elementary statistical mechanics. If Stirling's approximation is used for $\ln N!$, the Helmholtz free energy is

$$\frac{F^{id}}{N} = k_B T (\ln \Lambda^3 \rho - 1) \tag{2.3.16}$$

and the chemical potential is

$$\mu^{id} = k_B T \ln \Lambda^3 \rho \tag{2.3.17}$$

The partition function of a system of interacting particles is conveniently written in the form

$$Q_N = Q_N^{id} \frac{Z_N}{V^N} \tag{2.3.18}$$

Then, on taking the logarithm of both sides, the Helmholtz free energy separates naturally into 'ideal' and 'excess' parts:

$$F = F^{id} + F^{ex} \tag{2.3.19}$$

where F^{id} is given by (2.3.16) and the excess part is

$$F^{ex} = -k_B T \ln \frac{Z_N}{V^N} \tag{2.3.20}$$

The excess part contains the contributions to the free energy that arise from interactions between particles; in the case of an inhomogeneous fluid there will also be a contribution that depends explicitly on the external potential. A similar division into ideal and excess parts can be made of any thermodynamic function obtained by differentiation of F with respect to either V or T. For example, the internal energy derived from (2.3.10) and (2.3.18) is

$$U = U^{id} + U^{ex} \tag{2.3.21}$$

where $U^{id} = \frac{3}{2} N k_B T$ and

$$U^{ex} = \langle V_N \rangle = \frac{1}{Z_N} \int V_N \exp(-\beta V_N) \, d\mathbf{r}^N \tag{2.3.22}$$

Note the simplification compared with the expression for U given by the first equality in (2.3.10); because V_N is a function only of the particle coordinates, the integrations over momenta cancel between numerator and denominator.

In the *isothermal–isobaric* ensemble pressure rather than volume is a fixed parameter. The thermodynamic potential for a system having specified values of N, P and T is the Gibbs free energy, G, defined as

$$G = F + PV \tag{2.3.23}$$

and other state functions are obtained by differentiation of G with respect to the independent variables. The link with statistical mechanics is now made through the relation

$$G = -k_B T \ln \Delta_N \qquad (2.3.24)$$

where the isothermal–isobaric partition function Δ_N is generally written[5] as a Laplace transform of the canonical partition function:

$$\Delta_N = \frac{1}{h^{3N} N!} \frac{1}{V_0} \int_0^\infty dV \iint \exp[-\beta(\mathcal{H} + PV)] \, dr^N \, dp^N$$

$$= \frac{1}{V_0} \int_0^\infty \exp(-\beta PV) Q_N \, dV \qquad (2.3.25)$$

where V_0 is a reference volume, inclusion of which makes Δ_N dimensionless. The form of (2.3.25) implies that the process of forming the ensemble average involves first calculating the canonical ensemble average at a volume V and then averaging over V with a weight factor $\exp(-\beta PV)$.

2.4   THE GRAND CANONICAL ENSEMBLE AND CHEMICAL POTENTIAL

The discussion of ensembles has thus far been restricted to uniform systems containing a fixed number of particles ('closed' systems). We now extend the argument to situations in which the number of particles may vary by interchange with the surroundings, but retain the assumption that the system is homogeneous. The thermodynamic state of an 'open' system is defined by specifying the values of μ, V and T and the corresponding thermodynamic potential is the grand potential, Ω, defined in terms of the Helmholtz free energy by

$$\Omega = F - N\mu \qquad (2.4.1)$$

When the internal energy is given by (2.3.6), the grand potential reduces to

$$\Omega = -PV \qquad (2.4.2)$$

and the differential form of (2.4.1) is

$$d\Omega = -S \, dT - P \, dV - N \, d\mu \qquad (2.4.3)$$

The thermodynamic functions S, P and N are therefore given as derivatives of Ω by

$$S = -\left(\frac{\partial \Omega}{\partial T}\right)_{V,\mu}, \quad P = -\left(\frac{\partial \Omega}{\partial V}\right)_{T,\mu}, \quad N = -\left(\frac{\partial \Omega}{\partial \mu}\right)_{T,V} \qquad (2.4.4)$$

An ensemble of systems having the same values of μ, V and T is called a *grand canonical* ensemble. The phase space of the grand canonical ensemble is the union of phase spaces corresponding to all values of the variable N for given values of V and T. The ensemble probability density is therefore a function of N as well as of the phase space variables $\mathbf{r}^N, \mathbf{p}^N$; at equilibrium it takes the form

$$f_0(\mathbf{r}^N, \mathbf{p}^N; N) = \frac{\exp[-\beta(\mathcal{H} - N\mu)]}{\Xi} \tag{2.4.5}$$

where

$$\Xi = \sum_{N=0}^{\infty} \frac{\exp(N\beta\mu)}{h^{3N} N!} \iint \exp(-\beta\mathcal{H}) \, d\mathbf{r}^N \, d\mathbf{p}^N = \sum_{N=0}^{\infty} \frac{z^N}{N!} Z_N \tag{2.4.6}$$

is the grand partition function and

$$z = \frac{\exp(\beta\mu)}{\Lambda^3} \tag{2.4.7}$$

is the *activity*. The definition (2.4.5) means that f_0 is normalised such that

$$\sum_{N=0}^{\infty} \frac{1}{h^{3N} N!} \iint f_0(\mathbf{r}^N, \mathbf{p}^N; N) d\mathbf{r}^N d\mathbf{p}^N = 1 \tag{2.4.8}$$

and the ensemble average of a microscopic variable $B(\mathbf{r}^N, \mathbf{p}^N)$ is

$$\langle B \rangle = \sum_{N=0}^{\infty} \frac{1}{h^{3N} N!} \iint B(\mathbf{r}^N, \mathbf{p}^N) f_0(\mathbf{r}^N, \mathbf{p}^N; N) \, d\mathbf{r}^N d\mathbf{p}^N \tag{2.4.9}$$

The link with thermodynamics is established through the relation

$$\Omega = -k_B T \ln \Xi \tag{2.4.10}$$

Equation (2.3.17) shows that $z = \rho$ for a uniform, ideal gas and in that case (2.4.6) reduces to

$$\Xi^{\mathrm{id}} = \sum_{N=0}^{\infty} \frac{\rho^N V^N}{N!} = \exp(\rho V) \tag{2.4.11}$$

which, together with (2.4.2), yields the equation of state in the form $\beta P = \rho$.

The probability $p(N)$ that at equilibrium a system of the ensemble contains precisely N particles, irrespective of their coordinates and momenta, is

$$p(N) = \frac{1}{h^{3N} N!} \iint f_0 \, d\mathbf{r}^N \, d\mathbf{p}^N = \frac{1}{\Xi} \frac{z^N}{N!} Z_N \tag{2.4.12}$$

The average number of particles in the system is

$$\langle N \rangle = \sum_{N=0}^{\infty} N p(N) = \frac{1}{\Xi} \sum_{N=0}^{\infty} N \frac{z^N}{N!} Z_N = \frac{\partial \ln \Xi}{\partial \ln z} \tag{2.4.13}$$

which is equivalent to the last of the thermodynamic relations (2.4.4). A measure of the fluctuation in particle number about its average value is provided by the mean-square deviation, for which an expression is obtained if (2.4.13) is differentiated with respect to $\ln z$:

$$
\begin{aligned}
\frac{\partial \langle N \rangle}{\partial \ln z} &= z \frac{\partial}{\partial z} \left(\frac{1}{\Xi} \sum_{N=0}^{\infty} N \frac{z^N}{N!} Z_N \right) \\
&= \frac{1}{\Xi} \sum_{N=0}^{\infty} N^2 \frac{z^N}{N!} Z_N - \left(\frac{1}{\Xi} \sum_{N=0}^{\infty} N \frac{z^N}{N!} Z_N \right)^2 \\
&= \langle N^2 \rangle - \langle N \rangle^2 \equiv \langle (\Delta N)^2 \rangle
\end{aligned}
\tag{2.4.14}
$$

or

$$
\frac{\langle (\Delta N)^2 \rangle}{\langle N \rangle} = \frac{k_B T}{\langle N \rangle} \frac{\partial \langle N \rangle}{\partial \mu}
\tag{2.4.15}
$$

The right-hand side of this equation is an intensive quantity and the same must therefore be true of the left-hand side. Hence the relative root-mean-square deviation, $\langle (\Delta N)^2 \rangle^{1/2} / \langle N \rangle$, tends to zero as $\langle N \rangle \to \infty$. In the *thermodynamic limit*, i.e. the limit $\langle N \rangle \to \infty$, $V \to \infty$ with $\rho = \langle N \rangle / V$ held constant, the number of particles in the system of interest (the thermodynamic variable N) may be identified with the grand canonical average, $\langle N \rangle$. More generally, in the same limit, thermodynamic properties calculated in different ensembles become identical.

The intensive ratio (2.4.15) is related to the isothermal compressibility χ_T, defined as

$$
\chi_T = -\frac{1}{V} \left(\frac{\partial V}{\partial P} \right)_T
\tag{2.4.16}
$$

To show this we note first that because the Helmholtz free energy is an extensive property it must be expressible in the form

$$
F = N \phi(\rho, T)
\tag{2.4.17}
$$

where ϕ, the free energy per particle, is a function of the intensive variables ρ and T. From (2.3.8) we find that

$$
\mu = \phi + \rho \left(\frac{\partial \phi}{\partial \rho} \right)_T
\tag{2.4.18}
$$

$$
\left(\frac{\partial \mu}{\partial \rho} \right)_T = 2 \left(\frac{\partial \phi}{\partial \rho} \right)_T + \rho \left(\frac{\partial^2 \phi}{\partial \rho^2} \right)_T
\tag{2.4.19}
$$

while

$$P = \rho^2 \left(\frac{\partial \phi}{\partial \rho} \right)_T \tag{2.4.20}$$

$$\left(\frac{\partial P}{\partial \rho} \right)_T = 2\rho \left(\frac{\partial \phi}{\partial \rho} \right)_T + \rho^2 \left(\frac{\partial^2 \phi}{\partial \rho^2} \right)_T = \rho \left(\frac{\partial \mu}{\partial \rho} \right)_T \tag{2.4.21}$$

Because $(\partial P/\partial \rho)_T = -(V^2/N)(\partial P/\partial V)_{N,T} = 1/\rho \chi_T$ and $(\partial \mu/\partial \rho)_T = V(\partial \mu/\partial N)_{V,T}$ it follows that

$$N \left(\frac{\partial \mu}{\partial N} \right)_{V,T} = \frac{1}{\rho \chi_T} \tag{2.4.22}$$

and hence, from (2.4.15), that

$$\frac{\langle (\Delta N)^2 \rangle}{\langle N \rangle} = \rho k_B T \chi_T \tag{2.4.23}$$

Thus the compressibility cannot be negative, since $\langle N^2 \rangle$ is always greater than or equal to $\langle N \rangle^2$.

Equation (2.4.23) and other fluctuation formulae of similar type can also be derived by purely thermodynamic arguments. In the thermodynamic theory of fluctuations described in Appendix A the quantity N in (2.4.23) is interpreted as the number of particles in a subsystem of macroscopic dimensions that forms part of a much larger thermodynamic system. If the system as a whole is isolated from its surroundings, the probability of a fluctuation within the subsystem is proportional to $\exp(\Delta S_t/k_B)$, where ΔS_t is the total entropy change resulting from the fluctuation. Since ΔS_t can in turn be related to changes in the properties of the subsystem, it becomes possible to calculate the mean-square fluctuations in those properties; the results thereby obtained are identical to their statistical mechanical counterparts. Because the subsystems are of macroscopic size, fluctuations in neighbouring subsystems will in general be uncorrelated. Strong correlations can, however, be expected under certain conditions. In particular, number fluctuations in two infinitesimal volume elements will be highly correlated if the separation of the elements is comparable with the range of the interparticle forces. A quantitative measure of these correlations is provided by the equilibrium distribution functions to be introduced later in Sections 2.5 and 2.6.

The definitions (2.3.1) and (2.4.5), together with (2.4.12), show that the canonical and grand canonical ensemble probability densities are related by

$$\frac{1}{h^{3N} N!} f_0(\mathbf{r}^N, \mathbf{p}^N; N) = p(N) f_0^{[N]}(\mathbf{r}^N, \mathbf{p}^N) \tag{2.4.24}$$

The grand canonical ensemble average of any microscopic variable is therefore given by a weighted sum of averages of the same variable in the canonical ensemble, the weighting factor being the probability $p(N)$ that the system contains precisely N particles.

In addition to its significance as a fixed parameter of the grand canonical ensemble, the chemical potential can also be expressed as a canonical ensemble average. This result, due to Widom,[6] provides some useful insight into the meaning of chemical potential. From (2.3.8) and (2.3.20) we see that

$$\mu^{ex} = F^{ex}(N+1, V, T) - F^{ex}(N, V, T) = k_B T \ln \frac{V Z_N}{Z_{N+1}} \qquad (2.4.25)$$

or

$$\frac{V Z_N}{Z_{N+1}} = \exp(\beta \mu^{ex}) \qquad (2.4.26)$$

where Z_N, Z_{N+1} are the configuration integrals for systems containing N or $(N+1)$ particles, respectively. The ratio Z_{N+1}/Z_N is

$$\frac{Z_{N+1}}{Z_N} = \frac{\int \exp[-\beta V_{N+1}(\mathbf{r}^{N+1})] d\mathbf{r}^{N+1}}{\int \exp[-\beta V_N(\mathbf{r}^N)] d\mathbf{r}^N} \qquad (2.4.27)$$

If the total potential energy of the system of $(N+1)$ particles is written as

$$V_{N+1}(\mathbf{r}^{N+1}) = V_N(\mathbf{r}^N) + \epsilon \qquad (2.4.28)$$

where ϵ is the energy of interaction of particle $(N+1)$ with all others, (2.4.27) can be re-expressed as

$$\frac{Z_{N+1}}{Z_N} = \frac{\int \exp(-\beta \epsilon) \exp[-\beta V_N(\mathbf{r}^N)] d\mathbf{r}^{N+1}}{\int \exp[-\beta V_N(\mathbf{r}^N)] d\mathbf{r}^N} \qquad (2.4.29)$$

If the system is homogeneous, translational invariance allows us to take $\mathbf{r}_{N+1}$ as origin for the remaining N position vectors and integrate over $\mathbf{r}_{N+1}$; this yields a factor V and (2.4.29) becomes

$$\frac{Z_{N+1}}{Z_N} = \frac{V \int \exp(-\beta \epsilon) \exp(-\beta V_N) d\mathbf{r}^N}{\int \exp(-\beta V_N) d\mathbf{r}^N} = V \langle \exp(-\beta \epsilon) \rangle \qquad (2.4.30)$$

where the angular brackets denote a canonical ensemble average for the system of N particles. Substitution of (2.4.30) in (2.4.25) gives

$$\mu^{ex} = -k_B T \ln \langle \exp(-\beta \epsilon) \rangle \qquad (2.4.31)$$

Hence the excess chemical potential is proportional to the logarithm of the mean Boltzmann factor of a test particle introduced randomly into the system.

Equation (2.4.31) is commonly referred to as the Widom insertion formula, particularly in connection with its use in computer simulations, where it provides a powerful and easily implemented method of determining the chemical potential of a fluid. It is also called the potential distribution theorem, since it may be written in the form

$$\beta\mu^{\mathrm{ex}} = -\ln \int \exp(-\beta\epsilon)p(\epsilon)\,d\epsilon \qquad (2.4.32)$$

where the quantity $p(\epsilon)\,d\epsilon$ is the probability that the potential energy of the test particle lies in the range $\epsilon \to \epsilon + d\epsilon$. Given a microscopic model of the distribution function $p(\epsilon)$, use of (2.4.32) provides a possible route to the calculation of the chemical potential of, say, a solute molecule in a liquid solvent. This forms the basis of what is called a 'quasi-chemical' theory of solutions.[7]

Equation (2.4.31) has a particularly simple interpretation for a system of hard spheres. Insertion of a test hard sphere can have one of two possible outcomes: either the sphere that is added overlaps with one or more of the spheres already present, in which case ϵ is infinite and the Boltzmann factor in (2.4.31) is zero, or there is no overlap, in which case $\epsilon = 0$ and the Boltzmann factor is unity. The excess chemical potential may therefore be written as

$$\mu^{\mathrm{ex}} = -k_{\mathrm{B}}T \ln p_0 \qquad (2.4.33)$$

where p_0 is the probability that a hard sphere can be introduced at a randomly chosen point in the system without creating an overlap. Calculation of p_0 poses a straightforward problem provided the density is low. As Figure 2.2 illustrates, centred on each particle of the system is a sphere of radius d and volume $v_{\mathrm{x}} = \frac{4}{3}\pi d^3$, or eight times the hard-sphere volume, from which the centre of the test particle is excluded if overlap is to be avoided. If the density is sufficiently low, the total excluded volume in a system of N hard spheres is to a good approximation N times that of a single sphere. It follows that

$$p_0 \approx \frac{V - N v_{\mathrm{x}}}{V} = 1 - \frac{4}{3}\pi\rho d^3 \qquad (2.4.34)$$

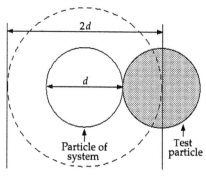

FIGURE 2.2 Widom's method for determining the excess chemical potential of a hard-sphere fluid. The broken line shows the sphere centred on a particle of the system into which the centre of a test hard sphere cannot penetrate without creating an overlap.

and hence, from (2.4.33), that at low densities:

$$\beta\mu^{\text{ex}} \approx \frac{4}{3}\pi\rho d^3 \tag{2.4.35}$$

As we shall see in Section 3.9, this is the correct result for the leading term in the density expansion of the excess chemical potential of the hard-sphere fluid. However, the argument used here breaks down as the density increases, because overlaps between the exclusion spheres around neighbouring particles can no longer be ignored. Use of the approximation represented by (2.4.34) therefore overestimates the coefficients of all higher-order terms in the expansion.

2.5 PARTICLE DENSITIES AND DISTRIBUTION FUNCTIONS

It was shown in Section 2.3 that a factorisation of the equilibrium phase space probability density $f_0^{[N]}(\mathbf{r}^N, \mathbf{p}^N)$ into kinetic and potential terms leads naturally to a separation of thermodynamic properties into ideal and excess parts. A similar factorisation can be made of the reduced phase space distribution functions $f_0^{(n)}(\mathbf{r}^n, \mathbf{p}^n)$ defined in Section 2.1. We assume again that there is no external field and hence that the hamiltonian is $\mathcal{H} = K_N + V_N$, where K_N is a sum of independent terms. For a system of fixed N, V and T, $f_0^{[N]}$ is given by the canonical distribution (2.3.1). If we recall from Section 2.3 that integration over each component of momentum yields a factor $(2\pi m k_B T)^{1/2}$, we see that $f_0^{(n)}$ can be written as

$$f_0^{(n)}(\mathbf{r}^n, \mathbf{p}^n) = \rho_N^{(n)}(\mathbf{r}^n) f_M^{(n)}(\mathbf{p}^n) \tag{2.5.1}$$

where

$$f_M^{(n)}(\mathbf{p}^n) = \frac{1}{(2\pi m k_B T)^{3n/2}} \exp\left(-\beta \sum_{i=1}^{n} \frac{|\mathbf{p}_i|^2}{2m}\right) \tag{2.5.2}$$

is the product of n independent Maxwell distributions of the form defined by (2.1.26) and $\rho_N^{(n)}$, the equilibrium *n-particle density* is

$$\rho_N^{(n)}(\mathbf{r}^n) = \frac{N!}{(N-n)!} \frac{1}{Q_N} \iint \exp(-\beta\mathcal{H}) \, d\mathbf{r}^{(N-n)} \, d\mathbf{p}^N$$

$$= \frac{N!}{(N-n)!} \frac{1}{Z_N} \int \exp(-\beta V_N) \, d\mathbf{r}^{(N-n)} \tag{2.5.3}$$

The quantity $\rho_N^{(n)}(\mathbf{r}^n) \, d\mathbf{r}^n$ determines the probability of finding n particles of the system with coordinates in the volume element $d\mathbf{r}^n$ irrespective of the positions of the remaining particles and irrespective of all momenta. The particle densities and the closely related, equilibrium *particle distribution functions*, defined below, provide a complete description of the structure of a fluid, while knowledge of the low-order particle distribution functions, in particular of the pair density $\rho_N^{(2)}(\mathbf{r}_1, \mathbf{r}_2)$, is often sufficient to calculate the equation of state and other thermodynamic properties of the system.

The definition of the n-particle density means that

$$\int \rho_N^{(n)}(\mathbf{r}^n)\, d\mathbf{r}^n = \frac{N!}{(N-n)!} \tag{2.5.4}$$

and in particular that

$$\int \rho_N^{(1)}(\mathbf{r})\, d\mathbf{r} = N \tag{2.5.5}$$

The single-particle density of a uniform fluid is therefore equal to the overall number density:

$$\rho_N^{(1)}(\mathbf{r}) = N/V = \rho \quad \text{(uniform fluid)} \tag{2.5.6}$$

In the special case of a uniform, ideal gas we know from (2.3.14) that $Z_N = V^N$. Hence the pair density is

$$\rho_N^{(2)} = \rho^2 \left(1 - \frac{1}{N} \right) \quad \text{(uniform ideal gas)} \tag{2.5.7}$$

The appearance of the term $1/N$ in (2.5.7) reflects the fact that in a system containing a fixed number of particles the probability of finding a particle in the volume element $d\mathbf{r}_1$, given that another particle is in the element $d\mathbf{r}_2$, is proportional to $(N-1)/V$ rather than ρ.

The n-particle distribution function $g_N^{(n)}(\mathbf{r}^n)$ is defined in terms of the corresponding particle densities by

$$g_N^{(n)}(\mathbf{r}^n) = \frac{\rho_N^{(n)}(\mathbf{r}_1, \dots, \mathbf{r}_n)}{\prod_{i=1}^{n} \rho_N^{(1)}(\mathbf{r}_i)} \tag{2.5.8}$$

which for a homogeneous system reduces to

$$\rho^n g_N^{(n)}(\mathbf{r}^n) = \rho_N^{(n)}(\mathbf{r}^n) \tag{2.5.9}$$

The particle distribution functions measure the extent to which the structure of a fluid deviates from complete randomness. If the system is also isotropic, the pair distribution function $g_N^{(2)}(\mathbf{r}_1, \mathbf{r}_2)$ is a function only of the separation $r_{12} = |\mathbf{r}_2 - \mathbf{r}_1|$; it is then usually called the *radial distribution function* and written simply as $g(r)$. When r is much larger than the range of the interparticle potential, the radial distribution function approaches the ideal gas limit; from (2.5.7) this limit can be identified as $(1 - 1/N) \approx 1$.

The particle densities defined by (2.5.3) are also expressible in terms of δ-functions of position in a form that is very convenient for later purposes. From the definition of a δ-function it follows that

$$\langle \delta(\mathbf{r} - \mathbf{r}_1) \rangle = \frac{1}{Z_N} \int \delta(\mathbf{r} - \mathbf{r}_1) \exp[-\beta V_N(\mathbf{r}_1, \mathbf{r}_2, \ldots, \mathbf{r}_N)] \, d\mathbf{r}^N$$

$$= \frac{1}{Z_N} \int \cdots \int \exp[-\beta V_N(\mathbf{r}, \mathbf{r}_2, \ldots, \mathbf{r}_N)] \, d\mathbf{r}_2 \cdots d\mathbf{r}_N$$

$$(2.5.10)$$

The ensemble average in (2.5.10) is a function of the coordinate $\mathbf{r}$ but is independent of the particle label (here taken to be 1). A sum over all particle labels is therefore equal to N times the contribution from any one particle. Comparison with the definition (2.5.3) then shows that

$$\rho_N^{(1)}(\mathbf{r}) = \left\langle \sum_{i=1}^{N} \delta(\mathbf{r} - \mathbf{r}_i) \right\rangle \tag{2.5.11}$$

which represents the ensemble average of a microscopic particle density $\rho(\mathbf{r})$. Similarly, the average of a product of two δ-functions is

$$\langle \delta(\mathbf{r} - \mathbf{r}_1)\delta(\mathbf{r}' - \mathbf{r}_2) \rangle = \frac{1}{Z_N} \int \delta(\mathbf{r} - \mathbf{r}_1)\delta(\mathbf{r}' - \mathbf{r}_2)$$

$$\exp[-\beta V_N(\mathbf{r}_1, \mathbf{r}_2, \ldots, \mathbf{r}_N)] \, d\mathbf{r}^N$$

$$= \frac{1}{Z_N} \int \cdots \int \exp[-\beta V_N(\mathbf{r}, \mathbf{r}', \mathbf{r}_3, \ldots, \mathbf{r}_N]$$

$$d\mathbf{r}_3 \cdots d\mathbf{r}_N \tag{2.5.12}$$

which implies that

$$\rho_N^{(2)}(\mathbf{r}, \mathbf{r}') = \left\langle \sum_{i=1}^{N} \sum_{j=1}^{N} {}' \delta(\mathbf{r} - \mathbf{r}_i)\delta(\mathbf{r}' - \mathbf{r}_j) \right\rangle \tag{2.5.13}$$

where the prime on the summation sign indicates that terms for which $i = j$ must be omitted. Finally, a useful δ-function representation can be obtained for the radial distribution function. It follows straightforwardly that

$$\left\langle \frac{1}{N} \sum_{i=1}^{N} \sum_{j=1}^{N} {}' \delta(\mathbf{r} - \mathbf{r}_j + \mathbf{r}_i) \right\rangle = \left\langle \frac{1}{N} \int \sum_{i=1}^{N} \sum_{j=1}^{N} {}' \delta(\mathbf{r}' + \mathbf{r} - \mathbf{r}_j)\delta(\mathbf{r}' - \mathbf{r}_i) \, d\mathbf{r}' \right\rangle$$

$$= \frac{1}{N} \int \rho_N^{(2)}(\mathbf{r}' + \mathbf{r}, \mathbf{r}') \, d\mathbf{r}' \tag{2.5.14}$$

Hence, if the system is both homogeneous and isotropic:

$$\left\langle \frac{1}{N} \sum_{i=1}^{N} \sum_{j=1}^{N} {}' \delta(\mathbf{r} - \mathbf{r}_j + \mathbf{r}_i) \right\rangle = \frac{\rho^2}{N} \int g_N^{(2)}(\mathbf{r}, \mathbf{r}') \, d\mathbf{r}' = \rho g(r) \tag{2.5.15}$$

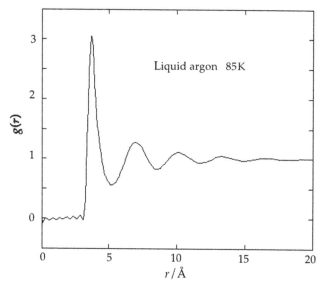

FIGURE 2.3 Results of neutron scattering experiments for the radial distribution function of argon near the triple point. The ripples at small r are artefacts of the data analysis. After Yarnell et al.[8]

The radial distribution function plays a key role in the physics of monatomic liquids. There are several reasons for this. First, $g(r)$ is measurable by radiation scattering experiments. The results of such an experiment on liquid argon are pictured in Figure 2.3; $g(r)$ shows a pattern of peaks and troughs that is typical of all monatomic liquids, tends to unity at large r, and vanishes as $r \to 0$ as a consequence of the strongly repulsive forces that act at small particle separations. Secondly, the form of $g(r)$ provides considerable insight into what is meant by the structure of a liquid, at least at the level of pair correlations. The definition of $g(r)$ implies that on average the number of particles lying within the range r to $r+dr$ from a reference particle is $4\pi r^2 \rho g(r)\, dr$ and the peaks in $g(r)$ represent 'shells' of neighbours around the reference particle. Integration of $4\pi r^2 \rho g(r)$ up to the position of the first minimum therefore provides an estimate of the nearest-neighbour 'coordination number'. The concepts of a 'shell' of neighbours and a 'coordination number' are obviously more appropriate to solids than to liquids, but they provide useful measures of the structure of a liquid provided the analogy with solids is not taken too far. The coordination number (≈ 12.2) calculated from the distribution function shown in the figure is in fact very close to the number (12) of nearest neighbours in the face-centred cubic structure into which argon crystallises. Finally, if the atoms interact through pairwise-additive forces, thermodynamic properties can be expressed in terms of integrals over $g(r)$, as we shall now show.

Consider a uniform fluid for which the total potential energy is given by a sum of pair terms:

$$V_N(\mathbf{r}^N) = \sum_{i=1}^{N} \sum_{j>i}^{N} v(r_{ij}) \qquad (2.5.16)$$

According to (2.3.22), the excess internal energy is

$$U^{\text{ex}} = \frac{N(N-1)}{2} \iint v(\mathbf{r}_{12}) \left(\frac{1}{Z_N} \int \cdots \int \exp(-\beta V_N) \, d\mathbf{r}_3 \cdots d\mathbf{r}_N \right) d\mathbf{r}_1 \, d\mathbf{r}_2 \qquad (2.5.17)$$

because the double sum over i, j in (2.5.16) gives rise to $\frac{1}{2}N(N-1)$ terms, each of which leads to the same result after integration. Use of (2.5.3) and (2.5.9) allows (2.5.17) to be rewritten as

$$U^{\text{ex}} = \frac{N^2}{2V^2} \iint v(r_{12}) g_N^{(2)}(\mathbf{r}_1, \mathbf{r}_2) \, d\mathbf{r}_1 d\mathbf{r}_2 \qquad (2.5.18)$$

We now take the position of particle 1 as the origin of coordinates, set $\mathbf{r}_{12} = \mathbf{r}_2 - \mathbf{r}_1$ and integrate over the coordinate $\mathbf{r}_1$ (which yields a factor V) to give

$$U^{\text{ex}} = \frac{N^2}{2V^2} \iint v(r_{12}) g(r_{21}) \, d\mathbf{r}_1 \, d\mathbf{r}_{12} = \frac{N^2}{2V} \int v(r) g(r) \, d\mathbf{r} \qquad (2.5.19)$$

or

$$\frac{U^{\text{ex}}}{N} = 2\pi\rho \int_0^{\infty} v(r) g(r) r^2 \, dr \qquad (2.5.20)$$

This result, usually referred to as the *energy equation*, can also be derived in a more intuitive way. The mean number of particles at a distance between r and $r+dr$ from a reference particle is $n(r) \, dr = 4\pi r^2 \rho g(r) \, dr$ and the total energy of interaction with the reference particle is $v(r)n(r) \, dr$. The excess internal energy per particle is then obtained by integrating $v(r)n(r)$ between $r = 0$ and $r = \infty$ and dividing the result by two to avoid counting each interaction twice.

It is also possible to express the equation of state (2.2.10) as an integral over $g(r)$. Given the assumption of pairwise additivity of the interparticle forces, the internal contribution to the virial function can be written, with the help of Newton's Third Law, as

$$\mathcal{V}_{\text{int}} = \sum_{i=1}^{N} \sum_{j>i}^{N} \mathbf{r}_i \cdot \mathbf{F}_{ij} = -\sum_{i=1}^{N} \sum_{j>i}^{N} r_{ij} v'(r_{ij}) \qquad (2.5.21)$$

where $v'(r) \equiv dv(r)/dr$. Then, starting from (2.2.10) and following the steps involved in the derivation of (2.5.20) but with $v(r_{ij})$ replaced by $r_{ij} v'(r_{ij})$:

$$\frac{\beta P}{\rho} = 1 - \frac{2\pi\beta\rho}{3} \int_0^{\infty} v'(r) g(r) r^3 \, dr \qquad (2.5.22)$$

Equation (2.5.22) is called either the *pressure equation* or, in common with (2.2.10), the *virial equation*.

Equations (2.5.20) and (2.5.22) are superficially simpler in form than (2.3.22) and (2.2.10), but the difficulty has merely shifted to that of determining the radial distribution function from the pair potential via (2.5.3) and (2.5.8). The problem is yet more complicated if there are many-body forces acting between particles or if the pair potential is not spherically symmetric. The presence of three-body forces, for example, leads to the appearance in expressions for the internal energy and pressure of integrals over the triplet distribution function $g_N^{(3)}(\mathbf{r}_1, \mathbf{r}_2, \mathbf{r}_3)$. We shall not pursue this matter further, since no new point of principle is involved, but the generalisation to systems of non-spherical particles is treated in detail in Chapter 11.

Because the pressure equation involves the derivative of the pair potential, it is not directly applicable in the calculation of the equation of state of hard spheres, or of other systems for which the pair potential contains a discontinuity. The problem can be overcome by rewriting (2.5.22) in terms of a function $y(r)$ defined as

$$y(r) = \exp[\beta v(r)]g(r) \qquad (2.5.23)$$

We show in Chapter 4 that $y(r)$ is a continuous function of r even when there are discontinuities in $v(r)$ and hence in $g(r)$; $y(r)$ is called the *cavity distribution function* for reasons that will become clear in Section 4.6. On introducing the definition of $y(r)$ into (2.5.22) we find that

$$\frac{\beta P}{\rho} = 1 - \frac{2\pi\beta\rho}{3} \int_0^\infty v'(r)e(r)y(r)r^3 \, dr$$

$$= 1 + \frac{2\pi\rho}{3} \int_0^\infty e'(r)y(r)r^3 \, dr \qquad (2.5.24)$$

where

$$e(r) = \exp[-\beta v(r)] \qquad (2.5.25)$$

is the Boltzmann factor for a pair of particles separated by a distance r and $e'(r) \equiv de(r)/dr$. In the case of hard spheres, $e(r)$ is a unit step function, the derivative of which is a δ-function, i.e. $e(r) = 0$ for $r < d$, $e(r) = 1$ for $r > d$ and $e'(r) = \delta(r - d)$, where d is the hard-sphere diameter. Thus

$$\frac{\beta P}{\rho} = 1 + \frac{2\pi\rho}{3} \int_0^\infty r^3 y(r)\delta(r - d) \, dr$$

$$= 1 + \frac{2\pi\rho}{3} \lim_{r \to d^+} r^3 y(r) = 1 + \frac{2\pi\rho}{3}d^3 g(d) \qquad (2.5.26)$$

The pressure of the hard-sphere fluid is therefore determined by the value of the radial distribution function at contact of the spheres, where $g(r)$ goes discontinuously to zero. We show in the next section that $g(r) \approx e(r)$ and hence that $g(d) \to 1$ in the limit $\rho \to 0$. Thus, at low densities:

$$\frac{\beta P}{\rho} \approx 1 + \frac{2}{3}\pi\rho d^3 \qquad (2.5.27)$$

This expression represents the first two terms in the *virial expansion* of the equation of state in powers of the density, which we derive in a systematic way in Section 3.9.

The contact value of $g(r)$ also appears in the theory of transport processes in gases. Elementary kinetic theory[9] shows that at low densities the mean time between collisions suffered by a given particle is $\lambda/\bar{u}$, where $\bar{u} = (8k_B T/\pi m)^{1/2}$ is the mean speed appropriate to a Maxwell distribution of momenta and λ is the mean free path. If the gas particles are treated as hard spheres of diameter d, the mean free path is $\lambda = 1/\sqrt{2}\pi\rho d^2$. Thus the collision rate in the dilute gas is

$$\Gamma_0 = \bar{u}/\lambda = 4\rho d^2 (\pi k_B T/m)^{1/2} \qquad (2.5.28)$$

At higher densities the collision rate is enhanced by the interactions between particles. Since the 'forces' between hard spheres act only at collisions, the collision rate is proportional to the non-ideal contribution to the pressure, as given by the hard-sphere equation of state (2.5.26). It follows that $\Gamma_E = g(d)\Gamma_0$ where Γ_E, the collision rate in the dense gas, is the quantity that arises in the Enskog theory discussed in Section 2.1. This enhancement of the collision rate leads to a corresponding reduction in the self-diffusion coefficient relative to the value obtained from the Boltzmann equation by a factor $1/g(d)$.

2.6   PARTICLE DENSITIES IN THE GRAND CANONICAL ENSEMBLE

The fact that in the canonical ensemble the pair distribution function behaves asymptotically as $(1 - 1/N)$ rather than tending strictly to unity is often irrelevant since the term of order N^{-1} vanishes in the thermodynamic limit. On the other hand, if a term of that order is integrated over the volume of the system, a result of order V/N is obtained, which usually cannot be ignored. The difficulties that this situation sometimes creates can be avoided by working in the grand canonical ensemble. As we shall see in later chapters, the grand canonical ensemble also provides a convenient framework for the derivation of density expansions of the particle distribution functions and, more generally, for the development of the theory of inhomogeneous fluids.

In the grand canonical ensemble the n-particle density is defined in terms of its canonical ensemble counterparts as the sum

$$\rho^{(n)}(\mathbf{r}^n) = \sum_{N \geq n}^{\infty} p(N)\rho_N^{(n)}(\mathbf{r}^n)$$

$$= \frac{1}{\Xi} \sum_{N=n}^{\infty} \frac{z^N}{(N-n)!} \int \exp(-\beta V_N)d\mathbf{r}^{(N-n)} \qquad (2.6.1)$$

where $p(N)$ is the probability (2.4.12). Integration of (2.6.1) over the coordinates $\mathbf{r}_1, \ldots, \mathbf{r}_n$ shows that $\rho^{(n)}$ is normalised such that

$$\int \rho^{(n)}(\mathbf{r}^n) d\mathbf{r}^n = \left\langle \frac{N!}{(N-n)!} \right\rangle \tag{2.6.2}$$

In particular:

$$\int \rho^{(1)} d\mathbf{r} = \langle N \rangle \tag{2.6.3}$$

and

$$\iint \rho^{(2)}(\mathbf{r}_1, \mathbf{r}_2) d\mathbf{r}_1 d\mathbf{r}_2 = \left\langle N^2 \right\rangle - \langle N \rangle \tag{2.6.4}$$

Equation (2.6.3) confirms that the single-particle density in a homogeneous system is

$$\rho^{(1)} = \langle N \rangle / V \equiv \rho \quad \text{(uniform fluid)} \tag{2.6.5}$$

We know from Section 2.4 that for a homogeneous, ideal gas the activity z is equal to ρ, while the integral in (2.6.1) is equal to $V^{(N-n)}$. Hence the particle densities of the ideal gas are

$$\rho^{(n)} = \rho^n \quad \text{(uniform ideal gas)} \tag{2.6.6}$$

The relation between the grand canonical n-particle density and the corresponding distribution function is the same as in the canonical ensemble, i.e.

$$g^{(n)}(\mathbf{r}^n) = \frac{\rho^{(n)}(\mathbf{r}_1, \ldots, \mathbf{r}_n)}{\prod_{i=1}^n \rho^{(1)}(\mathbf{r}_i)} \tag{2.6.7}$$

or $\rho^{(n)}(\mathbf{r}^n) = \rho^n g^{(n)}(\mathbf{r}^n)$ if the system is homogeneous, but now $g^{(n)}(\mathbf{r}^n) \to 1$ for all n as the mutual separations of all pairs of particles becomes sufficiently large. In particular, the *pair correlation function*, defined as

$$h^{(2)}(\mathbf{r}_1, \mathbf{r}_2) = g^{(2)}(\mathbf{r}_1, \mathbf{r}_2) - 1 \tag{2.6.8}$$

vanishes in the limit $|\mathbf{r}_2 - \mathbf{r}_1| \to \infty$. If we insert the definition (2.6.1) into (2.6.7) we obtain an expansion of the n-particle distribution function of a uniform fluid as a power series in z, which starts as

$$\Xi \left(\frac{\rho}{z} \right)^n g^{(n)}(\mathbf{r}^n) = \exp[-\beta V_n(\mathbf{r}^n)] + \mathcal{O}(z) \tag{2.6.9}$$

The first term on the right-hand side is the one corresponding to the case $N = n$ in (2.6.1). As $\rho \to 0$, it follows from earlier definitions that $z \to 0$, $\rho/z \to 1$ and $\Xi \to 1$. Hence, taking $n = 2$, we find that the low-density limit of the radial distribution function is equal to the Boltzmann factor of the pair potential:

$$\lim_{\rho \to 0} g(r) = \exp[-\beta v(r)] \tag{2.6.10}$$

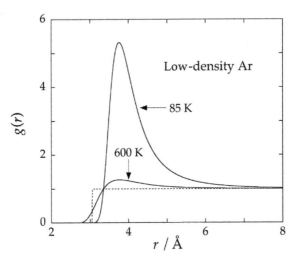

FIGURE 2.4 Low-density limit of the radial distribution function for argon at two temperatures, calculated from (2.6.10) for the accurate pair potential pictured in Figure 1.3. The dotted line shows the low-density distribution function for a system of hard spheres of diameter 3.23 Å (see text).

Figure 2.4 shows the low-density limit of the radial distribution function of argon for the accurate, gas phase potential pictured in Figure 1.3. Results are shown for two temperatures, one some four times greater than the experimental critical temperature (150.7 K) and one close to the experimental triple point (83.8 K). The strong peak seen at 85 K in the region of the minimum in the pair potential is indicative of the known tendency for argon to form weakly bound, van der Waals dimers at low temperatures.[10] At 600 K, by contrast, the distribution function closely resembles that of a hard-sphere gas, with the attractive part of the potential playing only a minor role. The hard-sphere results are for hard spheres of diameter equal to 3.23 Å, corresponding to the pair separation at which the gas phase potential is equal to $k_B T$; this is known to provide a realistic estimate of the effective 'size' of an atom as a function of temperature.

The δ-function representations of $\rho_N^{(1)}(\mathbf{r})$, $\rho_N^{(2)}(\mathbf{r}, \mathbf{r}')$ and $g(r)$ provided by (2.5.11), (2.5.13) and (2.5.15), respectively, are also valid (without the subscript N) in the grand canonical ensemble, as are the energy and pressure equations, (2.5.20) and (2.5.22). On the other hand, the *compressibility equation*, which expresses χ_T as an integral over $g(r)$, can be derived only in the grand canonical ensemble because the compressibility is related to fluctuations in an open system via (2.4.23). The normalisations (2.6.3) and (2.6.4) show that

$$\iint \left[\rho^{(2)}(\mathbf{r}_1, \mathbf{r}_2) - \rho^{(1)}(\mathbf{r}_1)\rho^{(1)}(\mathbf{r}_2) \right] d\mathbf{r}_1 \, d\mathbf{r}_2 = \left\langle N^2 \right\rangle - \langle N \rangle - \langle N \rangle^2$$

$$(2.6.11)$$

In the homogeneous case it follows immediately that

$$1 + \rho \int [g(r) - 1]\mathrm{d}\mathbf{r} = \frac{\langle N^2 \rangle - \langle N \rangle^2}{\langle N \rangle} = \rho k_\mathrm{B} T \chi_T \qquad (2.6.12)$$

Unlike the energy and pressure equations, the applicability of this relation does not rely on the assumption of pairwise additivity of the interparticle forces. For an ideal gas in the grand canonical ensemble, $g(r) = 1$ for all r; it follows from (2.6.12) that $\chi_T^{\mathrm{id}} = \beta/\rho$, in agreement with the result obtained by differentiation of the ideal gas equation of state.

2.7 MOLECULAR DYNAMICS SIMULATION

As we briefly mentioned at the end of Chapter 1, the behaviour of liquids, solids and dense gases at the microscopic level can be simulated in one of two ways: by the method of molecular dynamics or by a Monte Carlo method. The importance of computer simulation from the standpoint of liquid state theory is the fact that it provides essentially exact, quasi-experimental data on well-defined models, particularly on those that are prototypical models of simple liquids. In this section we give a brief account of how classical computer simulations are carried out. Excellent books exist that provide much fuller descriptions of the principles underlying the large variety of techniques that are now available and of the computer codes needed for their implementation.[1,11]

We begin by considering the method of molecular dynamics. In a conventional molecular dynamics simulation of a bulk fluid a system of N particles is allocated a set of initial coordinates within a cell of fixed volume, most commonly a cube. A set of velocities is also assigned, usually drawn from a Maxwell distribution appropriate to the temperature of interest and selected in such a way that the net linear momentum of the system is zero. The subsequent calculation tracks the motion of the particles through space by integration of the classical equations of motion. Equilibrium properties are obtained as time averages over the dynamical history of the system in the manner outlined in Section 2.2 and correspond to averages over a microcanonical ensemble. In modern work N is typically of order 10^3 or 10^4, though much larger systems have occasionally been studied. To minimise surface effects, and thereby simulate more closely the behaviour expected of a macroscopic system, it is customary to use a periodic boundary condition. The way in which the periodic boundary condition is applied is illustrated for the two-dimensional case in Figure 2.5. The system as a whole is divided into cells. Each cell is surrounded on all sides by periodic images of itself and particles that are images of each other have the same relative positions within their respective cells and the same momenta. When a particle enters or leaves a cell, the move is balanced by an image of that particle leaving or entering through the opposite face.

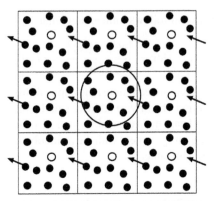

FIGURE 2.5 Periodic boundary conditions used in computer simulations. The circle represents the truncation sphere around a white particle in the central cell. When a particle leaves a cell it is replaced by an image of that particle entering through the opposite face.

A key question that arises in both the molecular dynamics and Monte Carlo methods is whether the properties of an infinite, periodic fluid with a unit cell containing, typically, of order 10^3 particles are representative of the properties of the macroscopic system that the calculation is designed to simulate. There is no easy or general answer to this,[12] but broadly speaking it appears that bulk properties are only weakly dependent on sample size beyond $N \approx 500$, and that the remaining errors, relative to the $N \to \infty$ limit, are no larger than the inevitable statistical uncertainties. Nonetheless, the restriction on sample size does have some drawbacks. For example, it is impossible to study collective, spatial fluctuations of wavelength greater than L, the length of the cell. Use of a periodic boundary condition also has an effect on time correlations. In a molecular dynamics simulation a local disturbance will move through the periodic system and reappear at the same place, albeit in attenuated form, after a recurrence time of order L/c, where c is a speed of propagation that can be roughly equated to the speed of sound. The effects of periodicity will manifest themselves in spurious contributions to time correlations calculated over time intervals greater than this. Another difficulty, which is particularly acute for small samples, is the so-called quasi-ergodic problem. In the context of a computer simulation the term refers to the possibility that the system may become trapped in some region of phase space. Near the melting temperature, for example, an initial, lattice-type arrangement of particles may persist for very long times unless the density is appreciably less than the freezing density of the fluid. Whatever the starting conditions, time must be allowed for the system to equilibrate before the 'production' stage of the calculation begins, while throughout the simulation it is important to monitor the properties of the system in such a way as to detect any tendency towards a long-time drift. Non-ergodic behaviour is also observed in simulations in which a liquid is quenched below a 'glass transition' temperature into a disordered, glassy state which is metastable with respect to the equilibrium, crystalline phase. Such states are characterised by very slow relaxation processes of the type to be discussed in Section 8.8.

The interactions between particles can be of any form but in the great majority of cases they are assumed to be pairwise additive. For economy in computing time it is customary to truncate the interaction at a separation $r_c \leq \frac{1}{2}L$, where the cut-off radius r_c is typically a few particle diameters. When a truncation sphere is used, the interaction of a particle with its neighbours is calculated with a 'nearest-neighbour' convention. The principle of this convention is illustrated in Figure 2.5: a particle i lying within a given cell is assumed to interact only with the nearest image of any other particle j (including j itself), the interaction being set equal to zero if the distance from the nearest image is greater than r_c. The upper limit imposed on r_c ensures that interactions with other images of j are automatically ignored. Use of such a cut-off is inappropriate when the interparticle forces are long ranged, particularly for ionic systems, since there is no guarantee that the truncation sphere would be electrically neutral. One way to overcome this difficulty is to calculate the coulombic interaction of a particle not only with all other particles in the same cell but with all images in other cells. An infinite lattice sum of this type can be evaluated by the method of Ewald, the essence of which is to convert the slowly convergent sum in r^{-1} into two series that are separately rapidly convergent. One series is a sum in real space of a short-range potential that may safely be truncated, and the other is a sum over reciprocal-lattice vectors of the periodic array of cells. Strongly polar systems also require special treatment.

The earliest applications of the molecular dynamics method were those of Alder and Wainwright[13] to systems of hard spheres and other hard-core particles. A feature of hard-sphere dynamics is that the velocities of the particles change only as the result of collisions; between collisions, the particles move in straight lines at constant speeds. The time evolution of a many particle, hard-sphere system may therefore be treated as a sequence of strictly binary, elastic collisions. Thus the algorithm for calculation of the trajectories consists of first advancing the coordinates of all particles until such a time as a collision occurs somewhere in the system, and then of exploiting the fact that both energy and momentum are conserved to calculate the changes in velocities of the colliding particles. Since that calculation is exact, the trajectories of the particles can be computed with a precision limited only by round-off errors. The instantaneous temperature of the system remains constant because the total kinetic energy is conserved.

When the potentials are continuous, the trajectories of the particles, unlike those of hard spheres, can no longer be calculated exactly. In the case of spherically symmetric potentials the equations of motion are the $3N$ coupled, second-order differential equations (2.2.1). These equations must be solved numerically by finite difference methods, which leads unavoidably to errors in the particle trajectories. One of the simplest but also most successful algorithms is that first used by Verlet[14] in studies of the properties of the Lennard-Jones fluid. Let the coordinates of particle i at time t be $\mathbf{r}_i(t)$. The coordinates at times $t \pm \Delta t$ are given by Taylor expansions forwards and backwards in time around $\mathbf{r}_i(t)$:

$$\mathbf{r}_i(t \pm \Delta t) = \mathbf{r}_i(t) \pm \Delta t\, \dot{\mathbf{r}}_i(t) + \frac{1}{2}\Delta t^2 \ddot{\mathbf{r}}_i(t) \pm \mathcal{O}(\Delta t^3) \qquad (2.7.1)$$

By adding together the two expansions in (2.7.1), we obtain an estimate for the particle coordinates at time $t + \Delta t$:

$$\mathbf{r}_i(t + \Delta t) \approx -\mathbf{r}_i(t - \Delta t) + 2\mathbf{r}_i(t) + \frac{\Delta t^2}{m}\mathbf{F}_i(t) \qquad (2.7.2)$$

where $\mathbf{F}_i(t)$ is the total force acting on particle i at time t. The error in the predicted coordinates is of order Δt^4. If we subtract the two expansions in (2.7.1), we obtain an estimate of the velocity of particle i at time t:

$$\dot{\mathbf{r}}_i(t) \approx \frac{1}{2\Delta t}[\mathbf{r}_i(t + \Delta t) - \mathbf{r}_i(t - \Delta t)] \qquad (2.7.3)$$

The error now is of order Δt^2, but velocities play no part in the integration scheme and the particle trajectories are therefore unaffected. In one of a number of variants of the Verlet algorithm, the 'velocity' version, the predicted coordinates are obtained solely from the forward expansion in (2.7.1), i.e.

$$\mathbf{r}_i(t + \Delta t) \approx \mathbf{r}_i(t) + \Delta t\, \dot{\mathbf{r}}_i(t) + \frac{1}{2}\Delta t^2\, \ddot{\mathbf{r}}_i(t) \qquad (2.7.4)$$

and the velocity is calculated as

$$\dot{\mathbf{r}}_i(t + \Delta t) \approx \dot{\mathbf{r}}_i(t) + \frac{1}{2}\Delta t[\ddot{\mathbf{r}}_i(t + \Delta t) + \ddot{\mathbf{r}}_i(t)] \qquad (2.7.5)$$

Taken together, (2.7.4) and (2.7.5) are equivalent to (2.7.2). In other words, the particle trajectories in configuration space are identical in the two versions of the algorithm, but different estimates are obtained for the velocities.

Although simple in form, the original Verlet algorithm and its modifications are at least as satisfactory as higher-order schemes that make use of derivatives of the particle coordinates beyond $\ddot{\mathbf{r}}_i(t)$. It may be less accurate than others at short times but, more importantly, it conserves energy well even over very long times; it is also time reversible, as it should be for consistency with the equations of motion. Some understanding of the reasons for the stability of the algorithm may be obtained in the following way.[15]

The true dynamics of a system of particles is described by the action of the operator $\exp(i\mathcal{L}t)$ on the phase space coordinates $\mathbf{r}^N$, $\mathbf{p}^N$ in the manner described by (2.1.14). Let the time interval t be divided into P equal intervals of length Δt. Then

$$\exp(i\mathcal{L}t) = [\exp(i\mathcal{L}\Delta t)]^P \qquad (2.7.6)$$

If the Liouville operator is divided in the form

$$i\mathcal{L} = i\mathcal{L}_{\mathbf{r}} + i\mathcal{L}_{\mathbf{p}} \qquad (2.7.7)$$

where

$$i\mathcal{L}_{\mathbf{r}} \equiv \sum_{i=1}^{N} \dot{\mathbf{r}}_i \cdot \frac{\partial}{\partial \mathbf{r}_i}, \quad i\mathcal{L}_{\mathbf{p}} = \sum_{i=1}^{N} \mathbf{F}_i \cdot \frac{\partial}{\partial \mathbf{p}_i} \qquad (2.7.8)$$

and if Δt is sufficiently small, the operator $\exp(i\mathcal{L}\Delta t)$ can be written as[16]

$$\exp(i\mathcal{L}\Delta t) \approx \exp\left(i\frac{1}{2}\mathcal{L}_{\mathbf{p}}\Delta t\right)\exp(i\mathcal{L}_{\mathbf{r}}\Delta t)\exp\left(i\frac{1}{2}\mathcal{L}_{\mathbf{p}}\Delta t\right) \qquad (2.7.9)$$

This relationship is only approximate, since the operators $\mathcal{L}_{\mathbf{r}}$ and $\mathcal{L}_{\mathbf{p}}$ do not commute; the error involved is of order Δt^3. The action of an exponential operator of the type appearing in (2.7.9) is

$$\exp\left(a\frac{\partial}{\partial x}\right)f(x) \equiv 1 + a\frac{\partial f}{\partial x} + \frac{1}{2}a^2\frac{\partial^2 f}{\partial x^2} + \cdots = f(x + a) \qquad (2.7.10)$$

The effect of operating with $\exp(i\mathcal{L}_{\mathbf{r}}\Delta t)$ or $\exp(i\mathcal{L}_{\mathbf{p}}\Delta t)$ on $\mathbf{r}^N, \mathbf{p}^N$ is therefore to displace the position or momentum, respectively, of each particle according to the rules

$$\mathbf{r}_i \rightarrow \mathbf{r}_i + \Delta t\,\dot{\mathbf{r}}_i = \mathbf{r}_i + (\Delta t/m)\mathbf{p}_i$$
$$\mathbf{p}_i \rightarrow \mathbf{p}_i + \Delta t\,\dot{\mathbf{p}}_i = \mathbf{p}_i + \Delta t\,\mathbf{F}_i \qquad (2.7.11)$$

The three operations involved in (2.7.9) may be regarded as successive steps in a simple, predictor–corrector scheme. The first step yields an estimate of the momentum of the particle at time $t + \Delta t/2$:

$$\mathbf{p}_i(t + \Delta t/2) = \mathbf{p}_i(t) + \frac{1}{2}\Delta t\,\dot{\mathbf{p}}_i(t) = \mathbf{p}_i(t) + \frac{1}{2}\Delta t\,\mathbf{F}_i(t) \qquad (2.7.12)$$

In the second step this estimate of the momentum is used to predict the coordinates of the particle at time $t + \Delta t$:

$$\mathbf{r}_i(t + \Delta t) = \mathbf{r}_i(t) + (\Delta t/m)\mathbf{p}_i(t + \Delta t/2)$$
$$= \mathbf{r}_i(t) + \Delta t\,\dot{\mathbf{r}}_i(t) + (\Delta t^2/2m)\mathbf{F}_i(t) \qquad (2.7.13)$$

Finally, an improved estimate is obtained for the momentum, based on the value of the force acting on the particle at its predicted position:

$$\mathbf{p}_i(t + \Delta t) = \mathbf{p}_i(t + \Delta t/2) + \frac{1}{2}\Delta t\,\dot{\mathbf{p}}_i(t + \Delta t)$$
$$= \mathbf{p}_i(t) + \frac{1}{2}\Delta t\,[\mathbf{F}_i(t) + \mathbf{F}_i(t + \Delta t)] \qquad (2.7.14)$$

The results thereby obtained for $\mathbf{r}_i(t + \Delta t)$, $\mathbf{p}_i(t + \Delta t)$ are precisely those that appear in the velocity version of the Verlet algorithm, (2.7.4) and (2.7.5). It is remarkable that a practical and widely used algorithm can be derived from a well-defined approximation for the propagator $\exp(i\mathcal{L}t)$. What is more significant, however, is that each of the three steps implied by use of (2.7.9) is time reversible and conserves volume in phase space in the sense of Section 2.1; the same is therefore true of the algorithm overall. The fact that the Verlet algorithm preserves these key features of hamiltonian dynamics is almost certainly the reason why it is numerically so stable. Other time-reversible algorithms can be derived by dividing the Liouville operator in ways different from that adopted in (2.7.9).

A molecular dynamics calculation is organised as a loop over time. At each step, the time is incremented by Δt, the total force acting on each particle is computed and the particles are advanced to their new positions. In the early stages of the simulation it is normal for the temperature to move away from the value at which it was set and some occasional rescaling of particle velocities is therefore needed. Once equilibrium is reached, the system is allowed to evolve undisturbed, with both potential and kinetic energies fluctuating around steady, mean values; the temperature of the system is calculated from the time-averaged kinetic energy, as in (2.2.4). The choice of the time step Δt is made on the basis of how well total energy is conserved. In the case of a model of liquid argon, for example, an acceptable level of energy conservation is achieved with a time step of 10^{-14} s, and a moderately long run would be one lasting about 10^5 time steps, corresponding to a real time span of the order of a nanosecond. By treating argon atoms as hard spheres of diameter 3.4 Å, the mean 'collision' time in liquid argon near its triple point can be estimated as roughly 10^{-13} s. Hence the criterion for the choice of time step based on energy conservation leads to the physically reasonable result that Δt should be roughly an order of magnitude smaller than the typical time between 'collisions'. As the time step is increased, the fluctuations in total energy become larger, until eventually an overall, upward drift in energy develops. Even when a small time step is used, deviations from the true dynamics are inevitable, and the phase space trajectory of the system can be expected to diverge exponentially from that given by the exact solution of the equations of motion. In this respect an error in the algorithm plays a similar role to a small change in initial conditions. Any such change is known to lead to a divergence in phase space that grows with time as $\exp(\lambda t)$, where λ is a 'Lyapunov exponent'; the consequences in terms of loss of correlation between trajectories can be dramatic.[17]

The methods outlined above are easily extended to models of molecular fluids in which the molecules consist of independent atoms bound together by continuous intramolecular forces, but small molecules are in general more efficiently treated as rigid particles. One approach to the solution of the equations of motion of a rigid body involves a separation of internal and centre-of-mass coordinates. Another is based on the method of 'constraints', in which

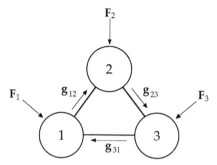

FIGURE 2.6 The method of constraints applied to a triatomic molecule: $\mathbf{F}_i$ is the total intermolecular force on atom i and $\mathbf{g}_{ij}$ is the force of constraint that maintains the rigidity of the bond between i and j.

the equations of motion are solved in Cartesian form.[18] As an illustration of the use of constraint dynamics, consider the example of the triatomic molecule shown in Figure 2.6, in which each internuclear bond is of length L and each atom (labelled 1–3) is of mass m. The geometry of the molecule is described by three constraints, $\sigma_{ij}(\mathbf{r}_1, \mathbf{r}_2, \mathbf{r}_3)$, such that

$$\sigma_{ij} = \frac{1}{2}(\mathbf{r}_{ij} \cdot \mathbf{r}_{ij} - L^2) = 0 \qquad (2.7.15)$$

where $\mathbf{r}_{ij} = \mathbf{r}_j - \mathbf{r}_i$. The total force acting on atom 1, say, at time t is the sum of three terms: $\mathbf{F}_1(t)$, the force due to interactions with other molecules; a force of constraint, $\mathbf{g}_{12}(t)$, which ensures that the bond vector $\mathbf{r}_{12}$ remains of fixed length; and a second force of constraint, $\mathbf{g}_{13}(t)$, which preserves the bond length between atoms 1 and 3. Similar considerations apply to the other atoms. The forces of constraint are directed along the corresponding bond vectors and the law of action and reaction requires that $\mathbf{g}_{ij} = -\mathbf{g}_{ji}$. Thus $\mathbf{g}_{ij} = \lambda_{ij}\mathbf{r}_{ij}$, where λ_{ij} is a time-dependent scalar quantity, with $\lambda_{ij} = \lambda_{ji}$. The newtonian equations of motion are therefore of the form

$$m\ddot{\mathbf{r}}_1(t) = \mathbf{F}_1(t) + \lambda_{12}\mathbf{r}_{12}(t) + \lambda_{13}\mathbf{r}_{13}(t)$$
$$m\ddot{\mathbf{r}}_2(t) = \mathbf{F}_2(t) - \lambda_{12}\mathbf{r}_{12}(t) + \lambda_{23}\mathbf{r}_{23}(t) \qquad (2.7.16)$$
$$m\ddot{\mathbf{r}}_3(t) = \mathbf{F}_3(t) - \lambda_{13}\mathbf{r}_{13}(t) - \lambda_{23}\mathbf{r}_{23}(t)$$

Comparison with (2.7.15) shows that the total force of constraint on atom i, $\mathbf{G}_i$, can be written as

$$\mathbf{G}_i = -\sum_{j \neq i} \lambda_{ij}\frac{\partial \sigma_{ij}}{\partial \mathbf{r}_i} \qquad (2.7.17)$$

As is to be expected, the sum of the forces of constraint is zero: $\sum_i \mathbf{G}_i = 0$.

It is possible to eliminate the unknown quantities λ_{12}, λ_{13} and λ_{23} from (2.7.16) by requiring the second time derivative of the constraint conditions

(2.7.15) to vanish, i.e. by setting $\ddot{\sigma}_{ij} = \dot{\mathbf{r}}_{ij} \cdot \dot{\mathbf{r}}_{ij} + \mathbf{r}_{ij} \cdot \ddot{\mathbf{r}}_{ij} = 0$ and replacing $\ddot{\mathbf{r}}_i$ by $(\mathbf{F}_i + \mathbf{G}_i)/m$. The resulting system of equations for the constrained coordinates can then be integrated numerically. In practice this procedure does not work; the errors inherent in any approximate algorithm cause the bond lengths to drift away rapidly from their initial values. What is done instead is to require the constraints to be satisfied exactly after each time step in a manner dictated by the chosen integration scheme. If the original Verlet algorithm is used, for example, we find that

$$\mathbf{r}_1(t + \Delta t) = \mathbf{r}_1'(t + \Delta t) + (\Delta t^2/m)[\lambda_{12}\mathbf{r}_{12}(t) + \lambda_{13}\mathbf{r}_{13}(t)]$$
$$\mathbf{r}_2(t + \Delta t) = \mathbf{r}_2'(t + \Delta t) + (\Delta t^2/m)[-\lambda_{12}\mathbf{r}_{12}(t) + \lambda_{23}\mathbf{r}_{23}(t)] \qquad (2.7.18)$$
$$\mathbf{r}_3(t + \Delta t) = \mathbf{r}_3'(t + \Delta t) + (\Delta t^2/m)[-\lambda_{13}\mathbf{r}_{13}(t) - \lambda_{23}\mathbf{r}_{23}(t)]$$

where $\mathbf{r}_i'(t + \Delta t)$ are the predicted coordinates of atom i in the absence of constraints, given by (2.7.4). Equations (2.7.18) must be solved subject to the requirement that $|\mathbf{r}_{ij}(t+\Delta t)|^2 = L^2$ for all i, j. This leads to three simultaneous equations for the quantities $\lambda_{ij}(t)$, to which a solution can be obtained by an iterative method; three to four iterations per molecule are normally sufficient to maintain the bond lengths constant to within one part in 10^4.

Apart from its simplicity, a particular merit of the method of constraints is the fact that it can be used for both rigid and flexible molecules. A partially flexible chain molecule, for example, can be treated by employing a suitable mixture of constraints on bond angles and bond lengths in a way that allows for torsional motion and bending but freezes the fast vibrations.[19]

The algorithms discussed thus far are limited to the calculation of the trajectories of particles moving solely under the influence of the interparticle forces. However, some of the most interesting applications of the molecular dynamics method have involved the incorporation into the dynamics of one or more additional degrees of freedom that describe, for example, a coupling between the physical system of interest and its surroundings or some fluctuating molecular property such as an induced dipole moment. The equations of motion of the resulting 'extended' system are most easily derived within the framework of Lagrangian mechanics. As an example of this approach we shall briefly describe the scheme, developed in a classic paper by Andersen,[20] which allows a molecular dynamics simulation to be carried out for state conditions corresponding to constant pressure rather than constant volume.

The Lagrangian of a mechanical system is defined as the difference between the kinetic and potential energies taken as functions of a set of generalised coordinates, one for each degree of freedom of the system, and a corresponding set of generalised velocities. In the case of an atomic fluid there are $3N$ degrees of freedom and the generalised coordinates are simply the Cartesian coordinates, denoted collectively by $\mathbf{r}^N$, with generalised velocities similarly denoted by $\dot{\mathbf{r}}^N$.

The Lagrangian for a one-component system is therefore

$$L(\mathbf{r}^N, \dot{\mathbf{r}}^N) = \frac{1}{2}m \sum_{i=1}^{N} |\dot{\mathbf{r}}_i|^2 - V_N(\mathbf{r}^N) \qquad (2.7.19)$$

and the equations of motion of the particles are given by

$$\frac{\partial}{\partial t}\left(\frac{\partial L}{\partial \dot{\mathbf{r}}_i}\right) = \frac{\partial L}{\partial \mathbf{r}_i} \qquad (2.7.20)$$

The generalised momenta are

$$\mathbf{p}_i = \frac{\partial L}{\partial \dot{\mathbf{r}}_i} \qquad (2.7.21)$$

and the link with the hamiltonian description of the system is provided by the relation

$$\mathcal{H}(\mathbf{r}^N, \mathbf{p}^N) = \sum_{i=1}^{N} \mathbf{p}_i \cdot \dot{\mathbf{r}}_i - L\left(\mathbf{r}^N, \dot{\mathbf{r}}^N\right) \qquad (2.7.22)$$

In the simple case just described, use of (2.7.20) leads to the Newtonian equations (2.2.1) and the hamiltonian is that given by (2.1.1), with the contribution from the external field omitted.

Consider a system of structureless particles enclosed in a cube of volume V. The 'extended Lagrangian' proposed by Andersen differs from (2.7.19) in two ways. First, the Cartesian coordinates and associated velocities are replaced by the scaled variables $\boldsymbol{\tau}_i$ and $\dot{\boldsymbol{\tau}}_i$, defined as

$$\boldsymbol{\tau}_i = V^{-1/3}\mathbf{r}_i, \quad \dot{\boldsymbol{\tau}}_i = V^{-1/3}\dot{\mathbf{r}}_i \qquad (2.7.23)$$

Secondly, V itself is treated as an additional, generalised coordinate. The extended system may then be visualised as one that fluctuates in volume against a fixed external pressure equal to P_{ext}. With that picture in mind the Lagrangian is assumed to have the form

$$L(\boldsymbol{\tau}^N, \dot{\boldsymbol{\tau}}^N, V, \dot{V}) = \frac{1}{2}m \sum_{i=1}^{N} |\dot{\boldsymbol{\tau}}_i|^2 + \frac{1}{2}W\dot{V}^2 - V_N(V^{1/3}\boldsymbol{\tau}^N) - P_{\text{ext}}V \quad (2.7.24)$$

where the quantities $\frac{1}{2}W\dot{V}^2$ and $P_{\text{ext}}V$ are respectively the kinetic and potential energies associated with the coordinate V; W is an inertial factor which plays the role of a 'mass' in the kinetic energy term. The equations of motion derived from the analogues of (2.7.20) for the scaled variables are

$$\ddot{\boldsymbol{\tau}}_i = \frac{\mathbf{F}_i}{mV^{1/3}} - \frac{2}{3V}\dot{V}\dot{\boldsymbol{\tau}}_i \qquad (2.7.25)$$

where $\mathbf{F}_i$ is the total force on particle i, and

$$W\ddot{V} = \frac{m}{3V^{1/3}} \sum_{i=1}^{N} |\dot{\boldsymbol{\tau}}_i|^2 + \frac{1}{3V^{2/3}} \sum_{i=1}^{N} \boldsymbol{\tau}_i \cdot \mathbf{F}_i - P_{\text{ext}} \qquad (2.7.26)$$

Equation (2.7.25) shows that the motion of the particles is now coupled to the motion of the coordinate V, while the meaning of (2.7.26) is most easily grasped by rewriting it in terms of the unscaled variables:

$$W\ddot{V} = \frac{m}{3V}\sum_{i=1}^{N}|\dot{\mathbf{r}}_i|^2 + \frac{1}{3V}\sum_{i=1}^{N}\mathbf{r}_i \cdot \mathbf{F}_i - P_{\text{ext}} \qquad (2.7.27)$$

Comparison with (2.2.9) shows that the sum of the first two terms on the right-hand side is the instantaneous value of the internal pressure P of the system and hence that

$$\ddot{V} = \frac{1}{W}\left(P - P_{\text{ext}}\right) \qquad (2.7.28)$$

Thus the difference between internal and external pressures represents the 'force' that causes the volume of the system to change. When averaged over a sufficiently long time, the 'force' must vanish, and the mean value of the internal pressure will be equal to the pre-set value P_{ext}.

The generalised momenta conjugate to the generalised coordinates $\dot{\boldsymbol{\tau}}_i$ and V, respectively, are

$$\boldsymbol{\pi}_i = \frac{\partial \mathrm{L}}{\partial \dot{\boldsymbol{\tau}}_i} = mV^{2/3}\dot{\boldsymbol{\tau}}_i, \quad \pi_V = \frac{\partial \mathrm{L}}{\partial \dot{V}} = W\dot{V} \qquad (2.7.29)$$

The hamiltonian of the extended system, which is conserved by the equations of motion, is therefore

$$\mathcal{H}(\boldsymbol{\tau}^N, \boldsymbol{\pi}^N; V, \pi_V) = \sum_{i=1}^{N} \boldsymbol{\pi}_i \cdot \dot{\boldsymbol{\tau}}_i + \pi_V \dot{V} - \mathrm{L}(\boldsymbol{\tau}^N, \dot{\boldsymbol{\tau}}^N; V, \dot{V})$$

$$= \frac{1}{2mV^{2/3}}\sum_{i=1}^{N}|\boldsymbol{\pi}_i|^2 + V_N(V^{1/3}\boldsymbol{\tau}^N)$$

$$+ \frac{\pi_V^2}{2W} + P_{\text{ext}}V \qquad (2.7.30)$$

This is equal to the enthalpy, H, of the physical system apart from the presence of the fictitious kinetic energy term $\pi_V^2/2W$. Since the extra term is a quadratic function of momentum its average value is

$$\left\langle \frac{\pi_V^2}{2W} \right\rangle = \frac{1}{2}k_{\mathrm{B}}T \qquad (2.7.31)$$

which, relative to the remaining terms in (2.7.30), becomes negligibly small in the limit $N \to \infty$. Thus, to a good approximation, time averages over the trajectories of the particles correspond to averages in the constant N, P and H or isobaric–isoenthalpic ensemble. This is true irrespective of the value chosen for the inertial parameter W. However, if the value used is too small, the motion of the coordinate V is effectively decoupled from that of the particles; if it is too large, the phase space of the extended system is inefficiently sampled. In the limit $W \to \infty$, and assuming that $\dot{V}$ is initially zero, the equations of motion reduce to those of conventional molecular dynamics at constant N, V and E.

Andersen's paper also describes a method for controlling the temperature of the system by adding a stochastic collision term to the equations of motion. Later work by Nosé[21] showed that the same effect could be achieved by use of the equations of motion derived from an extended Lagrangian in which a variable is introduced that scales the velocities of the particles; this mimics the interaction between the system and a reservoir of fixed temperature. Nosé's method was later reformulated by Hoover[22] in a way that made it easier to implement and the so-called Nosé–Hoover 'thermostat' is now very widely used in molecular dynamics calculations at constant N, V and T or N, P and T.

2.8 MONTE CARLO METHODS

Given a set of initial conditions, a conventional molecular dynamics simulation is, in principle, entirely deterministic in nature. By contrast, as the name suggests, a stochastic element is an essential part of any Monte Carlo calculation. In a Monte Carlo simulation a system of N particles, subject to the same boundary condition used in molecular dynamics calculations and interacting through some known potentials, is again assigned a set of arbitrarily chosen, initial coordinates. A sequence of configurations is then generated, which in the simplest case would occur by random displacements of randomly chosen particles, usually of one particle at a time. Not all configurations that are generated are added to the sequence. The decision whether to 'accept' or 'reject' a trial configuration is made in such a way that asymptotically configuration space is sampled according to the probability density corresponding to a particular statistical mechanical ensemble. The ensemble average of any function of the particle coordinates, such as the total potential energy, is then obtained as an unweighted average over the resulting set of configurations. The particle momenta do not enter the calculation, there is no time scale involved, and the order in which the configurations occur has no special significance. The method is therefore limited to the calculation of static properties.

The problem of devising a scheme for sampling configuration space according to a specific probability distribution is most easily formulated in terms of the theory of Markov processes.[23] Suppose we have a sequence of random variables. Here the 'variable' consists of the coordinates of the particles, and possibly also the volume of the system or the number of particles it contains, while its range is the set of all accessible states of the system. Hence, instead of speaking of the 'value' of the variable at a given point in the sequence, it is more natural to say that at that point the system occupies a particular state.

If the probability of finding the system in a state n at 'time' $(t+1)$ is dependent only on the state it occupied at the previous time, t, the sequence of states constitutes a Markov chain. Note that the concept of 'time' is introduced merely for descriptive purposes; there is no connection with any physical time scale.

Let $q_n(t)$ be the probability that the system is in a state n at time t. A Markov process is one for which

$$q_n(t) = \sum_m p_{n \leftarrow m} q_m(t-1) \tag{2.8.1}$$

where $p_{n \leftarrow m}$ is a transition probability, with $\sum_n p_{n \leftarrow m} = 1$. If we regard the probabilities $\{q_n(t)\}$ as the components of a column vector $\mathbf{q}(t)$ and the quantities $\{p_{n \leftarrow m}\}$ as the elements of a square *transition matrix* $\mathbf{p}$, (2.8.1) may be rewritten in more compact form as

$$\mathbf{q}(t) = \mathbf{p} \cdot \mathbf{q}(t-1) \tag{2.8.2}$$

Equation (2.8.2) can be immediately generalised to yield the probability distribution at time t given an initial distribution $\mathbf{q}(0)$:

$$\mathbf{q}(t) = \overbrace{\mathbf{p} \cdots \mathbf{p}}^{t \text{ times}} \cdot \mathbf{q}(0) \equiv \mathbf{p}^t \cdot \mathbf{q}(0) \tag{2.8.3}$$

where $\mathbf{p}^t \equiv \{p_{n \leftarrow m}^{(t)}\}$ is the t-fold product of $\mathbf{p}$ with itself. If all elements of the matrix $\mathbf{p}^t$ are non-zero for some finite t, each state of the system can be reached from any other state in a finite number of steps (or finite 'time'), and the Markov chain is said to be ergodic; it is clear that this usage of the term 'ergodic' is closely related to its meaning in statistical mechanics. When the chain is ergodic, it can be shown that the limits

$$Q_n = \lim_{t \to \infty} p_{n \leftarrow m}^{(t)} q_m(0) \tag{2.8.4}$$

exist and are the same for all m. In other words there exists a limiting probability distribution $\mathbf{Q} \equiv \{Q_n\}$ that is independent of the initial distribution $\mathbf{q}(0)$. When the limiting distribution is reached, it persists, because $\mathbf{p} \cdot \mathbf{Q} = \mathbf{Q}$ or, in component form:

$$Q_n = \sum_m p_{n \leftarrow m} Q_m \tag{2.8.5}$$

This result is called the *steady-state condition*. In the context of statistical mechanics the limiting distribution is determined by the appropriate equilibrium probability density, which in the case of the canonical ensemble, for example, is proportional to the Boltzmann factor, so the desired limits are $Q_n \propto \exp[-\beta V_N(n)]$. The task of finding a set of transition probabilities consistent with the known, limiting distribution is greatly simplified by seeking a transition matrix that satisfies *microscopic reversibility*, i.e. one for which

$$Q_n p_{m \leftarrow n} = Q_m p_{n \leftarrow m} \tag{2.8.6}$$

If this relation holds, the steady-state condition is automatically satisfied.[24]

Let us suppose that the system is in state m at a given time and that a trial state n is generated in some way. If the probability of choosing n as the trial state is the same as that of choosing m when n is the current state, a choice of transition probabilities that satisfies (2.8.6) is

$$\begin{aligned} p_{n \leftarrow m} &= 1, \quad \text{if } Q_n \geq Q_m \\ &= \frac{Q_n}{Q_m}, \quad \text{if } Q_n < Q_m \end{aligned} \tag{2.8.7}$$

with $p_{m \leftarrow m} = 1 - p_{n \leftarrow m}$. The transition matrix defined by (2.8.7) is the one proposed in the pioneering work of Metropolis et al.[25] and remains much the most commonly used prescription for $\mathbf{p}$. In practice, in the case of the canonical ensemble, the trial state is normally generated by selecting a particle i at random and giving it a small, random displacement, $\mathbf{r}_i \rightarrow \mathbf{r}_i + \Delta \mathbf{r}$, where $\Delta \mathbf{r}$ is chosen uniformly within prescribed limits. If the difference in potential energy of the two states is $\Delta U = V_N(n) - V_N(m)$, the trial state is accepted unconditionally when $\Delta U \leq 0$ and with a probability $\exp(-\beta \Delta U)$ when $\Delta U > 0$, i.e.

$$p_{n \leftarrow m} = \min \{1, \exp(-\beta \Delta U)\} \tag{2.8.8}$$

The procedure takes a particularly simple form for a system of hard spheres: trial configurations in which two or more spheres overlap are rejected, but all others are accepted. One important point to note about the Metropolis scheme is that the system remains in its current state if the trial state n is rejected. In that case, state m appears a further time in the Markov chain, and the contribution it makes to any ensemble average must be counted again.

Monte Carlo methods similar to that outlined above are easily devised for use in other ensembles. All that changes are the form of the equilibrium probability density and the way in which trial states are generated. In the case of the isothermal–isobaric ensemble random displacements of the particles must be combined with random changes in volume. The corresponding probability density can be deduced from the form of the partition function (2.3.25), but allowance needs to be made for the fact that a change in volume alters the range of integration over particle coordinates. That can be done, in the case where the periodic cell is cubic, by switching to the scaled coordinates $\boldsymbol{\tau}_i$ defined by (2.7.23). This has the effect of transforming the integral over the region V into an integral over the unit cube ω:

$$\int_V \cdots \mathrm{d}\mathbf{r}^N \rightarrow V^N \int_\omega \cdots \mathrm{d}\boldsymbol{\tau}^N \tag{2.8.9}$$

and the partition function, after integration over momenta, takes the form

$$\Delta_N = \frac{1}{N!} \frac{V_0^{-1}}{\Lambda^{3N}} \int_0^\infty dV \, V^N \int \exp[-\beta(U_N + PV)] d\boldsymbol{\tau}^N \qquad (2.8.10)$$

where, to avoid confusion, we use the symbol U_N rather than V_N to denote the total potential energy. The required probability density is therefore proportional to $V^N \exp[-\beta(U_N + PV)]$. Thus the selection rule for displacements is the same as in the canonical ensemble while that for a change in volume from V to $V + \Delta V$ is

$$p_{n \leftarrow m} = \min \left\{ 1, \exp \left[-\beta(\Delta U + P\Delta V) + N \ln \left(1 + \frac{\Delta V}{V} \right) \right] \right\} \qquad (2.8.11)$$

where ΔU is the change in potential energy brought about by the change in volume. As in the case of particle displacements, the choice of ΔV must be made within prescribed limits.

In simulations in the grand canonical ensemble displacements are combined with random attempts to insert or delete particles, a choice that must be made randomly but with equal probabilities. By switching to scaled coordinates and integrating over momenta in the definition of the grand partition function (2.4.6) we find that the equilibrium probability density is

$$\frac{(zV)^N}{N!} \exp(-\beta U_N)$$

The acceptance rule for displacements is again given by (2.8.8) and those for insertion and deletion of particles by

$$p_{n \leftarrow m} = \min \left\{ 1, \frac{zV}{N+1} \exp(-\beta \Delta U) \right\}, \quad \text{insertion}, \ N \to N+1 \quad (2.8.12)$$

and

$$p_{n \leftarrow m} = \min \left\{ 1, \frac{N}{zV} \exp(-\beta \Delta U) \right\}, \quad \text{deletion}, \ N \to N-1 \quad (2.8.13)$$

where ΔU is the change in potential energy associated with the gain or loss of a particle.

The extension to molecular systems is straightforward. Interactions between particles are now dependent on their mutual orientation and 'displacements' are either random translational moves or random reorientations. The choice of which type of move is to be attempted at any given stage should be made randomly to guarantee that microscopic reversibility is preserved.

Monte Carlo methods are widely used in the study of phase equilibria for model systems, particularly that of equilibrium between liquid and vapour. The liquid – vapour coexistence curve of a one-component system can be determined

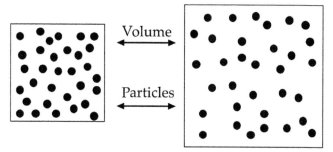

FIGURE 2.7 The Gibbs ensemble. The system of interest consists of two sub-systems A and B held at a constant temperature and between which volume and particles can be exchanged while keeping the total volume $V = V_A + V_B$ and total number of particles $N = N_A + N_B$ constant.

if the chemical potential is known as a function of density and temperature over the relevant region of the phase diagram. The necessary data may be obtained by working either in the grand canonical ensemble, where the chemical potential is an input parameter, or in the isobaric–isothermal ensemble[26] if supplemented by calculation of the chemical potential by the particle insertion method.[27] A more direct approach to the problem of liquid–vapour coexistence is provided by the 'Gibbs ensemble' methodology developed by Panagiotopoulos.[28] Consider a system held at a constant temperature T and divided into two sub-systems, A and B, which represent the two phases, as pictured in Figure 2.7. The equilibrium properties of the composite system can be determined from a Monte Carlo simulation involving particle displacements within each sub-system and exchanges of volume and particles between them, while keeping both the total volume V and total number of particles N constant. If the temperature and overall density are well chosen, the ensemble averages will be those corresponding to phase equilibrium in which subsystem A, say, has a density equal to that of the vapour, and B has a density equal to that of the liquid, while the pressure in the two subsystems will be the same and equal to the vapour pressure. The coexistence curve in the density – temperature plane can therefore be determined without measurement of the chemical potential, which should, however, be the same for A and B; this can checked by use of a test particle method[29] to ensure that a true equilibrium state has been reached.

The Gibbs ensemble approach is straightforward to implement and requires only modest computing resources. If high accuracy is required, however, other methods must be used, of which the most powerful is based on calculations in the grand canonical ensemble combined with a 'histogram reweighting' scheme. Such schemes are ones in which data obtained from multiple simulations for the same values of μ and V but different temperatures are pooled in such a way as to minimise the statistical uncertainties in the results of the individual simulations. The principle involved can be readily understood by focusing on the simpler problem of the calculation of the internal energy from data obtained in canonical ensemble simulations.

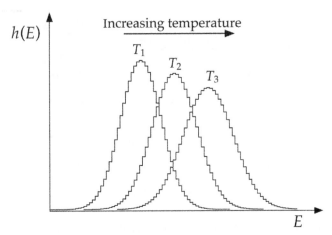

FIGURE 2.8 Histograms showing the number of times a system is found to have a total potential energy lying in an interval ΔE around a value E in typical Monte Carlo calculations in the canonical ensemble.

If a Monte Carlo run is carried out at constant N, V and T it is a trivial matter to construct a histogram, $h(E)$, incremented at each step in the calculation, which records the number of times that the potential energy $V_N(\mathbf{r}^N)$ of the system is found to lie in a narrow interval ΔE around a value E. The histogram typically has the form pictured in Figure 2.8, where results are plotted for three different temperatures; as the temperature increases, the histogram broadens and the peak shifts to higher energies. The excess internal energy is the mean value of E, given in terms of histogram entries by

$$U^{\text{ex}}(T) = \frac{\sum_E E h(E)}{\sum_E h(E)} \tag{2.8.14}$$

Let $p(E)\,dE$ be the probability of finding the system in a state of potential energy in the range E to $E + dE$. An estimate of the probability density $p(E)$ is provided by the quantity

$$p(E) = \frac{h(E)}{\mathcal{N}\Delta E} \tag{2.8.15}$$

where $\mathcal{N}$ is the total number of steps in the Monte Carlo run. The probability density is the product of an energy density of states, $W(N,V,E)$, and a Boltzmann factor, $\exp(-\beta E)$:

$$p(E) = \frac{1}{\mathcal{Z}} W(N,V,E) \exp(-\beta E) \tag{2.8.16}$$

where

$$\mathcal{Z} = \sum_E W(N,V,E) \exp(-\beta E) \tag{2.8.17}$$

is an unknown, run-dependent, normalisation constant.[30] Since the density of states is independent of temperature, an estimate of the excess internal energy at a temperature T' can be derived from (2.8.14) if $h(E)$ is replaced by a histogram $h'(E)$, reweighted to match the target temperature:

$$h(E) \to h'(E) = h(E) \exp[(-\beta' + \beta)E] \qquad (2.8.18)$$

This makes it possible to determine the excess internal energy over a limited range of temperature from data accumulated in a single simulation, while other physical properties can be determined by extensions of the basic method.[31] As the temperature difference $|T' - T|$ increases, however, greater weight is placed on the contributions to $\langle E \rangle$ from the wings of the histogram, which correspond to energies rarely sampled in the simulation. This leads to a rapidly growing loss in accuracy.

A more efficient method is one based on multiple histograms[32] obtained from independent simulations at the same values of N and V but different temperatures, T_m say. The temperatures should be sufficiently closely spaced to ensure a significant degree of overlap between neighbouring histograms, as exemplified in Figure 2.8. Equations (2.8.15) and (2.8.16) together show that each simulation provides an estimate of the density of states in the form

$$W_m(N, V, E) \approx \mathcal{Z}_m \frac{h_m(E)}{\mathcal{N} \Delta E} \exp(\beta_m E) \qquad (2.8.19)$$

where $\mathcal{N}$ and ΔE are assumed to be the same in each case. The results for different temperatures can then then be combined to provide an estimate of the density of states over the full range of energies sampled by the simulations in the form of a weighted sum:

$$W(N, V, E) \approx \frac{\sum_m c_m W_m(N, V, E)}{\sum_m c_m} \qquad (2.8.20)$$

How are the coefficients c_m to be chosen? Let us imagine that not one but n simulations are carried out at a temperature T_m, where n is a very large number and let $\langle h_m(E) \rangle$ be the histogram obtained by averaging over the n sets of results, which in the limit $n \to \infty$ is related to the exact density of states by

$$\mathcal{Z}_m \frac{\langle h_m(E) \rangle}{\mathcal{N} \Delta E} \exp(\beta_m E) \to W(N, V, E), \quad n \to \infty \qquad (2.8.21)$$

The limiting value of the quantity $\langle h_m(E) \rangle$ provides a natural choice of weight factor in (2.8.20). Though the limiting value cannot be computed, it is sufficient to know that the limit exists; justification[33] for its use as a weight factor rests ultimately on the fact that the error associated with an individual histogram is proportional to $\langle h_m(E) \rangle^{-1/2}$. That given, it follows from (2.8.19) and (2.8.21) that the best estimate of the density of states is

$$W(N, V, E) \approx \frac{\sum_m \langle h_m(E) \rangle W_m(N, V, E)}{\sum_m \langle h_m(E) \rangle}$$

$$= \frac{1}{\mathcal{N} \Delta E} \frac{\sum_m h_m(E)}{\sum_m [\mathcal{Z}_m]^{-1} \exp(-\beta_m E)} \tag{2.8.22}$$

This expression cannot be used as it stands, since the normalisation constants remain unknown, but substitution of (2.8.22) in (2.8.17) shows that $\mathcal{Z}_{m'}$, corresponding to a temperature $T_{m'}$, is given in closed form by

$$\mathcal{Z}_{m'} = \sum_E W(N, V, E) \exp(-\beta_{m'} E)$$

$$= \sum_E \frac{\sum_m h_m(E)}{\mathcal{N} \Delta E \sum_m [\mathcal{Z}_m]^{-1} \exp[(\beta_{m'} - \beta_m) E]} \tag{2.8.23}$$

The set of equations represented by (2.8.23) can be solved self-consistently to yield values of the normalisation constants relative to that at one, arbitrarily chosen temperature. The density of states given by (2.8.22) can then be constructed, from which the excess internal energy is easily computed at any temperature in the range originally chosen.

Application of histogram reweighting to the grand canonical ensemble involves the accumulation of data on both potential energy and particle number in a two-dimensional histogram.[34] This allows reweighting to be made to other values of chemical potential as well as temperature. The method has proved particularly valuable in studies of the critical region, where high precision in the calculation of physical properties is needed but is also difficult to achieve. In the case of Gibbs ensemble simulations, for example, fluctuations in density in the two subsystems at temperatures close to T_c become comparable with the difference between the equilibrium densities of liquid and vapour, making accurate measurement of the individual densities impossible.

REFERENCES

[1] Frenkel, D. and Smit, B., 'Understanding Molecular Simulation', 2nd edn. Academic Press, San Diego, 2002, p. 492.

[2] Résibois, P. and DeLeener, M., 'Classical Kinetic Theory of Fluids'. John Wiley, New York, 1977.

[3] Maitland, G.C., Rigby, M., Smith, E.B. and Wakeham, W.A., 'Intermolecular Forces'. Clarendon Press, Oxford, 1981, Chap. 5.

[4] Hill, T.L., 'Statistical Mechanics'. McGraw-Hill, New York, 1956, p. 189.

[5] For a different formulation of the constant pressure ensemble, see Koper, G.J.M. and Reiss, H., *J. Phys. Chem. B* **100**, 422 (1996) and Corti, D.S. and Soto-Campos, D., *J. Chem. Phys.* **108**, 7959 (1998).

[6]   Widom, B., *J. Chem. Phys.* **39**, 2808 (1963). See also Shing, K.S. and Gubbins, K.E., *Mol. Phys.* **46**, 1109 (1982).

[7]   Beck, T.L., Paulitis, M.E. and Pratt, L.R., 'The Potential Distribution Theorem and Models of Molecular Solutions'. Cambridge University Press, New York, 2006.

[8]   Yarnell, J.L., Katz, M.J., Wenzel, R.G. and Koenig, S.H., *Phys. Rev. A* **7**, 2130 (1973).

[9]   See, e.g., Reif, F., 'Fundamentals of Statistical and Thermal Physics'. McGraw-Hill, New York, 1965, pp. 268 and 471.

[10] Ref. 3, Sec. 7.5.

[11] (a) Allen, M.P. and Tildesley, D.J., 'Computer Simulation of Liquids'. Clarendon Press, Oxford, 1987. (b) Newman, E.J. and Barkema, G.T., 'Monte Carlo Methods in Statistical Physics'. Oxford University Press, New York, 1999. (c) Landau, D.P. and Binder, K., 'A Guide to Monte Carlo Simulations in Statistical Physics', 3rd edn. Cambridge University Press, Cambridge, 2009. (d) Tuckerman, M.E., 'Statistical Mechanics and Molecular Simulation'. Oxford University Press, Oxford, 2010.

[12] Theoretical results have been obtained for the N-dependence of the chemical potential: see Siepmann, J.I., McDonald, I.R. and Frenkel, D., *J. Phys.: Condensed Matter* **4**, 679 (1992).

[13] Alder, B.J. and Wainwright, T.E., *J. Chem. Phys.* **31**, 459 (1959).

[14] Verlet, L., *Phys. Rev.* **159**, 98 (1967).

[15] Tuckerman, M.E., Berne, B.J. and Martyna, G., *J. Chem. Phys.* **97**, 1990 (1992).

[16] This follows from the so-called Trotter expansion: see ref. 11(d), App. C.

[17] See, e.g., ref. 11(c), p. 81.

[18] (a) Ryckaert, J.P., Ciccotti, G. and Berendsen, H.J.C., *J. Comp. Phys.* **23**, 237 (1977). (b) Ciccotti, G., Ferrario, M. and Ryckaert, J.P., *Mol. Phys.* **47**, 1253 (1982).

[19] Ryckaert, J.P., *Mol. Phys.* **55**, 549 (1985).

[20] Andersen, H.C., *J. Chem. Phys.* **72**, 2384 (1980).

[21] (a) Nosé, S., *Mol. Phys.* **52**, 255 (1984). (b) Nosé, S., *J. Chem. Phys.* **81**, 511 (1984).

[22] Hoover, W.G., *Phys. Rev. A* **31**, 1695 (1985).

[23] Wood, W.W. and Parker, F.R., *J. Chem. Phys.* **27**, 720 (1957).

[24] Microscopic reversibility is a sufficient but not necessary condition: see Manousiouthakis, V.I. and Deem, M.W., *J. Chem. Phys.* **94**, 2753 (1999).

[25] Metropolis, N., Rosenbluth, A.W., Rosenbluth, M.N., Teller, A.N. and Teller, E., *J. Chem. Phys.* **21**, 1087 (1953). This was the first paper to be published in the field of molecular simulation.

[26] In this case the molecular dynamics method can also be used: see, e.g., Lotfi, A., Vrabec, J. and Fischer, J., *Mol. Phys.* **76**, 1319 (1992).

[27] The insertion formula takes a different form in the isobaric – isothermal ensemble: see ref. 11(c), pp. 176-7.

[28] (a) Panagiotopoulos, A.Z., *Mol. Phys.* **61**, 813 (1987). (b) Panagiotopoulos, A.Z., Quirke, N., Stapleton, M., and Tildesley, D.J., *Mol. Phys.* **63**, 527 (1988).

[29] Smit, B. and Frenkel, D., *Mol. Phys.* **68**, 951 (1989).

[30] The quantity $\mathcal{Z}$ is essentially a discretised form of the canonical partition function.

[31] For an early application, see McDonald, I.R. and Singer, K., *Disc. Faraday Soc.* **43**, 40 (1967).

[32] Ferrenberg, A.M. and Swendsen, R.H., *Phys. Rev. Lett.* **63**, 1195 (1989).

[33] See the very clear account given in ref. 11(b), pp. 222–3.

[34] For a discussion of reweighting techniques with particular reference to phase equilibria see Wilding, N.B., *Am. J. Phys.* **69**, 1147 (2001).

Chapter 3

Static Properties of Liquids: Thermodynamics and Structure

Liquids are homogeneous in the bulk but inhomogeneities appear close to the confining walls or other physical boundaries and wherever different phases coexist. Although it might seem natural to develop the theory of uniform fluids first, it turns out to be equally convenient and in many ways more illuminating to treat uniform and non-uniform systems simultaneously from the outset. In the first six sections of this chapter we describe a general approach to the study of inhomogeneous fluids based on the formalism of the grand canonical ensemble.[1] The starting point is a hamiltonian that includes a term representing the interaction of the particles with some spatially varying, external field. The effect of this term is to break the translational symmetry of the system, but results for uniform fluids can be recovered by taking the limit in which the external field vanishes. A key component of the theory is a variational principle for the grand potential, which is a classical version of a principle originally derived for the interacting electron gas.[2] The last three sections provide an introduction to the use of diagrammatic methods in the theory of liquids, with examples chosen to complement the work discussed in earlier parts of the chapter.

3.1 A FLUID IN AN EXTERNAL FIELD

We consider again a system of identical, spherical particles in a volume V. The hamiltonian of the system in the presence of an external potential $\phi(\mathbf{r})$ is that given already by (2.1.1) but repeated here for ease of reference:

$$\mathcal{H}(\mathbf{r}^N, \mathbf{p}^N) = K_N(\mathbf{p}^N) + V_N(\mathbf{r}^N) + \Phi_N(\mathbf{r}^N) \tag{3.1.1}$$

The external field is assumed to couple to the microscopic particle density $\rho(\mathbf{r})$, defined as a sum of δ-functions in the form already introduced implicitly in (2.5.11), i.e.

$$\rho(\mathbf{r}) = \sum_{i=1}^{N} \delta(\mathbf{r} - \mathbf{r}_i) \tag{3.1.2}$$

Theory of Simple Liquids, Fourth Edition. http://dx.doi.org/10.1016/B978-0-12-387032-2.00003-9
61

Thus the total potential energy due to the field is

$$\Phi_N(\mathbf{r}^N) = \sum_{i=1}^{N} \phi(\mathbf{r}_i) = \int \rho(\mathbf{r})\phi(\mathbf{r})d\mathbf{r} \tag{3.1.3}$$

The average density at a point $\mathbf{r}$ is the single-particle density, or *density profile*, $\rho^{(1)}(\mathbf{r})$:

$$\langle \rho(\mathbf{r}) \rangle = \rho^{(1)}(\mathbf{r}) \tag{3.1.4}$$

where the angular brackets denote an average over a grand canonical ensemble. Thus the average value of Φ_N is

$$\langle \Phi_N \rangle = \int \rho^{(1)}(\mathbf{r})\phi(\mathbf{r})d\mathbf{r} \tag{3.1.5}$$

Fluctuations in the local density about its average value are described by a density-density correlation function, $H^{(2)}(\mathbf{r}, \mathbf{r}')$, defined as

$$\begin{aligned} H^{(2)}(\mathbf{r}, \mathbf{r}') &= \langle [\rho(\mathbf{r}) - \langle \rho(\mathbf{r}) \rangle][\rho(\mathbf{r}') - \langle \rho(\mathbf{r}') \rangle] \rangle \\ &= \rho^{(2)}(\mathbf{r}, \mathbf{r}') + \rho^{(1)}(\mathbf{r})\delta(\mathbf{r} - \mathbf{r}') - \rho^{(1)}(\mathbf{r})\rho^{(1)}(\mathbf{r}') \\ &= \rho^{(1)}(\mathbf{r})\rho^{(1)}(\mathbf{r}')h^{(2)}(\mathbf{r}, \mathbf{r}') + \rho^{(1)}(\mathbf{r})\delta(\mathbf{r} - \mathbf{r}') \end{aligned} \tag{3.1.6}$$

where $\rho^{(2)}(\mathbf{r}, \mathbf{r}')$ is given by the analogue of (2.5.13) in the grand canonical ensemble and $h^{(2)}(\mathbf{r}, \mathbf{r}')$ is the pair correlation function (2.6.8). The function $H^{(2)}(\mathbf{r}, \mathbf{r}')$ represents the first in a hierarchy of density correlation functions having the general form

$$H^{(n)}(\mathbf{r}_1, \ldots, \mathbf{r}_n) = \left\langle [\rho(\mathbf{r}_1) - \rho^{(1)}(\mathbf{r}_1)] \cdots [\rho(\mathbf{r}_n) - \rho^{(1)}(\mathbf{r}_n)] \right\rangle \tag{3.1.7}$$

for $n \geq 2$. Each function $H^{(n)}$ is a linear combination of all particle densities up to and including $\rho^{(n)}$.

Inclusion of the external field term in the hamiltonian requires some modification of earlier definitions. As before, the grand partition function is related to the grand potential by $\Xi = \exp(-\beta \Omega)$, but now has the form

$$\Xi = \sum_{N=0}^{\infty} \frac{1}{N!} \int \exp(-\beta V_N) \left(\prod_{i=1}^{N} z \exp[-\beta\phi(\mathbf{r}_i)] \right) d\mathbf{r}^N \tag{3.1.8}$$

and the definition of the particle densities in (2.6.1) is replaced by

$$\rho^{(n)}(\mathbf{r}^n) = \frac{1}{\Xi} \sum_{N=n}^{\infty} \frac{1}{(N-n)!} \int \exp(-\beta V_N) \left(\prod_{i=1}^{N} z \exp[-\beta\phi(\mathbf{r}_i)] \right) d\mathbf{r}^{(N-n)} \tag{3.1.9}$$

Equation (3.1.8) may be recast as

$$\Xi = \sum_{N=0}^{\infty} \frac{1}{N!} \int \cdots \int \exp(-\beta V_N) \left(\prod_{i=1}^{N} \Lambda^{-3} \exp[\beta \psi(\mathbf{r}_i)] \right) d\mathbf{r}_1 \cdots d\mathbf{r}_N$$

(3.1.10)

where

$$\psi(\mathbf{r}) = \mu - \phi(\mathbf{r})$$

(3.1.11)

The quantity $\psi(\mathbf{r})$ is called the *intrinsic chemical potential*. It is the contribution to μ that is not explicitly dependent on $\phi(\mathbf{r})$.

The intrinsic chemical potential arises naturally in a thermodynamic description of the system. We suppose that the definition of $\phi(\mathbf{r})$ includes the confining potential, i.e. the interaction between the particles and the containing walls.[3] The usual thermodynamic variable V may then be replaced by $\phi(\mathbf{r})$, the volume accessible to the particles being that region of space in which $\phi(\mathbf{r})$ is finite. The change in U resulting from an infinitesimal change in equilibrium state is now

$$\delta U = T \, \delta S + \int \rho^{(1)}(\mathbf{r}) \delta \phi(\mathbf{r}) d\mathbf{r} + \mu \, \delta N$$

(3.1.12)

(cf. (2.3.5)), where the integral extends over all space rather than over a large but finite volume. The definition of the Helmholtz free energy remains $F = U - TS$ and the change in F in an infinitesimal process is therefore

$$\delta F = -S \, \delta T + \int \rho^{(1)}(\mathbf{r}) \delta \phi(\mathbf{r}) d\mathbf{r} + \mu \, \delta N$$

(3.1.13)

By analogy with (3.1.11), we can also define an *intrinsic free energy*, $\mathcal{F}$, as

$$\mathcal{F} = F - \int \rho^{(1)}(\mathbf{r}) \phi(\mathbf{r}) d\mathbf{r}$$

(3.1.14)

with

$$\delta \mathcal{F} = -S \, \delta T - \int \delta \rho^{(1)}(\mathbf{r}) \phi(\mathbf{r}) d\mathbf{r} + \mu \, \delta N$$

$$= -S \, \delta T + \int \delta \rho^{(1)}(\mathbf{r}) \psi(\mathbf{r}) d\mathbf{r}$$

(3.1.15)

Thus $\psi(\mathbf{r})$ appears as the field variable conjugate to $\rho^{(1)}(\mathbf{r})$. Finally, the grand potential $\Omega = F - N\mu$, when expressed in terms of $\mathcal{F}$, is

$$\Omega = \mathcal{F} + \int \rho^{(1)}(\mathbf{r}) \phi(\mathbf{r}) d\mathbf{r} - N\mu$$

(3.1.16)

with a differential given by

$$\delta \Omega = -S \, \delta T + \int \rho^{(1)}(\mathbf{r}) \delta \phi(\mathbf{r}) - N \, \delta \mu$$

$$= -S \, \delta T - \int \rho^{(1)}(\mathbf{r}) \delta \psi(\mathbf{r}) d\mathbf{r}$$

(3.1.17)

We see from (3.1.15) and (3.1.17) that it is natural to take $\mathcal{F}$ and Ω as functions of T and *functionals*, respectively, of $\rho^{(1)}$ and ψ; the functional relationships are expressed by use of the notation $\mathcal{F}[\rho^{(1)}]$ and $\Omega[\psi]$. Hence the change in $\mathcal{F}$, say, created by a change in $\rho^{(1)}(\mathbf{r})$ is determined by the *functional derivative* of $\mathcal{F}$ with respect to $\rho^{(1)}$. Some familiarity is therefore required with the rules of functional differentiation, a summary of which is given in the section that follows.

The intrinsic free energy can also be written as an ensemble average. The definition (2.4.5) of the grand canonical probability density $f_0(\mathbf{r}^N, \mathbf{p}^N; N)$ shows that in the presence of an external field

$$\ln f_0 = \beta\Omega - \beta K_N - \beta V_N - \beta\Phi_N + N\beta\mu \tag{3.1.18}$$

Thus

$$\langle K_N + V_N + k_{\mathrm{B}}T \ln f_0 \rangle = \Omega + \int \rho^{(1)}(\mathbf{r})\psi(\mathbf{r})\mathrm{d}\mathbf{r} = \mathcal{F} \tag{3.1.19}$$

If there are no correlations between particles, the intrinsic chemical potential at a point $\mathbf{r}$ is given by the usual expression (2.3.17) for the chemical potential of a system of non-interacting particles, but with the overall number density ρ replaced by $\rho^{(1)}(\mathbf{r})$. Thus the chemical potential of an inhomogeneous, ideal gas is

$$\mu^{\mathrm{id}} = k_{\mathrm{B}}T \ln[\Lambda^3 \rho^{(1)}(\mathbf{r})] + \phi(\mathbf{r}) \tag{3.1.20}$$

where the first term on the right-hand side is the intrinsic part. Equation (3.1.20) can be rearranged to give the well-known *barometric law*:

$$\rho^{(1)}(\mathbf{r}) = z^{\mathrm{id}} \exp[-\beta\phi(\mathbf{r})] \tag{3.1.21}$$

where the activity $z^{\mathrm{id}} = \Lambda^{-3} \exp(\beta\mu^{\mathrm{id}})$ is equal to the number density of the uniform gas at the same chemical potential. The intrinsic free energy of an ideal gas also has a purely 'local' form, given by an integral over $\mathbf{r}$ of the free energy per unit volume of a non-interacting system of density $\rho^{(1)}(\mathbf{r})$:

$$\mathcal{F}^{\mathrm{id}} = k_{\mathrm{B}}T \int \rho^{(1)}(\mathbf{r}) \left(\ln[\Lambda^3 \rho^{(1)}(\mathbf{r})] - 1 \right) \mathrm{d}\mathbf{r} \tag{3.1.22}$$

This expression reduces to (2.3.16) in the uniform case.

3.2   FUNCTIONALS AND FUNCTIONAL DIFFERENTIATION

A functional is a natural extension of the familiar mathematical concept of a function. The meaning of a function is that of a mapping from points in n-space to a real or complex number, n being the number of variables on which the function depends. A functional, by contrast, depends on all values of a function $u(x)$, say, in a range $a \leq x \leq b$. It can therefore be interpreted as a mapping from ∞-space to a real or complex number, the points in ∞-space being the values of $u(x)$ at the infinite number of points in the relevant range of the variable x. Functions of several variables and functionals are therefore conveniently treated as discrete and continuous versions of the same mathematical concept, making it possible to construct the rules of functional differentiation by analogy with those of elementary calculus. As usual, a sum in the discrete case is replaced by an integral in the limit in which the distribution of variables becomes continuous.

If f is a function of the n variables $\mathbf{z} \equiv z_1, \ldots, z_N$ the change in f due to an infinitesimal change in $\mathbf{z}$ is

$$\mathrm{d}f = f(\mathbf{z} + \mathrm{d}\mathbf{z}) - f(\mathbf{z}) = \sum_{i=1}^{n} A_i(\mathbf{z})\mathrm{d}z_i \qquad (3.2.1)$$

where

$$A_i(\mathbf{z}) \equiv \frac{\partial f}{\partial z_i} \qquad (3.2.2)$$

Similarly, if F is a functional of $u(x)$, then

$$\delta F = F[u + \delta u] - F[u] = \int_a^b A[u; x]\delta u(x)\mathrm{d}x \qquad (3.2.3)$$

and the functional derivative

$$A[u; x] \equiv \frac{\delta F}{\delta u(x)} \qquad (3.2.4)$$

is a functional of u and a function of x. The functional derivative determines the change in F resulting from a change in u at a particular value of x; to calculate the change in F due to a variation in $u(x)$ throughout the range of x it is necessary to integrate over x, as in (3.2.3).

The rules of functional differentiation are most easily grasped by considering some specific examples. If f is a linear function of n variables we know that

$$f(\mathbf{z}) = \sum_{i=1}^{n} a_i z_i, \quad \mathrm{d}f = \sum_{i=1}^{n} a_i\,\mathrm{d}z_i \qquad (3.2.5)$$

and

$$\frac{\partial f}{\partial z_i} = a_i \qquad (3.2.6)$$

The analogue of (3.2.5) for a linear functional is

$$F[u] = \int a(x)u(x)\mathrm{d}x, \quad \delta F = \int a(x)\delta u(x)\mathrm{d}x \qquad (3.2.7)$$

and comparison with (3.2.3) shows that

$$\frac{\delta F}{\delta u(x)} = a(x) \tag{3.2.8}$$

A more general example of the same type is when

$$F = \int \cdots \int a(x_1, \ldots, x_N) u(x_1) u(x_2) \cdots u(x_N) dx_1 \cdots dx_N \tag{3.2.9}$$

where the function $a(x_1, \ldots, x_N)$ is symmetric with respect to permutation of the labels $1, \ldots, N$. Then

$$\delta F = \int \cdots \int a(x_1, \ldots, x_N) \delta u(x_1) u(x_2) \cdots u(x_N) dx_1 \cdots dx_N$$
$$+ (N - 1) \text{ other terms} \tag{3.2.10}$$

The N terms on the right-hand side are all equivalent, so the change in F is N times the value of any one term. Thus

$$\frac{\delta F}{\delta u(x_1)} = N \int \cdots \int a(x_1, \ldots, x_N) u(x_2) \cdots u(x_N) dx_2 \cdots dx_N \tag{3.2.11}$$

As a slightly more complicated example, consider the non-linear functional

$$F[u] = \int u(x) \ln u(x) dx \tag{3.2.12}$$

for which

$$\delta F = \int [\delta u(x) \ln u(x) + u(x) \delta \ln u(x)] dx$$
$$= \int [\ln u(x) + 1] \delta u(x) dx \tag{3.2.13}$$

and hence

$$\frac{\delta F}{\delta u(x)} = \ln u(x) + 1 \tag{3.2.14}$$

This example shows how functional derivatives can be evaluated with the help of rules appropriate to ordinary differentiation.

An important special case is when

$$F[u] = u(x') = \int \delta(x - x') u(x) dx \tag{3.2.15}$$

Then

$$\delta F = \int \delta(x - x') \delta u(x) dx = \delta u(x') \tag{3.2.16}$$

and

$$\frac{\delta u(x')}{\delta u(x)} = \delta(x - x') \tag{3.2.17}$$

When u is a function of two variables the functional derivative is defined through the relation

$$\delta F = \iint \frac{\delta F}{\delta u(x_1, x_2)} \delta u(x_1, x_2) dx_1\, dx_2 \tag{3.2.18}$$

In applications in statistical mechanics symmetry often leads to a simplification similar to that seen in the example (3.2.9). Consider the functional defined as

$$F[u] = \iiint a(x_1, x_2, x_3) u(x_1, x_2) u(x_2, x_3) u(x_3, x_1) dx_1\, dx_2\, dx_3 \tag{3.2.19}$$

where $a(x_1, x_2, x_3)$ is symmetrical with respect to permutation of the labels 1, 2 and 3. The change in F due to an infinitesimal change in the function u is now

$$\delta F = \iiint a(x_1, x_2, x_3) \delta u(x_1, x_2) u(x_2, x_3) u(x_3, x_1) dx_1\, dx_2\, dx_3$$
$$+ \text{ two equivalent terms} \tag{3.2.20}$$

Thus

$$\frac{\delta F}{\delta u(x_1, x_2)} = 3 \int a(x_1, x_2, x_3) u(x_2, x_3) u(x_3, x_1) dx_3 \tag{3.2.21}$$

Higher-order derivatives are defined in a manner similar to (3.2.3). In particular, the second derivative is defined through the relation

$$\delta A[u; x] = \int \frac{\delta A[u; x]}{\delta u(x')} \delta u(x') dx' \tag{3.2.22}$$

The second derivative of the functional (3.2.9), for example, is

$$\frac{\delta^2 F}{\delta u(x_1) \delta u(x_2)} = N(N-1) \int \cdots \int a(x_1, \ldots, x_N) u(x_3) \cdots u(x_N) dx_3 \cdots dx_N \tag{3.2.23}$$

and is a functional of u and a function of both x_1 and x_2. If the derivatives exist, a functional $F[u]$ can be expanded in a Taylor series around a function u_0:

$$F[u] = F[u_0] + \int \frac{\delta F}{\delta u(x)}\bigg|_{u=u_0} [u(x) - u_0(x)] dx$$
$$+ \frac{1}{2!} \iint \frac{\delta^2 F}{\delta u(x) \delta u(x')}\bigg|_{u=u_0} [u(x) - u_0(x)]$$
$$\times [u(x') - u_0(x')] dx\, dx' + \cdots \tag{3.2.24}$$

Finally, the equivalent of the chain rule of ordinary differentiation is

$$\frac{\delta F}{\delta u(x)} = \int \frac{\delta F}{\delta v(x')} \frac{\delta v(x')}{\delta u(x)} dx' \tag{3.2.25}$$

3.3 FUNCTIONAL DERIVATIVES OF THE GRAND POTENTIAL

The methods of the previous section can be used to derive some important results involving derivatives of the grand potential. We saw in Section 3.1 that it is natural to treat the intrinsic free energy as a functional of the single-particle density. The manner in which the functional $\mathcal{F}[\rho^{(1)}]$ varies with $\rho^{(1)}$ is described by (3.1.15) and from that result, given the definition of a functional derivative, it follows immediately that

$$\frac{\delta \mathcal{F}}{\delta \rho^{(1)}(\mathbf{r})} = \psi(\mathbf{r}) \tag{3.3.1}$$

where the derivative is taken at constant T. The intrinsic free energy can be divided into ideal and excess parts in the form

$$\mathcal{F}[\rho^{(1)}] = \mathcal{F}^{\mathrm{id}}[\rho^{(1)}] + \mathcal{F}^{\mathrm{ex}}[\rho^{(1)}] \tag{3.3.2}$$

where the ideal part is given by (3.1.22). Use of example (3.2.14) confirms that the functional derivative of $\mathcal{F}^{\mathrm{id}}$ is

$$\frac{\delta \mathcal{F}^{\mathrm{id}}}{\delta \rho^{(1)}(\mathbf{r})} = k_{\mathrm{B}} T \ln[\Lambda^3 \rho^{(1)}(\mathbf{r})] \tag{3.3.3}$$

in agreement with (3.1.20). In the same way it follows from (3.1.17) that the functional derivative of $\Omega[\psi]$ with respect to ψ is

$$\frac{\delta \Omega}{\delta \psi(\mathbf{r})} = -\rho^{(1)}(\mathbf{r}) \tag{3.3.4}$$

Taken together, (3.3.1) and (3.3.4) show that the functionals $\Omega[\psi]$ and $\mathcal{F}[\rho^{(1)}]$ are related by a generalised Legendre transformation,[4] i.e.

$$\Omega[\psi] - \int \psi(\mathbf{r}) \frac{\delta \Omega}{\delta \psi(\mathbf{r})} \mathrm{d}\mathbf{r} \to \Omega[\psi] + \int \psi(\mathbf{r}) \rho^{(1)}(\mathbf{r}) \mathrm{d}\mathbf{r} = \mathcal{F}[\rho^{(1)}] \tag{3.3.5}$$

In the limit $\phi \to 0$, ψ and $\rho^{(1)}$ can be replaced by μ and $\langle N \rangle / V$, respectively, and (3.3.1) and (3.3.4) reduce to standard thermodynamic results, $\partial F / \partial N = \mu$ and $\partial \Omega / \partial \mu = -N$.

The relationship that exists between Ω and Ξ means that it must also be possible to obtain (3.3.4) by differentiation of $\ln \Xi$. We already know the outcome of this calculation, but the exercise is nonetheless a useful one, since it points the way towards the calculation of higher-order derivatives. In carrying out the differentiation it proves helpful to introduce a *local activity*, z^*, defined as

$$z^*(\mathbf{r}) = \frac{\exp[\beta \psi(\mathbf{r})]}{\Lambda^3} = z \exp[-\beta \phi(\mathbf{r})] \tag{3.3.6}$$

If we also adopt a simplified notation in which a position vector $\mathbf{r}_i$ is denoted by i, the grand partition function (3.1.10) can be rewritten in the form

$$\varXi = \sum_{N=0}^{\infty} \frac{1}{N!} \int \cdots \int \exp\left(-\beta V_N\right) \prod_{i=1}^{N} z^*(i)\, \mathrm{d}1 \cdots \mathrm{d}N \qquad (3.3.7)$$

The derivative we require is

$$\frac{\delta\Omega}{\delta\psi(1)} = -k_{\mathrm{B}}T\frac{\delta\ln\varXi}{\delta\psi(1)} = -\frac{z^*(1)}{\varXi}\frac{\delta\varXi}{\delta z^*(1)} \qquad (3.3.8)$$

The term for $N = 0$ in (3.3.7) vanishes on differentiation. Higher-order terms are of the general form considered in example (3.2.9) and differentiation of each term therefore yields a factor N. Thus

$$\frac{\delta\varXi}{\delta z^*(1)} = \sum_{N=1}^{\infty} \frac{1}{(N-1)!} \int \cdots \int \exp\left(-\beta V_N\right) \prod_{i=2}^{N} z^*(i)\, \mathrm{d}2 \cdots \mathrm{d}N \qquad (3.3.9)$$

and combination of (3.3.8) and (3.3.9) with the definition of the particle densities in (3.1.9) leads back to (3.3.4). Further differentiation of $\varXi$ shows that

$$\rho^{(n)}(1,\ldots,n) = \frac{z^*(1)\cdots z^*(n)}{\varXi}\frac{\delta^n\varXi}{\delta z^*(1)\cdots\delta z^*(n)} \qquad (3.3.10)$$

The grand partition function is said to be the *generating functional* for the particle densities.

Calculation of the second derivative of Ω with respect to ψ is only slightly more complicated. The quantity to be determined is now

$$\frac{\delta^2\Omega}{\delta\psi(1)\delta\psi(2)} = -\beta z^*(2)\frac{\delta}{\delta z^*(2)}\left(\frac{1}{\varXi}z^*(1)\frac{\delta\varXi}{\delta z^*(1)}\right) \qquad (3.3.11)$$

Differentiation of successive factors in the product in brackets gives rise, respectively, to a term in $\rho^{(1)}(2)$, a term in $\delta(1,2)$ (as in example (3.2.17)) and a term in $\rho^{(2)}(1,2)$ (from (3.3.10)). On combining these results we find that

$$\frac{\delta^2\Omega}{\delta\psi(1)\delta\psi(2)} = \beta[\rho^{(1)}(1)\rho^{(1)}(2) - \rho^{(1)}(1)\delta(1,2) - \rho^{(2)}(1,2)]$$

$$= -\beta H^{(2)}(1,2) \qquad (3.3.12)$$

where $H^{(2)}(1,2)$ is the density-density correlation function defined by (3.1.6). The process of differentiation can again be extended; although the algebra becomes increasingly tedious, the general result has a simple form:

$$\frac{\delta^n\beta\Omega}{\delta\beta\psi(1)\cdots\delta\beta\psi(n)} = -H^{(n)}(1,\ldots,n), \quad n \geq 2 \qquad (3.3.13)$$

The grand potential is therefore the generating functional for the n-fold density correlation functions.

3.4   DENSITY FUNCTIONAL THEORY

The grand potential has temperature and intrinsic chemical potential as its natural variables. However, it turns out to be more profitable to treat $\rho^{(1)}$ rather than ψ as the fundamental field variable. The definition (3.1.9) shows that $\rho^{(1)}$ is a functional of ϕ. What is not obvious is the fact that for a given interparticle potential energy function V_N and fixed values of T and μ, there is only one external potential that gives rise to a specific density profile. This result, the proof of which is given in Appendix B, has far-reaching implications. The grand canonical probability density f_0 defined by (2.4.5) is a functional of $\phi(\mathbf{r})$. Hence any quantity which, for given V_N, T and μ, is wholly determined by f_0 is necessarily a functional of $\rho^{(1)}$ and its functional dependence on $\rho^{(1)}$ is independent of the external potential. In particular, because the intrinsic free energy is the ensemble average of $(K_N + V_N + k_B T \ln f_0)$ (see (3.1.19)), it follows that $\mathcal{F}[\rho^{(1)}]$ is a unique functional of $\rho^{(1)}$.

Let $n(\mathbf{r})$ be some average of the microscopic density, not necessarily the equilibrium one, and let $\Omega_\phi[n]$ be a functional of n, defined for fixed external potential by

$$\Omega_\phi[n] = \mathcal{F}[n] + \int n(\mathbf{r})\phi(\mathbf{r})d\mathbf{r} - \mu \int n(\mathbf{r})d\mathbf{r} \qquad (3.4.1)$$

At equilibrium, $n(\mathbf{r}) = \rho^{(1)}(\mathbf{r})$, and Ω_ϕ reduces to the grand potential, i.e.

$$\Omega_\phi[\rho^{(1)}] = \Omega \qquad (3.4.2)$$

while differentiation of (3.4.1) with respect to $n(\mathbf{r})$ gives

$$\left.\frac{\delta\Omega_\phi}{\delta n(\mathbf{r})}\right|_{n=\rho^{(1)}} = \left.\frac{\delta\mathcal{F}[n]}{\delta n(\mathbf{r})}\right|_{n=\rho^{(1)}} - \mu + \phi(\mathbf{r})$$
$$= 0 \qquad (3.4.3)$$

where the right-hand side vanishes by virtue of (3.3.1). Thus Ω_ϕ is stationary with respect to variations in $n(\mathbf{r})$ around the equilibrium density. It is also straightforward to show that

$$\Omega_\phi[n] \geq \Omega \qquad (3.4.4)$$

where the equality applies only when $n(\mathbf{r}) = \rho^{(1)}(\mathbf{r})$. In other words, the functional Ω_ϕ has a lower bound equal to the exact grand potential of the system. A proof of (3.4.4) is also given in Appendix B.

Equations (3.4.3) and (3.4.4) provide the ingredients for a variational calculation of the density profile and grand potential of an inhomogeneous fluid. What is required in order to make the theory tractable is a parameterisation of the free energy functional $\mathcal{F}[n]$ in terms of $n(\mathbf{r})$. Since the ideal part is known exactly, the difficulty lies in finding a suitable form for $\mathcal{F}^{ex}[n]$. The best estimates of $\rho^{(1)}$ and Ω are then obtained by minimising the functional $\Omega_\phi[n]$ with respect to variations in $n(\mathbf{r})$. Minimisation of such a functional is the central problem in the calculus of variations and normally requires the solution to a differential equation called the Euler or Euler–Lagrange equation. Computational schemes of this type are grouped together under the title *density functional theory*. The theory has found application to a very wide range of problems, some of which are discussed in later chapters. As in any variational calculation, the success achieved depends on the skill with which the trial functional is constructed. Because $\mathcal{F}$ is a unique functional of $\rho^{(1)}$, a good approximation would be one that was suitable for widely differing choices of external potential, but in practice most approximations are designed for use in specific physical situations.

If V_N is a sum of pair potentials, it is possible to derive an exact expression for $\mathcal{F}^{ex}$ in terms of the pair density in a form that lends itself readily to approximation. The grand partition function can be written as

$$\Xi = \sum_{N=0}^{\infty} \frac{1}{N!} \int \cdots \int \prod_{i<j}^{N} e(i,j) \prod_{i=1}^{N} z^*(i) \, \mathrm{d}1 \cdots \mathrm{d}N \qquad (3.4.5)$$

where $e(i,j) \equiv \exp[(-\beta v(i,j)]$. Then the functional derivative of Ω with respect to v at constant T and ψ is

$$
\frac{\delta \Omega}{\delta v(1,2)} = \frac{\delta \ln \Xi}{\delta \ln e(1,2)} = \frac{e(1,2)}{\Xi} \frac{\delta \Xi}{\delta e(1,2)}
$$

$$
= \frac{1}{\Xi} \sum_{N=2}^{\infty} \frac{N(N-1)}{2N!} \int \cdots \int \prod_{i<j}^{N} e(i,j) \prod_{i=1}^{N} z^*(i) \, \mathrm{d}3 \cdots \mathrm{d}N \quad (3.4.6)
$$

where the factor $\frac{1}{2}N(N-1)$ is the number of equivalent terms resulting from the differentiation (cf. (3.2.20)). Comparison with the definition of $\rho^{(n)}$ in (3.1.9) shows that

$$\rho^{(2)}(\mathbf{r}, \mathbf{r}') = 2\frac{\delta \Omega}{\delta v(\mathbf{r}, \mathbf{r}')} \qquad (3.4.7)$$

and hence that

$$\rho^{(2)}(\mathbf{r}, \mathbf{r}') = 2\frac{\delta \mathcal{F}^{ex}[\rho^{(1)}]}{\delta v(\mathbf{r}, \mathbf{r}')} \qquad (3.4.8)$$

We now suppose that the pair potential can be expressed as the sum of a 'reference' part, $v_0(\mathbf{r}, \mathbf{r}')$, and a 'perturbation', $w(\mathbf{r}, \mathbf{r}')$, and define a family of intermediate potentials by

$$v_\lambda(\mathbf{r}, \mathbf{r}') = v_0(\mathbf{r}, \mathbf{r}') + \lambda w(\mathbf{r}, \mathbf{r}'), \quad 0 \le \lambda \le 1 \qquad (3.4.9)$$

The reference potential could, for example, be the hard-sphere interaction and the perturbation could be a weak, attractive tail, while the increase in λ from 0 to 1 would correspond to a gradual 'switching on' of the perturbation. It follows from integration of (3.4.8) at constant single-particle density that the free energy functional for the system of interest, characterised by the full potential $v(\mathbf{r}, \mathbf{r}')$, is related to that of the reference system by

$$
\begin{aligned}
\mathcal{F}^{\text{ex}}[\rho^{(1)}] &= \mathcal{F}_0^{\text{ex}}[\rho^{(1)}] + \frac{1}{2} \int_0^1 d\lambda \iint \rho^{(2)}[\mathbf{r}, \mathbf{r}'; \lambda] w(\mathbf{r}, \mathbf{r}') d\mathbf{r} \, d\mathbf{r}' \\
&= \mathcal{F}_0^{\text{ex}}[\rho^{(1)}] + \frac{1}{2} \iint \rho^{(1)}(\mathbf{r}) \rho^{(1)}(\mathbf{r}') w(\mathbf{r}, \mathbf{r}') d\mathbf{r} \, d\mathbf{r}' \\
&\quad + \mathcal{F}_{\text{corr}}[\rho^{(1)}]
\end{aligned}
\tag{3.4.10}
$$

where $\rho^{(2)}(\mathbf{r}, \mathbf{r}'; \lambda)$ is the pair density for the system with potential v_λ and

$$
\mathcal{F}_{\text{corr}}[\rho^{(1)}] = \frac{1}{2} \int_0^1 d\lambda \iint \rho^{(1)}(\mathbf{r}) \rho^{(1)}(\mathbf{r}') h^{(2)}(\mathbf{r}, \mathbf{r}'; \lambda) w(\mathbf{r}, \mathbf{r}') d\mathbf{r} d\mathbf{r}'
\tag{3.4.11}
$$

is the contribution to $\mathcal{F}^{\text{ex}}$ due to correlations induced by the perturbation. Equation (3.4.10) provides a basis for the perturbation theories of uniform fluids discussed in Chapter 5.

3.5   DIRECT CORRELATION FUNCTIONS

We saw in Section 3.3 that the grand potential is a generating functional for the density correlation functions $H^{(n)}(\mathbf{r}^n)$. In a similar way, the excess part of the free energy functional acts as a generating functional for a parallel hierarchy of *direct correlation functions*, $c^{(n)}(\mathbf{r}^n)$. The single-particle function is defined as the first functional derivative of $\mathcal{F}^{\text{ex}}$ with respect to $\rho^{(1)}$:

$$
c^{(1)}(\mathbf{r}) = -\beta \frac{\delta \mathcal{F}^{\text{ex}}[\rho^{(1)}]}{\delta \rho^{(1)}(\mathbf{r})}
\tag{3.5.1}
$$

The pair function is defined as the functional derivative of $c^{(1)}$:

$$
c^{(2)}(\mathbf{r}, \mathbf{r}') = \frac{\delta c^{(1)}(\mathbf{r})}{\delta \rho^{(1)}(\mathbf{r}')} = -\beta \frac{\delta^2 \mathcal{F}^{\text{ex}}[\rho^{(1)}]}{\delta \rho^{(1)}(\mathbf{r}) \delta \rho^{(1)}(\mathbf{r}')}
\tag{3.5.2}
$$

and similarly for higher-order functions: $c^{(n+1)}(\mathbf{r}^{n+1})$ is the derivative of $c^{(n)}(\mathbf{r}^n)$. It follows from (3.3.1), (3.3.3) and (3.5.1) that

$$
\beta \psi(\mathbf{r}) = \beta \frac{\delta \mathcal{F}[\rho^{(1)}]}{\delta \rho^{(1)}(\mathbf{r})} = \ln[\Lambda^3 \rho^{(1)}(\mathbf{r})] - c^{(1)}(\mathbf{r})
\tag{3.5.3}
$$

or, given that $\psi = \mu - \phi$ and $z = \exp(\beta\mu)/\Lambda^3$:

$$\rho^{(1)}(\mathbf{r}) = z \exp[-\beta\phi(\mathbf{r}) + c^{(1)}(\mathbf{r})] \qquad (3.5.4)$$

Comparison with the corresponding ideal gas result in (3.1.21) (the barometric law) shows that the effects of particle interactions on the density profile are wholly contained in the function $c^{(1)}(\mathbf{r})$. It is also clear from (3.5.3) that the quantity $-k_B T c^{(1)}(\mathbf{r})$, which acts in (3.5.4) as a self-consistent addition to the external potential, is the excess part of the intrinsic chemical potential. By appropriately adapting the argument of Section 2.4 it can be shown that $-k_B T c^{(1)}(\mathbf{r})$ is given by an expression identical to that on the right-hand side of (2.4.31), but where ϵ is now the energy of a test particle placed at $\mathbf{r}$ that interacts with particles of the system but not with the external field.[5] If there is no external field, (3.5.4) can be rearranged to give

$$-k_B T c^{(1)} = \mu - k_B T \ln \Lambda^3 \rho = \mu^{\text{ex}} \qquad (3.5.5)$$

To obtain a useful expression for $c^{(2)}(\mathbf{r}, \mathbf{r}')$ we must return to some earlier results. Equations (3.3.4) and (3.3.12) show that, apart from a constant factor, the density-density correlation function is the functional derivative of $\rho^{(1)}$ with respect to ψ:

$$H(\mathbf{r}, \mathbf{r}') = k_B T \frac{\delta \rho^{(1)}(\mathbf{r})}{\delta \psi(\mathbf{r}')} \qquad (3.5.6)$$

where, for notational simplicity, we have temporarily omitted the superscript (2). It therefore follows from (3.2.17) and (3.2.25) that the functional inverse of H, defined through the relation

$$\int H(\mathbf{r}, \mathbf{r}'') H^{-1}(\mathbf{r}'', \mathbf{r}') d\mathbf{r}'' = \delta(\mathbf{r} - \mathbf{r}') \qquad (3.5.7)$$

is

$$H^{-1}(\mathbf{r}, \mathbf{r}') = \beta \frac{\delta \psi(\mathbf{r})}{\delta \rho^{(1)}(\mathbf{r}')} \qquad (3.5.8)$$

Functional differentiation of the expression for ψ in (3.5.3) gives

$$\beta \frac{\delta \psi(\mathbf{r})}{\delta \rho^{(1)}(\mathbf{r}')} = \frac{1}{\rho^{(1)}(\mathbf{r})} \delta(\mathbf{r} - \mathbf{r}') - c^{(2)}(\mathbf{r}, \mathbf{r}') = H^{-1}(\mathbf{r}, \mathbf{r}') \qquad (3.5.9)$$

If we now substitute for H and H^{-1} in (3.5.7), integrate over $\mathbf{r}''$ and introduce the pair correlation function defined by (3.1.6), we obtain the *Ornstein–Zernike relation*:

$$h^{(2)}(\mathbf{r}, \mathbf{r}') = c^{(2)}(\mathbf{r}, \mathbf{r}') + \int c^{(2)}(\mathbf{r}, \mathbf{r}'') \rho^{(1)}(\mathbf{r}'') h^{(2)}(\mathbf{r}'', \mathbf{r}') d\mathbf{r}'' \qquad (3.5.10)$$

This relation is often taken as the definition of $c^{(2)}$, but the definition as a derivative of the intrinsic free energy gives the function greater physical meaning. It can be solved recursively to give

$$h^{(2)}(1,2) = c^{(2)}(1,2) + \int c^{(2)}(1,3)\rho^{(1)}(3)c^{(2)}(3,2)\mathrm{d}3$$
$$+ \iint c^{(2)}(1,3)\rho^{(1)}(3)c^{(2)}(3,4)\rho^{(1)}(4)c^{(2)}(4,2)\mathrm{d}3\,\mathrm{d}4$$
$$+ \cdots \tag{3.5.11}$$

Equation (3.5.11) has an obvious physical interpretation: the 'total' correlation between particles 1 and 2, represented by $h^{(2)}(1,2)$, is due in part to the 'direct' correlation between 1 and 2 but also to the 'indirect' correlation propagated via increasingly large numbers of intermediate particles. With this physical picture in mind it is plausible to suppose that the range of $c^{(2)}(1,2)$ is comparable with that of the pair potential $v(1,2)$ and to ascribe the fact that $h^{(2)}(1,2)$ is generally much longer ranged than $v(1,2)$ to the effects of indirect correlation. The differences between the two functions for the Lennard-Jones fluid at high density and low temperature are illustrated in Figure 3.1; $c(r)$ is not only shorter ranged than $h(r)$ but also simpler in structure.

If the fluid is uniform and isotropic, the Ornstein–Zernike relation becomes

$$h(r) = c(r) + \rho \int c(|\mathbf{r} - \mathbf{r}'|)h(r')\mathrm{d}\mathbf{r}' \tag{3.5.12}$$

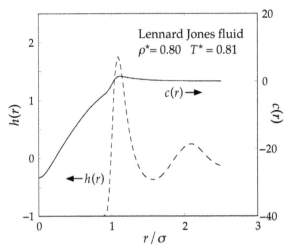

FIGURE 3.1 The pair functions $h(r)$ and $c(r)$ obtained by Monte Carlo calculations for the Lennard-Jones potential at a high density and low temperature. After Llano–Restrepo and Chapman.[6]

where the term representing the indirect correlation now appears as a convolution integral. We have again followed the convention adopted earlier for $g(r)$ by omitting the superscripts (2) when the system is homogeneous and shall continue to do so in circumstances where there is no risk of ambiguity. On taking the Fourier transform of both sides of (3.5.12) we obtain an algebraic relation between $\hat{h}(k)$ and $\hat{c}(k)$:

$$\hat{h}(k) = \frac{\hat{c}(k)}{1 - \rho\hat{c}(k)} \tag{3.5.13}$$

Equation (3.5.13) provides a link with thermodynamics via the compressibility equation (2.6.12). Since $h(r) = g(r) - 1$, it follows from (2.6.12) that the isothermal compressibility can be written in either of the two equivalent forms:

$$\rho k_B T \chi_T = 1 + \rho\hat{h}(0) \tag{3.5.14}$$

or

$$\frac{1}{\rho k_B T \chi_T} = 1 - \rho\hat{c}(0) \tag{3.5.15}$$

These results illustrate very clearly the inverse relationship that exists between h and c.

The definitions of $c^{(1)}$ and $c^{(2)}$ in (3.5.1) and (3.5.2) are useful in characterising the nature of an approximate free energy functional. As a simple example, consider the functional derived from the exact result (3.4.10) by discarding the term $\mathcal{F}_{corr}$, which amounts to treating the effects of the perturbation $w(\mathbf{r}, \mathbf{r}')$ in a mean field approximation. Then

$$c^{(1)}(\mathbf{r}) \approx c_0^{(1)}(\mathbf{r}) - \beta \int \rho^{(1)}(\mathbf{r}')w(\mathbf{r}, \mathbf{r}')d\mathbf{r}' \tag{3.5.16}$$

and

$$c^{(2)}(\mathbf{r}, \mathbf{r}') \approx c_0^{(2)}(\mathbf{r}, \mathbf{r}') - \beta w(\mathbf{r}, \mathbf{r}') \tag{3.5.17}$$

where $c_0^{(1)}$ and $c_0^{(2)}$ are the direct correlation functions of the reference system. Substitution of (3.5.16) in (3.5.4) yields an integral equation for $\rho^{(1)}(\mathbf{r})$, which can be solved iteratively if the properties of the reference system are known or if some further approximation is made for $c_0^{(1)}$. Equation (3.5.17) is a well-known approximation in the theory of uniform fluids[7]; for historical reasons it is called the *random phase approximation* or RPA. It is generally accepted that $c^{(2)}(\mathbf{r}, \mathbf{r}')$ behaves asymptotically as $-\beta v(\mathbf{r}, \mathbf{r}')$. The RPA should therefore be exact when $|\mathbf{r} - \mathbf{r}'|$ is sufficiently large; this assumes that the perturbation contains the long-range part of the potential, which is almost invariably the case.

The formally exact expression for the intrinsic free energy given by (3.4.10) was obtained by thermodynamic integration with respect to the interparticle potential. Another exact expression can be derived from the definitions of $c^{(1)}$ and $c^{(2)}$ by integration with respect to the single-particle density. Let $\rho_0^{(1)}(\mathbf{r})$

and $c_0^{(1)}(\mathbf{r})$ be the single-particle density and single-particle direct correlation function, respectively, in a reference state of the system of interest. We choose a linear integration path between the reference state and the final state of density $\rho^{(1)}(\mathbf{r})$ such that

$$\rho^{(1)}(\mathbf{r}; \lambda) = \rho_0^{(1)}(\mathbf{r}) + \lambda \Delta \rho^{(1)}(\mathbf{r}) \tag{3.5.18}$$

where $\Delta \rho^{(1)} = \rho^{(1)} - \rho_0^{(1)}$. Then integration of (3.5.1) gives

$$\mathcal{F}^{\text{ex}}[\rho^{(1)}] = \mathcal{F}_0^{\text{ex}}[\rho_0^{(1)}] - k_B T \int_0^1 d\lambda \int \frac{\partial \rho^{(1)}(\mathbf{r}; \lambda)}{\partial \lambda} c^{(1)}(\mathbf{r}; \lambda) d\mathbf{r}$$

$$= \mathcal{F}_0^{\text{ex}}[\rho_0^{(1)}] - k_B T \int_0^1 d\lambda \int \Delta \rho^{(1)}(\mathbf{r}) c^{(1)}(\mathbf{r}; \lambda) d\mathbf{r} \tag{3.5.19}$$

Similarly, from integration of (3.5.2):

$$c^{(1)}(\mathbf{r}; \lambda) = c_0^{(1)}(\mathbf{r}) + \int_0^\lambda d\lambda' \int \Delta \rho^{(1)}(\mathbf{r}') c^{(2)}(\mathbf{r}, \mathbf{r}'; \lambda') d\mathbf{r}' \tag{3.5.20}$$

and hence, after substitution of (3.5.20) in (3.5.19):

$$\mathcal{F}^{\text{ex}}[\rho^{(1)}] = \mathcal{F}_0^{\text{ex}}[\rho_0^{(1)}] - k_B T \int \Delta \rho^{(1)}(\mathbf{r}) c_0^{(1)}(\mathbf{r}) d\mathbf{r}$$

$$- k_B T \int_0^1 d\lambda \int_0^\lambda d\lambda' \iint \Delta \rho^{(1)}(\mathbf{r}) \Delta \rho^{(1)}(\mathbf{r}') c^{(2)}(\mathbf{r}, \mathbf{r}'; \lambda) d\mathbf{r} \, d\mathbf{r}'$$

$$\tag{3.5.21}$$

The integration path defined by (3.5.18) is chosen for mathematical convenience, but the final result is independent of path, since $\mathcal{F}^{\text{ex}}$ is a unique functional of $\rho^{(1)}$.

Some simplification of (3.5.21) is possible. An integration by parts shows that

$$\int_0^1 d\lambda \int_0^\lambda y(\lambda') d\lambda' = \int_0^1 (1 - \lambda) y(\lambda) d\lambda \tag{3.5.22}$$

for any function $y(\lambda)$. Thus

$$\mathcal{F}^{\text{ex}}[\rho^{(1)}] = \mathcal{F}_0^{\text{ex}}[\rho_0^{(1)}] - k_B T \int \Delta \rho^{(1)}(\mathbf{r}) c_0^{(1)}(r) d\mathbf{r}$$

$$- k_B T \int_0^1 d\lambda (1 - \lambda) \iint \Delta \rho^{(1)}(\mathbf{r}) \Delta \rho^{(1)}(\mathbf{r}') c^{(2)}(\mathbf{r}, \mathbf{r}'; \lambda) d\mathbf{r} \, d\mathbf{r}'$$

$$\tag{3.5.23}$$

In contrast to (3.4.10), use of this result in constructing a trial functional requires an approximation for $c^{(2)}(\mathbf{r}', \mathbf{r}'; \lambda)$ rather than $h^{(2)}(\mathbf{r}', \mathbf{r}'; \lambda)$, while its derivation does not rely on the assumption of pairwise additivity of the particle interactions.

If we assume that the final state is homogeneous and that the initial state is one of zero density, (3.5.23) yields an expression for the excess free energy of a uniform fluid of density ρ:

$$F^{\text{ex}}(\rho) = \rho^2 k_B T \int_0^1 d\lambda (\lambda - 1) \int d\mathbf{r} \int c(|\mathbf{r}' - \mathbf{r}|; \lambda\rho) d(\mathbf{r}' - \mathbf{r}) \quad (3.5.24)$$

or, after integration over $\mathbf{r}$:

$$\frac{\beta F^{\text{ex}}(\rho)}{N} = \rho \int_0^1 d\lambda (\lambda - 1) \int c(r; \lambda\rho) d\mathbf{r} \quad (3.5.25)$$

3.6 THE DENSITY RESPONSE FUNCTION

Let us suppose that a uniform fluid of number density ρ_0 is exposed to a weak, external potential $\delta\phi(\mathbf{r})$. The hamiltonian of the system is

$$\mathcal{H} = \mathcal{H}_0 + \sum_{i=1}^{N} \delta\phi(\mathbf{r}_i) \quad (3.6.1)$$

where $\mathcal{H}_0$ is the hamiltonian of the uniform fluid. The external potential acts as a perturbation on the system and creates an inhomogeneity, measured by the deviation $\delta\rho^{(1)}(\mathbf{r})$ of the single-particle density from its value in the uniform state:

$$\delta\rho^{(1)}(\mathbf{r}) = \rho^{(1)}(\mathbf{r}) - \rho_0 \quad (3.6.2)$$

Because the perturbation is weak, it can be assumed that the response is a linear but non-local function of $\delta\phi(\mathbf{r})$, expressible in terms of a *linear response function* $\chi(\mathbf{r}, \mathbf{r}')$ in the form

$$\delta\rho^{(1)}(\mathbf{r}) = \int \chi(\mathbf{r}, \mathbf{r}') \delta\phi(\mathbf{r}') d\mathbf{r}' \quad (3.6.3)$$

It follows from the definition of a functional derivative that

$$\chi(\mathbf{r}, \mathbf{r}') = \left. \frac{\delta\rho^{(1)}(\mathbf{r})}{\delta\phi(\mathbf{r}')} \right|_{\phi=0} = -\left. \frac{\delta\rho^{(1)}(\mathbf{r})}{\delta\psi(\mathbf{r}')} \right|_{\phi=0} \quad (3.6.4)$$

and hence, from (3.5.6), that

$$\chi(\mathbf{r}, \mathbf{r}') = -\beta H^{(2)}(\mathbf{r}, \mathbf{r}') \quad (3.6.5)$$

where $H^{(2)}(\mathbf{r}, \mathbf{r}')$ is the density-density correlation function of the unperturbed system. Because the unperturbed system is homogeneous, the response function can be written as

$$\chi(|\mathbf{r} - \mathbf{r}'|) = -\beta[\rho_0^2 h(|\mathbf{r} - \mathbf{r}'|) + \rho_0 \delta(|\mathbf{r} - \mathbf{r}'|)] \quad (3.6.6)$$

and the change in density due to the perturbation divides into local and non-local terms:

$$\delta\rho^{(1)}(\mathbf{r}) = -\beta\rho_0\delta\phi(\mathbf{r}) - \beta\rho_0^2 \int h(|\mathbf{r} - \mathbf{r}'|)\delta\phi(\mathbf{r}')d\mathbf{r}' \qquad (3.6.7)$$

This result is called the Yvon equation; it is equivalent to a first-order Taylor expansion of $\rho^{(1)}$ in powers of $\delta\phi$.

We now take the Fourier transform of (3.6.3) and relate the response $\delta\hat{\rho}^{(1)}(\mathbf{k})$ to the Fourier components of the external potential, defined as

$$\delta\hat{\phi}(\mathbf{k}) = \int \exp(-i\mathbf{k} \cdot \mathbf{r})\delta\phi(\mathbf{r})d\mathbf{r} \qquad (3.6.8)$$

The result is

$$\delta\hat{\rho}^{(1)}(\mathbf{k}) = \chi(\mathbf{k})\hat{\phi}(\mathbf{k}) = -\beta\rho_0 S(\mathbf{k})\delta\hat{\phi}(\mathbf{k}) \qquad (3.6.9)$$

where

$$S(\mathbf{k}) = 1 + \rho_0\hat{h}(\mathbf{k}) = \frac{1}{1 - \rho_0\hat{c}(k)} \qquad (3.6.10)$$

is the *static structure factor* of the uniform fluid; the second equality in (3.6.10) follows from (3.5.13). The structure factor appears in (3.6.9) as a generalised response function, akin to the magnetic susceptibility of a spin system. The linear density response to an external field is therefore determined by the density-density correlation function in the absence of the field; this is an example of the *fluctuation–dissipation theorem*. More specifically, $S(\mathbf{k})$ is a measure of the density response of a system, initially in equilibrium, to a weak, external perturbation of wavelength $2\pi/k$. When the probe is a beam of neutrons, $S(\mathbf{k})$ is proportional to the total scattered intensity in a direction determined by the momentum transfer $\hbar\mathbf{k}$ between beam and sample. Use of such a probe provides an experimental means of determining the radial distribution function of a liquid, as in the example shown in Figure 2.3. Equations (3.5.14) and (3.6.10) together show that at long wavelengths $S(\mathbf{k})$ behaves as

$$\lim_{\mathbf{k}\to 0} S(\mathbf{k}) = \rho k_B T \chi_T \qquad (3.6.11)$$

and is therefore a measure of the response in one macroscopic quantity (the number density) to a change in another (the applied pressure). If the system is isotropic, the structure factor is a function only of the wavenumber k.

An example of an experimentally determined structure factor for liquid sodium near the triple point is pictured in Figure 3.2; the dominant feature is a pronounced peak at a wavenumber approximately equal to $2\pi/\Delta r$, where Δr is the spacing of the peaks in $g(r)$. As the figure shows, the experimental structure factor is very well fitted by Monte Carlo results for a purely repulsive potential that varies as r^{-4}. Since the r^{-4} potential is only a crude representation of the effective potential for liquid sodium, the good agreement seen in the figure

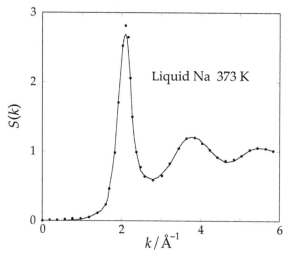

FIGURE 3.2   Structure factor of liquid sodium near the normal melting temperature. The points are experimental X-ray scattering results[8] and the curve is obtained from a Monte Carlo calculation[9] for the r^{-4} potential.

strongly suggests that the structure factor is insensitive to details of the atomic interactions.

The discussion until now has been limited to one-component systems, but the ideas developed in this section and the preceding one can be extended to mixtures without major complications. Consider a system containing N_ν particles of species ν, with $\nu = 1$ to n. If $N = \sum_\nu N_\nu$ is the total number of particles, the number concentration of species ν is $x_\nu = N_\nu/N$. The partial microscopic density $\rho_\nu(\mathbf{r})$ and its average value $\rho_\nu^{(1)}(\mathbf{r})$ (the single-particle density of species ν) are defined in a manner identical to (3.1.2) and (3.1.4), except that the sum on i is limited to particles of species ν. At the pair level, the structure of the fluid is described by $\frac{1}{2}n(n+1)$ partial pair correlation functions $h_{\nu\mu}^{(2)}(\mathbf{r},\mathbf{r}')$ and $\frac{1}{2}n(n+1)$ direct correlation functions $c_{\nu\mu}^{(2)}(\mathbf{r},\mathbf{r}')$. The two sets of functions are linked by a set of coupled equations, representing a generalisation of the Ornstein–Zernike relation (3.5.10), which in the homogeneous case becomes

$$h_{\nu\mu}(r) = c_{\nu\mu}(r) + \rho \sum_\lambda x_\lambda \int c_{\nu\lambda}(|\mathbf{r}-\mathbf{r}'|)h_{\lambda\mu}(r')\mathrm{d}\mathbf{r}' \qquad (3.6.12)$$

The change in single-particle density of species ν induced by a weak, external potential $\delta\phi_\mu(\mathbf{r})$ which couples to the density of species μ is given by a straightforward generalisation of (3.6.7):

$$\delta\rho_\nu^{(1)}(\mathbf{r}) = -x_\nu\delta_{\nu\mu}\beta\rho\,\delta\phi_\mu(\mathbf{r}) - x_\nu x_\mu\beta\rho^2 \int h_{\nu\mu}(|\mathbf{r}-\mathbf{r}'|)\delta\phi_\mu(\mathbf{r}')\mathrm{d}\mathbf{r}' \quad (3.6.13)$$

or, after Fourier transformation:

$$\delta\hat{\rho}_v^{(1)}(\mathbf{k}) = \chi_{v\mu}(\mathbf{k})\delta\hat{\phi}_\mu(\mathbf{k}) = -\beta\rho S_{v\mu}(\mathbf{k})\delta\hat{\phi}_\mu(\mathbf{k}) \tag{3.6.14}$$

where $\chi_{v\mu}(\mathbf{k})$ is a linear response function and

$$S_{v\mu}(\mathbf{k}) = x_v\delta_{v\mu} + x_v x_\mu \rho \hat{h}_{v\mu}(\mathbf{k}) \tag{3.6.15}$$

is a partial structure factor of the uniform fluid. Note that the local contribution to $\delta\rho_v^{(1)}(\mathbf{r})$ in (3.6.13) disappears unless the labels v, μ refer to the same species. Finally, the generalisation to mixtures of the expression for the compressibility given by (3.5.15) is

$$\frac{1}{\rho k_B T \chi_T} = 1 - \rho \sum_v \sum_\mu \hat{c}_{v\mu}(0) \tag{3.6.16}$$

If the partial structure factors are represented as a matrix, $\mathbf{S}(k)$, combination of (3.6.12) and (3.6.15), together with a matrix inversion, shows that the corresponding generalisation of (3.6.11) is

$$\rho k_B T \chi_T = \frac{|\mathbf{S}(0)|}{\sum_v \sum_\mu x_v x_\mu |\mathbf{S}(0)|_{v\mu}} \tag{3.6.17}$$

where $|\mathbf{S}(0)|_{v\mu}$ is the cofactor of $S_{v\mu}(0)$ in the determinant $|\mathbf{S}(0)|$. Equation (3.6.17) is called the Kirkwood–Buff formula.[10]

3.7 DIAGRAMMATIC METHODS

The grand partition function and particle densities are defined as many-dimensional integrals over particle coordinates. Such integrals are conveniently represented by *diagrams* or *graphs*, which in turn can be manipulated by graph theoretical methods. These methods include simple prescriptions for the evaluation of functional derivatives of the type encountered in earlier sections of this chapter. As we shall see, the diagrammatic approach leads naturally to expansions of thermodynamic properties and particle distribution functions in powers of either the activity or density. While such expansions are in general more appropriate to gases than to liquids, diagrammatic methods have played a prominent role in the development of the modern theory of dense fluids. The statistical mechanics of non-uniform fluids, for example, was originally formulated in diagrammatic terms.[11] The introductory account given here is based largely on the work of Morita and Hiroike,[12] de Dominicis[13] and Stell.[14] Although the discussion is self-contained, it is limited in scope, and no attempt is made at mathematical rigour.

We consider again the case when the interparticle potential energy is a sum of pair terms. As we shall see later, it is sometimes convenient to replace the Boltzmann factor $\exp(-\beta V_N)$ by a sum of products of *Mayer functions*, $f(i, j)$, defined as

$$f(i, j) = \exp[-\beta v(i, j)] - 1 \equiv e(i, j) - 1 \qquad (3.7.1)$$

Then, for example, in the definition of $\rho^{(1)}(1)$ given by (3.1.9) the term for $N = 4$ involves an integral of the form

$$I = \iiint \left(\prod_{i=1}^{4} z^*(i) \right) f(1, 2) f(1, 4) f(2, 3) f(3, 4) \, \mathrm{d}2 \, \mathrm{d}3 \, \mathrm{d}4 \qquad (3.7.2)$$

To each such integral there corresponds a *labelled diagram* consisting of a number of *circles* linked by *bonds*. Circles represent particle coordinates and carry an appropriate label; for that reason the diagrams are sometimes called 'cluster' diagrams. The circles are of two types: *white circles* (or 'root points'), which correspond to coordinates held constant in the integration, and *black circles* (or 'field points'), which represent the variables of integration. With a circle labelled i we associate a function of coordinates, $\gamma(i)$, say. The circle is then referred to as a white or black γ-circle; a 1-circle is a circle for which $\gamma(i) = 1$. Bonds are drawn as lines between circles. With a bond between circles i and j we associate a function $\eta(i, j)$, say, and refer to it as an η-bond; a *simple* diagram is one in which no pair of circles is linked by more than one bond. The *value* of a labelled diagram is the value of the integral that the diagram represents; it is a function of the coordinates attached to the white circles and a functional of the functions associated with the black circles and bonds. Thus the integral in (3.7.2) is represented by a simple, labelled diagram consisting of z^*-circles (both white and black) and f-bonds:

$$I \ = \quad$$

The black circles in a diagram correspond to the dummy variables of integration. The manner in which the black circles are labelled is therefore irrelevant and the labels may conveniently be omitted altogether. The value of the resulting *unlabelled diagram* involves a combinatorial factor related to the topological structure of the diagram. Consider a labelled diagram containing m black γ-circles and any number of white circles. Each of the $m!$ possible permutations of labels of the black circles leaves the value of the diagram unchanged. There is, however, a subgroup of permutations that give rise to diagrams which are *topologically equivalent*. Two labelled diagrams are said to be topologically equivalent if they are characterised by the same set of *connections*, meaning that circles labelled i and j in one diagram are linked by an η-bond if and only if they are similarly linked in the other. In the case when all black circles are associated with the same function, the *symmetry number* of a simple diagram is the order of the subgroup of permutations that leave the connections unaltered. We adopt the convention that where the word 'diagram' or the symbol for a diagram appears in an equation, the quantity to be inserted is the value of that diagram. Then the value of a simple diagram Γ consisting of n white circles labelled 1 to n and m unlabelled black circles is

$$\Gamma = (1/m!)[\text{the sum of all topologically inequivalent diagrams obtained}$$
$$\text{by labelling the black circles}] \qquad (3.7.3)$$

The number of labelled diagrams appearing on the right-hand side of this equation is equal to $m!/S$, where S is the symmetry number, and each of the diagrams has a value equal to that of the integral it represents. The definition (3.7.3) may therefore be reformulated as

$$\Gamma = (1/S)[\text{any diagram obtained by labelling the black circles}]$$
$$= (1/S)[\text{the value of the corresponding integral}] \qquad (3.7.4)$$

In the example already pictured the symmetry number of the diagram is equal to two, since the connections are unaltered by interchange of the labels 2 and 4. Thus the unlabelled diagram obtained by removing the labels 2, 3 and 4 has a value equal to $\frac{1}{2}I$.

The definition of the value of a diagram can be extended to a wider class of diagrams than those we have discussed but the definition of symmetry number may have to be modified. For example, if a diagram is *composite* rather than simple, the symmetry number is increased by a factor $n!$ for every pair of circles linked by n bonds of the same *species*. On the other hand, if the functions associated with the black circles are not all the same, the symmetry number is reduced.

The difference in value of labelled and unlabelled diagrams is important because the greater ease with which unlabelled diagrams are manipulated is due precisely to the inclusion of the combinatorial factor S. In all that follows, use of the word 'diagram' without qualification should be taken as referring to the unlabelled type, though the distinction will often be irrelevant. Two unlabelled diagrams are *topologically distinct* if it is impossible to find a permutation that converts a labelled version of one diagram into a labelled version of the other. Diagrams that are topologically distinct represent different integrals. Statistical mechanical quantities usefully discussed in diagrammatic terms are frequently obtained as 'the sum of all topologically distinct diagrams' having certain properties. To avoid undue repetition we shall always replace the cumbersome phrase in quotation marks by the expression 'all diagrams'. We also adopt the convention that any diagrams we discuss are simple unless they are otherwise described.

Two circles are *adjacent* if they are linked by a bond. A sequence of adjacent circles and the bonds that link them is called a *path*. Two paths between a given pair of circles are *independent* if they have no intermediate circle in common. A *connected* diagram is either *simply* or *multiply* connected; if there exist (at least) n independent paths between any pair of circles the diagram is (at least) *n-tuply* connected. In the examples shown below, diagram (a) is simply connected, (b) is triply connected and (c) is a *disconnected* diagram with two doubly connected *components*.

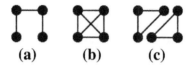

(a)   (b)   (c)

A bond is said to *intersect* the circles that it links. *Removal* of a circle from a diagram means that the circle and the bonds that intersect it are erased. A *connecting* circle is a circle whose removal from a connected diagram causes the diagram to become disconnected; the *multiplicity* of a connecting circle is the number of components into which the diagram separates when the circle is removed. Removal of an *articulation* circle from a connected diagram causes the diagram to separate into two or more components, of which at least one contains no white circle; an *articulation pair* is a pair of circles whose removal has the same effect. A diagram that is free of articulation circles is said to be *irreducible*; the absence of articulation pairs implies irreducibility but not vice versa. If a diagram contains at least two white circles, a *nodal* circle is one through which all paths between two particular white circles pass. Clearly there can be no nodal circle associated with a pair of white circles linked by a bond. A nodal circle is necessarily also a connecting circle and may also be an articulation circle if its multiplicity is three or more. In the examples given below the arrows point (a) to an articulation circle, (b) to an articulation pair and (c) to a nodal circle

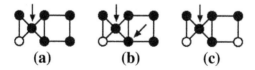

(a)   (b)   (c)

A *sub-diagram* of a diagram Γ is any diagram that can be obtained from Γ by some combination of the removal of circles and erasure of bonds. A sub-diagram is *maximal* with respect to a given property if it is not embedded in another sub-diagram with the same property; a particularly important class of maximal sub-diagrams are those that are irreducible. The *star product* of two connected diagrams Γ_1, Γ_2 is the diagram Γ_3 obtained by linking together the two diagrams in such a way that white circles carrying the same labels are superimposed, as in the example below:

The two diagrams are said to be *connected in parallel* at the white circles having labels that are common to both Γ_1 and Γ_2; if the two diagrams are connected in parallel at white γ-circles, the corresponding circles in Γ_3 are γ^2-circles. If Γ_1 and Γ_2 have no white circles in common, or if one or both contain only black circles, the star product is a disconnected diagram having Γ_1 and Γ_2 as its components.

Star-irreducible diagrams are connected diagrams that cannot be expressed as the star product of two other diagrams except when one of the two is the diagram consisting of a single white circle. This definition of star-irreducibility excludes all diagrams containing white connecting circles or connecting subsets of white circles, all diagrams with adjacent white circles and, by convention, the diagram consisting of a single white circle. The star product of two star-irreducible diagrams can be uniquely decomposed into the factors that form the product; thus the properties of star-irreducible diagrams are analogous to those of prime numbers.

Diagrammatic expressions are manipulated with the aid of certain rules, the most important of which are contained in a series of lemmas derived by Morita and Hiroike.[12] The lemmas are stated here without proof and illustrated by simple examples.[15] Some details of the proofs are given in Appendix C.

Lemma 1 *Let G be a set of topologically distinct, star-irreducible diagrams and let H be the set of all diagrams in G and all possible star products of diagrams in G. Then*

$$[\text{all diagrams in } H] = \exp{[\text{all diagrams in } G]} - 1$$

Illustration. If G consists of a single diagram, Γ, where

$$\Gamma = \text{O}\!\!-\!\!\bullet$$

then

Lemma 1 is called the 'exponentiation theorem'. If the diagrams in G consist solely of black circles and bonds, use of the lemma makes it possible to express a sum of connected and disconnected diagrams in terms of the connected subset.

Lemmas 2 and 3 contain the diagrammatic prescriptions for the evaluation of two important types of functional derivative.

Lemma 2 *Let Γ be a diagram consisting of black γ-circles and bonds. Then*

$$\delta\Gamma/\delta\gamma(\mathbf{r}) = [\text{all diagrams obtained by replacing a black } \gamma\text{-circle of } \Gamma$$
$$\text{by a white 1-circle labelled } \mathbf{r}]$$

Illustration.

$$\Gamma = \quad \longrightarrow \quad \delta\Gamma/\delta\gamma(\mathbf{r}) = \quad + \quad$$

Lemma 3 *Let Γ be a diagram consisting of black circles and η-bonds. Then*

$$\delta\Gamma/\delta\eta(\mathbf{r}, \mathbf{r}') = \tfrac{1}{2}[\text{all diagrams obtained by erasing an } \eta\text{-bond of } \Gamma,$$
$$\text{whitening the circles that it linked and labelling those}$$
$$\text{circles } \mathbf{r} \text{ and } \mathbf{r}']$$

Illustration.

$$\Gamma = \quad \longrightarrow \quad \delta\Gamma/\delta\eta(\mathbf{r},\mathbf{r}') = \tfrac{1}{2}$$

This is the diagrammatic representation of example (3.2.21) for the case when $a = 1$. The numerical factor present in (3.2.21) is taken care of by the different symmetry numbers before ($S = 6$) and after ($S = 1$) differentiation.

Lemmas 4 and 5 are useful in the process of *topological reduction*.

Lemma 4 *Let G be a set of topologically distinct, connected diagrams consisting of a white circle labelled $\mathbf{r}$, black γ-circles and bonds, and let $\mathcal{G}(\mathbf{r})$ be the sum of all diagrams in G. If Γ is a connected diagram, if H is the set of all topologically distinct diagrams obtained by decorating all black circles of Γ with diagrams in G, and if each diagram in H is uniquely decomposable, then*

$$[\text{all diagrams in } H] = [\text{the diagram obtained from } \Gamma \text{ by replacing the}$$
$$\text{black } \gamma\text{-circles by } \mathcal{G}\text{-circles}]$$

The process of *decorating* the diagram Γ consists of attaching one of the elements in G in such a way that its white circle is superimposed on a black circle of Γ and then blackened. For the diagrams in H to be *uniquely decomposable* it must be possible, given the structure of Γ, to determine by inspection which diagram in G has been used to decorate each black circle of Γ; this is always possible if Γ is free of black articulation circles.

Illustration. If the set G consists of the two diagrams:

and if

$$\Gamma = \quad$$

then the set H consists of the three diagrams

Although the example is a simple one, it illustrates the key ingredient of a topological reduction: the sum of a number of diagrams (here the diagrams in H, where the black circles are γ-circles) is replaced by a single diagram of simpler structure (here Γ, where the black circles are $\mathcal{G}$-circles).

Lemma 5 *Let G be a set of topologically distinct, connected diagrams consisting of two white circles labelled* **r** *and* **r**′, *black circles and η-bonds, and let* $\mathcal{G}(\mathbf{r}, \mathbf{r}')$ *be the sum of all diagrams in G. If Γ is a connected diagram, if H is the set of all topologically distinct diagrams obtained by replacing all bonds of Γ by diagrams in G, and if each diagram in H is uniquely decomposable, then*

$$[\text{all diagrams in } H] = [\text{the diagram obtained from } \Gamma \text{ by replacing the}$$
$$\eta\text{-bonds by } \mathcal{G}\text{-bonds}]$$

Replacement of bonds in Γ involves superimposing the two white circles of the diagram drawn from G onto the circles of Γ and erasing the bond between them. The circles take the same colour and, if white, the same label as the corresponding circle in Γ.

Illustration. If the set G consists of the two diagrams:

and if

$$\Gamma = \quad$$

then the set H consists of the three diagrams

3.8   DIAGRAMMATIC EXPANSIONS OF THE DIRECT CORRELATION FUNCTIONS

We now give examples of how the definitions and lemmas of the previous section can be used to obtain results of physical interest. The examples we choose are ones that lead to series expansions of the direct correlation functions $c^{(1)}(\mathbf{r})$ and $c^{(2)}(\mathbf{r}, \mathbf{r}')$ introduced in Section 3.5. We assume again that the interparticle forces are pairwise additive and take as our starting point the expression for Ξ given by (3.4.5), from which it follows immediately that Ξ can be represented diagrammatically as

$$\Xi = 1 + [\text{all diagrams consisting of black } z^*\text{-circles with an } e\text{-bond}$$
$$\text{linking each pair}]$$

$$= \quad 1 \;+\; \bullet \;+\; \bullet\!\!-\!\!\bullet \;+\; \triangle \;+\; \boxtimes \;+\cdots$$

$$(3.8.1)$$

Note that the definition of the value of a diagram takes care of the factors $1/N!$ in (3.4.5).

Because $e(i, j) \to 1$ as $|\mathbf{r}_j - \mathbf{r}_i| \to \infty$, the contribution from the Nth term in (3.8.1) is of order V^N, and problems arise in the thermodynamic limit. It is therefore better to reformulate the series in terms of Mayer functions by making the substitution $f(i, j) = e(i, j) - 1$, as in example (3.7.2). The series then becomes

$$\Xi = 1 + [\text{all diagrams consisting of black } z^*\text{-circles and } f\text{-bonds}]$$

$$= \quad 1 \;+\; \bullet \;+\; \bullet\ \ \bullet \;+\; \bullet\!\!-\!\!\bullet \;+\; \overset{\bullet}{\underset{\bullet\ \ \bullet}{}} \;+\; \overset{\bullet}{\underset{\bullet\!-\!\bullet}{}} \;+\; \overset{\bullet}{\underset{\bullet\!-\!\bullet}{\diagup}} \;+\; \triangle \;+\cdots$$

$$(3.8.2)$$

The disconnected diagrams in (3.8.2) can be eliminated by taking the logarithm of Ξ and applying Lemma 1. This yields an expansion of the grand potential in the form

$$-\beta\Omega = [\text{all connected diagrams consisting of black } z^*\text{-circles and } f\text{-bonds}]$$

$$= \quad \bullet \;+\; \bullet\!\!-\!\!\bullet \;+\; \bigwedge \;+\; \triangle \;+\cdots$$

$$(3.8.3)$$

Since there is no need to consider disconnected diagrams again, the requirement that diagrams must be connected will from now on be omitted.

At each order in z^* beyond the second, many of the diagrams in the series (3.8.3) contain articulation circles; those contributing at third and fourth orders are shown below, with the articulation circles marked by arrows:

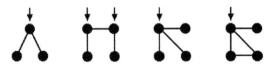

If the system were translationally invariant, the articulation circles could be chosen as the origin of coordinates in the corresponding integrals. The integrals would then factorise as products of integrals that already appear at lower order in the expansion. While this is not possible in the general case, diagrams that contain articulation circles can be eliminated by switching from an activity to a density expansion. This requires, as an intermediate step, the activity expansion of $\rho^{(1)}(\mathbf{r})$. The single-particle density at a point $\mathbf{r}$ is the functional derivative of the grand potential with respect to either $\psi(\mathbf{r})$ or, equivalently, $\ln z^*(\mathbf{r})$. From (3.3.10) and Lemma 2 it follows that

$$\rho^{(1)}(\mathbf{r})/z^*(\mathbf{r}) = 1 + \text{[all diagrams consisting of a white 1-circle labelled}$$
$$\mathbf{r}, \text{ at least one black } z^*\text{-circle and } f\text{-bonds]} \quad (3.8.4)$$

The diagrams in (3.8.4) fall into two classes: those in which the articulation circle is a white circle and those in which it is not and are therefore star-irreducible. The first of these classes is just the set of all diagrams that can be expressed as star products of diagrams in the second class. Use of Lemma 1 therefore eliminates the diagrams with white articulation circles to give an expansion of $\ln[\rho^{(1)}(\mathbf{r})/z^*(\mathbf{r})]$ which, from (3.5.3), is equal to $c^{(1)}(\mathbf{r})$:

$$c^{(1)}(\mathbf{r}) = \text{[all diagrams consisting of a white 1-circle labelled } \mathbf{r}, \text{ at least}$$
$$\text{one black } z^*\text{-circle and } f\text{-bonds, such that the white circle}$$
$$\text{is not an articulation circle]} \quad (3.8.5)$$

The diagrams in (3.8.5) are all star-irreducible, but some contain black articulation circles. To eliminate the latter, we proceed as follows. For each diagram Γ in (3.8.5) we identify a maximal, irreducible sub-diagram Γ_m that contains the single white circle.

Illustration.

$$\Gamma = \quad \longrightarrow \quad \Gamma_m =$$

In the example shown there is one articulation circle (marked by an arrow) and there are two maximal, irreducible sub-diagrams, one of which contains the white circle. It can be shown[15] that for each Γ there is a unique choice of Γ_m; if Γ itself is irreducible, Γ and Γ_m are the same. The set $\{\Gamma_m\}$ is a subset of the diagrams in (3.8.5). Given any Γ_m, the diagram from which it derives can be reconstructed by decorating the black circles with diagrams taken from the set defined in (3.8.4). Lemma 4 can therefore be used in a topological reduction whereby the z^*-circles in (3.8.5) are replaced by $\rho^{(1)}$-circles and diagrams with black articulation circles disappear. Thus

$$c^{(1)}(\mathbf{r}) = \text{[all irreducible diagrams consisting of a white 1-circle labelled } \mathbf{r},$$
$$\text{at least one black } \rho^{(1)}\text{-circle and } f\text{-bonds]}$$

$$(3.8.6)$$

The final step is to exploit the definition (3.5.2) of the two-particle direct correlation function as a functional derivative of the one-particle function by applying Lemma 2 to the series (3.8.6). The diagrams in (3.8.6) are irreducible; since they contain only one white circle this is equivalent to saying that they are free of connecting circles. Clearly they remain free of connecting circles when a second black circle is whitened as a result of the functional differentiation. It follows that $c^{(2)}(\mathbf{r},\mathbf{r}')$ can be expressed diagrammatically as

$$c^{(1)}(\mathbf{r},\mathbf{r}') = \text{[all diagrams consisting of two white 1-circles labelled } \mathbf{r}$$
$$\text{and } \mathbf{r}', \text{black } \rho^{(1)}\text{-circles and } f\text{-bonds, and which are free}$$
$$\text{of connecting circles]}$$

$$(3.8.7)$$

In the absence of an external field, (3.8.7) becomes an expansion of $c(r)$ in powers of the number density.[16]

The form of (3.8.7) suggests that the range of the direct correlation function should be roughly the range of the pair potential, as anticipated in Section 3.5. To lowest order in ρ, $c(r) \approx f(r)$ or, at large r, $c(r) \approx -\beta v(r)$. Since all higher-order diagrams in (3.8.7) are at least doubly connected, the contributions they make to $c(r)$ decay at least as fast as $[f(r)]^2$, and are therefore negligible in comparison with the leading term in the limit $r \to \infty$. However, the effects of indirect correlations are such that $h(r)$ may be significantly different from zero even for distances at which the potential is very weak. The contrast in behaviour between $c(r)$ and $h(r)$ is particularly evident close to the critical point. As the critical point is approached the compressibility χ_T becomes very large. It follows from (3.5.14) that $\hat{h}(k)$, the Fourier transform of $h(r)$, acquires a strong peak at the origin, eventually diverging as $T \to T_c$, which implies that $h(r)$ becomes very long ranged. On the other hand, (3.5.15) shows that

$$\rho\hat{c}(0) = 1 - \beta/\rho\chi_T \tag{3.8.8}$$

Close to the critical point $\rho\hat{c}(0) \approx 1$; $c(r)$ therefore remains short ranged.

The argument concerning the relative ranges of $h(r)$ and $c(r)$ does not apply to ionic fluids. The effect of screening in ionic systems is to cause $h(r)$ to decay exponentially at large r, whereas $c(r)$ still has the range of the potential and therefore decays as r^{-1}. In that situation $c(r)$ is of longer range than $h(r)$.

3.9 VIRIAL EXPANSION OF THE EQUATION OF STATE

The derivation of the series expansion of $c^{(1)}(\mathbf{r})$ yields as a valuable by-product the *virial expansion* of the equation of state of a homogeneous fluid. If there is no external field, $c^{(1)}$ can be replaced by $-\beta\mu^{ex}$ and $\rho^{(1)}$ by ρ. Equation (3.8.6) then becomes

$$\beta\mu = \beta\mu^{id} - \sum_{i=1}^{\infty} \beta_i \rho^i \tag{3.9.1}$$

where the coefficients β_i are the irreducible *cluster integrals*. The quantity $\beta_i \rho^i$ is the sum of all diagrams in (3.8.6) that contain precisely i black circles but with $\rho^{(1)}$ replaced by ρ, the so-called[17] *Mayer diagrams*. The first two coefficients are

$$\beta_1 = \int f(0,1)\mathrm{d}1 \tag{3.9.2}$$

$$\beta_2 = \frac{1}{2} \iint f(0,1)f(0,2)f(1,2)\mathrm{d}1\,\mathrm{d}2 \tag{3.9.3}$$

where, in each case, the white circle is labelled 0. Substitution of (3.9.1) in (2.4.21) and integration with respect to ρ gives

$$\beta P = \rho - \sum_{i=1}^{\infty} \frac{i}{i+1} \beta_i \rho^{i+1} \tag{3.9.4}$$

If the *virial coefficients* are defined as

$$B_{i+1} = -\frac{i}{i+1}\beta_i, \quad i \geq 1 \tag{3.9.5}$$

we recover the virial expansion in its standard form:

$$\frac{\beta P}{\rho} = 1 + \sum_{i=2}^{\infty} B_i(T)\rho^{i-1} \tag{3.9.6}$$

The coefficients B_2 and B_3 are given by

$$B_2 = -\frac{1}{2}\beta_1 = -\frac{1}{2}\int f(r)d\mathbf{r} \qquad (3.9.7)$$

$$B_3 = -\frac{2}{3}\beta_2 = -\frac{1}{3}\iint f(r)f(r')f(|\mathbf{r}-\mathbf{r}'|)d\mathbf{r}\,d\mathbf{r}' \qquad (3.9.8)$$

where the coordinates of the white circle have been taken as origin.

The expression for the second virial coefficient is more easily obtained by inserting in the virial equation (2.5.22) the low-density limit of $g(r)$ given by (2.6.10), i.e.

$$g(r) \approx e(r) = f(r) + 1 \qquad (3.9.9)$$

Then

$$\frac{\beta P}{\rho} \approx 1 - \frac{2\pi\beta\rho}{3}\int_0^\infty v'(r)e(r)r^3\,dr \qquad (3.9.10)$$

If the pair potential decays faster than r^{-3} at large r, (3.9.10) can be integrated by parts to give

$$\frac{\beta P}{\rho} \approx 1 - 2\pi\rho\int_0^\infty f(r)r^2\,dr \qquad (3.9.11)$$

in agreement with (3.9.7). For some simple potential models the second virial coefficient can be determined analytically. In the simplest case, that of hard spheres, the Mayer function $f(r)$ is equal to -1 for $r < d$ and vanishes for $r > d$. It follows from (3.9.7) that

$$B_2 = -2\pi\int_0^d (-1)r^2\,dr = \frac{2}{3}\pi d^3 = 4v_0 \qquad (3.9.12)$$

where v_0 is the volume of a sphere. Given the relation provided by (3.9.5) between the coefficients in the expansions (3.9.1) and (3.9.4) we see that the excess chemical potential at low densities behaves as

$$\beta\mu^{\text{ex}} \approx \frac{4}{3}\pi\rho d^3 = 8v_0\rho \qquad (3.9.13)$$

in agreement with the result derived from the Widom insertion formula in Section 2.4.

For more realistic, continuous potentials numerical integration of (3.9.7) is usually required and, unlike the case of hard spheres, B_2 and all higher-order virial coefficients are temperature dependent. Measurement of the extent to which the equation of state of a dilute gas deviates from the ideal gas law allows the second virial coefficient to be determined experimentally as a function of temperature and such measurements have played an important role in the development of accurate pair potentials for atoms and small molecules. Figure 3.3 shows the experimental results obtained for argon over a wide

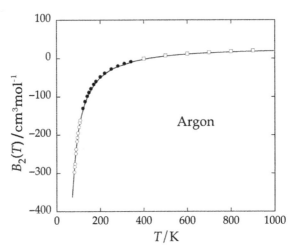

FIGURE 3.3 The second virial coefficient of argon as a function of temperature. The curve is calculated from the accurate pair potential[18] plotted in Figure 1.3 and the circles, points and squares are the experimental results[19] at low, intermediate and high temperatures, respectively.

temperature range together with those calculated[18] from the accurate pair potential for argon pictured in Figure 1.3. The shape of the curve reflects the way in which the limiting, low-density form of the pair distribution function changes with temperature, as illustrated by the examples plotted in Figure 2.4. At high temperatures the distribution function resembles that of a hard-sphere fluid. The virial coefficient is therefore positive and varies only slowly with temperature, its numerical value providing a rough measure of the size of the repulsive core of the potential. At low temperatures the rapid decrease towards increasingly negative values is related to the depth and width of the potential well, which determine the magnitude and shape of the pronounced peak that develops in $g(r)$ as the temperature is reduced. The good agreement between theory and experiment in Figure 3.3 is not unexpected, since experimental values of B_2 formed part of the data used in parameterising the potential.

The number of diagrams that contribute to the ith virial coefficient grows very rapidly with i and the associated integrals become increasingly more complicated, as Figure 3.4 reveals. The number of diagrams arising at each order can be roughly halved by reformulating the diagrammatic expansion in (3.9.1) in terms of both f-bonds and e-bonds rather than f-bonds alone;[20] this leads to a significant degree of cancellation between diagrams. Even with this simplification, however, the computational challenge for large values of i remains severe.[21] Not surprisingly, therefore, explicit calculations have for the most part been confined to the low-order coefficients. Hard spheres are an exception. In addition to B_2, both B_3 and B_4 are known analytically, and the coefficients B_5 to B_{10} have been evaluated numerically. If we define the *packing*

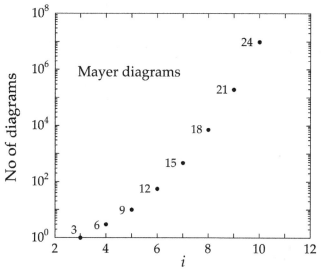

FIGURE 3.4 The points show the number of Mayer diagrams that contribute to the ith virial coefficient for hard spheres; the numbers give the dimensions of the associated integrals.

fraction, η, as the ratio of the volume occupied by the spheres to the total volume in which they are enclosed, i.e.

$$\eta = \frac{Nv_0}{V} = \frac{1}{6}\pi\rho d^3 \tag{3.9.14}$$

the virial expansion for hard spheres can be rewritten as

$$\frac{\beta P}{\rho} = 1 + \sum_{i=1}^{\infty} \mathcal{B}_i \eta^i \tag{3.9.15}$$

with

$$\mathcal{B}_i = \left(\frac{6}{\pi d^3}\right)^i B_{i+1} \tag{3.9.16}$$

The ten-term series, based on tabulated values[22] of the coefficients $\mathcal{B}_i$, is now

$$\frac{\beta P}{\rho} = 1 + 4\eta + 10\eta^2 + 18.365\eta^3 + 28.224\eta^4 + 39.82\eta^5$$
$$+ 53.34\eta^6 + 68.54\eta^7 + 85.8\eta^8 + 105.8\eta^9 + \cdots \tag{3.9.17}$$

The uncertainty[22b] in the numerical values is largest for $\mathcal{B}_{10}$, but even there it is no more than ± 0.4. Figure 3.5 shows that the pressures calculated from the truncated, ten-term series are in very good agreement with the results of computer simulations[23]; it is only at densities close to the fluid-solid transition that differences become detectable.

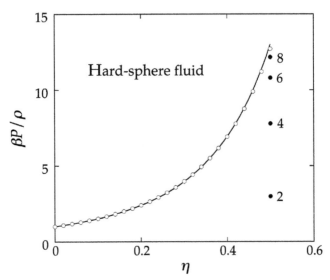

FIGURE 3.5 Virial expansion of the equation of state of the hard-sphere fluid. The curves are the results obtained from the truncated, ten-term series (3.9.17), the numbered points mark the values at $\eta = 0.50$ when the series is truncated after 2, 4, 6 or 8 terms, and the circles are the results of molecular dynamics calculations.[23]

Guided by the form of (3.9.17), Carnahan and Starling[24] were able to construct a simple but accurate hard-sphere equation of state. Noting that $\mathcal{B}_1$ and $\mathcal{B}_2$ are both integers, they chose to replace $\mathcal{B}_3$ by the nearest integer, 18, and supposed that $\mathcal{B}_i$ for all i is given by

$$\mathcal{B}_i = a_1 i^2 + a_2 i + a_3 \qquad (3.9.18)$$

With $\mathcal{B}_1 = 4$, $\mathcal{B}_2 = 10$ and $\mathcal{B}_3 = 18$, the solution to (3.9.15) is $a_1 = 1, a_2 = 3$ and $a_3 = 0$. The formula then predicts that the coefficients $\mathcal{B}_4$–$\mathcal{B}_9$ are 28, 40, 54,70, 88 and 108, respectively, in close agreement with the those in (3.9.17). The expression

$$\frac{\beta P}{\rho} = 1 + \sum_{i=1}^{\infty} (i^2 + 3i)\eta^i \qquad (3.9.19)$$

may be written as a linear combination of the first and second derivatives of the geometric series $\sum_i \eta^i$. It can therefore be summed explicitly to give

$$\frac{\beta P}{\rho} = \frac{1 + \eta + \eta^2 - \eta^3}{(1 - \eta)^3} \qquad (3.9.20)$$

Equation (3.9.20) provides an excellent fit to the results of computer simulations over the entire fluid range. It systematically underestimates the pressure but the discrepancies are never greater than 0.3%. A large number of other hard-sphere equations of state have been devised,[24] but the simple form of the

Carnahan–Starling equation makes it very convenient for use in thermodynamic calculations. In particular, a closed expression for the excess Helmholtz free energy is obtained by combining (3.9.17) with the second of the thermodynamic relations (2.3.8):

$$\frac{\beta F^{\text{ex}}}{N} = \int_0^\eta \left(\frac{\beta P}{\rho} - 1 \right) \frac{d\eta'}{\eta'} = \frac{\eta(4 - 3\eta)}{(1 - \eta)^2} \tag{3.9.21}$$

differentiation of which with respect to N yields an expression for the excess chemical potential:

$$\beta \mu^{\text{ex}} = \frac{8\eta - 9\eta^2 + 3\eta^3}{(1 - \eta)^3} \tag{3.9.22}$$

Expansion of (3.9.21) in powers of η gives

$$\frac{\beta F^{\text{ex}}}{N} = 4\eta + 5\eta^2 + 6\eta^3 + 7\eta^4 + \cdots \tag{3.9.23}$$

Thus (3.9.20) can be derived from the simple assumption, suggested by the known, exact values of the two leading terms in (3.9.23), that the coefficients in the expansion of the excess free energy form an arithmetic progression.

Barboy and Gelbart[25] have shown that a series that is much more rapidly convergent than the virial series (3.9.17) is obtained if the equation of state is expanded in powers, not of η, but of the quantity

$$y = \frac{\eta}{1 - \eta} \tag{3.9.24}$$

to give

$$\frac{\beta P}{\rho} = \frac{1}{\eta} \sum_{n=1}^\infty c_n y^n$$
$$= c_1 \frac{1}{1 - \eta} + c_2 \frac{\eta}{(1 - \eta)^2} + c_3 \frac{\eta^2}{(1 - \eta)^3} + \mathcal{O}(y^4) \tag{3.9.25}$$

The coefficients c_n are then determined by expanding the right-hand side in powers of η and equating the coefficients of successive powers to the virial coefficients in (3.9.17). A simple calculation shows that $c_1 = 1, c_2 = B_1 - 1 = 3$ and $c_3 = B_2 - 2c_2 - 1 = 3$. Truncation at order y^3 – the 'Y_3 approximation' – therefore leads to an equation of state of the form

$$\frac{\beta P}{\rho} = \frac{1}{1 - \eta} + 3\frac{\eta}{(1 - \eta)^2} + 3\frac{\eta^2}{(1 - \eta)^3} \tag{3.9.26}$$

or, equivalently:

$$\frac{\beta P}{\rho} = \frac{1 + \eta + \eta^2}{(1 - \eta)^3} \tag{3.9.27}$$

The same result is obtained from 'scaled particle' theory, a short account of which is given in Appendix E. The predicted pressures are systematically higher than those given by the Carnahan–Starling equation; at $\eta = 0.4$, for example, the difference is about 4%. In the higher-order, $Y4$ approximation, the discrepancy is of opposite sign and more than an order of magnitude smaller, though it remains less accurate than the Carnahan–Starling equation. Use of the y-expansion does, however, have the great merit of yielding an equation of state for any system of hard particles that is both simple in form and reasonably accurate, while requiring as input only the values of the low-order virial coefficients.

Note that the Carnahan–Starling equation can be recovered from the Y_3 approximation by replacing the coefficient of the last term in (3.9.26) by $(3 - \eta)$. We shall see in the next section that the analogous modification of the Y_3 approximation for mixtures of hard spheres leads again to a very accurate equation of state.

3.10 BINARY SYSTEMS

The two chapters that follow describe some of the more important methods that have been developed for the calculation of thermodynamic properties of simple, one-component fluids, but which can be extended without undue complication to the case of mixtures. In this section we deal briefly with applications to mixtures that lie outside the scope of those chapters, limiting ourselves for sake of simplicity to the case of binary systems. Prominent among these are the methods peculiar to mixtures that are grouped together under the heading of *conformal solution theory* and are designed primarily for the calculation of the changes in thermodynamic properties that occur on mixing.[26] Properties of mixing can be defined in a variety of ways, but for liquids the most important are those that refer to mixing at constant pressure and temperature.

An ideal mixture is one formed from particles that are labelled but are otherwise identical. In that case mixing leads to changes only in the entropy and free energy. For real, non-ideal mixtures the quantities of primary interest are the 'excess' properties of mixing, defined as the differences between the observed changes in thermodynamic properties and those of an ideal mixture at the same pressure and composition. It is found experimentally that the excess Gibbs free energy and enthalpy are positive for most mixtures of simple liquids, but the excess volume may be of either sign. For simple liquids the excess properties are always small in comparison with the properties of the mixture itself. In the case of argon and krypton, for example, mixing in equal proportions at the triple point temperature of krypton leads to a net decrease in volume of about 2%.

The use of conformal solution theory is restricted to mixtures for which the pair potentials, and those of the pure components, are all of the form

$$v_{\nu\mu} = \epsilon_{\nu\mu} u(r/\sigma_{\nu\mu}) \tag{3.10.1}$$

where $\epsilon_{\nu\mu}$ and $\sigma_{\nu\mu}$ are, respectively, a characteristic energy and a characteristic length and the function u is the same for all potentials. Lennard-Jones fluids are an obvious example and the one on which we shall focus. The principle of corresponding states applies rigorously to any family of pure substances whose potentials are conformal in the sense of (3.10.1) and whose thermodynamic properties are therefore described by a single equation of state. As a prerequisite for implementation of conformal solution theory that equation of state must be known; in the case of the Lennard-Jones fluid this information is provided by a multi-parameter function fitted to the results of Monte Carlo simulations.[27] It is normally assumed that the cross-interaction parameters ($\nu \neq \mu$) are related to those for like particles by the Lorentz and Berthelot combining rules:

$$\sigma_{12} = \tfrac{1}{2}(\sigma_{11} + \sigma_{22}), \quad \text{Lorentz rule}$$
$$\epsilon_{12} = (\epsilon_{11}\epsilon_{22})^{1/2}, \quad \text{Berthelot rule} \tag{3.10.2}$$

Because the changes in thermodynamic properties on mixing are small, a few percent deviation from the Lorentz–Berthelot rules can result in appreciable changes in magnitude, and even a change in sign, of the calculated excess properties. Simulations of systems designed to model a variety of real liquid mixtures[28] show that agreement with experiment is usually much improved if the value for ϵ_{12} given by the Berthelot rule is reduced by a few percent.

The simplest form of conformal solution theory is that provided by a 'one-fluid' approximation. This is a zeroth-order theory in which the properties of the mixture, apart from the ideal terms, are taken to be those of a hypothetical fluid of the same conformal family as the pure components and characterised by potential parameters σ_0 and ϵ_0. The best known and most successful approximation of this type is the van der Waals one-fluid model (vdW1),[29] so called because it represents the equivalent in modern terms of the rules used by van der Waals to calculate the constants in his equation of state of a mixture in terms of the corresponding constants for the individual components. To obtain expressions[30] for σ_0 and ϵ_0 let us suppose that the radial distribution functions $g_{\nu\mu}(r)$ in the mixture scale with $\sigma_{\nu\mu}$, and hence that

$$g_{\nu\mu}(r/\sigma_{\nu\mu}) = g_0(r/\sigma_0), \quad \text{say} \tag{3.10.3}$$

for all ν, μ. The generalisation to mixtures of the energy equation (2.5.20):

$$\frac{U^{\text{ex}}}{N} = 2\pi\rho \sum_\nu \sum_\mu x_\nu x_\mu \int_0^\infty v_{\nu\mu}(r) g_{\nu\mu}(r) r^2 \, dr \tag{3.10.4}$$

can then be rewritten, after a change of variable, as

$$\frac{U^{\text{ex}}}{N} = 2\pi\rho \sum_\nu \sum_\mu x_\nu x_\mu \epsilon_{\nu\mu} \sigma_{\nu\mu}^3 \int_0^\infty u(s) g_0(s) s^2 \, ds$$
$$= 2\pi\rho \epsilon_0 \sigma_0^3 \int_0^\infty u(s) g_0(s) s^2 \, ds \tag{3.10.5}$$

This result has the form of the energy equation for a one-component fluid having a radial distribution function $g_0(r)$ and potential parameters σ_0, ϵ_0 such that

$$\epsilon_0 \sigma_0^3 = \sum_\nu \sum_\mu x_\nu x_\mu \epsilon_{\nu\mu} \sigma_{\nu\mu}^3 \qquad (3.10.6)$$

For the hypothetical fluid to be defined uniquely this result must be supplemented by a second, independent expression for σ_0 or ϵ_0. Nothing is gained by substitution of (3.10.1) in the multicomponent form of the virial equation (2.5.22), since this leads again to (3.10.6). One possibility[31] is to force agreement between the compressibility of the hypothetical fluid and that of the mixture by supposing that

$$\rho \sum_\nu \sum_\mu x_\nu x_\mu \int [g_{\nu\mu}(r) - 1] \mathrm{d}\mathbf{r} = \rho \int [g_0(r) - 1] \mathrm{d}\mathbf{r} \qquad (3.10.7)$$

Substitution of (3.10.3) in (3.10.7), combined with a change of variable, yields a second relation in the form

$$\sigma_0^3 = \sum_\nu \sum_\mu x_\nu x_\mu \sigma_{\nu\mu}^3 \qquad (3.10.8)$$

The vdW1 approximation is a strikingly simple one and very easy to implement.[32] It has nonetheless proved successful in predicting the excess thermodynamic properties, chemical potentials and phase diagrams of Lorentz–Berthelot mixtures, at least in cases where the interaction parameters are not very different. In particular, the approximation becomes rapidly less satisfactory as the size difference between components increases. Elaborations of conformal solution theory have been proposed in which the properties of the system of interest are identified with those of an ideal mixture of two hypothetical pure fluids, but the results are often inferior to those obtained by the one-fluid approach. Corrections to the vdW1 approximation have been worked out in certain cases by expansion of the free energy of the system of interest about that of the hypothetical fluid, in the spirit of the perturbation theories which are discussed in Chapter 5, but the simplicity of the method, which is its main attraction, is thereby lost.

Much attention has also been given to mixtures of hard spheres of different diameters, since these serve as simple models of a very wide range of physical systems. The potentials in such a mixture are conformal with each other, though in a trivial sense insofar as there is no scale of energy. The vdW1 approximation now corresponds to equating the properties of the mixture to those of a system of hard spheres of diameter d_0 such that

$$d_0^3 = \sum_\nu \sum_\mu x_\nu x_\mu d_{\nu\mu}^3 \qquad (3.10.9)$$

the equation of state of which is given very accurately by the Carnahan–Starling equation. However, the approximation is expected to work well only when the diameter ratio is close to unity, which is frequently not the case for those physical systems that can be satisfactorily modelled by a mixture of hard spheres. In general it is more profitable to exploit the fact that the absence of an energy scale means that the virial coefficients are functions only of density at a given composition, which offers the possibility of building an equation of state for a mixture based on known values of the low-order coefficients. The y-expansion introduced in the previous section provides a systematic method of achieving this goal.

We first consider the case of additive hard spheres, for which, in a mixture of spheres of diameter d_{11} and d_{22}, the quantity d_{12} is given by

$$d_{12} = \tfrac{1}{2}(d_{11} + d_{22}) \tag{3.10.10}$$

The equation of state is now expanded in powers of

$$y_\nu = \frac{x_\nu \eta}{1 - \eta} \tag{3.10.11}$$

where $\eta_\nu = x_\nu \eta$ is the volume fraction of component ν. The coefficients in this expansion, if truncated at third order, are related to the second and third virial coefficients of additive hard spheres, which are known analytically as functions of composition.[25] The analogue of the Y_3 approximation (3.9.26) derived in this way can be written in compact form as

$$\frac{\beta P}{\rho} = \frac{1}{1 - \eta} + 3\frac{\langle d^1 \rangle \langle d^2 \rangle}{\langle d^3 \rangle} \frac{\eta}{(1 - \eta)^2} + 3\frac{\langle d^2 \rangle^3}{\langle d^3 \rangle^2} \frac{\eta^2}{(1 - \eta)^3} \tag{3.10.12}$$

where $\langle d^n \rangle \equiv x_1 d_{11}^n + x_2 d_{22}^n$ and $\eta = \pi\rho \langle d^3 \rangle /6$ is the packing fraction in the mixture. As in the one-component case, the same result emerges from scaled particle theory, and it is easy to see that (3.10.12) reduces to (3.9.26) when $d_{11} = d_{22}$. In addition, significant improvement is again obtained by replacing the numerical coefficient in the last term on the right-hand side by $(3 - \eta)$, thereby yielding a generalisation of the Carnahan–Starling equation to the case of mixtures that was proposed independently by Boublík[33] and Mansoori et al.[34] The result is commonly referred to as the BMCSL equation. Figure 3.6 shows a comparison between Monte Carlo results and the predictions of the vdW1, Y_3 and BMCSL approximations for mixtures of hard spheres with diameter ratios $R = d_{11}/d_{22} = 2$ and $R = 10$ at a fractional concentration x_1 of the larger particles equal to 0.5 and 0.1, respectively. These are systems for which the one-fluid model is poor, as the figure reveals. The fit achieved with the BMCSL equation is by contrast very good; the Y_3 approximation also works well at all except at the highest densities, where the trend is similar to that found for the one-component fluid in the previous section.

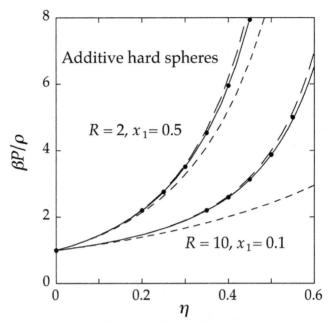

FIGURE 3.6 Equation of state of a system of additive hard spheres for two choices of the diameter ratio $R = d_{11}/d_{22}$ and concentration x_1 of the larger particles. The curves show the predictions of the BMCSL equation (full curve), the Y_3 approximation (3.10.12) (long dashes) and the vdW1 model (short dashes). The points are the results of simulations.[35,36]

The arguments originally used to obtain the BMCSL equation were basically *ad hoc* in nature. However, the same result was subsequently shown to arise naturally within the framework of a theory[37] based on the assumption, inspired by scaled particle theory, that the Helmholtz free energy can be written as a sum of terms in powers of the quantities $\rho \langle d^n \rangle$ with n limited to the values 1, 2 and 3. This approach leads at different levels of approximation to a sequence of three equations of state which improve successively on the scaled particle result. Of these approximations, the first leads to the BMCSL equation, the second to the 'extended' Carnahan–Starling equation of state[38], and the third gives a virtually exact fit to the results of simulations over wide ranges of diameter ratio and concentration. Both the BMSCL equation and the Y_3 approximation predict that hard spheres are miscible in all proportions whatever their relative size may be. There is, however, some theoretical evidence, based on certain of the integral equation approximations to be discussed in Chapter 4, which suggests that demixing may occur for highly asymmetric systems at low concentrations of the larger species.[39]

Mixtures of additive hard spheres provide useful models of many colloidal dispersions, but a wider range of physical phenomena can be described by non-additive systems, for which

$$d_{12} = \tfrac{1}{2}(d_{11} + d_{22})(1 + \Delta) \qquad (3.10.13)$$

where Δ is a non-additivity parameter. Negative non-additivity $(-1 \leq \Delta < 0)$ favours hetero-coordination, typified by the short-range, chemical ordering seen in certain liquid alloys. Positive non-additivity $(\Delta > 0)$ favours homo-coordination and therefore leads, as simulations have confirmed, to phase separation at a density determined by the value of Δ; it plays an important role in the theoretical description of colloid-polymer mixtures. An extreme example of positive non-additivity is that provided by the 'penetrable sphere' model of Widom and Rowlinson,[40] devised for the study of the liquid-vapour transition, in which $d_{11} = d_{22} = 0$ but $d_{12} = d$.

The equation of state of a mixture of non-additive hard spheres in the Y_3 approximation can be obtained, like that of an additive system, by expansion to third order in powers of $\eta_v/(1 - \eta)$. However, in the case of a symmetric mixture, i.e. one for which $d_{11} = d_{22} = d$, say, the problem is equivalent to that of a one-component system of hard spheres of diameter d but having virial coefficients that are functions of the non-additivity parameter and composition. The known, exact expressions for B_2 and B_3 lead[41] to values of the coefficients in (3.9.25) given by $c_1 = 1$, $c_2 = 3 + 8x_1x_2(3\Delta + 3\Delta^2 + \Delta^3)$ and $c_3 = 3 + 12x_1x_2(6\Delta + 9\Delta^2 + 4\Delta^3)$.

Figure 3.7 shows the variation of $\beta P/\rho$ with Δ predicted by the Y_3 approximation at two values of reduced density and two compositions, together with the results of Monte Carlo calculations for the same state conditions. Overall the agreement between theory and simulation is good, but at the higher density the predicted pressures are once more systematically too high; the vdW1 model also gives satisfactory results, at least for negative Δ. Given the Y_3 equation of state it is possible[43] to determine the density $\rho_c d^3$ above which demixing would occur for a given value of Δ. The critical density is found to decrease smoothly with increasing non-additivity, as intuition would suggest, and lies close to the value found by Gibbs ensemble Monte Carlo calculations[44] for $\Delta = 0.2$; the arrows in the figure mark the values of Δ at which phase separation is predicted to occur at equal concentrations of the two components. The Y_3 approximation does, however, suffer from a defect that becomes apparent when Δ is large and negative. As the results for $\rho^* = 0.6$ show, the predicted pressure begins to increase weakly with decreasing Δ below $\Delta \approx -0.6$. This behaviour, which becomes much more pronounced at higher densities, is physically unrealistic. It is linked to an inconsistency in the approximation which is most easily understood in the case of equal concentrations. In the limit $\Delta \to -1$ there is no interaction between particles of different species and the system reduces to that of two, identical, pure fluids confined to the same volume at a total packing fraction η. Under these conditions the equation of state of a mixture for which $x_1 = x_2$ should be the same as that of a one-component fluid

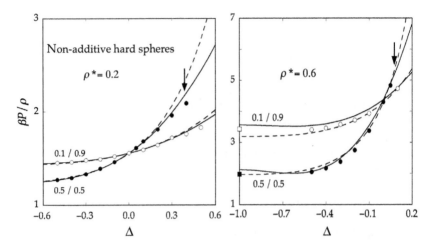

FIGURE 3.7 Equation of state of a symmetric mixture of non-additive hard spheres as a function of the non-additivity parameter at two values of the reduced density $\rho^* = \rho d^3$ and two compositions, $x_1 = 0.1$, $x_2 = 0.9$ and $x_1 = x_2 = 0.5$. The continuous and broken curves show the predictions of the Y_3 approximation and the vdW1 model, respectively, the points are the results of Monte Carlo calculations,[41,42] and the arrows mark the values of Δ at which phase separation is predicted to occur in the Y_3 approximation for the case of equal concentrations. The squares in the figure for $\rho = 0.6$ show the correct limiting behaviour as $\Delta \to -1$ as given by the Y_3 approximation for a one-component system (see text for details).

at a packing fraction equal to $\frac{1}{2}\eta$, which, within the Y_3 approximation, is

$$\frac{\beta P}{\rho} = \frac{1 + \frac{1}{2}\eta + \frac{1}{4}\eta^2}{\left(1 - \frac{1}{2}\eta\right)^3} = 1 + 2\eta + \frac{5}{2}\eta^2 + \frac{37}{24}\eta^3 + \cdots \qquad (3.10.14)$$

On the other hand, when $\Delta = -1$ and $x_1 = x_2$, the coefficients in (3.9.25) are $c_1 = c_2 = 1$ and $c_3 = 0$. This leads to an equation of state of the form

$$\frac{\beta P}{\rho} = \frac{1}{(1 - \eta)^2} = 1 + 2\eta + 3\eta^2 + 2\eta^3 + \cdots \qquad (3.10.15)$$

for which the calculated pressure is always larger than that given by (3.10.14). Similar considerations apply at other compositions. The correct limits at $\Delta = -1$ for the case when $\rho^* = 0.6$ are shown in the figure and are more consistent with the trends in the Monte Carlo results than those provided by (3.10.15). No such difficulty arises in the case of the vdW1 model, since the diameter of the equivalent hard spheres decreases monotonically with decreasing Δ. The limiting value as $\Delta \to -1$ is nearly exact, its accuracy being limited only by the use of the Carnahan–Starling equation of state for the hypothetical pure fluid. However, although the model is moderately successful at the state conditions of Figure 3.7, the agreement with the results of simulations of asymmetric mixtures deteriorates rapidly as the diameter ratio increases.[45]

REFERENCES

[1] Our treatment draws freely on the classic review article by Evans, R., *Adv. Phys.* **28**, 143 (1979). See also Evans R., *In* 'Fundamentals of Inhomogeneous Fluids' (D. Henderson ed.). (Marcel Decker, New York, 1991).

[2] See Section 3.4. The principle was established for the ground state of the electron gas by Hohenberg, P. and Kohn, W., *Phys. Rev.* **136**, B864 (1964) and extended to finite temperatures by Mermin, N.D., *Phys. Rev.* **137**, A1441 (1964). It was first applied to classical systems by Ebner, C., Saam, W.F. and Stroud, D., *Phys. Rev. A* **14**, 226 (1976).

[3] There may also be a contribution from an external source such as an electric or gravitational field.

[4] See, e.g. D. Chandler, 'Modern Statistical Mechanics'. Oxford University Press, New York, 1987, p. 16.

[5] Widom, B., *J. Stat. Phys.* **19**, 563 (1978). See also Widom, B., *J. Phys. Chem.* **86**, 869 (1982).

[6] Llano-Restrepo, M. and Chapman, W.G., *J. Chem. Phys.* **97**, 2046 (1992).

[7] See Chapter 5, Section 5.5.

[8] Greenfield, A.J., Wellendorf, J. and Wiser, N., *Phys. Rev. A* **4**, 1607 (1971).

[9] Hansen, J.P. and Schiff, D., *Mol. Phys.* **25**, 1281 (1973).

[10] Kirkwood, J.G. and Buff, F.P., *J. Chem. Phys.* **19**, 774 (1951).

[11] (a) Buff, F.P. and Stillinger, F.H., *J. Chem. Phys.* **25**, 312 (1956). (b) Stillinger, F.H. and Buff, F.P., *J. Chem. Phys.* **37**, 1 (1962).

[12] Morita, T. and Hiroike, K., *Prog. Theor. Phys.* **25**, 537 (1961).

[13] (a) de Dominicis, C., *J. Math. Phys.* **3**, 983 (1962). (b) de Dominicis, C., *J. Math. Phys.* **4**, 255 (1963).

[14] Stell, G., *In* 'The Equilibrium Theory of Classical Fluids' (H.L. Frisch and J.L. Lebowitz, eds). W.A. Benjamin, New York, 1964.

[15] See also McDonald, I.R. and O'Gorman, S.P., *Phys. Chem. Liq.* **8**, 57 (1978).

[16] This result was first obtained by Rushbrooke, G.S. and Scoins, H.L., *Proc. Roy. Soc. A* **216**, 203 (1953).

[17] Mayer, J.E. and Mayer, M.G., 'Statistical Mechanics'. Wiley, New York, 1940, Chap. 13.

[18] Maitland, G.C., Rigby, M., Smith, E.B. and Wakeham, W.A., 'Intermolecular Forces'. Clarendon Press, Oxford, 1981, pp. 497-8 and 562.

[19] (a) Weir, R.D., Wynn Jones, I., Rowlinson, J.S. and Saville, G., *Trans. Faraday Soc.* **63**, 1320 (1967). (b) Tegeler, Ch., Span, R. and Wagner, W., *J. Phys. Chem. Ref. Data* **28**, 779 (1999). (c) Ref. 18, p. 570.

[20] Ree, F.H. and Hoover, W.G., *J. Chem. Phys.* **40**, 939 (1964); *J. Chem. Phys.* **41**, 1635 (1964); *J. Chem. Phys.* **46**, 4181 (1967).

[21] For a comprehensive description of the computational techniques employed, see McCoy, B.M., '*Advanced Statistical Mechanics*'. Oxford University Press, New York, 2010, Chaps. 6 and 7.

[22] (a) Labík, S., Kolafa, J. and Malijevský, A., *Phys. Rev. E* **71** (2005). (b) Clisby, N. and McCoy, B.M., *J. Stat. Phys.* **122**, 15 (2006).

[23] Kolafa, J., Labík, S. and Malijevský, A., *Phys. Chem. Chem. Phys.* **6**, 2335 (2004).

[24] Carnhan, N.F. and Starling, K.E., *J. Chem. Phys.* **51**, 635 (1969). For a listing of other hard-sphere equations of state and an assessment of their relative merits, see Mulero, A., Galán, C.A., Parra, M.I. and Cuadros, F., *Lect. Notes Phys.* **753**, 37 (2008).

[25] Barboy, B. and Gelbart, W.M., *J. Chem. Phys.* **71**, 3053 (1979).

[26] Rowlinson, J.S. and Swinton, F.L., *Liquids and Liquid Mixtures*, 3rd edn. Butterworth, London, 1982, Chap. 8.

[27] Johnson, K., Zollweg, J.A. and Gubbins, K.E., *Mol. Phys.* **78**, 591 (1993).

[28] McDonald, I.R., *Mol. Phys.* **23**, 41 (1972).

[29] Leland, T.W., Rowlinson, J.S. and Sather, G.A., *Trans. Faraday Soc.* **64**, 1447 (1968).

[30] Henderson, D. and Leonard, P.J., *In* 'Physical Chemistry: An Advanced Treatise', (H. Eyring, D. Henderson and W. Jost, eds), Vol. VIIIB, Chap. 7. Academic Press, New York, 1971.

[31] MacGowan, D., Lebowitz, J.L. and Waisman, E.M., *Chem. Phys. Lett.* **114**, 321 (1985).

[32] For representative applications, see Shing, K.S. and Gubbins, K.E., *Mol. Phys.* **65**, 1235 (1988); Georgoulaki, A.M., Ntouros, I.V., Tassios, D.P. and Panagiotopoulos, A.Z., *Fluid Phase Equilib.* **100**, 153 (1994); and Blas, F.J. and Fujihara, I., *Mol. Phys.* **100**, 2823 (2002).

[33] Boublík, T., *J. Chem. Phys.* **53**, 471 (1970).

[34] Mansoori, G.A., Carnahan, N.F., Starling, K.E. and Leland, T., *J. Chem. Phys.* **54**, 1523 (1971).

[35] Lue, L. and Woodcock, L.V., *Mol. Phys.* **96**, 1453 (1999).

[36] Barrio, C. and Solana, J.R., *Physica A* **351**, 387 (2005).

[37] This is the fundamental measure theory of Rosenfeld, Y., *Phys. Rev. Lett.* **63**, 980 (1989), which is described in detail in Section 6.4. The application referred to here is due to Hansen-Goos, H. and Roth, R., *J. Chem. Phys.* **124**, 154506 (2006).

[38] Santos, A., Yuste, S.B. and López de Haro, M., *Mol. Phys.* **96**, 1 (1999).

[39] Biben, T. and Hansen, J.P., *Phys. Rev. Lett.* **66**, 2215 (1991).

[40] Widom, B. and Rowlinson, J.S., *Molecular Theory of Capillarity*. Oxford University Press, New York, 1982, Chap. 5.

[41] Jung, J., Jhon, M.S. and Ree, F.H., *J. Chem. Phys.* **100**, 9064 (1994).

[42] Jung, J., Jhon, M.S. and Ree, F.H., *J. Chem. Phys.* **100**, 528 (1994).

[43] Biben, T. and Hansen, J.P., *Physica A* **235**, 142 (1997). For related calculations, see Dijkstra, M., *Phys. Rev. E.* **58**, 7523 (1998) and Pellicane, G., Caccamo. C., Giaquinta, P.V. and Saija, F., *J. Phys. Chem. B* **111**, 4503 (2007).

[44] Amar, J.G., *Mol. Phys.* **67**, 739 (1989).

[45] Paricaud, P., *Phys. Rev. E* **78**, 021202 (2008).

Chapter 4

Distribution Function Theories

In this chapter we describe some of the more important theoretical methods available for calculation of the pair distribution function of a uniform fluid. If the pair distribution function is known, thermodynamic properties of the system can be obtained by a number of different routes. We begin, however, by describing the way in which the distribution function is measured in radiation scattering experiments.

4.1 THE STATIC STRUCTURE FACTOR

The structure factor of a uniform fluid was defined in Section 3.6 in terms of the Fourier transform of the pair correlation function, $h(r)$. It can be defined more generally as

$$S(\mathbf{k}) = \left\langle \frac{1}{N} \rho_{\mathbf{k}} \rho_{-\mathbf{k}} \right\rangle \tag{4.1.1}$$

where $\rho_{\mathbf{k}}$ is a Fourier component of the microscopic density (3.1.2):

$$\rho_{\mathbf{k}} = \int \rho(\mathbf{r}) \exp(-i\mathbf{k} \cdot \mathbf{r}) \, d\mathbf{r} = \sum_{i=1}^{N} \exp(-i\mathbf{k} \cdot \mathbf{r}_i) \tag{4.1.2}$$

Given the δ-function representation of the pair density in (2.5.13), the definition (4.1.1) implies that in the homogeneous case:

$$S(\mathbf{k}) = \left\langle \frac{1}{N} \sum_{i=1}^{N} \sum_{j=1}^{N} \exp(-i\mathbf{k} \cdot \mathbf{r}_i) \exp(i\mathbf{k}.\mathbf{r}_j) \right\rangle$$

$$= 1 + \left\langle \frac{1}{N} \sum_{i=1}^{N} \sum_{j \neq i}^{N} \exp[-i\mathbf{k} \cdot (\mathbf{r}_i - \mathbf{r}_j)] \right\rangle$$

$$= 1 + \left\langle \frac{1}{N} \sum_{i=1}^{N} \sum_{j \neq i}^{N} \iint \exp[-i\mathbf{k} \cdot (\mathbf{r} - \mathbf{r}')] \delta(\mathbf{r} - \mathbf{r}_i) \delta(\mathbf{r}' - \mathbf{r}_j) \, d\mathbf{r} \, d\mathbf{r}' \right\rangle$$

$$= 1 + \frac{1}{N} \iint \exp[-i\mathbf{k} \cdot (\mathbf{r} - \mathbf{r}')] \rho_N^{(2)}(\mathbf{r} - \mathbf{r}') \, d\mathbf{r} \, d\mathbf{r}'$$

$$= 1 + \rho \int g(r) \exp(-i\mathbf{k} \cdot \mathbf{r}) \, d\mathbf{r} \tag{4.1.3}$$

Theory of Simple Liquids, Fourth Edition. http://dx.doi.org/10.1016/B978-0-12-387032-2.00004-0

In the last step we have used the definition (2.5.8) of the pair distribution function and exploited the fact that the system is translationally invariant in order to integrate over $\mathbf{r}'$. Conversely, $g(r)$ is given by the inverse transform

$$\rho g(\mathbf{r}) = (2\pi)^{-3} \int [S(\mathbf{k}) - 1] \exp(i\mathbf{k} \cdot \mathbf{r}) \, d\mathbf{k} \qquad (4.1.4)$$

The final result in (4.1.3) can also be written as

$$S(\mathbf{k}) = 1 + (2\pi)^3 \rho \delta(\mathbf{k}) + \rho \hat{h}(\mathbf{k}) \qquad (4.1.5)$$

The definitions (3.6.10) and (4.1.1) are therefore equivalent apart from a δ-function term, which henceforth we shall ignore. Experimentally (see below) that term corresponds to radiation which passes through the sample unscattered.

The structure factor of a fluid can be determined experimentally from measurements of the cross-section for scattering of neutrons or X-rays by the fluid as a function of scattering angle. Here we give a simplified treatment of the calculation of the neutron cross-section in terms of $S(\mathbf{k})$.

Let us suppose that an incident neutron is scattered by the sample through an angle θ. The incoming neutron can be represented as a plane wave:

$$\psi_1(\mathbf{r}) = \exp(i\mathbf{k}_1 \cdot \mathbf{r}) \qquad (4.1.6)$$

while at sufficiently large distances from the sample the scattered neutron can be represented as a spherical wave:

$$\psi_2(\mathbf{r}) \sim \frac{\exp(ik_2 r)}{r} \qquad (4.1.7)$$

Thus, asymptotically ($r \to \infty$), the wave function of the neutron behaves as

$$\psi(\mathbf{r}) \sim \exp(i\mathbf{k}_1 \cdot \mathbf{r}) + f(\theta) \frac{\exp(ik_2 r)}{r} \qquad (4.1.8)$$

and the amplitude $f(\theta)$ of the scattered component is related to the differential cross-section $d\sigma/d\Omega$ for scattering into a solid angle $d\Omega$ in the direction θ, ϕ by

$$\frac{d\sigma}{d\Omega} = |f(\theta)|^2 \qquad (4.1.9)$$

The geometry of a scattering event is illustrated in Figure 4.1. The momentum transferred from neutron to sample in units of $\hbar$ is

$$\mathbf{k} = \mathbf{k}_1 - \mathbf{k}_2 \qquad (4.1.10)$$

To simplify the calculation we assume that the scattering is elastic. Then $|\mathbf{k}_1| = |\mathbf{k}_2|$ and

$$k = 2k_1 \sin \frac{1}{2}\theta = \frac{4\pi}{\lambda} \sin \frac{1}{2}\theta \qquad (4.1.11)$$

where λ is the neutron wavelength.

The scattering of the neutron occurs as the result of interactions with the atomic nuclei. These interactions are very short ranged, and the total

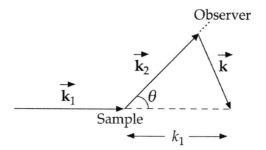

FIGURE 4.1 Geometry of an elastic scattering event.

scattering potential $V(\mathbf{r})$ may therefore be approximated by a sum of δ-function pseudopotentials in the form

$$V(\mathbf{r}) = \frac{2\pi \hbar^2}{m} \sum_{i=1}^{N} b_i \delta(\mathbf{r} - \mathbf{r}_i) \qquad (4.1.12)$$

where b_i is the *scattering length* of the ith nucleus and m is the neutron mass. For most nuclei, b_i is positive, but it may also be negative and even complex; it varies both with isotopic species and with the spin state of the nucleus.

The wave function $\psi(\mathbf{r})$ must be a solution of the Schrödinger equation:

$$\left(-\frac{\hbar^2}{2m} \nabla^2 + V(\mathbf{r}) \right) \psi(\mathbf{r}) = E\psi(\mathbf{r}) \qquad (4.1.13)$$

The general solution having the correct asymptotic behaviour is

$$\psi(\mathbf{r}) = \exp(i\mathbf{k}_1 \cdot \mathbf{r}) - \frac{m}{2\pi \hbar^2} \int \frac{\exp(ik_1|\mathbf{r} - \mathbf{r}'|)}{|\mathbf{r} - \mathbf{r}'|} V(\mathbf{r}')\psi(\mathbf{r}')\, d\mathbf{r}' \qquad (4.1.14)$$

where the second term on the right-hand side represents a superposition of spherical waves emanating from each point in the sample.

Equation (4.1.14) is an integral equation for $\psi(\mathbf{r})$. The solution in the case when the interaction $V(\mathbf{r})$ is weak is obtained by setting $\psi(\mathbf{r}) \approx \exp(i\mathbf{k}_1 \cdot \mathbf{r})$ inside the integral sign. This substitution yields the so-called first Born approximation to $\psi(\mathbf{r})$:

$$\psi(\mathbf{r}) \approx \exp(i\mathbf{k}_1 \cdot \mathbf{r}) - \frac{m}{2\pi \hbar^2} \int \frac{\exp(ik_1|\mathbf{r} - \mathbf{r}'|)}{|\mathbf{r} - \mathbf{r}'|} V(\mathbf{r}') \exp(i\mathbf{k}_1 \cdot \mathbf{r}')\, d\mathbf{r}'$$

$$(4.1.15)$$

from which an expression for $f(\theta)$ is obtained by taking the $r \to \infty$ limit and matching the result to the known, asymptotic form of $\psi(\mathbf{r})$ given by (4.1.8). If $|\mathbf{r}| \gg |\mathbf{r}'|$, then

$$|\mathbf{r} - \mathbf{r}'| = (r^2 + r'^2 - 2\mathbf{r} \cdot \mathbf{r}')^{1/2} \approx r - \hat{\mathbf{r}} \cdot \mathbf{r}' \qquad (4.1.16)$$

where $\hat{\mathbf{r}}$ is a unit vector in the direction of $\mathbf{r}$. Since we have assumed that the scattering is elastic, $k_1\hat{\mathbf{r}} = \mathbf{k}_2$. Thus, as $r \to \infty$:

$$\psi(\mathbf{r}) \sim \exp(i\mathbf{k}_1 \cdot \mathbf{r}) - \frac{\exp(ik_1r)}{r} \frac{m}{2\pi\hbar^2} \int \exp(-i\mathbf{k}_2 \cdot \mathbf{r}')V(\mathbf{r}')$$

$$\times \exp(i\mathbf{k}_1 \cdot \mathbf{r}')\,d\mathbf{r}' \qquad (4.1.17)$$

By comparing (4.1.17) with (4.1.8), and remembering that $k_1 = k_2$, we find that

$$f(\theta) = -\frac{m}{2\pi\hbar^2} \int \exp(-i\mathbf{k}_2 \cdot \mathbf{r})V(\mathbf{r})\exp(i\mathbf{k}_1 \cdot \mathbf{r})\,d\mathbf{r}$$

$$= -\frac{m}{2\pi\hbar^2} \int V(\mathbf{r})\exp(i\mathbf{k} \cdot \mathbf{r})\,d\mathbf{r} \qquad (4.1.18)$$

Hence the amplitude of the scattered component is proportional to the Fourier transform of the scattering potential. The first line of (4.1.18) also shows that $f(\theta)$ is expressible as a matrix element of the interaction $V(\mathbf{r})$ between initial and final plane-wave states of the neutron. Use of the first Born approximation is therefore equivalent to calculating the cross-section $d\sigma/d\Omega$ by the 'golden rule' of quantum mechanical perturbation theory.

An expression for $d\sigma/d\Omega$ can now be derived by substituting for $V(\mathbf{r})$ in (4.1.18), inserting the result in (4.1.9) and taking the thermal average. This yields the expression

$$\frac{d\sigma}{d\Omega} = \left\langle \left| \sum_{i=1}^{N} b_i \exp(-i\mathbf{k} \cdot \mathbf{r}_i) \right|^2 \right\rangle$$

$$= \left\langle \sum_{i=1}^{N} \sum_{j=1}^{N} b_i b_j \exp[-i\mathbf{k} \cdot (\mathbf{r}_j - \mathbf{r}_i)] \right\rangle \qquad (4.1.19)$$

A more useful result is obtained by taking an average of the scattering lengths over isotopes and nuclear spin states, which can be done independently of the thermal averaging over coordinates. We therefore introduce the notation

$$\left\langle b_i^2 \right\rangle = \left\langle b^2 \right\rangle, \quad \left\langle b_i b_j \right\rangle = \left\langle b_i \right\rangle \left\langle b_j \right\rangle \equiv \left\langle b \right\rangle^2$$

$$\left\langle b \right\rangle^2 \equiv b_{coh}^2, \quad \left(\left\langle b^2 \right\rangle - \left\langle b \right\rangle^2 \right) \equiv b_{inc}^2 \qquad (4.1.20)$$

and rewrite (4.1.19) as

$$\frac{d\sigma}{d\Omega} = N\left\langle b^2 \right\rangle + \left\langle b \right\rangle^2 \left\langle \sum_{i=1}^{N} \sum_{j\neq i}^{N} \exp[-i\mathbf{k} \cdot (\mathbf{r}_i - \mathbf{r}_j)] \right\rangle$$

$$= N\left(\left\langle b^2 \right\rangle - \left\langle b \right\rangle^2 \right) + \left\langle b \right\rangle^2 \left\langle \left| \sum_{i=1}^{N} \exp(-i\mathbf{k} \cdot \mathbf{r}_i) \right|^2 \right\rangle$$

$$= Nb_{inc}^2 + Nb_{coh}^2 S(\mathbf{k}) \qquad (4.1.21)$$

The subscripts 'coh' and 'inc' refer, respectively, to *coherent* and *incoherent* scattering. Information about the structure of the fluid is contained entirely within the coherent contribution to the cross-section; there is no incoherent contribution if the sample consists of one isotopic species of zero nuclear spin. The amplitude of the wave scattered by a single, fixed nucleus is

$$f(\theta) = -b \int \delta(\mathbf{r}) \exp(i\mathbf{k} \cdot \mathbf{r}) \, d\mathbf{r} = -b \qquad (4.1.22)$$

In the absence of incoherent scattering the cross-section for scattering by a liquid is

$$\frac{d\sigma}{d\Omega} = Nb^2 S(\mathbf{k}) \qquad (4.1.23)$$

where Nb^2 is the cross-section for a system of N independent nuclei and $S(\mathbf{k})$ represents the effects of spatial correlations.

A similar calculation can be made of the cross-section for elastic scattering of X-rays. There is now no separation into coherent and incoherent parts, but the expression for the differential cross-section has the same general form as in (4.1.23). One important difference is that X-rays are scattered by interaction with the atomic electrons and the analogue of the neutron scattering length is the atomic *form factor*, $f(\mathbf{k})$. The latter, unlike b, is a function of $\mathbf{k}$ and defined as

$$f(\mathbf{k}) = \left\langle \sum_{n=1}^{Z} \exp\left[i\mathbf{k} \cdot \left(\mathbf{r}_i^{(n)} - \mathbf{r}_i\right)\right] \right\rangle_Q \qquad (4.1.24)$$

where the subscript Q denotes a quantum mechanical expectation value, $\mathbf{r}_i^{(n)}$ represents the coordinates of the nth electron of the ith atom (with nuclear coordinates $\mathbf{r}_i$) and Z is the atomic number; for large atoms, $f(\mathbf{k}) \approx Z$ over the range of $\mathbf{k}$ in which $S(\mathbf{k})$ displays a significant degree of structure.

The pair distribution function is derived from a measured structure factor, such as that pictured in Figure 3.2, by numerically transforming the experimental data according to (4.1.4). Difficulties arise in practice because measurements of $S(\mathbf{k})$ necessarily introduce a cut-off at large values of k. These difficulties are the source of the unphysical ripples seen at small r in the distribution function for liquid argon shown in Figure 2.1.

The definition of the structure factor given by (4.1.1) is easily extended to systems of more than one component. As in Section 3.6, we consider an n-component system in which the number concentration of species ν is x_ν. The microscopic partial density $\rho^\nu(\mathbf{r})$ and its Fourier components $\rho_\mathbf{k}^\nu$ are defined in a manner analogous to (3.1.2) and (4.1.2), except that the sums run only over the particles of species ν. Thus

$$\rho_\mathbf{k}^\nu = \sum_{i=1}^{N_\nu} \exp(-i\mathbf{k} \cdot \mathbf{r}_i) \qquad (4.1.25)$$

If the fluid is homogeneous, the partial pair distribution function

$$g_{\nu\mu}(\mathbf{r}) = h_{\nu\mu}(\mathbf{r}) + 1 \tag{4.1.26}$$

has a δ-function representation given by

$$x_\nu x_\mu \rho g_{\nu\mu}(\mathbf{r}) = \left\langle \frac{1}{N} \sum_{i=1}^{N_\nu} \sum_{j=1}^{N_\mu} \delta(\mathbf{r} + \mathbf{r}_i - \mathbf{r}_j) \right\rangle \tag{4.1.27}$$

The partial structure factor defined by a generalisation of (4.1.1) as

$$S_{\nu\mu}(\mathbf{k}) = \left\langle \frac{1}{N} \rho_\mathbf{k}^\nu \rho_{-\mathbf{k}}^\mu \right\rangle \tag{4.1.28}$$

is related to $g_{\nu\mu}(r)$ by

$$S_{\nu\mu}(\mathbf{k}) = x_\nu \delta_{\nu\mu} + x_\nu x_\mu \rho \int g_{\nu\mu}(r) \exp(-i\mathbf{k} \cdot \mathbf{r}) \, d\mathbf{r} \tag{4.1.29}$$

which again differs from the earlier definition (3.6.15) by an unimportant δ-function term.

4.2   THE YBG HIERARCHY AND THE BORN–GREEN EQUATION

It was shown in Section 2.1 that the non-equilibrium phase-space distribution functions $f^{(n)}(\mathbf{r}^n, \mathbf{p}^n; t)$ are coupled together by a set of equations called the BBGKY hierarchy. A similar hierarchy exists for the equilibrium particle densities, assuming again that the forces between particles are pairwise additive; this is generally known as the Yvon–Born–Green or YBG hierarchy.

Consider first the case when $n = 1$. At equilibrium (2.1.20) becomes

$$\left(\frac{\mathbf{p}_1}{m} \cdot \frac{\partial}{\partial \mathbf{r}_1} + \mathbf{X}_1 \cdot \frac{\partial}{\partial \mathbf{p}_1} \right) f_0^{(1)}(\mathbf{r}_1, \mathbf{p}_1)$$

$$= -\iint \mathbf{F}_{12} \cdot \frac{\partial}{\partial \mathbf{p}_1} f_0^{(2)}(\mathbf{r}_1, \mathbf{p}_1, \mathbf{r}_2, \mathbf{p}_2) \, d\mathbf{r}_2 \, d\mathbf{p}_2 \tag{4.2.1}$$

where, from the expression for $f_0^{(n)}$ given by (2.5.1) with the subscript N omitted:

$$f_0^{(1)}(\mathbf{r}_1, \mathbf{p}_1) = \rho^{(1)}(\mathbf{r}_1) f_M(\mathbf{p}_1) \tag{4.2.2}$$

and

$$f_0^{(2)}(\mathbf{r}_1, \mathbf{p}_1, \mathbf{r}_2, \mathbf{p}_2) = \rho^{(2)}(\mathbf{r}_1, \mathbf{r}_2) f_M(\mathbf{p}_1) f_M(\mathbf{p}_2) \tag{4.2.3}$$

On inserting (4.2.2) and (4.2.3) into (4.2.1), exploiting the normalisation (2.1.27) and the fact that $(\partial/\partial\mathbf{p}) f_M(\mathbf{p}) = -(\beta/m)\mathbf{p} f_M(\mathbf{p})$, and finally dividing through by $(\beta/m) f_M(\mathbf{p}_1)$, we obtain a relation between the single-particle $(n = 1)$ and pair $(n = 2)$ densities:

$$(k_B T \mathbf{p}_1 \cdot \nabla_1 - \mathbf{p}_1 \cdot \mathbf{X}_1) \rho^{(1)}(\mathbf{r}_1) = \int (\mathbf{p}_1 \cdot \mathbf{F}_{12}) \rho^{(2)}(\mathbf{r}_1, \mathbf{r}_2) \, d\mathbf{r}_2 \qquad (4.2.4)$$

Equation (4.2.4) may be cast in the form $\mathbf{p}_i \cdot \mathbf{Q} = 0$ where $i = 1$, but because this result would be true for any choice of $\mathbf{p}_i$ it follows that $\mathbf{Q} = 0$. Thus, replacing the forces $\mathbf{X}_1$ and $\mathbf{F}_{12}$ in (4.2.4) by the negative gradients of the external potential $\phi(\mathbf{r}_1)$ and interparticle potential $v(\mathbf{r}_1, \mathbf{r}_2)$, respectively, and dividing through by $\rho^{(1)}(\mathbf{r}_1)$, we find that

$$-k_B T \nabla_1 \ln \rho^{(1)}(\mathbf{r}_1) = \nabla_1 \phi(\mathbf{r}_1) + \int \nabla_1 v(\mathbf{r}_1, \mathbf{r}_2) \rho^{(1)}(\mathbf{r}_2) g^{(2)}(\mathbf{r}_1, \mathbf{r}_2) \, d\mathbf{r}_2$$
$$(4.2.5)$$

This expression provides a possible starting point for the calculation of the density profile of a fluid in an external field, while if there are no interactions between particles it reduces to the usual barometric law, $\rho^{(1)}(\mathbf{r}) \propto \exp[-\beta\phi(\mathbf{r})]$.

Similar manipulations for the case when $n = 2$ yield a relationship between the pair and triplet distribution functions which, in the absence of an external field, takes the form

$$-k_B T \nabla_1 \ln g^{(2)}(\mathbf{r}_1, \mathbf{r}_2)$$
$$= \nabla_1 v(\mathbf{r}_1, \mathbf{r}_2) + \rho \int \nabla_1 v(\mathbf{r}_1, \mathbf{r}_3) \left(\frac{g^{(3)}(\mathbf{r}_1, \mathbf{r}_2, \mathbf{r}_3)}{g^{(2)}(\mathbf{r}_1, \mathbf{r}_2)} - g^{(2)}(\mathbf{r}_1, \mathbf{r}_3) \right) d\mathbf{r}_3$$
$$(4.2.6)$$

where on the right-hand side we have subtracted a term that vanishes in the isotropic case. We now eliminate the triplet distribution function by use of Kirkwood's *superposition approximation*,[1] i.e.

$$g^{(3)}(\mathbf{r}_1, \mathbf{r}_2, \mathbf{r}_3) \approx g^{(2)}(\mathbf{r}_1, \mathbf{r}_2) g^{(2)}(\mathbf{r}_2, \mathbf{r}_3) g^{(2)}(\mathbf{r}_3, \mathbf{r}_1) \qquad (4.2.7)$$

which becomes exact in the limit $\rho \to 0$. When this approximation is introduced into (4.2.6) the result is a non-linear integro–differential equation for the pair distribution function in terms of the pair potential:

$$-k_B T \nabla_1 [\ln g(\mathbf{r}_1, \mathbf{r}_2) + \beta v(\mathbf{r}_1, \mathbf{r}_2)]$$
$$= \rho \int \nabla_1 v(\mathbf{r}_1, \mathbf{r}_3) g(\mathbf{r}_2, \mathbf{r}_3) [g(\mathbf{r}_3, \mathbf{r}_1) - 1] \, d\mathbf{r}_3 \qquad (4.2.8)$$

This is the Born–Green equation.[2] Given $v(r)$, (4.2.8) can be solved numerically to yield $g(r)$, from which in turn all thermodynamic properties can be derived via the energy, pressure and compressibility equations. The work of Born and Green represented one of the earliest attempts to determine the structure and thermodynamics of a classical fluid by following a well-defined statistical mechanical route, but the results obtained are satisfactory only at low densities.[3] As we shall see later, other approximate integral equations have subsequently been proposed that work well even at high densities.

By construction, the superposition approximation satisfies the so-called core condition for hard-core systems, meaning that $g^{(3)}(\mathbf{r}_1, \mathbf{r}_2, \mathbf{r}_3)$ vanishes when any of the interparticle distances r_{12}, r_{13}, or r_{23} is less than the hard-core diameter. However, it violates the sum rule

$$g^{(2)}(\mathbf{r}_1, \mathbf{r}_2) = \frac{\rho}{N-2} \int g^{(3)}(\mathbf{r}_1, \mathbf{r}_2, \mathbf{r}_3)\, d\mathbf{r}_3 \qquad (4.2.9)$$

which follows directly from the definitions (2.5.3) and (2.5.9). An alternative to (4.2.7) is provided by the 'convolution' approximation,[4] which has the merit of satisfying (4.2.9) exactly. The approximation is most easily expressed in **k**-space, where it takes the form

$$S^{(3)}(\mathbf{k}, \mathbf{k}') \equiv \left\langle \frac{1}{N} \rho_{\mathbf{k}} \rho_{\mathbf{k}'} \rho_{-\mathbf{k}-\mathbf{k}'} \right\rangle \approx S(k) S(k') S(|\mathbf{k}+\mathbf{k}'|) \qquad (4.2.10)$$

The product of structure factors in (4.2.10) transforms in **r**-space into a convolution product of pair distribution functions, but this fails to satisfy the core condition and in practice is rarely used. The convolution approximation can be derived[5] by setting the triplet function $\hat{c}^{(3)}(\mathbf{k}, \mathbf{k}')$ equal to zero in the three-particle analogue of the Ornstein–Zernike relation (3.5.10).

4.3 FUNCTIONAL EXPANSIONS AND INTEGRAL EQUATIONS

A series of approximate integral equations for the pair distribution function of a uniform fluid in which the particles interact through pairwise–additive forces can be derived systematically by an elegant method due to Percus.[6] The basis of the method is the interpretation of the quantity $\rho g(r)$ as the single-particle density at a point **r** in the fluid when a particle of the system is known to be located at the origin, $\mathbf{r} = 0$. The particle at the origin, labelled 0, is assumed to be fixed in space, while the other particles move in the force field of particle 0. Then the total potential energy of the remaining particles in the 'external' field due to particle 0 is of the form (3.1.3), with

$$\phi(i) = v(0, i) \qquad (4.3.1)$$

Let $\Xi[\phi]$, as given by (3.1.8), be the grand partition function in the presence of the external field. In that expression, V_N is the total interatomic potential energy of particles $1, \ldots, N$. Alternatively, we may treat the particle at the origin as an $(N + 1)$th particle. Then

$$V_N + \sum_{i=1}^{N} \phi(i) = \sum_{i=1}^{N}\sum_{j>i}^{N} v(i, j) + \sum_{i=1}^{N} v(0, i) = V_{N+1} \qquad (4.3.2)$$

If we denote the partition function in the absence of the field by Ξ_0, (3.1.8) can be rewritten as

$$\begin{aligned}
\Xi[\phi] &= \sum_{N=0}^{\infty} \frac{z^N}{N!} \int \cdots \int \exp(-\beta V_{N+1}) \, d1 \cdots dN \\
&= \frac{\Xi_0}{z} \sum_{N=0}^{\infty} \frac{1}{\Xi_0} \frac{z^{N+1}}{N!} \int \cdots \int \exp(-\beta V_{N+1}) \, d1 \cdots dN \\
&= \frac{\Xi_0}{z} \sum_{N=1}^{\infty} \frac{1}{\Xi_0} \frac{z^N}{(N-1)!} \int \cdots \int \exp(-\beta V_N) \, d1 \cdots d(N-1)
\end{aligned}$$

$$(4.3.3)$$

Equation (2.5.3) shows that the sum on N in (4.3.3) is the definition of the single-particle density in a homogeneous system. Thus

$$\Xi[\phi] = \frac{\rho \Xi_0}{z} \qquad (4.3.4)$$

The physical content of this result is closely related to that of (2.4.30). By a similar manipulation, but starting from (3.1.9), it can be shown that the single-particle density in the presence of the external field is related to the two-particle density in the absence of the field by

$$\rho^{(1)}(1|\phi) = \frac{\rho^{(2)}(0, 1|\phi = 0)}{\rho} \qquad (4.3.5)$$

Because the system is spatially uniform in the absence of the field, (2.6.7) and (4.3.5) together yield the relation

$$\rho^{(1)}(1|\phi) = \rho g(0, 1) \qquad (4.3.6)$$

which is the mathematical expression of Percus's idea. The effect of switching on the force field of particle 0 is to change the potential $\phi(1)$ from zero to $\Delta\phi = v(0, 1)$; the response, measured by the change in the single-particle density, is

$$\Delta\rho^{(1)}(1) = \rho^{(1)}(1|\phi) - \rho^{(1)}(1|\phi = 0) = \rho g(0, 1) - \rho = \rho h(0, 1) \quad (4.3.7)$$

If the field due to particle 0 is regarded as a perturbation it is natural to consider functional Taylor expansions of various functionals of ϕ or $\rho^{(1)}$ with respect to $\Delta\phi$. One obvious choice is to expand $\Delta\rho^{(1)}$ itself in powers of $\Delta\phi$. The first-order result is simply the Yvon equation (3.6.7), with the infinitesimal quantities $\delta\rho^{(1)}, \delta\phi$ replaced by $\Delta\rho^{(1)}, \Delta\phi$. On combining this expression with (4.3.1) and (4.3.7) we find that

$$h(0, 1) = -\beta v(0, 1) + \rho \int h(1, 2)[-\beta v(0, 2)]\, d2 \tag{4.3.8}$$

Comparison with the Ornstein–Zernike relation (3.5.12) shows that in this approximation

$$c(0, 1) \approx -\beta v(0, 1) \tag{4.3.9}$$

When the potential is steeply repulsive at short range, (4.3.8) and (4.3.9) are very poor approximations, because $\Delta\rho^{(1)}$ is then a highly non-linear functional of ϕ. The approach is more successful in the case of the Coulomb potential; as we shall see in Section 4.5, (4.3.9) is equivalent to the Debye–Hückel approximation.

Better results are obtained for short-ranged potentials by expansion in powers of $\Delta\rho^{(1)}$. In combination with the Ornstein–Zernike relation, each choice of functional to be expanded yields a different integral equation for the pair distribution function. Here we consider the effect of expanding the intrinsic free energy. Equation (3.5.23) is an exact relation for $\mathcal{F}^{ex}[\rho^{(1)}]$ relative to the free energy of a reference system at the same temperature and chemical potential. If we take the reference system to be a uniform fluid of density ρ_0 and chemical potential μ_0, the quantities $c_0^{(1)}, \mathcal{F}^{ex}$ can be replaced by $-\beta\mu_0^{ex}$, F_0^{ex} and (3.5.23) becomes

$$\mathcal{F}^{ex}[\rho^{(1)}] = F_0^{ex} + \mu_0^{ex} \int \Delta\rho^{(1)}(\mathbf{r})\, d\mathbf{r}$$
$$- k_B T \int_0^1 d\lambda (1 - \lambda) \iint \Delta\rho^{(1)}(\mathbf{r}) c^{(2)}(\mathbf{r}, \mathbf{r}'; \lambda) \Delta\rho^{(1)}(\mathbf{r}')\, d\mathbf{r}\, d\mathbf{r}' \tag{4.3.10}$$

This result is still exact, but if we make the approximation of setting $c^{(2)}(\mathbf{r}, \mathbf{r}'; \lambda)$ equal to the direct correlation function of the reference system, $c_0^{(2)}(\mathbf{r}, \mathbf{r}')$, for all values of λ, we obtain an expansion of $\mathcal{F}^{ex}[\rho^{(1)}]$ correct to second order in $\Delta\rho^{(1)} \equiv \rho^{(1)} - \rho_0$:

$$\mathcal{F}^{ex} \approx F_0^{ex} + \mu_0^{ex} \int \Delta\rho^{(1)}(\mathbf{r})\, d\mathbf{r}$$
$$- \frac{1}{2} k_B T \iint \Delta\rho^{(1)}(\mathbf{r}) c_0^{(2)}(\mathbf{r}, \mathbf{r}') \Delta\rho^{(1)}(\mathbf{r}')\, d\mathbf{r}\, d\mathbf{r}' \tag{4.3.11}$$

or, after adding the ideal part (3.1.22) and replacing μ_0^{ex} by $\mu_0 - k_B T \ln \Lambda^3 \rho_0$:

$$
\mathcal{F}[\rho^{(1)}] \approx F_0 + (\mu_0 - k_B T) \int \Delta \rho^{(1)}(\mathbf{r}) d\mathbf{r} + k_B T \int \rho^{(1)}(\mathbf{r}) \ln \frac{\rho^{(1)}(\mathbf{r})}{\rho_0} d\mathbf{r}
$$
$$
- \frac{1}{2} k_B T \iint \Delta \rho^{(1)}(\mathbf{r}) c_0^{(2)}(|\mathbf{r} - \mathbf{r}'|) \Delta \rho^{(1)}(\mathbf{r}') d\mathbf{r} \, d\mathbf{r}' \qquad (4.3.12)
$$

The grand potential functional $\Omega_\phi[\rho^{(1)}]$ defined by (3.4.1) is

$$
\Omega_\phi[\rho^{(1)}] = \mathcal{F}[\rho^{(1)}] + \int \rho^{(1)}(\mathbf{r}) \phi(\mathbf{r}) d\mathbf{r} - \mu \int \rho^{(1)}(\mathbf{r}) d\mathbf{r} \qquad (4.3.13)
$$

or, after substitution for $\mathcal{F}$ from (4.3.12):

$$
\Omega_\phi[\rho^{(1)}] \approx \Omega_0 + \int \rho^{(1)}(\mathbf{r}) \phi(\mathbf{r}) d\mathbf{r}
$$
$$
+ k_B T \int \left(\rho^{(1)}(\mathbf{r}) \ln \frac{\rho^{(1)}(\mathbf{r})}{\rho_0} - \Delta \rho^{(1)}(\mathbf{r}) \right) d\mathbf{r}
$$
$$
- \frac{1}{2} k_B T \iint \Delta \rho^{(1)}(\mathbf{r}) c_0^{(2)}(|\mathbf{r} - \mathbf{r}'|) \Delta \rho^{(1)}(\mathbf{r}') d\mathbf{r} \, d\mathbf{r}' \qquad (4.3.14)
$$

where

$$
\Omega_0 = F_0 - \mu_0 \int \rho_0 \, d\mathbf{r} \qquad (4.3.15)
$$

is the grand potential of the reference system. At equilibrium, Ω_ϕ is a minimum with respect to variations in the single-particle density, and it is straightforward to show that the density that minimises (4.3.14) is

$$
\rho^{(1)}(\mathbf{r}) = \rho_0 \exp \left(-\beta \phi(\mathbf{r}) + \int \Delta \rho^{(1)}(\mathbf{r}') c_0^{(2)}(|\mathbf{r} - \mathbf{r}'|) d\mathbf{r}' \right) \qquad (4.3.16)
$$

The same result is obtained by minimising the total free-energy functional obtained by adding the external field term to (4.3.12), but subject now to the constraint that the total number of particles must remain constant, i.e.

$$
\int \Delta \rho^{(1)}(\mathbf{r}) \, d\mathbf{r} = 0 \qquad (4.3.17)
$$

Equation (4.3.16) may be interpreted either as an expression for the density profile of a fluid in a true external field or, following Percus, as an expression for the pair distribution function of a uniform fluid of density ρ_0, when $\phi(r)$ can be identified with the pair potential. In the uniform case it follows from (4.3.7) that

$$
g(\mathbf{r}) = \exp \left(-\beta v(\mathbf{r}) + \rho \int c(|\mathbf{r} - \mathbf{r}'|) h(\mathbf{r}') d\mathbf{r}' \right) \qquad (4.3.18)
$$

or, from the Ornstein–Zernike relation (3.5.12):

$$g(\mathbf{r}) = \exp[-\beta v(\mathbf{r})] \exp[h(\mathbf{r}) - c(\mathbf{r})] \tag{4.3.19}$$

This is the *hypernetted chain* or HNC approximation.[7] The corresponding expression for the grand potential is obtained by substituting (4.3.16) for $\rho^{(1)}(\mathbf{r})$ in (4.3.14). The result, after some rearrangement and use of the Ornstein–Zernike relation and of (4.3.7), is

$$\Omega = \Omega_0 + \frac{1}{2}\rho k_B T \int h(\mathbf{r})[h(\mathbf{r}) - c(\mathbf{r})]\,d\mathbf{r} - \rho k_B T \int c(\mathbf{r})\,d\mathbf{r} \tag{4.3.20}$$

The quantity $\Delta\Omega = \Omega - \Omega_0$ is the change in grand potential arising from the introduction of a particle that acts as the source of the external field. Since that particle is fixed in space, it makes no contribution to the ideal free energy, and the change in grand potential is therefore equal to the excess chemical potential. Thus, in the HNC approximation:

$$\beta\mu^{\text{ex}} = \frac{1}{2}\rho \int h(\mathbf{r})[h(\mathbf{r}) - c(\mathbf{r})]\,d\mathbf{r} - \rho \int c(\mathbf{r})\,d\mathbf{r} \tag{4.3.21}$$

Equation (4.3.19) represents an approximate closure of the Ornstein–Zernike relation, since it provides a second, independent relation between $h(r)$ and $c(r)$. Elimination of $c(r)$ between the two relations yields the HNC integral equation:

$$\ln g(\mathbf{r}) + \beta v(\mathbf{r}) = \rho \int [g(\mathbf{r} - \mathbf{r}') - 1][g(\mathbf{r}') - 1 - \ln g(\mathbf{r}') - \beta v(\mathbf{r}')]\,d\mathbf{r}' \tag{4.3.22}$$

Equation (4.3.22) and other integral equations of a similar type can be solved numerically by an iterative approach, starting with a guess for either of the functions h or c. Perhaps the easiest method is to use the relation (3.5.13) between the Fourier transforms of h and c. An initial guess, $c_{(0)}(r)$ say, is made and its Fourier transform inserted in (3.5.13); an inverse transformation yields a first approximation for $h(r)$. The closure relation between h and c is then used to obtain an improved guess, $c_{(1)}(r)$ say. The process is repeated, with $c_{(1)}(r)$ replacing $c_{(0)}(r)$ as input, and the iteration continues until convergence is achieved.[8] To ensure convergence it is generally necessary to mix successive approximations to $c(r)$ before they are used at the next level of iteration. A variety of elaborations of this basic scheme have been worked out, based on a decomposition of $h - c$ into coarse and fine parts and use of the Newton–Raphson algorithm to solve the integral equation on the coarse grid.[8]

Use of (4.3.19) is equivalent to setting

$$c(\mathbf{r}) = h(\mathbf{r}) - \ln[h(\mathbf{r}) + 1] - \beta v(\mathbf{r}) \tag{4.3.23}$$

For sufficiently large r, $h(r) \ll 1$; if we expand the logarithmic term in (4.3.23), we find that $c(r) \approx -\beta v(r)$. As we shall see in Chapter 10, the r^{-1} decay of $c(r)$ at large r is crucial in determining the properties of ionic fluids. For such systems we must expect the HNC approximation to be superior to those approximations in which $c(r)$ has a different asymptotic behaviour.

4.4 THE PERCUS–YEVICK EQUATION

The derivation of (4.3.19) has a strong appeal, since it shows that the HNC closure of the Ornstein–Zernike relation corresponds to minimising a well-defined grand potential (or free energy) functional, albeit an approximate one. It also leads naturally to an expression for the chemical potential of a uniform fluid expressed solely in terms of the functions $h(r)$ and $c(r)$. The HNC equation can, however, be derived in a simpler way by expanding the single-particle direct correlation function $c^{(1)}(\mathbf{r})$ of an inhomogeneous fluid about that of a uniform reference system in powers of $\Delta\rho^{(1)}$ where, as before, we follow Percus's idea by supposing that the inhomogeneity is induced by 'switching on' the interaction $\phi(\mathbf{r})$ with a particle fixed at the origin. To first order in $\Delta\rho^{(1)}$ the result is

$$
\begin{aligned}
c^{(1)}(\mathbf{r}) &\approx c_0^{(1)} + \int \Delta\rho^{(1)}(\mathbf{r}') \left. \frac{\delta c^{(1)}(\mathbf{r})}{\delta\rho^{(1)}(\mathbf{r}')} \right|_{\phi=0} d\mathbf{r}' \\
&= -\beta\mu_0^{\mathrm{ex}} + \int \Delta\rho^{(1)}(\mathbf{r}') c_0^{(2)}(\mathbf{r}, \mathbf{r}') \, d\mathbf{r}'
\end{aligned}
\tag{4.4.1}
$$

where the subscript 0 again denotes a property of the reference system. When taken together with the relation (3.5.4) between $c^{(1)}(\mathbf{r})$ and $\rho^{(1)}(\mathbf{r})$, it is easy to show that (4.4.1) is equivalent to (4.3.16), and therefore leads again to the HNC expression (4.3.19). This method of approach is also suggestive of routes to other integral equation approximations, since there are many functionals that could be expanded to yield a possibly useful closure of the Ornstein–Zernike relation. We can, for example, choose to expand $\exp[c^{(1)}(\mathbf{r})]$ in powers of $\Delta\rho^{(1)}$. The first-order result is now

$$
\begin{aligned}
\exp[c^{(1)}(\mathbf{r})] &\approx \exp\left(-\beta\mu_0^{\mathrm{ex}}\right) + \int \Delta\rho^{(1)}(\mathbf{r}') \left. \frac{\delta \exp[c^{(1)}(\mathbf{r})]}{\delta\rho^{(1)}(\mathbf{r}')} \right|_{\phi=0} d\mathbf{r}' \\
&= \exp\left(-\beta\mu_0^{\mathrm{ex}}\right) \left(1 + \int \Delta\rho^{(1)}(\mathbf{r}') c_0^{(2)}(\mathbf{r}, \mathbf{r}') \, d\mathbf{r}'\right)
\end{aligned}
\tag{4.4.2}
$$

which leads, via (3.5.4), to an expression for the pair distribution function of a uniform fluid:

$$
\begin{aligned}
g(\mathbf{r}) &= \exp[-\beta v(\mathbf{r})]\left(1 + \rho \int c(|\mathbf{r} - \mathbf{r}'|)h(\mathbf{r}') \, d\mathbf{r}'\right) \\
&= \exp[-\beta v(\mathbf{r})][1 + h(\mathbf{r}) - c(\mathbf{r})]
\end{aligned}
\tag{4.4.3}
$$

This is the Percus–Yevick or PY approximation.[9] The integral equation that results from using the Ornstein–Zernike relation to eliminate $c(\mathbf{r})$ from (4.4.3) is

$$\exp[\beta v(\mathbf{r})]g(\mathbf{r}) = 1 + \rho \int [g(\mathbf{r}-\mathbf{r}') - 1]g(\mathbf{r}')\left(1 - \exp[\beta v(\mathbf{r}')]\right) d\mathbf{r}' \quad (4.4.4)$$

The approximation (4.4.3) is equivalent to taking

$$c(\mathbf{r}) \approx \left(1 - \exp[\beta v(\mathbf{r})]\right) g(\mathbf{r}) = g(\mathbf{r}) - y(\mathbf{r}) \qquad (4.4.5)$$

where $y(r)$ is the cavity distribution function defined by (2.5.23). It follows that $c(\mathbf{r})$ is assumed to vanish wherever the potential is zero. The PY equation has proved to be more successful than the HNC approximation when the potential is strongly repulsive and short ranged. From comparison of (4.4.3) with (4.3.19) we see that the PY approximation is recovered by linearisation of the HNC result with respect to $(h-c)$, while a diagrammatic analysis shows that the PY equation corresponds to summing a smaller class of diagrams in the density expansion of $h(r)$. To some extent, therefore, the greater success of the PY equation in the case of short-range potentials must be due to a cancellation of errors.

The HNC and PY equations are the classic integral equation approximations of liquid state theory. We shall deal shortly with the question of their quantitative reliability, but it is useful initially to note some general features of the two approximations. Both equations predict, correctly, that $g(r)$ behaves as $\exp[-\beta v(r)]$ in the limit $\rho \to 0$. As we shall see in Section 4.6, they also yield the correct expression for the term of order ρ in the density expansion of $g(r)$. It follows that they both give the correct second and third virial coefficients in the density expansion of the equation of state. At order ρ^2 and beyond, each approximation neglects a certain number (different for each theory) of the diagrams appearing in the exact expansion of $g(r)$. Once a solution for the pair distribution function has been obtained, the internal energy, pressure and compressibility can be calculated from (2.5.20), (2.5.22) and (2.6.12), respectively. The pressure may also be determined in two other ways. First, the inverse compressibility can be integrated numerically with respect to density to yield the so-called compressibility equation of state. Secondly, the internal energy can be integrated with respect to inverse temperature to give the Helmholtz free energy (see (2.3.9)); the latter can in turn be differentiated numerically with respect to volume to give the 'energy' equation of state. The results obtained via the three routes (virial, compressibility and energy) are in general different, sometimes greatly so. This lack of thermodynamic consistency is a common feature of approximate theories. The HNC equation is a special case insofar as it corresponds to a well-defined free energy functional, and differentiation of that free energy with respect to volume can be shown[10] to give the same result as the virial equation. The energy and virial routes to the equation of state are therefore equivalent.

The PY equation (4.4.4) is of particular interest in the theory of simple liquids because it is soluble analytically in the important case of the hard-sphere fluid. Written in terms of the function $y(r)$, the PY approximation (4.4.5) is

$$c(r) = y(r)f(r) \tag{4.4.6}$$

For hard spheres of diameter d, (4.4.6) is equivalent to setting

$$
\begin{aligned}
c(r) &= -y(r), & r &< d \\
&= 0, & r &> d
\end{aligned}
\tag{4.4.7}
$$

It follows that $c(r)$ has a discontinuity at $r = d$, since $y(r)$ is continuous everywhere (see below in Section 4.6). The solution is further restricted by the fact that $g(r)$ must vanish inside the hard core, i.e.

$$g(r) = 0, \quad r < d \tag{4.4.8}$$

Given (4.4.7) and (4.4.8) it is possible to rewrite the PY equation as an integral equation for $y(r)$ in the form

$$y(r) = 1 + \rho \int_{r' < d} y(r') \, d\mathbf{r}' - \rho \int_{\substack{r' < d \\ |\mathbf{r} - \mathbf{r}'| > d}} y(r')y(|\mathbf{r} - \mathbf{r}'|) \, d\mathbf{r}' \tag{4.4.9}$$

which was solved independently by Thiele and Wertheim by use of Laplace transform methods.[11] The final result for $c(r)$ is

$$
\begin{aligned}
c(x) &= -\lambda_1 - 6\eta\lambda_2 x - \frac{1}{2}\eta\lambda_1 x^3, & x &< 1 \\
&= 0, & x &> 1
\end{aligned}
\tag{4.4.10}
$$

where $x = r/d$, η is the packing fraction and

$$\lambda_1 = (1 + 2\eta)^2/(1 - \eta)^4, \qquad \lambda_2 = -(2 + \eta)^2/4(1 - \eta)^4 \tag{4.4.11}$$

Appendix D describes a different method of solution, due to Baxter[12]; this has the advantage of being easily generalised to cases where the potential consists of a hard-sphere core and a tail.

The compressibility of the hard-sphere fluid is obtained by substitution of (4.4.10) in (3.5.15), and integration with respect to η yields the compressibility equation of state:

$$\frac{\beta P^c}{\rho} = \frac{1 + \eta + \eta^2}{(1 - \eta)^3} \tag{4.4.12}$$

Alternatively, substitution of

$$\lim_{r \to d^+} g(r) = y(d) = -\lim_{r \to d^-} c(r) \tag{4.4.13}$$

in (2.5.26) leads to the virial equation of state:

$$\frac{\beta P^{\mathrm{v}}}{\rho} = \frac{1 + 2\eta + 3\eta^2}{(1 - \eta)^2} \tag{4.4.14}$$

The difference between P^{c} and P^{v} increases with increasing density. The general expressions for the nth virial coefficient, obtained by expanding the two equations in powers of η, are

$$B_n^{\mathrm{c}}/b^{n-1} = 2[2 + 3n(n - 1)]/4^n$$
$$B_n^{\mathrm{v}}/b^{n-1} = 8[3n - 4]/4^n \tag{4.4.15}$$

where $b \equiv B_2 = (2\pi/3)d^3$. Both equations yield the exact values of B_2 and B_3 but give incorrect (and different) values for the higher-order coefficients.

The full equations of state are plotted in Figure 4.2 for comparison with results predicted by the Carnahan–Starling formula (3.9.20), which is nearly exact. The pressures calculated from the compressibility equation lie systematically closer to and above the Carnahan–Starling results at all densities, while the virial pressures lie below them. It appears that the Carnahan–Starling formula interpolates accurately between the two PY expressions; in fact

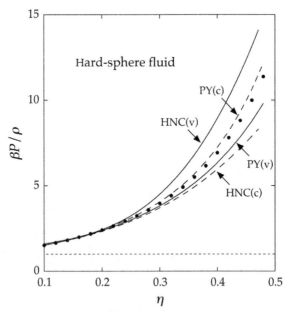

FIGURE 4.2 Equation of state of the hard-sphere fluid in the PY and HNC approximations. The full curves and dashes show results from the virial and compressibility equations, respectively, and the points are results obtained from the Carnahan–Starling equation (3.9.20).

(3.9.20) is recovered if (4.4.12) and (4.4.14) are added together with weights, respectively, of two-thirds and one-third:

$$\frac{\beta P}{\rho} = \frac{\beta}{3\rho}(2P^c + P^v) = \frac{1 + \eta + \eta^2 - \eta^3}{(1 - \eta)^3} \tag{4.4.16}$$

The compressibility and virial equations of state obtained by numerical solution of the HNC equation are also shown in Figure 4.2. They are clearly inferior to their PY counterparts.

The Thiele–Wertheim solution of the PY equation was later extended to the case of binary mixtures of additive hard spheres by Lebowitz and Rowlinson.[13] Their results show that the two components should be miscible in all proportions irrespective of diameter ratio. It is therefore understandable that the same conclusion follows from the BMLCS equation (3.10.12), since this can be derived by weighting the PY expressions for the compressibility and virial equations of state of the mixture in a manner identical to the first equality in (4.4.16).

The PY approximation to the pair distribution function is obtained by substitution of (4.4.10) into the Ornstein–Zernike relation; as a consequence of the discontinuity in $c(r)$ at $r = d$, $g(r)$ is only a piecewise–analytical function.[14] A comparison of the calculated distribution function with the results of a Monte Carlo simulation of the hard-sphere fluid at a density ($\eta = 0.49$) close to the fluid–solid transition is shown in Figure 4.3. Although the general agreement is good, the theoretical curve shows two significant defects. First, the value at contact is too low. Secondly, the oscillations are slightly out of phase with the Monte Carlo results. In addition, the amplitude of the oscillations decreases too slowly with increasing distance, with the consequence that the main peak in the structure factor is too high, reaching a maximum value of 3.05 rather than the value 2.85 obtained by simulation. An accurate representation of the pair distribution function of the hard-sphere fluid is an important ingredient of many theories. To meet that need, a simple, semi-empirical modification of the PY result has been devised in which the faults seen in Figure 4.3 are corrected.[15]

An analytical solution of the PY equation has also been derived for the 'sticky sphere' model of Baxter[16] along the lines followed for hard spheres in Appendix D. The model is one that corresponds to the square-well potential of Figure 1.2 in the limit of vanishing range of attraction ($\gamma \to 1^+$) and divergent well depth ($\epsilon \to \infty$):

$$\beta v(r) = \infty, \qquad\qquad r < d$$

$$= \ln\left[\frac{12\tau(\gamma - 1)}{\gamma d}\right], \quad d \le r < \gamma d, \, \gamma \to 1^+$$

$$= 0, \qquad\qquad\qquad r > \gamma d \tag{4.4.17}$$

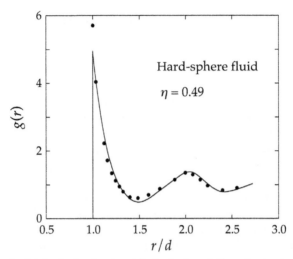

FIGURE 4.3 Radial distribution function of the hard-sphere fluid at a density close to the fluid–solid transition. The curve shows the PY solution and the points are the results of Monte Carlo calculations.

The quantity τ is a dimensionless measure of the temperature that increases monotonically with T, while the form of the attractive term ensures that the second virial coefficent (3.9.7) remains finite:

$$B_2(\tau) = \frac{\pi d^3}{6}\left(4 - \frac{1}{\tau}\right) \qquad (4.4.18)$$

The PY solution shows that the model undergoes a first-order, gas–liquid transition below a critical point at $\tau_c = (2 - \sqrt{2})/6 \approx 0.0976$ and $\eta_c = (3\sqrt{2} - 4)/2 \approx 0.1213$. Sticky spheres provide a useful model of colloidal systems, where the attractive interactions are frequently both strong and very short ranged compared with the particle dimensions.

Solutions to the PY and HNC equations have been obtained for a variety of other pair potentials over wide ranges of temperature and density. Comparison of results for the Lennard-Jones potential with those of computer simulations shows that the PY approximation is superior at all thermodynamic states for which calculations have been made.[3] At high temperatures the agreement with simulations is excellent both for internal energy and for pressure, but it worsens rapidly as the temperature is reduced. Figure 4.4 shows results for the virial and energy equations of state along the isotherm $T^* = 1.35$, which corresponds to a near-critical temperature. Although the pressures calculated by the energy route are in good agreement with those obtained by simulation,[18] the more significant feature of the results is the serious thermodynamic inconsistency that they reveal, which becomes more severe as the temperature is lowered further.

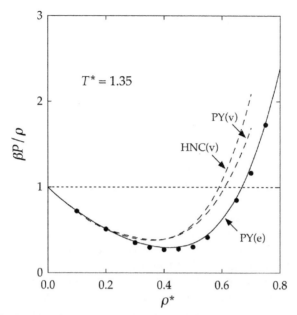

FIGURE 4.4 Equation of state of the Lennard-Jones fluid along the isotherm $T^* = 1.35$. The curves show results obtained from the PY and HNC equations via the virial (v) and energy (e) routes and the points are the results of Monte Carlo calculations.[17]

The deficiencies in the PY approximation at low temperatures are also evident in the behaviour of the pair distribution function. The main peak in $g(r)$ has too great a height and occurs at too small a value of r, while the later oscillations are out of phase with the results of simulations; in the latter respect, the situation is markedly worse than it is for hard spheres. These weaknesses show that the PY approximation cannot be regarded as a quantitatively satisfactory theory of the liquid state.

4.5   THE MEAN SPHERICAL APPROXIMATION

There are a variety of model fluids of interest in the theory of liquids for which the pair potential consists of a hard-sphere interaction plus a tail. The tail is normally attractive, but not necessarily spherically symmetric. Such systems have been widely studied in the *mean spherical approximation* or MSA. The name comes from the fact that the approximation was first proposed as a generalisation of the mean spherical model of Ising spin systems. The general form of the potential in the spherically symmetric case is

$$v(r) = \infty, \qquad r < d$$
$$= v_1(r), \quad r > d \tag{4.5.1}$$

where d is the hard-sphere diameter. The MSA is defined in terms of the pair distribution and direct correlation functions by

$$g(r) = 0, \qquad r < d$$
$$c(r) = -\beta v_1(r), \quad r > d \qquad (4.5.2)$$

When supplemented by the Ornstein–Zernike relation, these two expressions combine to yield an integral equation for $g(r)$. The first expression is exact, while the second extends the asymptotic behaviour of $c(r)$ to all $r > d$ and is clearly an approximation. Despite the crude form assumed for $c(r)$, the MSA gives good results in many cases. For example, it provides a much better description of the properties of the square-well fluid[19] than is given by either the PY or HNC approximation. However, the most attractive feature of the MSA is the fact that the integral equation can be solved analytically for a number of potential models of physical interest, including the hard-core Yukawa fluid defined by (1.2.2) as well as simple models of electrolyte solutions (discussed in Chapter 10) and polar liquids (Chapter 11).

The PY equation for hard spheres is the special case of the MSA when the tail in the potential is absent and the analytical solution of the MSA for certain pair potentials is closely linked to the method of solution of the PY hard-sphere problem. The two theories also have a common diagrammatic structure,[20] but the connection between them can be established more easily in the following way. The basic PY approximation (4.4.3) may be expressed in the form

$$c(r) = f(r) + f(r)[h(r) - c(r)] \qquad (4.5.3)$$

where $f(r)$ is the Mayer function for the potential $v(r)$. In the low-density limit, $h(r)$ and $c(r)$ become the same, and the right-hand side of (4.5.3) reduces to $f(r)$. Equation (4.5.3) can therefore be rewritten as

$$c(r) = c_0(r) + f(r)[h(r) - c(r)] \qquad (4.5.4)$$

where $c_0(r)$, the limiting value of $c(r)$ at low density, is equal to $f(r)$ both in an exact theory and in the PY approximation. If we choose another form for $c_0(r)$ in (4.5.4), we generate a different theory. For a potential of the type defined by (4.5.1) the exact $c_0(r)$ is

$$c_0(r) = \exp[-\beta v(r)] - 1 = [1 + f_d(r)]\exp[-\beta v_1(r)] - 1 \qquad (4.5.5)$$

where $f_d(r)$ is the Mayer function for hard spheres. The MSA is equivalent to linearising (4.5.5) with respect to $v_1(r)$ by setting

$$c_0(r) \approx [1 + f_d(r)][1 - \beta v_1(r)] - 1 = f_d(r) - \beta v_1(r)[1 + f_d(r)] \qquad (4.5.6)$$

and at the same time replacing f by f_d in (4.5.4). Taken together, these two approximations give rise to the expression

$$f_d(r)[1 + h(r)] = [c(r) + \beta v_1(r)][1 + f_d(r)] \qquad (4.5.7)$$

which is equivalent to the closure relation (4.5.2). This characterisation of the MSA shows that it involves approximations additional to those underlying the PY equation. One would therefore not expect the MSA to be of comparable accuracy to the PY approximation. In practice, as the results for the square-well fluid show, this is not always true.

The structure of (4.5.7) suggests a natural way in which the MSA can be extended to a class of pair potentials wider than that defined by (4.5.1).[21] Let us suppose that the potential $v(r)$ is divided in the form

$$v(r) = v_0(r) + v_1(r) \tag{4.5.8}$$

The conventional MSA applies only when v_0 is the hard-sphere potential. When $v_0(r)$ is strongly repulsive but continuous the natural generalisation of the closure relation (4.5.7) is obtained by replacing f_d by f_0, the Mayer function for the potential v_0. The resulting equation can then be rearranged to give

$$g(r) = \exp[-\beta v_0(r)][1 + h(r) - c(r) - \beta v_1(r)] \tag{4.5.9}$$

which reduces to the PY approximation (4.4.3) when $v_1(r)$ is very weak. When applied to the Lennard-Jones fluid the 'soft-core' MSA gives good results when the potential is divided at its minimum in the manner that has also proved very successful when used in thermodynamic perturbation theory (see Section 5.4).

4.6 DIAGRAMMATIC EXPANSIONS OF THE PAIR FUNCTIONS

In Section 3.8 we derived the density expansion of the two-particle direct correlation function $c^{(2)}(1, 2)$. We now wish to do the same for other pair functions. One of our main goals is to obtain a precise, diagrammatic characterisation of the HNC approximation of Section 4.3. The simplest way to proceed is to take as starting point the iterative solution of the Ornstein–Zernike relation in (3.5.11). That solution can be expressed in diagrammatic terms as

$h(1, 2) = $ [all chain diagrams consisting of two terminal white 1-circles

labelled 1 and 2, black $\rho^{(1)}$-circles and c-bonds]

$$= \underset{1 \quad 2}{\circ\!\!-\!\!\circ} + \underset{1 \qquad 2}{\circ\!\!-\!\!\bullet\!\!-\!\!\circ} + \underset{1 \qquad\quad 2}{\circ\!\!-\!\!\bullet\!\!-\!\!\bullet\!\!-\!\!\circ} + \cdots \tag{4.6.1}$$

where the meaning of the terms 'chain' diagram and 'terminal' circle is self-evident. We now replace the c-bonds in (4.6.1) by their series expansion. The first term on the right-hand side of (4.6.1) yields the complete set of diagrams that contribute to $c(1, 2)$ and are therefore free of connecting circles, which means they contain neither articulation circles nor nodal circles. The black circles appearing at higher order are all nodal circles; they remain nodal circles when the c-bonds are replaced by diagrams drawn from the series (3.8.7), but no articulation circles appear. The topology of the resulting diagrams is therefore similar to that of the diagrams in the series for $c(1, 2)$ except that nodal circles are now permitted. Thus[22]

$h(1, 2) =$ [all irreducible diagrams consisting of two white 1-circles

labelled 1 and 2, black $\rho^{(1)}$-circles and f-bonds]   (4.6.2)

Equation (4.6.2) contains more diagrams than (3.8.7) at each order in density beyond the zeroth-order term; the additional diagrams contain at least one nodal circle. For example, of the two second-order terms shown below, (a) appears in both expansions but (b) appears only in (4.6.2), because in (b) the black circles are nodal circles:

(a)   (b)

Diagrams (a) and (b) differ only by the presence in (a) of an f-bond between the white circles. If we recall that $e(1, 2) = f(1, 2) + 1$, we see that the sum of (a) and (b) is given by a single diagram in which the white circles are linked by an e-bond. All diagrams in (4.6.2) can be paired uniquely in this way, except that the lowest-order diagram

appears alone. We therefore add to (4.6.2) the disconnected diagram consisting of two white 1-circles:

$= 1$

and obtain an expansion of $g(1, 2) = h(1, 2) + 1$ in terms of diagrams in which the white circles are linked by an e-bond and all other bonds are f-bonds. Alternatively, on dividing through by $e(1, 2)$, we find that the cavity distribution function $y(1, 2) = g(1, 2)/e(1, 2)$ can be expressed in the form

$y(1, 2) =$ [all irreducible diagrams consisting of two non-adjacent white

1-circles labelled 1 and 2, black $\rho^{(1)}$-circles and f-bonds]

(4.6.3)

If the system is homogeneous and the factor $e(1, 2)$ is restored, (4.6.3) becomes an expansion of $g(1, 2)$ in powers of ρ with coefficients $g_n(r)$ such that

$$g(r) = \exp[-\beta v(r)] \left(1 + \sum_{n=1}^{\infty} \rho^n g_n(r) \right) \qquad (4.6.4)$$

Both $g_1(r)$ and $g_2(r)$ have been evaluated analytically for hard spheres.[23]

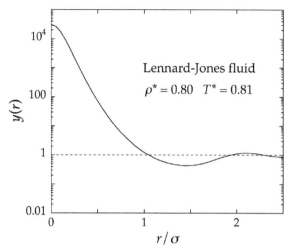

FIGURE 4.5 Monte Carlo results for the cavity distribution function of the Lennard-Jones fluid. After Llano–Restrepo and Chapman.[24]

The form of the series (4.6.4) leads immediately to two important results. First, $g(r)$ behaves as $\exp[-\beta v(r)]$ as $\rho \to 0$, as we proved in a different way in Section 2.6. Secondly, $y(r)$ is a continuous function of r even for hard spheres, for which the discontinuity in $g(r)$ at $r = d$ is wholly contained in the factor $\exp[-\beta v(r)]$. This useful property has already been exploited in the derivation of the hard-sphere equation of state (2.5.26). It is also clear from (4.6.3) that $y(1, 2)$ can be interpreted as the distribution function for a pair $1, 2$ in a 'mixed' system in which the interaction between those particles is suppressed (and hence $e(1, 2) = 1$) but other interactions remain the same. For a system of hard spheres, two such particles would correspond to spheres that can overlap each other, but not other particles, and therefore play a role equivalent to that of spherical cavities of volume equal to that of a hard sphere. Figure 4.5 shows the calculated cavity distribution function for the Lennard-Jones fluid in a high-density, low-temperature thermodynamic state. The very rapid increase in $y(r)$ as $r \to 0$ implies that there is a high probability of finding the two 'cavity' particles at very small separations.[25]

The pair distribution function is sometimes written as

$$g(1, 2) = \exp[-\beta \psi(1, 2)] \tag{4.6.5}$$

where $\psi(1, 2)$ is the *potential of mean force*. The name is justified by the fact that the quantity $-\nabla_1 \psi(1, 2)$ is the force on particle 1, averaged over all positions of particles $3, 4, \ldots$, with particles 1 and 2 held at $\mathbf{r}_1$ and $\mathbf{r}_2$, respectively. This can be proved[26] by taking the logarithm of both sides of the definition of $g(1, 2)$ provided by (2.5.3) and (2.5.8) and differentiating with respect to the coordinates of particle 1. In thermodynamic terms the potential of mean force

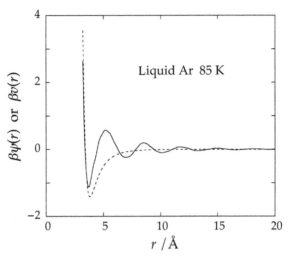

FIGURE 4.6 The full curve shows the potential of mean force for liquid argon at 85 K, derived from the experimental results of Figure 2.3; the dashed curve shows the Lennard-Jones potential with parameters chosen to fit experimental properties of the liquid.

is the reversible work that must be done on the system to bring together at a separation r two particles that initially were infinitely separated. The example plotted in Figure 4.6 is for liquid argon at 85 K, calculated from the experimental neutron scattering results for $g(r)$ shown in Figure 2.3. Depending on the final separation of the particles, the reversible work required may be either positive or negative, with fluctuations that reflect the structure of the liquid.

It is clear from the behaviour of the pair distribution function at low density that $\psi(1,2) \to v(1,2)$ as $\rho \to 0$. If we define a function $\omega(1,2)$ by

$$\omega(1,2) = \beta[v(1,2) - \psi(1,2)] \qquad (4.6.6)$$

then

$$g(1,2) = e(1,2)\exp[\omega(1,2)] \qquad (4.6.7)$$

and therefore

$$\omega(1,2) = \ln y(1,2) \qquad (4.6.8)$$

An application of Lemma 1 of Section 3.7 to the diagrams in (4.6.3) shows that

$$\omega(1,2) = \text{[all diagrams consisting of two non-adjacent white 1-circles}$$
$$\text{labelled 1 and 2, black } \rho^{(1)}\text{-circles and } f\text{-bonds, such that}$$
$$\text{the white circles are not an articulation pair]} \qquad (4.6.9)$$

The effect of this operation is to eliminate those diagrams in the expansion of $y(1,2)$ that are star products of other diagrams in the same expansion. For example, it eliminates the penultimate diagram pictured in (4.6.3), since this is the star product of the first diagram with itself:

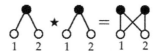

The fact that the white circles in (4.6.9) are not an articulation pair means that there exists at least one path between each pair of black circles which does not pass through either white circle.

From the earlier discussion we know that $c(1,2)$ is the sum of all diagrams in $h(1,2)$ that are free of nodal circles. We therefore define a function $s(1,2)$ such that

$$s(1,2) = h(1,2) - c(1,2) \qquad (4.6.10)$$

where

$s(1,2) =$ [all irreducible diagrams consisting of two white 1-circles labelled

1 and 2, black $\rho^{(1)}$-circles and f-bonds, and which contain

at least one nodal circle]

$$= \text{[diagrams]} + \cdots \qquad (4.6.11)$$

Diagrams belonging to the set (4.6.11) are called the *series* diagrams; the function $s(1,2)$ is given by the convolution integral on the right-hand side of the Ornstein–Zernike relation (3.5.10) and is therefore termed the *indirect correlation function*.

All series diagrams are also members of the set (4.6.9). The function $\omega(1,2)$ can therefore be re-expressed as

$$\omega(1,2) = s(1,2) + b(1,2) \qquad (4.6.12)$$

where $b(1,2)$ is the sum of the diagrams in (4.6.9) that are free of nodal circles; these are called the *bridge* or *elementary* diagrams and $b(1,2)$ is called the *bridge function*. To second order in density the only bridge diagram is

On combining (4.6.7), (4.6.10) and (4.6.12), we obtain the following, exact relation:

$$\ln[h(1,2) + 1] = -\beta v(1,2) + b(1,2) + h(1,2) - c(1,2) \qquad (4.6.13)$$

Since $h(1,2)$ and $c(1,2)$ are linked by the Ornstein–Zernike relation, (4.6.13) would be transformed into an integral equation for h (or c) if the unknown function $b(1, 2)$ were replaced by some function of h (or c). For example, the f-bond expansion of $b(1, 2)$ can be rewritten as an h-bond expansion[27] and inserted in (4.6.13). The result, together with the Ornstein–Zernike relation, constitutes an exact integral equation for $h(1, 2)$, but because the h-bond expansion introduces an infinite series of many-dimensional integrals of products of h, the equation is intractable. If instead we set $b(1, 2) = 0$, we recover the HNC approximation, which was arrived at in a very different way in Section 4.3. By rewriting the exact relation (4.6.13) as

$$y(1, 2) = \exp[s(1, 2) + b(1, 2)] \quad \text{(exact)} \tag{4.6.14}$$

we see that the HNC and PY approximations are equivalent to taking either

$$y(1, 2) \approx \exp[s(1, 2)] \quad \text{(HNC)} \tag{4.6.15}$$

or

$$y(1, 2) \approx s(1, 2) + 1 \quad \text{(PY)} \tag{4.6.16}$$

In each case differences with respect to the exact result arise initially only at second order in density. From comparison of (4.6.14) with (4.6.16) it also follows that the PY approximation may be viewed as one for which the bridge function is approximated by

$$b(1, 2) \approx \ln[s(1, 2) + 1] - s(1, 2) \quad \text{(PY)} \tag{4.6.17}$$

While this interpretation is certainly correct it is important not to misunderstand its meaning. In particular, it does not imply that the PY approximation represents a partial summation of the diagrammatic expansion of $b(r)$. On the contrary, the diagrammatic effect of (4.6.17) is to replace the bridge diagrams by star products of series diagrams.

The results just given make it possible to understand, at least for low densities, why the PY results for hard spheres, and more generally for potentials with a strongly repulsive core, are superior to those obtained from the HNC equation. The coefficient of the term of order ρ^n in the density expansion (4.6.4) of the pair distribution function of a homogeneous fluid is given by the sum of all diagrams in (4.6.3) that contain precisely n black circles. Thus $g_1(r)$ is represented by the single diagram

$$g_1(r) = \underset{1 \quad 2}{\overset{}{\diagup\hspace{-0.3em}\diagdown}}$$

and $g_2(r)$ is the sum of five diagrams:

$$g_2(r) = \underset{1 \quad 2}{\| \|} + \underset{1 \quad 2}{\boxtimes} + \underset{1 \quad 2}{\boxtimes} + \underset{1 \quad 2}{\boxtimes} + \underset{1 \quad 2}{\boxtimes}$$

where all black circles are now 1-circles and the second and third diagrams in the expression for $g_2(r)$ are equal in value. The diagram representing $g_1(r)$ is also the first-order diagram in the expansion of $s(r)$, showing that (4.6.15) and (4.6.16) are both exact to order ρ. At second order the HNC approximation is obtained by discarding the bridge diagram in the exact expression. Thus

If the bridge function is written as a power series in density:

$$b(r) = b^{(2)}(r)\rho^2 + b^{(3)}(r)\rho^3 + \cdots \qquad (4.6.18)$$

then in the PY approximation the coefficient of the second-order term is

$$b^{(2)}(r) = -\frac{1}{2}[g_1(r)]^2 \qquad (4.6.19)$$

or, diagrammatically:

where the factor $\frac{1}{2}$ is taken care of by the symmetry number of the product diagram. The contributions from the bridge and product diagrams in the exact expression for $g_2(r)$ therefore cancel each other to give

The same result follows directly from (4.6.16).

The relative merits of the two approximations can be tested numerically in the case of hard spheres, since analytical expressions are available[23b] for the different contributions to the exact result for $g_2(r)$. The results are shown in Figure 4.7, from which it is clear that the cancellation on which the PY approximation for $g_2(r)$ rests is nearly complete; it becomes exact for $r \geq \sqrt{3}d$. Discarding both the bridge and product diagrams is therefore an improvement on omission of the bridge diagram alone. Complete cancellation would be achieved if the f-bond linking the two black circles of the bridge diagram were set equal to -1. This is an approximation that is clearly most appropriate for hard spheres, for which $f(r)$ takes only the values -1 or zero depending on whether r is less than or greater than d; as the potential softens it becomes more difficult to justify. Numerically the effect is small because the value of the bridge diagram is largely determined by the contribution from regions in which the coordinates associated with the black circles are separated by distances shorter than d. Similar considerations apply at higher densities.[28]

The derivation of the Debye–Hückel expression for the radial distribution function of a system of charged particles provides a simple but useful example

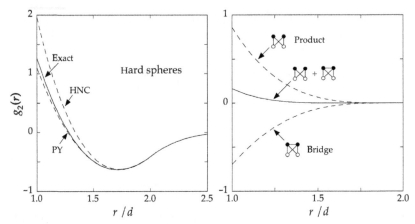

FIGURE 4.7 Left-hand panel: The coefficient $g_2(r)$ in the density expansion of the pair distribution function of the homogeneous hard-sphere fluid; the full curve gives the exact result, and the dashed curves show the results obtained from the HNC and PY approximations. Right-hand panel: The dashed curves are the contributions to the exact result for $g_2(r)$ from the product diagram (above) and the bridge diagram (below), and the full curve is the sum of the two.

of the application of diagrammatic techniques to the calculation of pair functions. Consider a homogeneous, one-component plasma of point charges q, immersed in a neutralising, uniform background of opposite charge, for which the pair potential[29] is

$$v(r) = q^2/r \qquad (4.6.20)$$

Use of (4.6.20) in expansions of the pair functions leads to divergent integrals but convergent results can be obtained if entire classes of diagrams are summed. The most strongly divergent integrals in the expansion of $\omega(1, 2)$ are those associated with the most weakly connected diagrams, namely the chain diagrams. If the chain diagrams are summed to all orders in ρ, but all other diagrams are ignored, the result is an approximation for $\omega(1, 2)$ of the form

$\omega(1, 2) \approx$ [all chain diagrams consisting of two terminal white1-circles

labelled 1 and 2, one or more black ρ-circles and f-bonds]

$$= \underset{1}{\circ}\!\!-\!\!\bullet\!\!-\!\!\underset{2}{\circ} + \underset{1}{\circ}\!\!-\!\!\bullet\!\!-\!\!\bullet\!\!-\!\!\underset{2}{\circ} + \underset{1}{\circ}\!\!-\!\!\bullet\!\!-\!\!\bullet\!\!-\!\!\bullet\!\!-\!\!\underset{2}{\circ} + \cdots \qquad (4.6.21)$$

By analogy with (3.5.10) and (4.6.1), $\omega(1, 2)$ is given by

$$\omega(1, 2) = \rho \int f(1, 3)[f(3, 2) + \omega(3, 2)]\,d3 \qquad (4.6.22)$$

On taking Fourier transforms (4.6.22) becomes

$$\hat{\omega}(k) = \frac{\rho[\hat{f}(k)]^2}{1 - \rho\hat{f}(k)} \qquad (4.6.23)$$

with

$$\rho \hat{f}(k) = \rho \int \exp(-i\mathbf{k} \cdot \mathbf{r}) f(r) \, d\mathbf{r}$$

$$\approx -\beta \rho q^2 \int \frac{\exp(-i\mathbf{k} \cdot \mathbf{r})}{r} \, d\mathbf{r} = -\frac{k_D^2}{k^2} \tag{4.6.24}$$

where

$$k_D = (4\pi \beta \rho q^2)^{1/2} \tag{4.6.25}$$

is the Debye wavenumber. We now substitute for $\rho \hat{f}(k)$ in (4.6.23) and find that

$$\rho[\hat{\omega}(k) - \beta \hat{v}(k)] = \frac{k_D^2}{k_D^2 + k^2} \tag{4.6.26}$$

or

$$\omega(r) - \beta v(r) = -\beta \psi(r) = -\frac{\beta q^2}{r} \exp(-k_D r) \tag{4.6.27}$$

We see that summing the chain diagrams leads to a potential of mean force or 'renormalised' potential equal to $v(r) \exp(-k_D r)$. This damping of the Coulomb potential by the factor $\exp(-k_D r)$ is familiar from elementary Debye–Hückel theory and corresponds physically to the effects of screening. It follows from (4.6.5) that the corresponding approximation for the radial distribution function is

$$g(r) = \exp\left(-\frac{\beta q^2}{r} \exp(-k_D r)\right) \tag{4.6.28}$$

Equation (4.6.28) is more familiar in its linearised form, valid for $k_D r \gg 1$, i.e.

$$g(r) \approx 1 - \frac{\beta q^2}{r} \exp(-k_D r) \tag{4.6.29}$$

This result could have been obtained more directly by replacing $c(r)$ by $-\beta v(r)$ in (4.6.1). A serious weakness of the linearised approximation is the fact that it allows $g(r)$ to become negative at small r; this failing is rectified in the non-linear version (4.6.28).

The pair functions discussed in this section, together with their definitions, are summarised in Table 4.1.

4.7   EXTENSIONS OF INTEGRAL EQUATIONS

We saw in the previous section that the development of an accurate integral equation for $g(r)$ can be reduced to the problem of devising a satisfactory approximation for the bridge function $b(r)$. The HNC approximations consists in setting $b(r) = 0$. Hence the integral equations to which some other

TABLE 4.1 Selected pair functions and their definitions.

Function	Symbol	Definition	
Pair distribution function	$g(r)$	(2.5.15)	
Pair correlation function	$h(r)$	$g(r) - 1$	(4.6.2)[a]
Direct correlation function	$c(r)$	(3.5.2), (3.5.10)	(3.8.7)[a]
Cavity distribution function	$y(r)$	$\exp[\beta v(r)]\, g(r)$	(4.6.3)[a]
Potential of mean force	$\psi(r)$	$-k_B T \ln g(r)$	
[Unnamed]	$\omega(r)$	$\ln y(r)$	(4.6.9)[a]
Indirect correlation function	$s(r)$	$h(r) - c(r)$	(4.6.11)[a]
Bridge function	$b(r)$	$\omega(r) - s(r)$	(4.6.12)[a]

[a]*Diagrammatic expansion.*

approximation, $b(r) \approx b_0(r)$ say, gives rise can be regarded as a modified HNC equation in which the exact relation (4.6.13) is replaced by

$$\ln g(r) = -\beta[v(r) - k_B T b_0(r)] + h(r) - c(r) \qquad (4.7.1)$$

The task of solving the modified equation is therefore equivalent to finding the solution to the HNC equation for an effective potential $v_{\mathrm{eff}}(r)$ defined as

$$v_{\mathrm{eff}}(r) = v(r) - k_B T b_0(r) \qquad (4.7.2)$$

It is possible to improve the HNC approximation systematically by including successively higher-order terms in the series expansion of the bridge function, but the calculations are computationally demanding and the slow convergence of the series means that in general only modest improvement is achieved.[30]

The true bridge function for a given system can be calculated from (4.6.14) if $c(r), h(r)$ and $y(r)$ are known. A conventional simulation provides values of $h(r)$ at separations where $g(r)$ is non-zero, from which $c(r)$ for all r can be obtained via the Ornstein–Zernike relation; in this range of r the calculation of $y(r)$ from $h(r)$ is a trivial task. To determine $b(r)$ at smaller separations, where $h(r) \approx -1$, an independent calculation of $y(r)$ is required. This can be achieved by simulation of the mixed system, described in the previous section, in which the particles labelled 1 and 2 do not interact with each other. The calculation is straightforward in principle, but the very rapid rise in $y(r)$ as $r \to 0$ means that special techniques are needed to ensure that the full range of r is adequately sampled.[24,31]

Figure 4.8 shows the bridge function derived from Monte Carlo calculations for the Lennard-Jones fluid in a thermodynamic state not far from the triple point and compares the results with those given by the PY approximation (4.6.17). In the example illustrated, the bridge function makes a contribution to the effective potential (4.7.2) that is both short ranged and predominantly repulsive, but the

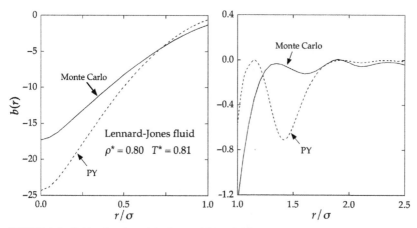

FIGURE 4.8 Bridge function of the Lennard-Jones fluid obtained by Monte Carlo calculations for $r < \sigma$ (left) and $r > \sigma$ (right). The PY results are those given by (4.6.17). After Llano–Restrepo and Chapman.[24]

same is true for the Lennard-Jones fluid at other thermodynamic states and for other model fluids. The PY approximation is poor at small values of r, but in that region the pair potential is so strongly repulsive that errors in the effective potential are unimportant for many purposes. So far as the calculation of thermodynamic properties is concerned, the most serious deficiencies in the PY approximation occur in the region of the main peak in $g(r)$.

Alternatives to the PY approximation have been proposed[32] that resemble (4.6.17) insofar as $b(r)$ is written as a function of $s(r)$. These approximations give results for the hard-sphere fluid that improve on those obtained from the PY equation and they have also been applied, though with generally less success, to systems having an attractive term in the potential. There is no reason to suppose, however, that the functional relationship between $b(r)$ and $s(r)$ is the same for all potentials, or even for different thermodynamic states of a given system.[24,33] To improve on the PY or PY-like approximations it seems necessary to make the assumed form of $b(r)$ explicitly dependent on $v(r)$. The soft-core MSA (SMSA) discussed in Section 4.5 provides an example of how this can be done. From the SMSA expression for $g(r)$ given by (4.5.9) it follows that

$$y(r) \equiv \exp\left[\beta v(r)\right] g(r) = \exp\left[\beta v_1(r)\right]\left[1 + s(r) - \beta v_1(r)\right] \qquad (4.7.3)$$

where $v_1(r)$ is the tail in the potential. Comparison with (4.6.13) shows that this is equivalent to replacing the bridge function by

$$b(r) \approx \ln[1 + s^*(r)] - s^*(r) \quad \text{(SMSA)} \qquad (4.7.4)$$

where

$$s^*(r) = s(r) - \beta v_1(r) \qquad (4.7.5)$$

Equation (4.7.4) is identical to its PY counterpart (4.6.17) but with $s(r)$ replaced by $s^*(r)$. The result, as we have seen, is a marked improvement relative to the PY approximation in the case of the Lennard-Jones fluid.

We showed in Section 4.3 that the HNC approximation can be derived by minimising the grand potential functional obtained from a functional Taylor expansion of the intrinsic free energy truncated at second order. The question therefore arises as to whether any significant improvement is obtained when the third-order term is included.[5] Equation (4.3.10) again provides the starting point of the calculation, but $c^{(2)}(\mathbf{r}, \mathbf{r}'; \lambda)$ is now replaced, not by $c_0^{(2)}(\mathbf{r}, \mathbf{r}')$, but by

$$c^{(2)}(\mathbf{r}, \mathbf{r}'; \lambda) \approx c_0^{(2)}(\mathbf{r}, \mathbf{r}') + \lambda \int \Delta \rho^{(1)}(\mathbf{r}'') \frac{\delta c_0^{(2)}(\mathbf{r}, \mathbf{r}')}{\delta \rho^{(1)}(\mathbf{r}'')} d\mathbf{r}''$$

$$= c_0^{(2)}(\mathbf{r}, \mathbf{r}') + \lambda \int \Delta \rho^{(1)}(\mathbf{r}'') c_0^{(3)}(\mathbf{r}, \mathbf{r}', \mathbf{r}'') d\mathbf{r}'' \quad (4.7.6)$$

where $c_0^{(3)}(\mathbf{r}, \mathbf{r}', \mathbf{r}'')$ is the three-particle direct correlation function of the reference fluid. The effect is to add to the functional (4.3.14) the term

$$-\frac{1}{6}k_B T \iiint \Delta \rho^{(1)}(\mathbf{r}) \Delta \rho^{(1)}(\mathbf{r}') \Delta \rho^{(1)}(\mathbf{r}'') c_0^{(3)}(\mathbf{r}, \mathbf{r}', \mathbf{r}'') d\mathbf{r} \, d\mathbf{r}' \, d\mathbf{r}''$$

If we now follow the steps that previously led to the HNC approximation (4.3.19), we obtain an expression for the pair distribution function of a uniform fluid having the form (4.7.1), with

$$b_0(r) = 12\rho^2 \iint c^{(3)}(\mathbf{r} - \mathbf{r}', \mathbf{r} - \mathbf{r}'') h(r') h(r'') d\mathbf{r}' \, d\mathbf{r}'' \quad (4.7.7)$$

Solution of the integral equation for $g(r)$ requires some further approximation[5] to be made for the triplet function $c^{(3)}$. Equation (4.7.7) is equivalent to the lengthier expression in terms of $g^{(3)}$ obtained from an expansion of $c^{(1)}(\mathbf{r})$ taken to second order, the so-called HNC2 approximation.[34]

Results based on (4.7.7) show a clear improvement over the HNC approximation for a number of model fluids but the method is computationally demanding. The HNC equation can more easily and successfully be extended by identifying $b_0(r)$ with the bridge function of a suitable reference system, a step that leads to the 'reference' HNC (RHNC) approximation.[35] The obvious choice of reference system is a fluid of hard spheres, since this is the only potential model for which the bridge function is known with sufficient accuracy over the full range of state conditions.[36] Equation (4.7.1) then represents a one-parameter theory in which the only unknown quantity is the hard-sphere diameter d. It was originally argued that the bridge function was likely to be highly insensitive to details of the potential and that its representation by a hard-sphere function should therefore be a good approximation. Although it is now recognised that the bridge function does not have a genuinely 'universal' character,[37] this approach

TABLE 4.2 Thermodynamic properties of the Lennard-Jones fluid: comparison between molecular dynamics results (MD) and calculations based on the RHNC approximation. After Lado et al.[35c]

ρ^*	T^*	$\beta P/\rho$		$\beta U^{ex}/N$	
		MD	RHNC	MD	RHNC
0.85	0.719	0.36	0.424	−6.12	−6.116
0.85	2.889	4.36	4.364	−4.25	−4.240
0.75	1.071	0.89	0.852	−5.17	−5.166
0.65	1.036	−0.11	−0.155	−4.52	−4.522
0.65	2.557	2.14	2.136	−3.78	−3.786
0.45	1.552	0.57	0.552	−2.98	−2.982
0.45	2.935	1.38	1.377	−2.60	−2.608
0.40	1.424	0.38	0.382	−2.73	−2.728

has been applied successfully in calculations for a variety of different systems. The overall agreement with the results of simulations is very good, as illustrated by the results for thermodynamic properties of the Lennard-Jones fluid given in Table 4.2; the errors in the corresponding pair distribution functions are barely discernible, even under conditions close to the triple point. In the work on which Table 4.2 is based, the hard-sphere diameter was chosen in such a way as to minimise an approximate free energy functional. So far as internal consistency of the theory is concerned, use of this procedure gives the RHNC approximation a status comparable with that of the HNC equation. The method has also been applied to mixtures of Lennard-Jones fluids, again with very good results.[35e]

A number of attempts have been made to combine different closure relations in hybrid schemes that ensure a degree of thermodynamic consistency. For example, whereas the HNC approximation is correct at large separations, the PY approximation, being much superior for strongly repulsive potentials, is presumably more accurate at short distances. It is therefore plausible to mix the two closures in such a way[38] that the function $y(r)$ in (4.6.14) reduces to its PY value as $r \to 0$ and to its HNC value as $r \to \infty$. The parameter that determines the proportions in which the two approximations are mixed at intermediate values of r can then be chosen to force consistency between the compressibility and virial equations of state. The method works well for systems of particles interacting through purely repulsive potentials, but breaks down for the Lennard-Jones potential for which, at low temperatures, it is impossible to find a value of the mixing parameter that provides thermodynamic consistency. Where successful, the method relies heavily on the fact that the HNC and PY approximations in some sense bracket the exact solution for the system of interest. The difficulty in the case of the Lennard-Jones fluid lies in the fact that the PY approximation is poor at low temperatures. The problem can be overcome[39] by interpolating instead between the HNC approximation and the soft-core MSA, an approach – called the HMSA – that yields results comparable in quality with those obtained by the RHNC approximation.

A more ambitious method of building thermodynamic consistency into an integral equation theory is to write the direct correlation function in a form that can be adjusted so as to satisfy some consistency criterion. This is the basis of the self-consistent Ornstein–Zernike approximation or SCOZA developed by Stell and coworkers[40] for application to potentials consisting of a hard core and a tail, $v_1(r)$ say, as in (4.5.1). Since $g(r)$ vanishes inside the hard core, closure of the Ornstein–Zernike is achieved by making some approximation for $c(r)$ in the range $r > d$; this is typically of the form

$$c(r) = c_d(r) - \alpha(\rho, T)v_1(r), \quad r > d \qquad (4.7.8)$$

where $c_d(r)$ is the direct correlation function of the hard-sphere fluid. The quantity $\alpha(\rho, T)$, which plays the role of an effective, density-dependent, inverse temperature, can then be chosen in such a way as to enforce consistency between the compressibility and energy routes to the equation of state. Equation (4.7.8) resembles certain other closure relations insofar as the range of $c(r)$ is the same as that of the pair potential, but in contrast, say, to the MSA, its amplitude is now density-dependent. If the compressibility and internal energy are to be consistent with each other, they must come from the same free energy, and hence must satisfy the relation[41]

$$-\frac{\partial \hat{c}(k = 0)}{\partial \beta} = \frac{\partial^2 u}{\partial \rho^2} \qquad (4.7.9)$$

thereby providing a partial differential equation for $\alpha(\rho, T)$; here $u \equiv U^{\mathrm{ex}}/V$ while $\hat{c}(k = 0)$ is related to the compressibility by (3.5.15).

Most of the published calculations based on the SCOZA are concerned with the hard-core Yukawa model (1.2.2), a system for which the analytical solution to the MSA is known.[42] A major simplification of the problem is then possible. If $c_d(r)$ for $r > d$ is represented by a second Yukawa term, $\hat{c}(k = 0)$ can be related analytically to u and (4.7.9) becomes a partial differential equation for the variable $u(\rho, T)$, which can be solved numerically; the two free parameters in the second Yukawa term are chosen so as to reproduce the Carnahan–Starling equation of state in the limit $T \rightarrow \infty$. The same simplification applies when the long-range contribution to the potential is represented by a linear combination of Yukawa terms, a strategy that makes it possible to mimic a variety of pair potentials of physical interest.[43] For other choices of $v_1(r)$, such as that provided by the square-well potential,[43,44] a fully numerical solution is required, thereby substantially increasing the computational effort involved. The SCOZA gives good results for the structure and thermodynamics of the Yukawa and square-well fluids over a range of state conditions and choices of the Yukawa

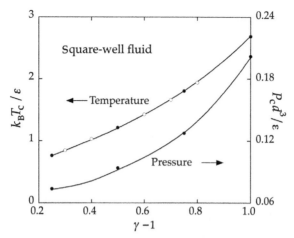

FIGURE 4.9 Critical temperature and critical pressure of the square-well fluid as a function of the well width ($\gamma - 1$) in units of d. The curves are calculated from the SCOZA[44a]; the symbols show the results of Monte Carlo simulations.[45]

inverse-range parameter γ or the well-width parameter in the square-well potential (see Figure 1.2), but its chief merit is the fact that it remains accurate in the critical region, where the performance of other integral equation theories is mostly poor. The success of the SCOZA in the case of the square-well fluid is illustrated in Figure 4.9, which shows the behaviour of the reduced critical temperature and critical pressure as functions of γ. Agreement with the results of Monte Carlo calculations is excellent for both properties. There are, however, some differences between theory and simulation in the results for the critical density; these discrepancies increase as the range of the attractive interaction is reduced, a trend that is also apparent in calculations for the Yukawa fluid.[46]

4.8 ASYMPTOTIC DECAY OF THE PAIR CORRELATION FUNCTION

Results from simulations, integral equation approximations and radiation scattering experiments invariably show that in the liquid range the pair correlation function decays to zero in the damped, oscillatory manner exemplified in Figures 2.3 and 3.2. At low densities, by contrast, it decays monotonically. The oscillatory decay is associated with packing-induced layering of neighbours around a central particle, while at low density the result in (2.6.10) implies that the decay of $h(r)$ is governed by the behaviour of the pair potential at large r. Working on the basis of a one-dimensional model, Fisher and Widom[47] predicted that at least for short-range potentials there should be a sharp cross-over from monotonic to oscillatory decay along a locus of points in the density–temperature plane, now termed the Fisher–Widom line.

We take as a starting point the diagrammatic expansion of the direct correlation function in (3.8.7) and the discussion that follows, which tell us that $c(r)$ behaves as $-\beta v(r)$ as $r \to \infty$. Consider first the case of short-range potentials,[48] either of finite range, vanishing beyond some cut-off value, or decaying exponentially at large r. The Fourier transform $\hat{c}(k)$ of such potentials can be expanded in even powers of k:

$$\hat{c}(k) = c_0 + c_2 k^2 + c_4 k^4 + \mathcal{O}(k^6) \tag{4.8.1}$$

The Ornstein–Zernike relation (3.5.13) expresses $\hat{h}(k)$ in terms of $\hat{c}(k)$ and an inverse Fourier transform yields an expression for $h(r)$ that can be written in two equivalent forms, either

$$rh(r) = \frac{1}{4\pi^2 i} \int_{-\infty}^{\infty} \exp{(ikr)} \frac{\hat{c}(k)}{1 - \rho\hat{c}(k)} k \, dk \tag{4.8.2}$$

or

$$rh(r) = \frac{1}{2\pi^2} \operatorname{Im} \int_{0}^{\infty} \exp{(ikr)} \frac{\hat{c}(k)}{1 - \rho\hat{c}(k)} k \, dk \tag{4.8.3}$$

If $\hat{c}(k)$ is a known function, the integrals on the right-hand side of these equations can be evaluated by contour integration in the plane of complex wavenumbers, $k = k_1 + ik_2$, pictured schematically in Figure 4.10. The poles of the integrand correspond to zeros of the denominator, given by the complex solutions of the equation

$$1 - \rho\hat{c}(k) = 0 \tag{4.8.4}$$

To calculate the integral in (4.8.2) the contour must be closed by an infinite semi-circle in the upper half-plane. The value of the integral is the sum of the

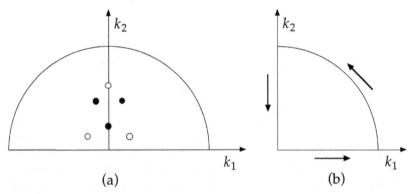

FIGURE 4.10 The complex wavenumber plane. (a) The open and closed circles show two possible distributions of the poles of $\hat{h}(k)$ lying closest to the real axis; see text for details. (b) Contour used in the evaluation of the integral in (4.8.3).

residues at the poles, and a pair of equations that together determine the real and imaginary parts of the poles is obtained by taking the real and imaginary parts of (4.8.4):

$$4\pi\rho \int_0^\infty c(r)\frac{\sinh(k_2 r)}{k_2 r}\cos(k_1 r)r^2\,dr = 1$$

$$4\pi\rho \int_0^\infty c(r)\cosh(k_2 r)\frac{\sin(k_1 r)}{k_1 r}r^2\,dr = 1$$

(4.8.5)

If R_n is the residue of $\hat{c}(k)/[1 - \rho\hat{c}(k)]$ at the nth pole, $k = k^{(n)}$, the integral in (4.8.2) reduces to

$$rh(r) = \frac{1}{2\pi}\sum_n R_n \exp\left(k^{(n)}r\right) = \frac{1}{2\pi}\sum_n R_n \exp\left(-ik_2^{(n)}r\right)\exp\left(ik_1^{(n)}r\right)$$

(4.8.6)

The poles may lie on the imaginary axis, $k_1^{(n)} = 0$, or may form a conjugate pair, $k^{(n)} = \pm k_1^{(n)} + ik_2^{(n)}$. In the first case the contribution to the decay of $rh(r)$ from the single pole is purely exponential; in the second case there is a damped, oscillatory contribution from the conjugate pair. There could in principle be an infinite number of such terms but the presence of the exponential factors in (4.8.6) ensures that asymptotically the dominant contribution will come from the pole or poles nearest the real axis. Two scenarios are therefore possible. If the nearest pole is purely imaginary, corresponding to the black circles in Figure 4.10, then

$$\lim_{r\to\infty} h(r) = \frac{A}{r}\exp(-k_2 r)$$

(4.8.7)

where the amplitude $A = R/2\pi$, with R being the residue at the pole. If all poles are simple, the residue theorem implies that

$$A = \frac{-ik_2}{2\pi\rho^2\hat{c}'(ik_2)}$$

(4.8.8)

where the prime denotes a derivative with respect to the argument; then differentiation of the Fourier transform of $c(r)$ shows that for $k = ik_2$:

$$\hat{c}'(ik_2) = \frac{4\pi}{ik_2}\int_0^\infty c(r)\left(\cosh(k_2 r) - \frac{\sinh(k_2 r)}{k_2 r}\right)r^2\,dr$$

(4.8.9)

Alternatively, if the poles closest to the real axis form a conjugate pair, corresponding to the white circles in the figure, the asymptotic behaviour is oscillatory:

$$\lim_{r\to\infty} h(r) = \frac{2|A|}{r}\exp(-k_2 r)\cos(k_1 r - \theta)$$

(4.8.10)

where the amplitude $|A|$ and phase angle θ are related by

$$|A|\exp(-i\theta) = -\frac{k_1 + ik_2}{2\pi\rho^2\hat{c}'(k_1 + ik_2)}$$

(4.8.11)

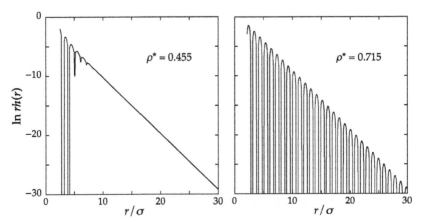

FIGURE 4.11 Asymptotic behaviour of the function $\ln rh(r)$ predicted by the pole analysis described in the text for the truncated Lennard-Jones fluid at $T^* = 1.2$ and two densities. From R.J.F. Leote de Carvalho et al., 'The decay of the pair correlation function in simple fluids: long- versus short-ranged potentials', *J. Phys. Condens. Matter* **6**, 9275–9294 (1994).© *IOP Publishing 1994. Reproduced by permission of IOP Publishing. All rights reserved.*

Calculations that use as input the direct correlation functions derived from integral equation approximations show that the relative positions of the lowest-lying imaginary and complex conjugate poles change as the density increases along an isotherm.[48] At low densities, the purely imaginary pole lies below the conjugate pair and $h(r)$ is found to decay monotonically; at high densities the situation is reversed, leading to an oscillatory decay. The cross-over in relative positions of the poles defines a point on the Fisher–Widom line. The curves of the function $\ln rh(r)$ plotted in Figure 4.11 illustrate the striking difference in asymptotic behaviour at densities on different sides of the Fisher–Widom line in the case of the Lennard-Jones potential truncated[49] at $r = 2.5\sigma$. The results shown are the contributions to the expansion (4.8.6) from the poles pictured in Figure 4.10, calculated from input provided by numerical solution of the HMSA equation of Section 4.9, which is known to be very accurate. Beyond $r \approx 2\sigma$ they are indistinguishable on the scale of the figure from the results derived directly from the HMSA values of $h(r)$. Some oscillations are seen at intermediate values of r even at low density, but these rapidly merge into an exponential decay; at high density the oscillations are exponentially damped but persisting. By repeating the calculations for a large number of points in the density–temperature plane it possible to map out the Fisher–Widom line for the potential, with the results shown in Figure 4.12. The line intersects the liquid–vapour coexistence curve on the liquid side at $T/T_c \approx 0.9$ and $\rho/\rho_c \approx 1.8$, numbers that are very close to those obtained in similar calculations for the square-well fluid.

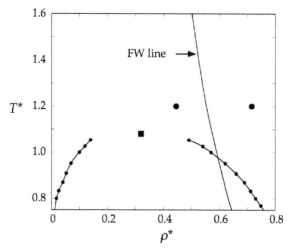

FIGURE 4.12 The Fisher–Widom line for the truncated Lennard-Jones fluid calculated from the HMSA. The small black circles show the results of Monte Carlo calculations[50] of the liquid–vapour coexistence curve, with lines drawn as a guide to the eye. The square is the Monte Carlo estimate of the critical point and the large black circles mark the state points for which the results shown in Figure 4.11 were calculated. From R.J.F. Leote de Carvalho et al., 'The decay of the pair correlation function in simple fluids: long- versus short-ranged potentials', *J. Phys. Condens. Matter* **6**, 9275–9294 (1994).© *IOP Publishing 1994. Reproduced by permission of IOP Publishing. All rights reserved.*

The asymptotic analysis is more complicated for potentials that decay as a power law, as is the case for dispersion forces, where the dominant interaction at large r is $v(r) \approx -a_6/r^6$, with a Fourier transform given by

$$\hat{v}(k) = -\frac{\pi^2 a_6}{12}k^3 \tag{4.8.12}$$

The dependence on k^3 means that the $\hat{c}(k)$ can no longer be expanded purely in terms of even powers of k. Instead we can write

$$\hat{c}(k) = \hat{c}^{\text{sr}}(k) + ak^3 \tag{4.8.13}$$

where the short-range part $\hat{c}^{\text{sr}}(k)$ can be expanded in the manner of (4.8.1) and a is the coefficient (apart from the negative sign) of k^3 in (4.8.12). In this case the function $rh(r)$ is evaluated by contour integration of (4.8.3) with the contour taken around the upper-right quadrant of the complex plane (see Figure 4.10). The contribution from the circular part vanishes. Hence, from the residue theorem, the integral is given by

$$rh(r) = \frac{1}{2\pi^2}\text{Im}\left(2\pi i \sum_n \exp{(ik^{(n)}r)}R_n + \int_0^\infty ik_2 \exp{(-k_2 r)}\frac{\hat{c}(ik_2)}{1 - \rho\hat{c}(ik_2)}\,\mathrm{d}(ik_2)\right) \tag{4.8.14}$$

where R_n is the residue at a pole $k^{(n)}$ in the upper-right quadrant. The poles are again the roots of (4.8.4), which now has the form

$$1 - \rho\left[\hat{c}^{\text{sr}}(k) + ak^3\right] = 0 \qquad (4.8.15)$$

The presence of the term in k^3 means that there are no purely imaginary solutions to this equation and hence no poles on the imaginary axis. In fact, since $\hat{c}^{\text{sr}}(k)$ is a real function, the imaginary part of (4.8.15) implies that $k_2 = 0$. It is precisely the absence of poles on the imaginary axis that allows the use of the contour shown in Figure 4.10.

The final task is to determine the long-range behaviour of the integral, $I(r)$, on the right-hand side of (4.8.14), which can be rewritten in the form

$$I(r) = \frac{1}{2\pi^2\rho}\text{Im}\int_0^\infty k_2 \exp\left(-k_2 r\right)\left(1 - \frac{1}{1 - \rho\hat{c}(ik_2)}\right)dk_2 \qquad (4.8.16)$$

The first term in large brackets leads to a real integral; what remains is

$$I(r) = -\frac{1}{2\pi^2\rho}\text{Im}\int_0^\infty k_2 \exp\left(-k_2 r\right)\frac{1}{1 - \rho\left[\hat{c}^{\text{sr}}(ik_2) - iak_2^3\right]}dk_2 \qquad (4.8.17)$$

The presence of the exponential factor means that the integral is dominated by the contribution from small k_2. Use of the expansion (4.8.1) for $\hat{c}^{\text{sr}}(k)$ and Taylor expansion of the integrand to order k_2^4 leads, after taking the imaginary part, to

$$I(r) = \frac{a}{2\pi^2}\frac{1}{\left[1 - \rho\hat{c}^{\text{sr}}(k = 0)\right]^2}\int_0^\infty \exp\left(-k_2 r\right)\left[k_2^4 + \mathcal{O}(k_2^6)\right]dk_2$$
$$= \beta a_6 S(0)^2\frac{1}{r^5} + \mathcal{O}(r^{-7}) \qquad (4.8.18)$$

where $S(0)$ is the long-wavelength limit of the static structure factor and (3.6.10), (4.8.12) and (4.8.13) have been used. Thus, from (4.8.14), and by analogy with (4.8.10), we find that

$$h(r) = [S(0)]^2\frac{\beta a_6}{r^6} + \sum_n |A_n| \exp\left(-k_2^{(n)}r\right)\cos\left(k_1^{(n)}r - \theta_n\right) \qquad (4.8.19)$$

where the second term is the contribution from all poles within the upper-right quadrant of the complex plane. The absence of the factor 2 in front of the sum compared with (4.8.10) comes from the fact that the conjugate poles in the upper-left quadrant make no contribution. The conclusion, therefore, is that at large r, $h(r)$ behaves in the same manner as $c(r)$ but with a prefactor which is small in dense, weakly compressible liquids. The same result had been arrived at earlier and via a different route by Enderby et al.[51]

The picture of the asymptotic behaviour of $h(r)$ presented by (4.8.19) is that of oscillations which eventually merge into a power law decay. The effect of increasing density at constant temperature is simply to increase the value of r at which the oscillations disappear. Hence there is no sharp cross-over between different regimes of the type found for short-range potentials, for which there is a pure exponential contribution to the decay. Efforts have been made[52] to redefine the Fisher–Widom line to cater for such a situation, based on a more detailed study of the pole structure for potentials that behave as r^{-6}. This has revealed that although there can be no purely imaginary pole, there is a 'pseudo-exponential' pole that lies off the imaginary axis but very close to it, the contribution from which substantially modifies the asymptotic decay.

The extension of the asymptotic analysis to binary mixtures is straightforward but it leads to some surprising results.[48a] The k-space representation of the Ornstein–Zernike relations (3.6.12) is

$$\hat{h}_{\nu\mu}(k) = \hat{c}_{\nu\mu}(k) + \sum_{\lambda} \rho_{\lambda} \hat{c}_{\nu\lambda}(k)\hat{h}_{\lambda\mu}(k) \qquad (4.8.20)$$

where $\rho_{\lambda} = x_{\lambda}\rho$. These coupled equations can be solved for $\hat{h}_{\nu\mu}(k)$ in the form of ratios of k-space functions, a key feature of which is the fact that the denominator is the same for all ν, μ. The poles of $\hat{h}_{\nu\mu}(k)$ are given by the zeros of this common denominator and are therefore the same for all pairs. The functions $h_{\nu\mu}(r)$ can again be calculated by contour integration with a result given by a generalisation of (4.8.6);

$$rh_{\nu\mu}(r) = \frac{1}{2\pi} \sum_{n} R_n^{\nu\mu} \exp\left(ik^{(n)}r\right) \qquad (4.8.21)$$

which implies that asymptotically all pair correlation functions decay with the same characteristic length, $2\pi/k_2$, and the same oscillatory period, $2\pi/k_1$, where k_1 and k_2 are the real and imaginary parts of the pole or poles nearest to the real axis, conclusions that are somewhat counter–intuitive. The amplitude and phase of oscillation will, however, be different. Explicit calculations for highly size-asymmetric, binary mixtures of hard spheres show that the period of oscillation is close to the diameter of the larger species.

The decay of the density profile at a planar, fluid–fluid or wall–fluid interface can also be analysed along the same lines as those we have described. Calculation of the asymptotic behaviour close to the critical point or in ionic liquids[53] introduces new problems, discussion of which is deferred until Sections 5.7 and 10.3, respectively.

REFERENCES

[1] Kirkwood, J.G., *J. Chem. Phys.* **3**, 300 (1935).
[2] Born, M. and Green M.S., 'A General Kinetic Theory of Liquids'. Cambridge University Press, Cambridge, 1949.

[3]   Levesque, D., *Physica* **32**, 1985 (1966).

[4]   Ichimaru, S., *Phys. Rev. A* **2**, 494 (1970).

[5]   Barrat, J.L., Hansen, J.P. and Pastore, G., *Mol. Phys.* **63**, 747 (1988).

[6]   (a) Percus, J.K., *Phys. Rev. Lett.* **8**, 462 (1962).
      (b) Percus, J.K., *In* 'The Equilibrium Theory of Classical Fluids' (H.L. Frisch and J.L. Lebowitz, eds). W.A. Benjamin, New York, 1964.

[7]   The HNC approximation was developed independently by several workers. For some historical background, see Rowlinson, J.S., *Rep. Prog. Phys.* **28**, 169 (1965).

[8]   (a) Gillan, M.J, *Mol. Phys.* **38**, 1781 (1979). (b) Labik, S., Malijevski, A. and Vonka, P., *Mol. Phys.* **56**, 709 (1985).

[9]   The PY equation was originally derived in a very different way by Percus, J.K. and Yevick, G.J., *Phys. Rev.* **110**, 1 (1958).

[10] (a) Morita, T., *Prog. Theor. Phys.* **23**, 423, 829 (1960). (b) See also Schlijper, A.G., Telo da Gama, M.M. and Ferreira, P.G., *J. Chem. Phys.* **98**, 1534 (1993).

[11] (a) Thiele, E., *J. Chem. Phys.* **39**, 474 (1963). (b) Wertheim, M.S., *Phys. Rev. Lett.* **10**, 321 (1963). (c) Wertheim, M.S., *J. Math. Phys.* **5**, 643 (1964).

[12] Baxter, R.J. *Aust. J. Phys.* **21**, 563 (1968).

[13] (a) Lebowitz, J.L., *Phys. Rev.* **133**, A895 (1964). (b) Lebowitz, J.L. and Rowlinson, J.S., *J. Chem. Phys.* **41**, 133 (1964).

[14] (a) Analytical expressions covering the range $x = 1$ to 5 are given by Smith, W.R. and Henderson, D., *Mol. Phys.* **19**, 411 (1970). (b) See also Chang, J. and Sandler, S.I., *Mol. Phys.* **81**, 735 (1994).

[15] (a) Verlet, L. and Weis, J.J., *Phys. Rev. A* **5**, 939 (1972). (b) The implementation of certain theories also requires a knowledge of the hard-sphere $y(r)$ inside the hard core, a parameterisation of which is given by Grundke, E.W. and Henderson, D., *Mol. Phys.* **24**, 269 (1972).

[16] Baxter, R.J., *J. Chem. Phys.* **49**, 2770 (1968).

[17] Hansen, J.P. and Verlet, L., *Phys. Rev.* **184**, 151 (1969).

[18] Chen, M., Henderson, D. and Barker, J.A., *Can. J. Phys.* **54**, 703 (1969).

[19] Smith, W.R., Henderson, D. and Tago Y., *J. Chem. Phys.* **67**, 5308 (1977).

[20] Madden, W.G., *J. Chem. Phys.* **75**, 1984 (1981).

[21] Madden, W.G. and Rice, S.A., *J. Chem. Phys.* **72**, 4208 (1980).

[22] This result was first derived by Mayer, J.E. and Montroll, E., *J. Chem. Phys.* **9**, 2 (1941). We recall from Section 3.7 that an irreducible diagram is one without articulation circles.

[23] (a) Kirkwood, J.G., *J. Chem. Phys.* **3**, 300 (1935). (b) Nijboer, B.R.A. and van Hove, L., *Phys. Rev.* **85**, 177 (1952).

[24] Llano-Restrepo, M. and Chapman, W.G., *J. Chem. Phys.* **97**, 2046 (1992).

[25] (a) In the case of hard spheres the limiting value as $r \to 0$ is equal to $\exp(\beta\mu^{ex})$. See Hoover, W.G. and Poirier, J.C., *J. Chem. Phys.* **37**, 1041 (1962). (b) Widom, B., *J. Chem. Phys.* **39**, 2808 (1963).

[26] Hill, T.L., 'Introduction to Statistical Thermodynamics'. Addison-Wesley, Reading, 1960, p. 313.

[27] This transformation is achieved by a topological reduction based on Lemma 5 of Section 3.7.

[28] (a) Stell, G., *Physica* **29**, 517 (1963). (b) See also Attard, P., "Thermodynamics and Statistical Mechanics". Academic Press, London, 2002, Section 9.3.2.

[29] We use electrostatic units (esu).

[30] See, e.g. Perkyns, J.S., Dyer, K.M. and Pettitt, B.M., *J. Chem. Phys.* **116**, 9404 (2002).

[31] Torrie, G. and Patey, G.N., *Mol. Phys.* **34**, 1623 (1977).

[32] (a) For examples of this approach, see Caccamo, C., *Phys. Rep.* **274**, 1 (1996). (b) Bomont, J.M., *Adv. Chem. Phys.* **139**, 1 (2008).

[33] Fantoni, R. and Pastore, G. *J. Chem. Phys.* **120**, 10681 (2004).

[34] (a) Verlet, L., *Physica* **30**, 95 (1964). (b) Verlet, L., *Physica* **31**, 959 (1965).

[35] (a) Lado, F., *Phys. Rev. A* **8**, 2548 (1973). (b) Rosenfeld, Y. and Ashcroft, N.W., *Phys. Rev. A* **20**, 1208 (1979). (c) Lado, F., Foiles, S.M. and Ashcroft, N.W., *Phys. Rev. A* **28**, 2374 (1983). (d) Foiles, S.M., Ashcroft, N.W. and Reatto, L., *J. Chem. Phys.* **80**, 4441 (1984). (e) Enciso, E., Lado, F., Lombardero, M., Abascal, J.L.F. and Lago, S., *J. Chem. Phys.* **87**, 2249 (1987). Note that the bridge function in these and related papers is sometimes defined, in the present notation, as $-b(r)$.

[36] See Note 14. The hard-sphere bridge function has also been parameterised as a function of packing fraction by Malijevský, A. and Labík, S., *Mol. Phys.* **60**, 663 (1987).

[37] See, e.g. Kambayashi, S. and Chihara, J., *Phys. Rev. E* **50**, 1317 (1994).

[38] Rogers, F.J. and Young, D.A., *Phys. Rev. A* **30**, 999 (1984).

[39] Zerah, G. and Hansen, J.P., *J. Chem. Phys.* **84**, 2336 (1986).

[40] (a) Pini, D., Stell, G. and Høye, J.S., *Int. J. Thermophys.* **19**, 1029 (1998). (b) Pini, D., Stell, G. and Wilding, N.B., *Mol. Phys.* **95**, 483 (1998). (c) Caccamo, C., Pellicane, G., Costa, D., Pini, D. and Stell, G., *Phys. Rev. E* **60**, 5533 (1999). (d) Kahl, G., Schöll-Paschinger, E. and Stell, G., *J. Phys. Condensed Matter* **14**, 9153 (2002).

[41] This result follows straightforwardly from the thermodynamic relations (2.3.8) and (2.3.9).

[42] (a) Waisman, E., *Mol. Phys.* **25**, 45 (1973). (b) For later developments see, e.g. Høye, J.S. and Blum, L., *J. Stat. Phys.* **16**, 399 (1977). (c) Ginoza, M., *Mol. Phys.* **71**, 145 (1990). (d) Tang, Y., *J. Chem. Phys.* **118**, 4140 (2003).

[43] See, e.g. Pini, D., Stell, G. and Wilding, N.B., *J. Chem. Phys.* **115**, 2702 (2001).

[44] (a) Schöll-Paschinger, E., Benavides, A.L. and Castañeda-Priego, R., *J. Chem. Phys.* **123**, 234513 (2005). (b) Pini, D., Parola, A., Colombo, J. and Reatto, L., *Mol. Phys.* **109**, 1343 (2011).

[45] (a) Del Río, F., Avalos, E., Espíndola, R., Rull, L.F., Jackson, G. and Lago, S., *Mol. Phys.* **100**, 2531 (2002) (closed circles). (b) Largo, J., Miller, M.A. and Sciortino, F., *J. Chem. Phys.* **128**, 134513 (2008) (open circles).

[46] Orea, P., Tapia-Medina, C., Pini, D. and Reiner, A., *J. Chem. Phys.* **132**, 114108 (2010).

[47] Fisher, M.E. and Widom, B., *J. Chem. Phys.* **50**, 3756 (1969).

[48] (a) Evans, R., Leote de Carvalho, R.J.F., Henderson, J.R. and Hoyle, D.C., *J. Chem. Phys.* **100**, 591 (1994). (b) Leote de Carvalho, R.J.F., Evans, R., Hoyle, D.C. and Henderson, J.R., *J. Phys. Condensed Matter* **6**, 9275 (1994). In these papers the notation $\alpha_1 \equiv k_1$, $\alpha_0 \equiv k_2$ is used.

[49] The potential is also shifted upwards to make it vanish at the truncation distance.

[50] Smit, B., *J. Chem. Phys.* **96**, 8639 (1992).

[51] Enderby, J.E., Gaskell, T. and March, N.H., *Proc. Phys. Soc.* **85**, 217 (1965).

[52] Evans, R. and Henderson, J.R., *J. Phys.: Condensed Matter* **21**, 474220 (2009).

[53] Leote de Carvalho, R.J.F. and Evans, R., *Mol. Phys.* **83**, 619 (1994).

Chapter 5

Perturbation Theory

5.1 INTRODUCTION: THE VAN DER WAALS MODEL

The intermolecular pair potential often separates in a natural way into two parts: a harsh, short-range repulsion and a smoothly varying, long-range attraction. A separation of this type is an explicit ingredient of many empirical representations of the intermolecular forces, including the Lennard-Jones potential. It is now generally accepted that the structure of most simple liquids, at least at high density, is largely determined by the way in which the molecular hard cores pack together. By contrast, the attractive interactions may, in a first approximation, be regarded as giving rise to a uniform background potential that provides the cohesive energy of the liquid but has little effect on its structure. A further plausible approximation consists in modelling the short-range forces by the infinitely steep repulsion of the hard-sphere potential. The properties of the liquid of interest can in this way be related to those of a hard-sphere reference system, the attractive part of the potential being treated as a perturbation. The choice of the hard-sphere fluid as a reference system is an obvious one, since its thermodynamic and structural properties are well known.

The idea of representing a liquid as a system of hard spheres moving in a uniform, attractive potential is an old one, providing as it does the physical basis for the famous van der Waals equation of state. At the time of van der Waals little was known of the properties of the dense, hard-sphere fluid. The approximation that van der Waals made was to set the excluded volume per sphere of diameter d equal to $\frac{2}{3}\pi d^3$ (or four times the hard-sphere volume), which leads to an equation of state of the form

$$\frac{\beta P_0}{\rho} = \frac{1}{1 - 4\eta} \tag{5.1.1}$$

where, as before, η is the packing fraction. Equation (5.1.1) gives the second virial coefficient correctly (see (3.9.14)) but it fails badly at high densities. In particular, the pressure diverges as $\eta \to 0.25$, a packing fraction lying well below that of the fluid–solid transition ($\eta \approx 0.49$).

Theory of Simple Liquids, Fourth Edition. http://dx.doi.org/10.1016/B978-0-12-387032-2.00005-2

Considerations of thermodynamic consistency[1] show that the equation of state compatible with the hypothesis of a uniform, attractive background is necessarily of the form

$$\frac{\beta P}{\rho} = \frac{\beta P_0}{\rho} - \beta \rho a \qquad (5.1.2)$$

where a is a positive constant; this is equivalent to supposing that the chemical potential is lowered with respect to that of the hard spheres by an amount proportional to the density and equal to $2a\rho$. The classic van der Waals equation is then recovered by substituting for P_0 from (5.1.1). It is clear that a first step towards improving on van der Waals's result is to replace (5.1.1) by a more accurate hard-sphere equation of state, such as that of Carnahan and Starling, (3.9.20). A calculation along these lines was first carried out by Longuet–Higgins and Widom,[2] who thereby were able to account successfully for the melting properties of rare-gas solids.

The sections that follow are devoted to perturbation methods that may be regarded as attempts to improve the theory of van der Waals in a systematic fashion. The methods we describe have as a main ingredient the assumption that the structure of a dense, monatomic fluid resembles that of an assembly of hard spheres. Justification for this intuitively appealing idea is provided by the great success of the perturbation theories to which it gives rise, and which mostly reduce to (5.1.2) in some well-defined limit, but more direct evidence exists to support it. For example, it has long been known[3] that the experimental structure factors of a variety of liquid metals near their normal melting points can to a good approximation be superimposed on the structure factor of an 'equivalent' hard-sphere fluid, and Figure 5.1 shows the results of a similar but more elaborate analysis of data obtained by molecular dynamics calculations for the Lennard-Jones fluid. The fact that the attractive forces play such an apparently minor role in these examples is understandable through the following argument.[4] Equation (3.6.9) shows that the structure factor determines the density response of the fluid to a weak, external field. If the external potential is identified with the potential due to a test particle placed at the origin, the long-range part of that potential gives rise to a long-wavelength response in the density. In the long-wavelength limit ($k \to 0$), the response is proportional to $S(k = 0)$ and hence, through (3.6.11), to the isothermal compressibility. Under triple-point conditions the compressibility of a liquid is very small: typically the ratio of χ_T to its ideal-gas value is approximately 0.02. The effects of long-wavelength perturbations are therefore greatly reduced. At lower densities, particularly in the critical region, the compressibility can become very large. The role of the attractive forces is then important and the simple van der Waals model no longer has a sound physical basis.

We shall assume throughout this chapter that the interactions between particles are spherically symmetric and pairwise additive, though there is no difficulty in principle in extending the treatment to include three-body and higher-order forces. We also suppose that the system of interest is homogeneous.

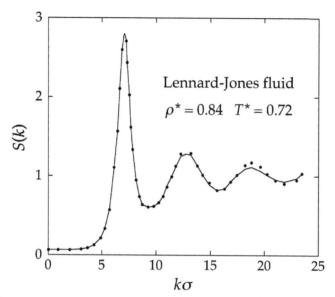

FIGURE 5.1 Structure factor of the Lennard-Jones fluid close to the triple point (curve) and its representation by a hard-sphere model (points). *Redrawn with permission from Ref. 4 © 1968 American Physical Society.*

The basis of all the perturbation theories we discuss is a division of the pair potential of the form

$$v(1,2) = v_0(1,2) + w(1,2) \qquad (5.1.3)$$

where $v_0(1,2)$ is the pair potential of the reference system and $w(1,2)$ is the perturbation. The calculation then usually proceeds in two stages. The first step is to compute the effects of the perturbation on the thermodynamic properties and pair distribution function of the reference system. This can be done systematically via an expansion in powers either of inverse temperature (the 'λ-expansion') or of a parameter that measures the range of the perturbation (the 'γ-expansion'). When hard spheres themselves are the reference system, this completes the calculation, but in the more general situation the properties of some 'soft-core' reference system must in turn be related to those of the hard-sphere fluid.

5.2 THE λ-EXPANSION

Consider a pair potential $v_\lambda(1,2)$ of the form

$$v_\lambda(1,2) = v_{\lambda_0}(1,2) + w_\lambda(1,2) \qquad (5.2.1)$$

where λ is a parameter that varies between λ_0 and λ_1. When $\lambda = \lambda_0$, w_λ vanishes and the potential $v_{\lambda_0} \equiv v_0$ reduces to that of a reference system, the properties of which are assumed to be known, whereas for $\lambda = \lambda_1$ the potential $v_{\lambda_0} \equiv v$ is the one that characterises the system of interest. The quantity λ has the meaning of a coupling parameter: the effect of varying λ continuously from λ_0 to λ_1 is that of gradually increasing the perturbation $w_\lambda(1,2)$. The commonest example of such a potential is

$$v_\lambda(1,2) = v_0(1,2) + \lambda w(1,2) \tag{5.2.2}$$

with $\lambda_0 = 0$ and $\lambda_1 = 1$; when $\lambda = 1$, the potential is the same as that introduced in (5.1.3).

Let $V_N(\lambda)$, given by

$$V_N(\lambda) = \sum_{i=1}^{N} \sum_{j>i}^{N} v_\lambda(i,j) \tag{5.2.3}$$

be the total potential energy of a system of particles interacting through the potential (5.2.1). From the definitions of the configuration integral, (2.3.13), and the excess free energy (here denoted simply by F), (2.3.20), it follows immediately that the derivative of $F(\lambda)$ with respect to the coupling parameter is

$$\beta \frac{\partial F(\lambda)}{\partial \lambda} = \frac{1}{Z_N(\lambda)} \int \exp\left[-\beta V_N(\lambda)\right] \beta V_N'(\lambda) \mathrm{d}\mathbf{r}^N = \beta \langle V_N'(\lambda) \rangle_\lambda \tag{5.2.4}$$

where $V_N'(\lambda) \equiv \partial V_N(\lambda)/\partial \lambda$ and $\langle \cdots \rangle_\lambda$ denotes a canonical ensemble average for the system characterised by the potential $v_\lambda(1,2)$. Integration of (5.2.4) gives

$$\beta F(\lambda_1) = \beta F_0 + \beta \int_{\lambda_0}^{\lambda_1} \langle V_N'(\lambda) \rangle_\lambda \, \mathrm{d}\lambda \tag{5.2.5}$$

where $F_0 \equiv F_{\lambda_0}$ is the excess free energy of the reference system. A series expansion of the ensemble average $\langle V_N'(\lambda) \rangle_\lambda$ can now be made around its value for $\lambda = \lambda_0$:

$$\langle V_N'(\lambda) \rangle_\lambda = \langle V_N'(\lambda) \rangle_{\lambda_0} + (\lambda - \lambda_0) \frac{\partial}{\partial \lambda} \langle V_N'(\lambda) \rangle_\lambda \bigg|_{\lambda=\lambda_0} + \mathcal{O}(\lambda - \lambda_0)^2 \tag{5.2.6}$$

The derivative with respect to λ in (5.2.6) is

$$\frac{\partial}{\partial \lambda} \langle V_N'(\lambda) \rangle_\lambda = \langle V_N''(\lambda) \rangle_\lambda - \beta \left(\langle [V_N'(\lambda)]^2 \rangle_\lambda - \langle V_N'(\lambda) \rangle_\lambda^2 \right) \tag{5.2.7}$$

and insertion of this result in (5.2.5) yields an expansion of the free energy in powers of $(\lambda_1 - \lambda_0)$:

$$\beta F(\lambda_1) = \beta F_0 + (\lambda_1 - \lambda_0)\beta \langle V_N'(\lambda_0) \rangle_{\lambda_0}$$
$$+ \frac{1}{2}(\lambda_1 - \lambda_0)^2 \left(\beta \langle V_N''(\lambda_0) \rangle_{\lambda_0} - \beta^2 \left(\langle [V_N'(\lambda_0)]^2 \rangle_{\lambda_0} \right. \right.$$
$$\left. \left. - \langle V_N'(\lambda_0) \rangle_{\lambda_0}^2 \right) \right) + \mathcal{O}(\lambda_1 - \lambda_0)^3 \tag{5.2.8}$$

We now restrict ourselves to the important special case when $v_\lambda(1,2)$ is given by (5.2.2). If we define the total perturbation energy for $\lambda = 1$ as

$$W_N = \sum_{i=1}^{N}\sum_{j>i}^{N} w(i,j) \qquad (5.2.9)$$

then $V'_N = W_N$, $V''_N = 0$ and (5.2.8) simplifies to give

$$\beta F = \beta F_0 + \beta \langle W_N \rangle_0 - \frac{1}{2}\beta^2 \left(\langle W_N^2 \rangle_0 - \langle W_N \rangle_0^2 \right) + \mathcal{O}(\beta^3) \qquad (5.2.10)$$

The series (5.2.10) is called the *high-temperature expansion*, but the name is not entirely appropriate. Although successive terms are multiplied by increasing powers of β, the ensemble averages are also, in general, functions of temperature. However, when the reference system is a hard-sphere fluid, the averages depend only on density and the λ-expansion reduces to a Taylor series in T^{-1}. Equation (5.2.10) was first derived by Zwanzig,[5] who showed that the nth term in the series can be written in terms of the mean fluctuations $\langle [(W_N - \langle W_N \rangle_0)]^\nu \rangle_0$, with $\nu \leq n$. Thus every term in the expansion corresponds to a statistical average evaluated in the reference system ensemble. The third and fourth-order terms, for example, are

$$\beta F_3 = \frac{\beta^3}{3!} \langle [W_N - \langle W_N \rangle_0]^3 \rangle_0$$

$$\beta F_4 = -\frac{\beta^4}{4!} \left(\langle [W_N - \langle W_N \rangle_0]^4 \rangle_0 - 3 \langle [W_N - \langle W_N \rangle_0]^2 \rangle_0^2 \right) \qquad (5.2.11)$$

The assumption of pairwise additivity of the potential means that (5.2.5) can be written as

$$\frac{\beta F}{N} = \frac{\beta F_0}{N} + \frac{\beta}{2N} \int_0^1 d\lambda \iint \rho_\lambda^{(2)}(1,2)w(1,2)d1\,d2 \qquad (5.2.12)$$

where $\rho_\lambda^{(2)}(1,2)$ is the pair density for the system with potential $v_\lambda(1,2)$; this is a special case of the general result contained in (3.4.10). The pair density can then be expanded in powers of λ:

$$\rho_\lambda^{(2)}(1,2) = \rho_0^{(2)}(1,2) + \lambda \left. \frac{\partial \rho_\lambda^{(2)}(1,2)}{\partial \lambda} \right|_{\lambda=0} + \mathcal{O}(\lambda^2) \qquad (5.2.13)$$

When this result is inserted in (5.2.12) the term of zeroth order in λ yields the first-order term in the high-temperature expansion of the free energy:

$$\frac{\beta F_1}{N} = \frac{\beta}{2N} \iint \rho_0^{(2)}(1,2)w(1,2)d1\,d2 = \frac{\beta\rho}{2} \int g_0(1,2)w(1,2)d\mathbf{r}_{12} \qquad (5.2.14)$$

In this approximation the structure of the fluid is unaltered by the perturbation. At second order in λ, however, calculation of the free energy involves the derivative $\partial \rho_\lambda^{(2)} / \partial \lambda$. Care is needed in passing to the thermodynamic limit and the differentiation is easier to perform in the grand canonical ensemble. The final result for a closed system[6] is

$$
\begin{aligned}
\frac{\beta F_2}{N} = -\frac{1}{2} \beta^2 \Bigg(& \frac{1}{2} \rho \int g_0(1,2)[w(1,2)]^2 \mathrm{d}2 \\
& + \rho^2 \iint g_0^{(3)}(1,2,3) w(1,2) w(1,3) \mathrm{d}2\, \mathrm{d}3 \\
& + \tfrac{1}{4} \rho^3 \iiint [g_0^{(4)}(1,2,3,4) - g_0^{(2)}(1,2) g_0^{(2)}(3,4)] \\
& \times w(1,2) w(3,4) \mathrm{d}2\, \mathrm{d}3\, \mathrm{d}4 \Bigg) \\
& + \tfrac{1}{4} \beta^2 S(0) \left(\frac{\partial}{\partial \rho} \left(\rho^2 \int g_0(1,2) w(1,2) \mathrm{d}2 \right) \right)^2
\end{aligned} \qquad (5.2.15)
$$

where $S_0(k)$ is the structure factor of the reference system.

We see from (5.2.15) that a rigorous calculation of the second-order term requires a knowledge of the three- and four-particle distribution functions of the reference system. The situation is even more complicated for higher-order terms, since the expression obtained for the term of order n involves the distribution functions of all orders up to and including $2n$. By the same rule, calculation of the first-order term requires only the pair distribution function of the reference system. If ϵ defines the energy scale of the perturbation, truncation at first order is likely to be justified whenever $\beta \epsilon \ll 1$. The fact that the second and higher-order terms are determined by fluctuations in the total perturbation energy suggests that they should be small, relative to F_1, whenever the perturbing potential is a very smoothly varying function of particle separation.

The question of whether or not a first-order treatment is adequate depends on the thermodynamic state, the form of the potential $v(1,2)$, and the manner in which $v(1,2)$ is divided into a reference system potential and a perturbation. It is clear that the high-temperature expansion is easiest to apply when terms beyond first order are negligible, but approximate schemes that simplify the calculation of F_2 have also been devised, the best known of which are the two 'compressibility' approximations of Barker and Henderson.[7] The argument on which these are based is a semi-macroscopic one. Let the range of interparticle distances be divided into equal intervals of length r_m to $r_{m+1} = \Delta r$, with $m = 0, 1, 2, \dots$. Now imagine that two concentric spheres of radius r_m and r_{m+1} are drawn around each particle of the reference system. On average, the number of neighbours lying in the spherical shells between two successive spheres will be

$$
\langle N_m \rangle_0 = 2\pi N \rho \int_{r_m}^{r_{m+1}} g_0(r) r^2 \, \mathrm{d}r \qquad (5.2.16)
$$

If Δr is sufficiently small, the perturbation $w(r)$ will have essentially the same value, w_m say, at all points within the shell. By repeating the same exercise for other values of m, (5.2.10) can be rewritten in terms of the numbers $\langle N_m \rangle_0$ in the form

$$
\begin{aligned}
\beta(F - F_0) = {} & \beta \sum_m \langle N_m \rangle_0 \, w_m \\
& - \frac{1}{2}\beta^2 \sum_m \sum_n \big(\langle N_m N_n \rangle_0 - \langle N_m \rangle_0 \langle N_n \rangle_0 \big) w_m w_n \\
& + \mathcal{O}(\beta^3)
\end{aligned}
\tag{5.2.17}
$$

If the shells were of macroscopic volume, there would be no correlation between the numbers of particles in different shells, so that $\langle N_m N_n \rangle_0 = \langle N_m \rangle_0 \langle N_n \rangle_0$ for $m \neq n$. The second-order term in (5.2.17) would then reduce to

$$
\beta F_2 = -\frac{1}{2} \sum_m \langle \Delta N_m^2 \rangle_0 (\beta w_m)^2
\tag{5.2.18}
$$

where $\langle \Delta N_m^2 \rangle_0 \equiv \langle N_m^2 \rangle_0 - \langle N_m \rangle_0^2$. In addition, the fluctuation in the number of particles in any given shell would be related to the compressibility of the reference system by the macroscopic expression (2.4.23):

$$
\langle \Delta N_m^2 \rangle_0 = \langle N_m \rangle_0 k_B T \rho \chi_T^0 = k_B T \left(\frac{\partial \rho}{\partial P} \right)_0
\tag{5.2.19}
$$

With these assumptions, and replacement of the sum on m by an integral, (5.2.18) becomes

$$
\frac{\beta F_2}{N} = -\pi \rho k_B T \int_0^\infty [\beta w(r)]^2 \left(\frac{\partial \rho}{\partial P} \right)_0 g_0(r) r^2 \, dr
\tag{5.2.20}
$$

Alternatively, it can be argued that the derivative of the bulk density with respect to pressure in (5.2.20) should be replaced by the derivative of a local density $\rho g_0(r)$, thereby yielding a second approximation in the form

$$
\frac{\beta F_2}{N} = -\pi \rho k_B T \int_0^\infty [\beta w(r)]^2 \left(\frac{\partial [\rho g_0(r)]}{\partial P} \right)_0 r^2 \, dr
\tag{5.2.21}
$$

The rationale for this is that the fluctuations involved are of microscopic rather than macroscopic character. Equations (5.2.20) and (5.2.21) are called, respectively, the 'macroscopic' and 'local' compressibility approximations. The two methods lead to similar results[7] but the macroscopic version is somewhat easier to implement.

If the reference system is the hard-sphere fluid and the perturbation potential $w(1,2)$ is very long ranged, the high-temperature expansion limited to first order reduces to the van der Waals equation (5.1.2). It is necessary only that the range of $w(1, 2)$ be large compared with the range of interparticle separations over which $g_0(1, 2)$ is significantly different from its asymptotic value. Then, to a good approximation:

$$\frac{\beta F_1}{N} \approx \frac{1}{2}\beta\rho \int w(\mathbf{r})d\mathbf{r} = -\beta\rho a \qquad (5.2.22)$$

where a is positive when the perturbing potential is attractive. On differentiating with respect to density we recover (5.1.2):

$$\frac{\beta P}{\rho} = \rho\frac{\partial}{\partial\rho}\left(\frac{\beta F_0}{N} + \frac{\beta F_1}{N}\right) = \frac{\beta P_0}{\rho} - \beta\rho a \qquad (5.2.23)$$

A further important feature of the high-temperature expansion is the fact that the first-order approximation yields a rigorous upper bound on the free energy of the system of interest, irrespective of the choice of reference system. This result is a further consequence of the Gibbs–Bogoliubov inequalities discussed in Appendix B in connection with the density functional theory of Section 3.4. Consider two integrable, non-negative but otherwise arbitrary configuration space functions $A(\mathbf{r}^N)$ and $B(\mathbf{r}^N)$, defined such that[8]

$$\int A(\mathbf{r}^N)d\mathbf{r}^N = \int B(\mathbf{r}^N)d\mathbf{r}^N \qquad (5.2.24)$$

The argument in Appendix B shows that the two functions satisfy the inequality

$$\int A(\mathbf{r}^N)\ln A(\mathbf{r}^N)d\mathbf{r}^N \geq \int A(\mathbf{r}^N)\ln B(\mathbf{r}^N)d\mathbf{r}^N \qquad (5.2.25)$$

We now make two particular choices for A and B. First, let

$$A(\mathbf{r}^N) = \exp\left(\beta[F_0 - V_N(0)]\right), \quad B(\mathbf{r}^N) = \exp\left(\beta[F_0 - V_N(1)]\right) \quad (5.2.26)$$

It follows from (5.2.19) that

$$F \leq F_0 + \left[\langle V_N(1)\rangle_0 - \langle V_N(0)\rangle_0\right] = F_0 + \langle W_N\rangle_0 \qquad (5.2.27)$$

This is precisely the inequality announced earlier. If we interchange the definitions of A and B, i.e. if we set

$$A(\mathbf{r}^N) = \exp\left(\beta[F_0 - V_N(1)]\right), \quad B(\mathbf{r}^N) = \exp\left(\beta[F_0 - V_N(0)]\right) \qquad (5.2.28)$$

then we find from (5.2.19) that

$$F \geq F_0 + \langle W_N\rangle_1 \qquad (5.2.29)$$

where the average of the perturbation energy is now taken over the ensemble of the system of interest. This result is less useful than (5.2.21) because the properties of the system of interest are in general unknown. With the assumption of pairwise additivity, (5.2.21) and (5.2.23) may be combined in the form

$$\frac{\beta F_0}{N} + \frac{1}{2}\beta\rho \int g(\mathbf{r})w(\mathbf{r})d\mathbf{r} \leq \frac{\beta F}{N} \leq \frac{\beta F_0}{N} + \frac{1}{2}\beta\rho \int g_0(\mathbf{r})w(\mathbf{r})d\mathbf{r} \quad (5.2.30)$$

The second of the inequalities in (5.2.24) can be used as the basis for a variational approach to the calculation of thermodynamic properties.[9] The variational procedure consists in choosing a reference system potential that depends on one or more parameters and then of minimising the last term on the right-hand side of (5.2.24) with respect to those parameters. As we shall see in the next section, the method has been applied with particular success[10] to systems of particles interacting through an inverse power or 'soft-sphere' potential of the form

$$v(r) = \epsilon(\sigma/r)^n \quad (5.2.31)$$

In these calculations the reference system is taken to be a fluid of hard spheres and the hard-sphere diameter is treated as the single variational parameter. Some modest improvement is achieved if a correction is made for the fact that the configuration space accessible to the hard-sphere and soft-sphere fluids is different for the two systems. The effect of this correction is to add to the right-hand side of (5.2.14) a term[11] involving a ratio of configuration integrals:

$$\frac{\beta\Delta F}{N} = \frac{1}{N}\ln\frac{\int_{\Omega_d}\exp\left[-\beta V_N(\mathbf{r}^N)\right]d\mathbf{r}^N}{\int_\Omega\exp\left[-\beta V_N(\mathbf{r}^N)\right]d\mathbf{r}^N} \quad (5.2.32)$$

where $V_N(\mathbf{r}^N)$ is the total potential energy of the system of interest (the soft-sphere fluid), Ω represents the full configuration space and Ω_d represents that part of configuration space in which there is no overlap between hard spheres of diameter d. Since Ω_d is smaller than Ω, the correction is always negative, thereby lowering the upper bound on the free energy provided by the inequality (5.2.21). The correction term can be evaluated numerically by a Monte Carlo method[11b] and an approximate but accurate expression for the term has been derived[9] that involves only the pair distribution function of the hard-sphere fluid.

5.3 SINGULAR PERTURBATIONS: THE f-EXPANSION

The form of perturbation theory described in Section 5.2 is well suited to deal with weak, smoothly varying perturbations but serious or even insurmountable difficulties appear when a short-range, repulsive, singular or rapidly varying perturbation is combined with a hard-sphere reference potential. Such a situation arises in the case of the square-shoulder potential pictured in Figure 5.2. This resembles the more widely studied square-well potential of Figure 1.2a, but

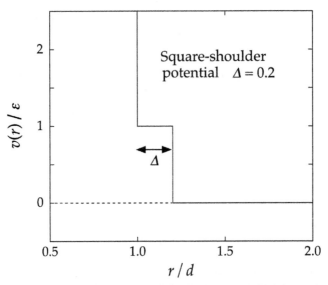

FIGURE 5.2 A square-shoulder potential with a repulsive barrier of height ϵ and width Δd, where $\Delta = 0.2$.

with the attractive well replaced by a repulsive barrier or 'shoulder' of height ϵ and width Δd, where d is the hard-sphere diameter. The square-shoulder potential has been adopted as a crude model of the interaction between metal ions of high atomic number such as Cs^+, which undergo electronic transitions at high pressures, and of the interactions in certain colloidal systems. It is also the simplest member of a class of 'core-softened' potentials that give rise to a rich variety of phase diagrams.

For given state conditions there will be ranges of ϵ and Δ for which the theory of Section 5.2 is adequate[12] but it will fail, in particular, when $\epsilon \gg k_B T$. The λ-expansion can be adapted to handle the more extreme situations by shifting the focus away from the perturbing potential $w(r)$ to the corresponding Mayer function, given by

$$f_w(r) = \exp\left[-\beta w(r)\right] - 1 \tag{5.3.1}$$

which remains finite for any repulsive potential.[13,14] The total perturbation energy for a given value of λ is now taken as

$$W_N(\lambda) = -k_B T \sum_{i=1}^{N} \sum_{j>i}^{N} \ln\left[1 + \lambda f_w(i, j)\right], \quad 0 \leq \lambda \leq 1 \tag{5.3.2}$$

and the total potential energy is therefore

$$V_N(\lambda) = V_N(0) + W_N(\lambda) \tag{5.3.3}$$

where $V_N(0)$ is the potential energy of the reference system. The expression for the excess Helmholtz free energy given by (5.2.8) remains valid, with $\lambda_0 = 0$ and $\lambda_1 = 1$, but the derivatives of $V_N(\lambda)$ or, equivalently, of $W_N(\lambda)$ with respect to λ are now

$$\beta W_N^{(n)} = \beta \left. \frac{\partial^n W_N(\lambda)}{\partial \lambda^n} \right|_{\lambda=0} = (-1)^n (n-1)! \sum_{i=1}^{N} \sum_{j>i}^{N} [f_w(i,j)]^n \qquad (5.3.4)$$

Substitution of (5.3.4) in (5.2.8) leads to an expansion of the free energy, usually called the f-expansion, which starts as

$$\beta F = \beta F_0 + \langle \beta W_N' \rangle_0$$
$$+ \frac{1}{2} \left(\langle \beta W_N'' \rangle_0 - \left\langle \left[\beta W_N' - \langle \beta W_N' \rangle_0 \right]^2 \right\rangle_0 \right) + \cdots \qquad (5.3.5)$$

Evaluation of the first-order correction is again given by an integral over the pair distribution function of the reference system:

$$\frac{\beta F_1}{N} = -\frac{1}{2} \rho \int g_0(1,2) f_w(1,2) \, \mathrm{d}\mathbf{r}_{12} \qquad (5.3.6)$$

while the second-order term can be recast in the form

$$\frac{\beta F_2}{N} = \frac{1}{4} \rho \int g_0(1,2) [f_w(1,2)]^2 \, \mathrm{d}\mathbf{r}_{12} - \frac{1}{2} \left\langle \left[\beta W_N' - \langle \beta W_N' \rangle_0 \right]^2 \right\rangle_0 \qquad (5.3.7)$$

The fluctuation term in this expression is given by the sum of the last three terms on the right-hand side of (5.2.15) with $\beta w(i,j)$ replaced everywhere by $f_w(i,j)$. A more useful result is provided by one of the compressibility approximations (5.2.20) or (5.2.21), with $\beta w(i,j)$ again replaced by $f_w(i,j)$.

A conceptually simple but challenging test of the f-expansion is provided by the following problem. Consider a mixture of equisized hard spheres of diameter d, labelled A and B, in which the interaction between differently labelled spheres is given by a hard-shoulder potential:

$$v_{AB}(r) = \infty, \quad r < d$$
$$= \epsilon, \quad d < r < d(1+\Delta)$$
$$= 0, \quad r > d(1+\Delta) \qquad (5.3.8)$$

We now take the limit $\epsilon \to \infty$, which transforms the system into a symmetrical, non-additive mixture of hard spheres with $d_{AB} = d(1+\Delta)$. The non-additivity can then be treated as a perturbation on a reference system corresponding to an ideal mixture of labelled but physically identical, hard spheres of diameter d; this brings the calculation close in spirit to that of the conformal solution theory described in Section 3.10. The perturbation associated with the non-additivity is simply

$$f_w(r) = -1, \quad d < r < d(1+\Delta)$$
$$= 0, \quad r > d(1+\Delta) \qquad (5.3.9)$$

and the first-order correction to the excess free energy provided by (5.3.6) therefore reduces to

$$\frac{\beta F_1}{N} = 4\pi \rho x_A x_B \int_d^{d(1+\Delta)} g_0(r) r^2 \, dR \qquad (5.3.10)$$

where $g_0(r)$ is the pair distribution function of a one-component hard-sphere fluid at a packing fraction $\eta = \pi d^3 N / 6V$, and $x_A, x_B = 1 - x_A$ are the fractions of particles labelled A and B, respectively. An additional factor $2x_A x_B$ appears compared with (5.3.6) because the perturbation affects only the A-B interaction.

As we saw in Section 3.10, positive non-additivity in mixtures of hard spheres is expected to drive a fluid–fluid phase separation above a critical density ρ_c. This has been confirmed by computer simulations, including Gibbs ensemble Monte Carlo calculations[15] for a binary mixture with $x_A = x_B$ and $\Delta = 0.2$. Figure 5.3 shows the Monte Carlo results for the phase diagram in the concentration-density plane together those predicted by first-order perturbation theory.[14] Given the severity of the test, the agreement between simulation and theory is good. In particular, the two estimates of the critical density $(\rho_c d^3 \approx 0.41)$ differ by only about 1%. The same theory shows that the critical density should decrease with increasing non-additivity, reaching a value $\rho_c d^3 \approx 0.08$ for $\Delta = 1$, in broad agreement with the predictions of other theoretical approaches and the results of other simulations[16]. An expression

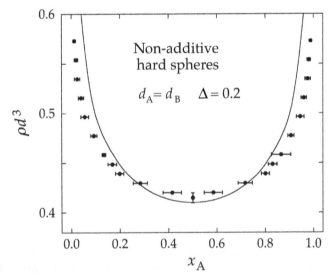

FIGURE 5.3 Phase diagram in the concentration-density plane for a binary mixture of non-additive hard spheres with $\Delta = 0.2$. The curve is calculated from first-order perturbation theory and the points with error bars show the results of Monte Carlo calculations.[15] *Redrawn with permission from Ref. 14 © Taylor & Francis Limited.*

for the first-order correction to the pair distribution function of the reference system has also been derived.[17]

5.4 SOFT-CORE REFERENCE SYSTEMS

Perturbation theories are useful only if they relate the properties of the system of interest to those of a well-understood reference system. Hard spheres are an obvious choice of reference system, for the reasons discussed in Section 5.1. On the other hand, realistic intermolecular potentials do not have an infinitely steep repulsive core, and there is no natural separation into a hard-sphere part and a weak perturbation. Instead, an arbitrary division of the potential is made, as in (5.2.1), and the properties of the reference system, with potential $v_0(r)$, are then usually related to those of hard spheres in a manner independent of the way in which the perturbation is treated. In this section we discuss how the relation between the reference system and the system of hard spheres can be established, postponing the question of how best to separate the potential until Section 5.5. We describe in detail only the 'blip function' method of Andersen, Weeks and Chandler,[18] but we also show how results obtained earlier by Rowlinson[19] and by Barker and Henderson[20] can be recovered from the same analysis. In each case the free energy of the reference system is equated to that of a hard-sphere fluid at the same temperature and density. The hard-sphere diameter is, in general, a functional of $v_0(r)$ and a function of ρ and T, and the various methods of treating the reference system differ from each other primarily in the prescription used to determine d.

If the reference system potential is harshly repulsive but continuous, the Boltzmann factor $e_0(r) = \exp[-\beta v_0(r)]$ typically has the appearance shown in Figure 5.4 and is not very different from the Boltzmann factor $e_d(r)$ of a hard-sphere potential. Thus, for a well-chosen value of d, the function

$$\Delta e(r) = e_0(r) - e_d(r) \tag{5.4.1}$$

is effectively non-zero only over a small range of r, which we denote by ξd. The behaviour of $\Delta e(r)$ as a function of r is sketched in the figure; the significance of the name 'blip function' given to it is obvious.

When ξ is small it is natural to seek an expansion of the properties of the reference system about those of hard spheres in powers of ξ. Such a series can be derived by making a functional Taylor expansion of the reduced free energy density $\phi = -\beta F^{\text{ex}}/V$ in powers of $\Delta e(r)$, i.e.

$$\phi = \phi_d + \int \left.\frac{\delta\phi}{\delta e(\mathbf{r})}\right|_{e=e_d} \Delta e(\mathbf{r})\, d\mathbf{r} + \frac{1}{2}\iint \left.\frac{\delta^2\phi}{\delta e(\mathbf{r})\delta e(\mathbf{r}')}\right|_{e=e_d} \Delta e(\mathbf{r})\Delta e(\mathbf{r}')\, d\mathbf{r}\, d\mathbf{r}' + \cdots$$

$$\tag{5.4.2}$$

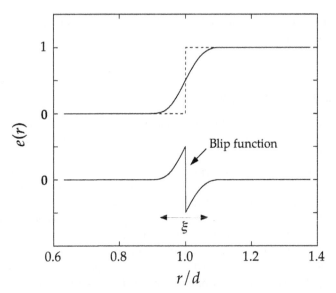

FIGURE 5.4   The blip function. The upper part of the figure shows the Boltzmann factors $e_0(r)$ and $e_d(r)$ for soft-core (full curve) and hard-sphere (dashes) potentials, respectively; the lower part shows the blip function, $\Delta e(r) = e_0(r) - e_d(r)$.

where ϕ_d is the free energy density of the hard-sphere fluid. We know from (2.5.23) and (3.4.8) that the functional derivative of ϕ with respect to $e(\mathbf{r})$ is

$$\frac{\delta\phi}{\delta e(\mathbf{r})} = \frac{1}{2}\rho^2 y(\mathbf{r}) \tag{5.4.3}$$

Equation (5.4.2) can therefore be rewritten as

$$\phi = \phi_d + \frac{1}{2}\rho^2 \int y_d(\mathbf{r})\Delta e(\mathbf{r})\,d\mathbf{r} + \cdots \tag{5.4.4}$$

The expression for the second-order term involves the three- and four-particle distribution functions of the hard-sphere system, but terms beyond first order are not needed for sufficiently steep potentials.

Since the range of $\Delta e(r)$ is ξd, the first-order term in the expansion (5.4.2) is of order ξ. A natural choice of d is one that causes the first-order term to vanish; d is then determined by the implicit relation

$$\int y_d(r)\Delta e(r)\,d\mathbf{r} = 0 \tag{5.4.5}$$

With this choice of d, the second order term in (5.4.2), which would normally be of order ξ^2, becomes of order ξ^4. Thus the free energy density of the reference system is related to that of the hard-sphere fluid by

$$\phi_0 = \phi_d + \mathcal{O}(\xi^4) \tag{5.4.6}$$

where d is defined by (5.4.5).

Equation (5.4.5) represents just one of many possible prescriptions for calculating the diameter of the 'equivalent' hard spheres. Because $\Delta e(r)$ is non-zero only in a narrow range of r, the factor $r^2 y_d(r)$ in (5.4.5) can be expanded in a Taylor series about $r = d$ in the form

$$r^2 y_d(r) = \sigma_0 + \sigma_1 \left(\frac{r}{d} - 1 \right) + \sigma_2 \left(\frac{r}{d} - 1 \right)^2 + \cdots \tag{5.4.7}$$

with

$$\frac{\sigma_m}{d^m} = \frac{d^m}{dr^m} r^2 y_d(r) \Big|_{r=d} \tag{5.4.8}$$

Substitution of the expansion (5.4.7) in (5.4.5) gives

$$\sum_{m=0}^{\infty} \frac{\sigma_m}{m!} I_m = 0 \tag{5.4.9}$$

where

$$I_m = \int_0^\infty \left(\frac{r}{d} - 1 \right)^m \Delta e(r) \, d(r/d)$$

$$= -\frac{1}{m+1} \int_0^\infty \left(\frac{r}{d} - 1 \right)^{m+1} \frac{d}{dr} \exp[-\beta v_0(r)] dr \tag{5.4.10}$$

If $v_0(r)$ varies rapidly with r, the derivative in (5.4.10) is approximately a δ-function at $r = d$ and the series (5.4.9) is rapidly convergent. If only the first term is retained, then $I_0 = 0$, and a straightforward integration shows that

$$d = \int_0^\infty \left(1 - \exp[-\beta v_0(r)] \right) dr \tag{5.4.11}$$

This expression is identical to one derived in a different way by Barker and Henderson.[20] In the case when $v_0(r)$ is a soft-sphere potential of the form (5.2.31), the integral in (5.4.11) can be evaluated explicitly in terms of the Γ-function to give

$$d = \sigma(\epsilon/k_B T)^{1/n} \Gamma\left(\frac{n-1}{n} \right) = \sigma(\epsilon/k_B T)^{1/n}(1 + \gamma/n) + \mathcal{O}(1/n^2) \tag{5.4.12}$$

where $\gamma = 0.5772\ldots$ is Euler's constant. On discarding terms of order $1/n^2$ we recover an expression due to Rowlinson.[19] Rowlinson's calculation is based on an expansion of the free energy in powers of the inverse steepness parameter $\lambda = 1/n$ about $\lambda = 0$ (hard spheres); the work of Barker and Henderson may be regarded as a generalisation of Rowlinson's method to a repulsive potential of arbitrary form.

The main difference between (5.4.5) and (5.4.11) is the fact that the former yields a hard-sphere diameter which is a function of both density and temperature, whereas the Barker–Henderson diameter is dependent only on temperature. The potentially greater flexibility that use of (5.4.5) thereby provides is irrelevant, however, in the case of the soft-sphere potential (5.2.31), for which the excess thermodynamic properties are determined by the single, dimensionless parameter Γ defined as

$$\Gamma = \rho\sigma^3 \left(\frac{\epsilon}{k_B T} \right)^{3/n} \tag{5.4.13}$$

Figure 5.5 show results[10] obtained by the two methods for the free energy and pressure of the soft-sphere fluid with $n = 12$ (the r^{-12} fluid) and makes comparison with the results of Monte Carlo calculations. The blip function approach is clearly superior but the differences become smaller as the potential becomes steeper.

Blip function theory also yields a very simple expression for the pair distribution function of the reference system. It follows from (5.4.3) and (5.4.4) that

$$y_0(r) = y_d(r) + \text{higher-order terms} \tag{5.4.14}$$

where the higher-order terms are of order ξ^2 or smaller if d is chosen to satisfy (5.4.5). Thus

$$g_0(r) = \exp[-\beta v_0(r)]y_0(r) \approx \exp[-\beta v_0(r)]y_d(r) \tag{5.4.15}$$

This result can now be used, in combination with (5.2.14), to compute the correction to the free energy which results from a perturbing potential $w(r)$. It also allows us to rewrite (5.4.5) in terms of the $k \to 0$ limits of the reference system and hard-sphere structure factors in the form $S_0(0) = S_d(0)$. Use of the hard-sphere diameter defined by (5.4.5) therefore has the effect of setting the compressibility of the reference system equal to that of the underlying hard-sphere fluid. Equation (5.4.15) is expected to be less accurate than the expression for the free energy, (5.4.6), because the neglected terms are now of order ξ^2 rather than ξ^4. This is borne out by the calculations made for the r^{-12}-fluid; the approximate $g_0(r)$ is in only moderate agreement with the results of simulations[21b] whereas the agreement obtained for the free energy is good, as illustrated in Figure 5.5. The situation improves markedly as n increases.

Although the blip function method yields satisfactory results for the thermodynamic properties of the r^{-12} fluid it is clear from Figure 5.5 that there is scope for improvement in the results obtained for the equation of state at large values of Γ, i.e. at high densities or low temperatures. There is, in addition, a lack of internal consistency in the theory. The results shown in the figure were obtained by numerical differentiation of the free energy and differ significantly from those obtained from the virial equation (2.5.22). Results derived

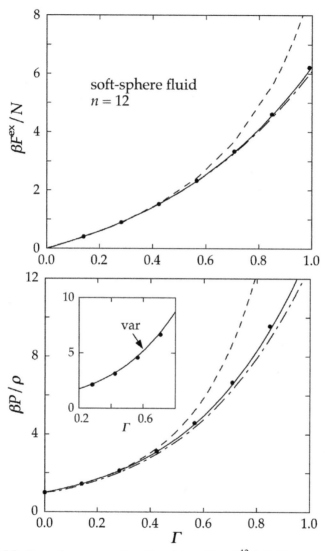

FIGURE 5.5 Excess free energy and equation of state of the r^{-12} fluid. The curves show the predictions of perturbation theory for four different choices of the diameter of the 'equivalent' hard spheres; in the main parts of the figure the chain, dashed and full curves are obtained from use of (5.4.5), (5.4.11) or (5.4.16), respectively, and the curve in the inset shows the results of the variational approach based on (5.2.30). The points give the results of Monte Carlo calculations.[21a] *Redrawn with permission from Ref. 10 © 2003 American Institute of Physics.*

from the free energy are the more reliable, but they are also more troublesome to compute. Equivalence of the two routes to the equation of state is guaranteed, however, if the hard-sphere diameter is calculated, not from (5.4.5), but from the relation[22]

$$\int \frac{\partial y_d(r)}{\partial d} \Delta e(r) d\mathbf{r} = 0 \tag{5.4.16}$$

Equation (5.4.16) is derived by requiring that the free energy of the system of interest be a minimum with respect to variations in the hard-sphere function $y_d(r)$. As Figure 5.5 shows, the calculated free energy and pressure of the r^{-12}-fluid are thereby brought into very close agreement with the Monte Carlo results.

The blip function expansion was designed specifically to treat the case of strongly repulsive potentials. This is true for the Lennard-Jones fluid, which we discuss in the next section, and the accuracy of the blip function method in such circumstances could hardly be improved upon. The method is less satisfactory for the softer repulsions relevant to liquid metals, because truncation of the expansion (5.4.2) after the first-order term is no longer justified. By contrast, though we see from Figure 5.6 that the hard-sphere variational approach described in Section 5.2 is comparable in accuracy with blip function theory for $n = 12$, it also retains its accuracy even for $n = 4$ while the first-order blip

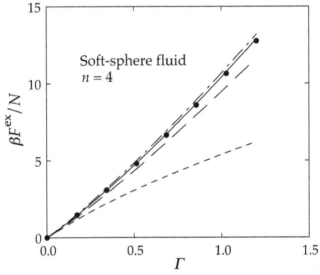

FIGURE 5.6 Excess free energy of the r^{-4} fluid. The points are Monte Carlo results and the curves show the predictions of different theories: blip function method based on (5.4.5) (short dashes) or (5.4.16) (long dashes), and variational theory based on a hard-sphere reference system with (full curve) or without (chain curve) the correction represented by (5.2.32). *Redrawn with permission from Ref. 23 © 2004 American Institute of Physics.*

function method does not.[23] We also see that within blip function theory the two prescriptions for the hard-sphere diameter, (5.4.5) and (5.4.16), give rise to significantly larger differences in free energy as the potential is softened. The correction (5.2.32) to the variational calculation is small but not negligible.

A soft-sphere, inverse-power potential also serves as a 'reference' system in a different sense to the one we have so far described. The underlying idea emerged from a study of the pressure-energy fluctuations observed in molecular dynamics simulations. In the case of an inverse-power potential the virial function and potential energy of every configuration are trivially related in the form

$$\mathcal{V} = \frac{n}{3} V_N \qquad (5.4.17)$$

The fluctuations in $\mathcal{V}$ and V_N relative to their equilibrium averages are therefore completely correlated, meaning that the correlation coefficient

$$\mathcal{R} = \frac{\langle \Delta \mathcal{V} \Delta V_N \rangle}{\left(\langle (\Delta \mathcal{V})^2 \rangle \langle (\Delta V_N)^2 \rangle \right)^{1/2}} \qquad (5.4.18)$$

is equal to one. It has been found, however, that in simulations of a number of models of simple liquids the correlation, though not complete, is very strong, and that $\mathcal{R}$ exceeds 0.9 over a large region of the phase diagram.[24] Thus for the Lennard-Jones fluid at near triple point conditions, $R \approx 0.94$, while for a model of water at room temperature the fluctuations are virtually uncorrelated, with $R < 0.001$. If the instantaneous values of $\mathcal{V}$ are plotted against those of V_N the result for a strongly correlating liquid is a scatter diagram in which the individual points are closely grouped around a straight line of slope q, while for an inverse-power potential all points would lie exactly on a line of slope $n/3$. The fluctuations in a strongly correlating system are therefore very similar to those in a soft-sphere fluid with an exponent $n = 3q$. In the case of the Lennard-Jones fluid, simulation shows that $q \approx 6$ and hence that the effective value of n lies in the range 18–19; as we shall see in the next section, that choice of n provides a good representation of the repulsive wall of the Lennard-Jones potential.

The matching of the behaviour of a strongly correlating liquid to that of a soft-sphere fluid has a remarkable and not easily anticipated consequence.[25] Equation (5.4.13) shows that properties of a soft-sphere fluid depend on density and temperature through the combination $\rho^{n/3}/T$. Similarly, it is found that along a line of constant ρ^q/T in the phase diagram of a strongly correlating liquid many properties are almost constant if all quantities involved are expressed in reduced units; such a line is therefore termed an 'isomorph'. The properties concerned include, among others, the excess entropy and heat capacity at constant volume, the pair distribution function and static structure factor, and the coefficients of self-diffusion and shear viscosity. The appearance of the excess entropy and diffusion constant in the list of near-invariants has a precedent in work, carried out much earlier, in which these two quantities were shown to be related in the same way for both inverse-power and Lennard-Jones fluids.[26]

5.5 AN EXAMPLE: THE LENNARD-JONES FLUID

The λ-expansion described in Section 5.2 is suitable for treating perturbations that vary slowly in space, while the blip function expansion and related methods of Section 5.3 provide a good description of reference systems for which the potential is rapidly varying but localised. In this section we show how the two approaches can be combined in a case where the pair potential has both a steep but continuous, repulsive part and a weak, longer ranged attraction. The example we choose is that of the Lennard-Jones fluid, a system for which sufficient data are available from computer simulations to allow a complete test to be made of different perturbation schemes.[27].

At first sight it might appear that the complications due to softness of the core would make it more difficult to obtain satisfactory results by perturbation theory than in situations where the potential consists of a hard-sphere interaction and a tail. This is not necessarily true, however, because there is now the extra flexibility provided by the arbitrary separation of the potential into a reference part, $v_0(r)$, and a perturbation, $w(r)$. A judicious choice of separation can significantly enhance the rate of convergence of the resulting perturbation series.

In the early work of McQuarrie and Katz[28] the r^{-12} term was chosen as the reference system potential and the r^{-6} term was treated as a perturbation. Given a scheme in which the properties of the reference system are calculated accurately, the method works well at temperatures above $T^* \approx 3$. At lower temperatures, however, the results are much less satisfactory. This is understandable, since the reference system potential is considerably softer than the full potential in the region close to the minimum in $v(r)$. In that region, as Figure 5.7 shows, the Lennard-Jones interaction is better described by an inverse-power potential with n in the range 18–20 rather than 12 and the choice of reference potential needs to reflect this behaviour if the resulting theory is to be successful over a wide range of state conditions.

The most commonly used divisions of the potential are those illustrated in Figure 5.8. In the separation used by Barker and Henderson[20] the reference system is defined by that part of the full potential which is positive $(r < \sigma)$ and the perturbation consists of the part that is negative $(r > \sigma)$. The reference system properties are then related to those of hard spheres of diameter d given by (5.4.11). In contrast to the case of the r^{-12} potential (see Figure 5.5), this treatment of the reference system yields very accurate results. The corrections due to the perturbation are handled in the framework of the λ-expansion; the first-order term is calculated from (5.2.14), with $g_0(r)$ taken to be the pair distribution function of the equivalent hard-sphere fluid. At $T^* = 0.72$ and

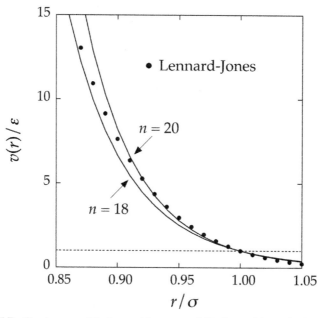

FIGURE 5.7 The steepness of the Lennard-Jones potential in the repulsive region in comparison with the behaviour of two inverse-power potentials. The Lennard-Jones energies have been shifted upwards by ϵ to make the three potentials coincide at $r = \sigma$.

$\rho^* = 0.85$, which is close to the triple point of the Lennard-Jones fluid, the results are $\beta F_0/N = 3.37$ and $\beta F_1/N = -7.79$. Thus the sum of the two leading terms is equal to -4.42, whereas the result obtained for the total excess free energy from Monte Carlo calculations is $\beta F/N = -4.87$. The sum of all higher-order terms in the λ-expansion is therefore far from negligible; detailed calculations show that the second-order term accounts for most of the remainder.[27a] The origin of the large second-order term lies in the way in which the potential is separated. As Figure 5.8 reveals, the effect of dividing $v(r)$ at $r = \sigma$ is to include in the perturbation the rapidly varying part of the potential between $r = \sigma$ and the minimum at $r = r_m \approx 1.122\,\sigma$. Since the pair distribution function has its maximum value in the same range of r, fluctuations in the total perturbation energy W_N, and hence the numerical values of F_2, are large.

The work of Barker and Henderson remains a landmark in the development of liquid state theory, since it demonstrated for the first time that thermodynamic perturbation theory is capable of yielding quantitatively reliable results even for states close to the triple point of the system of interest. A drawback to their method is the fact that its successful implementation requires a careful evaluation of the second-order term in the λ-expansion. The calculation of F_2 from (5.2.15) requires further approximations to be made and the theory is

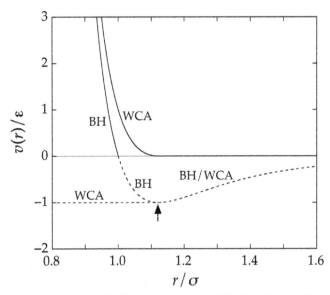

FIGURE 5.8 Two separations of the Lennard-Jones potential that have been used in perturbation theory calculations: BH, by Barker and Henderson;[20] WCA, by Weeks, Chandler and Andersen.[29] Full curves: the reference system potential; dashes: the perturbation. The arrow marks the position of the minimum in the full pair potential; at larger values of r the Barker–Henderson and WCA choices of perturbation are the same.

inevitably more awkward to handle than in the case when a first-order treatment is adequate.

The problem of the second-order term can be largely overcome by dividing the potential in the manner of Weeks, Chandler and Andersen,[29] usually called the WCA separation. In that method the potential is split at $r = r_m$ into its purely repulsive $(r < r_m)$ and purely attractive $(r > r_m)$ parts; the former defines the reference system and the latter constitutes the perturbation. To avoid a discontinuity at $r = r_m$, $w(r)$ is set equal to $-\epsilon$ for $r < r_m$, and $v_0(r)$ is shifted upwards by a compensating amount. Compared with the Barker–Henderson separation, the perturbation now varies more slowly over the range of r corresponding to the first peak in $g(r)$, and the perturbation series is therefore more rapidly convergent. For example, at $T^* = 0.72$, $\rho^* = 0.85$, the free energy of the reference system is $\beta F_0/N = 4.49$ and the first-order correction in the λ-expansion is -9.33; the sum of the two terms is -4.84, which differs by less than 1% from the Monte Carlo result for the full potential.[27b] Agreement of the same order is found throughout the high-density region and the perturbation series may confidently be truncated after the first-order term. The difficulties associated with the calculation of the second and higher-order terms are thereby avoided. At very high densities, on the other hand, the hard-sphere diameter calculated for the WCA separation may correspond to a packing fraction lying in the metastable region beyond the fluid–solid transition. This limits the range of applicability of the theory at supercritical temperatures.[30]

In most calculations based on the WCA separation the free energy of the reference system is related to that of hard spheres through (5.4.5) and (5.4.6). At high densities, the error (of order ξ^4) thereby introduced is very small. Under the same conditions, use of the approximate relation (5.4.5) to calculate the first-order correction from (5.2.14) also introduces only a very small error. In addition, when the hard-sphere diameter is calculated from (5.4.15), a simplification occurs, since it ensures that the compressibilities of the hard-sphere and reference systems are equal. The integral over the pair distribution function in (2.6.12) must therefore be the same for both $g_0(r)$ and $g_d(r)$. Since $g_0(r)$ and $g_d(r)$ are identical for $r > r_m$, it follows that the quantity

$$\int_0^{r_m} \left[g_0(r) - g_d(r)\right] r^2 \, dr$$

must vanish. But the perturbation is a constant (equal to $-\epsilon$) for $r < r_m$, so $g_0(r)$ can be replaced by $g_d(r)$ for all r in the evaluation of the first-order term. Thus

$$\frac{\beta A_1}{N} = 2\pi \int_d^\infty w(r) g_d(r) r^2 \, dr \qquad (5.5.1)$$

This argument does not apply for other choices of hard-sphere diameter, including that given by (5.4.16).

Equation (5.5.1) has precisely the same form as that of the first-order term in the Barker–Henderson approach, in which the hard-sphere fluid is identified as the reference system from the outset. The two methods can be brought even closer together by combining the choice of hard-sphere diameter made by Barker and Henderson with the WCA division of the potential. This leads to two first-order theories that differ only in the prescription used for the hard-sphere diameter. Results[10] obtained for the equation of state by the two methods are shown in Figure 5.9. The level of agreement with the results of computer simulations is good for both methods at densities below $\rho^* \approx 0.6$, but overall the WCA approach is clearly the more successful. However, the range of state conditions covered by the figure is very large. With a choice of Lennard-Jones parameters appropriate to argon, for example, the pressure reached at the highest density and temperature is of order 10 kbar.[33] Discrepancies appear when the results are plotted on an expanded scale and are particularly marked in the region close to the critical point, where the role played by fluctuations cannot be ignored. This is illustrated for the case of the isotherm $T^* = 1.4$ in the inset to the figure. The best estimate of the critical temperature of the Lennard-Jones fluid is $T_c^* \approx 1.31$ but the results obtained by first-order theory with either choice of hard-sphere diameter clearly correspond to a subcritical isotherm; it is evident that the critical temperature is significantly overestimated. Under these conditions we can expect a second-order theory to be more successful.

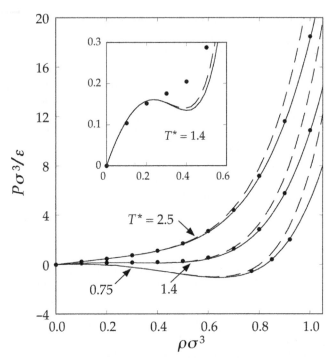

FIGURE 5.9 Equation of state of the Lennard-Jones fluid along three isotherms. The points are the results of simulations[31] and the curves show the predictions of first-order perturbation theory for two choices of the hard-sphere diameter when the WCA separation of the potential is used. Full curves: based on (5.4.5); chain curve: based on (5.4.11). The critical point temperature and density of the Lennard-Jones fluid are[32] $T^* \approx 1.31$ and $\rho^* \approx 0.32$. *Redrawn with permission from Ref. 10 © 2003 American Institute of Physics.*

This is confirmed by the results in Figure 5.10 for the isotherm $T^* = 1.35$, which shows the results of second-order Barker–Henderson theory[34] based on division of the potential at $r = \sigma$ and use of the macroscopic compressibility approximation (5.2.20) for F_2. The predicted critical temperature is now much closer to the true value but quantitative accuracy in the estimation of T_c cannot be expected from a truncated perturbation expansion. That would, in principle, require the summation of the expansion to all orders in the perturbation, a goal which can be reached within 'hierarchical reference theory', a description of which we defer until Section 5.8.

Ben-Amotz and Stell[35] have shown that a theory based on the WCA division of the potential can be formulated in a way that retains the accuracy of the original version but is easier to apply. This involves, first, the use of a hard-sphere system as the reference system rather than a soft repulsive system, the properties of which must be related to those of a hard-sphere fluid in a separate step. Secondly, the hard-sphere diameter is taken as the separation r at which

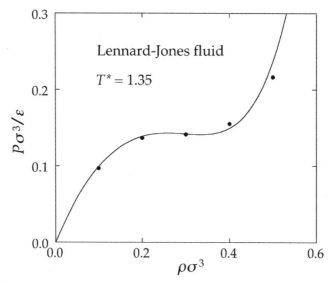

FIGURE 5.10 Equation of state of the Lennard-Jones fluid along the isotherm $T^* = 1.35$. The points are the results of simulations[31] and the curve shows the predictions of second-order perturbation theory.[34]

$v_0(r) = k_B T$, which makes the determination of d a trivial task. The resulting theory, which the authors refer to as HS-WCA, gives results that are virtually identical to those of the original method, with the added advantage of being less sensitive to the precise choice of hard-sphere diameter.

At low densities the attractive forces play an important role in determining the structure and the key assumption of a first-order theory, namely that $g(r) \approx g_0(r)$, is no longer valid. New methods are therefore required if the calculation of higher-order terms is to be avoided.

5.6 TREATMENT OF ATTRACTIVE FORCES

Situations in which the influence of the attractive forces on the structure cannot be ignored may be treated by methods similar to those used when the perturbation is both weak and very long ranged relative to the reference system potential. In such cases the natural expansion parameter is the inverse range rather than the strength of the perturbation; this leads to the so-called γ-expansion,[36] the nature of which differs significantly from that of the λ-expansion described in Section 5.2. Early work on the γ-expansion was motivated by the fact that the exact solution was known for the one-dimensional model of hard rods of length d which attract each other via the potential

$$w_\gamma(x) = -a\gamma \exp(-\gamma x), \quad a\gamma > 0 \tag{5.6.1}$$

where γ is an inverse-range parameter; the integral of $w_\gamma(x)$ over all one-dimensional space is independent of γ and equal to $-a$. Kac, Uhlenbeck and Hemmer[37] have shown that in the limit $\gamma \to 0$, taken after the thermodynamic limit, the pressure is given by the one-dimensional van der Waals equation, i.e.

$$\lim_{\gamma \to 0} \frac{\beta P}{\rho} = \frac{1}{1 - \rho d} - \beta \rho a \qquad (5.6.2)$$

where the first term on the right-hand side is the exact equation of state of the hard-rod reference system or 'Tonks gas'.[38] This result was later extended to three dimensions and it was proved rigorously that in the limit where the perturbation is both infinitesimally weak and infinitely long ranged, the equation of state is given exactly by the generalised van der Waals equation (5.1.2).

The γ-expansion is obtained by considering perturbations of the general form

$$w_\gamma(r) = -\gamma^3 f(\gamma r) \qquad (5.6.3)$$

and expanding the properties of the system of interest in powers of γ. If R is the range of the reference system potential (e.g. the hard-sphere diameter), the dimensionless parameter of the expansion is $\delta = (\gamma R)^3$; δ is roughly the ratio of the reference system interaction volume (e.g. the volume of a hard sphere) to the total interaction volume. In most simple liquids the attractive forces are not truly long ranged in the sense of (5.6.3), but many of the results of the γ-expansion may usefully be carried over to such systems by setting $\gamma = 1$. However, rather than following the original derivation of the γ-expansion, we describe instead the closely related but simpler method of Andersen and Chandler.[39] In doing so we make use of the diagrammatic definitions and lemmas of Section 3.7. We assume throughout that the pair potential has the general form given by (5.1.3).

We first require the diagrammatic expansion of the excess Helmholtz free energy. This can be derived from the corresponding expansion of the single-particle direct correlation function given by (3.8.6), taken for the case of zero external field. By comparison of (3.8.6) with the definition of $c^{(1)}(\mathbf{r})$ in (3.5.1) it can be deduced that the reduced free energy density $\phi = -\beta F^{ex}/V$ introduced in Section 5.4 is expressible diagrammatically as

$V\phi = $ [all irreducible diagrams consisting of two or more black
    ρ-circles and f-bonds]

$$\qquad (5.6.4)$$

If (5.6.4) is inserted in (3.5.1), a simple application of Lemma 2 leads back to (3.8.6).

The separation of the pair potential in (5.1.3) means that the Mayer function $f(1,2)$ can be factorised as

$$f(1,2) = f_0(1,2) + [1 + f_0(1,2)] \left(\exp[\Psi(1,2)] - 1\right) \qquad (5.6.5)$$

where $f_0(1,2)$ is the Mayer function of the reference system and

$$\Psi(1,2) = -\beta w(1,2) \qquad (5.6.6)$$

Since the perturbation is weak, the exponential term in (5.6.5) can be expanded to give

$$f(1,2) = f_0(1,2) + [1 + f_0(1,2)] \sum_{n=1}^{\infty} \frac{[\Psi(1,2)]^n}{n!} \qquad (5.6.7)$$

The form of (5.6.7) suggests the introduction of two different types of bond: short-range f_0-bonds and long-range Ψ-bonds. The presence of two types of bond transforms the simple diagrams in (5.6.4) into composite diagrams in which two circles are linked by at most one f_0-bond but an arbitrary number of Ψ-bonds. We recall from Section 3.7 that if two circles in a diagram are linked by n bonds of a given species, the symmetry number of the diagram is increased, and its value decreased, by a factor $n!$; this takes care of the factors $1/n!$ in (5.6.7). The full expansion of ϕ in terms of composite diagrams is

$V\phi = $ [all irreducible diagrams consisting of two or more black

$\qquad \rho$-circles, f_0-bonds and Ψ-bonds, where each pair of circles is

$\qquad$ linked by any number of Ψ-bonds but at most one f_0-bond] (5.6.8)

The corresponding expansion of the pair distribution function can be obtained from (3.4.8). Written in the notation of the present section the latter becomes

$$\rho^2 g(1,2) = 2V \frac{\delta\phi}{\delta\Psi(1,2)} \qquad (5.6.9)$$

and the diagrammatic prescription for $g(1,2)$ follows immediately from an application of Lemma 3.

The sum of all diagrams in (5.6.8) in which only f_0-bonds appear yields the free energy density ϕ_0 of the reference system. The f_0-bonds in the other diagrams can be replaced in favour of h_0-bonds by a process of topological reduction based on Lemma 5. This leads to the elimination of diagrams containing 'reference articulation pairs', which are pairs of circles linked by one or more independent paths consisting exclusively of black circles linked by reference system bonds.[40] Of the diagrams that remain after the topological reduction there are two of order ρ^2 that contain only a single Ψ-bond. The sum of the two is written as

$$
\begin{aligned}
V\phi_{\text{HTA}} &= \;\; \bullet\!\!-\!\!\bullet \;+\; \bullet\cdots\bullet \\
&= \frac{1}{2}\rho^2 \iint [\Psi(1,2) + h_0(1,2)\Psi(1,2)]\,d1\,d2 \\
&= -\frac{1}{2}V\beta\rho^2 \int g_0(\mathbf{r})w(\mathbf{r})d\mathbf{r}
\end{aligned}
\tag{5.6.10}
$$

where a broken line represents an h_0-bond, a solid line represents a Ψ-bond and HTA stands for 'high-temperature approximation'. Comparison of (5.6.10) with (5.2.14) shows that the HTA is equivalent to truncation of the λ-expansion after the first-order term, with

$$
\phi_{\text{HTA}} = -\frac{\beta F_1}{V}
\tag{5.6.11}
$$

The corresponding approximation to $g(1,2)$ is given by a trivial application of Lemma 3. If $\phi \approx \phi_{\text{HTA}}$, we find from (5.6.10) that

$$
\begin{aligned}
\rho^2 g(1,2) &\approx 2V\frac{\delta\phi_{\text{HTA}}}{\delta\Psi(1,2)} \\
&= \rho^2 + \rho^2 h_0(1,2) = \rho^2 g_0(1,2)
\end{aligned}
\tag{5.6.12}
$$

in agreement with the results of Section 5.4.

To proceed beyond the HTA it is necessary to sum a larger class of diagrams in the expansion of ϕ. An approximation similar in spirit to the Debye–Hückel theory of ionic fluids is

$$
\phi \approx \phi_0 + \phi_{\text{HTA}} + \phi_{\text{R}}
\tag{5.6.13}
$$

where

$$
V\phi_{\text{R}} = \;\bullet\!\!\frown\!\!\bullet\; + \;\triangle\; + \;\triangle\; + \;\square\; + \;\square\; + \;\square\; + \cdots
\tag{5.6.14}
$$

is the sum of all simple 'ring' diagrams plus the diagram consisting of two black circles linked by two Ψ-bonds; the absence of reference articulation pairs means that none of the ring diagrams in (5.6.14) contains two successive h_0-bonds. The approximation to $g(1,2)$ obtained by applying Lemma 3 is now

$$
g(1,2) \approx g_0(1,2) + C(1,2)
\tag{5.6.15}
$$

where the function $C(1,2)$ is given by

$$
\begin{aligned}
\rho^2 C(1,2) = \;[&\text{all chain diagrams consisting of two terminal} \\
&\text{white } \rho\text{-circles labelled 1 and 2, black } \rho\text{-circles,} \\
&\Psi\text{-bonds and } h_0\text{-bonds, but in which there are} \\
&\text{never two successive } h_0\text{-bonds}]
\end{aligned}
\tag{5.6.16}
$$

When the reference system is the ideal gas and $w(r)$ is the Coulomb potential, $-k_B T C(1,2)$ is the screened potential $\psi(r)$ of (4.6.27) and (5.6.15) reduces to the linearised Debye–Hückel result (4.6.29). For the systems of interest here, $-k_B T C(1,2)$ is a renormalised potential in which the perturbation is screened by the order imposed on the fluid by the short-range interaction between particles.

The function $C(1,2)$ can be evaluated by Fourier transform techniques similar to those used in the derivation of the Debye–Hückel result. We first group the chain diagrams according to the number of Ψ-bonds they contain. Let $C^{(n)}(1,2)$ be the sum of all chain diagrams with precisely n Ψ-bonds. Then

$$\rho^2 C(1,2) = \rho^2 \sum_{n=1}^{\infty} C^{(n)}(1,2) \tag{5.6.17}$$

where, for example:

$$\rho^2 C^{(1)}(1,2) = \underset{1\quad 2}{\text{O—O}} + \underset{1\qquad 2}{\text{O····●—O}} + \underset{1\qquad 2}{\text{O—●····O}} + \underset{1\qquad\qquad 2}{\text{O····●—●····O}}$$

$$\tag{5.6.18}$$

Any diagram that contributes to $C^{(n)}$ contains at most $(n+1)$ h_0-bonds and $C^{(n)}$ consists of 2^{n+1} topologically distinct diagrams.

The sum of all diagrams in $C^{(n)}(1,2)$ can be represented by a single 'generalised chain' in which circles are replaced by *hypervertices*. A hypervertex of order n is associated with a function of n coordinates, $\Sigma(1,\ldots,n)$, and is pictured as a large circle surrounded by n white circles; the latter correspond, as usual, to the coordinates $\mathbf{r}_1,\ldots,\mathbf{r}_n$. For present purposes we need consider only the hypervertex of order two associated with the reference system function $\Sigma_0(1,2)$ defined as

$$\Sigma_0(1,2) = \rho\delta(1,2) + \rho^2 h_0(1,2)$$

$$= \underset{1\qquad 2}{\text{O◯O}} \tag{5.6.19}$$

We can then re-express $C^{(n)}(1,2)$ for $n=1$ and $n=2$ in the form

$$\rho^2 C^{(1)}(1,2) = \iint \Sigma_0(1,3)\Psi(3,4)\Sigma_0(4,2)\,\mathrm{d}3\,\mathrm{d}4$$

$$= \underset{1\qquad\qquad\qquad 2}{\text{O◯●—●◯O}} \tag{5.6.20}$$

$$\rho^2 C^{(2)}(1,2) = \iiiint \Sigma_0(1,3)\Psi(3,4)\Sigma_0(4,5)\Psi(5,6)\Sigma_0(6,2)\,d3\,d4\,d5\,d6$$

$$\text{(5.6.21)}$$

and so on; the quantity $\rho^2 C^{(n)}(1,2)$ for any n is represented by a generalised chain consisting of n Ψ-bonds and $(n+1)$ Σ_0-hypervertices. Each generalised chain corresponds to a convolution integral with a Fourier transform given by

$$\rho^2 \hat{C}^{(n)}(\mathbf{k}) = [\hat{\Sigma}_0(\mathbf{k})\hat{\Psi}(\mathbf{k})]^n \hat{\Sigma}_0(\mathbf{k}) \tag{5.6.22}$$

where $\hat{\Sigma}_0(\mathbf{k})$ is related to the structure factor of the reference system by $\hat{\Sigma}_0(\mathbf{k}) = \rho S_0(\mathbf{k})$ and $\hat{\Psi}(\mathbf{k}) = -\beta \hat{w}(\mathbf{k})$. If $|\hat{\Sigma}_0(\mathbf{k})\hat{\Psi}(\mathbf{k})| < 1$, the Fourier transform of the function $C(1,2)$ is obtained as the sum of a geometric series:

$$\rho^2 \hat{C}(\mathbf{k}) = \sum_{n=1}^{\infty} \rho^2 \hat{C}^{(n)}(\mathbf{k}) = \frac{[\hat{\Sigma}_0(\mathbf{k})]^2 \hat{\Psi}(\mathbf{k})}{1 - \hat{\Sigma}_0(\mathbf{k})\hat{\Psi}(\mathbf{k})}$$

$$= -\frac{\rho^2 [S_0(\mathbf{k})]^2 \beta \hat{w}(\mathbf{k})}{1 + \rho S_0(\mathbf{k})\beta \hat{w}(\mathbf{k})} \tag{5.6.23}$$

The derivation of (5.6.23) tends to obscure the basic simplicity of the theory. If (4.1.5), (5.6.15) and (5.6.23) are combined, we find that the structure factor of the system of interest is related to that of the reference fluid by

$$S(\mathbf{k}) = S_0(\mathbf{k}) - \frac{\rho [S_0(\mathbf{k})]^2 \beta \hat{w}(\mathbf{k})}{1 + \rho S_0(\mathbf{k})\beta \hat{w}(\mathbf{k})} = \frac{S_0(\mathbf{k})}{1 + \rho S_0(\mathbf{k})\beta \hat{w}(\mathbf{k})} \tag{5.6.24}$$

On the other hand, we find with the help of (3.6.10) that the exact relation between the two structure factors is given in terms of the corresponding direct correlation functions by

$$S(\mathbf{k}) = \frac{S_0(\mathbf{k})}{1 - \rho[\hat{c}(\mathbf{k}) - \hat{c}_0(\mathbf{k})]S_0(\mathbf{k})} \tag{5.6.25}$$

Use of (5.6.24) is therefore equivalent to replacing the true direct correlation function by the random-phase approximation (RPA) of (3.5.17), i.e.

$$c(r) \approx c_0(r) - \beta w(r) \tag{5.6.26}$$

which is asymptotically correct if the perturbation contains the long-range part of the potential. The Debye–Hückel approximation corresponds to writing $c(r) \approx -\beta w(r)$; (5.6.26) improves on this by building in the exact form of the direct correlation function of the reference system.

The RPA approximation for the free energy is obtained by combining (5.6.10), (5.6.13) and (5.6.14). When functionally differentiated with respect to $\Psi(1,2)$ according to the rule (5.6.12), the total ring diagram contribution to ϕ yields the function $C(1,2)$. It follows that $V\phi_R$ may be expressed diagrammatically as

$$V\phi_R = \sum_{n=2}^{\infty} R^{(n)} \tag{5.6.27}$$

where $R^{(n)}$ is a 'generalised ring' consisting of Σ_0-hypervertices and Ψ-bonds. A generalised ring is derived from a generalised chain by inserting a Ψ-bond between the white circles and integrating over the coordinates associated with those circles. Thus

$$\begin{aligned} R^{(n)} &= \frac{\rho^2}{2n} \iint C^{(n-1)}(1,2)\Psi(1,2)\, d1\, d2 \\ &= \frac{V\rho^2}{2n} \int C^{(n-1)}(\mathbf{r})\Psi(\mathbf{r})\, d\mathbf{r} \\ &= \frac{V\rho^2}{2n} (2\pi)^{-3} \int \hat{C}^{(n-1)}(\mathbf{k})\hat{\Psi}(\mathbf{k})\, d\mathbf{k} \end{aligned} \tag{5.6.28}$$

where the factor $1/2n$ comes from the symmetry number of the generalised ring. If we now substitute for $\hat{C}^{(n-1)}(\mathbf{k})$ from (5.6.22) and assume again that $|\hat{\Sigma}_0(\mathbf{k})\hat{\Psi}(\mathbf{k})| < 1$, we find that the contribution to ϕ from the ring diagrams is

$$\begin{aligned} \phi_R &= \left(\frac{1}{2\pi}\right)^3 \int \sum_{n=2}^{\infty} \frac{1}{2n} [\hat{\Sigma}_0(\mathbf{k})\hat{\Psi}(\mathbf{k})]^n\, d\mathbf{k} \\ &= -\frac{1}{2}(2\pi)^{-3} \int \left(\hat{\Sigma}_0(\mathbf{k})\hat{\Psi}(\mathbf{k}) + \ln[1 - \hat{\Sigma}_0(\mathbf{k})\hat{\Psi}(\mathbf{k})]\right) d\mathbf{k} \end{aligned} \tag{5.6.29}$$

This result is used in the discussion of hierarchical reference theory in Section 5.8.

We saw in Section 4.6 that a defect of the linearised Debye–Hückel approximation is the fact that it yields a pair distribution function which behaves in an unphysical way at small separations. A similar problem arises here. Consider, for simplicity, the case in which the reference system is a fluid of hard spheres of diameter d. In an exact theory, $g(r)$ necessarily vanishes for $r < d$, but in the approximation represented by (5.6.15) there is no guarantee that this will be so, since in general $C(r)$ will be non-zero in that region. There is, however, some flexibility in the choice of $C(r)$ and this fact can be usefully exploited. Although $C(r)$ is a functional of $w(r)$ it is obvious on physical grounds that the true properties of the fluid must be independent of the choice of perturbation for $r < d$. The unphysical behaviour of the RPA can therefore be eliminated by choosing $w(r)$ for $r < d$ in such a way that

$$C(r) = 0, \quad r < d \tag{5.6.30}$$

Comparison of (5.6.15) with the general rule (5.6.9) shows that this condition is equivalent to requiring the free energy to be stationary with respect to variations in the perturbing potential within the hard core. The RPA together with the condition (5.6.30) is called the 'optimised' random-phase approximation or ORPA. The ORPA may also be regarded as a solution to the Ornstein–Zernike relation that satisfies both the closure relation (5.6.26) and the restriction that $g(r) = 0$ for $r < d$. It is therefore similar in spirit to the MSA of Section 4.5, the difference being that the treatment of the hard-sphere system is exact in the ORPA.

The derivation of (5.6.24) did not involve any assumption about the range of the potential $w(r)$. However, as we have seen in Section 3.5, the RPA can also be derived by treating the effects of the perturbation in mean field fashion, an approximation that is likely to work best when the perturbation is both weak and long ranged. In practice, the optimised version of the theory gives good results for systems such as the Lennard-Jones fluid.[41] Not surprisingly, however, it is less successful when the attractive well in the potential is both deep and narrow.[42] In that case better results are obtained by replacing $-\beta w(r)$ in (5.6.26) by the corresponding Mayer function; this modification also ensures that $c(r)$ behaves correctly in the low-density limit.

A different method of remedying the unphysical behaviour of the RPA pair distribution function can be developed by extending the analogy with Debye–Hückel theory. If the reference system is the ideal gas, the RPA reduces to

$$g(1,2) \approx 1 + C(1,2) \tag{5.6.31}$$

When $w(r)$ is the Coulomb potential, this result is equivalent to the linearised Debye–Hückel approximation (4.6.27). If we add to the right-hand side of (5.6.28) the sum of all diagrams in the exact expansion of $h(1,2)$ that can be expressed as star products of the diagram $C(1,2)$ with itself, and then apply Lemma 1, we obtain an improved approximation in the form

$$g(1,2) \approx \exp C(1,2)$$

$$\tag{5.6.32}$$

which is equivalent to the non-linear equation (4.6.28). In the present case a generalisation of the same approach replaces the RPA of (5.6.15) by the approximation

$$g(1,2) \approx g_0(1,2) \exp C(1,2) \tag{5.6.33}$$

This is called the 'exponential' or EXP approximation. At low density the renormalised potential behaves as $C(r) \approx \Psi(r) = -\beta w(r)$. In the same limit, $g_0(r) \approx \exp[-\beta v_0(r)]$. Thus, from (5.6.33):

$$\lim_{\rho \to 0} g(1,2) = \exp[-\beta v_0(r)] \exp[-\beta w(r)] = \exp[-\beta v(r)] \tag{5.6.34}$$

The EXP approximation, unlike either the HTA or the ORPA, is therefore exact in the low-density limit. Andersen and Chandler[39] give arguments to show that the contribution from diagrams neglected in the EXP approximation is minimised if the optimised $C(1,2)$ is used in the evaluation of (5.6.33) and the related expression for the free energy.

The ORPA and the EXP approximation with optimised $C(1,2)$ both correspond to a truncation of the diagrammatic expansion of the free energy in terms of ρ-circles, h_0-bonds and Ψ-bonds in which the perturbation inside the hard core is chosen so as to increase the rate of convergence. Each is therefore an approximation within a general theoretical framework called 'optimised cluster theory'. The optimised cluster expansion is not in any obvious way a systematic expansion in powers of a small parameter, but it has the great advantage of yielding successive approximations that are easy to evaluate if the pair distribution function of the reference system is known. The γ-expansion provides a natural ordering of the perturbation terms in powers of γ^3, but it leads to more complicated expressions for properties of the system of interest. If the perturbation is of the form of (5.6.3), the terms of order γ^3 in the expansion of the free energy consist of the second of the two diagrams in (5.6.10) (the HTA) and the sum of all diagrams in (5.6.14) (the ring diagrams). There is, in addition, a term of zeroth order in γ, given by the first of the two diagrams in (5.6.10), which in this case has the value

$$\frac{1}{2}\beta\rho^2\gamma^3 \int f(\gamma^3 \mathbf{r}) \, d\mathbf{r} = V\beta\rho^2 a \qquad (5.6.35)$$

where a is the constant introduced in (5.2.16). We see that the effect of the volume integration is to reduce the apparent order of the term from γ^3 to γ^0. As a consequence, the free energy does not reduce to that of the reference system in zeroth order. It yields instead the van der Waals approximation; the latter is therefore exact in the limit $\gamma \to 0$. Through order γ^3 the free energy (with $\gamma = 1$) is the same as in the RPA. On the other hand, the sum of all terms of order γ^3 in the expansion of $g(1,2)$ contains diagrams additional to the chain diagrams included in (5.6.15).[43]

Results obtained by the optimised cluster approach for a potential model consisting of a hard-sphere core plus a Lennard-Jones tail at two different thermodynamic states are compared with the results of Monte Carlo calculations in Figure 5.11. In the lower-density state, the HTA, ORPA and EXP pair distribution functions represent successively improved approximations to the 'exact' results. At the higher density, where the perturbation is heavily screened and the renormalised potential is correspondingly weak, the HTA is already very satisfactory. The difference in behaviour between the two states reflects the diminishing role of the attractive forces on the structure of the fluid as the density increases. Similar conclusions have been reached for other model fluids. Overall the results obtained by optimised cluster methods are comparable in accuracy with those of conventional perturbation theory taken to second order.

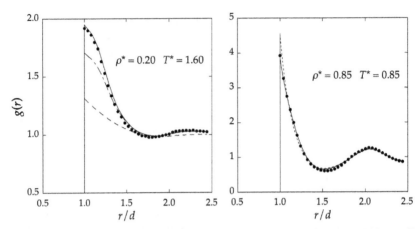

FIGURE 5.11   Radial distribution function for a fluid of hard spheres with a Lennard-Jones tail at two different thermodynamic states. The points are Monte Carlo results and the curves show the predictions of perturbation theory. Dashes: HTA; chain curve; ORPA; full curves: EXP. After Stell and Weis.[44]

5.7   MEAN FIELD THEORY OF LIQUID–VAPOUR COEXISTENCE

Coexistence of liquid and vapour arises from a balance between repulsive and attractive intermolecular forces. In the absence of any attractive interactions there is no liquid–vapour transition and only one fluid phase appears. Since perturbation theory is based explicitly on a division of the pair potential into repulsive and attractive parts, it is a natural choice for the description of phenomena associated with condensation. The integral equation approximations described in Chapter 4 provide another possible approach, but for the most part they either lead to spurious solutions or do not converge numerically in the thermodynamic region of interest.[45] These failings are a consequence of the underlying singularities in thermodynamic properties, in particular the divergence of the isothermal compressibility at the critical point.

For a two-phase system to be in equilibrium, each phase must be at the same pressure (for mechanical equilibrium) and temperature (for thermal equilibrium). However, the pressure and temperature of a two-phase system are not independent variables, since equality of the chemical potentials is also required. Thus, at equilibrium between liquid (L) and gas (G) in a one-component system:

$$\mu_L(P, T) = \mu_G(P, T) \qquad (5.7.1)$$

If μ_L and μ_G are known from some approximate theory, (5.7.1) can be solved for P as a function of T to yield the phase coexistence curve in the pressure–temperature plane. Condensation is a first-order phase transition, since it coincides with discontinuities in the first-order thermodynamic derivatives of the Gibbs free energy. The volume change, ΔV, corresponds to a discontinuity in $(\partial G/\partial P)_T$, while the change in entropy, ΔS, corresponds to a discontinuity in $(\partial G/\partial T)_P$; ΔS is related to the latent heat of the transition by $L = T\Delta S$. Differentiation of the equilibrium condition (5.7.1) with respect to temperature leads to the Clapeyron equation:

$$\frac{dP}{dT} = \frac{\Delta S}{\Delta V} = \frac{L}{T\Delta V} \tag{5.7.2}$$

Since V and S both increase on vaporisation, it follows that the slope of the coexistence curve is always positive.

We consider again a system for which the pair potential $v(r)$ consists of a hard-sphere repulsion supplemented by an attractive term, $w(r)$, for $r > d$, where, as usual, d is the hard-sphere diameter. If $w(r)$ is sufficiently long ranged, the free energy may be approximated by the first two terms of the λ-expansion of Section 5.2 or, within the mean field approximation (5.2.22), by

$$\frac{\beta F}{N} = \frac{\beta F_0}{N} - \beta\rho a \tag{5.7.3}$$

where F_0, the free energy of the hard-sphere reference system, is a function only of the packing fraction η. The equation of state is then given by (5.2.23) which, when combined with the Carnahan–Starling expression for P_0, takes the form

$$\frac{\beta P}{\rho} = \frac{1 + \eta + \eta^2 - \eta^3}{(1-\eta)^3} - \beta\rho a \tag{5.7.4}$$

Equation (5.7.4) is an example of what is commonly termed an 'augmented' van der Waals theory.

Above a critical temperature T_c, to be determined below, the pressure isotherms calculated from (5.7.4) are single-valued, increasing functions of ρ, as sketched in Figure 5.12. Below T_c, however, so-called van der Waals loops appear, which contain an unphysical section between their maxima and minima where the isothermal compressibility would be negative, thereby violating one of the conditions necessary for stability of the system against fluctuations (see Appendix A). The unstable states are eliminated by replacing the loops by horizontal portions between points on the isotherm determined via the Maxwell equal areas construction in the $P - V$ plane. The Maxwell construction is a graphical formulation of the requirement for equality of the pressures and chemical potentials of the two phases; it is equivalent[46] to the double-tangent construction on a plot of free energy versus volume, which ensures that F is always a convex function, i.e. that $(\partial^2 F/\partial V^2)_T > 0$. The end-points of the horizontal portions lie on the coexistence curve, while the locus of maxima and minima of the van der Waals loops, which separates the density-pressure plane into stable and unstable regions, forms the *spinodal curve*. States lying between the coexistence and spinodal curves are metastable,

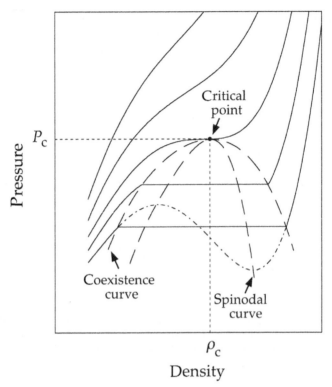

FIGURE 5.12   Isotherms of a simple fluid in the $\rho - P$ plane. The chain curve shows a van der Waals loop. Note that the Maxwell construction applies in the $V - P$, not the $\rho - P$ plane.

but can be reached experimentally if sufficient care is taken to prevent formation of the thermodynamically stable phase. As the temperature increases towards the critical value, the horizontal portion of the isotherm shrinks, eventually reducing to a point of inflection with a horizontal tangent. The critical parameters T_c and ρ_c are therefore determined by the conditions

$$\left(\frac{\partial P}{\partial \rho} \right)_{T=T_c} = 0, \qquad \left(\frac{\partial^2 P}{\partial \rho^2} \right)_{T=T_c} = 0 \qquad (5.7.5)$$

The first of these conditions confirms that the compressibility diverges at the critical point; it also diverges everywhere along the spinodal curve, the apex of which coincides with the critical point. The two coexisting phases, liquid and vapour, merge at the critical point, so the transition, which is of first order below T_c, becomes of second order. Second-order transitions are characterised by discontinuities in the second derivatives of the free energy, of which the compressibility is one.

Equations (5.7.4) and (5.7.5) can be solved numerically for the unknown quantities ρ_c, T_c and P_c (the critical pressure) to give[47]

$$\rho_c d^3 \approx 0.249, \qquad k_B T_c \approx 0.180 a/d^3, \qquad Z_c = \frac{P_c}{\rho_c k_B T_c} \approx 0.359 \qquad (5.7.6)$$

where, as usual, d is the hard-sphere diameter. Both the critical density and the critical compressibility factor Z_c are independent of the strength of the interparticle attraction as measured by the value of the quantity a. We now suppose that the perturbation is given by an inverse-power interaction of the general form

$$w(r) = -\epsilon \left(\frac{d}{r}\right)^{3+\alpha}, \qquad \alpha > 0 \qquad (5.7.7)$$

which becomes increasingly longer ranged as $\alpha \to 0$. In this case the quantity a in (5.2.22) can be identified as

$$a = 2\pi \frac{\epsilon d^3}{\alpha} \qquad (5.7.8)$$

The reduced critical temperature is therefore

$$\alpha T_c^* \approx 1.132 \qquad (5.7.9)$$

This relationship implies that the liquid–vapour coexistence curve plane for different values of the parameter α can be obtained by simple rescaling of temperature; the density scale remains the same.

Figure 5.13 makes comparison between Monte Carlo results[48] for the liquid–vapour existence curve and those obtained in the mean field approximation[47] for the case when $\alpha = 0.1$. Here the agreement between simulation and theory is very good; the Monte Carlo estimates of ρ_c^* and T_c^* and the values predicted by (5.7.6) and (5.7.9) differ by about 1%. However, the agreement deteriorates very rapidly with increasing values of α and when $\alpha = 3$ (the r^{-6}-fluid) liquid–vapour coexistence is found to be metastable with respect to freezing. These failings of the mean field approach can be ascribed in part to the approximation (5.2.22) used for the first-order term in the high-temperature expansion but also to the neglect of higher-order terms. Although the fluctuations corresponding to the higher-order terms are small for liquids close to freezing, they become much larger as the density is lowered. The figure also includes results from second-order perturbation theory, which for this value of α is essentially exact and remains moderately accurate even for $\alpha = 3$.

The deficiencies of mean field theory are also evident in the predictions to which it leads for the behaviour of thermodynamic properties in the immediate vicinity of the critical point. In the approximation represented by (5.7.4) the pressure is an analytic function of ρ and T over a range of packing fraction that extends well beyond the value corresponding to close packing,

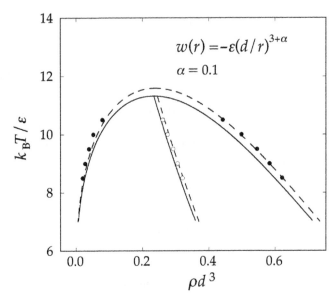

FIGURE 5.13 Liquid–vapour coexistence curve for a fluid of hard spheres with an attractive tail varying as $r^{-(3+\alpha)}$ with $\alpha = 0.1$. The circles show the results of Monte Carlo calculations[48] and the curves are the predictions of mean field theory (full curve) and second-order perturbation theory (dashes). The central part shows the results obtained for the mean density of liquid and vapour. *Redrawn with permission from Ref. 47 © 2003 American Physical Society.*

i.e. $\eta = \pi\sqrt{2}/6 \approx 0.74$. It is therefore legitimate to expand P around P_c in powers of the deviations $\Delta\rho = \rho - \rho_c$ and $\Delta T = T - T_c$. Expansion up to third order gives

$$P = P_c + P_{10}\Delta T + P_{11}\Delta T \Delta\rho + P_{03}(\Delta\rho)^3 + \cdots \qquad (5.7.10)$$

where the coefficients P_{ij} are

$$P_{ij} = \left(\frac{\partial^{i+j} P}{\partial T^i \partial\rho^j}\right)_{\rho=\rho_c, T=T_c} \qquad (5.7.11)$$

Terms in $\Delta\rho$ and $(\Delta\rho)^2$ are zero by virtue of the conditions (5.7.5) and other omitted terms play no role in the derivation that follows. Along the critical isotherm, $\Delta T = 0$, and (5.7.10) simplifies to give

$$\Delta P = P - P_c \sim (\Delta\rho)^3, \quad T = T_c \qquad (5.7.12)$$

Thus the critical isotherm is predicted to have an antisymmetric, cubic form. Division of both sides of (5.7.10) by $\Delta\rho$ gives

$$P_{03}(\Delta\rho)^2 = \frac{\Delta P}{\Delta\rho} - P_{10}\frac{\Delta T}{\Delta\rho} - P_{11}\Delta T \qquad (5.7.13)$$

On taking the limit $\Delta T \to 0$, we find that

$$\frac{\Delta P}{\Delta \rho} \to \left(\frac{\partial P}{\partial \rho}\right)_{T=T_c} = 0$$

$$\frac{\Delta T}{\Delta \rho} \to \left(\frac{\partial T}{\partial \rho}\right)_P = -\frac{(\partial P/\partial \rho)_{T=T_c}}{(\partial P/\partial T)_{\rho=\rho_c}} = 0 \qquad (5.7.14)$$

where the second result follows from the fact that $(\partial P/\partial T)_\rho > 0$ whatever the density. Thus (5.7.13) reduces to

$$\Delta \rho = \pm B |\Delta T|^{1/2}, \quad T < T_c \qquad (5.7.15)$$

where $B^2 = P_{11}/P_{03} > 0$. The coexistence curve close to the critical point should therefore be symmetrical about $\rho = \rho_c$, i.e. $(\rho_G - \rho_c) = -(\rho_L - \rho_c)$ and $\rho_L + \rho_G = 2\rho_c$. This is a special case of the empirical law of 'rectilinear diameters', according to which $\rho_L + \rho_G$ is a linear function of temperature, a relationship that is well satisfied by the results plotted in Figure 5.13.

Next we consider the behaviour of the isothermal compressibility. From (5.7.10) we see that near the critical point:

$$\left(\frac{\partial P}{\partial \rho}\right)_T \approx P_{11}\Delta T + P_{03}(\Delta \rho)^2 \qquad (5.7.16)$$

Along the critical isochore, where $\Delta \rho = 0$, we find that

$$\chi_T = \frac{1}{\rho}\left(\frac{\partial \rho}{\partial P}\right)_T \approx \frac{1}{P_{11}\rho_c}(\Delta T)^{-1}, \quad T \to T_c^+ \qquad (5.7.17)$$

Along the coexistence curve, (5.7.15) applies. Thus

$$\chi_T \approx \frac{1}{2P_{11}\rho_c}|\Delta T|^{-1}, \quad T \to T_c^- \qquad (5.7.18)$$

The behaviour of the inverse compressibility close to the critical point as $\chi_T \to \infty$ is illustrated schematically in Figure 5.14. Finally, it is easy to show that the specific heat c_V exhibits a finite discontinuity as the critical point is approached along either the critical isochore or the coexistence curve.

Equations (5.7.12), (5.7.15), (5.7.17) and (5.7.18) are examples of the *scaling laws* that characterise the behaviour of a fluid close to the critical point, some of which are summarised in Table 5.1. Scaling laws are expressed in terms of certain, experimentally measurable *critical exponents* $(\alpha, \beta, \gamma, \text{etc.})$, which have the same values for all fluids, irrespective of their chemical nature.[50] This *universality* extends to the behaviour of the Ising model and other magnetic systems near the paramagnetic-ferromagnetic transition. By comparing the definitions of the scaling laws in Table 5.1 with the results of mean field

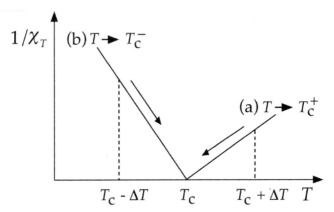

FIGURE 5.14 Divergence of the isothermal compressibility as the critical point is approached (a) from above T_c along the critical isochore and (b) from below along the liquid–vapour coexistence curve. Note the difference in slope in the two cases.

TABLE 5.1 Definitions of the critical scaling laws and numerical values of the exponents.

	Definition	$T - T_c$	$\rho - \rho_c$	Expt[46]	Classical
α	$c_V = A(T - T_c)^{-\alpha}$	>0	0	0.10 ± 0.05	0^a
α'	$c_V = A'(T - T_c)^{-\alpha'}$	<0	$\neq 0$		0^a
β	$\rho_L - \rho_G = B|T - T_c|^{\beta}$	<0	$\neq 0$	0.32 ± 0.01	0.5
γ	$\chi_T = C(T - T_c)^{-\gamma}$	>0	0	1.24 ± 0.1	1
γ'	$\chi_T = C|T - T_c|^{-\gamma'}$	<0	$\neq 0$		1
δ	$|P - P_c| = D|\rho - \rho_c|^{\delta}$	0	$\neq 0$	4.8 ± 0.2	3
ν	$\xi = \xi_0(T - T_c)^{-\nu}$	>0	0	0.63 ± 0.04	0.5
ν'	$\xi = \xi_0(T - T_c)^{-\nu'}$	<0	$\neq 0$		0.5

^{a} *Finite discontinuity.*

calculations, we can identify the so-called classical values of some of the critical exponents: $\alpha = \alpha' = 0$ (a finite discontinuity), $\beta = \frac{1}{2}, \gamma = \gamma' = 1$ and $\delta = 3$. These results differ significantly from the experimental values listed in the table. The classical values are independent of the explicit form of the equation of state. They follow solely from the assumption that the pressure or, equivalently, the free energy is an analytic function of ρ and T close to the critical point and can therefore be expanded in a Taylor series.[51] Analyticity also implies that the classical exponents should be independent of the spatial dimensionality, which is in contradiction both with experimental findings and with exact, theoretical results for the Ising model. The hypothesis of analyticity at the critical point, inherent in mean field theory, must therefore be rejected. The presence of mathematical singularities in the free energy, reflected in the fact that the true critical exponents are neither integers nor simple, rational numbers, can be traced back to the appearance of large-scale density fluctuations near the critical point. For any finite system the partition function and free energy are analytic functions of the independent thermodynamic variables. Singularities appear only in the thermodynamic limit, where fluctuations of very long wavelength become possible. Finite systems therefore behave classically, as the results of computer simulations have shown. Extrapolation techniques based on 'finite-size scaling' ideas are needed if non-classical values of the exponents are to be obtained by simulation.[52]

On approaching the critical point the amplitude of density fluctuations increases and local fluctuations become correlated over increasingly long distances. The compressibility equation (2.6.12) shows that the divergence of the compressibility must be linked to a divergence in the range of the pair correlation function $h(r)$; the range of $h(r)$ is called the *correlation length*, ξ. The behaviour of ξ for $T \approx T_c$ is described by critical exponents ν (along the critical isochore as $T \to T_c^+$) and ν' (along the coexistence curve as $T \to T_c^-$). These exponents are measurable by light and X-ray scattering experiments. Anomalies in the intensity of light scattered from a fluid near its critical point, particularly the phenomenon known as critical opalescence, were first studied theoretically by Ornstein and Zernike as far back as 1914; it was in the course of this work that the direct correlation function was introduced. Equation (3.5.15) shows that close to the critical point $\hat{c}(k)$ is of order $1/\rho$ at $k = 0$. Thus the range of $c(r)$ remains finite, which is consistent with the conjecture that $c(r) \to -\beta v(r)$ as $r \to \infty$ (see the discussion following (3.8.7)). If we also assume that $\hat{c}(k)$ has no singularities, it can be expanded in a Taylor series about $k = 0$ in the form

$$\rho\hat{c}(k) = c_0(\rho, T) + c_2(\rho, T)k^2 + \mathcal{O}(k^4) \tag{5.7.19}$$

where the coefficients of the two leading terms are

$$c_0(\rho, T) = \rho\hat{c}(0) = 1 - 1/\rho k_B T \chi_T$$
$$c_2(\rho, T) = -\frac{1}{6}\rho \int c(r)r^2 \, d\mathbf{r} \equiv -R^2 \tag{5.7.20}$$

The characteristic length R is sometimes called the Debye persistence length. Note that the conjecture regarding the asymptotic behaviour of $c(r)$ means that c_2 and higher-order coefficients in (5.7.20) are strictly defined only for pair potentials of sufficiently short range.

The key assumption of Ornstein–Zernike theory is that R remains finite at the critical point. Equations (3.6.12) and (5.7.19) then imply that

$$\frac{1}{S(k)} = 1 - \rho\hat{c}(k) \approx 1 - c_0(\rho, T) - c_2(\rho, T)k^2 \tag{5.7.21}$$

or, from (5.7.20):

$$S(k) = 1 + \rho \hat{h}(k) \approx \frac{R^{-2}}{K^2 + k^2} \tag{5.7.22}$$

where $K^2 = (1 - c_0)R^{-2} = R^{-2}/\rho k_B T \chi_T$. The asymptotic form of the pair correlation function is obtained by taking the Fourier transform of (5.7.22):

$$h(r) \sim \frac{1}{4\pi \rho R^2} \frac{\exp(-Kr)}{r}, \qquad r \to \infty \tag{5.7.23}$$

The form of this expression makes it natural to identify K with the inverse range of $h(r)$, i.e. with the inverse correlation length:

$$\xi = K^{-1} = R(\rho k_B T \chi_T)^{1/2} \tag{5.7.24}$$

From (5.7.24) and Table 5.1 it is obvious that within the Ornstein–Zernike approximation the critical exponents for ξ and χ_T are related by $\nu = \frac{1}{2}\gamma$. There are indications, however, that the theory is not entirely correct at the critical point. First, it breaks down in two dimensions, where it predicts that $h(r) \sim \ln r$ for large r, which is clearly absurd. Secondly, careful study of plots of $1/S(k)$ versus k^2 shows that the experimental data are not strictly linear, as suggested by (5.7.21), but curve slightly downwards in the limit $k^2 \to 0$. These difficulties can be circumvented[51] by the introduction of another exponent, η, which allows $h(r)$ for large r to be written as

$$h(r) \sim \frac{A \exp(-r/\xi)}{r^{\mathcal{D}-2+\eta}} \tag{5.7.25}$$

where $\mathcal{D}$ is the dimensionality; the Ornstein–Zernike approximation is recovered by putting $\eta = 0$. In the limit $\xi \to \infty$, the Fourier transform of (5.7.24) is

$$\hat{h}(k) \sim \frac{A}{k^{2-\eta}} \tag{5.7.26}$$

and a non-zero value of η can account for the non-linearity of the plots of $1/S(k)$ versus k^2. Substitution of (5.7.26) in the compressibility equation (2.6.12) yields a relation between the exponents γ, ν and η:

$$\nu(2 - \eta) = \gamma \tag{5.7.27}$$

This result is independent of dimensionality. The value of η is difficult to determine experimentally, but the available evidence suggests that it is a small, positive number, approximately equal to 0.05.

5.8 SCALING CONCEPTS AND HIERARCHICAL REFERENCE THEORY

The shortcomings of mean field theory in the critical region are linked to its inability to describe the onset of large-scale density fluctuations close to the critical point, where the correlation length ξ diverges. The scaling concepts introduced by Widom[53] and Kadanoff[54] in the 1960s, and later formalised by Wilson[55] within renormalisation group theory, are ultimately based on the recognition that ξ is the only relevant length scale near criticality. The divergence of ξ as $T \to T_c$ causes the fluid to become 'scale invariant', meaning that fluctuations on all length scales are self-similar; this in turn implies that critical behaviour is universal.

Scaling laws follow from an explicit assumption concerning the functional form of thermodynamic potentials near the critical point. The basic idea is perhaps most easily illustrated in the case of the chemical potential, which is the 'ordering field' conjugate to the 'order parameter' $(\rho_L - \rho_G)$. These two variables play roles analogous to the external field and magnetisation in the Ising model, which belongs to the same universality class as simple fluids. At the critical point we see from (2.4.21) and (5.7.5) that the chemical potential satisfies the conditions

$$\left(\frac{\partial \mu}{\partial \rho}\right)_{T=T_c} = \left(\frac{\partial^2 \mu}{\partial \rho^2}\right)_{T=T_c} = 0 \qquad (5.8.1)$$

If μ is assumed to be an analytic function of ρ and T at the critical point, a Taylor expansion similar to (5.7.10) can be made. By introducing the reduced variables

$$\mu^* = \frac{\mu \rho_c}{P_c}, \quad \Delta\rho^* = \frac{\rho - \rho_c}{\rho_c}, \quad \Delta T^* = \frac{T - T_c}{T_c} \qquad (5.8.2)$$

and taking account of (5.8.1), the result to first order in ΔT^* becomes

$$\Delta\mu^* = \mu^*(\rho, T) - \mu^*(\rho_c, T)$$
$$\approx \frac{(\mu - \mu_c)\rho_c}{P_c} - \mu_{10}^* \Delta T^* \approx \mu_{11}^* \Delta\rho^* \Delta T^* + \mu_{03}^* (\Delta\rho^*)^3 \qquad (5.8.3)$$

where

$$\mu_{ij}^* = \left(\frac{\partial^{i+j} \mu^*}{\partial \Delta T^{*i} \partial \Delta\rho^{*j}}\right)_{\rho=\rho_c, T=T_c} \qquad (5.8.4)$$

The classical values of the critical exponents are now easily recovered. In particular, since ΔT^* is zero along the critical isotherm:

$$\Delta\mu^* = \pm D^* |\Delta\rho^*|^\delta = D^* \Delta\rho^* |\Delta\rho^*|^{\delta-1} \qquad (5.8.5)$$

where $\delta = 3$ and $D^* = \mu_{03}^*$. Similarly, because $\Delta\mu^*$ vanishes along the coexistence curve:

$$\Delta\rho^* = \pm B^* |\Delta T^*|^\beta \qquad (5.8.6)$$

where $\beta = \frac{1}{2}$ and $B^* = (\mu_{11}^*/\mu_{03}^*)^\beta$.

We now introduce a dimensionless *scaling parameter*, defined as

$$x = \Delta T^* / |\Delta\rho^*|^{1/\beta} \qquad (5.8.7)$$

Clearly x is zero along the critical isotherm and is infinite along the critical isochore, while along the coexistence curve $x = -x_0 = -(1/B^*)^{-1/\beta}$. Equation (5.8.3) can therefore be rewritten in generic form as

$$\Delta\mu^* = \Delta\rho^*|\Delta\rho^*|^{\delta-1}h(x) \qquad (5.8.8)$$

where, in the classical theory:

$$h(x) = \mu_{03}^*(1 + x/x_0) \qquad (5.8.9)$$

One way of formulating the scaling hypothesis is to postulate that non-classical critical behaviour still yields a result having the general form of (5.8.8), but with non-classical values of the exponents β and δ and a different (but unspecified) expression for $h(x)$, assumed to be an analytic function of x for $-x_0 < x < \infty$ and to vanish as $x \to x_0$.[56]

The scaling hypothesis leads to relations between the critical exponents, from which the values of all exponents can be obtained once two are specified. Consider, for example, the exponent γ', which describes the behaviour of the isothermal compressibility along the coexistence curve. Given that $x = -x_0$ and $h(x) = 0$, it follows from (5.8.6) and (5.8.8) that

$$\left(\frac{\partial\Delta\mu^*}{\partial\Delta\rho^*}\right)_{\Delta T^*} = -\frac{1}{\beta}|\Delta\rho^*|^{\delta-1-1/\beta}\Delta T^* h'(-x_0) \sim |\Delta T^*|^{\beta(\delta-1)} \qquad (5.8.10)$$

where $h'(x) \equiv \mathrm{d}h(x)/\mathrm{d}x$. Then, since $\chi_T^{-1} = \rho^2\left(\partial\mu/\partial\rho\right)_T$ (see (2.4.22)), comparison with the definition of the exponent γ' in Table 5.1 shows that

$$\gamma' = \beta(\delta - 1) \qquad (5.8.11)$$

In a similar way it is possible to establish the relations

$$\gamma = \gamma', \quad \alpha' + 2\beta + \gamma' = 2, \quad \alpha' + \beta(1 + \delta) = 2 \qquad (5.8.12)$$

However, since this analysis rests on a hypothesis that refers only to thermodynamic quantities, it yields no information about the correlation-length exponents ν, ν' and η. Relations involving those quantities can be derived by exploiting scale invariance near the critical point within Kadanoff's 'block spin' construction for magnetic systems.[54] That approach leads back to the exponent relation (5.7.27) and to the 'hyperscaling' relation, which involves the dimensionality $\mathcal{D}$ of the system:

$$\nu\mathcal{D} = 2 - \alpha \qquad (5.8.13)$$

Although scaling arguments lead to relations between the critical exponents, they cannot be used to derive numerical values of the exponents given only the hamiltonian of the system. That goal can be reached within renormalisation group theory, which is basically an iterative scheme whereby the total number of degrees of freedom contained in a volume of order $\xi^{\mathcal{D}}$ is systematically reduced to a smaller set of effective degrees of freedom. The reduction is brought about by successive elimination of fluctuations of wavelength $\lambda < L$, where the length L is progressively allowed to approach ξ. Scaling laws turn out to be a natural consequence of the theory. The set of transformations τ associated with the progressive reduction in the numbers of degrees of freedom gradually transforms a given initial hamiltonian, belonging to some universality class, into a fixed point of τ, i.e. a hamiltonian that is invariant under the transformation; the existence of a fixed point is equivalent to the principle of universality. The theory shows that for dimensionality $\mathcal{D} > 4$, fluctuations of wavelength λ become negligible as λ increases, and mean field theory is therefore exact. Deviations from classical behaviour for $\mathcal{D} < 4$ can be expanded in powers of $\epsilon = 4 - \mathcal{D}$ by the use of techniques of field theory; this allows the calculation of the non-classical exponents in three dimensions.[57]

Renormalisation group ideas have been combined with those of thermodynamic perturbation theory in the hierarchical reference theory or HRT of Parola, Reatto and coworkers,[58] which leads to a non-classical description of criticality. The starting point of HRT is closely related to the treatment of long-range interactions in Section 5.6. We assume again that the total pair potential is divided into a repulsive, reference part, $v_0(r)$, and an attractive perturbation, $w(r)$. Then, in the random-phase approximation (5.6.13) and (5.6.29), the reduced free energy density $\phi = -\beta F^{\text{ex}}/V$ is given by

$$\phi = \phi_0 + \frac{1}{2}\rho^2 \int g_0(r)\Psi(r)\mathrm{d}\mathbf{r}$$
$$-\frac{1}{2}(2\pi)^{-3}\int \left(\hat{\Sigma}_0(k)\hat{\Psi}(k) + \ln[1 - \hat{\Sigma}_0(k)\hat{\Psi}(k)]\right)\mathrm{d}\mathbf{k} \quad (5.8.14)$$

where a subscript 0 denotes a property of the reference system, $\Psi(r) = -\beta w(r)$ and $\hat{\Sigma}_0(k) = \rho S_0(k) = \rho + \rho^2\hat{h}_0(k)$. Use of Parseval's relation shows that

$$(2\pi)^{-3}\int \hat{\Sigma}_0(k)\hat{\Psi}(k)\mathrm{d}\mathbf{k} = (2\pi)^{-3}\int \rho\hat{\Psi}(k)\mathrm{d}\mathbf{k} + \rho^2\int h_0(r)\Psi(r)\,\mathrm{d}\mathbf{r}$$
$$(5.8.15)$$

Equation (5.8.14) may therefore be rewritten as

$$\phi = \phi_0 + \frac{1}{2}\rho^2 \int \Psi(r)\mathrm{d}\mathbf{r} - \frac{1}{2}(2\pi)^{-3}\int \left(\rho\hat{\Psi}(k) + \ln[1 - \hat{\Sigma}_0(k)\hat{\Psi}(k)]\right)\mathrm{d}\mathbf{k}$$
$$(5.8.16)$$

where the first two terms on the right-hand side correspond to the mean field approximation (5.7.3) and the final term is the contribution made by fluctuations. The non-analyticities in the free energy that characterise the critical region mean, however, that a straightforward perturbative treatment of the effect of fluctuations is bound to fail. The renormalisation group approach provides a hint of how to go beyond conventional perturbation theory. Density fluctuations must be introduced selectively and recursively, starting from short-wavelength fluctuations, which modify the local structure of the reference fluid, up to longer wavelengths, which eventually lead to condensation. The gradual switching on of fluctuations is brought about by passing from the pair potential of the reference system to the full potential via an infinite sequence of intermediate potentials

$$v^{(Q)}(r) = v_0(r) + w^{(Q)}(r) \tag{5.8.17}$$

where the perturbation $w^{(Q)}(r)$ contains only those Fourier components of $w(r)$ corresponding to wavenumbers $k > Q$. In other words:

$$\hat{w}^{(Q)} = \hat{w}(k), \quad k > Q$$
$$= 0, \qquad k < Q \tag{5.8.18}$$

and the reference system and full potentials are recovered in the limits $Q \to \infty$ and $Q \to 0$, respectively:

$$\lim_{Q \to \infty} v^{(Q)}(r) = v_0(r), \quad \lim_{Q \to 0} v^{(Q)}(r) = v(r) \tag{5.8.19}$$

The 'Q-system', i.e. the fluid with pair potential $v^{(Q)}(r)$, serves as the reference system for a system of particles interacting through the potential $v^{(Q-\delta Q)}(r)$, corresponding to an infinitesimally lower cutoff in $\mathbf{k}$-space. The parameter Q, like the inverse-range parameter γ in (5.6.3), has no microscopic significance; its role, as we shall see, is merely to generate a sequence of approximations that interpolate between the mean field result and the exact solution for the fully interacting system.

The cutoff in $\hat{w}(k)$ at $k = Q$ leads to discontinuities in the free energy and pair functions of the Q-system. To avoid the difficulties that this would create, a modified free energy density $\bar{\phi}^{(Q)}$ is introduced, defined as

$$\bar{\phi}^{(Q)} = \phi^{(Q)} + \frac{1}{2}\rho^2[\hat{\Psi}(0) - \hat{\Psi}^{(Q)}(0)] - \frac{1}{2}\rho[\Psi(0) - \Psi^{(Q)}(0)] \tag{5.8.20}$$

together with a modified direct correlation function $\hat{C}^{(Q)}$, given by

$$\hat{C}^{(Q)}(k) = \hat{c}^{(Q)}(k) - 1/\rho + \hat{\Psi}(k) - \hat{\Psi}^{(Q)}(k) \tag{5.8.21}$$

where $c^{(Q)}(k)$ is the direct correlation function of the Q-system, defined in the usual way, and $\Psi^{(Q)}(r) = -\beta w^{(Q)}(r)$. Inclusion of the last two terms[59] on the right-hand side of (5.8.21) compensates for the discontinuity, equal to $\beta\hat{w}(k)$, that appears in the function $\hat{c}^{(Q)}(k)$ at $k = Q$. Thus

$$\hat{C}^{(Q)}(k) = -\frac{1}{\hat{\Sigma}^{(Q)}(k)}, \quad k > Q$$

$$= -\frac{1}{\hat{\Sigma}^{(Q)}(k)} + \hat{\Psi}(k), \quad k < Q \tag{5.8.22}$$

with $\hat{\Sigma}^{(Q)}(k) = \rho S^{(Q)}(k)$. With these definitions, the expression derived from (5.8.16) for $\bar{\phi}^{(Q-\delta Q)}$ in terms of $\bar{\phi}^{(Q)}$ can be written as

$$\bar{\phi}^{(Q-\delta Q)} = \bar{\phi}^{(Q)} + \frac{1}{2}(2\pi)^{-3} \int \ln\left(1 - \frac{\hat{\Psi}(k)}{\hat{C}^{(Q)}(k)}\right) d\mathbf{k} \qquad (5.8.23)$$

where the integration is confined to the interval $Q - \delta Q < k < Q$. By taking the limit $\delta Q \to 0$ we arrive at an exact, differential equation for $\bar{\phi}^{(Q)}$ which describes the evolution of the free energy with Q:

$$-\frac{d\bar{\phi}^{(Q)}}{dQ} = \frac{Q^2}{4\pi^2} \ln\left(1 - \frac{\hat{\Psi}(Q)}{\hat{C}^{(Q)}(Q)}\right) \qquad (5.8.24)$$

The initial condition is imposed at $Q = \infty$, where the free energy takes its mean field value, i.e.

$$\phi^{(\infty)} = \phi_0 + \frac{1}{2}\rho^2\hat{\Psi}(0) - \frac{1}{2}\rho\Psi(0) \qquad (5.8.25)$$

or, equivalently, $\bar{\phi}^{(\infty)} = \phi_0$.

Methods similar to those sketched above can be used to derive a formally exact, infinite hierarchy of differential equations that link the pair function $C^{(Q)}(k)$ to all higher-order direct correlation functions $\hat{c}_n^{(Q)}(\mathbf{r}_1, \ldots, \mathbf{r}_n), n \geq 3$. Close to the critical point some simplification occurs at small values of Q, i.e. when critical fluctuations begin to make a contribution to the free energy. The definitions (5.8.20) and (5.8.21) imply that a generalisation of the compressibility relation (3.5.15):

$$\hat{C}^{(Q)}(k = 0) = -\frac{\partial^2 \bar{\phi}^{(Q)}}{\partial \rho^2} \qquad (5.8.26)$$

applies for all Q. The resulting divergence of $1/\hat{C}^{(Q)}(k)$ in the limit $k \to 0$ means that the argument of the logarithmic function in (5.8.23) is dominated by the term describing pair correlations. Thus the evolution of the free energy with Q in its final stages has a universal character, being essentially independent of the interaction term $\hat{\Psi}(k)$. Similar simplifications appear at all levels of the hierarchy, and the distinctive features of renormalisation group theory, such as scaling laws and the expansion in powers of $\epsilon = 4 - \mathcal{D}$, emerge from the formalism without recourse to field theoretical models.

Away from the critical region some approximate closure of the hierarchy is required if numerical results are to be obtained. In practice this is achieved at the level of the free energy by approximating the function $\hat{C}^{(Q)}(k)$ in a form that is consistent both with (5.8.26) and with the Ornstein–Zernike assumption that $\hat{C}^{(Q)}(k)$ is analytic in k^2 (see (5.7.19)). The first equation of the hierarchy is thereby transformed into a partial differential equation in the variables Q

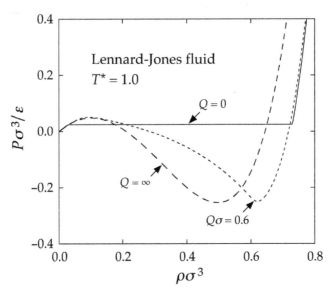

FIGURE 5.15 An isotherm of the Lennard-Jones fluid in the pressure-density plane, calculated at three different stages in the integration of (5.8.24). The limits $Q = \infty$ and $Q = 0$ correspond, respectively, to the mean field and final solutions. For $Q = 0$ the theory yields an isotherm which is rigorously flat in the two-phase region, while at finite Q van der Waals loops are obtained.

and ρ. Closures of this general type, having features in common with other approximate theories, have been used in calculations for a variety of simple fluids[60,61]. Overall the theory yields a very satisfactory description of liquid–vapour coexistence. Non-classical values are obtained for the critical exponents, though these differ somewhat from the nearly exact results derived from the ϵ expansion[58b]. For example, within HRT, $\beta \approx 0.345$, while the ϵ expansion gives $\beta \approx 0.327$. Below T_c the theory leads to rigorously flat isotherms in the two-phase region, illustrated by the results for the Lennard-Jones fluid shown in Figure 5.15. This significantly simplifies the task of locating the densities at coexistence compared with other theories, which rely on use of the Maxwell construction.

The method as described so far is called the 'sharp cutoff' formulation of HRT, in reference to the way in which the intermediate potentials are defined, and is the version employed in early applications of HRT to fluids. Its main deficiency is the fact that if the coexistence curve in the density-temperature plane is approached along a sub-critical isotherm the inverse isothermal compressibility decreases continuously to reach zero at coexistence. The compressibility therefore diverges at all points along the coexistence curve, which now coincides with the spinodal everywhere, not only at the critical point (see Figure 5.12). This makes it impossible to study states in the region of metastability that lies between those two curves. A potentially more serious feature of the method lies in the pathological nature of the intermediate potentials. The discontinuity imposed at $k = Q$ gives rise to a long range, oscillatory tail in $w^{(Q)}(r)$ and it is at least questionable whether use of the simple closure relations of conventional liquid state theory is justified under such conditions. Both difficulties can be overcome by use of a smooth cutoff.[62] This is achieved by introduction of a sequence of intermediate potentials dependent on a parameter t which varies between zero and infinity:

$$v^{(t)}(r) = v_0(r) + w^{(t)}(r) \tag{5.8.27}$$

such that

$$w^{(t)}(r) = w(r) - \psi(t)w(re^{-t}) \tag{5.8.28}$$

where $\psi(t)$ is a monotonically decreasing function which has an initial value $\psi(0) = 1$ and decays exponentially as $t \to \infty$.[63] Thus the limits analogous to (5.8.19) are now

$$\lim_{t \to 0} v^{(t)}(r) = v_0(r), \quad \lim_{t \to \infty} v^{(t)}(r) = v(r) \tag{5.8.29}$$

The difference here, in contrast to (5.8.18), is that the perturbation represented by $w^{(t)}(r)$ is a monotonically varying function of r and the potential $v^{(t)}(r)$ therefore remains well behaved at all t. As t increases, both the amplitude and, crucially, the range of the potential increase, thereby meeting a fundamental requirement of HRT. Figure 5.16 shows examples of the changing form of $v^{(t)}(r)$ for the case of a Yukawa potential with $\lambda = 1.8$, a choice of the inverse-range parameter that provides a fair approximation to the form of the Lennard-Jones potential. It is clear from the figure that the potential is near its asymptotic limit when $t \approx 1.5$, but the small difference that remains has an important role to play, since it is only in the limit $t \to \infty$ that a proper description of criticality can emerge.[64]

Use of first-order perturbation theory leads to an exact expression for the variation with t of the reduced free energy density of the 't-system' in the form

$$\frac{\partial \beta \phi^{(t)}}{\partial t} = \frac{1}{2}\rho^2 \int g^{(t)}(r) \frac{\partial \beta w^{(t)}(r)}{\partial t} \, d\mathbf{r} \tag{5.8.30}$$

where $g^{(t)}(r)$ is the corresponding radial distribution function. For a Yukawa potential $g^{(t)}(r)$ vanishes for r less than the hard-core diameter d and the Ornstein–Zernike relation can be closed by use of a simple, MSA-like approximation for the direct correlation function $c^{(t)}(r)$ outside the core:

$$c^{(t)}(r) = -\beta[w^{(t)}(r) + \alpha_t w(r)], \quad r > d \tag{5.8.31}$$

which, in combination with (5.8.30), corresponds again to truncation at first order of an exact, infinite hierarchy of equations. The value of the parameter α_t is chosen to satisfy the compressibility sum rule (3.5.15); in particular, its value

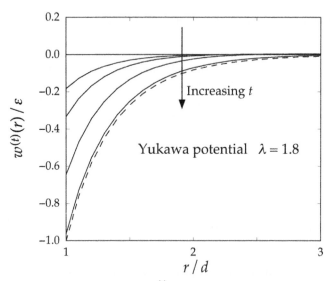

FIGURE 5.16 Growth of the perturbation $w^{(t)}(r)$ in the smooth cutoff formulation of HRT for the case of a Yukawa potential with $\lambda = 0.8$. The full curves correspond to $t = 0, 0.1, 0.2, 0.5$ and 1.5 and the dashed curve shows the result obtained in the limit $t \to \infty$, corresponding to the attractive term in the full potential.

at $t = 0$ can be used to force agreement with the Carnahan–Starling result for the reference system. There is a parallel here with the procedure adopted in the self-consistent Ornstein–Zernike approximation (SCOZA) described in Section 4.7. In fact the SCOZA is equivalent to a smooth cutoff version of HRT in which the perturbation increases linearly with t, i.e. when $w^{(t)}(r) = tw(r), t = 0 \to 1$. But this changes only the amplitude of the perturbation whereas in HRT, as emphasised already, it is the increase in range of the potential that forges the link with renormalisation group ideas and leads to success in the description of critical properties. The two theories can, however, be expected to yield very similar results for thermodynamic properties, including the liquid–vapour coexistence curve.[65]

The way in which the intermediate potentials are defined in the smooth cutoff approach has allowed the theory to be cast in more familiar, real-space terms than is the case for the sharp cutoff discussed earlier. The apparent simplicity of the resulting equations is, however, illusory. Their numerical solution poses formidable computational problems,[66] alleviated only partly by the fact that the MSA has an analytical solution for the Yukawa potential. The main difference between the results and those previously obtained by the sharp cutoff route lies in the behaviour of the isothermal compressibility close to coexistence, which in the limit $t \to \infty$ now changes discontinuously as the coexistence line is crossed. Below that line, but bounded by it, the compressibility may be either positive or negative, corresponding respectively to regions of metastability or

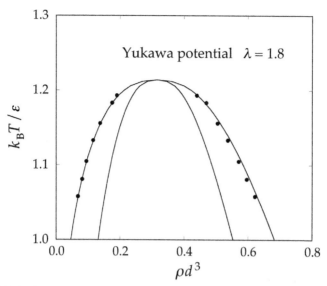

FIGURE 5.17 Liquid–vapour coexistence curve and spinodal line for a Yukawa fluid with $\lambda = 1.8$. The curves are those predicted by the smooth cutoff version of HRT and the points show the results of Monte Carlo calculations. *Redrawn with permission from Ref. 62 © 2008 American Physical Society.*

instability. The spinodal line can then be identified with the boundary of the region in which the compressibility remains negative for all t. Figure 5.17 shows the predicted coexistence curve and spinodal of a Yukawa fluid, again for $\lambda = 1.8$. Agreement with the results of simulations is very good, though somewhat better on the vapour side of the coexistence curve than on the liquid side. The SCOZA results for the coexistence curve are almost indistinguishable from those of HRT except in the region close to the critical point, where the HRT curve is flatter, and are therefore omitted for sake of clarity. Discrepancies between theory and simulation increase with λ. This is not unexpected, since the closure relation (5.8.31) becomes increasingly less accurate as the range of the potential is reduced. In contrast to the case of a sharp cutoff, truncation of the hierarchy at first order now causes some memory of the microscopic model to persist even in the limit $t \to \infty$, and strict universality is thereby lost. For the exponent β the numerical evidence suggests that the truncation error is small, with β lying in the range[62,66] 0.330–0.335; this represents a clear improvement over the result obtained by the sharp cutoff method.

REFERENCES

[1] Widom, B., *J. Chem. Phys.* **39**, 2808 (1963).

[2] Longuet-Higgins, H.C. and Widom, B., *Mol. Phys.* **8**, 549 (1964).

[3] Ashcroft, N.W. and Lekner, J., *Phys. Rev.* **145**, 83 (1966).

[4] Verlet, L., *Phys. Rev.* **165**, 201 (1968).

[5] Zwanzig, R., *J. Chem. Phys.* **22**, 1420 (1954).

[6] Henderson, D. and Barker, J.A., *In* 'Physical Chemistry: An Advanced Treatise', vol. VIIIA (H. Eyring, D. Henderson and W. Jost, eds). Academic Press, New York, 1971.

[7] Barker, J.A. and Henderson, D., *J. Chem. Phys.* **47**, 2856 (1967).

[8] Isihara, A., *J. Phys. A* **1**, 539 (1968).

[9] (a) Mansoori, G.A. and Canfield, F.B., *J. Chem. Phys.* **51**, 4958 (1969). (b) Rasaiah, J.C. and Stell, G., *Mol. Phys.* **18**, 249 (1970).

[10] Ben-Amotz, D. and Stell, G., *J. Chem. Phys.* **119**, 10777 (2003).

[11] (a) Mon, K.K., *J. Chem. Phys.* **112**, 3245 (2000). (b) Mon, K.K., *J. Chem. Phys.* **115**, 4766 (2001).

[12] Ryzhov, V.N. and Stishov, S.M., *Phys. Rev. E* **67**, 010201 (2003).

[13] Kincaid, J.M., Stell, G. and Hall, C.K., *J. Chem. Phys.* **65**, 2161 (1976).

[14] Sillrén, P. and Hansen, J.P., *Mol. Phys.* **105**, 1803 (2007).

[15] Amar, J.G., *Mol. Phys.* **67**, 739 (1989).

[16] Góźdź, W.T., *J. Chem. Phys.* **119**, 3309 (2003).

[17] Molina, J.J., Thèse de doctorat, Université Pierre et Marie Curie (2011).

[18] Andersen, H.C., Weeks, J.D. and Chandler, D., *Phys. Rev. A* **4**, 1597 (1971).

[19] Rowlinson, J.S., *Mol. Phys.* **8**, 107 (1964).

[20] Barker, J.A. and Henderson, D., *J. Chem. Phys.* **47**, 4714 (1967).

[21] (a) Hansen, J.P., *Phys. Rev. A* **2**, 221 (1970). (b) Hansen, J.P. and Weis, J.J., *Mol. Phys.* **23**, 853 (1972).

[22] Lado, F., *Mol. Phys.* **52**, 871 (1984). See also Ben-Amotz, D. and Stell, G., *J. Phys. Chem. B* **108**, 6877 (2004).

[23] Ben-Amotz, D. and Stell, G., *J. Chem. Phys.* **120**, 4844 (2004).

[24] Bailey, N.P., Pedersen, U.R., Gnan, N., Schrøder, T.R. and Dyre, J.C., *J. Chem. Phys.* **129**, 184507 (2008).

[25] Gnan, N., Schrøder, T.B., Pedersen, U.R., Bailey, N.P. and Dyre, J.C., *J. Chem. Phys.* **131**, 234504 (2009). This paper is one of a series; for a review, see Ingebrigtsen, T.S., Schrøder, T.B. and Dyre, J.C., *Phys. Rev. X* **2**, 011011 (2012) and references therein.

[26] Rosenfeld, Y., *Phys. Rev. A* **15**, 2545 (1977). The fact that the functional relationship is the same for different potentials hints at a possible quasiuniversality in the properties of simple liquids. For a detailed discussion, see Dyre, J.C., *Phys. Rev. E* **87**, 022106 (2013).

[27] (a) Levesque, D. and Verlet, L. *Phys. Rev.* **182**, 307 (1969). (b) Verlet, L. and Weis, J.J. *Mol. Phys.* **24**, 1013 (1972).

[28] McQuarrie, D.A. and Katz, J.L., *J. Chem. Phys.* **44**, 2398 (1966).

[29] Weeks, J.D., Chandler, D. and Andersen, H.C., *J. Chem. Phys.* **54**, 5237 (1971).

[30] Talbot J., Lebowitz, J.L., Waisman, E.M., Levesque, D. and Weis, J.J., *J. Chem. Phys.* **85**, 2187 (1986).

[31] Johnson, J.K., Zollweg, J.A. and Gubbins, K.E., *Mol. Phys.* **78**, 591 (1993).

[32] For a survey of the results obtained by simulation, see Pérez-Pellitero, J., Ungerer, P., Orkoulas, G. and Mackie, A.D., *J. Chem. Phys.* **125**, 054515 (2006).

[33] With $\epsilon/k_B = 120\ K$, $\sigma = 3.405$ Å, values appropriate to liquid argon, $P\sigma^3/\epsilon = 1$ corresponds to a pressure $P \approx 4.2 \times 10^7$ Pa (0.42 kbar).

[34] Mansoori, G.A., Provine, J.A. and Canfield, F.B., *J. Chem. Phys.* **51**, 5295 (1969).

[35] Ben-Amotz, D. and Stell, G., *J. Phys. Chem. B.* **108**, 6877 (2004).

[36] (a) Hemmer, P.C., *J. Math. Phys.* **5**, 75 (1964). (b) Lebowitz, J.L., Stell, G. and Baer, S., *J. Math. Phys.* **6**, 1282 (1965).

[37] Kac, M., Uhlenbeck, G.E. and Hemmer, P.C., *J. Math. Phys.* **4**, 216 (1963). For the extension to three dimensions, see van Kampen, N.G., *Phys. Rev.* **135**, 362 (1964). Lebowitz, J.L. and Penrose, O., *J. Math. Phys.* **7**, 98 (1966).

[38] Tonks, L., *Phys. Rev.* **50**, 955 (1936).

[39] Andersen, H.C. and Chandler, D., *J. Chem. Phys.* **57**, 1918 (1972).

[40] Except in trivial cases, removal of a reference articulation pair causes a diagram to separate into two or more components, of which at least one contains only reference-system bonds.

[41] Barker, J.A. and Henderson, D., *Rev. Mod. Phys.* **48**, 587 (1976), Tables VIII and IX.

[42] Pini, D., Parola, A. and Reatto, L., *Mol. Phys.* **100**, 1507 (2002).

[43] For a discussion of the relationship between the two expansions, see Stell, G., *J. Chem. Phys.* **55**, 1485 (1971).

[44] Stell, G. and Weis, J.J., *Phys. Rev. A* **21**, 645 (1980).

[45] See, e.g., Caccamo, C., *Phys. Rep.* **274**, 1 (1996).

[46] Huang, K. 'Statistical Mechanics', 2nd edn. John Wiley, New York, 1987, p. 41.

[47] Camp, P.J., *Phys. Rev. E* **67**, 011503 (2003).

[48] Camp, P.J. and Patey, G.N., *J. Chem. Phys.* **114**, 399 (2001).

[49] (a) Rowlinson, J.S. and Swinton, F.L., 'Liquids and Liquid Mixtures', 3rd edn. Butterworth, London, 1982. (b) Chaikin, P.M. and Lubensky T.C., 'Principles of Condensed Matter Physics'. Cambridge University Press, Cambridge, 1995.

[50] Note that the prefactors (the critical amplitudes) that appear in the scaling laws are not universal quantities.

[51] Fisher, M.E., *J. Math. Phys.* **5**, 944 (1964).

[52] See, e.g., Ferrenberg, A.M. and Landau, D.P., *Phys. Rev B* **44**, 5081 (1991) (Ising model). Wilding, N.B., *Phys. Rev. E* **52**, 602 (1995) (Lennard-Jones fluid). For an introductory treatment, see Wilding, N.B., *Am. J. Phys.* **69**, 1147 (2001).

[53] Widom, B., *J. Chem. Phys.* **43**, 3898 (1965).

[54] (a) Kadanoff, L.P., *Physics* **2**, 263 (1966). (b) Kadanoff, L.P., 'Statistical Physics: Statics, Dynamics and Renormalization'. World Scientific, Singapore, 1999.

[55] (a) Wilson, K.G., *Phys. Rev. B* **4**, 3174 and 3184 (1971). (b) Fisher, M.E., *Rev. Mod. Phys.* **70**, 653 (1998). For introductory treatments, see the book by Huang, Ref. 46, and Plischke, M. and Bergersen, B., 'Equilibrium Statistical Physics', 2nd edn. World Scientific, Singapore, 1994.

[56] Griffiths, R.B. *Phys. Rev.* **158**, 176 (1967).

[57] (a) Wilson, K.G. and Kogut, J., *Phys. Rep.* **12**, 75 (1974). (b) Amit, D.J., 'Field Theory, the Renormalization Group and Critical Phenomena', 2nd edn. World Scientific, Singapore, 1993.

[58] (a) Parola, A. and Reatto, L., *Phys. Rev. A* **31**, 3309 (1985). (b) Parola, A., Pini, D. and Reatto, L., *Phys. Rev. E.* **48**, 3321 (1993). (c) Parola, A. and Reatto, L., *Adv. Phys.* **44**, 211 (1995).

[59] The second term is introduced in order to simplify the resulting expressions.

[60] See, e.g., Tau, M., Parola, A., Pini, D. and Reatto, L., *Phys. Rev. E* **52**, 2644 (1995).

[61] HRT has also been successfully generalised to binary systems. See Pini, D., Tau, M., Parola, A. and Reatto, L., *Phys. Rev. E* **67**, 046116 (2003).

[62] Parola, A., Pini, D. and Reatto, L., *Phys. Rev. Lett.* **100**, 165704 (2008). This paper contains the first report of explicit calculations for a microscopic model of a fluid but the smooth cutoff procedure itself had been proposed much earlier: see Parola, A., *J. Phys. C* **19**, 5071 (1986).

[63] More specifically, $\psi(t) = \exp(-dt)\bar{\psi}(t)$, where d is the dimensionality and $\bar{\psi}(t)$ is assumed to decay as $\exp(-2t)$ at large t, but the final results are insensitive to the precise form of $\bar{\psi}(t)$.

[64] See, e.g., Fig. 1 of Ref. 62.

[65] For an intercomparison of HRT and SCOZA in the critical region, see Høye, J.S. and Lomba, E., *Mol. Phys.* **109**, 2773 (2011).

[66] Parola, A., Pini, D. and Reatto, L., *Mol. Phys.* **107**, 503 (2009).

Chapter 6

Inhomogeneous Fluids

Chapters 4 and 5 were concerned with theories designed primarily for the calculation of thermodynamic and structural properties of bulk, uniform fluids. We now turn our attention to non-uniform systems. The translational invariance characteristic of a homogeneous fluid is broken by exposure to an external force field, in the vicinity of a confining surface (which may be regarded as the source of an external field), or in the presence of an interface between coexisting phases. Static properties of inhomogeneous fluids are most effectively studied within the framework of density functional theory, the foundations of which were laid in Sections 3.1 and 3.4. As we saw there, use of the theory requires as a starting point some approximate expression for the intrinsic free energy as a functional of the single-particle density, or density profile, $\rho^{(1)}(\mathbf{r})$. In this chapter we show how useful approximations can be devised and describe their application to a variety of physical problems.

6.1   LIQUIDS AT INTERFACES

Molecular interactions at fluid interfaces are responsible for many familiar, physical processes, from lubrication and bubble formation to the wetting of solids and the capillary rise of liquids in narrow tubes. Questions of a fundamental character that a theory needs to address include the nature of the interface that arises spontaneously between, say, a liquid and its vapour or between two immiscible liquids; the layering of dense fluids near a solid substrate; the properties of liquids confined to narrow pores; the formation of electric double layers in electrolyte solutions; and the factors that control interfacial phase transitions, such as the capillary condensation of undersaturated vapour in porous media. In all these situations surface contributions to the thermodynamic potentials (proportional to the surface area) are no longer negligible compared with the contributions from the bulk (proportional to the volume). The equilibrium values of the potentials are therefore determined by the competition between bulk and surface effects.[1]

The change in grand potential associated with an infinitesimal change in thermodynamic state of a system containing an interface is given by a

Theory of Simple Liquids, Fourth Edition. http://dx.doi.org/10.1016/B978-0-12-387032-2.00006-4

generalisation of (2.4.3):

$$d\Omega = -S\,dT - P\,dV - N\,d\mu + \gamma\,d\mathcal{A} \qquad (6.1.1)$$

or, in the case of a mixture:

$$d\Omega = -S\,dT - P\,dV - \sum_\nu N_\nu\,d\mu_\nu + \gamma\,d\mathcal{A} \qquad (6.1.2)$$

where ν labels a species, $\mathcal{A}$ is the interfacial area and γ, the variable conjugate to $\mathcal{A}$, is the surface tension. The corresponding change in Helmholtz free energy is

$$dF = -S\,dT - P\,dV + \sum_\nu \mu_\nu\,dN_\nu + \gamma\,d\mathcal{A} \qquad (6.1.3)$$

The surface tension is the work required to increase the interface by unit area. It is positive for any real liquid, since intermolecular forces tend to reduce the interfacial area. Hence, in the absence of gravity, formation of a spherical interface is always favoured. From (6.1.2) and (6.1.3) it follows that γ may be written as a thermodynamic derivative in either of two ways:

$$\gamma = \left(\frac{\partial\Omega}{\partial\mathcal{A}}\right)_{V,T,\{\mu_\nu\}} = \left(\frac{\partial F}{\partial\mathcal{A}}\right)_{V,T,\{N_\nu\}} \qquad (6.1.4)$$

Since Ω is a homogeneous function of first order in V and $\mathcal{A}$, (6.1.2) can be integrated at constant μ_ν and T to give

$$\Omega = -PV + \gamma\mathcal{A} \qquad (6.1.5)$$

which is the generalisation to interfacial systems of the thermodynamic relation (2.4.2). Thus the surface tension may also be written as:

$$\gamma = \frac{1}{\mathcal{A}}(\Omega + PV) \equiv \frac{\Omega^{(s)}}{\mathcal{A}} \qquad (6.1.6)$$

where $\Omega^{(s)}$ is the *surface excess* grand potential.

The concept of a surface excess property can be extended to other thermodynamic quantities. Consider, for example, the interface between a one-component liquid and its vapour. Under the influence of gravity, the interface is planar and horizontal, and the density profile depends only on the vertical coordinate, z. Macroscopically the interface appears sharp, but on the molecular scale it varies smoothly over a few molecular diameters. A typical density profile, $\rho^{(1)}(z)$, is shown schematically in Figure 6.1, where the z-axis is drawn perpendicular to the interface. The physical interface is divided into two parts by an imaginary plane located at $z = z_0$, called the *Gibbs dividing surface*. The liquid phase extends below $z = z_0$, where $\rho^{(1)}(z)$ rapidly approaches its bulk liquid value, ρ_L, while for $z > z_0$, $\rho^{(1)}(z)$ tends towards the bulk gas value, ρ_G.

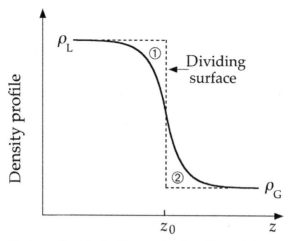

FIGURE 6.1 Density profile at the liquid–vapour interface. The z-axis is perpendicular to the interface and the Gibbs dividing surface is located at $z = z_0$; ρ_L and ρ_G are the bulk densities of liquid and gas, respectively. The customary choice of z_0 is one that makes the regions labelled 1 and 2 equal in area.

The liquid and gas *adsorptions*, Γ_L and Γ_G, are defined as integrals over the regions labelled 1 and 2 in the figure:

$$\Gamma_L = \int_{-\infty}^{z_0} [\rho^{(1)}(z) - \rho_L]dz < 0, \quad \Gamma_G = \int_{z_0}^{\infty} [\rho^{(1)}(z) - \rho_G]dz > 0 \quad (6.1.7)$$

Though the location of the dividing surface is arbitrary, it is commonly positioned so as to make the two labelled regions equal in area, in which case the total adsorption, $\Gamma = \Gamma_L + \Gamma_G$, is zero. We shall follow this convention. If the interface were infinitely sharp, with the two bulk phases meeting discontinuously at the dividing surface, the total number of particles would be

$$N_L + N_G = V_L \rho_L + V_G \rho_G \qquad (6.1.8)$$

where V_L, V_G are the volumes occupied by the two phases. The total number of particles in the inhomogeneous system contained in the volume $V = V_L + V_G$ may therefore be written as

$$N = N_L + N_G + N^{(s)} \qquad (6.1.9)$$

where $N^{(s)}$ is the surface excess number of particles, and the total adsorption is $\Gamma = N^{(s)}/\mathcal{A}$. With the conventional choice of z_0, $N^{(s)} = 0$. In a solution, z_0 may be chosen such that the adsorption of the solvent vanishes, but the adsorptions of the solutes will then in general be non-zero. Expressions analogous to (6.1.9) serve as definitions of the other surface excess quantities.

The surface excess grand potential is related to the surface tension by (6.1.6). When that relation is combined with (6.1.2) and the corresponding expressions for the two bulk phases we find that

$$d\Omega^{(s)} = \gamma \, d\mathcal{A} + \mathcal{A} \, d\gamma = -S^{(s)} \, dT - \sum_\nu N_\nu^{(s)} \, d\mu_\nu + \gamma \, d\mathcal{A} \qquad (6.1.10)$$

which leads, after division by $\mathcal{A}$, to

$$s^{(s)} \, dT + \sum_\nu \Gamma_\nu \, d\mu_\nu + d\gamma = 0 \qquad (6.1.11)$$

where $s^{(s)} \equiv S^{(s)}/\mathcal{A}$ is the surface excess entropy per unit area. Equation (6.1.11) is called the *Gibbs adsorption equation*. This is the surface equivalent of the Gibbs-Duhem relation in the bulk phase and shows that the adsorptions of the different species are related to the surface tension by

$$\Gamma_\nu = -\left(\frac{\partial \gamma}{\partial \mu_\nu}\right)_{T,\{\mu_{\nu'\neq\nu}\}} \qquad (6.1.12)$$

Equations (6.1.11) and (6.1.12) have been derived with the example of a liquid–gas interface in mind, but their applicability is more general. They hold also in the case of a fluid in contact with a solid surface. There, depending on the nature of the solid-fluid interaction, the adsorptions may be either positive or negative.

Thus far we have assumed that the system contains a single, planar (or weakly curved) interface, well separated from any other surface. When a fluid is narrowly confined, an additional control variable comes into play, namely the quantity that characterises the spacing between the bounding surfaces. In the simplest situation, that of a liquid confined to a slit-like pore between two parallel plates of area $\mathcal{A}$, the new variable is the spacing L of the plates. The necessary generalisation of (6.1.2) is

$$d\Omega = -S \, dT - P \, dV - \sum_\nu N_\nu \, d\mu_\nu + 2\gamma \, d\mathcal{A} - f_S \mathcal{A} \, dL \qquad (6.1.13)$$

where $\gamma = \frac{1}{2}(\partial\Omega/\partial\mathcal{A})_{V,T,\{\mu_\nu\},L}$ is the substrate-fluid interfacial tension. The quantity $-f_S$ is the variable per unit area conjugate to L; f_S has the dimensions of pressure, but is commonly referred to as the 'solvation force'. Physically, f_S is the force over and above any direct interaction between the plates that must be exerted on the plates in order to maintain them at a separation L; when $f_S > 0$, the force is repulsive. If Γ_ν, $\rho_\nu^{(1)}(z)$ and $\rho_{\nu L}$ are, respectively, the total adsorption, density profile and bulk liquid density of species ν, then

$$\Gamma_\nu = \int_0^L [\rho_\nu^{(1)}(z) - \rho_{\nu L}] dz \qquad (6.1.14)$$

and the differential of the surface excess grand potential is

$$d\Omega^{(s)} = -2s^{(s)}\mathcal{A}\,dT - \mathcal{A}\sum_\nu \Gamma_\nu\,d\mu_\nu + 2\gamma\,d\mathcal{A} - f_S\mathcal{A}\,dL \qquad (6.1.15)$$

The interfacial tension is again the surface excess grand potential per unit area, i.e. $\gamma = \Omega^{(s)}(\mu, T, L)/2\mathcal{A}$, and the solvation force is

$$f_S = -2\left(\frac{\partial\gamma}{\partial L}\right)_{T,\{\mu_\nu\}} = -\frac{1}{\mathcal{A}}\left(\frac{\partial\Omega}{\partial L}\right)_{\mathcal{A},T,\{\mu_\nu\}} - P \qquad (6.1.16)$$

since $dV = \mathcal{A}\,dL$. In the limit $L \to \infty$, the first term on the right-hand side of (6.1.16) becomes equal to the bulk pressure and the solvation force vanishes. In the same limit, the total adsorptions Γ_ν become equal to the sum of the adsorptions at each plate 1, 2 considered separately, i.e. $\Gamma_\nu \to \Gamma_\nu^{(1)} + \Gamma_\nu^{(2)}$, and $2\gamma \to \gamma^{(1)} + \gamma^{(2)}$. The 'solvation potential' per unit area is defined as

$$W(L) = \frac{1}{\mathcal{A}}[\Omega^{(s)}(L) - \Omega^{(s)}(L \to \infty)] = (2\gamma - \gamma^{(1)} - \gamma^{(2)}) - f_S L \qquad (6.1.17)$$

with $f_S = -\partial W(L)/\partial L$. In the limit $L \to 0$, the confined fluid is completely expelled and $\gamma \to 0$. Thus $W(L = 0) = -\gamma^{(1)} - \gamma^{(2)}$.

6.2 APPROXIMATE FREE ENERGY FUNCTIONALS

We saw in Chapter 3 that the grand potential of an inhomogeneous fluid is a functional of the intrinsic chemical potential $\psi(\mathbf{r}) = \mu - \phi(\mathbf{r})$, where $\phi(\mathbf{r})$ is the external potential. Equation (3.3.13) shows that Ω is also the generating functional for the set of n-particle correlation functions $H^{(n)}(\mathbf{r}_1, \ldots, \mathbf{r}_n)$. Similarly, the Helmholtz free energy is a functional of the single-particle density, and its excess (non-ideal) part is the generating functional for the set of n-particle direct correlation functions $c^{(n)}(\mathbf{r}_1, \ldots, \mathbf{r}_n)$. Implementation of density functional theory is based on the variational principle embodied in (3.4.3), according to which the functional $\Omega_\phi[n] = \mathcal{F}[n] - \int n(\mathbf{r})\psi(\mathbf{r})d\mathbf{r}$ reaches its minimum value when the trial density $n(\mathbf{r})$ coincides with the equilibrium density, while the minimum value itself is the grand potential of the system. This in turn requires the construction of an intrinsic free energy functional $\mathcal{F}$ in a form appropriate to the physical problem of interest. While the ideal part is given exactly by (3.1.22), the non-trivial, excess part is in general unknown, and some approximation must be invoked.

We consider first the case of a small-amplitude modulation of the single-particle density of the form $\delta\rho^{(1)}(\mathbf{r}) = \rho^{(1)}(\mathbf{r}) - \rho_0$, where ρ_0 is the number density of the uniform reference fluid. If the modulation is produced by a weak, external potential $\delta\phi(\mathbf{r})$, the Fourier components of $\delta\rho^{(1)}$ are related to those of $\delta\phi$ by the linear response formula (3.6.9), the constant of proportionality being the density response function $\chi(k)$. A similar result emerges if $\mathcal{F}$ is assumed to be a quadratic functional of the density modulation, i.e.

$$\mathcal{F}[\rho^{(1)}] = Vf_0 + \frac{1}{2}\int d\mathbf{r} \int d\mathbf{r}'\, \delta\rho^{(1)}(\mathbf{r})X_0(\mathbf{r},\mathbf{r}')\delta\rho^{(1)}(\mathbf{r}') + \mathcal{O}\left((\delta\rho^{(1)})^3\right)$$

(6.2.1)

where f_0 is the free energy per unit volume of the reference system; the function $X_0(\mathbf{r},\mathbf{r}')$ is also a property of the reference system and therefore dependent only on the separation $\mathbf{r} - \mathbf{r}'$. The absence from (6.2.1) of a term linear in $\delta\rho^{(1)}$ is explained by the fact that when $\phi(\mathbf{r}) = 0$, $\mathcal{F}[\rho^{(1)}]$ has its minimum value for a uniform density. When written in terms of Fourier components, (6.2.1) becomes

$$\mathcal{F}[\rho^{(1)}] = Vf_0 + \frac{1}{2V}\sum_{\mathbf{k}} \delta\hat{\rho}^{(1)}(\mathbf{k})\hat{X}_0(\mathbf{k})\delta\hat{\rho}^{(1)}(-\mathbf{k}) + \mathcal{O}(\delta\rho^{(1)})^3 \quad (6.2.2)$$

Then, on applying the variational formula (3.4.3), where the derivative is now taken with respect to $\delta\hat{\rho}^{(1)}(\mathbf{k})$, we find that $\delta\hat{\rho}^{(1)}(\mathbf{k})$ and $\delta\hat{\phi}(\mathbf{k})$ are linearly related in the form

$$\hat{X}_0(\mathbf{k})\delta\hat{\rho}^{(1)}(\mathbf{k}) = -\delta\hat{\phi}(\mathbf{k}) \quad (6.2.3)$$

Comparison of (6.2.3) with the linear response expression (3.6.9) shows that

$$\hat{X}_0(\mathbf{k}) \equiv -\frac{1}{\chi(k)} = \frac{k_B T}{\rho_0 S(\mathbf{k})} \quad (6.2.4)$$

where $S(\mathbf{k})$ is the static structure factor of the uniform fluid. The cost in free energy of creating a weak density modulation of wavevector $\mathbf{k}$ is therefore proportional to $1/S(\mathbf{k})$.

Next we consider the slow modulation limit, corresponding to the case of an inhomogeneity of wavelength such that $|\nabla\rho^{(1)}(\mathbf{r})|/\rho_0 = 1/\xi \ll 1/\xi_0$, where ξ_0 is a typical correlation length in the bulk system. The simplest assumption to make is that macroscopic thermodynamics applies locally, i.e. within volume elements of order ξ^3, and hence that a local free energy can be defined at each point in the fluid. In this *local density approximation* the intrinsic free energy is written as

$$\mathcal{F}[\rho^{(1)}] = \int f_0(\rho^{(1)})d\mathbf{r} \quad (6.2.5)$$

where $f_0(\rho^{(1)})$ is the free energy per unit volume of the homogeneous fluid at a density $\rho^{(1)}(\mathbf{r})$. Because the ideal contribution to the free energy functional is precisely of the local form represented by (6.2.5), the approximation is needed only for the excess part, $\mathcal{F}^{ex}$. The Euler–Lagrange formula that results from substitution of (6.2.5) in the variational formula (3.4.3) is

$$f_0'(\rho^{(1)}) = \mu - \phi(\mathbf{r}) \quad (6.2.6)$$

where, here and below, the prime denotes a derivative of a function with respect to its argument, in this case $\rho^{(1)}(\mathbf{r})$. If we now take the gradient of both sides of

(6.2.6) and use the second of the thermodynamic relations (2.3.8), we find that (6.2.6) is equivalent to the macroscopic condition of mechanical equilibrium:

$$\nabla P(\mathbf{r}) = -\rho^{(1)}(\mathbf{r})\nabla\phi(\mathbf{r}) \tag{6.2.7}$$

The local density approximation has proved successful in predicting the concentration profiles of colloidal dispersions in sedimentation equilibrium, where the external potential is either gravity or a centrifugal potential and the slow modulation criterion is therefore well satisfied.[2]

To go beyond the local density approximation we suppose initially that the inhomogeneity extends in only one direction, as is true, for example, of the interface pictured in Figure 6.1. The density profile is then a function of a single coordinate, which we take to be z. The free energy functional can be formally expanded in powers of $1/\xi$, the inverse range of the inhomogeneity. Thus, since $d\rho^{(1)}(z)/dz$ is of order $1/\xi$, a natural generalisation of (6.2.5) is one in which the free energy density f is taken to be a function not only of $\rho^{(1)}(z)$ but also of its low-order derivatives, i.e.

$$\mathcal{F}[\rho^{(1)}] = \int_{-\infty}^{\infty} f\left(\rho^{(1)}(z), \frac{d\rho^{(1)}(z)}{dz}, \frac{d^2\rho^{(1)}(z)}{dz^2}\right) dz \tag{6.2.8}$$

with

$$f = f_0 + f_1\frac{d\rho^{(1)}(z)}{dz} + f_{2'}\left(\frac{d\rho^{(1)}(z)}{dz}\right)^2 + f_{2''}\frac{d^2\rho^{(1)}(z)}{dz^2} + \mathcal{O}(1/\xi^4) \tag{6.2.9}$$

where the coefficients f_n on the right-hand side are all functions of $\rho^{(1)}(z)$. Terms beyond f_0 in (6.2.9) represent successive 'gradient' corrections to the local density approximation. However, the coefficient f_1 is zero, since the functional must be invariant under reflections. Indeed, if $\rho^{(1)}(z)$ is a solution of (3.4.3), the mirror-image profile $\rho^{(1)}(-z)$ must also be a solution. A change of variable from z to $z' = -z$ in the integral (6.2.8) proves that this is possible only if $f_1 = 0$; a similar argument shows that all odd coefficients must also vanish. When (6.2.9) is substituted in (6.2.8), the term involving $d^2\rho^{(1)}(z)/dz^2$ can be transformed into one proportional to $[d\rho^{(1)}(z)/dz]^2$ through an integration by parts. The resulting expression for $\mathcal{F}$ is called the *square-gradient* functional:

$$\mathcal{F}[\rho^{(1)}] = \int_{-\infty}^{\infty} \left(f_0 + f_2\left(\frac{d\rho^{(1)}(z)}{dz}\right)^2\right) dz \tag{6.2.10}$$

Substitution of (6.2.10) in (3.4.3) yields a differential equation for $\rho^{(1)}(z)$ of the form

$$f_0' - f_2'\left(\frac{d\rho^{(1)}(z)}{dz}\right)^2 - 2f_2\frac{d^2\rho^{(1)}(z)}{dz^2} = \mu - \phi(z) \tag{6.2.11}$$

The generalisation of these results to the three-dimensional case is straightforward, requiring only the replacement of $d\rho^{(1)}(z)/dz$ by $\nabla\rho^{(1)}(\mathbf{r})$. Thus (6.2.10) becomes

$$\mathcal{F}[\rho^{(1)}] = \int \left(f_0 + f_2|\nabla\rho^{(1)}(\mathbf{r})|^2 \right) d\mathbf{r} \qquad (6.2.12)$$

where f_0 and f_2 are functions of $\rho^{(1)}(\mathbf{r})$.

The coefficient f_2 can be determined by considering again the case of a slowly varying, small-amplitude inhomogeneity $\delta\rho^{(1)}(\mathbf{r})$ around a bulk density ρ_0. If the integrand in (6.2.12) is expanded to second order in $\delta\rho^{(1)}(\mathbf{r})$ and the result expressed in terms of Fourier components, we find that

$$\mathcal{F}[\rho^{(1)}] \approx \int \left(f_0 + \frac{1}{2}f_0''(\delta\rho^{(1)})^2 + f_2\nabla\delta\rho^{(1)}(\mathbf{r}) \cdot \nabla\delta\rho^{(1)}(\mathbf{r}) \right) d\mathbf{r}$$
$$= Vf_0 + \frac{1}{2V}\sum_{\mathbf{k}} \left(f_0'' + 2f_2k^2 \right) \delta\hat{\rho}^{(1)}(\mathbf{k})\delta\hat{\rho}^{(1)}(-\mathbf{k}) \qquad (6.2.13)$$

where f_0 and f_2 are now functions of ρ_0. This result should be compared with the quadratic functional (6.2.2). Both approximations assume that the inhomogeneity is small in amplitude, but whereas (6.2.2) is valid for any $\mathbf{k}$, (6.2.13) holds only in the long-wavelength limit. The structure factor and two-particle direct correlation function of the reference fluid are related by (3.6.10). If $\hat{c}(\mathbf{k})$ is expanded in even powers of k in the manner of (5.7.19), the quantity $\hat{X}_0(\mathbf{k})$ in (6.2.2) can be replaced by

$$\hat{X}_0(\mathbf{k}) = \frac{k_BT}{\rho_0}\left(1 - \rho_0\hat{c}(\mathbf{k})\right) = \frac{k_BT}{\rho_0}\left(1 - c_0 - c_2k^2 + \mathcal{O}(k^4)\right) \qquad (6.2.14)$$

where the coefficients c_0 and c_2 are given by (5.7.20). Then, on identifying the resulting expression with (6.2.13), we find that

$$f_0''(\rho_0) = k_BT \int c(r)d\mathbf{r} \qquad (6.2.15)$$

and

$$f_2(\rho_0) = \frac{1}{12}k_BT \int c(r)r^2 \, d\mathbf{r} \qquad (6.2.16)$$

Equation (6.2.15) is merely a restatement of the compressibility relation (3.5.15) while (6.2.16) shows that the coefficient f_2 is determined by the second moment of the direct correlation function of the homogeneous system.

The form of the results obtained for f_0 and f_2 suggests that terms of order higher than quadratic are likely to involve still higher-order moments of $c(r)$, thereby exposing a limitation inherent in an expansion in powers of the density profile gradient (or powers of $1/\xi$). Because $c(r)$ decays as $v(r)$ at large r, moments of any given order will diverge for sufficiently long ranged potentials. For example, if the potential contains a contribution from dispersion forces, $c(r)$ will decay as r^{-6}, leading to a divergence of the fourth- and higher-order moments and hence of the coefficients f_n for $n \geq 4$. Even within the square-gradient approximation there is the further difficulty that in the presence of attractive interactions the equilibrium state of the reference system may be one in which liquid and vapour coexist, and neither f_0 nor f_2 is properly defined in the two-phase region. The square-gradient functional has nonetheless proved extremely useful in studies of the liquid–gas interface, as the work described in the next section will illustrate.[3] Long-range interactions can be treated by dividing the pair potential into a short-range reference part and long-range perturbation in the spirit of the perturbation theories of Chapter 5. This separation leads to the formally exact expression for the excess part of the free energy functional given by (3.4.10), from which an approximate, mean field functional is obtained if the correlation term is ignored. The mean field approach provides the basis for the Poisson–Boltzmann theory of the electric double layer described in Section 10.6.

The local density and square-gradient functionals are both designed for use in cases where the inhomogeneity is weak and slowly varying. Two different strategies have been devised to deal with situations in which these conditions are not met. The first, already discussed in a different context in Section 4.3, is based on a functional Taylor expansion of $\mathcal{F}^{\text{ex}}$ in powers of the deviation from the bulk density. Truncation of the expansion at second order, and replacement of the direct correlation function by that of the reference system, leads to the expression for the density profile given by (4.3.16); the quadratic functional (6.2.1) is then recovered if the ideal contribution to the free energy is also expanded to second order. Equation (4.3.16) provides the starting point for a theory of freezing described in Section 6.8. The alternative approach involves the concept of a weighted or coarse-grained local density. There are some circumstances in which the local density may reach values greater than that corresponding to close packing. This is true, for example, of a dense, hard-sphere fluid close to a solid surface. In such cases the local density approximation becomes meaningless. However, a non-local approximation with a structure not unlike (6.2.5) can be devised by introducing a coarse-grained density $\bar{\rho}(\mathbf{r})$, defined as a weighted average of $\rho^{(1)}(\mathbf{r})$ over a volume comparable with the volume of a particle, i.e.

$$\bar{\rho}(\mathbf{r}) = \int w(|\mathbf{r} - \mathbf{r}'|)\rho^{(1)}(\mathbf{r}')d\mathbf{r}' \qquad (6.2.17)$$

where $w(|\mathbf{r}|)$ is some suitable weight function, normalised such that

$$\int w(|\mathbf{r}|)d\mathbf{r} = 1 \qquad (6.2.18)$$

The excess part of the free energy functional is then taken to be

$$\mathcal{F}^{\mathrm{ex}}[\rho^{(1)}] = \int \phi^{\mathrm{ex}}(\bar{\rho})\rho^{(1)}(\mathbf{r})\mathrm{d}\mathbf{r} \qquad (6.2.19)$$

where $\phi^{\mathrm{ex}}(\bar{\rho}) = f^{\mathrm{ex}}(\bar{\rho})/\bar{\rho}$ is the excess free energy per particle of the homogeneous fluid at a density $\bar{\rho}(\mathbf{r})$; the exact form (3.1.22) is retained for the ideal part. Equation (6.2.19) represents a *weighted density approximation*.

The difficulty in implementing a weighted density approximation lies in making an appropriate choice of weight function.[4] A useful guide is obtained by considering the low-density limit. The virial expansion developed in Section 3.9 shows that to lowest order in density the excess free energy per particle of a homogeneous fluid of density ρ_0 is $\phi^{\mathrm{ex}}(\rho_0) = k_B T \rho_0 B_2$, where B_2 is the second virial coefficient (3.9.7). In the case of hard spheres, B_2 is given by the integral

$$B_2 = \frac{1}{2} \int \Theta(|\mathbf{r}| - d)\mathrm{d}\mathbf{r} \qquad (6.2.20)$$

where d is the hard-sphere diameter and $\Theta(x)$ is a unit step function: $\Theta(x) = 1$, $x < 0$; $\Theta(x) = 0$, $x > 0$. The total excess free energy of the homogeneous fluid may therefore be written as

$$\beta F^{\mathrm{ex}} = \beta \int \rho_0 \phi^{\mathrm{ex}}(\rho_0)\mathrm{d}\mathbf{r} = \frac{1}{2} \int \mathrm{d}\mathbf{r} \int \mathrm{d}\mathbf{r}' \, \rho_0^2 \Theta(|\mathbf{r} - \mathbf{r}'| - d) \qquad (6.2.21)$$

This result can be immediately generalised to the inhomogeneous case in the form

$$\beta \mathcal{F}^{\mathrm{ex}}[\rho^{(1)}] = \frac{1}{2} \int \mathrm{d}\mathbf{r} \int \mathrm{d}\mathbf{r}' \, \rho^{(1)}(\mathbf{r})\Theta(|\mathbf{r} - \mathbf{r}'| - d)\rho^{(1)}(\mathbf{r}')$$

$$= \frac{1}{2}\beta \int \phi^{\mathrm{ex}}(\bar{\rho})\rho^{(1)}(\mathbf{r})\mathrm{d}\mathbf{r} \qquad (6.2.22)$$

where $\bar{\rho}(\mathbf{r})$ is the weighted density defined by (6.2.17), with a weight function given by

$$w(|\mathbf{r}|) = \frac{1}{2B_2}\Theta(|\mathbf{r}| - d) = \frac{3}{4\pi d^3}\Theta(|\mathbf{r}| - d) \qquad (6.2.23)$$

which corresponds to averaging the density uniformly over a sphere of radius d. The same approximation may be used at higher densities if combined with a suitable expression for $\phi^{\mathrm{ex}}(\bar{\rho})$, such as that derived from the Carnahan–Starling equation of state. This leads to qualitatively satisfactory results for the oscillatory density profiles of hard spheres near hard, planar walls[5]; an example is shown later in Figure 6.5, from which the quantitative deficiencies in the approximation are evident. Significant improvement is achievable, at the cost of greater computational effort, if the weight function itself is made dependent on the weighted density.[6] For example, we can retain (6.2.19) but replace (6.2.17) by

$$\bar{\rho}(\mathbf{r}) = \int w(|\mathbf{r} - \mathbf{r}'|, \bar{\rho})\rho^{(1)}(\mathbf{r}')d\mathbf{r}' \qquad (6.2.24)$$

Alternatively, we can write the free energy functional in the form

$$\mathcal{F}^{ex}[\rho^{(1)}] = N\phi^{ex}(\bar{\rho}) \qquad (6.2.25)$$

where $\bar{\rho}$ is a position independent, weighted density given by

$$\bar{\rho} = \frac{1}{N} \int d\mathbf{r}\, \rho^{(1)}(\mathbf{r}) \int d\mathbf{r}'\, w(|\mathbf{r} - \mathbf{r}'|, \bar{\rho})\rho^{(1)}(\mathbf{r}') \qquad (6.2.26)$$

In each case a solution for $w(|\mathbf{r}|, \bar{\rho})$ can be obtained by functionally differentiating $\mathcal{F}^{ex}$ twice with respect to $\rho^{(1)}$ to give $c(\mathbf{r})$ (see (3.5.2)) and matching the results to those for the reference system. Numerical calculations therefore require as input not only the free energy of the uniform fluid but also the direct correlation function, which would normally be obtained from some approximate integral equation. For many purposes, however, these methods has been superseded by the *fundamental measure theory* of Rosenfeld,[7] a discussion of which we defer until Section 6.5.

6.3 THE LIQUID–VAPOUR INTERFACE

An interface between bulk phases will form spontaneously whenever the thermodynamic conditions necessary for phase coexistence are met. The most familiar example is the interface that forms between a liquid and its coexisting vapour, for which the density profile $\rho^{(1)}(z)$ varies smoothly with the single coordinate z in the manner illustrated schematically in Figure 6.1. At low temperatures the width of the interface is of the order of a few particle diameters, but since the distinction between the two phases vanishes continuously at the critical temperature the width is expected to increase rapidly as the critical point is approached and the densities ρ_L and ρ_G merge towards a common value, the critical density ρ_c. The smoothness of the profile makes this a problem to which the square-gradient approximation is well suited. Such a calculation was first carried out by van der Waals, whose work is the earliest known example of the use in statistical mechanics of what are now called density functional methods. The Euler-Lagrange equation to be solved is (6.2.11) in the limit in which the gravitational potential $\phi(z) = mgz$ becomes vanishingly small. So long as the inhomogeneity is of small amplitude, i.e. $(\rho_L - \rho_G) \ll \rho_c$, the coefficient f_2 of the square-gradient term is related by (6.2.16) to the direct correlation function of the bulk, reference system.

For condensation to occur, the interparticle potential must contain an attractive term, $w(r)$ say. Within the random phase approximation, $c(r) \approx c_0(r) - \beta w(r)$ (see (3.5.17)), but the presence of a factor r^4 in the integrand means that the contribution to the integral in (6.2.16) from the short-range function $c_0(r)$ can be ignored. Thus

$$f_2 \approx -\frac{1}{3}\pi \int_0^\infty w(r)r^4 \, dr = \frac{1}{2}m \tag{6.3.1}$$

where m is a positive constant which is independent of density. Equation (6.2.11) then takes the simpler form

$$m\frac{d^2\rho^{(1)}(z)}{dz^2} = -\frac{dW(\rho^{(1)})}{d\rho^{(1)}} \tag{6.3.2}$$

where $W(\rho^{(1)}) = -f_0(\rho^{(1)}) + \mu\rho^{(1)}$. The analogy between this expression and Newton's equation of motion is obvious, with m, z, $\rho^{(1)}(z)$ and $W(\rho^{(1)})$ playing the roles of mass, time, position and potential energy, respectively. Equation (6.3.2) is a non-linear differential equation that must be solved subject to the boundary conditions $\lim_{z\to\pm\infty} W(\rho^{(1)}) = W(\rho_B) = -f_0(\rho_B) + \mu\rho_B = P$, where ρ_B is the bulk density of either liquid (as $z \to -\infty$) or gas (as $z \to +\infty$) and P is the bulk pressure. When integrated, (6.3.2) becomes

$$W(\rho^{(1)}) + \frac{1}{2}m\left(\frac{d\rho^{(1)}(z)}{dz}\right)^2 = P \tag{6.3.3}$$

which is analogous to the conservation of mechanical energy, while a second integration yields a parametric representation of the density profile in the form of a quadrature:

$$z = -\left(\frac{1}{2}m\right)^{1/2} \int_{\rho^{(1)}(0)}^{\rho^{(1)}(z)} [P - W(\rho)]^{-1/2} \, d\rho \tag{6.3.4}$$

By definition, $W(\rho) = -\omega(\rho)$, where $\omega = \Omega/V$ is the grand potential per unit volume of the fluid at a density $\rho = \rho^{(1)}(z)$. At liquid–gas coexistence, the function $\omega(\rho)$ has two minima of equal depth, situated at $\rho = \rho_L$ and $\rho = \rho_G$, with $\omega(\rho_L) = \omega(\rho_G) = -P$. A simple parameterisation of $\omega(\rho)$, valid near the critical point is

$$\omega(\rho) = \frac{1}{2}C(\rho - \rho_L)^2(\rho - \rho_G)^2 - P \tag{6.3.5}$$

where both C and the pressure at coexistence, P, are functions of temperature. Substitution of (6.3.5) in (6.3.4) gives

$$z = -\left(\frac{m}{C}\right)^{1/2} \int_{\rho^{(1)}(0)}^{\rho^{(1)}(z)} \frac{d\rho}{(\rho_L - \rho)(\rho - \rho_G)} = -\zeta \ln\left(\frac{\rho^{(1)}(z) - \rho_G}{\rho_L - \rho^{(1)}(z)}\right) \tag{6.3.6}$$

where $\zeta = (m/C)^{1/2}/(\rho_L - \rho_G)$ is a characteristic length that provides a measure of the interfacial width. Equation (6.3.6) is easily solved to give $\rho^{(1)}$ as a function of z:

$$\rho^{(1)}(z) = \frac{\rho_G}{1 + \exp(-z/\zeta)} + \frac{\rho_L}{1 + \exp(z/\zeta)}$$

$$= \frac{1}{2}(\rho_L + \rho_G) - \frac{1}{2}(\rho_L - \rho_G)\tanh\left(\frac{z}{2\zeta}\right) \qquad (6.3.7)$$

which has the general shape pictured in Figure 6.1. The predicted profile is therefore antisymmetric with respect to the mid-point, a result consequent on the symmetric form assumed for the grand potential in (6.3.5) and the neglect of the density dependence of the coefficient f_2. In reality, the profile is steeper on the liquid than on the vapour side. Equation (6.3.7) also implies that the width of the interface diverges at the critical point. Within the mean field theory of phase transitions, $(\rho_L - \rho_G)$ behaves as $(T_c - T)^{1/2}$ as the critical temperature is approached from below,[8] so ζ diverges as $(T_c - T)^{-1/2}$. Note, however, that density functional theory provides only an 'intrinsic' or averaged description of the density profile. The physical interface is a fluctuating object; these 'capillary' fluctuations lead to a thermal broadening of the interface that can be comparable with the theoretical, intrinsic width.

The surface tension is defined thermodynamically as the additional free energy per unit area due to the presence of an interface. Accordingly, within the square-gradient approximation:

$$\gamma = \int_{-\infty}^{\infty} \left(f_0(\rho^{(1)}) + \frac{1}{2}m\left(d\rho^{(1)}/dz\right)^2 - f_B \right) dz \qquad (6.3.8)$$

where f_B is the bulk free energy density, equal to f_L for $z < z_0$ and to f_G for $z > z_0$. Now $f_0(\rho) = -W(\rho) + \mu\rho$ and $W(\rho)$ is given by (6.3.3), from which the bulk pressure can be eliminated by use of the thermodynamic relation $P = f_B - \mu\rho_B$. Equation (6.3.8) therefore reduces to

$$\gamma = \int_{-\infty}^{\infty} \left(-P + \mu\rho^{(1)}(z) + m\left(d\rho^{(1)}/dz\right)^2 - f_B \right) dz$$

$$= \int_{-\infty}^{\infty} \left(\mu[\rho^{(1)}(z) - \rho_B] + m\left(d\rho^{(1)}/dz\right)^2 \right) dz$$

$$= m\int_{-\infty}^{\infty} \left(d\rho^{(1)}/dz\right)^2 dz \qquad (6.3.9)$$

Use of (6.3.3) and (6.3.5) allows (6.3.9) to be recast in the equivalent form:

$$\gamma = m\int_{-\infty}^{\infty} \frac{d\rho^{(1)}}{dz}d\rho^{(1)} = (2m)^{1/2}\int_{\rho_L}^{\rho_G} [P + \omega(\rho)]^{1/2}d\rho$$

$$= -(mC)^{1/2}\int_{\rho_L}^{\rho_G} (\rho_L - \rho)(\rho - \rho_G)d\rho = \frac{1}{6}(mC)^{1/2}(\rho_L - \rho_G)^3$$

$$(6.3.10)$$

which shows that close to the critical point the surface tension is expected to behave as $\gamma \sim (T_c - T)^{3/2}$. Experimentally, however, the critical exponent is found to be somewhat smaller than the predicted value of $\frac{3}{2}$.

6.4 A MICROSCOPIC EXPRESSION FOR THE SURFACE TENSION

Thus far surface tension has been defined only in thermodynamic terms. In this section we show that the surface tension at a fluid–fluid interface can also be expressed microscopically[9] in terms of the interfacial density profile and the direct correlation function of an inhomogeneous fluid, $c^{(2)}(\mathbf{r}_1, \mathbf{r}_2)$. We take as an example a planar interface between liquid (L) and vapour (G). At equilibrium the interface has a density profile $\rho_0^{(1)}(z)$ and a Gibbs dividing surface located at $z = z_G = 0$, as shown in Figure 6.1; the origin $z = 0$ is chosen such that

$$\int_{-\infty}^{0} \left[\rho_0^{(1)}(z) - \rho_L \right] dz + \int_{0}^{\infty} \left[\rho_0^{(1)}(z) - \rho_G \right] dz = 0 \qquad (6.4.1)$$

Capillary wave fluctuations within the interface will cause the instantaneous Gibbs dividing surface to deviate from its average, planar form. Before discussing that problem we need to consider briefly the way in which a surface can be described geometrically. Let S be some arbitrarily chosen surface within the interface, pictured schematically in Figure 6.2. If the surface deviates only weakly from the x–y plane,[10] the vertical displacement of the surface with respect to the equilibrium dividing surface will be a single-valued function $h(x, y)$ of the coordinates $(x, y) \equiv \mathbf{R}$ of a point in the $z = 0$ plane. The position of any point M on S is then uniquely specified by the coordinates[11]

$$\mathbf{r} \equiv \left(x, y, h(x, y) \right) \qquad (6.4.2)$$

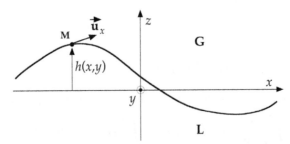

FIGURE 6.2 The curve shows a cut through the x–z plane of a surface S within a liquid–vapour interface. The vector $\mathbf{u}_x$ is one of the two vectors defined by (6.4.3); the vector $\mathbf{u}_y$ lies in the y–z plane. The horizontal plane at $z = 0$ is the Gibbs dividing surface at equilibrium; if S is the instantaneous dividing surface, $h(x, y) = z_G(\mathbf{R})$.

with two vectors tangent to the surface at M given by

$$\mathbf{u}_x = \frac{d\mathbf{r}}{dx} = \left(1, 0, h_x(x, y)\right), \quad \mathbf{u}_y = \frac{d\mathbf{r}}{dy} = \left(0, 1, h_y(x, y)\right) \qquad (6.4.3)$$

The significance of these vectors for our purposes is the fact that an area element $d^2 A'$ of S which is underpinned by an element of area in the x–y plane, $d^2 A = dx dy$, is proportional to the modulus of the vector product $\mathbf{u}_x \wedge \mathbf{u}_y$:

$$d^2 A' = |\mathbf{u}_x \wedge \mathbf{u}_y| dx\, dy = \left(1 + h_x^2 + h_y^2\right)^{1/2} d^2 A \qquad (6.4.4)$$

Now consider the effect of a low-amplitude, long-wavelength fluctuation in the local density $\rho(\mathbf{r})$ around its equilibrium value:

$$\rho(\mathbf{r}) = \rho_0^{(1)}(z) + \Delta\rho(\mathbf{r}) \qquad (6.4.5)$$

The displacement $z_G(\mathbf{R})$ of the Gibbs dividing surface resulting from the fluctuation is equal to $h(x, y)$ in the coordinate system defined above. Its value is determined implicitly by a generalisation of the equilibrium relation (6.4.1):

$$\int_{-\infty}^{z_G} \left[\rho(\mathbf{R}, z) - \rho_L\right] dz + \int_{z_G}^{\infty} \left[\rho(\mathbf{R}, z) - \rho_G\right] dz = 0 \qquad (6.4.6)$$

and an explicit expression follows by subtraction of (6.4.1) from (6.4.6):

$$z_G(\mathbf{R}) = \frac{1}{\rho_L - \rho_G} \int_{-\infty}^{\infty} \Delta\rho(\mathbf{R}, z) dz \qquad (6.4.7)$$

In the case of a long-wavelength modulation of the interface along $\mathbf{R}$, the local density at a point $(\mathbf{R}, z)$ corresponds to a uniform shift of the equilibrium profile by $z_G(\mathbf{R})$:

$$\rho(\mathbf{R}, z) = \rho_0^{(1)}\left((z - z_G(\mathbf{R}))\right) \approx \rho_0^{(1)}(z) - z_G(\mathbf{R})\frac{d\rho_0^{(1)}(z)}{dz} \qquad (6.4.8)$$

which clearly satisfies (6.4.7).

The change in grand potential at fixed chemical potential associated with the fluctuation can be obtained by expansion of the functional

$$\Omega\left[\rho(\mathbf{r})\right] = F\left[\rho(\mathbf{r})\right] - \mu \int \rho(\mathbf{r}) d\mathbf{r} \qquad (6.4.9)$$

in powers of $\Delta\rho(\mathbf{r})$. The first-order term vanishes by virtue of (3.4.3). To second order:

$$\begin{aligned}
\Delta\Omega &= \Omega\left[\rho_0^{(1)}(z) + \Delta\rho(\mathbf{r})\right] - \Omega\left[\rho_0^{(1)}(z)\right] \\
&= \frac{1}{2} \int d\mathbf{r}_1 \int d\mathbf{r}_2 \frac{\delta^2 F}{\delta\rho^{(1)}(\mathbf{r}_1)\delta\rho^{(1)}(\mathbf{r}_2)}\bigg|_{\rho_0^{(1)}(z)} \Delta\rho(\mathbf{r}_1)\Delta\rho(\mathbf{r}_2) \\
&= \frac{1}{2} k_B T \int d\mathbf{r}_1 \int d\mathbf{r}_2 C^{(2)}(\mathbf{r}_1, \mathbf{r}_2)\Delta\rho(\mathbf{r}_1)\Delta\rho(\mathbf{r}_2) \qquad (6.4.10)
\end{aligned}$$

where

$$C^{(2)}(\mathbf{r}_1, \mathbf{r}_2) = \frac{1}{\rho_0^{(1)}(z_1)} \delta(\mathbf{r}_2 - \mathbf{r}_1) - c^{(2)}(\mathbf{r}_1, \mathbf{r}_2) \qquad (6.4.11)$$

The first term on the right-hand side of (6.4.11) comes from the ideal contribution to the free energy (see (3.1.22)) while the second term follows from (3.5.2). The planar geometry at equilibrium and the isotropy in the x–y plane together imply that

$$C^{(2)}(\mathbf{r}_1, \mathbf{r}_2) = C^{(2)} \left(|\mathbf{R}_2 - \mathbf{R}_1|, z_1, z_2 \right) \qquad (6.4.12)$$

We now take two-dimensional Fourier transforms with respect to $\mathbf{R} = \mathbf{R}_2 - \mathbf{R}_1$:

$$\Delta\hat{\rho}(\mathbf{K}, z) = \int \exp(i\mathbf{K} \cdot \mathbf{R}) \Delta\rho(\mathbf{R}, z) d\mathbf{R}$$
$$\hat{C}^{(2)}(\mathbf{K}, z_1, z_2) = \int \exp(i\mathbf{K} \cdot \mathbf{R}) C^{(2)}(\mathbf{R}, z_1, z_2) d\mathbf{R} \qquad (6.4.13)$$

Use of Parseval's relation shows that (6.4.10) may now be written as

$$\Delta\Omega = \frac{k_B T}{2A} \int dz_1 \int dz_2 \sum_{\mathbf{K}} \hat{C}^{(2)}(\mathbf{K}, z_1, z_2) \Delta\hat{\rho}(\mathbf{K}, z_1) \Delta\hat{\rho}(-\mathbf{K}, z_2) \qquad (6.4.14)$$

where A is the total area of the equilibrium interface. In the limit $A \to \infty$ the sum on $\mathbf{K}$-vectors goes over to an integral.

It follows from the general relation (6.4.4) that the change in interfacial area due to the fluctuation is

$$\Delta A = A' - A = \int_A \left[\left(1 + |\nabla_{\mathbf{R}} z_G(\mathbf{R})|^2\right)^{1/2} - 1 \right] d\mathbf{R}$$
$$\approx \frac{1}{2} \int_A |\nabla_{\mathbf{R}} z_G(\mathbf{R})|^2 d\mathbf{R} \qquad (6.4.15)$$

Since the fluctuation has only long-wavelength components parallel to the equilibrium surface, it is sufficient to replace the function $C^{(2)}(K, z_1, z_2)$ in (6.4.14) by its small-K expansion, i.e.

$$\hat{C}^{(2)}(K, z_1, z_2) = C_0(z_1, z_2) + K^2 C_2(z_1, z_2) + \mathcal{O}(K^4) \qquad (6.4.16)$$

On substitution of (6.4.8) and (6.4.16) in (6.4.14), and returning to $\mathbf{R}$-space via Parseval's relation, we find that

$$
\begin{aligned}
\Delta\Omega &= \frac{1}{2}k_B T \int dz_1 \int dz_2 \left\{ C_0(z_1, z_2) \int d\mathbf{R}\,\Delta\rho(\mathbf{R}, z_1)\Delta\rho(\mathbf{R}, z_2) \right. \\
&\quad \left. + C_2(z_1, z_2) \int d\mathbf{R}\,(\nabla_{\mathbf{R}}\,\Delta\rho(\mathbf{R}, z_1)) \cdot (\nabla_{\mathbf{R}}\,\Delta\rho(\mathbf{R}, z_2)) \right\} \\
&= \frac{1}{2}k_B T \int dz_1 \int dz_2 \frac{d\rho_0^{(1)}(z_1)}{dz_1}\frac{d\rho_0^{(1)}(z_2)}{dz_2} \times \left\{ C_0(z_1, z_2) \int d\mathbf{R}\,|z_G(\mathbf{R})|^2 \right. \\
&\quad \left. + C_2(z_1, z_2) \int d\mathbf{R}\,|\nabla_{\mathbf{R}}\, z_G(\mathbf{R})|^2 \right\}
\end{aligned}
\tag{6.4.17}
$$

Equation (F.5) of Appendix F shows that the contribution from terms involving $C_0(z_1, z_2)$ must vanish. It follows finally, by comparison with the thermodynamic relation (6.1.4) at constant μ, V and T, that the surface tension $\gamma = \Delta\Omega/\Delta A$ is given by the microscopic expression

$$
\gamma = k_B T \int_{-\infty}^{\infty} dz_1 \int_{-\infty}^{\infty} dz_2\, C_2(z_1, z_2)\frac{d\rho_0^{(1)}(z_1)}{dz_1}\frac{d\rho_0^{(1)}(z_2)}{dz_2}
\tag{6.4.18}
$$

Taken together, (6.4.11), (6.4.12) and (6.4.16) show that $C^{(2)}(z_1, z_2)$ is related to the inhomogeneous two-particle direct correlation function by

$$
C_2(z_1, z_2) = \frac{1}{2}\pi \int_0^{\infty} R^3 c^{(2)}(R, z_1, z_2) dR
\tag{6.4.19}
$$

A parallel exists between the result for γ provided by combination of (6.4.18) and (6.4.19), and the expression for the isothermal compressibility in terms of the direct correlation function of the bulk fluid, given by (3.5.15); in each case no explicit reference is made to the interactions between particles. An alternative, 'mechanical' expression for γ in terms of the pair density (2.5.13), which involves the pair potential explicitly and is therefore restricted to hamiltonians of the form (2.5.16), had been derived earlier by Kirkwood and Buff[12]; this is akin to the virial relation (2.5.22) for the bulk pressure.

6.5 FUNDAMENTAL MEASURE THEORY

Fundamental measure theory is a generalised form of weighted density approximation for a fluid consisting of hard particles. In contrast to similar approximations discussed in Section 6.2, the free energy density is taken to be a function not just of one but of several different weighted densities, defined by weight functions that emphasise the geometrical characteristics of the particles. The theory was originally formulated for hard-sphere mixtures, but for the sake of simplicity we consider in detail only the one-component case. Its development[7] was inspired by the link that exists between scaled particle theory[13] (described in Appendix E) and the Percus – Yevick approximation for hard spheres. Scaled particle theory provides only thermodynamic properties, while the PY approximation is a theory of pair structure, but the PY equation of state obtained via the compressibility route is identical to the scaled-particle result; the same is true for binary mixtures.

The development of the theory starts from the observation that the PY expression (4.4.10) for the two-particle direct correlation function of the hard-sphere fluid may be rewritten in terms of quantities that characterise the geometry of two intersecting spheres of radius $R(=\frac{1}{2}d)$ and separated by a distance $r < 2R$, as pictured in Figure 6.3. The quantities involved are the overlap volume $\Delta V(r)$, the overlap surface area $\Delta S(r)$ and the 'overlap radius' $\Delta R(r) = 2R - \bar{R}$, where $\bar{R} = R + \frac{1}{4}r$ is the mean radius of the convex envelope surrounding the spheres. Written in this way, (4.4.10) becomes

$$- c(r) = \chi^{(3)} \Delta V(r) + \chi^{(2)} \Delta S(r) + \chi^{(1)} \Delta R(r) + \chi^{(0)} \Theta(|\mathbf{r}| - 2R) \quad (6.5.1)$$

where the step function $\Theta(|\mathbf{r}| - 2R)$, defined in the previous section, is the 'characteristic' volume function of the exclusion sphere shown in the figure. The density-dependent coefficients $\chi^{(\alpha)}$ can be expressed in the form

$$\chi^{(0)} = \frac{1}{1 - \xi_3}, \quad \chi^{(1)} = \frac{\xi_2}{(1 - \xi_3)^2}$$

$$\chi^{(2)} = \frac{\xi_1}{(1 - \xi_3)^2} + \frac{\xi_2^2}{4\pi(1 - \xi_3)^3} \quad (6.5.2)$$

$$\chi^{(3)} = \frac{\xi_0}{(1 - \xi_3)^2} + \frac{2\xi_1\xi_2}{(1 - \xi_3)^3} + \frac{\xi_2^3}{4\pi(1 - \xi_3)^4}$$

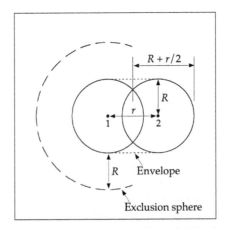

FIGURE 6.3 Geometry of two overlapping hard spheres of radius R and separation r. The exclusion sphere of radius $2R$ drawn around sphere 1 defines the region into which the centre of sphere 2 cannot enter without creating an overlap.

with $\xi_\alpha = \rho \mathcal{R}^{(\alpha)}$, where the quantities $\mathcal{R}^{(\alpha)}$ are the 'fundamental geometric measures' of a sphere:

$$\mathcal{R}^{(3)} = \frac{4}{3}\pi R^3 \text{(volume)}, \quad \mathcal{R}^{(2)} = 4\pi R^2 \text{(surface area)}$$

$$\mathcal{R}^{(1)} = R \text{(radius)}, \quad \mathcal{R}^{(0)} = 1 \tag{6.5.3}$$

The variables ξ_α also arise naturally in scaled particle theory. In particular, the scaled particle free energy density (see Appendix E) can be written as

$$\frac{\beta F^{\text{ex}}}{V} = -\xi_0 \ln(1 - \xi_3) + \frac{\xi_1 \xi_2}{1 - \xi_3} + \frac{\xi_2^3}{24\pi(1 - \xi_3)^2} \tag{6.5.4}$$

The same result applies to mixtures if the scaled particle variables are replaced by their multi-component generalisations,[14] i.e. $\xi_\alpha = \sum_\nu \rho_\nu \mathcal{R}_\nu^{(\alpha)}$, where ρ_ν is the number density of spheres of species ν and fundamental measures $\mathcal{R}_\nu^{(\alpha)}$.

The overlap volume, surface and radius are geometric measures associated with a pair of overlapping spheres, but they are also expressible in terms of convolutions of the characteristic volume and surface functions of individual spheres:

$$\omega^{(3)}(\mathbf{r}) = \Theta(|\mathbf{r}| - R)\text{(volume)}, \quad \omega^{(2)}(\mathbf{r}) = \delta(|\mathbf{r}| - R)\text{(surface)} \tag{6.5.5}$$

via the relations

$$\Delta V(r) = \omega^{(3)} \otimes \omega^{(3)} = \int \Theta(|\mathbf{r}'| - R)\Theta(|\mathbf{r} - \mathbf{r}'| - R)d\mathbf{r}'$$

$$= \frac{2}{3}\pi \left(2R^3 - 3R^2 r + r^3\right) \Theta(|\mathbf{r}| - 2R)$$

$$\Delta S(r) = 2\omega^{(3)} \otimes \omega^{(2)} = 2\int \Theta(|\mathbf{r}'| - R)\delta(|\mathbf{r} - \mathbf{r}'| - R)d\mathbf{r}' \tag{6.5.6}$$

$$= 4\pi R^2 \left(1 - r/2R\right) \Theta(|\mathbf{r}| - 2R)$$

$$\Delta R(r) = \frac{\Delta S(r)}{8\pi R} + \frac{1}{2}R\Theta(|\mathbf{r}| - 2R) = \left(R - r/4\right)\Theta(|\mathbf{r}| - 2R)$$

When results are brought together, it is straightforward to show that (6.5.1) is identical to (4.4.10); in particular, $c(r)$ is strictly zero for $r > 2R$ and $c(r) \to -\Theta(|\mathbf{r}| - 2R)$ as $\rho \to 0$. In addition, it is clear that if $c(r)$ is to be written solely in terms of functions characteristic of individual spheres, the pair function $\Theta(|\mathbf{r}| - 2R)$ must be replaced by some convolution of single-sphere functions; this can be achieved with a basis set consisting of the two scalar functions (6.5.5), a vector function

$$\boldsymbol{\omega}^{(2)}(\mathbf{r}) = \nabla \omega^{(3)}(\mathbf{r}) = \frac{\mathbf{r}}{r}\delta(|\mathbf{r}| - R) \tag{6.5.7}$$

and three further functions proportional to either $\omega^{(2)}(\mathbf{r})$ or $\boldsymbol{\omega}^{(2)}(\mathbf{r})$:

$$\boldsymbol{\omega}^{(1)}(\mathbf{r}) = \frac{\boldsymbol{\omega}^{(2)}(\mathbf{r})}{4\pi R}, \quad \omega^{(0)}(\mathbf{r}) = \frac{\omega^{(2)}(\mathbf{r})}{4\pi R^2}, \quad \omega^{(1)}(\mathbf{r}) = \frac{\omega^{(2)}(\mathbf{r})}{4\pi R} \qquad (6.5.8)$$

The vector functions are needed to account for the discontinuity in the step function. Then

$$\Theta(|\mathbf{r}| - 2R) = 2\left(\omega^{(3)} \otimes \omega^{(0)} + \boldsymbol{\omega}^{(2)} \otimes \boldsymbol{\omega}^{(1)} + \omega^{(2)} \otimes \omega^{(1)}\right) \qquad (6.5.9)$$

where the convolution of two vector functions also implies a scalar product; this result is most easily verified by taking Fourier transforms. In the limit $\mathbf{k} \to 0$, the transforms of the scalar characteristic functions are related to the scaled particle variables by

$$\rho\hat{\omega}^{(\alpha)}(\mathbf{k} = 0) = \xi_\alpha, \quad \alpha = 0 \text{ to } 3 \qquad (6.5.10)$$

while the transforms of the vector functions vanish:

$$\hat{\boldsymbol{\omega}}^{(\alpha')}(\mathbf{k} = 0) = 0, \quad \alpha' = 1, 2 \qquad (6.5.11)$$

Use of the characteristic functions (6.5.5), (6.5.7) and (6.5.8) as a basis therefore allows the PY direct correlation function to be expressed as a linear combination of convolutions in the form

$$c(r) = \sum_\alpha \sum_\beta c_{\alpha\beta}\omega^{(\alpha)} \otimes \omega^{(\beta)} \qquad (6.5.12)$$

where a simplified notation has been adopted in which the sums on α and β run over both scalar and vector functions; the density-dependent coefficients $c_{\alpha\beta}$ are proportional[7] to the functions $\chi^{(\alpha)}$ defined by (6.5.2). A different set of basis functions that does not involve vector functions has been proposed, but turns out to be equivalent to the one we have described in the sense that it leads ultimately to the same free energy functional.[15]

The key assumption of fundamental measure theory is that the excess free energy functional has the form

$$\beta\mathcal{F}^{\text{ex}}[\rho^{(1)}] = \int \Phi^{\text{ex}}(\{\bar{\rho}_\alpha(\mathbf{r}')\})\mathrm{d}\mathbf{r}' \qquad (6.5.13)$$

where the free energy density Φ^{ex} (in units of $k_\mathrm{B}T$) is a function of a set of weighted densities, each defined in the manner of (6.2.17), i.e.

$$\bar{\rho}_\alpha(\mathbf{r}) = \int w_\alpha(|\mathbf{r} - \mathbf{r}'|)\rho^{(1)}(\mathbf{r}')\mathrm{d}\mathbf{r}' \qquad (6.5.14)$$

It follows from (3.5.2) that if the scheme contained in (6.5.13) and (6.5.14) is adopted the direct correlation function of the uniform fluid is of the form

$$c(r) = -\sum_\alpha \sum_\beta \frac{\partial^2 \Phi^{\text{ex}}}{\partial\bar{\rho}_\alpha \partial\bar{\rho}_\beta} w_\alpha \otimes w_\beta \qquad (6.5.15)$$

Comparison of (6.5.15) with (6.5.12) suggests immediately that the appropriate choice of weight functions in (6.5.14) are the characteristic functions $\omega^{(\alpha)}(\mathbf{r})$ and $\omega^{(\alpha')}(\mathbf{r})$, and hence that the set $\{\bar{\rho}_\alpha\}$ is one consisting of four scalar and two vector densities:

$$\bar{\rho}_\alpha(\mathbf{r}) = \int \omega^{(\alpha)}(|\mathbf{r}-\mathbf{r}'|)\rho^{(1)}(\mathbf{r}')d\mathbf{r}', \quad \alpha = 0 \text{ to } 3$$

$$\bar{\boldsymbol{\rho}}_{\alpha'}(\mathbf{r}) = \int \boldsymbol{\omega}^{(\alpha')}(\mathbf{r}-\mathbf{r}')\rho^{(1)}(\mathbf{r}')d\mathbf{r}', \quad \alpha' = 1,2 \tag{6.5.16}$$

If the system is homogeneous, the scalar weighted densities reduce to the scaled particle variables (6.5.2) and the vector densities vanish. The scalar densities have the dimensions of the corresponding ξ_α, i.e. $[\mathrm{L}]^{\alpha-3}$; $\bar{\boldsymbol{\rho}}_1$ and $\bar{\boldsymbol{\rho}}_2$ have the same dimensions as $\bar{\rho}_1$ and $\bar{\rho}_2$, respectively.

The precise functional form of the free energy density remains to be specified. One obvious possibility, in the spirit of a virial expansion, is to write Φ^{ex} as a linear combination of the lowest powers of the weighted densities and their products. In that case, since Φ^{ex} is a scalar quantity with the dimensions of density, it can only be a sum of terms in $\bar{\rho}_0$, $\bar{\rho}_1\bar{\rho}_2$, $\bar{\rho}_2^3$, $\bar{\boldsymbol{\rho}}_1 \cdot \boldsymbol{\rho}_2$ and $\bar{\rho}_2(\bar{\boldsymbol{\rho}}_2 \cdot \bar{\boldsymbol{\rho}}_2)$, with coefficients ϕ_i that are functions of the dimensionless density $\bar{\rho}_3$. Thus

$$\Phi^{\mathrm{ex}}(\{\bar{\rho}_\alpha\}) = \phi_0\bar{\rho}_0 + \phi_1\bar{\rho}_1\bar{\rho}_2 + \phi_2\bar{\rho}_2^3 + \phi_3\bar{\boldsymbol{\rho}}_1 \cdot \bar{\boldsymbol{\rho}}_2 + \phi_4\bar{\rho}_2(\bar{\boldsymbol{\rho}}_2 \cdot \bar{\boldsymbol{\rho}}_2) \tag{6.5.17}$$

or, in the case of a uniform fluid:

$$\Phi^{\mathrm{ex}}(\{\xi_\alpha\}) = \phi_0\xi_0 + \phi_1\xi_1\xi_2 + \phi_2\xi_2^3 \tag{6.5.18}$$

The excess free energy functional follows from (6.5.13) and the corresponding excess grand potential is

$$\Omega^{\mathrm{ex}}[\rho^{(1)}] = -\int P^{\mathrm{ex}}[\rho^{(1)}]d\mathbf{r} = \mathcal{F}^{\mathrm{ex}}[\rho^{(1)}] - \int \rho^{(1)}(\mathbf{r})\frac{\delta\mathcal{F}^{\mathrm{ex}}}{\delta\rho^{(1)}(\mathbf{r})}d\mathbf{r} \tag{6.5.19}$$

Hence the excess pressure P^{ex} (a functional of $\rho^{(1)}$) is given by the expression

$$\beta P^{\mathrm{ex}}[\rho^{(1)}] = -\Phi^{\mathrm{ex}} + \sum_\alpha \bar{\rho}_\alpha(\mathbf{r})\frac{\partial\Phi^{\mathrm{ex}}}{\partial\bar{\rho}_\alpha} \tag{6.5.20}$$

where the sum runs over all densities in the set $\{\bar{\rho}_\alpha\}$.

Now consider the problem from the point of view of scaled particle theory, which provides an approximation for the excess chemical potential μ_v^{ex} of a solute particle of radius R_v in a uniform fluid of hard spheres. It is shown in Appendix E that in the limit $R_v \to \infty$, $\mu_v^{\mathrm{ex}} \to PV_v$, where V_v is the volume of the particle and P is the bulk pressure. But it follows from (6.5.18), as applied to a mixture, that the chemical potential of the solute, $\mu_v^{\mathrm{ex}} = k_B T(\partial\Phi^{\mathrm{ex}}/\partial\rho_v)$, must also satisfy the relation

$$\beta\mu_v^{\mathrm{ex}} = \sum_\alpha \frac{\partial\Phi^{\mathrm{ex}}}{\partial\xi_\alpha}\frac{\partial\xi_\alpha}{\partial\rho_v} = \frac{\partial\Phi^{\mathrm{ex}}}{\partial\xi_3}V_v + \mathcal{O}(R_v^2) \tag{6.5.21}$$

Thus the derivative $\partial \Phi^{\text{ex}}/\partial \xi_3$ can be identified as βP. Within fundamental measure theory the further assumption is now made that the analogous relation is valid for the inhomogeneous fluid, i.e. that

$$\frac{\partial \Phi^{\text{ex}}}{\partial \bar{\rho}_3} = \beta P^{\text{ex}}[\rho^{(1)}] + \bar{\rho}_0 \qquad (6.5.22)$$

and combination of (6.5.20) and (6.5.22) yields a differential equation for the free energy density in the form

$$-\Phi^{\text{ex}} + \sum_{\alpha} \bar{\rho}_{\alpha} \frac{\partial \Phi^{\text{ex}}}{\partial \bar{\rho}_{\alpha}} + \bar{\rho}_0 = \frac{\partial \Phi^{\text{ex}}}{\partial \bar{\rho}_3} \qquad (6.5.23)$$

Substitution of (6.5.17) into (6.5.23), and identification of the coefficients of the basis functions in the expansion (6.5.17), leads to five, first-order differential equations, one for each of the coefficients ϕ_i; these equations are easily solved to give

$$\phi_0 = -\ln(1 - \bar{\rho}_3) + c_0, \quad \phi_1 = \frac{c_1}{1 - \bar{\rho}_3}$$

$$\phi_2 = \frac{c_2}{(1 - \bar{\rho}_3)^2}, \quad \phi_3 = \frac{c_3}{1 - \bar{\rho}_3}, \quad \phi_4 = \frac{c_4}{(1 - \bar{\rho}_3)^2} \qquad (6.5.24)$$

The constants of integration c_i in (6.5.24) are chosen to ensure that both the free energy and its second functional derivative, i.e. the two-particle direct correlation function (see (3.5.2)), go over correctly to their known, low-density limits in the case of a uniform fluid.[16] These constraints give $c_0 = 0$, $c_1 = 1$, $c_2 = 1/24\pi$, $c_3 = -1$ and $c_4 = -1/8\pi$.[17] The excess free energy density is thereby completely determined and may be written in the form

$$\Phi^{\text{ex}}(\{\bar{\rho}_{\alpha}\}) = \Phi_1 + \Phi_2 + \Phi_3 \qquad (6.5.25)$$

with

$$\Phi_1 = -\bar{\rho}_0 \ln(1 - \bar{\rho}_3), \quad \Phi_2 = \frac{\bar{\rho}_1 \bar{\rho}_2 - \bar{\boldsymbol{\rho}}_1 \cdot \bar{\boldsymbol{\rho}}_2}{1 - \bar{\rho}_3}$$

$$\Phi_3 = \frac{\bar{\rho}_2^3 - 3\bar{\rho}_2(\bar{\boldsymbol{\rho}}_2 \cdot \bar{\boldsymbol{\rho}}_2)}{24\pi(1 - \bar{\rho}_3)^2} \qquad (6.5.26)$$

which reduces to the scaled particle result (6.5.4) for a uniform fluid. The two-particle direct correlation function obtained by differentiation of the free energy reduces in turn to the PY expression (6.5.1), while the third functional derivative yields a three-particle function in good agreement with the results of Monte Carlo calculations.[18] As Figure 4.2 shows, the scaled particle (or PY compressibility) equation of state slightly overestimates the pressure of the hard-sphere fluid. Some improvement in performance may therefore be expected if the assumed form of the free energy density is modified in such a way as to recover the Carnahan–Starling equation of state (3.9.20) in the uniform-fluid limit.[19]

The theory can be generalised to the case of hard-sphere mixtures in a straightforward way.[20] Scalar and vector characteristic functions $\omega_\nu^{(\alpha)}(\mathbf{r})$ and $\boldsymbol{\omega}_\nu^{(\alpha)}$ are defined for each species ν in a manner completely analogous to the one-component case, with R_ν replacing R. The characteristic functions are then used as weight functions in the definition of a set of global weighted densities:

$$
\begin{aligned}
\bar{\rho}_\alpha(\mathbf{r}) &= \sum_\nu \int \omega_\nu^{(\alpha)}(|\mathbf{r} - \mathbf{r}'|)\rho_\nu^{(1)}(\mathbf{r}')d\mathbf{r}', \quad \alpha = 0 \text{ to } 3 \\
\bar{\boldsymbol{\rho}}_{\alpha'}(\mathbf{r}) &= \sum_\nu \int \boldsymbol{\omega}_\nu^{(\alpha')}(|\mathbf{r} - \mathbf{r}'|)\rho_\nu^{(1)}(\mathbf{r}')d\mathbf{r}', \quad \alpha' = 1, 2
\end{aligned}
$$

$$(6.5.27)$$

where $\rho_\nu^{(1)}$ is the density profile of species ν, and the free energy density of the mixture is again given by (6.5.26), or some other, improved form.

The same general approach[21] can be used to derive free energy functionals for hard-core systems in dimensions $\mathcal{D} = 1$ (hard rods) or $\mathcal{D} = 2$ (hard disks). For $\mathcal{D} = 1$, where only two weight functions are required, this leads to the exact hard-rod functional due to Percus.[22] For $\mathcal{D} = 2$, the procedure is less straightforward, since the decomposition of the Mayer function analogous to (6.5.9) is not achievable with any finite set of basis functions and the PY equation does not have an analytical solution. One and two-dimensional hard-core systems may be regarded as special cases of a hard-sphere fluid confined to a cylindrical pore ($\mathcal{D} = 1$) or a narrow slit ($\mathcal{D} = 2$), for which the diameter of the cylinder or width of the slit is equal to the hard-sphere diameter. Narrow confinement therefore corresponds to a reduction in effective dimensionality or 'dimensional crossover', the most extreme example of which ($\mathcal{D} = 0$) occurs when a hard sphere is confined to a spherical cavity large enough to accommodate at most one particle. If the $\mathcal{D} = 3$ functional is to be used in studies of highly confined fluids, it is clearly desirable that it should reduce to the appropriate one or two-dimensional functional for density profiles of the form $\rho^{(1)}(\mathbf{r}) = \rho^{(1)}(x)\delta(y)\delta(z)$ (for $\mathcal{D} = 1$) or $\rho^{(1)}(\mathbf{r}) = \rho^{(1)}(x, y)\delta(z)$ (for $\mathcal{D} = 2$). This turns out not to be the case. The exact results for $\mathcal{D} = 0$ and $\mathcal{D} = 1$ are recovered if the term Φ_3 in (6.5.25) is omitted, but that leads to a considerable deterioration in the results for $\mathcal{D} = 3$. A good compromise is achieved[23] if Φ_3 is replaced by

$$
\Phi_3' = \frac{\bar{\rho}_2^3}{24\pi(1 - \bar{\rho}_3)^2}(1 - \xi^2)^3 \tag{6.5.28}
$$

where $\xi(\mathbf{r}) = |\bar{\boldsymbol{\rho}}_2(\mathbf{r})/\bar{\rho}_2(\mathbf{r})|$. The modified term vanishes for $\mathcal{D} = 0$ and is numerically small, except at the highest densities, for $\mathcal{D} = 1$. In addition, since Φ_3' differs from Φ_3 only by terms of order ξ^4, differentiation of the resulting functional still leads to the PY result for the direct correlation function of the uniform fluid. However, the modification is essentially empirical in nature.

A more systematic method of constructing free energy functionals with the correct dimensional crossover properties is to start from the exact result for $\mathcal{D} = 0$ and build in successively the additional terms needed in higher dimensions.[24] That the functional should have at least the correct qualitative behaviour for $\mathcal{D} = 0$ is essential for application to the solid phase, where each particle is confined to the nearly spherical cage formed by its nearest neighbours. The contribution from Φ_3 diverges to negative infinity in the zero-dimensional limit. Thus the theory in its unmodified form cannot account for solid-fluid coexistence, since the solid is always the stable phase.

6.6 CONFINED FLUIDS

The density functional formalism has been successfully applied to a wide range of physical problems involving inhomogeneous fluids. In this section we describe some of the results obtained from calculations for fluids in confined geometries. The simplest example, illustrated in Figure 6.4, is that of a fluid near a hard, planar wall which confines the fluid strictly to a half-space $z \geq 0$, say, where the normal to the wall is taken as the z-axis. The particles of the fluid interact with the wall via a potential $\phi(z)$, which plays the role of the external potential in the theoretical treatment developed in earlier sections. For a hard wall the potential has a purely excluded-volume form, i.e. $\phi(z) = \infty$, $z < 0$, $\phi(z) = 0$, $z > 0$, but more generally it will contain a steeply repulsive term together with a longer ranged, attractive part. If the particles making up the wall are assumed to interact with those of the fluid through a Lennard-Jones potential with parameters ϵ and σ, integration over a continuous distribution of particles within the wall leads to a wall-fluid potential given by

$$\phi(z) = \frac{2}{3}\pi \rho_{\mathrm{W}} \sigma^3 \epsilon \left[\frac{2}{15}(\sigma/z)^9 - (\sigma/z)^3 \right] \qquad (6.6.1)$$

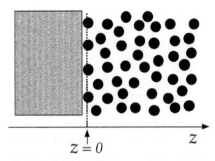

$$z = 0 \qquad\qquad z$$

FIGURE 6.4   A fluid confined by a hard wall; the centres of interaction of the particles are restricted to the region $z > 0$. For hard spheres of diameter d, the surface of the wall is at $z = -\frac{1}{2}d$.

where ρ_W is the density of particles in the wall; the surface of the wall is now at $z = 0$. This so-called 9-3 potential has been widely adopted as a model of the wall-fluid interaction.

The density profile of a fluid against a planar wall is a function of the single coordinate z. If the bulk density ρ_B (the density far from the wall) is sufficiently large, the profile has a pronounced layer structure that extends several particle diameters into the fluid. When all interactions are of hard-core type, $\rho^{(1)}(z)$ can be calculated by density functional theory with the boundary conditions:

$$\lim_{z \to \infty} \rho^{(1)}(z) = \rho_B \qquad (6.6.2)$$

and

$$\lim_{z \to 0+} \rho^{(1)}(z) = \beta P \qquad (6.6.3)$$

where P is the bulk pressure; these conditions must be supplemented by the requirement that $\rho^{(1)}(z) = 0$ for $z < 0$. Equation (6.6.3) is an expression of the *contact theorem*, $z = 0$ being the distance of closest approach of a hard sphere of diameter d to a hard wall with a surface at $z = -\frac{1}{2}d$ (see Figure 6.4). The proof of the contact theorem is similar to that of the relation (2.5.26) between the pressure of a uniform hard-sphere fluid and the value of the pair distribution function at contact. The density profile of a fluid against a hard wall is discontinuous at $z = 0$, but whatever the nature of the wall-fluid interaction the density profile can always be written in the form $\rho^{(1)}(z) = \exp[-\beta\phi(z)]y(z)$, where $y(z)$ is a continuous function of z, analogous to the cavity distribution function of a homogeneous fluid. The pressure exerted by the fluid on the wall must be balanced by the force per unit area exerted by the wall on the fluid, i.e.

$$P = -\int_0^\infty \frac{d\phi(z)}{dz} \rho^{(1)}(z)dz = k_B T \int_0^\infty \frac{d}{dz} \exp[-\beta\phi(z)]y(z)dz \qquad (6.6.4)$$

and hence, in the case of a hard wall:

$$P = k_B T \int_0^\infty \delta(z)y(z)dz = k_B T \rho^{(1)}(z = 0+) \qquad (6.6.5)$$

which is (6.6.3).

The layering of a high-density, hard-sphere fluid near a hard wall is illustrated in Figure 6.5, where comparison is made between the density profile derived from fundamental measure theory and results obtained by Monte Carlo calculations. Agreement between theory and simulation is excellent. The only significant discrepancies (not visible in the figure) occur close to contact, where the theoretical values are too high. The source of these small errors lies in the fact that in the theory as implemented here the value at contact is determined, via the boundary condition (6.6.3), by the pressure calculated from scaled particle theory. As discussed in Section 6.5, such errors can be largely eliminated by

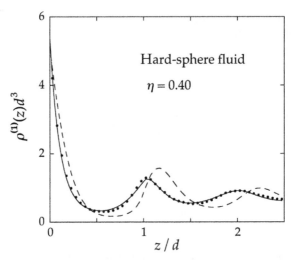

FIGURE 6.5 Density profile of a hard-sphere fluid close to a hard wall at a packing fraction $\eta = 0.40$. The full curve is calculated from fundamental measure theory and the points show the results of Monte Carlo calculations.[25] The dashed curve is calculated from the simpler weighted density approximation provided by (6.2.22).

tailoring the free energy functional to reproduce a more accurate equation of state.

Though designed for systems of hard particles, fundamental measure theory may also be used to calculate the density profiles and associated thermodynamic properties of a wider class of fluids if combined with the methods of perturbation theory described in Chapter 5. We suppose, as usual, that the pair potential $v(r)$ of the system of interest can be divided into a reference part, $v_0(r)$, and a perturbation, $w(r)$. Then (3.4.10) provides an exact relation between the free energy functional corresponding to the full potential, $\mathcal{F}[\rho^{(1)}]$, and that of the reference system, $\mathcal{F}_0[\rho^{(1)}]$. The obvious choice of reference system is again a fluid of hard spheres of diameter d given, say, by the Barker-Henderson prescription (5.3.11). If the perturbation is sufficiently weak to be treated in a mean field approximation, the correlation term in (3.4.10) can be ignored. The grand potential functional to be minimised is then of the form

$$\Omega_\phi[n] = \mathcal{F}_d[n] + \frac{1}{2} \iint n(\mathbf{r})w(\mathbf{r},\mathbf{r}')n(\mathbf{r}')\mathrm{d}\mathbf{r}\,\mathrm{d}\mathbf{r}' + \int n(\mathbf{r})[\phi(\mathbf{r}) - \mu]\mathrm{d}\mathbf{r}$$

$$(6.6.6)$$

where $\mathcal{F}_d[\rho^{(1)}]$ is the free energy functional of the hard-sphere system, taken to be of fundamental measure form, and $n(\mathbf{r})$ is a trial density. This approximation has been used in a variety of applications to confined fluids. An example of the results obtained for the density profile of a Lennard-Jones fluid confined to a slit formed by two parallel plates separated by a distance L is pictured in Figure 6.6; the wall-fluid potential has a form similar to (6.6.1). When $L/\sigma \approx 3$, the density

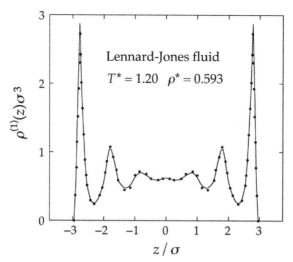

FIGURE 6.6 Density profile of a Lennard-Jones fluid in a slit of width $L = 7.5\sigma$. The curve is calculated from fundamental measure theory and the points show the results of a Monte Carlo simulation.[26] *Redrawn with permission from Ref. 27 © 1991 American Physical Society.*

profile displays a double-peaked structure, with maxima close to the walls of the slit. As the slit width increases, the number of layers of particles that can be accommodated also increases, with a third peak appearing initially mid-way between the walls. In the example shown, corresponding to $L/\sigma = 7.5$, six clearly defined layers can be detected, together with a weak maximum at the centre of the slit. The agreement with simulations is again outstandingly good. Figure 6.7 shows the solvation force as a function of L for the same system, calculated from the microscopic expression

$$f_S = -\int_0^L \frac{d\phi(z)}{dz} \rho^{(1)}(z)dz - P \qquad (6.6.7)$$

which is easily derived from the definition (6.1.16). The force is seen to oscillate around zero, its asymptotic value as $L \to \infty$. Oscillatory solvation forces are a direct consequence of the layering evident in Figure 6.6; they have been observed experimentally with the aid of 'surface force machines', which have a spatial resolution better than 1 Å. The amplitude of oscillation in the figure decreases rapidly with L, and is already negligible for $L = 7.5\sigma$ despite the high degree of layering still observed at this separation.

Functionals of the general form represented by (6.6.6), with various levels of approximation for the contribution from the reference system, have also been used extensively in studies of phenomena such as capillary condensation in a narrow pore and the wetting of solid substrates.[29] These two effects are closely related and each is strongly dependent on the nature of the interaction between the fluid and the confining surface. Capillary condensation is the phenomenon

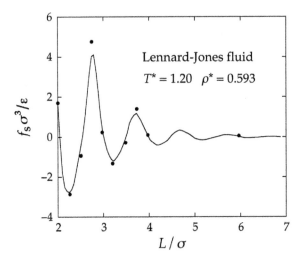

FIGURE 6.7 Solvation force for a Lennard-Jones fluid in a slit of width L. The curve is calculated from fundamental measure theory and the points show the results of a Monte Carlo simulation.[28] *Redrawn with permission from Ref. 27 © 1991 American Physical Society.*

whereby a confined gas condenses to a liquid at a chemical potential below that corresponding to liquid–vapour coexistence in the bulk; wetting is discussed in Section 6.7.

A different type of problem to which density functional theory has been successfully applied concerns the size selectivity of porous materials in which the pores have a confining length of molecular dimensions. As a simple example, consider an infinitely long, cylindrical pore of diameter D connecting two reservoirs which contain a three-component mixture of hard spheres under identical physical conditions (packing fraction and concentrations). The fluid in the reservoirs consists of a majority component – the 'solvent' S – at a packing fraction $\eta = 0.41$, and two 'solute' components, A and B, at concentrations of 0.05 M, with relative hard-sphere diameters $d_A : d_B : d_S$ appropriate to water (S) and the ions Na$^+$ (A) and K$^+$ (B).[30] Spheres of different diameters will permeate the pore to different extents, and at equilibrium the chemical potentials of each species will be the same inside the pore as in the reservoirs. The density profiles within the pore depend only on the radial distance r from the axis of the cylinder; they can be calculated by minimising a fundamental measure functional, modified in the manner represented by (6.5.28) to cater for the quasi-one-dimensional nature of the confinement. The degree of permeation (or 'absorbance') ζ_ν of species ν may be defined as the ratio of the mean density of particles of that species inside the pore to the density of the same species in the reservoirs. When the cylinder diameter D is comparable with the sphere diameters, the pore absorbs preferentially one of the two solutes. The selectivity

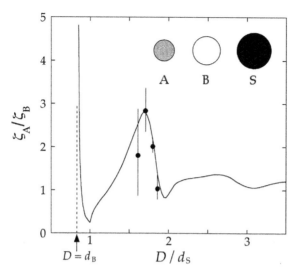

FIGURE 6.8 Selective absorption by a cylindrical pore of solute hard spheres (A, B) at low concentration in a solvent of larger spheres (S) as a function of the cylinder diameter. The curve is calculated from fundamental measure theory and the points with error bars show the results of Monte Carlo calculations. See text for details. *Redrawn with permission from Ref. 30 © 2001 The Royal Society of Chemistry.*

of the pore is measured by the relative absorbance ζ_A/ζ_B, plotted as a function of cylinder diameter in Figure 6.8. This varies with D by a factor of order 10, in fair agreement with calculations by a grand canonical Monte Carlo method, though the low concentrations of solute particle mean that the statistical uncertainties in the results of the simulations are large. When $d_B < d_S$, only A-particles can be absorbed. Thus, for cylinder diameters only slightly larger than d_B, the selectivity is initially very large but falls rapidly as D increases. When $D \approx d_S$, the larger solute is up to four times more likely to be adsorbed than the smaller one, a purely entropic effect that is somewhat counter-intuitive. However, when the cylinder diameter exceeds d_S and solvent particles can enter the pore, the selectivity rises, reaching a maximum value of about 2.8 at $D \approx 1.7d_S$. The degree of selectivity can be greatly enhanced by changes in the relative diameters of the species involved.

6.7   DENSITY FUNCTIONAL THEORY OF WETTING

Density functional theory has proved particularly valuable in its application to the study of three-phase equilibria. In this section we focus on the equilibrium between liquid and its vapour near a solid substrate but much of the theory we describe applies equally well to the situation when all three phases are fluids.

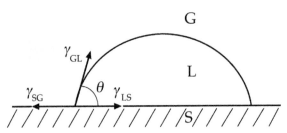

FIGURE 6.9 Schematic representation of a liquid drop (L) in equilibrium with a gas (G) and a planar, solid substrate (S). The balance of forces acting at the contact line leads to Young's equation (6.7.1).

In each case the competition between interfacial free energies, or surface tensions, gives rise to surface phase transitions that are distinct from those occurring in the bulk phases.

If a liquid drop in equilibrium with its vapour is deposited on a planar, solid substrate it will either remain localised or spread out to form a film that *wets* the substrate; what occurs in practice depends on the physical conditions. The case when the liquid remains localised is pictured in Figure 6.9; the line where the liquid (L), gas (G) and substrate (S) meet forms the *contact line* and the angle θ between the substrate and the plane tangent to the drop along the contact line is the *contact angle*. The three surface tensions, γ_{SG}, γ_{GL} and γ_{LS} are the forces per unit length acting at the contact line along each of the three interfaces. In equilibrium, when the liquid has come to rest, these forces must balance and their projections onto the substrate plane are related by Young's equation:

$$\gamma_{SG} = \gamma_{LS} + \gamma_{GL} \cos \theta \qquad (6.7.1)$$

It follows that the equilibrium *spreading coefficient* S, defined as

$$S = \gamma_{SG} - \left(\gamma_{LS} + \gamma_{GL}\right) \qquad (6.7.2)$$

is necessarily less than or equal to zero.

The physical significance of the spreading coefficient is clear: when $S = 0$, the contact angle is also zero, and the liquid spreads to form a macroscopically thick film. This is the phenomenon of *complete wetting*. There is no thermodynamic cost to the growth of a macroscopic layer of liquid between the substrate and the vapour, since the surface excess free energy (or grand potential) associated with the solid-gas interface is equal to the sum of the excess free energies of the liquid–solid and liquid–gas interfaces. If S is negative, the free energy per unit area of the solid-gas interface is lower than the sum of the other two, so there is no thermodynamic driving force that would cause the drop to spread. This corresponds to *partial wetting*, a situation characterised by a non-zero value of the contact angle. Ellipsometric measurements show that while the drop does not spread to form a macroscopic film, liquid is lost from the drop to form a film of thickness equal to a few molecular layers. The case when $\theta \to \pi$ corresponds to non-wetting or *drying*, where a layer of gas intrudes between solid and liquid; it is clear from (6.7.1) that there is no free energy cost involved in such a process.

Suppose one starts from a situation of partial wetting. What happens as the temperature increases towards the bulk critical temperature T_c? A simple argument, due to Cahn,[31] shows that one can expect to see a transition from partial to complete wetting at a *wetting temperature* $T_w < T_c$. This is a surface phase transition, for which the order parameter is the adsorption Γ, analogous to the liquid and gas adsorptions defined by (6.1.7):

$$\Gamma = \int_0^\infty \left[\rho^{(1)}(z) - \rho_G \right] dz \qquad (6.7.3)$$

where z is the vertical distance from the substrate surface placed at $z = 0$ and ρ_G is the bulk density of the gas phase. A useful definition of the *thickness* ξ of the adsorbed liquid film is provided by the reduced adsorption

$$\xi = \frac{\Gamma}{\rho_G} \qquad (6.7.4)$$

where ρ_G is the bulk density of the gas phase.

If the vapour–liquid coexistence curve is approached from the undersaturated vapour side at a temperature T below T_w, the adsorption remains finite, corresponding to partial wetting, whereas for $T > T_w$ the adsorption diverges as the chemical potential tends to its value, $\mu_0(T)$, at coexistence ($\mu \to \mu_0^-(T)$). If T_w is approached from below along the coexistence curve, i.e. for $\mu = \mu_0(T)$, Γ increases from a finite value towards infinity. In that case there are two possibilities. If Γ increases continuously, and diverges at $T = T_w$, the wetting transition is continuous or second order. Alternatively, Γ may change discontinuously from a finite value just below T_w to become infinite at $T = T_w$; this is a first-order transition. When the transition is first order there is a further twist predicted both by Cahn and later, for a microscopic model within density functional theory, by Ebner and Saam.[32] As coexistence is approached from the vapour side above T_w, an additional, *prewetting* transition occurs as a 'prewetting line' is crossed, marked by a discontinuity in Γ and shown on the schematic phase diagram pictured in Figure 6.10. The prewetting line starts on the coexistence curve at the wetting temperature, where the discontinuity is infinite, and moves into the undersaturated vapour region for $T > T_w$, where the jump in adsorption is finite. The amplitude of the discontinuity decreases as T increases, and vanishes at a prewetting critical temperature $T_{pwc} < T_c$, above which Γ increases continuously along an isotherm and diverges at coexistence. Thus, everywhere except at the wetting temperature itself, the discontinuous, prewetting transition is one between thin and thick films. The existence of different classes of wetting transitions has been confirmed experimentally

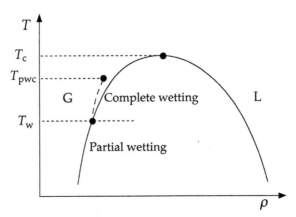

FIGURE 6.10 Schematic phase diagram corresponding to first-order wetting at or near liquid–gas coexistence. The prewetting line branches off from the coexistence line at the wetting temperature T_w and ends at the prewetting critical temperature T_{pwc}. See text for details.

by measurements of the contact angle as a function of temperature.[33] Most observed transitions are first order, signalled by a large hysteresis as T_w is approached from above or below along the coexistence curve. Observation of the prewetting transition between thin and thick films requires more sophisticated optical techniques.

In density functional theory the equilibrium density profile $\rho^{(1)}(z)$ is obtained by minimisation of the one-dimensional form of the grand potential functional (3.4.1) with respect to a trial profile. This leads to an expression for $\rho^{(1)}(z)$ of the generic form provided by (3.5.4):

$$\rho^{(1)}(z) = \rho_G \exp\left(-\beta\left[\psi^{ex}\left(\rho^{(1)}(z)\right) - \mu^{ex}\left(\rho_G\right) + \phi(z)\right]\right) \quad (6.7.5)$$

where $\phi(z)$ is the potential exerted on the fluid particles by the substrate, $\rho_G = \rho^{(1)}(z \to \infty)$, $\mu^{ex}(\rho_G)$ is the corresponding excess chemical potential and

$$\psi^{ex}\left(\rho^{(1)}(z)\right) = \frac{\delta\mathcal{F}^{ex}\left[\rho^{(1)}\right]}{\delta\rho^{(1)}(z)} \quad (6.7.6)$$

is the excess, intrinsic chemical potential, defined as the functional derivative of the excess, intrinsic free energy functional. Equation (6.7.5) must be solved iteratively for a given, approximate choice of $\mathcal{F}^{ex}[\rho^{(1)}]$. The adsorption can then be determined from its definition (6.7.3) and the surface excess grand potential, i.e. the surface tension, from the relation

$$\gamma\left[\rho^{(1)}\right] = \Omega^s\left[\rho^{(1)}(z)\right] = \Omega\left[\rho^{(1)}\right] - \Omega(\rho_G)$$

$$= \mathcal{F}\left[\rho^{(1)}\right] - F(\rho_G) + \int_0^\infty \rho^{(1)}(z)\phi(z)\mathrm{d}z - \mu\Gamma \quad (6.7.7)$$

The Gibbs–Duhem relation (6.1.12) provides a means of testing the thermo-dynamic self-consistency of results obtained in this way.

The phenomenological approach used by Cahn is a generalisation of van der Waals's square-gradient theory of the liquid–vapour interface, described in Section 6.3, in which allowance is made for the presence of a substrate. Combination of (3.4.1), (6.2.10) and (6.3.1) shows that the grand potential functional per unit area of substrate is

$$\Omega_\phi\left[\rho^{(1)}\right] = \int_0^\infty \left(\omega\left(\rho^{(1)}(z)\right) + \frac{1}{2}m\left(\frac{d\rho^{(1)}(z)}{dz}\right)^2 + \rho^{(1)}(z)\phi(z)\right) dz$$

(6.7.8)

where $\omega(\rho) = \Omega(\rho)/V = f_0(\rho) - \mu\rho$ is the bulk grand potential per unit volume at a density $\rho = \rho^{(1)}(z)$; near the critical point $\omega\left(\rho^{(1)}(z)\right)$ may again be represented in the parametric form given by (6.3.5). Use of the functional (6.7.8) is appropriate only for slowly varying profiles; it cannot describe the fluid layering discussed in Section 6.6 and seen, for example, in Figure 6.6. There is consequently no inconsistency involved in assuming that the substrate-fluid interaction acts only at contact, i.e.

$$\rho^{(1)}(z)\phi(z) = \phi_0\left(\rho^{(1)}(z)\right)\delta(z) = \phi_0(\rho_S)\delta(z)$$

(6.7.9)

where $\rho_S = \rho^{(1)}(z = 0)$ is the density of the fluid at contact with the substrate. The quantity ϕ_0 is usually taken to be of quadratic form:

$$\phi_0(\rho_S) = \gamma_0 - \gamma_1\rho_S + \frac{1}{2}\gamma_2\rho_S^2$$

(6.7.10)

The linear term, with a positive value for γ_1, represents the attraction between the particles of the liquid and the substrate, while choice of a positive value for γ_2 allows for the reduction of cohesion in the liquid in the immediate vicinity of the substrate, where particles have on average fewer neighbours than in the bulk.

Substitution of (6.7.9) and (6.7.10) in (6.7.8) shows that the grand potential functional divides into two parts:

$$\Omega_\phi\left[\rho^{(1)}\right] = \Omega\left[\rho^{(1)}\right] + \Omega(\rho_S)$$

(6.7.11)

where $\Omega(\rho_S) = \phi_0(\rho_S)$; the wetting behaviour is determined by the competition between the first (liquid film) and second (substrate) terms on the right-hand side of this expression. The minimisation of $\Omega_\phi\left[\rho^{(1)}\right]$ proceeds as in the case of the liquid–vapour interface except that the density profile now varies between ρ_S, at $z = 0$, and ρ_G, as $z \to \infty$. Equation (6.3.6) is therefore replaced by

$$z = -\zeta \ln\left\{\left(\frac{\rho^{(1)}(z) - \rho_G}{\rho_L - \rho^{(1)}(z)}\right)\left(\frac{\rho_L - \rho_S}{\rho_S - \rho_G}\right)\right\}$$

(6.7.12)

which can be rearranged to yield an expression for the density profile in the form

$$\rho^{(1)}(z) = \rho_G + \frac{\rho_L - \rho_G}{1 + \exp\left[(z - \xi)/\zeta\right]} \tag{6.7.13}$$

where ζ is the thickness of the liquid–vapour interface, introduced in (6.3.7), and ξ is the film thickness defined by (6.7.4) and now given explicitly by

$$\xi = -\zeta \ln \frac{\rho_L - \rho_S}{\rho_S - \rho_G} \equiv -\zeta \ln \Phi \tag{6.7.14}$$

Values of z calculated from (6.3.6) and (6.7.13) for given values of ρ_L and ρ_G therefore differ only by the thickness of the film. Substitution of (6.7.13) and (6.7.9) in (6.7.8) and integration over z shows that the surface excess grand potential per unit area is

$$\Omega^{(s)} = \Omega\left[\rho^{(1)}\right] + \Omega(\rho_S) - \Omega(\rho_G)$$
$$= \gamma \left[3\left(\frac{\rho_S - \rho_G}{\rho_L - \rho_G}\right)^2 - 2\left(\frac{\rho_S - \rho_G}{\rho_L - \rho_G}\right)^3\right] + \gamma_0 - \gamma_1 \rho_S + \frac{1}{2}\gamma_2 \rho_S^2 \tag{6.7.15}$$

where γ is the surface tension at the liquid–gas interface, given by (6.3.10).

For temperatures sufficiently far below T_c for ρ_G to be very much smaller than ρ_L, (6.7.15) may be written in an approximate but more convenient form as[34]

$$\Omega^{(s)} = \Omega_0^{(s)} + \gamma \left[\frac{3}{(1 + \Phi^2)} - \frac{2}{(1 + \Phi)^3} - 1 + 6(p_1 - p_2)\left(\frac{\Phi}{1 + \Phi}\right)\right.$$
$$\left. + 3p_2 \left(\frac{\Phi}{1 + \Phi}\right)^2\right] \tag{6.7.16}$$

where Φ is defined in (6.7.14) and $\Omega_0^{(s)} = \gamma + \gamma_0 - \gamma_1 \rho_L + \frac{1}{2}\gamma_2 \rho_L^2$ is the value of the surface excess grand potential at complete wetting, i.e. the limit in which $\rho_L \to \rho_S$, $\Phi \to 0$ and $\xi \to \infty$. The dimensionless quantities p_1 and p_2 are given by

$$p_1 = \frac{\gamma_1}{(mC)^{1/2}}(\rho_L - \rho_G)^{-2}, \quad p_2 = \frac{\gamma_2}{(mC)^{1/2}}(\rho_L - \rho_G)^{-1} \tag{6.7.17}$$

It is straightforward to show that (6.7.15) and (6.7.16) are equivalent in the limit $\rho_G \to 0$. The quantities p_1 and p_2 can be altered independently of ρ_S and therefore act as 'control parameters' in the manipulation of state conditions.

The final step is to minimise $\Omega^{(s)}$ with respect to either Φ or, equivalently, the contact density ρ_S, from which the equilibrium film thickness can be determined via (6.7.14). Depending on the values chosen for p_1 and p_2, the lowest minimum of $\Omega^{(s)}$ may appear either at $\Phi = 0$, corresponding to complete wetting, or at $\Phi = \Phi^*$, where

$$\Phi^* = \frac{\left[(p_2 + 1)^2 - 4p_1\right]^{1/2} + p_2 + 1}{2p_1} - 1 \qquad (6.7.18)$$

which corresponds to partial wetting by a film of finite thickness $\xi = -\zeta \ln \Phi^*$. If $p_2 < 1$, a discontinuous jump in film thickness from a microscopic value to infinity occurs at $p_1 = (p_2 + 3)(3p_2 + 1)/16$; this is a first-order transition with a wetting temperature implicitly determined by the relation between p_1 and p_2. If $p_2 \geq 1$, the equilibrium value of ξ diverges continuously as $p_1 \rightarrow p_2$, representing a second-order transition. The different possibilities are illustrated graphically in Figure 6.11, which shows schematic plots of the variation with film thickness of the quantity $\Delta\Omega^{(s)} = \Omega^{(s)} - \Omega_0^{(s)}$, the surface excess grand potential relative to that at complete wetting. The left-hand panel of the figure illustrates the behaviour typical of a first-order transition. Below T_w the lower of the two minima in $\Delta\Omega^{(s)}$ corresponds to a finite film thickness (partial wetting); above T_w the lower minimum occurs as $\xi \rightarrow \infty$ (complete wetting); and at $T = T_w$ the film thickness jumps discontinuously between a finite value and infinity. The two minima are separated by a thermodynamic potential barrier, giving rise to hysteresis in measurements of ξ as the temperature increases or decreases around T_w. The right-hand panel corresponds to a second-order transition. The global minimum now shifts continuously from a finite film thickness to infinity as T increases; in this case there is no potential barrier.

The theory requires only a minor extension to explain the onset of prewetting as the coexistence curve is approached from the undersaturated vapour side. Away from coexistence the two minima in the bulk grand potential no longer have the same depth; $\omega(\rho_G)$ now lies below $\omega(\rho_L)$ because the gas is the stable phase while the liquid is metastable. The parameterisation of $\omega(\rho)$ in (6.3.5) must therefore be generalised by inclusion of a contribution linear in $(\rho - \rho_G)$

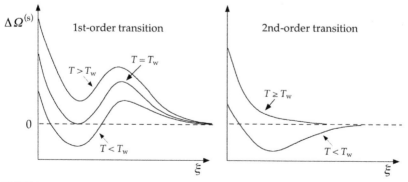

FIGURE 6.11 Schematic plots of the surface excess grand potential relative to that at complete wetting as a function of film thickness in both first and second-order transitions. See text for details.

and therefore proportional to the degree of undersaturation. The calculation of $\Omega^{(s)}$ as a function of ξ then proceeds in the manner already described and a first-order transition between microscopically thin and thick films is found to occur away from the coexistence curve at temperatures above T_w but below a limiting temperature T_{pwc}. As coexistence is approached, the minimum in $\Omega^{(s)}$ corresponding to the thick film moves continuously towards the limit of infinite thickness.

Cahn's theory is invaluable in the qualitative discussion of wetting phenomena. A limitation it has is the fact that its quantitative implementation requires as input the values of a large number of thermodynamic quantities or phenomenological parameters: $\rho_L(T)$, $\rho_G(T)$, m, C, γ_1 and γ_2. These values must be obtained from separate, theoretical treatments or from experiment. Versions of density functional theory that are microscopically more specific are needed if a direct link is to be established between a model hamiltonian and the interfacial properties and wetting behaviour of the physical system that the model represents. As we have seen in Section 6.6, the microscopic approach is usually based on the separation of the free energy functional into a part arising from the short-range, repulsive forces between particles, which is represented by the interaction between hard spheres of appropriately chosen diameter, and a long-range, attractive interaction, which is treated in a mean field manner.[35,36] Minimisation of the resulting grand potential function given by (6.6.6), adapted to the one-dimensional case, leads again to the implicit equation for the density profile provided by (6.7.5). The excess, intrinsic chemical potential in (6.7.5) is now the sum of hard-sphere and mean field terms, defined as the functional derivatives of the corresponding contributions to the free energy functional.

Density functional studies based on (6.7.5) have been made of the wetting of a solid substrate by a Yukawa hard-core fluid, for which the pair potential is given by (1.2.2). In this case the potential splits immediately into a true hard-sphere term and an attractive tail which serves as the perturbation:

$$w(r) = -\frac{\epsilon_F d}{r} \exp\left[-\lambda_F(r/d - 1)\right], \quad r > d \qquad (6.7.19)$$

where the dimensionless factor λ_F measures the range of the attraction. It is then natural to assume that the interaction between fluid and substrate is also of Yukawa form:

$$\phi(z) = \infty, \qquad\qquad z < \frac{1}{2}d$$

$$= -\epsilon_S \exp\left[-\lambda_S(z/d - 1)\right], \qquad z > \frac{1}{2}d \qquad (6.7.20)$$

where the range of the substrate-fluid interaction is governed by the parameter λ_S. The first numerical results were obtained by calculations based on a local density approximation for the hard-sphere free energy functional, from which the key conclusion to emerge was that the order of the wetting transition depends strongly on the relative ranges of the two Yukawa potentials.[36] If the range of the substrate-fluid attraction, which favours adsorption, is equal to or shorter than that of the fluid–fluid potential, i.e. if $\lambda_S \geq \lambda_F$, the transition is second order, but it may be first order if $\lambda_S < \lambda_F$. A prewetting transition at undersaturated vapour conditions above T_w was also identified at pressures very close to the coexistence curve. These findings were subsequently confirmed in calculations based on a more accurate, fundamental measure approximation for the hard-sphere functional.[37]

Typical results for the density profile of the microscopic model defined by (6.7.19) and (6.7.20) at gas densities lower than the value at coexistence, $\rho_0(T)$, are shown in Figure 6.12 for a case when $\lambda_F = \lambda_S$. Below the wetting temperature, estimated to occur at $T_w \approx 0.761 T_c$, the liquid film is restricted to two or three molecular layers and its thickness increases only modestly as coexistence is approached; above T_w the film grows continuously with no evidence of any discontinuity in thickness. This is consistent with the form of the adsorption isotherms plotted in the left-hand panel of Figure 6.13, where the degree of undersaturation is now measured in terms of the difference in chemical potential rather than density. When $T < T_w$ the adsorption, which is related to the film thickness by (6.7.4), levels off at a finite, microscopic value as coexistence is approached but diverges logarithmically when $T > T_w$. This is clearly a second-order transition. The right-hand panel reveals a very different behaviour at temperatures above T_w in a case where $\lambda_S < \lambda_F$ and $T_w \approx 0.783 T_c$. Discontinuous jumps in adsorption, characteristic of a first-order transition, followed by a continuous transition to complete wetting at coexistence are now visible. The amplitude of the jump decreases with increasing temperature and

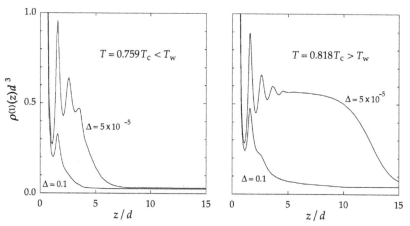

FIGURE 6.12 Density profiles at undersaturated vapour conditions for the microscopic model defined by (6.7.19) and (6.7.20), with $\epsilon_S = 1.75\epsilon_F$ and $\lambda_S = \lambda_F = 1.8$ at temperatures below (left) and above (right) the wetting temperature, $T_w \approx 0.761 T_c$. The quantity $\Delta = 1 - \rho/\rho_0(T)$ is a measure of the distance from coexistence. *Redrawn with permission from Ref. 37 © Taylor & Francis Limited.*

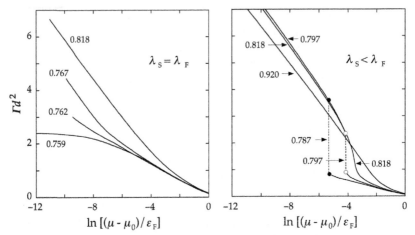

FIGURE 6.13 Adsorption isotherms at several values of the reduced temperature T/T_c for the microscopic model defined by (6.7.19) and (6.7.20), with $\epsilon_S = 1.75\epsilon_F(\lambda_S/\lambda_F)$; $\mu_0(T)$ is the chemical potential at coexistence. Left-hand panel: for $\lambda_S = \lambda_F = 1.8$ and $T_w \approx 0.761T_c$. Right-hand panel: for $\lambda_S = 1.2$, $\lambda_F = 1.8$ and $T_w \approx 0.783T_c$. *Redrawn with permission from Ref. 37 © Taylor & Francis Limited.*

disappears at a prewetting critical temperature $T_{pwc} \approx 0.81T_c$. Both the length of the prewetting line relative to the bulk critical temperature ($\approx 0.03T_c$) and its displacement from the coexistence curve, measured by the difference in chemical potential, are therefore very small. At temperatures beyond T_{pwc} the adsorption diverges logarithmically.

In a more realistic description of the intermolecular forces involved, account must be taken of the effect of dispersion interactions, which give rise to an attractive term in r^{-6} in the pair potential; the corresponding substrate-fluid attraction then varies as r^{-3}, as in the case represented by (6.6.1). It has been shown that when attractive interactions of such long range are present, the wetting transition should always be first order,[36] which is consistent with the fact that continuous wetting transitions are rarely observed experimentally.

A prewetting transition appears when the coexistence curve is approached along an isotherm from the vapour side. When approached from the liquid side ($\mu \to \mu_0^+(T)$) near a solid substrate that repels the fluid,[38] a drying transition occurs, sometimes called 'wetting by gas'. Well away from coexistence the density profile has a layered structure similar to that seen for a hard-sphere fluid near a hard wall in Figure 6.5, but as coexistence is neared the oscillations in the profile gradually disappear and the density at contact decreases. The changes in form of the density profile signal the growth of a layer of gas between the substrate and the liquid, the thickness of which diverges at coexistence. The effect is well reproduced[39] by a mean field version of density functional theory combined with a simple, weighted density approximation for the hard-sphere functional. Drying arises from a lack of cohesion in the fluid close to the substrate. It plays a key part in the mechanism whereby an effective, 'hydrophobic' attraction is created between large molecules or colloidal particles that repel the particles of the liquid (such as water) in which they are dissolved or suspended, leading ultimately to their aggregation or 'hydrophobic assembly'.[40]

6.8   DENSITY FUNCTIONAL THEORY OF FREEZING

If cooled or compressed sufficiently gently, a liquid will freeze into an ordered, solid phase. The transition is accompanied by a discontinuous change in volume, $\Delta V = V_{\rm L} - V_{\rm S}$, which is usually positive (water is a notable exception), and by a latent heat, $T \Delta S$, which is always positive. The discontinuities in V and S, both of which are first derivatives of the free energy, are the signatures of a first-order phase transition. Freezing of simple liquids is largely driven by entropic factors, a fact most obvious in the case of the hard-sphere fluid, since the nature of the hard-sphere interaction means that the difference in free energy of the solid and fluid phases at a given temperature is equal to $-T \Delta S$. One of the most significant findings to emerge from the earliest molecular simulations[41] was that the hard-sphere fluid freezes into a stable, face-centred-cubic crystal; accurate calculations[42] of the free energies of the fluid and solid as functions of density subsequently showed that the packing fractions at coexistence are $\eta_{\rm F} \approx 0.494$ and $\eta_{\rm S} \approx 0.545$. We can obtain a rough estimate of the difference in configurational entropy between the two phases by temporarily ignoring the correlations between particles brought about by excluded volume effects. If we treat the fluid as a system of non-interacting particles moving freely in a volume V and the solid as a system of localised (and hence distinguishable) particles in which each particle is confined by its neighbours to a region of order V/N around its lattice site, a simple calculation shows that the configurational entropy per particle of the solid lies below that of the fluid by an amount equal to $k_{\rm B}$. In reality, of course, correlations make a large contribution to the entropy, which at densities beyond $\eta \approx 0.5$ must be appreciably larger for the 'ordered' solid than for the 'disordered' fluid, since the solid is the stable phase. The explanation of this apparent paradox is the fact that the free volume available to a particle is larger in the solid than in the 'jammed' configurations that are generated when a fluid is overcompressed. This ties in with Bernal's observation[43] that the maximum density achievable by random packing of hard spheres ($\eta \approx 0.64$) lies well below that of the face-centred-cubic structure ($\eta \approx 0.74$).

The relative volume change on freezing of a hard-sphere fluid is $|\Delta V|/V \approx 0.10$ and the entropy change per particle is $\Delta S/Nk_{\rm B} \approx 1.16$. Simple perturbation theory shows that the effect of adding an attractive term to the hard-sphere interaction is to broaden the freezing transition, i.e. to increase the relative volume change, but the opposite effect occurs if the short-range repulsion is softened. In the case of the soft-sphere potentials defined by (5.2.31), for example, the relative volume change is found to decrease rapidly[44] with reduction in the exponent n, becoming strictly zero[45] in the limiting case of the one-component plasma ($n = 1$). The change in entropy also decreases with n, but much more slowly, and remains close to k_B per particle. Both experiments and simulations show that for a wide variety of systems consisting of spherical or nearly spherical particles the amplitude of the main peak in the static structure factor at freezing is approximately 2.85. This provides a useful criterion for freezing that appears to be independent of the crystal structure of the solid phase.[46] It applies, for example, to the family of soft-sphere fluids, for which the stable crystal phase is face-centred cubic at large values of n but body-centred cubic for softer potentials.

The lattice structure of a crystalline solid means that the density profile must be a periodic function of $\mathbf{r}$ such that

$$\rho^{(1)}(\mathbf{r} + \mathbf{R}_i) = \rho^{(1)}(\mathbf{r}) \tag{6.8.1}$$

where the set $\{\mathbf{R}_i\}$ represents the lattice coordinates of the particles in the perfectly ordered crystal. Let $\mathbf{u}_i = \mathbf{r}_i - \mathbf{R}_i$ be the displacement of particle i from its equilibrium position. Then the Fourier transform of the density profile can be written (see (3.1.4)) as

$$\hat{\rho}^{(1)}(\mathbf{k}) = \sum_{i=1}^{N} \langle \exp\left(-i\mathbf{k} \cdot \mathbf{r}_i\right)\rangle = \sum_{i=1}^{N} \exp\left(-i\mathbf{k} \cdot \mathbf{R}_i\right) \langle \exp\left(-i\mathbf{k} \cdot \mathbf{u}_i\right) \tag{6.8.2}$$

Away from any interface, all lattice sites are equivalent, and the second statistical average in (6.8.2) is therefore independent of i. Thus

$$\hat{\rho}^{(1)}(\mathbf{k}) = \langle \exp\left(-i\mathbf{k} \cdot \mathbf{u}\right)\rangle \sum_{i=1}^{N} \exp\left(-i\mathbf{k} \cdot \mathbf{R}_i\right) \tag{6.8.3}$$

The sum over lattice sites is non-zero only if $\mathbf{k}$ coincides with a reciprocal-lattice vector $\mathbf{G}$. Hence

$$\sum_{i=1}^{N} \exp\left(-i\mathbf{k} \cdot \mathbf{R}_i\right) = N\delta_{\mathbf{k},\mathbf{G}} \tag{6.8.4}$$

and the only non-zero Fourier components of the density are

$$\hat{\rho}^{(1)}(\mathbf{G}) = N \langle \exp\left(-i\mathbf{G} \cdot \mathbf{u}\right)\rangle \tag{6.8.5}$$

In the harmonic approximation, valid for small-amplitude vibrations of the particles around their lattice positions, the displacement vectors $\mathbf{u}$ have a gaussian distribution:

$$\langle \exp\left(-i\mathbf{G} \cdot \mathbf{u}\right)\rangle = \exp\left(-\frac{1}{6}G^2 \langle u^2 \rangle\right) \tag{6.8.6}$$

where $\langle u^2 \rangle$ is the mean-square displacement of a particle from its lattice site. If we substitute (6.8.6) in (6.8.5) and take the inverse transform, we find that

$$\rho^{(1)}(\mathbf{r}) = \frac{1}{V} \sum_{\mathbf{G}} \sum_{i=1}^{N} \exp\left(i\mathbf{G} \cdot (\mathbf{r} - \mathbf{R}_i)\right) \exp\left(-\frac{1}{6}G^2\langle u^2\rangle\right)$$

$$\approx \frac{1}{(2\pi)^3} \sum_{i=1}^{N} \int \exp\left(i\mathbf{G} \cdot (\mathbf{r} - \mathbf{R}_i)\right) \exp\left(-\frac{1}{6}G^2\langle u^2\rangle\right) d\mathbf{G}$$

$$= \left(\frac{\alpha}{\pi}\right)^{3/2} \sum_{i=1}^{N} \exp\left(-\alpha(\mathbf{r} - \mathbf{R}_i)^2\right) \qquad (6.8.7)$$

where $\alpha = 3/(2\langle u^2\rangle)$ is an inverse-width parameter. The density profile of the crystal therefore appears as the sum of N gaussian peaks, each centred on a lattice site $\mathbf{R}_i$. As α increases, the particles become more strongly localised and the peaks become narrower. The most general representation of $\rho^{(1)}(\mathbf{r})$ compatible with lattice periodicity is

$$\rho^{(1)}(\mathbf{r}) = \rho_S \left(1 + \sum_{\mathbf{G}\neq 0} \zeta(\mathbf{G}) \exp\left(i\mathbf{G} \cdot \mathbf{r}\right)\right) \qquad (6.8.8)$$

where ρ_S is the overall number density of the solid. In the harmonic approximation the coefficients of the 'density waves' $\exp(i\mathbf{G} \cdot \mathbf{r})$ are related to the parameter α by

$$\zeta(\mathbf{G}) = \exp\left(-G^2/4\alpha\right) \qquad (6.8.9)$$

The vibrational mean-square displacement $\langle u^2 \rangle$ can be determined by analysis of the lineshape of the Bragg peaks observed in X-ray or neutron-scattering experiments; it is found to decrease sharply as the crystal is cooled along an isochore or compressed along an isotherm. The quantity $L = \langle u^2\rangle^{1/2}/R_0$, where R_0 is the nearest-neighbour distance in the crystal, is called the Lindemann ratio. According to the 'Lindemann rule', melting should occur when L reaches a value that is only weakly material dependent and equal to about 0.15. Simulations have shown that for hard spheres the value at melting is approximately 0.13, but is slightly larger for softer potentials. That such a criterion exists is not surprising: instability of the solid can be expected once the vibrational amplitude of the particles becomes a significant fraction of the spacing between neighbouring lattice sites.

The idea that underpins much of the density functional approach to freezing goes back to the work of Kirkwood and Monroe.[47] While the periodic density profile is clearly very different from the uniform density of the fluid, it is reasonable to assume that the short-range pair correlations in the solid are similar to those of some effective, reference fluid. In other words, a crystal may be regarded as a highly inhomogeneous fluid, and different versions of the theory differ mostly in the choice made for the density of the reference fluid.[48]

We showed in Section 4.3 that expansion of the free energy functional in powers of $\delta\rho^{(1)}(\mathbf{r})$ around that of a homogeneous fluid of density ρ_0 leads, when truncated at second order, to the expression for the density profile given by (4.3.16). In the application to freezing there is no external field and (4.3.16) becomes

$$\rho^{(1)}(\mathbf{r}) = \rho_0 \exp\left(\int c_0^{(2)}(\mathbf{r} - \mathbf{r}')[\rho^{(1)}(\mathbf{r}) - \rho_0]d\mathbf{r}'\right) \tag{6.8.10}$$

Higher-order terms in the expansion can be derived, but explicit calculations become increasingly involved and are therefore rarely attempted. Equation (6.8.10) always has the trivial solution $\rho^{(1)}(\mathbf{r}) = \rho_0$, but at sufficiently high densities there exist, in addition, periodic solutions of the form (6.8.8). In order to decide whether the uniform or periodic solution corresponds to the stable phase it is necessary to compute the free energies of the two phases. The free energy of the solid phase is related to that of the reference fluid by (4.3.12), where the choice of ρ_0 remains open. It is clear, however, that ρ_0 should be comparable with ρ_S, the mean number density in the solid, since the density change on freezing is typically less than 10%. One obvious possibility is to set $\rho_0 = \rho_S$, which simplifies the problem because the linear term in (4.3.12) then vanishes, but other choices have been made.[49] If we substitute (6.8.8) (with $\rho_0 = \rho_S$) into the quadratic term in (4.3.12) and use the convolution theorem, we find that

$$\frac{\beta\Delta F}{N} = \frac{\beta\mathcal{F}[\rho^{(1)}]}{N} - \frac{\beta F_0(\rho_S)}{N}$$

$$= \int \rho^{(1)}(\mathbf{r}) \ln\left(\frac{\rho^{(1)}(\mathbf{r})}{\rho_S}\right) d\mathbf{r} - \frac{1}{2}\rho_S \sum_{\mathbf{G}\neq 0} \hat{c}_0^{(2)}(\mathbf{G})|\zeta\mathbf{G}|^2 \tag{6.8.11}$$

The difference in free energy, ΔF, must now be minimised with respect to $\rho^{(1)}(\mathbf{r})$, i.e. with respect to the order parameters $\zeta_{\mathbf{G}}$. In practice, most calculations are carried out using the gaussian form (6.8.9), in which case the inverse width α is the only variational parameter. The ideal contribution to the free energy favours the homogeneous phase; the quadratic, excess term favours the ordered phase provided the quantities $\hat{c}_0^{(2)}(\mathbf{G})$ are positive for the smallest reciprocal-lattice vectors, since the contributions thereafter decrease rapidly with increasing G. The competition between ideal and excess contributions leads to curves of ΔF versus α of the Landau type, shown schematically in Figure 6.14. When the density ρ_S is low (curves (a) and (b)), there is a single minimum at $\alpha = 0$, corresponding to a homogeneous, fluid phase. At higher densities (curve (c)), a minimum appears at a positive value of ΔF,

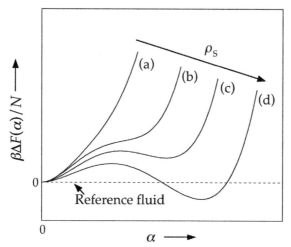

FIGURE 6.14   Typical behaviour of the free energy difference defined by (6.8.11) as a function of the variational parameter α for increasing values of the density ρ_S. Curve (d) corresponds to a density at which the ordered crystal is the stable phase. See text for details.

signalling the appearance of a metastable, crystalline phase. Further increase in density leads to a lowering of the value of ΔF at the second minimum, which eventually becomes negative (curve (d)); the ordered crystal is now the stable phase. Once the free energies of fluid and solid along a given isotherm are known, the densities of the coexisting phases can be determined from the Maxwell double-tangent construction, which ensures equality of the chemical potentials and pressures of the two phases.[8] The calculations are carried out for a given Bravais lattice and hence for a given set of reciprocal-lattice vectors. If the relative stability of different crystal structures is to be assessed, separate calculations are needed for each lattice.

The method we have outlined is essentially that of Ramakrishnan and Yussouff,[50] reformulated in the language of density functional theory.[51] It works satisfactorily in the case of hard spheres, but the quality of the results deteriorates for softer potentials, for which the stable solid has a body-centred-cubic structure. In that case, if the potential is sufficiently soft, the contribution to the sum over $\mathbf{G}$ in (6.8.11) from the second shell of reciprocal-lattice vectors is negative. The resulting contribution to ΔF is therefore positive and sufficiently large to destabilise the solid. This defect in the method can be overcome by inclusion of the third-order term in the expansion of the free energy functional, but that requires some approximation to be made for the three-particle direct correlation function of the reference system.[52] Other approaches to the problem of freezing have also been used. The most successful of these are variants of fundamental measure theory of the type discussed at the end of Section 6.5,

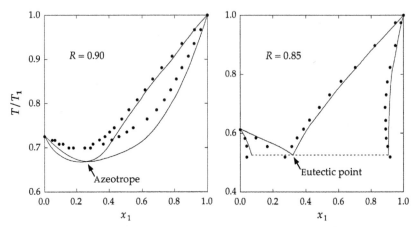

FIGURE 6.15 Phase diagrams of binary hard-sphere mixtures at a fixed pressure for two values of the diameter ratio R; x_1 and T_1 are, respectively, the number concentration of the larger spheres and the freezing temperature for $x_1 = 1$. Left-hand panel: an azeotropic-type diagram; right-hand panel: a eutectic-type diagram. The full curves are calculated from density functional theory and the points are the results of Monte Carlo calculations[54]; the broken line in the right-hand panel shows the miscibility gap at the eutectic temperature. *Redrawn with permission from Ref. 53 © American Institute of Physics.*

which lead to values for the densities at coexistence of the hard-sphere fluid and solid that agree with those obtained by simulation to within one percent.

The theory can be extended to mixtures and in that form has been used to study the freezing of binary hard-sphere mixtures into substitutionally disordered, face-centred-cubic structures, where the nature of the resulting phase diagram depends critically on the value of the diameter ratio, $R = d_1/d_2$. Figure 6.15 shows phase diagrams in the temperature-concentration plane obtained from a version of density functional theory[53] in which the free energy of the solid is determined by a generalisation of the weighted density approximation (6.2.25); earlier calculations based on a generalisation of (6.8.11) had led to qualitatively similar results.[55] When R is greater than approximately 0.94, the two species are miscible in all proportions in both phases, the concentration of large spheres being slightly higher in the solid. At lower values of R ($0.88 < R < 0.93$), the phase diagram has the form shown in the left-hand panel of the figure, in which we see the appearance of an azeotrope, i.e. a point where the coexistence curves pass through a minimum and solid and fluid have identical compositions. When R is reduced below 0.88, as in the right-hand panel, the azeotrope is replaced by a eutectic point. There is now a wide range of concentration over which the two species are immiscible in the solid; the solubilities of large spheres in a solid consisting mostly of small spheres or vice versa are each less than 10% and become rapidly smaller as R is further reduced. This behaviour is broadly consistent with the empirical

Hume–Rothery rule, according to which the disordered solid phases of metallic alloys become unstable for diameter ratios less than about 0.85. As the figure shows, there is good agreement with the results of simulations both here and in the azeotropic case. Other density functional calculations[56] have shown that ordered phases of AB_n-type structure remain stable at values of R below 0.8, which is consistent both with simulations of hard-sphere mixtures and with experimental studies of colloidal suspensions.[57]

6.9 FLUIDS ADSORBED IN POROUS MEDIA

In earlier sections of this chapter we discussed the properties of fluids confined either between two, parallel plates or within infinitely long cylinders. Simple pore geometries such as these, which give rise to one-dimensional inhomogeneities, lend themselves easily to density functional treatments of the structure and thermodynamics of the confined fluid. However, the great majority of natural and synthetic materials within which fluids can be adsorbed have a much more complex, usually random topology. Zeolites, for example, consist of periodic arrays of interconnected, parallel channels, while in other materials, including silica gels, clays and sintered powders, there is a random distribution of pores of various sizes and shapes, dispersed throughout a rigid matrix. The most general void topology consists of a single percolating void, or network of interconnected cavities, together with a large number of isolated cavities. The *porosity* ϕ of a material is defined as the fraction of the total volume which is accessible to molecules of the adsorbed fluid; in the case of soil clays, say, the porosity is typically in the range 0.5–0.6. A random distribution of voids causes the external field that the porous medium exerts on the fluid to vary randomly in space, thereby precluding the use of density functional theory in the form employed in Sections 6.1 and 6.5.

Let us consider the case of a rigid matrix consisting of N_0 spherical particles contained in a volume V and frozen in place in disordered configurations $\mathbf{q}^{N_0} \equiv \mathbf{q}_1, \mathbf{q}_2, \ldots, \mathbf{q}_{N_0}$ distributed according to a probability density $\mathcal{P}_0(\mathbf{q}^{N_0})$. A given *realisation* of the matrix corresponds to one such configuration. Adsorbed fluid particles (species 1) are confined to that fraction of the total volume not occupied by particles of the matrix (species 0) and interact both with the matrix particles and among themselves.[58] We suppose that the matrix configurations are obtained by quenching an initial, equilibrium configuration of the N_0 particles at a temperature T_0 to a temperature at which the particles remain fixed at their initial positions. If the total interaction energy of the matrix particles is V_{00} the distribution $\mathcal{P}_0(\mathbf{q}^{N_0})$ will be the normalised Boltzmann distribution:

$$\mathcal{P}_0(\mathbf{q}^{N_0}) = \frac{1}{Z_0} \exp\left[-\beta V_{00}(\mathbf{q}^{N_0})\right] \qquad (6.9.1)$$

The grand potential of the adsorbed fluid (with coordinates $\mathbf{r}^{N_1} \equiv \mathbf{r}_1, \mathbf{r}_2, \ldots, \mathbf{r}_{N_1}$) for a given realisation $\mathbf{q}^{N_0}$ of the matrix is

$$\beta \Omega_1(\mathbf{q}^{N_0}) = -\ln \Xi_1(\mathbf{q}^{N_0}) \tag{6.9.2}$$

where the grand partition function Ξ_1 is

$$\Xi_1 = \sum_{N_1=0}^{\infty} \frac{z_1^{N_1}}{N_1!} \int \exp\left\{-\beta\left[V_{01}(\mathbf{q}^{N_0}; \mathbf{r}^{N_1}) + V_{11}(\mathbf{r}^{N_1})\right]\right\} d\mathbf{r}^{N_1} \tag{6.9.3}$$

Here V_{11} and V_{01} are, respectively, the total fluid–fluid and fluid-matrix interaction energies, and z_1 is the activity of the adsorbed fluid, assumed to be in equilibrium with an external reservoir which fixes the chemical potential at a value μ_1. The grand potential must then be averaged over the matrix probability density to give

$$\beta\overline{\Omega_1} = -\int \mathcal{P}_0(\mathbf{q}^{N_0}) \ln\left[\Xi_1(\mathbf{q}^{N_0})\right] d\mathbf{q}^{N_0}$$

$$= -\frac{1}{Z_0} \int \exp\left[-\beta V_{00}(\mathbf{q}^{N_0})\right] \ln[\Xi_1(\mathbf{q}^{N_0})] d\mathbf{q}^{N_0} \tag{6.9.4}$$

where here and subsequently an overline denotes an average over matrix configurations. More generally, calculation of the value of a macroscopic property $\mathcal{A}$ of the adsorbed fluid at given values of T_1, V and μ_1 involves taking, first, the grand canonical average (denoted by $\langle \cdots \rangle$) over fluid variables of the corresponding dynamical variable $A(\mathbf{q}^{N_0}; N_1, \mathbf{r}^{N_1})$ for a fixed matrix configuration, followed by an average over the disordered matrix variables, weighted by $\mathcal{P}_0(\mathbf{q}^{N_0})$. Hence, when the probability density is given by (6.9.1):

$$\mathcal{A} \equiv \overline{\langle A \rangle} = \frac{1}{Z_0} \int d\mathbf{q}^{N_0} \frac{1}{\Xi_1(\mathbf{q}^{N_0})} \sum_{N_1=0}^{\infty} \frac{z_1^N}{N_1!} \int d\mathbf{r}^{N_1} A(\mathbf{q}^{N_0}; N_1, \mathbf{r}^{N_1})$$

$$\times \exp\left\{-\beta\left[V_{00}(\mathbf{q}^{N_0}) + V_{01}(\mathbf{q}^{N_0}; \mathbf{r}^{N_1}) + V_{11}(\mathbf{r}^{N_1})\right]\right\} \tag{6.9.5}$$

Both here and in (6.9.4) it is assumed implicitly that the temperature T_0 at which the particles of species 0 was equilibrated before the quench is the same as the temperature at which the adsorbed fluid is held and hence that $T_0 = T_1 = T$, say. This point is irrelevant for much of what follows, since the applications described are concerned mostly with systems in which all interactions are of hard-sphere form and all matrix realisations are given equal weight.

The statistical average in (6.9.5) is one appropriate to a 'quenched-annealed' or QA system. This is very different from the corresponding average for an equilibrium, binary mixture in an ensemble in which N_0 and μ_1 have fixed values. The frozen configurations of a QA system are independent of the configurations of the adsorbed fluid and provide only a random, external field that acts on the fluid particles. In an equilibrium mixture the configurations of the two species are fully correlated.

The evaluation of the grand potential (6.9.4) is complicated by the fact that the logarithm of the grand partition function of the adsorbed fluid, which appears under the integral sign, depends parametrically on the matrix coordinates $\mathbf{q}^{N_0}$. This difficulty can be circumvented by a continuum version[59,60] of the 'replica trick' used in the theory of spin glasses.[61,62] The trick is based on the identity,[63] valid for any positive Ξ:

$$\ln \Xi = \lim_{s \to 0} \frac{d}{ds} \Xi^s \tag{6.9.6}$$

and is implemented by supposing that s replicas of the fluid are adsorbed, all with the same activity z_1. The result is an equilibrium, $(s + 1)$-component mixture consisting of the N_0 matrix particles and s replicas of the fluid. Particles belonging to a replica labelled α, say, interact with each other and with the matrix particles, but not with particles in other replicas. The grand partition function of the replicated system is therefore

$$\Xi^{(\mathrm{r})}(s) = \frac{1}{N_0!} \sum_{N_1=0}^{\infty} \cdots \sum_{N_s=0}^{\infty} \frac{z_1^{N_1+\cdots+N_s}}{N_1! \cdots N_s!} \int d\mathbf{q}^{N_0} d\mathbf{r}^{N_1} \cdots d\mathbf{r}^{N_s}$$

$$\times \exp\left[-\beta\left(V_{00}(\mathbf{q}^{N_0}) + \sum_{\alpha=1}^{s} V_{0\alpha}(\mathbf{q}^{N_0}; \mathbf{r}^{N_\alpha}) + \sum_{\alpha=1}^{s} V_{\alpha\alpha}(\mathbf{r}^{N_\alpha})\right)\right] \tag{6.9.7}$$

where $V_{0\alpha}$ and $V_{\alpha\alpha}$ are the total matrix-fluid and fluid–fluid interaction energies for replica α. The corresponding grand potential is

$$\beta\Omega^{(\mathrm{r})}(s) = -\ln \Xi^{(\mathrm{r})}(s) \tag{6.9.8}$$

Combination of (6.9.4), (6.9.6) and (6.9.7), coupled with analytic continuation to $s = 0$, leads to an expression for the grand potential of the QA system in terms of the grand partition function of the replicated system:

$$\beta\overline{\Omega_1} = -\frac{1}{Z_0} \lim_{s \to 0} \frac{d}{ds} \Xi^{(\mathrm{r})}(s) = \beta \lim_{s \to 0} \frac{d}{ds} \Omega^{(\mathrm{r})}(s) \tag{6.9.9}$$

where we have used the fact that $\lim_{s \to 0} \Xi^{(\mathrm{r})}(s) = Z_0$, which follows immediately from (6.9.7).

Thermodynamic properties of the QA system, which is the system of interest, can now be derived from those of the equilibrium $(s + 1)$-component grand potential $\Omega^{(\mathrm{r})}(s)$. Since the matrix particles are treated in the canonical ensemble, the matrix density ρ_0 is constant and $dN_0 = d(N_0 V) = \rho_0 dV$. Given that all replicas of the fluid have the same chemical potential, the fundamental relation (2.4.3) becomes

$$d\Omega^{(r)}(s) = -S^{(r)}(s)dT - P^{(r)}(s)dV - sN_1^{(r)}(s)d\mu_1 + \mu_0^{(r)}(s)dN_0$$
$$= \left[-P^{(r)}(s) + \rho_0\mu_0^{(r)}(s)\right]dV - S^{(r)}(s)dT - sN_1^{(r)}d\mu_1 \quad (6.9.10)$$

On substituting (6.9.10) in (6.9.9) we find that

$$d\overline{\Omega_1} = -\lim_{s\to 0}\frac{dS^{(r)}(s)}{ds}dT + \lim_{s\to 0}\left(-\frac{dP^{(r)}(s)}{ds} + \rho_0\frac{d\mu_0^{(r)}(s)}{ds}\right)dV - N_1^{(r)}(s=0)d\mu_1$$
$$(6.9.11)$$

Thermodynamic properties of the confined fluid are therefore obtained by differentiation:

$$S_1 = -\left.\frac{\partial\overline{\Omega_1}}{\partial T}\right|_{V,\mu_1,\rho_0} = \lim_{s\to 0}\frac{dS^{(r)}(s)}{ds}$$

$$P_1 = -\left.\frac{\partial\overline{\Omega_1}}{\partial V}\right|_{T,\mu_1,\rho_0} = \lim_{s\to 0}\frac{d}{ds}\left[P^{(r)}(s) - \rho_0\mu_0^{(r)}(s)\right] \quad (6.9.12)$$

$$N_1 = -\left.\frac{d\overline{\Omega_1}}{d\mu_1}\right|_{V,T,\rho_0} = N_1^{(r)}(s=0)$$

In a similar way, the Gibbs–Duhem relation for the $(s+1)$-component mixture leads, via the relations (6.9.12), to the corresponding equation for the adsorbed fluid

$$-V\,dP_1 + S_1\,dT + N_1\,d\mu_1 = 0 \quad (6.9.13)$$

Combination of (6.9.12) and (6.9.13) shows that $d\overline{\Omega_1} = -d(P_1V)$, from which the thermodynamic relation (2.4.2) is recovered in the form

$$\overline{\Omega_1} = -P_1V \quad (6.9.14)$$

We turn now to a discussion of the pair correlation functions of the QA system, proceeding along the lines of Section 2.6. If $\rho_1^{(1)}(\mathbf{r}_1; \mathbf{q}^{N_0})$, $\rho_{11}^{(2)}(\mathbf{r}_1, \mathbf{r}_2; \mathbf{q}^{N_0})$ are, respectively, the single-particle and pair density of the adsorbed fluid for a given realisation of the matrix, and if the matrix structure is assumed to be statistically homogeneous, the fluid–fluid pair correlation function is

$$\rho_1^2 h_{11}(r_{12}) = \overline{\rho_{11}^{(2)}(\mathbf{r}_1, \mathbf{r}_2; \mathbf{q}_{N_0})} - \rho_1^2 \quad (6.9.15)$$

where $\rho_1 = \overline{N_1}/V$. Because pair correlations are mediated both through intervening fluid particles and via matrix particles it proves convenient for later purposes to introduce two auxiliary correlation functions by dividing the correlation function into 'connected' (c) and 'blocking' (b) parts[64] in the form

$$h_{11}(r_{12}) = h_c(r_{12}) + h_b(r_{12}) \quad (6.9.16)$$

where

$$\rho_1^2 h_c(r_{12}) = \overline{\rho_{11}^{(2)}\left(\mathbf{r}_1, \mathbf{r}_2; \mathbf{q}^{N_0}\right)} - \overline{\rho_1^{(1)}\left(\mathbf{r}_1; \mathbf{q}^{N_0}\right)\rho_1^{(1)}\left(\mathbf{r}_2; \mathbf{q}^{N_0}\right)} \qquad (6.9.17)$$

and

$$\rho_1^2 h_b(r_{12}) = \overline{\rho_1^{(1)}\left(\mathbf{r}_2; \mathbf{q}^{N_0}\right)\rho_1^{(1)}\left(\mathbf{r}_2; \mathbf{q}^{N_0}\right)} - \rho_1^2 \qquad (6.9.18)$$

The physical significance of this division will become clear later. For the present it is sufficient to recognise that for any given realisation of the matrix the single-particle and pair densities must satisfy the grand canonical normalisation conditions (2.6.3) and (2.6.4), which lead to the compressibility formula (2.6.12). Thus, when an average over matrix realisations is taken,[60] it follows from the definition (6.9.17) that the isothermal compressibility of the adsorbed fluid is determined solely by the connected part of $h_{11}(r)$:

$$1 + \rho_1 \int h_c(\mathbf{r})d\mathbf{r} = \frac{\overline{\left\langle [N_1(\mathbf{q}^{N_0})]^2 \right\rangle} - \overline{[\langle N_1(\mathbf{q}^{N_0}) \rangle]^2}}{\overline{\langle N_1(\mathbf{q}^{N_0}) \rangle}} = \rho_1 k_B T \chi_{T_1} \qquad (6.9.19)$$

The pair correlation functions of the QA system can be related to those of the replicated system by applying the replica relation (6.9.9) to the functional relations (3.4.7) for the mixture. For example:

$$h_{11}(r_{12}) = \lim_{s \to 0} h_{11}^{(r)}(r_{12}; s) \qquad (6.9.20)$$

Because the replicas are identical, it follows that $h_{0\alpha}^{(r)} \equiv h_{01}^{(r)}$ and $h_{\alpha\alpha}^{(r)} \equiv h_{11}^{(r)}$ for all $\alpha = 1$ to s, and $h_{\alpha\beta}^{(r)} \equiv h_{12}^{(r)}$ for $1 \le \alpha < \beta \le s$; the same is true for the corresponding set of direct correlation functions. The Ornstein–Zernike relations for the replicated system can then be obtained as the special case of those provided by (3.6.12) in which account is taken of these identities. Using the symbol $\otimes$ to denote a convolution product we find that

$$h_{00}^{(r)} = c_{00}^{(r)} + \rho_0 c_{00}^{(r)} \otimes h_{00}^{(r)} + s\rho_1 c_{01}^{(r)} \otimes h_{01}^{(r)}$$

$$h_{01}^{(r)} = c_{01}^{(r)} + \rho_0 c_{00}^{(r)} \otimes h_{01}^{(r)} + \rho_1 c_{01}^{(r)} \otimes h_{11}^{(r)} + (s-1)\rho_1 c_{01}^{(r)} \otimes h_{12}^{(r)}$$

$$h_{10}^{(r)} = c_{10}^{(r)} + \rho_0 c_{10}^{(r)} \otimes h_{00}^{(r)} + \rho_1 c_{11}^{(r)} \otimes h_{10}^{(r)} + (s-1)\rho_1 c_{12}^{(r)} \otimes h_{10}^{(r)}$$

$$h_{11}^{(r)} = c_{11}^{(r)} + \rho_0 c_{01}^{(r)} \otimes h_{01}^{(r)} + \rho_1 c_{11}^{(r)} \otimes h_{11}^{(r)} + (s-1)\rho_1 c_{12}^{(r)} \otimes h_{12}^{(r)} \qquad (6.9.21)$$

$$h_{12}^{(r)} = c_{12}^{(r)} + \rho_0 c_{01}^{(r)} \otimes h_{01}^{(r)} + \rho_1 c_{11}^{(r)} h_{12}^{(r)} + \rho_1 c_{12}^{(r)} \otimes h_{11}^{(r)}$$
$$+ (s-2)\rho_1 c_{12}^{(r)} \otimes h_{12}^{(r)}$$

Note that both $h_{12}^{(r)}$ and $c_{12}^{(r)}$ are non-zero. Although there is no interaction between particles in different replicas, they are spatially correlated through mutual interaction with particles of the matrix.

The correlation functions of the QA system satisfy the 'replica Ornstein–Zernike' or ROZ relations, obtained by taking the $s \to 0$ limit of the set (6.9.21):

$$h_{00} = c_{00} + \rho_0 c_{00} \otimes h_{00}$$

$$h_{01} = c_{01} + \rho_0 c_{00} \otimes h_{01} + \rho_1 c_{01} \otimes h_{11} - \rho_1 c_{01} \otimes h_{12}$$

$$h_{10} = c_{10} + \rho_0 c_{10} \otimes h_{00} + \rho_1 c_{11} h_{10} - \rho_1 c_{12} \otimes h_{10}$$

$$h_{11} = c_{11} + \rho_0 c_{01} \otimes h_{01} + \rho_1 c_{11} \otimes h_{11} - \rho_1 c_{12} \otimes h_{12}$$

$$h_{12} = c_{12} + \rho_0 c_{01} \otimes h_{01} + \rho_1 c_{11} \otimes h_{12} + \rho_1 c_{12} \otimes h_{11}$$
$$\qquad - 2\rho_1 c_{12} \otimes h_{12}$$

(6.9.22)

The first of these relations, which involves only matrix correlation functions, is disconnected from the others in the set, while symmetry implies that $h_{01} = h_{10}$ and $c_{01} = c_{10}$, so the second and third relations are equivalent. The appearance of h_{12} and c_{12} is at first sight surprising, since in the QA system there is only one fluid component. However, as we have seen, $h_{11}(r)$ divides naturally into connected and blocking parts. In the diagrammatic expansion (4.6.2) of $h_{11}(r)$, $h_b(r)$ corresponds to the subset of irreducible diagrams in which all paths between the two white circles pass through at least one black ρ_0-circle, which is precisely the definition of the correlation function $h_{12}^{(r)}$ in the replicated system.[64] Thus

$$h_b(r) = \lim_{s \to 0} h_{12}^{(r)}(r; s) = h_{12}(r) \qquad (6.9.23)$$

and, similarly, $c_b(r) = c_{12}(r) = c_{11}(r) - c_c(r)$. With these identifications the functions h_{12} and c_{12} can be eliminated from the ROZ equations to give

$$h_{00} = c_{00} + \rho_0 c_{00} \otimes h_{00}$$

$$h_{10} = c_{10} + \rho_0 c_{10} \otimes h_{00} + \rho_1 c_c \otimes h_{10}$$

$$h_{11} = c_{11} + \rho_0 c_{10} \otimes h_{01} + \rho_1 c_c \otimes h_{11} + \rho_1 c_b \otimes h_c$$

$$h_c = c_c + \rho_1 c_c \otimes h_c$$

(6.9.24)

As in the case of bulk fluids, solution of these equations requires the use of some approximate closure relation of the type discussed in Chapter 4.

An approximation introduced by Madden and Glandt,[58] expected to be valid for highly porous matrices, assumes that the blocking part of the direct correlation function $c_{11}(r)$ vanishes, i.e

$$c_b(r) = 0, \quad \text{(Madden–Glandt)} \qquad (6.9.25)$$

and hence that $c_{11} = c_c(r)$. The ROZ relations then reduce to the first three equations, since the last relation is no longer needed to determine the correlation functions for $00, 01$ and 11 interactions. The three remaining equations resemble the Ornstein–Zernike relations for a binary mixture given by (3.6.12), except that the equation for $h_{00}(r)$ is decoupled from the other two; a term $\rho_1 c_{01} \otimes h_{10}$ is missing. The Madden–Glandt approximation is compatible with both the PY and MSA closures, since in the underlying replicated system the pair potential and Mayer function are both zero for particles belonging to different replicas.

Two simple models of the rigid matrix have been studied in detail.[65] One consists of additive hard spheres of diameter d_{00} at a packing fraction $\eta_0 = \pi \rho_0 d_{00}^3/6$ and porosity $\phi = 1 - \eta_0$. The second is a random hard-sphere model, made up of overlapping hard spheres each centred on a sphere of diameter d_{01} from which fluid particles are excluded; in this case the porosity is[66] $\exp(-\eta_{01})$, where $\eta_{01} = \pi \rho_0 d_{01}^3/6$. The simplest model of the adsorbed fluid is one of hard spheres of diameter d_{11}; if fluid-matrix interactions are assumed to be additive, $d_{01} = \frac{1}{2}(d_{00} + d_{11})$. An extreme example of non-additivity is provided by a model in which $d_{00} = 0$ (the random model) and $d_{11} = 0$ (an adsorbed ideal gas), but $d_{01} = d > 0$; this is the quenched-annealed version of the Widom–Rowlinson model,[67] introduced in Section 3.10. Unlike the original Widom–Rowlinson model, however, its QA analogue can be solved analytically. The Mayer functions f_{00}, f_{11} are both zero, $f_{01}(r) = -\Theta(d - r)$ and, since the matrix particles are randomly distributed, $h_{00}(r)$ and $c_{00}(r)$ vanish for all r. The generalisation to binary systems of the diagrammatic expansion (3.8.7) of the direct correlation function shows that with these simplifications the only non-zero diagram which contributes to $c_{01}(r)$ is the first one, i..e.

$$c_{01}(r) = h_{01}(r) = f_{01}(r) \qquad (6.9.26)$$

The same expansion shows that $c_{11}(r)$ is defined diagrammatically as

$c_{11}(1, 2) =$ [all diagrams consisting of two white 1-circles labelled

        1 and 2 and linked by paths consisting of two f_{01}-bonds

        and one black ρ_0-circle]

$$(6.9.27)$$

Use of Lemma 1 of Section 3.7 and the definition of star-irreducible diagrams allows the density expansion to be summed in the form

$$c_{11}(1, 2) = \exp\left[\rho_0 \mathcal{O}_{01}(1, 2)\right] - \rho_0 \mathcal{O}_{01}(1, 2) - 1 \qquad (6.9.28)$$

where $\mathcal{O}_{01}$ is the overlap volume for two spheres of diameter d at a centre-to-centre separation equal to r:

$$\mathcal{O}_{01}(1, 2) = \int f_{01}(1, 3) f_{10}(3, 2) \mathrm{d}3 = \frac{4\pi}{3}\left(2d^3 - 3d^2 r + r^3\right) \qquad (6.9.29)$$

All diagrams in the expansion of $c_{11}(1, 2)$ contribute to the blocking function $c_b(r)$ and the connected function is identically zero; there are no paths between the two white circles that pass through a black ρ_1 circle, since $f_{11} = 0$. This is a situation in which the approximation (6.9.25) fails. Because $c_c(r) = 0$, it follows from (6.9.24) that in the Madden–Glandt approximation $h_c(r)$ also vanishes, whereas use of the third of the ROZ relations (6.9.24) shows that the exact result is

$$h_{11}(r) = c_{11}(r) + \rho_0 f_{10}(r) \otimes f_{01}(r) = c_{11}(r) + \rho_0 \mathcal{O}_{01}(r)$$
$$= \exp\left[\rho_0 \mathcal{O}_{01}(r)\right] - 1 \qquad (6.9.30)$$

The results in (6.9.26), (6.9.28) and (6.9.30) are trivially compatible with the HNC closure (4.3.23) for all pair correlation functions of the Widom–Rowlinson model. The link with the thermodynamics of the adsorbed fluid is most easily established via the compressibility relation (6.9.19). Since $h_c(r) = 0$ for all r, the equation of state is that of an ideal gas within the accessible volume ϕV, as one would expect. However, the pair correlation function $h_{11}(r)$ is non-zero due to the effect of blocking correlations induced by interactions with the matrix. More generally, while the compressibility route to the equation of state provides an operational link to the thermodynamics of a QA system the virial route involves complications not encountered in the case of a bulk fluid, which make it unusable in practical calculations.[60,68]

Extensive comparisons have been made[65,69] for both additive and random hard-sphere matrices between the results of Monte Carlo calculations and integral equation results based on a variety of approximate closures of the ROZ relations. As an example, Figure 6.16 shows the results obtained for the correlation functions $h_{11}(r)$ and $h_{01}(r)$ in the case of an additive hard-sphere matrix of porosity $\phi \approx 0.75$ for which $d_{00} = 3d_{11}$; the curves plotted are

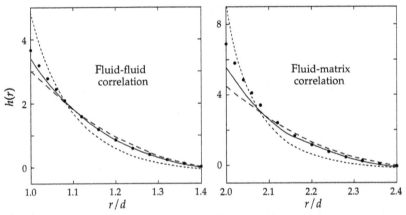

FIGURE 6.16 Fluid–fluid and fluid-matrix pair correlation functions for a fluid of hard spheres of density $\rho_1 d^3 = 0.481$ adsorbed in an additive hard-sphere matrix of density $\rho d^3 = 0.0179$, with $d_{00} = 3d, d_{01} = 2d$ and $d_{11} = d$. The full, long-dashed and short-dashed curves show the results of the RHNC, PY and HNC approximations, respectively, and the circles are the results of Monte Carlo calculations. *Redrawn with permission from Ref. 69 © American Institute of Physics.*

limited to the region of near contact between spheres, where the demands on a theory are greatest. Of the three approximations considered – PY, HNC and RHNC – the RHNC closure is clearly the most successful, though significant discrepancies between theory and simulation are apparent for the fluid-matrix function very close to contact. The fluid compressibilities derived from the RHNC approximation are also in very good agreement with the Monte Carlo data over a wide range of fluid and matrix densities.[69]

With the addition of attractive terms to the fluid–fluid and fluid-matrix potentials the replica formalism can be adapted to the study of the liquid–gas transition of adsorbed fluids within the framework of the perturbation theories described in Chapter 5.[70] The confinement of the fluid and the interactions between fluid and matrix lead to substantial changes in the phase diagram compared with that of the bulk fluid. Overall the trend is for significant decreases to occur in both the critical temperature and density as the porosity increases, while the ratio ρ_c/ϕ remains approximately constant, a pattern of behaviour which is broadly consistent with the available experimental data.[71]

6.10 THERMODYNAMICS OF GLASSES

When a liquid is slowly cooled at constant pressure it normally undergoes a first-order transition to an ordered, crystalline phase at a temperature lying on the equilibrium, liquid–solid coexistence curve. However, if the rate of cooling is sufficiently rapid, crystallisation can be by-passed; in that case the liquid is quenched into an amorphous solid or glass. Glassy materials have mechanical properties that are similar to those of crystalline solids. For example, they respond elastically to an applied shear stress. On the other hand their disordered, microscopic structure, as revealed by diffraction experiments, is very similar to that of a dense liquid and lacks the long-range periodicity of a crystal lattice. The glass transition temperature T_G is lower than the freezing temperature T_f of the liquid, but its value depends on factors such as the cooling rate and the experimental diagnostic used to locate the transition, including the time scale over which the observation is made. It is not an intrinsic property of the system.

A sketch[72] of the way in which the volume or enthalpy of a liquid varies with temperature at constant pressure is shown in Figure 6.17; at the freezing temperature both quantities change discontinuously and their rate of increase with temperature below the liquid–solid transition is less than in the liquid. If the liquid is quenched below T_f the slopes of both curves also change rapidly but continuously around a transition temperature T_G, which itself decreases as the rate of cooling slows. The rapid changes at temperatures close to T_G lead to sharp falls in the heat capacity $(\partial H/\partial T)_P$ and thermal expansion coefficient $(\partial \ln V/\partial T)_P$. As we shall see in Section 8.8, these thermodynamic signatures of a glass transition are strongly correlated with changes in dynamical properties; the structural relaxation time of a supercooled liquid rises rapidly

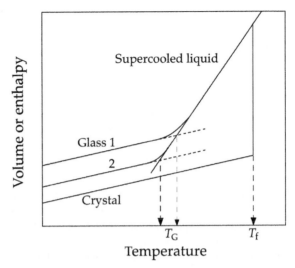

FIGURE 6.17 A schematic representation of the variation with temperature at constant pressure of either the volume or enthalpy of a liquid as it is passes through the glass transition. The labels 1 and 2 refer to two different cooling rates, which is slower for 2 than for 1; T_f is the freezing temperature. *Redrawn by permission from Ref. 72 © 2001 Macmillan Publishers Ltd.*

with decreasing temperature, reaching values of minutes or even hours at temperatures around T_G.

It was pointed out by Kauzmann[73] many years ago that because $(\partial S/\partial T)_P$ is greater for the liquid than the ordered solid, extrapolation of thermodynamic data for the liquid phase to temperatures below T_G would show that the entropy of a fully equilibrated liquid becomes equal to that of the equilibrium crystal at $T = T_K$, the Kauzmann temperature, and becomes negative at some sufficiently low temperature, contrary to physical law. One suggested way of avoiding this so-called entropy crisis involves a transition at the Kauzmann temperature to an 'ideal glass' state[74] having a unique, amorphous structure. A phase change of this type is called a random, first-order transition and is similar in character to the first-order transitions predicted by mean field theories of spin glasses.[75,76] This section is devoted to the question of the possible existence and nature of a transition between a hypothetical, supercooled and equilibrated liquid and an ideal glass phase. In structural glasses, by contrast with spin glass models, the randomness is self-generated, meaning that it does not arise from the hamiltonian and is usually associated with some form of frustration. The transition would not be observable experimentally, since the time for the system to reach equilibrium would become infinitely long at low temperatures.

A qualitative understanding of the thermodynamic behaviour of a supercooled liquid can be obtained from an energy landscape,[77] a hypersurface in $3N + 1$ dimensions that represents the total interaction energy $V_N(\mathbf{r}^N)$ of a system of N spherical particles in a given volume as a function of its

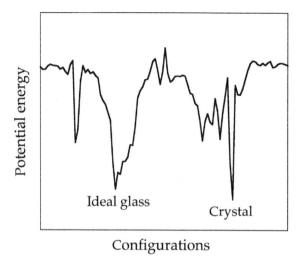

FIGURE 6.18 A schematic representation of a potential energy landscape for an ideal glass former: see text for details.

coordinates $\mathbf{r}^N$. A typical landscape consists of a sequence of potential energy valleys corresponding to local minima in $V_N(\mathbf{r}^N)$ separated by local maxima or saddle points. A schematic, two-dimensional representation of such a surface is pictured in Figure 6.18. At sufficiently high temperatures the system can explore all the energy minima by crossing the transition states, i.e. the local energy barriers; the system is ergodic and there is a single, equilibrium state corresponding to the uniform, liquid phase. As the temperature is lowered it becomes increasingly likely that the system will be found trapped in one of the local minima, including that corresponding to the hypothetical ideal glass. The lowest energy minimum corresponds to the crystalline phase, for which the microscopic, single-particle density is periodic, while the other low-lying minima will be populated by long-lived, metastable, glassy states. If the rare transitions between metastable states are ignored, each glassy state α is characterised by a local free energy per particle $f_\alpha = F_\alpha/N$ and an aperiodic density profile $\rho_\alpha(\mathbf{r})$. The latter plays the role of a multi-variate order parameter dependent on the disordered set of fixed sites $\mathbf{R}^N \equiv \mathbf{R}_1, \ldots, \mathbf{R}_N$ around which particles vibrate. The free energy landscape will be similar in form to the potential energy landscape in Figure 6.18 but differs from it by being temperature dependent; the two landscapes coincide only at zero temperature.

In the case of a hard-sphere system there is no potential energy and the glass or 'jamming' transition is driven by compression rather than temperature. An early attempt[78] to locate the supposed transition between the hard-sphere fluid and the ideal glass used a generalisation of the density functional theory of crystallisation described in the Section 6.8. The free energy was assumed to be a functional of an aperiodic single-particle density $\rho^{(1)}(\mathbf{r})$ of the generic form given in the last line of (6.8.7), with equilibrium positions $\mathbf{R}^N$ corresponding to a random, close-packed configuration generated by a cluster growth algorithm.[79] The only variational parameter is then the inverse width of the gaussian peaks in (6.8.7), as in the case of crystallisation; the resulting free energy curves for different packing fractions are qualitatively similar to those shown in Figure 6.8 and lead to a crossing of the curves for fluid and glass at $\eta \approx 0.6$. This approach was later made more general[80] by associating an inverse-width parameter α_i with each coordinate $\mathbf{R}_i$ and minimising the free energy functional with respect not only to the quantities α_i but also to variations in the coordinates compatible with the no-overlap constraint. In that way allowance is made for the fact that in a disordered glass the amplitude of vibration around a mean position will not be uniform. In practice a rather broad distribution of the local Debye–Waller factor $\langle|\mathbf{r} - \mathbf{R}_i|^2\rangle^{1/2}$ was obtained on global minimisation of the free energy. The scatter in values of the Debye-Waller factors is closely related to the dynamical heterogeneity of glasses to be discussed in Section 8.8.

The main weakness in the density functional approach is its reliance on use of the direct correlation function of the equilibrium fluid as determined by integral equations or simulations, results which are of proven accuracy only at densities well below that of the glass transition. It also fails to provide a complete thermodynamic description of the metastable glass phase in the range of temperatures between T_K and T_G, or the corresponding range of volume fractions in the case of hard spheres. The more general formulation sketched below is inspired by mean field theories of discontinuous spin glasses.[81]

The free energy landscape of a macroscopic system contains a very large number of minima within a temperature-dependent free energy interval $f_{\min}(T) < f = F/N < f_{\max}(T)$, where $f_{\min}$ is the lowest minimum. The density of free energy minima is given by

$$\mathcal{N}(f,T,N) = \exp[N s_c(f,T)/k_B] \tag{6.10.1}$$

where $s_c(f,T) = (1/N)k_B \ln \mathcal{N}(f,T,N)$ is the 'configurational entropy' per particle or 'complexity' of a glass of free energy per particle f at a temperature T. Since at low temperatures the free energy minima f_α are separated by energy barriers that are high relative to the thermal energy of the particles, the total partition function Q_N of the system may be approximated by the sum of the contributions from individual minima:

$$\begin{aligned} Q_N &\approx \sum_\alpha q_\alpha = \sum_\alpha \exp\left(-\beta N f_\alpha\right) \\ &\approx \int_{f_{\min}}^{f_{\max}} \mathcal{N}(f,T,N) \exp\left(-\beta N f\right) \mathrm{d}f \\ &= \int_{f_{\min}}^{f_{\max}} \exp\left[-N\beta\left(f - T s_c(f,T)\right)\right] \mathrm{d}f \end{aligned} \tag{6.10.2}$$

The total free energy per particle, $f_t = F_t/N$, is

$$f_t = -\frac{k_B T}{N} \ln Q_N \qquad (6.10.3)$$

Since N is very large, the integral in (6.10.2) may be evaluated by the saddle point method, in which the integral is replaced by the maximum value of the sharply peaked integrand. Equation (6.10.3) can then be combined with the saddle point estimate of (6.10.2) to give

$$f_t(T) = \min_f \left[f - T s_c(f, T) \right] = f^*(T) - T s_c(T) \qquad (6.10.4)$$

The free energy $f^*(T)$ is determined by the condition that

$$\left. \frac{\partial s_c(f, T)}{\partial f} \right|_{f=f^*} = \frac{1}{T} \qquad (6.10.5)$$

which implies that the rate at which the configurational entropy changes with f necessarily increases as the temperature falls, and hence that s_c is a convex function of f. Equation (6.10.4) resembles the standard thermodynamic relation, $F = U - TS$. Here, however, the internal energy is replaced by a local free energy, to which there is an entropic contribution additional to s_c and linked to the number of configurations explored within a free energy valley. For temperatures between T_K and T_G the free energy $f^*(T)$ in (6.10.4) is expected to lie in the interval $f_{min} < f^* < f_{max}$ and to decrease with T until it coincides with f_{min}. That stage is reached at the Kauzmann temperature, where the system will have become trapped within a single state, or possibly a small number of low-lying states, and $s_c \approx 0$. Below the Kauzmann temperature the state of the system is that of an ideal glass.

The discussion thus far has outlined a plausible scenario in which a random, first-order transition might occur. We are left, however, with the formidable task of computing the equilibrium thermodynamic properties, including $s_c(T)$, $f_t(T)$ and T_K. That can be done by use of the replica method[81] already encountered in a different context in Section 6.9. We suppose that m replicas of a system of N spherical particles are confined in a volume V, with coordinates $\mathbf{r}_1^a, \dots, \mathbf{r}_N^a$ for $1 \leq a \leq m$. Particles within the same replica interact via a pair potential $v(r)$ and those in different replicas interact via a weak, short-range, attractive potential $-\epsilon w(r)$. The role of the attractive potential is to ensure that the configurations of different replicas remain close to each other, since it favours the formation of clusters or 'molecules' composed of m particles, one from each replica; the m replicas therefore remain within the same free energy valley. The canonical partition function (or, more precisely, the configuration integral) of the replicated system consisting of Nm particles is

$$Q_{Nm} = \frac{1}{(N!)^m} \int \exp \left(-\beta \sum_{a=1}^{m} \sum_{i<j}^{N} v \left(\mathbf{r}_i^a - \mathbf{r}_j^a \right) + \beta \epsilon \sum_{a<b}^{m} \sum_{i,j}^{N} w \left(\mathbf{r}_i^a - \mathbf{r}_j^b \right) \right)$$

$$\times \prod_{a=1}^{m} \prod_{i=1}^{N} d\mathbf{r}_i^a \tag{6.10.6}$$

Consider first the case when $m = 2$. The replicated system is then a binary mixture of equal numbers of a and b particles, characterised by two distinct pair distribution functions, $g_{aa}(r) = g_{bb}(r)$ and $g_{ab}^{\epsilon}(r)$. The unlike pair distribution function $g_{ab}^{\epsilon}(r)$ serves as an order parameter for the transition from liquid to ideal glass. To see this we start by taking the thermodynamic limit of g_{ab}^{ϵ} followed by the limit $\epsilon \to 0$. In the ergodic liquid state there are no correlations between a and b and $\lim_{\epsilon \to 0} g_{ab}^{\epsilon}(r) = 1$ for all r. In the ideal glass state a and b particles are both constrained to remain close to the disordered but fixed positions $\mathbf{R}^N$ and therefore cannot drift away from their partners of opposite species as $\epsilon \to 0$. The function $g_{ab}^{0}(r)$ will therefore have a structure very similar to that of $g_{aa}(r)$ (or $g_{bb}(r)$) except for the appearance of an additional peak at $r = 0$, since an a and a b particle can overlap. Thus $g_{ab}^{0}(r)$ represents an order parameter which is discontinuous at the transition temperature, showing that in this respect the transition from liquid to ideal glass is one of first order. The same is true of spin glass models, where the magnetisation changes discontinuously.

The two-replica scheme gives no information about thermodynamic properties close to the transition. That problem can be overcome[81] by extending the method to non-integral values of m, with analytic continuation providing the definition of the partition function (6.10.6). This leads to generalisations of (6.10.2) and (6.10.4) in the form

$$Q_{Nm} \approx \int_{f_{min}}^{f_{max}} \exp \left[-N\beta \left(mf - T s_c(f, T) \right) \right] df \tag{6.10.7}$$

and

$$f_t(m, T) = -\frac{k_B T}{N} \ln Q_{Nm} \approx \min_f \left[mf - T s_c(f, T) \right] \tag{6.10.8}$$

where, as we shall see, m serves as a variational parameter that can be tuned to meet different situations. The saddle point value $f^*(m, T)$ is now given by

$$\left. \frac{\partial s_c(f, T)}{\partial f} \right|_{f=f^*} = \frac{m}{T} \tag{6.10.9}$$

The situation of special interest is that where m is less than one. In that case, provided m is sufficiently small, then at any given temperature $f^*(m, T)$ will be larger than $f_{\min}(T)$, and will reach and thereafter remain equal to $f_{\min}(T)$ when m exceeds a critical value $m^*(T)$. On introducing the free energy per particle of the replicated system:

$$\phi(m, T) = \frac{f_t(m, T)}{m} \tag{6.10.10}$$

we see from (6.10.8) and (6.10.10) that

$$f^*(m, T) = \frac{\partial[m\phi(m, T)]}{\partial m}, \quad s_c(f^*, T) = \frac{m^2}{T} \frac{\partial \phi(m, T)}{\partial m} \tag{6.10.11}$$

The configurational entropy vanishes in the ideal glass, so the second of these relations implies that the free energy of the glass is independent of m.

The replica method makes it possible to relate the free energy $\phi(m = 1, T)$ of the ideal glass phase for $T < T_K$ to the free energy $\phi(m, T)$ of a molecular liquid phase in which the 'molecules' consist of m 'atoms' with $m < 1$ and interact with each other via a potential effectively scaled down from $v(r)$ to $mv(r)$. Scaling of the potential is in turn equivalent to use of an effective temperature $T_m = T/m$ and substitution of T/T_m for m in (6.10.11) leads to expressions analogous to the fundamental relations (2.3.8) and (2.3.9) for S and U in terms of derivatives of F. If m is increased at constant T, the effective coupling between molecules becomes stronger until the critical value $m^*(T)$ is reached, at which stage the liquid would freeze into an ideal glass. Since $s_c \geq 0$, it follows from (6.10.11) that the free energy of the liquid must also increase with m. Thus the critical value of m can be identified as the point at which the free energy has its maximum value and s_c is therefore zero. The physical liquid corresponds to the case when $m = 1$, so T_K is the temperature at which $m^*(T) = 1$.

What remains is to calculate the free energy of the molecular liquid, where use can be made of techniques borrowed from standard liquid state theory. One method of doing so is based on an ingenious adaptation[81] of the hypernetted chain approximation to constrained, replicated systems described by the partition function (6.10.6). The excess free energy density of a homogeneous, one-component system can be expanded diagrammatically, as in (5.6.4), in terms of black ρ-circles and f-bonds. Topological reduction can then be used to derive a second expansion in which the f-bonds are replaced by h-bonds; the resulting expression for the free energy is therefore a functional of $h(r)$. The HNC approximation is unusual among approximate integral equations insofar as the excess free energy functional can be written in closed form[82] in terms of $h(r)$ and the pair potential $v(r)$, just as the chemical potential in (4.3.21) is expressed as a functional of $h(r)$ and $c(r)$, The equivalent free energy functional for a replicated system has a similar structure to that of a one-component system but involves both the like and unlike pair correlation functions. Optimisation of the functional with respect to $h_{aa}(r)$ and $h_{ab}(r)$ leads to a set of coupled closure relations which represent a generalisation of (4.3.19). These relations can be solved iteratively for a system consisting of equal numbers of a and b particles with the help of further approximations which exploit the fact that the amplitude of vibration of atoms around their molecular centre of mass will be small. The end result is an expression for the free energy which must be maximised with respect to m to determine $m^*(T)$, including the special case when $m^*(T) = 1$. Once the free energy is known the configurational entropy can be obtained by differentiation via (6.10.11).

Applications of the replica method have been made to soft-sphere[81] and Lennard-Jones[83] fluids and to the jamming transition of hard spheres.[84] The calculations for continuous potentials lead to values of T_K that lie well below the freezing temperature. They also lead to discontinuities at T_K, not only in the order parameter but also in the heat capacity rather than the internal energy, meaning that thermodynamically the transition is one of second order. Overall there is satisfactory agreement with the results of simulations but the remark made earlier in connection with the density functional calculations applies here as well; the integral equation approximations used to obtain numerical results are of unknown accuracy at the very low temperatures involved. It must also be kept in mind that the existence of an underlying, albeit unobservable transition to an ideal glass state remains a conjecture, but one that reveals deep seated analogies between the behaviour of intrinsically disordered systems, such as spin glasses, and that of structural glass formers

REFERENCES

[1] For a more detailed treatment of the material discussed in this section, see Evans, R., *In* 'Liquids at Interfaces: Les Houches, Session XLVIII' (J. Charvolin, J.F. Joanny and J. Zinn-Justin, eds). Elsevier, Amsterdam, 1990; and Rowlinson, J.S. and Widom, B., 'Molecular Theory of Capillarity'. Clarendon Press, Oxford, 1982.

[2] Biben, T., Hansen, J.P. and Barrat, J.L., *J. Chem. Phys.* **98**, 7330 (1993).

[3] It played a central role in early work on the wetting transition. See Cahn, J.W., *J. Chem. Phys.* **66**, 3667 (1977).

[4] For a critical survey of different approximations, see Evans, R., *In* 'Fundamentals of Inhomogeneous Fluids' (D. Henderson, ed). Marcel Decker, New York, 1991.

[5] Tarazona, P. and Evans, R., *Mol. Phys.* **52**, 847 (1984).

[6] (a) Tarazona, P., *Phys. Rev. A* **31**, 2672 (1985). (b) Curtin, W.A. and Ashcroft, N.W., *Phys. Rev. A* **32**, 2909 (1985). (c) Denton, A.R. and Ashcroft, N.W., *Phys. Rev. A* **39**, 4701 (1989).

[7] (a) Rosenfeld, Y., *Phys. Rev. Lett.* **63**, 980 (1989). (b) Rosenfeld, Y., Levesque, D. and Weis, J.J., *J. Chem. Phys.* **92**, 6818 (1990).

[8] See Chapter 5, Section 5.7.

[9] Triezenberg, D.G. and Zwanzg, R., *Phys. Rev. Lett.* **28**, 1183 (1972).

[10] There must be no 'overhangs' in the surface if $h(x, y)$ is to be single-valued.

[11] This choice of coordinates is called the Monge representation of the surface. See, e.g. Safran, S.F., 'Statistical Thermodynamics of Surfaces, Interfaces and Membranes'. Addison-Wesley, Reading, MA, 1994, Section 1.5.

[12] Kirkwood, J.G. and Buff, F.P., *J. Chem. Phys.* **65**, 338 (1949). The equivalence of the two expressions for γ in the case of pair potentials was demonstrated by Schofield, P., *Chem. Phys. Lett.* **62**, 413 (1979); see also Rowlinson and Widom, Ref. 1, Section 4.7.

[13] Reiss, H., Frisch, H.L. and Lebowitz, J.L., *J. Chem. Phys.* **31**, 369 (1959).

[14] The generalisations to mixtures are trivially related to the quantities $\langle d^n \rangle$ introduced in Section 3.10.

[15] (a) Kierlik, E. and Rosinberg, M.L., *Phys. Rev. A* **42**, 3382 (1990). (b) Phan, S., Kierlik, E., Rosinberg, M.L., Bildstein, B. and Kahl, G., *Phys. Rev. E* **48**, 618 (1993).

[16] The detailed calculation for a different but related system (parallel hard cubes) is given by Cuesta, J.A. and Martínez-Ratón, Y., *J. Chem. Phys.* **107**, 6379 (1997).

[17] Note that in Refs. 7 the signs of the coefficients c_3 and c_4 are incorrect.

[18] See Figure 1 in Ref. 15(a). The correct results are those given by the full curves: for an explanation, see Ref. 15(b).

[19] Roth, R., Evans, R., Lang, A. and Kahl, G., *J. Phys. Condens. Matter* **14**, 12063 (2002).

[20] Roth, R., *J. Phys. Cond. Matter* **22**, 063102 (2010).

[21] Rosenfeld, Y., *Phys. Rev. A* **42**, 5978 (1990).

[22] Percus, J.K., *J. Stat. Phys.* **15**, 505 (1976).

[23] Rosenfeld, Y., Schmidt, M., Löwen, H. and Tarazona P., *Phys. Rev. E* **55**, 4245 (1997).

[24] Tarazona, P. and Rosenfeld, Y., *Phys. Rev. E* **55**, R4873 (1997). See also González, A., White, J.A. and Evans, R., *J. Phys. Condens. Matter* **9**, 2375 (1997).

[25] Groot, R.D., Faber, N.M. and van der Eerden, J.P., *Mol. Phys.* **62**, 861 (1987).

[26] Snook, I.K. and van Megen, W., *J. Chem. Phys.* **72**, 2907 (1980).

[27] Kierlik, E. and Rosinberg, M.L., *Phys. Rev. A* **44**, 5025 (1991).

[28] Magda, J.J., Tirell, M. and Davis, H.T., *J. Chem. Phys.* **83**, 1888 (1985).

[29] (a) Evans, R., *J. Phys. Condens. Matter* **2**, 8989 (1990). (b) Gelb, L.V., Gubbins, K.E., Radhakrishnan, R. and Sliwinska-Bartkowiak, M., *Rep. Prog. Phys.* **62**, 1573 (1999).

[30] Goulding, D., Melchionna, S., and Hansen, J.P., *Phys. Chem. Chem. Phys.* **3**, 1644 (2001).

[31] Cahn, J.W., *J. Chem. Phys.* **66**, 3667 (1977).

[32] Ebner, C. and Saam, W.F., *Phys. Rev. Lett.* **38**, 1486 (1977).

[33] For a review of experimental results on wetting, see Bonn, D. and Ross, D., *Rep. Prog. Phys.* **64**, 1085 (2001).

[34] Denesyuk, N.A. and Hansen, J.P., *J. Chem. Phys.* **121**, 3613 (2004).

[35] Sullivan, D.E., *Phys. Rev. B* **20**, 3991 (1979).

[36] Tarazona, P. and Evans, R., *Mol. Phys.* **48**, 799 (1983).

[37] Oleksy, A. and Hansen, J.P., *Mol. Phys.* **107**, 2609 (2009).

[38] As, e.g. is the case for the model discussed in the text if $\epsilon_S = 0$.

[39] Tarazona, P. and Evans, R., *Mol. Phys.* **52**, 847 (1984).

[40] Chandler, D., *Nature* **437**. 640 (2005).

[41] (a) Alder, B.J. and Wainwright, T.E., *J. Chem. Phys.* **27**, 1208 (1957). (b) Wood, W.W. and Jacobson, J.D., *J. Chem. Phys.* **27**, 1207 (1957).

[42] Hoover, W.G. and Ree, F.H., *J. Chem. Phys.* **49**, 3609 (1968).

[43] Bernal, J.D. and Mason, J., *Nature* **188**, 910 (1960).

[44] (a) Hoover, W.G., Gray, S.G. and Johnson, K.W., *J. Chem. Phys.* **55**, 1128 (1971). (b) Agrawa, R. and Kofke, D.A., *Phys. Rev. Lett.* **74**, 122 (1995).

[45] Weeks, J.D., *Phys. Rev. B* **24**, 1530 (1981).

[46] (a) Hansen, J.P. and Verlet, L., *Phys. Rev.* **184**, 151 (1969). (b) Hansen, J.P. and Schiff, D., *Mol. Phys.* **25**, 1281 (1973).

[47] Kirkwood, J.G. and Monroe, E., *J. Chem. Phys.* **9**, 511 (1941).

[48] Fundamental measure theory is an exception to this rule.

[49] Baus, M., *J. Phys. Condens. Matter* **2**, 2111 (1990).

[50] Ramakrishnan, T.V. and Yussouff, M., *Phys. Rev. B* **19**, 2775 (1979).

[51] Haymet, A.D. and Oxtoby, D.W., *J. Chem. Phys.* **74**, 2559 (1981).

[52] See, e.g. Barrat, J.L., Hansen, J.P. and Pastore, G., *Mol. Phys.* **63**, 747 (1988).

[53] Zeng, X.C. and Oxtoby, D.W., *J. Chem. Phys.* **93**, 4357 (1990). See also Denton, A.R. and Ashcroft, N.W., *Phys. Rev.* **42**, 7312 (1990).

[54] Kranendonk, W.G.T. and Frenkel, D., *Mol. Phys.* **72**, 679 (1991).

[55] Barrat, J.L., Baus, M. and Hansen, J.P., *J. Phys. C* **20**, 1413 (1987).

[56] Xu, H. and Baus, M., *J. Phys. Condens. Matter* **4**, L663 (1992).

[57] Bartlett, P., Ottewill, R.H. and Pusey, P.N., *Phys. Rev. Lett.* **68**, 3801 (1992).

[58] Madden, W.G. and Glandt, E.D., *J. Stat. Phys.* **51**, 537 (1988).

[59] Given, J.A., *Phys. Rev. A* **45**, 816 (1992).

[60] Rosinberg, M.L., Tarjus, G. and Stell, G., *J. Chem. Phys.* **100**, 5172 (1994).

[61] (a) Edwards, S.F. and Anderson, P.W., *J. Phys. F* **5**, 965 (1975). (b) Edwards, S.F. and Jones, R.C., *J. Phys. A* **9**, 1595 (1976).

[62] Mézard, M., Parisi, G. and Virasoro, M.A., 'Spin Glass Theory and Beyond'. World Scientific, Singapore, 1987.

[63] If s is small, $\Xi^s = \exp(s \ln \Xi) \approx 1 + s \ln \Xi$.

[64] Given, J.A. and Stell, G., *Physica A* **209**, 495 (1994).

[65] Lomba, E., Given, J.A., Stell, G., Weis, J.J. and Levesque, D., *Phys. Rev. E* **48, 233 (1993)**.

[66] See, e.g. Reiss, H., *J. Phys. Chem.* **96**, 4736 (1992).

[67] Widom, B. and Rowlinson, J.S., *J. Chem. Phys.* **52**, 1670 (1970).

[68] Kierlik, E., Rosinberg, M.L., Tarjus, G. and Monson, P., *J. Chem. Phys.* **103**, 4256 (1995).

[69] Meroni, A., Levesque, D. and Weis, J.J., *J. Chem. Phys.* **105**, 1101 (1996).

[70] Kierlik, E., Rosinberg, M.L., Tarjus, G. and Monson, P.A., *J. Chem. Phys.* **106**, 264 (1997).

[71] Rosinberg, M.L., *In* 'New Approaches to Problems in Liquid State Theory' (C. Caccamo, J.P. Hansen and G. Stell, eds.). Kluwer, Dordrecht, 1999.

[72] Debenedetti, P.G. and Stillinger, F.H., *Nature* **410**, 259 (2001).

[73] Kauzmann, A.W., *Chem. Rev.* **43**, 219 (1948).

[74] The word 'ideal' as used here does not have the same implication as in 'ideal gas', where it means non-interacting. In an 'ideal glass' the interactions are very strong.

[75] Kirkpatrick, T.R., Thirumalai, D. and Wolynes, P.G., *Phys. Rev. A* **40**, 1045 (1989).

[76] (a) Berthier, L. and Biroli, G., *Rev. Mod. Phys.* **83**, 587, (2011). (b) Biroli, G. and Bouchaud, J.P., *In* 'Structural Glasses and Supercooled Liquids: Theory, Experiment and Applications' (P.G. Wolynes and V. Lubchenko, eds). John Wiley, Hoboken, NJ, 2012.

[77] Goldstein, M., *J. Chem. Phys.* **51**, 3728 (1969).

[78] Singh, Y., Stoessel, J.P. and Wolynes, P.G., *Phys. Rev. Lett.* **54**, 1059 (1985).

[79] Bennett, C.H., *J. Appl. Phys.* **43**, 2727 (1972).

[80] Chaudhuri, P., Karmakar, S. and Dasgupta, C., *Phys. Rev. Lett.* **100**, 125701 (2008).

[81] (a) Mézard, M. and Parisi, G., *J. Chem. Phys.* **111**, 1076 (1999). (b) Mézard, M. and Parisi, G., *J. Phys. Condens. Matter* **12**, 6655 (2000). (c) Mézard, M. and Parisi, G., *In* 'Structural Glasses and Supercooled Liquids: Theory, Experiment and Applications' (P.G. Wolynes and V. Lubchenko, eds). John Wiley, Hoboken, NJ, 2012.

[82] Morita, T. and Hiroike, K., *Prog. Theor. Phys.* **25**, 537 (1961).

[83] Coluzzi, B., Parisi, G. and Verrocchio, P., *J. Chem. Phys.* **112**, 2933 (2000).

[84] Parisi, G. and Zamponi, F., *Rev. Mod. Phys.* **82**, 789 (2010).

Chapter 7

Time-dependent Correlation and Response Functions

The next three chapters are devoted to a discussion of the transport properties and microscopic dynamics of simple, dense fluids.[1] The present chapter deals with the general formalism of time correlation functions and with linear response theory; Chapter 8 is concerned with the behaviour of time-dependent fluctuations in the long-wavelength, low-frequency limit, where contact can be made with the macroscopic equations of hydrodynamics; and Chapter 9 describes methods that allow the explicit calculation of time correlation functions.

7.1 GENERAL PROPERTIES OF TIME CORRELATION FUNCTIONS

A dynamical variable, $A(t)$ say, of a system consisting of N structureless particles is a function of some or all of the time-varying coordinates $\mathbf{r}_i$ and momenta $\mathbf{p}_i$, $i = 1$ to N. We recall from Section 2.1 that the time evolution of A is determined by the equation of motion $A(t) = \exp(i\mathcal{L}t)A(0)$, where $\mathcal{L}$ is the Liouville operator. It follows that A has the signature $\epsilon_A = +1$ or -1 under time reversal depending on whether or not it changes sign under the transformation $\mathbf{p}_i \rightarrow -\mathbf{p}_i$. Now consider two such variables, A and B, each of which may be either real or complex. Their equilibrium time correlation function is written as

$$C_{AB}(t', t'') = \left\langle A(t')B^*(t'') \right\rangle \tag{7.1.1}$$

with the convention that $t' \geq t''$. The superscript $*$ denotes a complex conjugate and the angular brackets represent either an average over time or an ensemble average over initial conditions. Thus $C_{AB}(t', t'')$ is defined either as

$$\left\langle A(t')B^*(t'') \right\rangle = \lim_{\tau \to \infty} \frac{1}{\tau} \int_0^\tau A(t' + t)B^*(t'' + t)\mathrm{d}t \tag{7.1.2}$$

Theory of Simple Liquids, Fourth Edition. http://dx.doi.org/10.1016/B978-0-12-387032-2.00007-6

or as

$$\langle A(t')B^*(t'')\rangle = \iint f_0^{[N]}(\mathbf{r}^N, \mathbf{p}^N) B^*(\mathbf{r}^N, \mathbf{p}^N)$$
$$\times \exp[i\mathcal{L}(t'-t'')]A(\mathbf{r}^N, \mathbf{p}^N)\,\mathrm{d}\mathbf{r}^N\mathrm{d}\mathbf{p}^N \qquad (7.1.3)$$

The average in (7.1.3) is taken over all possible states of the system at time t'', weighted by the equilibrium probability density $f_0^{[N]}$; for a system characterised by fixed values of N, V and T, $f_0^{[N]}$ is given by the canonical distribution (2.3.1). Equations (7.1.2) and (7.1.3) yield the same result in the thermodynamic limit if the system is ergodic. The most important class of time correlation functions are the *autocorrelation* functions $C_{AA}(t)$, for which A and B are the same variable.

Since the equilibrium probability density is independent of time, the ensemble average in (7.1.3) is independent of the choice of time origin t'' and the correlation function $C_{AB}(t', t'')$ is invariant under time translation. If we put $t'' = s$ and $t' = s + t$ the correlation function is a function only of the time difference t and is said to be *stationary* with respect to s. It is therefore customary to set $s = 0$ and use the more compact notation

$$C_{AB}(t) = \langle A(t)B^*\rangle \qquad (7.1.4)$$

where $B^* \equiv B^*(0)$. The stationary character of the correlation function means that

$$\frac{\mathrm{d}}{\mathrm{d}s}\langle A(t+s)B^*(s)\rangle = \langle \dot{A}(t+s)B^*(s)\rangle + \langle A(t+s)\dot{B}^*(s)\rangle = 0 \qquad (7.1.5)$$

and hence that

$$\langle \dot{A}(t)B^*\rangle = -\langle A(t)\dot{B}^*\rangle \qquad (7.1.6)$$

In particular:

$$\langle \dot{A}A^*\rangle = 0 \qquad (7.1.7)$$

Repeated differentiation with respect to s leads to a number of useful relations; these can also be deduced by exploiting the definition (7.1.2). For example:

$$\frac{\mathrm{d}^2}{\mathrm{d}t^2}\langle A(t)B^*\rangle = \langle \ddot{A}(t)B^*\rangle$$
$$= \lim_{\tau\to\infty}\frac{1}{\tau}\int_0^\tau \ddot{A}(t+t')B^*(t')\mathrm{d}t'$$
$$= -\lim_{\tau\to\infty}\frac{1}{\tau}\int_0^\tau \dot{A}(t+t')\dot{B}^*(t')\mathrm{d}t'$$
$$= -\langle \dot{A}(t)\dot{B}^*\rangle \qquad (7.1.8)$$

The invariance of correlation functions under time translation implies that

$$C_{AB}(t) = \epsilon_A\epsilon_B C_{AB}(-t) = \epsilon_A\epsilon_B\langle A(-t)B^*\rangle$$
$$= \epsilon_A\epsilon_B\langle AB^*(t)\rangle = \epsilon_A\epsilon_B C_{BA}^*(t) \qquad (7.1.9)$$

where ϵ_A, ϵ_B are the time-reversal signatures of the two variables. From this result it follows that autocorrelation functions are real functions of time.

It is clear that

$$\lim_{t \to 0} C_{AB}(t) = \langle AB^* \rangle \tag{7.1.10}$$

where $\langle AB^* \rangle$ is a static correlation function. In the limit $t \to \infty$ the variables $A(t)$ and B become uncorrelated and

$$\lim_{t \to \infty} C_{AB}(t) = \langle A \rangle \langle B^* \rangle \tag{7.1.11}$$

However, it is usually more convenient to define the dynamical variables in such a way as to exclude their average values and to consider only the time correlation of their fluctuating parts, i.e.

$$C_{AB}(t) = \langle [A(t) - \langle A \rangle][B^* - \langle B^* \rangle] \rangle \tag{7.1.12}$$

With this convention, $C_{AB}(t) \to 0$ as $t \to \infty$. Because

$$\langle [A(t) \pm A][A(t) \pm A]^* \rangle \geq 0 \tag{7.1.13}$$

it is also true that

$$-\langle AA^* \rangle \leq C_{AA}(t) \leq \langle AA^* \rangle \tag{7.1.14}$$

The magnitude of an autocorrelation function is therefore bounded above by its initial value. This is to be expected, since an autocorrelation function describes the averaged way in which spontaneous (thermal) fluctuations in a variable A decay in time.

If $C_{AB}(t)$ is defined as in (7.1.12), it is also possible to define its Fourier transform or *power spectrum*:

$$C_{AB}(\omega) = \frac{1}{2\pi} \int_{-\infty}^{\infty} C_{AB}(t) \exp(i\omega t) dt \tag{7.1.15}$$

and its Laplace transform:

$$\tilde{C}_{AB}(z) = \int_{0}^{\infty} C_{AB}(t) \exp(izt) dt \tag{7.1.16}$$

where z is a complex frequency. Since $C_{AB}(t)$ is bounded, $\tilde{C}_{AB}(z)$ is analytic in the upper half of the complex z plane (Im $z > 0$); it is also related to $C_{AB}(\omega)$ by a Hilbert transform, i.e.

$$\tilde{C}_{AB}(z) = \int_{0}^{\infty} dt \exp(izt) \int_{-\infty}^{\infty} C_{AB}(\omega) \exp(-i\omega t) d\omega$$

$$= i \int_{-\infty}^{\infty} \frac{C_{AB}(\omega)}{z - \omega} d\omega \tag{7.1.17}$$

An integral such as that in (7.1.17) can be evaluated with the help of a standard relation commonly written in short-hand form as

$$\lim_{\epsilon \to 0} \frac{1}{x \pm i\epsilon} \equiv \mathcal{P}\left(\frac{1}{x}\right) \mp i\pi\delta(x) \qquad (7.1.18)$$

where $\mathcal{P}$ denotes the principal value. Thus, since $C_{AA}(\omega)$ is necessarily real:

$$\lim_{\epsilon \to 0} \text{Re } \tilde{C}_{AA}(\omega + i\epsilon) = \lim_{\epsilon \to 0} \text{Re}\left(i \int_{-\infty}^{\infty} \frac{C_{AA}(\omega')}{\omega - \omega' + i\epsilon}\, d\omega'\right)$$
$$= \pi C_{AA}(\omega) \qquad (7.1.19)$$

It can also be shown that $C_{AA}(\omega) \geq 0$ for all ω. Consider an auxiliary variable, $A_T(\omega)$, defined as

$$A_T(\omega) = \frac{1}{\sqrt{2T}} \int_{-T}^{T} A(t) \exp(i\omega t)\, dt \qquad (7.1.20)$$

The statistical average of $\langle A_T(\omega)A_T^*(\omega)\rangle$ cannot be negative. Hence

$$\langle A_T(\omega)A_T^*(\omega)\rangle = \frac{1}{2T} \int_{-T}^{T} dt \int_{-T}^{T} dt' \langle A(t)A^*(t')\rangle \exp[i\omega(t - t')] \geq 0 \qquad (7.1.21)$$

If we now make a change of variable from t' to $\tau = t - t'$ and take the limit $T \to \infty$, we find that

$$\lim_{T \to \infty} \langle A_T(\omega)A_T^*(\omega)\rangle = \int_{-\infty}^{\infty} C_{AA}(\tau) \exp(i\omega\tau)\, d\tau$$
$$= C_{AA}(\omega) \geq 0 \qquad (7.1.22)$$

The experimental significance of time correlation functions lies in the fact that the spectra measured by various spectroscopic techniques are the power spectra of well-defined dynamical variables. This connection between theory and experiment will be made explicit in Section 7.5 for the special but important case of inelastic neutron scattering. In addition, as we shall see later, the linear transport coefficients of hydrodynamics are related to time integrals of certain autocorrelation functions. Finally, time correlation functions provide a quantitative description of the microscopic dynamics in liquids. Computer simulations play a key role here, since they give access to a large variety of correlation functions, many of which are not measurable by laboratory experiments.

Apart from the limitation to classical mechanics the properties of time correlation functions given thus far are completely general. We now restrict the discussion to systems of particles for which the interaction potential is continuous; the hamiltonian is therefore differentiable and the Liouville operator has the form given by (2.1.8). An autocorrelation function of such a system is an even function of time and can be expanded in a Taylor series in even powers of t around $t = 0$. Thus

$$C_{AA}(t) = \sum_{n=0}^{\infty} \frac{t^{2n}}{(2n)!} \left\langle A^{(2n)} A^* \right\rangle = \sum_{n=0}^{\infty} \frac{t^{2n}}{(2n)!} (-1)^n \left\langle A^{(n)} A^{(n)*} \right\rangle$$

$$= \sum_{n=0}^{\infty} \frac{t^{2n}}{(2n)!} (-1)^n \left\langle |(i\mathcal{L})^n A|^2 \right\rangle \tag{7.1.23}$$

where the superscript $(2n)$ denotes a $2n$-fold derivative and repeated use has been made of (7.1.8). Differentiation of the inverse Fourier transform of (7.1.15) $2n$ times with respect to t gives

$$\left\langle \omega^{2n} \right\rangle_{AA} \equiv \int_{-\infty}^{\infty} \omega^{2n} C_{AA}(\omega) \, d\omega = (-1)^n C_{AA}^{(2n)}(t = 0) \tag{7.1.24}$$

Thus, apart from a possible change of sign, the frequency moments of the power spectrum are equal to the derivatives of the autocorrelation function taken at $t = 0$; these derivatives are static correlation functions which are expressible as integrals over the particle distribution functions. On expanding the right-hand side of (7.1.17) in powers of $1/z$ it becomes clear that the frequency moments defined by (7.1.24) are also the coefficients in the high-frequency expansion of the Laplace transform:

$$\tilde{C}_{AA}(z) = \frac{i}{z} \sum_{n=0}^{\infty} \frac{\left\langle \omega^{2n} \right\rangle_{AA}}{z^{2n}} \tag{7.1.25}$$

Expansions of the type displayed in (7.1.23) cannot be used for systems such as the hard-sphere fluid. The impulsive nature of the forces between particles with hard cores means that the Liouville operator no longer has the form[2] shown in (2.1.8). As a result, the time correlation functions are non-analytic at $t = 0$, and their power spectra have frequency moments that are infinite.

The definition of a time correlation function provided by (7.1.3) has the form of an inner product of the 'vectors' $A(t)$ and B in the infinite-dimensional, Hilbert space of dynamical variables, usually called Liouville space. A useful notation based on this identification is one in which a time correlation function is written as

$$\left\langle A(t)B^* \right\rangle \equiv (B, A(t)) \tag{7.1.26}$$

where $(\cdots, \cdots)$ denotes an inner product. The usual requirements of an inner product are therefore satisfied. In particular, $(A, A) \geq 0$ and $(A, B) = (B, A)^*$. Formal properties of time correlation functions can then be deduced from the fact that the Liouville operator is hermitian (and hence $i\mathcal{L}$ is anti-hermitian) with respect to the inner product, i.e.

$$(B, \mathcal{L}A) = (A, \mathcal{L}B)^* = (\mathcal{L}B, A) \tag{7.1.27}$$

Because $\mathcal{L}$ is hermitian the propagator $\exp(i\mathcal{L}t)$ is a unitary operator with an hermitian conjugate given by $\exp(-i\mathcal{L}t)$. It follows that

$$
\begin{aligned}
\langle A(t)B^* \rangle &\equiv (B, \exp(i\mathcal{L}t)A) = (B, \exp(-i\mathcal{L}s)A(t+s)) \\
&= (A(t+s), \exp(i\mathcal{L}s)B)^* \equiv \langle A(t+s)B^*(s) \rangle
\end{aligned} \tag{7.1.28}
$$

thereby proving that the correlation function is stationary. Note that the effect of the operation $A(t) = \exp(i\mathcal{L}t)A$ is to 'rotate' A through an angle $\mathcal{L}t$ in Liouville space. By exploiting the fact that $i\mathcal{L}A = \dot{A}$, properties of time correlation functions which involve time derivatives of dynamical variables are also easily derived . For example:

$$
\begin{aligned}
\langle \dot{A}(t)B^* \rangle &\equiv (B, i\mathcal{L}A(t)) \\
&= -(A(t), i\mathcal{L}B)^* \equiv -\langle A(t)\dot{B}^* \rangle
\end{aligned} \tag{7.1.29}
$$

in agreement with (7.1.6).

The proof that the Liouville operator is hermitian requires an integration by parts of the derivatives appearing in the Poisson bracket representation (2.1.8). The inner product is sometimes defined without the weighting factor $f_0^{(N)}$, but the Liouville operator retains its hermitian character, since $\mathcal{L}f_0^{(N)} = 0$.

7.2   AN ILLUSTRATION: THE VELOCITY AUTOCORRELATION FUNCTION AND SELF-DIFFUSION

The ideas introduced in Section 7.1 can be usefully illustrated by considering one of the simplest but most important examples of a time correlation function, namely the autocorrelation function of the velocity $\mathbf{u} = \mathbf{p}/m$ of a tagged particle moving through a fluid. The velocity autocorrelation function, defined as

$$
Z(t) = \frac{1}{3} \langle \mathbf{u}(t) \cdot \mathbf{u} \rangle = \langle u_x(t)u_x \rangle \tag{7.2.1}
$$

is a measure of the projection of the particle velocity onto its initial value, averaged over initial conditions. Its value at $t = 0$ is given by the equipartition theorem:

$$
Z(0) = \frac{1}{3}\langle u^2 \rangle = \frac{k_B T}{m} \tag{7.2.2}
$$

At times long compared with any microscopic relaxation time the initial and final velocities will be completely uncorrelated. Thus $Z(t \to \infty) = 0$. The results of computer simulations of argon-like liquids show that the velocities are already largely decorrelated after times of order 10^{-12} s, but in general $Z(t)$ also has a weak, slowly decaying part. The detailed behaviour at long times varies with thermodynamic state, as is evident from the examples plotted

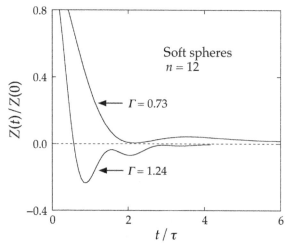

FIGURE 7.1 Normalised velocity autocorrelation function of the r^{-12}-fluid at two different values of the dimensionless coupling parameter Γ defined by (5.4.13). The larger value of Γ represents a thermodynamic state close to the fluid-solid transition and the unit of time is $\tau = (\sigma^2/48\epsilon)^{1/2}$. After Heyes *et al.*[3]

in Figure 7.1. We shall return later to a discussion of the main features of curves such as these, but first we show that there exists a general relationship between the self-diffusion coefficient D and the time integral of $Z(t)$.

Consider a set of identical, tagged particles having initial positions $\{\mathbf{r}_i(0)\}$. If the particles diffuse in time t to positions $\{\mathbf{r}_i(t)\}$, the self-diffusion coefficient is given by a well-known relation due to Einstein:

$$D = \lim_{t \to \infty} \frac{\langle |\mathbf{r}_i(t) - \mathbf{r}_i(0)|^2 \rangle}{6t} \qquad (7.2.3)$$

This result is a direct consequence of Fick's law of diffusion, as we shall see in Section 8.2. It is also a relation characteristic of a 'random walk', in which the mean-square displacement of the walker becomes a linear function of time after a sufficiently large number of random steps. The nature of the limiting process involved in (7.2.3) highlights the general importance of taking the thermodynamic limit before the limit $t \to \infty$. For a system of finite volume V, the diffusion coefficient defined by (7.2.3) is strictly zero, since the maximum achievable mean-square displacement is of order $V^{2/3}$. In practice, for a system of macroscopic dimensions, the ratio on the right-hand side of (7.2.3) will reach a plateau value at times much shorter than those required for the diffusing particles to reach the boundaries of the system; it is the plateau value that provides the definition of D for a finite system.

We now rewrite the Einstein relation in terms of the velocity autocorrelation function. The displacement in a time interval t of any tagged particle is

$$\mathbf{r}(t) - \mathbf{r}(0) = \int_0^t \mathbf{u}(t')dt' \tag{7.2.4}$$

When squared and averaged over initial conditions, (7.2.4) becomes

$$\left\langle |\mathbf{r}(t) - \mathbf{r}(0)|^2 \right\rangle = \left\langle \int_0^t \mathbf{u}(t')dt' \cdot \int_0^t \mathbf{u}(t'')dt'' \right\rangle$$

$$= 2\int_0^t dt' \int_0^{t'} dt'' \left\langle \mathbf{u}(t') \cdot \mathbf{u}(t'') \right\rangle$$

$$= 6\int_0^t dt' \int_0^{t'} dt'' Z(t' - t'') \tag{7.2.5}$$

A change of variable from t'' to $s = t' - t''$ followed by an integration by parts with respect to t' shows that

$$\left\langle |\mathbf{r}(t) - \mathbf{r}(0)|^2 \right\rangle = 6\int_0^t dt' \int_0^{t'} ds\, Z(s)$$

$$= 6t \int_0^t \left(1 - \frac{s}{t}\right) Z(s)\, ds \tag{7.2.6}$$

and substitution of (7.2.5) in (7.2.3) gives the required result:

$$D = \int_0^\infty Z(t)dt \tag{7.2.7}$$

Equation (7.2.7) is an example of a *Green–Kubo formula*, an important class of relations in which a macroscopic dynamical property is written as the time integral of a microscopic time correlation function.

If the interparticle potential is continuous, the short-time expansion of $Z(t)$ starts as

$$Z(t) = \frac{k_B T}{m}\left(1 - \Omega_0^2 \frac{t^2}{2} + \cdots\right) \tag{7.2.8}$$

Equation (7.1.23) shows that the coefficient of $\frac{1}{2}t^2$ is

$$\Omega_0^2 = \frac{m}{3k_B T}\langle \dot{\mathbf{u}} \cdot \dot{\mathbf{u}} \rangle = \frac{\langle |\mathbf{F}|^2 \rangle}{3mk_B T} \tag{7.2.9}$$

where $\mathbf{F}$ is the total force exerted on the diffusing particle by its neighbours. If the tagged particle is identical to all other particles in the fluid, $\mathbf{F} = -\nabla V_N$, where V_N is the total potential energy. When V_N is a sum of pair terms, Ω_0^2 can be expressed in terms of the equilibrium pair distribution function and the interparticle potential. To show this we first derive a useful, general result. Let $A(\mathbf{r}^N)$ be some function of the particle coordinates. Then

$$
\left\langle A(\mathbf{r}^N) \frac{\partial V_N}{\partial x_i} \right\rangle
$$

$$
= \frac{1}{Z_N} \int \cdots \int A(\mathbf{r}^N) \frac{\partial V_N}{\partial x_i} \exp(-\beta V_N) \mathrm{d}\mathbf{r}_1 \cdots \mathrm{d}x_i \, \mathrm{d}y_i \, \mathrm{d}z_i \cdots \mathrm{d}\mathbf{r}_N
$$

$$
= \frac{k_B T}{Z_N} \int \cdots \int \frac{\partial A(\mathbf{r}^N)}{\partial x_i} \exp(-\beta V_N) \mathrm{d}\mathbf{r}_1 \cdots \mathrm{d}x_i \, \mathrm{d}y_i \, \mathrm{d}z_i \cdots \mathrm{d}\mathbf{r}_N
$$

$$(7.2.10)$$

or

$$
\left\langle A(\mathbf{r}^N) \frac{\partial V_N}{\partial x_i} \right\rangle = k_B T \left\langle \frac{\partial A(\mathbf{r}^N)}{\partial x_i} \right\rangle \tag{7.2.11}
$$

The second equality in (7.2.10) follows from an integration by parts with respect to x_i.

Equation (7.2.11) is called the Yvon theorem. When applied to the current problem it shows that the mean-square force on a particle is

$$
\left\langle |\mathbf{F}|^2 \right\rangle = k_B T \left\langle \nabla^2 V_N \right\rangle \tag{7.2.12}
$$

With the assumption of pairwise additivity manipulations similar to those used in Section 2.5 now allow (7.2.9) to be rewritten in the form

$$
\Omega_0^2 = \frac{(N-1)}{3m} \left\langle \nabla^2 v(r) \right\rangle = \frac{\rho}{3m} \int \nabla^2 v(r) g(r) \mathrm{d}\mathbf{r} \tag{7.2.13}
$$

The quantity Ω_0 is called the Einstein frequency; it represents the frequency at which the tagged particle would vibrate if it were undergoing small oscillations in the potential well produced by the surrounding particles when maintained at their mean equilibrium positions around the tagged particle. Numerically, Ω_0 is of order 10^{13} s^{-1} for liquid argon near its triple point.

Equation (7.2.8) does not apply to systems of hard spheres because the hard-sphere potential is not differentiable.[4] The short-time behaviour of $Z(t)$ now takes the form

$$
\langle \mathbf{u}(t) \cdot \mathbf{u} \rangle = \left\langle u^2 \right\rangle + t \left(\frac{\mathrm{d}}{\mathrm{d}t} \langle \mathbf{u}(t) \cdot \mathbf{u} \rangle \right)_{t=0} + \cdots \tag{7.2.14}
$$

where the differentiation with respect to time must be carried out after the ensemble averaging. Thus

$$
Z(t) = \frac{1}{3} \left\langle u^2 \right\rangle (1 - \Omega_0' t + \cdots) \tag{7.2.15}
$$

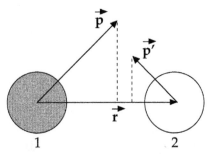

FIGURE 7.2 A collision between a tagged hard sphere, (1), and a sphere from the bath, (2). The projections of the momenta **p** and **p**$'$ of the colliding particles along the vector **r** are the quantities denoted by p and p' in the text.

where the frequency Ω_0' is

$$\Omega_0' = -\frac{1}{\langle u^2 \rangle} \lim_{\Delta t \to 0} \frac{\langle \Delta \mathbf{u} \cdot \mathbf{u} \rangle}{\Delta t} \tag{7.2.16}$$

Consider a tagged hard sphere of diameter d moving in a fluid of untagged but otherwise identical hard spheres.[5] Over a sufficiently short time interval the tagged sphere will suffer at most one collision with a sphere from the bath. To evaluate Ω_0' from its definition (7.2.16), let us suppose that the tagged sphere, of momentum **p**, collides with a sphere of momentum **p**$'$, as pictured in Figure 7.2. Because the collision is elastic, the momentum gained by the tagged particle is $\Delta \mathbf{p} = -(\mathbf{p} \cdot \hat{\mathbf{r}} - \mathbf{p}' \cdot \hat{\mathbf{r}})\hat{\mathbf{r}}$, where $\hat{\mathbf{r}} = \mathbf{r}/r$ is a unit vector along the line joining the two centres of mass. Thus $-\Delta \mathbf{p} \cdot \mathbf{p} = p(p - p')$ where p, p' are the components of **p** and **p**$'$, respectively, along $\hat{\mathbf{r}}$. If $p > p'$, the separation of the two spheres will decrease in a short time Δt by an amount $\Delta r = (p - p')\Delta t/m$. On average, given that Δr is small, the number of spheres that initially lie within a distance d to $d + \Delta r$ of the tagged sphere will be $n(\Delta r) \approx 4\pi \rho d^2 g(d)(p - p')\Delta t/m$, where $g(d)$ is the pair distribution function at contact, and the probability that the tagged sphere will suffer a collision with a sphere having a component of momentum p' along **r** is $P(p') = f_M(p')\mathrm{d}p'$, where f_M is the Maxwell distribution (2.1.26) in its component form, i.e.

$$f_M(p) = \frac{4\pi \exp\left(-\beta p^2/2m\right)}{(2\pi m k_B T)^{1/2}} \tag{7.2.17}$$

The statistical average of $-\Delta \mathbf{p} \cdot \mathbf{p}$ is therefore obtained by multiplying $P(p')$ by $n(\Delta r)p(p - p')f_M(p)\mathrm{d}p$ and integrating over p and p' subject to the constraint that $p > p'$. Bringing these results together we find that

$$\begin{aligned}
\Omega_0' &= -\frac{1}{3mk_B T} \lim_{\Delta t \to 0} \frac{\langle \Delta \mathbf{p} \cdot \mathbf{p} \rangle}{\Delta t} \\
&= \frac{4\pi \rho d^2 g(d)}{3m^2 k_B T} \iint_{p > p'} p(p - p')^2 f_M(p) f_M(p') \mathrm{d}p \, \mathrm{d}p' \tag{7.2.18}
\end{aligned}$$

or, on changing variables from p, p' to $p_+ = (p+p')/\sqrt{2}$, $p_- = (p-p')/\sqrt{2}$:

$$\Omega'_0 = \frac{4\sqrt{2}\pi\rho d^2 g(d)}{3m^2 k_B T} \int_{-\infty}^{\infty} dp_+ \int_{0}^{\infty} dp_- p_-^3 f_M(p_-) f_M(p_+) \qquad (7.2.19)$$

The double integral is now easily evaluated to give

$$\Omega'_0 = \frac{8\rho d^2 g(d)}{3} \left(\frac{\pi k_B T}{m}\right)^{1/2} = \frac{2}{3}\Gamma_E \qquad (7.2.20)$$

where Γ_E is the Enskog collision rate introduced in Section 2.5.

The derivation of (7.2.20) shows that the Enskog approximation makes allowance for static correlations in the fluid, but the key assumption underlying the Boltzmann equation (see Section 2.1) is retained, namely that successive collisions are completely uncorrelated. The velocity of a tagged particle immediately following a collision is therefore dependent on its velocity immediately prior to the collision, but not on its velocity at earlier times. Because collisions between hard spheres are instantaneous events, this is tantamount to saying that the 'memory' associated with the velocity of the tagged particle is of infinitesimally short duration, with the consequence, as we shall see in later sections, that the velocity autocorrelation function is exponential in time. By identifying the right-hand side of (7.2.15) with the leading terms in the expansion of an exponential function we recover Enskog's expression[6] for the velocity autocorrelation function of hard spheres:

$$Z_E(t) = \frac{k_B T}{m} \exp\left(-\frac{2}{3}\Gamma_E |t|\right) \qquad (7.2.21)$$

where the absolute value of t appears because $Z(t)$ must be an even function of t. The corresponding approximation for the diffusion coefficient is obtained by substitution of (7.2.21) in (7.2.7):

$$D_E = \frac{3k_B T}{2m\Gamma_E} = \frac{3}{8\rho d^2 g(d)} \left(\frac{k_B T}{\pi m}\right)^{1/2} \qquad (7.2.22)$$

Equation (7.2.22) is nearly exact in the low-density limit[7] while its applicability at higher densities has been thoroughly tested in molecular dynamics calculations.[8] From Figure 7.3 we see that the diffusion coefficient obtained by simulation exceeds the Enskog value at intermediate densities, but falls below it at densities close to crystallisation.[10] The high-density deviations arise from back-scattering effects, corresponding to the fact that collisions lead, on average, to the reversal of the velocity of a tagged particle into a comparatively narrow range of angles. This gives rise to an extended negative region in $Z(t)$; the same effect is seen for other potential models, as exemplified in Figure 7.1. The increase in the ratio D/D_E at intermediate densities is attributable in

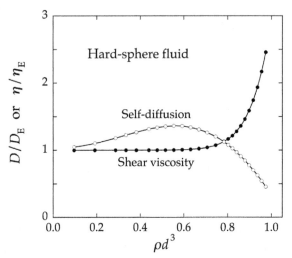

FIGURE 7.3 Molecular dynamics results[9] for the self-diffusion coefficient D and shear viscosity η of the hard-sphere fluid relative to their values in the Enskog approximation. The curves are shown as a guide to the eye. After Heyes.[8c]

large part to an enhancement of velocity correlations due to the excitation of slowly decaying, collective motions in the fluid. The motion of the tagged particle induces a backflow pattern in the surrounding fluid that reacts on the particle at later times, giving rise to persistence (or 'memory') effects and an unexpectedly slow ($\sim t^{-3/2}$) decay of $Z(t)$ at very long times; this behaviour is again not specific to hard spheres. We shall return to the question of the 'long-time tails' of correlation functions in Section 8.7. Figure 7.3 also shows the corresponding results for the shear viscosity of the hard-sphere fluid, but we postpone discussion of these until Section 8.4.

A treatment of self-diffusion by kinetic theory that goes beyond the Enskog approximation must take account of the correlated sequences of binary collisions that a tagged particle experiences. In such a sequence the tagged particle collides initially with a particle from the bath, then diffuses through the fluid, suffering collisions with other bath particles, before colliding either with the same particle it met initially or with another particle whose motion is correlated in some way with that of the initial collision partner. Examples of collision sequences are illustrated in Figure 7.4; in each case the tagged particle is labelled 1 and A, B represent two different space-time points. In example (a), the two collisions are uncorrelated. In (b) and (c), particles 1 and 2 first meet at A, then recollide at B; in (b) the recollision involves one intermediate collision between 2 and 3 (a three-body event) and in (c) it involves intermediate collisions between 1 and 4 and between 2 and 3 (a four-body event). Example (d) is a different type of four-body event in which the initial (at A) and final (at B) collision partners are different but the collisions suffered

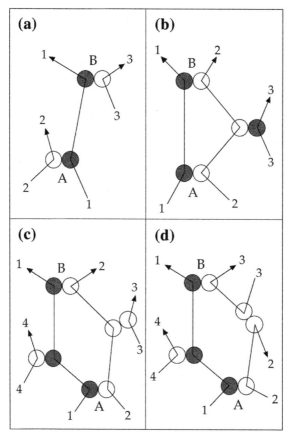

FIGURE 7.4 Examples of uncorrelated (a) and correlated (b–d) sequences of binary collisions suffered by a tagged particle labelled 1. In each case A and B represent two different space-time points with the first collision in the sequence occurring at A and the final collision at B. See text for details.

by 1 at A and B are nonetheless correlated. Sequences (b–d) are all examples of 'ring-collision' events.

7.3 BROWNIAN MOTION AND A GENERALISED LANGEVIN EQUATION

Calculations of the velocity autocorrelation function either by the Enskog method or by other, more sophisticated versions of kinetic theory are largely limited to hard-sphere systems, though efforts have been made to apply similar techniques in calculations for continuous potentials. In this section we describe a different approach that is more phenomenological in character but has found wide application in the theory of transport processes in liquids. Its basis is the stochastic theory used by Langevin to describe the brownian motion of a large and massive particle in a bath of particles that are much smaller and lighter than itself. The problem is characterised by two very different time scales, one associated with the slow relaxation of the initial velocity of the brownian particle and another linked to the frequent collisions that the brownian particle suffers with particles of the bath. Langevin assumed that the force acting on the brownian particle consists of two parts: a systematic, frictional force proportional to the velocity $\mathbf{u}(t)$, but acting in the opposite sense, and a randomly fluctuating force, $\mathbf{R}(t)$, which arises from collisions with surrounding particles. The equation of motion of a brownian particle of mass m is therefore written as

$$m\dot{\mathbf{u}}(t) = -m\xi\mathbf{u}(t) + \mathbf{R}(t) \qquad (7.3.1)$$

where ξ is the *friction coefficient*. The random force is assumed to vanish in the mean:

$$\langle \mathbf{R}(t) \rangle = 0 \qquad (7.3.2)$$

to be uncorrelated with the velocity at any earlier time:

$$\langle \mathbf{R}(t) \cdot \mathbf{u} \rangle = 0, \quad t > 0 \qquad (7.3.3)$$

and to have an infinitesimally short correlation time, i.e.

$$\langle \mathbf{R}(t+s) \cdot \mathbf{R}(s) \rangle = 2\pi R_0 \delta(t) \qquad (7.3.4)$$

which in turn means that the power spectrum of the random force is a constant, R_0 (a 'white' spectrum):

$$\frac{1}{2\pi} \int_{-\infty}^{\infty} \langle \mathbf{R}(t) \cdot \mathbf{R} \rangle \exp{(i\omega t)}\mathrm{d}t = R_0 \qquad (7.3.5)$$

These are reasonable assumptions when the brownian particle is much larger than its neighbours, since even on a short time scale its motion will be determined by a very large number of essentially uncorrelated collisions. When all particles are of the same size, the assumptions are less well justified, and a generalisation of a type to be described later is required.

The two terms on the right-hand side of the Langevin equation (7.3.1) are not independent. To see the connection between them we first write the solution to (7.3.1) in the form

$$m\mathbf{u}(t) = m\mathbf{u}(0) \exp{(-\xi t)} + \exp{(-\xi t)} \int_0^t \exp{(\xi s)}\mathbf{R}(s)\mathrm{d}s \qquad (7.3.6)$$

On squaring and taking the statistical average we find, given (7.3.3) and (7.3.4), that

$$m^2 \left\langle |\mathbf{u}(t)|^2 \right\rangle = m^2 \left\langle |\mathbf{u}(0)|^2 \right\rangle \exp(-2\xi t)$$

$$+ \exp(-2\xi t) \int_0^t \mathrm{d}s \int_0^t \mathrm{d}s' \exp[\xi(s+s')] 2\pi R_0 \delta(s-s')$$

$$= m^2 \left\langle |\mathbf{u}(0)|^2 \right\rangle \exp(-2\xi t) + \frac{\pi R_0}{\xi} [1 - \exp(-2\xi t)] \qquad (7.3.7)$$

We now take the limit $t \to \infty$; the brownian particle will then be in thermal equilibrium with the bath regardless of the initial conditions. Hence $\left\langle |\mathbf{u}(\infty)|^2 \right\rangle = 3k_B T/m$ and (7.3.7) can be rearranged to give an expression for the friction coefficient:

$$\xi = \frac{\pi \beta R_0}{3m} = \frac{\beta}{3m} \int_0^\infty \langle \mathbf{R}(t) \cdot \mathbf{R} \rangle \, \mathrm{d}t \qquad (7.3.8)$$

From a physical point of view it is not surprising to find a link between the frictional and random forces. If the brownian particle were to be drawn through the bath by an external field, random collisions suffered by the particle would give rise to a systematic retarding force proportional to the particle velocity. Equation (7.3.8) is an illustration of the fluctuation-dissipation theorem already discussed in Section 3.5 and which we shall establish more generally in Section 7.6.

The friction coefficient is also related to the diffusion coefficient. Consider the case when the brownian particle is initially ($t = 0$) situated at the origin ($\mathbf{r} = 0$). We wish to calculate the mean-square displacement of the particle after a time t. By multiplying through (7.3.1) by $\mathbf{r}(t)$ and using the results

$$\mathbf{r} \cdot \mathbf{u} = \mathbf{r} \cdot \dot{\mathbf{r}} = \frac{1}{2} \frac{\mathrm{d}}{\mathrm{d}t} r^2 \qquad (7.3.9)$$

and

$$\mathbf{r} \cdot \dot{\mathbf{u}} = \mathbf{r} \cdot \ddot{\mathbf{r}} = \frac{1}{2} \frac{\mathrm{d}^2}{\mathrm{d}t^2} r^2 - u^2 \qquad (7.3.10)$$

we find that

$$\frac{1}{2} m \frac{\mathrm{d}^2}{\mathrm{d}t^2} |\mathbf{r}(t)|^2 + \frac{1}{2} \xi m \frac{\mathrm{d}}{\mathrm{d}t} |\mathbf{r}(t)|^2 = m |\mathbf{u}(t)|^2 + \mathbf{r}(t) \cdot \mathbf{R}(t) \qquad (7.3.11)$$

In the statistical mean (7.3.11) becomes

$$\frac{\mathrm{d}^2}{\mathrm{d}t^2} \left\langle |\mathbf{r}(t)|^2 \right\rangle + \xi \frac{\mathrm{d}}{\mathrm{d}t} \left\langle |\mathbf{r}(t)|^2 \right\rangle = \frac{6k_B T}{m} \qquad (7.3.12)$$

The solution to (7.3.12) that satisfies the boundary conditions $\left\langle |\mathbf{r}(0)|^2 \right\rangle = 0$ and

$$\frac{\mathrm{d}}{\mathrm{d}t} \left\langle |\mathbf{r}(t)|^2 \right\rangle \Big|_{t=0} = 2 \langle \mathbf{r}(0) \cdot \mathbf{u}(0) \rangle = 0 \qquad (7.3.13)$$

is

$$\langle |\mathbf{r}(t)|^2 \rangle = \frac{6k_B T}{\xi m} \left(t - \frac{1}{\xi} + \frac{1}{\xi} \exp(-\xi t) \right) \qquad (7.3.14)$$

At very short times, such that $\xi t \ll 1$, the solution becomes

$$\langle |\mathbf{r}(t)|^2 \rangle \approx \left(\frac{3k_B T}{m} \right) t^2 = \langle u^2 \rangle t^2 \qquad (7.3.15)$$

which corresponds to free-particle (or 'ballistic') motion. At very large times ($\xi t \gg 1$), (7.3.14) reduces to

$$\langle |\mathbf{r}(t)|^2 \rangle \approx \left(\frac{6k_B T}{\xi m} \right) t \qquad (7.3.16)$$

and comparison with (7.2.3) leads to Einstein's expression for the diffusion coefficient:

$$D = \frac{k_B T}{\xi m} \qquad (7.3.17)$$

An estimate of ξ can be obtained from a hydrodynamic calculation of the frictional force on a sphere of diameter d moving with constant velocity $\mathbf{u}$ in a fluid of shear viscosity η. This leads to a famous result known as Stokes's law, the precise form of which depends on the assumptions made about the behaviour at the surface of the sphere of the velocity field created by the fluid. If the 'stick' boundary condition is used the fluid velocity at the surface is everywhere taken equal to $\mathbf{u}$; in the 'slip' approximation the normal component of the fluid velocity is set equal to the normal component of $\mathbf{u}$, thereby ensuring that no fluid can enter or leave the sphere, and the tangential force acting on the sphere is assumed to vanish. The stress tensor at the surface is then obtained by solving the linearised Navier–Stokes equation (see Section 8.3) subject to one of these boundary conditions, supplemented by the requirement that the fluid velocity field must vanish at infinite distance from the sphere. When the stress tensor is known the total frictional force $\mathbf{F}$ can be calculated by integration over the surface. The final result has the form $\mathbf{F} = -\xi \mathbf{u}$, with

$$\xi = \frac{3\pi \eta d}{m} \text{ (stick)}, \quad \xi = \frac{2\pi \eta d}{m} \text{ (slip)} \qquad (7.3.18)$$

Combination of (7.3.17) with (7.3.18) leads to two different forms of an expression for the product $D\eta$, known as the Stokes–Einstein relation:

$$D\eta = \frac{k_B T}{3\pi d} \text{ (stick)}, \quad D\eta = \frac{k_B T}{2\pi d} \text{ (slip)} \qquad (7.3.19)$$

It is a remarkable feature of the Stokes–Einstein relation that although it is derived from macroscopic considerations and is apparently limited to brownian particles, it also provides a good, empirical correlation of experimental data on simple liquids, use of the slip boundary condition generally leading to more reasonable values of the effective diameter d, at least for simple liquids.

The form of the velocity autocorrelation function of the brownian particle is easily deduced. If we multiply through (7.3.1) by $\mathbf{u}(0)$ and take the thermal average we find that

$$Z(t) = \frac{1}{3} \langle \mathbf{u}(t) \cdot \mathbf{u}(0) \rangle = \left(\frac{k_B T}{m} \right) \exp(-\xi t) \tag{7.3.20}$$

where $t \geq 0$. The expression for the diffusion coefficient given by (7.3.17) is then recovered by inserting (7.3.20) in (7.2.7). Note that the autocorrelation function is of the same, exponential form as the Enskog result for the hard-sphere fluid. This is to be expected, since a markovian hypothesis underlies both calculations. In practice, as is evident from Figure 7.1, the velocity autocorrelation function of a simple liquid may be very far from exponential. Moreover, the power spectrum of an exponential correlation function has an infinite second moment, which for continuous potentials is not consistent with the result shown in (7.2.8). The inconsistency arises because the applicability of (7.3.20) does not extend to very short times. In a time interval t such that $\xi t \ll 1$ the brownian particle experiences very few collisions and the basic assumptions of the Langevin theory are no longer valid.

When the dimensions of the diffusing particle are similar to those of its neighbours the weakest part of the theory is the markovian approximation whereby the frictional force on the particle at a given time is assumed to be proportional only to its velocity at the same time. The implication of this assumption is that the motion of the particle adjusts itself instantaneously to changes in the surrounding medium. It would obviously be more realistic to suppose that the frictional force acting on a particle reflects the previous history of the system. In other words, we should associate a certain memory with the motion of the particle. This can be achieved by introducing a friction coefficient $\xi(t-s)$ that is non-local in time and determines the contribution to the systematic force at time t coming from the velocity at earlier times s. Mathematically this amounts to writing the frictional force as a convolution in time, giving rise to a non-markovian generalisation of the Langevin equation, which we write as

$$m\dot{\mathbf{u}}(t) = -m \int_0^t \xi(t-s)\mathbf{u}(s)\mathrm{d}s + \mathbf{R}(t) \tag{7.3.21}$$

The properties of $\mathbf{R}(t)$ expressed by (7.3.2) and (7.3.3) are assumed to be unaltered. If, therefore, we multiply through (7.3.21) by $\mathbf{u}(0)$ and take the thermal average, we arrive at an equation for the velocity autocorrelation function in the form

$$\dot{Z}(t) = -\int_0^t \xi(t-s)Z(s)\mathrm{d}s \tag{7.3.22}$$

The quantity $\xi(t)$ is called the *memory function* for the autocorrelation function $Z(t)$. An equation analogous to (7.3.22) can be written down for the autocorrelation function of an arbitrary dynamical variable, A say. Such an expression may be regarded as a generalised Langevin equation in which the random 'force' is proportional to that part of $A(t)$ which is uncorrelated with $A(0)$ (cf. (7.3.3)). All that is lost in extending the use of the generalised Langevin equation to other dynamical variables is a feeling for the physical meaning of the 'friction' coefficient and random 'force'.

If we take the Laplace transform of (7.3.22), we obtain a simple, algebraic relation between $\tilde{Z}(z)$ and $\tilde{\xi}(z)$:

$$\tilde{Z}(z) = \frac{k_B T / m}{-iz + \tilde{\xi}(z)} \tag{7.3.23}$$

On replacing the frequency-dependent friction coefficient in (7.3.23) by a constant, ξ, and inverting the transform, we recover the exponential form of $Z(t)$ given by (7.3.20); this amounts to choosing a purely local (markovian) memory function, $\xi(t) = \xi \delta(t)$, which leads back to the original Langevin equation (7.3.1). Similarly, the Enskog approximation (7.2.20) corresponds to taking $\xi(t) = (3/2\Gamma_E)\delta(t)$. Equation (7.3.22) is exact, but it serves only as a definition of the unknown function $\xi(t)$. What is lacking at this stage is any statistical mechanical definition of either $\mathbf{R}(t)$ or $\xi(t)$, nor is it obvious that $\xi(t)$ is a simpler object to understand than $Z(t)$ itself. The interpretation of the generalised Langevin equation and the memory function equation in terms of statistical mechanics is described in detail in Chapter 9. Here it is sufficient to say that $\xi(t)$ is expected to decay much faster than $Z(t)$. If this is so, it suggests that a phenomenological model of a complicated dynamical process can be devised by postulating a rather simple form for the appropriate memory function that satisfies, in particular, the low-order sum rules on the autocorrelation function. For example, to describe the diffusion process, we could suppose that the memory function decays exponentially[12] with a characteristic time τ:

$$\xi(t) = \xi(0) \exp(-|t|/\tau) \tag{7.3.24}$$

If we now differentiate (7.3.22) with respect to time, set $t = 0$ and use (7.2.9), we find that

$$\xi(0) = -\frac{\ddot{Z}(0)}{Z(0)} = \Omega_0^2 \tag{7.3.25}$$

Then, by taking the Laplace transform of (7.3.24) and substituting the result in (7.3.23), we obtain the expression

$$\tilde{Z}(z) = \frac{k_B T / m}{-iz + \dfrac{\Omega_0^2}{-iz + \tau^{-1}}} \tag{7.3.26}$$

It follows from (7.2.7) that the diffusion coefficient is

$$D = \tilde{Z}(0) = \frac{k_B T}{m \Omega_0^2 \tau} \tag{7.3.27}$$

and inverse Laplace transformation of (7.3.26) shows that the velocity autocorrelation function is given by

$$Z(t) = \left(\frac{k_B T/m}{\alpha_+ - \alpha_-}\right)[\alpha_+ \exp(-\alpha_-|t|) - \alpha_- \exp(-\alpha_+|t|)] \qquad (7.3.28)$$

where α_+, α_- are the two poles of $\tilde{Z}(z = i\alpha)$:

$$\alpha_\pm = \frac{1}{2\tau}\left[1 \mp \left(1 - 4\Omega_0^2\tau^2\right)^{1/2}\right] \qquad (7.3.29)$$

If $\tau < 1/2\Omega_0$, the poles are real and positive and $Z(t)$ decays monotonically with the correct curvature $\left(\Omega_0^2\right)$ at the origin. On the other hand, if $\tau > 1/2\Omega_0$, which from (7.3.27) is equivalent to the condition

$$\frac{mD\Omega_0}{k_B T} < 2 \qquad (7.3.30)$$

then the poles are a complex conjugate pair and the velocity autocorrelation function behaves as

$$Z(t) = \left(\frac{k_B T}{m}\right)\exp(-|t|/2\tau)[\cos\Omega_1|t| + (1/2\Omega_1\tau)\sin\Omega_1|t|] \qquad (7.3.31)$$

where $\Omega_1^2 = \Omega_0^2 - 1/4\tau^2$. The function defined by (7.3.31) has the form of a damped oscillation that becomes negative at intermediate times, in qualitative agreement with simulation results on simple liquids at low temperatures and high densities (see Figure 7.1), where the condition (7.3.30) is indeed well satisfied. The argument that leads to (7.3.31) is nonetheless inadequate in certain respects. First, it provides no prescription for the relaxation time τ, though the value of τ can be derived from (7.3.27) if D is known. Secondly, Fourier transformation of (7.3.28) yields a frequency spectrum for which all even moments beyond the second are infinite. Both defects can be overcome by postulating a gaussian rather than an exponential memory function and forcing agreement with the fourth frequency moment of $Z(\omega)$, which in turn requires a knowledge of the equilibrium triplet distribution function. However, none of the phenomenological memory function approximations that use as their basic ingredients only the short-time behaviour of the correlation function are capable of reproducing the observed slow $(\sim t^{-3/2})$ decay at very long times $(\Omega_0 t \gg 1)$.

7.4 CORRELATIONS IN SPACE AND TIME

A detailed description of the time evolution of spatial correlations in liquids requires the introduction of time-dependent generalisations of the static distribution functions defined in Sections 2.5 and 2.6. The relevant dynamical variable is the microscopic particle density (3.1.2), where account must now be taken of the time-dependence of the particle coordinates $\mathbf{r}_i$. More generally, we define a microscopic dynamical variable as

$$A(\mathbf{r}, t) = \sum_{i=1}^{N} a_i(t)\delta[\mathbf{r} - \mathbf{r}_i(t)] \tag{7.4.1}$$

where a_i is some physical quantity such as the mass, velocity or energy of particle i. The spatial Fourier components of $A(\mathbf{r}, t)$ are

$$A_{\mathbf{k}}(t) = \int A(\mathbf{r}, t)\exp(-i\mathbf{k} \cdot \mathbf{r})d\mathbf{r} = \sum_{i=1}^{N} a_i(t)\exp[-i\mathbf{k} \cdot \mathbf{r}_i(t)] \tag{7.4.2}$$

A microscopic dynamical variable is said to be *conserved* if it satisfies a continuity equation of the form

$$\frac{\partial A(\mathbf{r}, t)}{\partial t} + \nabla \cdot \mathbf{j}^A(\mathbf{r}, t) = 0 \tag{7.4.3}$$

where $\mathbf{j}^A$ is the *current* associated with the variable A. Equation (7.4.3) is a local expression of the fact that the quantity $\int A(\mathbf{r}, t)d\mathbf{r} = \sum_i a_i(t)$ is independent of time; the corresponding equation for the Fourier components of A is

$$\frac{\partial A_{\mathbf{k}}(t)}{\partial t} + i\mathbf{k} \cdot \mathbf{j}_{\mathbf{k}}^A(t) = 0 \tag{7.4.4}$$

which shows that spontaneous fluctuations in a conserved variable decay very slowly at long wavelengths.

The time-dependent, microscopic particle density

$$\rho(\mathbf{r}, t) = \sum_{i=1}^{N} \delta[\mathbf{r} - \mathbf{r}_i(t)] \tag{7.4.5}$$

corresponds to the case when $a_i = 1$ and is a particularly important example of a conserved local variable. The associated particle current is

$$\mathbf{j}(\mathbf{r}, t) = \sum_{i=1}^{N} \mathbf{u}_i(t)\delta[\mathbf{r} - \mathbf{r}_i(t)] \tag{7.4.6}$$

with Fourier components

$$\mathbf{j}_{\mathbf{k}}(t) = \sum_{i=1}^{N} \mathbf{u}_i(t)\exp[-i\mathbf{k} \cdot \mathbf{r}_i(t)] \tag{7.4.7}$$

where $\mathbf{u}_i$ is the velocity of particle i. Each Fourier component may be separated into longitudinal (l) and transverse (t) parts, the two parts being parallel and perpendicular, respectively, to the wavevector $\mathbf{k}$. The longitudinal component, j_{kl}, is related to the microscopic density via the continuity equation (7.4.4).

The time correlation function of two space-dependent dynamical variables is defined as in (7.1.2) or (7.1.3) but is now, in general, non-local in space:

$$C_{AB}(\mathbf{r}', \mathbf{r}''; t', t'') = \langle A(\mathbf{r}', t') B^*(\mathbf{r}'', t'') \rangle \qquad (7.4.8)$$

while the correlation functions of the Fourier components are defined as

$$C_{AB}(\mathbf{k}', \mathbf{k}''; t', t'') = \langle A_{\mathbf{k}'}(t') B^*_{\mathbf{k}''}(t'') \rangle = \langle A_{\mathbf{k}'}(t') B_{-\mathbf{k}''}(t'') \rangle \qquad (7.4.9)$$

These correlation functions have all the properties given in Section 7.1, in particular those associated with stationarity. In addition, for homogeneous liquids, translational invariance in space means that the correlation function (7.4.8) depends only on the relative coordinates $\mathbf{r} = \mathbf{r}' - \mathbf{r}''$. Thus

$$C_{AB}(\mathbf{r}', \mathbf{r}''; t', t'') = C_{AB}(\mathbf{r}' - \mathbf{r}'', t' - t'') \qquad (7.4.10)$$

Translational invariance also implies that correlations between Fourier components $A_{\mathbf{k}'}(t')$ and $B_{\mathbf{k}''}(t'')$ are non-zero only if $\mathbf{k}' = \mathbf{k}''$, i.e.

$$C_{AB}(\mathbf{k}', \mathbf{k}''; t) = \langle A_{\mathbf{k}'}(t) B_{-\mathbf{k}''} \rangle \delta_{\mathbf{k}'\mathbf{k}''} \qquad (7.4.11)$$

Clearly $C_{AB}(\mathbf{k}, t)$ is the spatial Fourier transform of $C_{AB}(\mathbf{r}, t)$:

$$C_{AB}(\mathbf{k}, t) = \int C_{AB}(\mathbf{r}, t) \exp(-i\mathbf{k} \cdot \mathbf{r}) d\mathbf{r} \qquad (7.4.12)$$

If the fluid is also isotropic, the correlation functions (7.4.10) and (7.4.11) share with their static counterparts the property that they are functions, respectively, of the scalar quantities r and k. The frequency moments of the power spectrum of an autocorrelation function $C_{AA}(k, t)$ are again given by (7.1.24) but are now wavenumber-dependent. The continuity equation for conserved variables leads to simple expressions for the second frequency moments, called f-sum rules. From (7.1.24) and (7.4.4) it follows that

$$\langle \omega^2 \rangle_{AA} = \langle \dot{A}_{\mathbf{k}} \dot{A}_{-\mathbf{k}} \rangle = k^2 \left\langle \left| j^A_{kl} \right|^2 \right\rangle \qquad (7.4.13)$$

The memory function, M_{AA} say, of a space-dependent autocorrelation function C_{AA} must allow for non-local effects in space as well as in time. The memory function equation satisfied by C_{AA} is therefore written as

$$\dot{C}_{AA}(\mathbf{r}, t) + \int_0^t dt' \int d\mathbf{r}' M_{AA}(\mathbf{r} - \mathbf{r}', t - t') C_{AA}(\mathbf{r}', t') = 0 \qquad (7.4.14)$$

or, by exploiting the convolution theorem:

$$\dot{C}_{AA}(\mathbf{k}, t) + \int_0^t dt' M_{AA}(\mathbf{k}, t - t') C_{AA}(\mathbf{k}, t') = 0 \qquad (7.4.15)$$

We now focus specifically on the way in which time-dependent correlations in the microscopic density and particle current are described. A convenient starting point is provided by the space and time-dependent distribution function introduced by van Hove.[11] The van Hove function for a uniform fluid is defined as

$$G(\mathbf{r}, t) = \frac{1}{N} \left\langle \sum_{i=1}^{N} \sum_{j=1}^{N} \int \delta[\mathbf{r} - \mathbf{r}_j(t) + \mathbf{r}_i(0)] \right\rangle \tag{7.4.16}$$

which can be rewritten successively as

$$G(\mathbf{r}, t) = \frac{1}{N} \left\langle \int \sum_{i=1}^{N} \sum_{j=1}^{N} \delta[\mathbf{r}' + \mathbf{r} - \mathbf{r}_j(t)] \delta[\mathbf{r}' - \mathbf{r}_i(0)] d\mathbf{r}' \right\rangle$$

$$= \frac{1}{N} \left\langle \int \rho(\mathbf{r}' + \mathbf{r}, t) \rho(\mathbf{r}', 0) d\mathbf{r}' \right\rangle = \frac{1}{\rho} \langle \rho(\mathbf{r}, t) \rho(\mathbf{0}, 0) \rangle \tag{7.4.17}$$

The van Hove function therefore has the meaning of a density–density time correlation function which for $t = 0$ is closely related to the static correlation function (3.1.6). It separates naturally into two terms, usually called the 'self' (s) and 'distinct' (d) parts, i.e.

$$G(\mathbf{r}, t) = G_s(\mathbf{r}, t) + G_d(\mathbf{r}, t) \tag{7.4.18}$$

where

$$G_s(\mathbf{r}, t) = \frac{1}{N} \left\langle \sum_{i=1}^{N} \delta[\mathbf{r} - \mathbf{r}_i(t) + \mathbf{r}_i(0)] \right\rangle$$

$$G_d(\mathbf{r}, t) = \frac{1}{N} \left\langle \sum_{i=1}^{N} \sum_{j \neq i}^{N} \delta[\mathbf{r} - \mathbf{r}_j(t) + \mathbf{r}_i(0)] \right\rangle \tag{7.4.19}$$

Hence $G_s(\mathbf{r}, 0) = \delta(\mathbf{r})$ and (from (2.5.15)) $G_d(\mathbf{r}, 0) = \rho g(\mathbf{r})$. The physical interpretation of the van Hove function is that $G(\mathbf{r}, t)d\mathbf{r}$ is the number of particles j in a region $d\mathbf{r}$ around a point $\mathbf{r}$ at time t given that there was a particle i at the origin at time $t = 0$; the division into self and distinct parts corresponds to the possibilities that i and j may be the same particle or different ones. As t increases, the self part broadens into a bell-shaped curve, and the peaks initially present in the distinct part gradually disappear. In the limit $t \to \infty$, both functions become independent of r, with $G_s(\mathbf{r}, t \to \infty) \sim 1/V$ and $G_d(\mathbf{r}, t \to \infty) \sim \rho$.

Rather than considering the density–density correlation in real space it is often more convenient to focus attention on the correlation function of the Fourier components $\rho_{\mathbf{k}}$:

$$F(\mathbf{k}, t) = \frac{1}{N} \langle \rho_{\mathbf{k}}(t)\rho_{-\mathbf{k}} \rangle \tag{7.4.20}$$

The function $F(\mathbf{k}, t)$ is called the *intermediate scattering function*; as we shall see later, it is closely related to the cross-section measured in an inelastic scattering experiment. By following steps almost identical to those that establish the relation (4.1.3) between the static structure factor and the pair distribution function it can be shown that $F(\mathbf{k}, t)$ is the spatial Fourier transform of the van Hove function, i.e.

$$F(\mathbf{k}, t) = \int G(\mathbf{r}, t) \exp(-i\mathbf{k} \cdot \mathbf{r}) d\mathbf{r} \tag{7.4.21}$$

The power spectrum of the intermediate scattering function:

$$S(\mathbf{k}, \omega) = \frac{1}{2\pi} \int_{-\infty}^{\infty} F(\mathbf{k}, t) \exp(i\omega t) dt \tag{7.4.22}$$

is called the *dynamic structure factor*. Combination of (4.1.1) and (7.1.24) shows that the static and dynamic structure factors are related by

$$\int_{-\infty}^{\infty} S(\mathbf{k}, \omega) d\omega = F(\mathbf{k}, 0) = S(\mathbf{k}) \tag{7.4.23}$$

The physical significance of this sum rule will become clear in the next section. Finally, we define the autocorrelation function of the Fourier components (7.4.7) of the current associated with the microscopic density. Because $\mathbf{j}_{\mathbf{k}}$ is a vector, the corresponding correlation function is a second-rank tensor, but rotational invariance implies that the longitudinal and transverse projections of the particle current are uncorrelated if the fluid is isotropic. When that is so, the correlation function tensor has only two independent components and may therefore be written in the form

$$C_{\alpha\beta}(\mathbf{k}, t) = \left\langle \frac{k^2}{N} j_{\mathbf{k}}^{\alpha}(t) j_{-\mathbf{k}}^{\beta} \right\rangle$$
$$= \hat{k}_{\alpha}\hat{k}_{\beta}C_l(k, t) + (\delta_{\alpha\beta} - \hat{k}_{\alpha}\hat{k}_{\beta})C_t(k, t) \tag{7.4.24}$$

where $\alpha, \beta = x, y$ or z and $\hat{k}_{\alpha}, \hat{k}_{\beta}$ are cartesian components of the unit vector $\hat{\mathbf{k}} = \mathbf{k}/k$. If the z-axis is chosen parallel to $\mathbf{k}$, the longitudinal and transverse current autocorrelation functions are given by

$$C_l(k, t) = \left\langle \frac{k^2}{N} j_{\mathbf{k}}^{z}(t) j_{-\mathbf{k}}^{z} \right\rangle$$
$$C_t(k, t) = \left\langle \frac{k^2}{N} j_{\mathbf{k}}^{x}(t) j_{-\mathbf{k}}^{x} \right\rangle \tag{7.4.25}$$

The continuity equation (7.4.4) (with $A = \rho$) and the general property (7.1.8) show that the density and longitudinal current correlation functions are not independent, since

$$C_l(k, t) = \left\langle \frac{1}{N} \dot{\rho}_{\mathbf{k}}(t) \dot{\rho}_{\mathbf{k}} \right\rangle = -\frac{d^2}{dt^2} F(k, t) \tag{7.4.26}$$

Written in terms of Laplace transforms, (7.4.26) becomes

$$\tilde{C}_l(k, z) = z^2 \tilde{F}(k, z) - i z S(k) \tag{7.4.27}$$

or, on taking the real part and making use of (7.1.19):

$$C_l(k, \omega) = \omega^2 S(k, \omega) \tag{7.4.28}$$

The function $C_l(k, \omega)$ describes the spectrum of longitudinal current fluctuations in the liquid. Fluctuations in density are therefore intimately related to fluctuations in longitudinal current, but are independent of the transverse current.

In classical statistical mechanics positions and velocities at a given instant are uncorrelated. Thus the definitions of the current autocorrelation functions show that their zero-time values are the same and given by

$$C_{l,t}(k, 0) = k^2 \left(\frac{k_B T}{m} \right) = \omega_0^2, \quad \text{say} \tag{7.4.29}$$

From (7.4.26) and the sum rule (7.4.13) it follows that the second frequency moment of the dynamic structure factor is given by

$$\left\langle \omega^2 \right\rangle_{\rho\rho} = \int_{-\infty}^{\infty} \omega^2 S(k, \omega) d\omega = -\ddot{F}(k, 0) = \omega_0^2 \tag{7.4.30}$$

Since the sum rule is a consequence of the continuity equation, the second moment is purely kinetic in origin, but higher-order moments depend on the interparticle potential. If the potential is continuous, the general results contained in (7.1.23) and (7.1.24) imply that the odd frequency moments of $S(k, \omega)$ are all zero and the fourth moment is equal, by virtue of the relation (7.4.28), to the second moment of $C_l(k, \omega)$. We may therefore base a calculation of the fourth moment on the short-time expansion of $C_l(k, t)$, which we write as

$$C_l(k, t) = \omega_0^2 \left(1 - \omega_{1l}^2 \frac{t^2}{2!} + \cdots \right) \tag{7.4.31}$$

Then it follows from (7.1.8) that

$$\omega_0^2 \omega_{1l}^2 = -\frac{d^2}{dt^2} C_l(k, t) \bigg|_{t=0} = \frac{d^4}{dt^4} F(k, t) \bigg|_{t=0} = \left\langle \frac{1}{N} \ddot{\rho}_{\mathbf{k}} \ddot{\rho}_{-\mathbf{k}} \right\rangle \tag{7.4.32}$$

If we again take the z-axis along the direction of $\mathbf{k}$ and set $\dot{u}_{iz} = -(1/m)(\partial V_N/\partial z_i)$, (7.4.32) becomes

$$\omega_0^2\omega_{1l}^2 = k^4 \langle u_{iz}^4 \rangle + k^2 \left(\frac{k_B T}{m}\right) \left\langle \sum_{i=1}^{N}\sum_{j=1}^{N} \frac{\partial V_N}{\partial z_i}\frac{\partial V_N}{\partial z_j} \exp[ik(z_i - z_j)] \right\rangle$$

(7.4.33)

For a maxwellian distribution of velocities, $\langle u_{iz}^4 \rangle = 3\langle u_{iz}^2 \rangle^2$, and the statistical average in (7.4.33) can be simplified with the help of Yvon's theorem (7.2.11) to give

$$\left\langle \sum_{i=1}^{N}\sum_{j=1}^{N} \frac{\partial V_N}{\partial z_i}\frac{\partial V_N}{\partial z_j} \exp[ik(z_i - z_j)] \right\rangle$$
$$= k_B T \left\langle N\frac{\partial^2 V_N}{\partial z_1^2} + N(N-1)\frac{\partial^2 V_N}{\partial z_1 \partial z_2} \exp[ik(z_1 - z_2)] \right\rangle \quad (7.4.34)$$

where 1 and 2 are the labels of two, arbitrarily chosen particles. Hence, if V_N is a sum of pair terms:

$$\omega_{1l}^2 = 3\omega_0^2 + \frac{\rho}{m}\int (1 - \cos kz)\frac{\partial^2 v(r)}{\partial z^2}g(r)\mathrm{d}\mathbf{r} \quad (7.4.35)$$

where $v(r)$ is the pair potential. At large k, the kinetic contribution dominates, corresponding to free-particle behaviour. From (7.4.28) we see that ω_{1l}^2 is related to the second and fourth frequency moments of $S(k,\omega)$ by $\omega_{1l}^2 = \langle \omega^4 \rangle_{\rho\rho} / \langle \omega^2 \rangle_{\rho\rho}$.

A similar calculation can be made for the transverse current. The short-time expansion of the correlation function is now

$$C_t(k, t) = \omega_0^2 \left(1 - \omega_{1t}^2 \frac{t^2}{2!} + \cdots\right) \quad (7.4.36)$$

with

$$\omega_0^2\omega_{1t}^2 = -\frac{\mathrm{d}^2}{\mathrm{d}t^2}C_t(k, t)\bigg|_{t=0} \quad (7.4.37)$$

By pursuing the methods already used in the longitudinal case we find that the analogue of (7.4.35) is

$$\omega_{1t}^2 = \omega_0^2 + \frac{\rho}{m}\int (1 - \cos kz)\frac{\partial^2 v(r)}{\partial x^2}g(r)\mathrm{d}\mathbf{r} \quad (7.4.38)$$

Higher-order moments of $C_l(k, \omega)$ and $C_t(k, \omega)$ involve correlations between increasingly large numbers of particles and rapidly become very tedious to evaluate.

7.5 INELASTIC SCATTERING OF NEUTRONS AND X-RAYS

We now show how the Fourier transforms of the van Hove functions $G(\mathbf{r}, t)$ and $G_s(\mathbf{r}, t)$ are related to measurements of the inelastic scattering of slow (or 'thermal') neutrons. To do so we require a generalisation of the calculation of Section 4.1 that allows for the exchange of energy between the neutrons and the target.[13] Neutrons are particularly useful as probes of the microscopic dynamics of liquids because their momentum $\hbar\mathbf{k}$ and energy $E = \hbar\omega$ are related by $E = \hbar^2 k^2 / 2m$, where m is the neutron mass. It follows that when E is of order $k_B T$, and therefore comparable with the thermal energies of particles in the liquid, the wavelength $\lambda = 2\pi/k$ associated with the neutron is of the same order of magnitude as the distance between neighbouring particles.

In a typical scattering event a neutron of momentum $\hbar\mathbf{k}_1$ and energy $\hbar\omega_1$ is scattered into a solid angle $d\Omega$. Let the momentum and energy of the neutron after the event be $\hbar\mathbf{k}_2$ and $\hbar\omega_2$ and let the momentum and energy transfer from neutron to sample be $\hbar\mathbf{k}$ and $\hbar\omega$. The dynamical conservation laws require that

$$\hbar\omega = E_2 - E_1 \equiv \hbar\omega_{12} \qquad (7.5.1)$$

where E_1 and E_2 are the initial and final energies of the sample, and

$$\hbar\mathbf{k} = \hbar\mathbf{k}_1 - \hbar\mathbf{k}_2 \qquad (7.5.2)$$

Note that the frequency shift ω is positive when the neutron loses energy to the sample.

The probability per unit time, W_{12}, for the transition $|1, \mathbf{k}_1\rangle \rightarrow |2, \mathbf{k}_2\rangle$, where $|1\rangle$ and $|2\rangle$ denote the initial and final states of the sample, is given by Fermi's 'golden rule':

$$W_{12} = \frac{2\pi}{\hbar} |\langle 1, \mathbf{k}_1 | \mathcal{V} | 2, \mathbf{k}_2 \rangle|^2 \delta(\hbar\omega - \hbar\omega_{12}) \qquad (7.5.3)$$

where $\mathcal{V}$ represents the perturbation, i.e. the interaction between the neutron and the atomic nuclei. For the sake of simplicity we have ignored the spin state of the neutron. The partial differential cross-section for scattering into the solid angle $d\Omega$ in a range of energy transfer $\hbar d\omega$ is calculated by averaging W_{12} over all initial states $|1\rangle$ with their statistical weights $P_1 \propto \exp(-\beta E_1)$, summing over all final states $|2\rangle$ allowed by energy conservation, multiplying by the density of final states of the neutron, namely

$$\frac{d\mathbf{k}_2}{8\pi^3} = \frac{k_2^2 \, dk_2 \, d\Omega}{8\pi^3} = \frac{m}{8\pi^3\hbar} k_2 \, d\omega \, d\Omega \qquad (7.5.4)$$

and dividing by the flux $\hbar k_1/m$ of incident neutrons, with the final result having the form

$$\frac{d^2\sigma}{d\Omega \, d\omega} = \frac{k_2}{k_1} \left(\frac{m}{2\pi\hbar^2}\right)^2 \sum_{\{1\}}\sum_{\{2\}} P_1 |\langle 1, \mathbf{k}_1 | \mathcal{V} | 2, \mathbf{k}_2 \rangle|^2 \delta(\omega - \omega_{12}) \qquad (7.5.5)$$

The differential cross-section (4.1.9) is obtained by integrating over all energy transfers:

$$\frac{d\sigma}{d\Omega} = \int \frac{d^2\sigma}{d\Omega\, d\omega}\, d\omega \tag{7.5.6}$$

The structure and dynamics of the liquid enter the calculation through the interaction of the neutron with the atomic nuclei. We assume again that $\mathcal{V}$ is given by the sum (4.1.12) of δ-function pseudopotentials between a neutron located at $\mathbf{r}$ and nuclei at positions $\mathbf{r}_i$. If the initial and final states of the neutron are taken as plane-wave states of the form (4.1.6), the matrix element in (7.5.5) may be rewritten as

$$\langle 1, \mathbf{k}_1 | \mathcal{V} | 2, \mathbf{k}_2 \rangle = \frac{2\pi\hbar^2}{m} \sum_{i=1}^{N} \langle 1|b_i \exp\left(-i\mathbf{k}\cdot\mathbf{r}_i\right)|2\rangle \tag{7.5.7}$$

where $\hbar\mathbf{k}$ is the momentum transfer already defined and b_i is the scattering length of the nucleus labelled i.

Consider first the case when all nuclei in the sample have the same scattering length. By combining (7.5.5) and (7.5.7), exploiting the definition (4.1.2) and introducing the integral representation of the δ-function, we obtain an expression for the cross-section in terms of the Fourier components of the microscopic density:

$$\begin{aligned}
\frac{d^2\sigma}{d\Omega d\omega} &= b^2 \left(\frac{k_2}{k_1}\right) \sum_{\{1\}} \sum_{\{2\}} P_1 |\langle 1|\rho_\mathbf{k}|2\rangle|^2 \delta(\omega - \omega_{12}) \\
&= b^2 \left(\frac{k_2}{k_1}\right) \sum_{\{1\}} \sum_{\{2\}} P_1 \frac{1}{2\pi} \int_{-\infty}^{\infty} |\langle 1|\rho_\mathbf{k}|2\rangle|^2 \exp[i(\omega - \omega_{12})t]dt
\end{aligned} \tag{7.5.8}$$

Equation (7.5.8) can be simplified by recognising that

$$\begin{aligned}
\exp\left(-i\omega_{12}t\right) &|\langle 1|\rho_\mathbf{k}|2\rangle|^2 \\
&= \exp\left(-iE_2t/\hbar\right)\exp\left(iE_1t/\hbar\right)\langle 1|\rho_\mathbf{k}|2\rangle\langle 2|\rho_{-\mathbf{k}}|1\rangle \\
&= \langle 1|\exp\left(iE_1t/\hbar\right)\rho_\mathbf{k}\exp\left(-iE_2t/\hbar\right)|2\rangle\langle 2|\rho_{-\mathbf{k}}|1\rangle \\
&= \langle 1|\exp\left(i\mathcal{H}t/\hbar\right)\rho_\mathbf{k}\exp\left(-i\mathcal{H}t/\hbar\right)|2\rangle\langle 2|\rho_{-\mathbf{k}}|1\rangle \\
&= \langle 1|\rho_\mathbf{k}(t)|2\rangle\langle 2|\rho_{-\mathbf{k}}|1\rangle
\end{aligned} \tag{7.5.9}$$

where $\mathcal{H}$ is the hamiltonian of the sample.

It remains only to sum over the initial states of the sample, which is equivalent to taking an ensemble average, and over the final states, which is done by exploiting the closure property, $\sum_j |j\rangle\langle j| = 1$, of a complete set of quantum states $|j\rangle$. The final result for the cross-section is

$$\frac{d^2\sigma}{d\Omega d\omega} = b^2 \left(\frac{k_2}{k_1}\right) \frac{1}{2\pi} \int_{-\infty}^{\infty} \langle \rho_{\mathbf{k}}(t) \rho_{-\mathbf{k}} \rangle \exp{(i\omega t)} dt$$

$$= Nb^2 \left(\frac{k_2}{k_1}\right) S(\mathbf{k}, \omega) \qquad (7.5.10)$$

where $S(\mathbf{k}, \omega)$ is the dynamic structure factor defined by (7.4.22). Equation (7.5.10) shows that a measurement of the experimental cross-section as a function of $\mathbf{k}$ and ω is equivalent, at least in principle, to a determination of the van Hove correlation function $G(\mathbf{r}, t)$. The connection with the elastic cross-section is made via (7.5.6); comparison of (4.1.23) with (7.5.10), taken for the case $k_1 = k_2$, shows that (7.5.6) provides the physical content of the so-called 'elastic' sum rule (7.4.23).

By analogy with (7.4.21) and (7.4.22), it is customary to define a *self dynamic structure factor* $S_s(\mathbf{k}, \omega)$ as the double Fourier transform of the self part of the van Hove function, i.e.

$$S_s(\mathbf{k}, \omega) = \frac{1}{2\pi} \int_{-\infty}^{\infty} dt \exp{(i\omega t)} \int G_s(\mathbf{r}, t) \exp{(-i\mathbf{k} \cdot \mathbf{r})} d\mathbf{r} \qquad (7.5.11)$$

together with a self intermediate scattering function $F_s(\mathbf{k}, t)$, defined through the transform

$$S_s(\mathbf{k}, \omega) = \frac{1}{2\pi} \int_{-\infty}^{\infty} F_s(\mathbf{k}, t) \exp{(i\omega t)} dt \qquad (7.5.12)$$

with $F_s(\mathbf{k}, 0) = 1$. The self functions are important for the discussion of inelastic scattering in situations where more than one scattering length is involved. As in Section 4.1, the averaging over scattering lengths can be carried out independently of the thermal average over nuclear coordinates. A generalisation of the result in (4.1.21) allows the inelastic cross-section to be written as the sum of incoherent and coherent parts in the form

$$\frac{d^2\sigma}{d\Omega d\omega} = \left(\frac{d^2\sigma}{d\Omega d\omega}\right)_{inc} + \left(\frac{d^2\sigma}{d\Omega d\omega}\right)_{coh} \qquad (7.5.13)$$

with

$$\left(\frac{d^2\sigma}{d\Omega d\omega}\right)_{inc} = Nb_{inc}^2 \left(\frac{k_2}{k_1}\right) S_s(\mathbf{k}, \omega)$$

$$\left(\frac{d^2\sigma}{d\Omega d\omega}\right)_{coh} = Nb_{coh}^2 \left(\frac{k_2}{k_1}\right) S(\mathbf{k}, \omega) \qquad (7.5.14)$$

By varying the isotopic composition of the sample, or by using polarised neutrons, it is possible to measure separately the coherent and incoherent cross-sections and thereby, again in principle, to separate the van Hove function into its self and distinct parts.

For systems with inversion symmetry, which includes all fluids, the dynamic structure factor is invariant under a change of sign of $\mathbf{k}$. In the classical limit, $S(\mathbf{k}, \omega)$ is also an even function of ω, but a measured cross-section cannot be strictly even with respect to ω; if that were the case, thermal equilibrium between radiation and sample would never be reached. The principle of detailed balance requires that the cross-sections for the scattering processes $|\mathbf{k}_1, 1\rangle \rightarrow |\mathbf{k}_2, 2\rangle$ and $|\mathbf{k}_2, 2\rangle \rightarrow |\mathbf{k}_1, 1\rangle$ be equal to the ratio of the statistical weights of the states $|1\rangle$ and $|2\rangle$ and hence that

$$S(\mathbf{k}, \omega) = \exp(\beta\hbar\omega)S(\mathbf{k}, -\omega) \qquad (7.5.15)$$

Allowance must be made for this effect when comparing a classical calculation of $S(k, \omega)$ with the results of scattering experiments.

In the limit $\mathbf{r}, t \rightarrow 0$, particles in a fluid move freely at constant velocity. These conditions correspond to the limit $k, \omega \rightarrow \infty$, where $S(k, \omega)$ behaves in the manner appropriate to an ideal gas. The limiting form of $S(k, \omega)$ is easily derived, since positions of different particles are uncorrelated in an ideal gas ($G_d = \rho$); the calculation of $S(k, \omega)$ is therefore equivalent to a calculation of $G_s(r, t)$. The probability that an ideal-gas particle will move a distance r in a time t is equal to the probability, given by the Maxwell distribution (2.1.28), that the particle has a velocity in the range $\mathbf{u}$ to $\mathbf{u} + d\mathbf{u}$, where $\mathbf{u} = \mathbf{r}/t$. Thus

$$G_s(\mathbf{r}, t) = \left(\frac{\beta m}{2\pi t^2}\right)^{3/2} \exp(-\beta m r^2/2t^2) \qquad (7.5.16)$$

where the form of the pre-exponential factor ensures that $\int G_s(\mathbf{r}, t)d\mathbf{r} = 1$ for all t. The corresponding result for $S(k, \omega)$ is

$$S(k, \omega) = \left(\frac{\beta m}{2\pi k^2}\right)^{1/2} \exp(-\beta m\omega^2/2k^2) \qquad (7.5.17)$$

Equation (7.5.17) provides a reasonable fit to data on simple liquids at wavelengths significantly shorter than the spacing between particles, typically for k greater than about 10 Å^{-1}; small deviations from the free-particle result can be allowed for by calculating the correction to $S(k, \omega)$ due to a single, binary collision. At longer wavelengths correlations between particles become increasingly important and the ideal-gas model is no longer valid. Very small values of k correspond to the hydrodynamic regime, where thermodynamic equilibrium is brought by frequent collisions between particles. This is the opposite extreme to the free-particle limit represented by (7.5.17), a point discussed in more detail in Section 8.2.

Neutron scattering experiments designed for the study of both single-particle and collective dynamical properties have been carried out for a number of monatomic liquids.[14] These experiments have been complemented by simulations[15] of the Lennard-Jones and hard-sphere fluids and of models of the liquid alkali metals. The greatest interest lies in the behaviour of the dynamic structure factor as a function of k and all existing experiments and simulations reveal broadly the same features. At reduced wavenumbers $kd \approx 1$ or smaller where d is the atomic diameter, $S(k, \omega)$ has a sharp peak at zero frequency and two more or less well defined side peaks, one on each side of the central peak. As k increases, the peaks shift to higher frequencies with a dispersion that is approximately linear. We shall see in Chapter 8 that the side peaks observed at long wavelengths correspond to propagating sound waves; they are clearly visible in the results of neutron scattering experiments on liquid caesium,[14b] some of which are plotted in Figure 7.5. At shorter wavelengths the sound waves are strongly damped and disappear when $kd \approx 2$, leaving only a central, lorentzian-like peak. The width of the central peak first increases with k, but then shows a marked decrease at wavenumbers close to the peak in the static structure factor (see curve (d) in Figure 7.5). This last effect is called 'de Gennes narrowing'; it corresponds to a dramatic slowing down in the decay of the density autocorrelation function $F(k, t)$, which in turn has its origins in the strong spatial correlations existing at these wavelengths. At still larger values of k, the spectrum broadens again, going over finally to its free-particle limit. The behaviour of $S_s(k, \omega)$ is much simpler; this has only a single, central peak, the width of which increases smoothly with k.

For a number of years neutron scattering measurements were the main source of experimental information on the collective dynamics of liquids for wavenumbers and frequencies at which propagating modes are expected to appear. More recently, with the development of increasingly powerful synchrotron radiation sources, high-resolution, inelastic X-ray scattering studies of liquids have become feasible which by-pass two limitations of the neutron scattering approach.[16] First, in contrast to neutron scattering, X-ray scattering measures only the coherent cross-section. This significantly simplifies analysis of the experimental data, though it also precludes the use of X-ray scattering in the study of single-particle motion, which remains the preserve of neutron scattering experiments. A second and more fundamental difficulty associated with neutron scattering is the fact that the momentum–energy relation for neutrons limits the range of energy transfer which is accessible for given values of momentum transfer and initial energy of the neutron, $E = \hbar^2 k_0^2 / 2m$. From (7.5.1) and (7.5.2) we find that

$$\frac{\hbar\omega}{E} = 1 - \frac{k_1^2}{k_0^2} \tag{7.5.18}$$

and

$$k^2 = k_1^2 + k_2^2 - 2k_1 k_2 \cos\theta \tag{7.5.19}$$

where θ is the scattering angle. These two results can be combined to give

$$\frac{k^2}{k_1^2} = 2 - \frac{\hbar\omega}{E} - 2\left(1 - \frac{\hbar\omega}{E}\right)^{1/2} \cos\theta \tag{7.5.20}$$

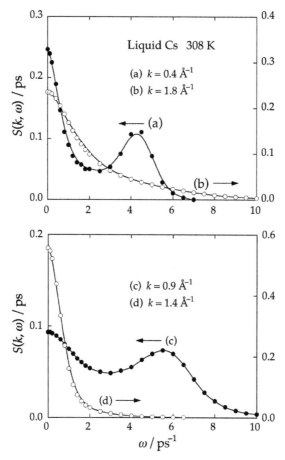

FIGURE 7.5 Results from neutron scattering experiments for the dynamic structure factor of liquid caesium. The spectra have been normalised to unit area and only the energy loss side is shown. The main peak in $S(k)$ is at $k \approx 1.4 \text{ Å}^{-1}$. *Redrawn with permission from Ref. 14b © 1992 American Physical Society.*

Since k_1^2 is proportional to E, this expression defines a line in the wavenumber–frequency plane for each choice of θ and the 'kinematically allowed region' is the area enclosed by the lines corresponding to $\theta = 0$ and $\theta = 180°$. Every point lying within that region is accessible at some value of the scattering angle but points lying outside it are not.

Part of the kinematically allowed region for neutrons having an initial energy equal to 50 meV is pictured in Figure 7.6, which shows that the lower boundary of the accessible range of energy transfer varies almost linearly with k when k is small; here the boundary is the line corresponding to $\theta = 0$. On expanding

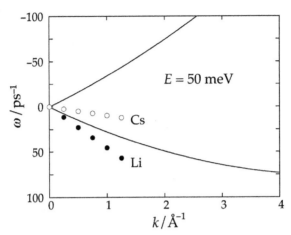

FIGURE 7.6 Part of the kinematically allowed region of wavenumber and frequency for the scattering of neutrons of initial energy $E = 50$ meV. The continuous line with a cusp at the origin forms the boundary of the allowed region; all points lying within that boundary are accessible for some value of the scattering angle. The points show the predicted values of the frequency shifts of the sound wave peaks in liquid caesium and lithium, calculated from the limiting, long-wavelength form of the dispersion relation; see text for details.

the square-root term in (7.5.20) we find that for $\theta = 0$:

$$k^2 = \frac{k_1^2 \hbar^2}{4E^2} \omega^2 + \mathcal{O}(\omega^4) \qquad (7.5.21)$$

The slope of the boundary as $k \to 0$ is therefore given by

$$\lim_{k \to 0} \frac{\partial \omega}{\partial k} = \pm \left(\frac{2E}{m}\right)^{1/2} = \pm u_0 \qquad (7.5.22)$$

where u_0 is the speed of the incident neutron. For the sound wave mode to be detectable, its dispersion curve must lie within the allowed region. This implies that the speed of sound in the liquid must be less than u_0, which for a 50 meV neutron is approximately 3100 m s^{-1}. We shall see in Chapter 8 that in the long-wavelength limit sound wave peaks appear at frequency shifts given by $\omega = \pm c_s k$, where c_s is the adiabatic speed of sound in the liquid. The same expression provides a fair guide to the dispersion of the peaks detected in the results of neutron scattering experiments, such as those pictured in Figure 7.5. We can therefore expect a propagating mode to be detectable for, say, caesium ($c_s \approx 1000$ m s^{-1}) but not for lithium ($c_s \approx 4500$ m s^{-1}), as the results shown in Figure 7.6 confirm. The kinematically allowed region can be expanded by an increase in energy of the incident neutrons, but at the cost of a loss in resolution.

The situation in the case of X-rays is very different. Equation (7.5.19) still applies, but the momentum–energy relation for a photon ($E = hck$, where c is the speed of light) means that (7.5.18) is replaced by

$$\frac{\hbar\omega}{E} = 1 - \frac{k_1}{k_0} \tag{7.5.23}$$

and the analogue of (7.5.20) is therefore

$$\frac{k}{k_1} = 1 + \left(1 - \frac{\hbar\omega}{E}\right)^2 - 2\left(1 - \frac{\hbar\omega}{E}\right)\cos\theta \tag{7.5.24}$$

Now, whereas the energy transfer remains of the order of tens of meV, the incident energy is typically three orders of magnitude larger than in the case of neutron scattering. The ratio of $\hbar\omega$ to E is consequently very small, the scattering is nearly elastic and (7.5.24) reduces to

$$k \approx 2k_1 \sin\frac{1}{2}\theta \tag{7.5.25}$$

Since k is almost independent of ω, it follows that the accessible range of energy transfer for a given momentum transfer is virtually unlimited. This allows measurements of $S(k, \omega)$ to be made over a very wide area in the wavenumber–frequency plane, including the region of special interest that lies below $k \approx 1 \, \text{Å}^{-1}$. The value of X-ray scattering in the study of collective motions in dense fluids has been demonstrated in experiments on liquid metals,[16] and a variety of more complicated systems, including hydrogen-bonded[17] and glass-forming[18] liquids.

Measurements of the dynamic structure factor can also be made by the inelastic scattering of light.[1a] As in the case of X-rays, light scattering measures only the coherent cross-section, but the wavelengths involved are much larger, of order 5000 Å. It is therefore possible to calculate the spectral distribution of scattered light from the macroscopic equations of hydrodynamics derived in Chapter 8. Light is scattered by fluctuations in the local dielectric constant, but for most liquids these are directly proportional to the fluctuations in density and the measured spectrum is proportional to $S(k, \omega)$.

7.6 LINEAR RESPONSE THEORY

We turn now to an investigation of the behaviour of a system under the perturbing influence of an external field to which the system is weakly coupled. As we shall see, the response of the system can be described entirely in terms of time correlation functions characteristic of the system at equilibrium, i.e. in the *absence* of the field; the expression already obtained for the neutron scattering cross-section in terms of the dynamic structure factor is an example of this relationship. The derivation of the general result requires only a straightforward calculation of the change produced in a dynamical variable B by an applied space and time-dependent field $\mathcal{F}$ conjugate to a variable A. Both A and B are to be regarded in general as functions of the coordinates and momenta of all particles in the system. The mean value of B in the equilibrium state is assumed to be zero.

The hamiltonian of the system in the presence of the external field is

$$\mathcal{H} = \mathcal{H}_0 + \mathcal{H}'(t) \tag{7.6.1}$$

where $\mathcal{H}_0$ characterises the unperturbed system and $\mathcal{H}'(t)$ represents the perturbation:

$$\mathcal{H}'(t) = -\int A(\mathbf{r}) \mathcal{F}(\mathbf{r}, t) \, d\mathbf{r} \tag{7.6.2}$$

The external field can always be treated as a superposition of monochromatic plane waves. Since we are interested in the linear response of the system, it is sufficient to consider a single plane wave:

$$\mathcal{F}(\mathbf{r}, t) = \frac{1}{V} \mathcal{F}_{\mathbf{k}} \exp[i(\mathbf{k} \cdot \mathbf{r} - \omega t)] \tag{7.6.3}$$

in which case (7.6.2) becomes

$$\mathcal{H}'(t) = -A_{-\mathbf{k}} \mathcal{F}_{\mathbf{k}} \exp(-i\omega t) \tag{7.6.4}$$

As a further simplification we shall temporarily suppose that the external field is spatially homogeneous and ignore the dependence on $\mathbf{k}$; the latter is trivially reintroduced at a later stage. We also assume that the system was in thermal equilibrium in the infinite past ($t \to -\infty$). Then $\mathcal{H}'(t)$ may be written as

$$\mathcal{H}'(t) = -A\mathcal{F}(t) = -A\mathcal{F}_0 \exp[-i(\omega + i\epsilon)t] \tag{7.6.5}$$

where A and B are now taken to be real. The factor $\exp(\epsilon t)$ ($\epsilon > 0$) is included to ensure that $\mathcal{F} \to 0$ as $t \to -\infty$; the limit $\epsilon \to 0$ is taken at the end of the calculation. The time evolution of the phase space probability density $f^{[N]}(t) \equiv f^{[N]}(\mathbf{r}^N, \mathbf{p}^N; t)$ in the presence of the perturbation is determined by the Liouville equation (2.1.9). Thus

$$\frac{\partial f^{[N]}(t)}{\partial t} = -i\mathcal{L} f^{[N]}(t) = \{\mathcal{H}_0 + \mathcal{H}', \, f^{[N]}(t)\}$$

$$= -i\mathcal{L}_0 f^{[N]}(t) - \{A, \, f^{[N]}(t)\} \mathcal{F}(t) \tag{7.6.6}$$

where $\mathcal{L}_0$ is the Liouville operator corresponding to the unperturbed hamiltonian. Equation (7.6.6) must be solved subject to the initial condition that $f^{[N]}(-\infty) = f_0^{[N]}$.

We are interested only in the response to a weak external field. We may therefore write the probability density as

$$f^{[N]}(t) = f_0^{[N]} + \Delta f^{[N]}(t) \tag{7.6.7}$$

and linearise (7.6.6) in the form

$$\frac{\partial \Delta f^{[N]}(t)}{\partial t} = -i \mathcal{L}_0 \Delta f^{[N]}(t) - \{A, f_0^{[N]}\} \mathcal{F}(t))$$ (7.6.8)

The solution to (7.6.8) is

$$\Delta f^{[N]}(t) = -\int_{-\infty}^{t} \exp[-i(t-s)\mathcal{L}_0] \left\{ A, f_0^{[N]} \right\} \mathcal{F}(s) ds$$ (7.6.9)

That this is the solution for all t is easily checked by differentiation, since it is obviously correct for $t = -\infty$. In the canonical ensemble $f_0^{[N]} \propto \exp(-\beta \mathcal{H}_0)$ and the Poisson bracket appearing in (7.6.9) can be re-expressed as

$$\left\{ A, f_0^{[N]} \right\} = \sum_{i=1}^{N} \left(\frac{\partial A}{\partial \mathbf{r}_i} \cdot \frac{\partial f_0^{[N]}}{\partial \mathbf{p}_i} - \frac{\partial A}{\partial \mathbf{p}_i} \cdot \frac{\partial f_0^{[N]}}{\partial \mathbf{r}_i} \right)$$

$$= -\beta \sum_{i=1}^{N} \left(\frac{\partial A}{\partial \mathbf{r}_i} \cdot \frac{\partial \mathcal{H}_0}{\partial \mathbf{p}_i} - \frac{\partial A}{\partial \mathbf{p}_i} \cdot \frac{\partial \mathcal{H}_0}{\partial \mathbf{r}_i} \right) f_0^{[N]}$$

$$= -\beta(i \mathcal{L}_0 A) f_0^{[N]} = -\beta \dot{A} f_0^{[N]}$$ (7.6.10)

The mean change in the variable $B(\mathbf{r}^N, \mathbf{p}^N)$ arising from the change in the distribution function is therefore

$$\langle \Delta B(t) \rangle = \iint B(\mathbf{r}^N, \mathbf{p}^N) \Delta f^{[N]}(t) d\mathbf{r}^N d\mathbf{p}^N$$

$$= \beta \int_{-\infty}^{t} \mathcal{F}(s) ds \iint f_0^{[N]} B \exp[-i(t-s)\mathcal{L}_0] \dot{A} \, d\mathbf{r}^N d\mathbf{p}^N$$

$$= \beta \int_{-\infty}^{t} \mathcal{F}(s) ds \iint f_0^{[N]} \dot{A} \exp[i(t-s)\mathcal{L}_0] B \, d\mathbf{r}^N d\mathbf{p}^N$$ (7.6.11)

where we have used a result contained in (7.1.27). The response of the system can therefore be written in the form

$$\langle \Delta B(t) \rangle = \int_{-\infty}^{t} \Phi_{BA}(t-s) \mathcal{F}(s) ds$$ (7.6.12)

in terms of an *after-effect function* $\Phi_{BA}(t)$, defined as

$$\Phi_{BA}(t) = \beta \langle B(t) \dot{A} \rangle = -\beta \langle \dot{B}(t) A \rangle$$ (7.6.13)

The thermal averages in (7.6.13) are taken over the unperturbed system because in the linear approximation represented by (7.6.11) the variable B evolves in time under the influence of the reference system propagator $\exp(i\mathcal{L}_0 t)$. It is sometimes convenient to use as an alternative definition of the after-effect function the expression

$$\theta_{BA}(t) = -\beta \langle \dot{B}(t)A \rangle \theta(t) \tag{7.6.14}$$

where $\theta(t)$ is the Heaviside step function. Since $\theta_{BA}(t) = 0$ for $t < 0$, the upper limit of the integral in (7.6.12) can then be extended to $+\infty$.

The physical meaning of (7.6.12) and (7.6.13) is that the response, i.e. the change in the variable B at time t, is a superposition of delayed effects and the response to a unit δ-function force applied at $t = 0$ is proportional to the after-effect function itself. The basic result of linear response theory embodied in these two equations can also be derived by calculating the changes in the phase space trajectories of the particles to first order in the applied force. That method of derivation emphasises the assumption of mechanical linearity which underlies linear response theory. Mechanical linearity cannot hold for macroscopic times, however, since it is known that the perturbed and unperturbed phase space trajectories diverge exponentially on a macroscopic time scale even when the external field is very weak. On the other hand, the corresponding deviation in the phase space distribution function is expected to behave smoothly as a function of the perturbation. Linearisation of the statistically averaged response should therefore be justified, in agreement with experimental observations. The apparent contradiction between mechanical non-linearity and statistical linearity is resolved by noting that the decay times of the relevant correlations, i.e. the times after which randomisation sets in, are generally quite short, and that use of a linear approximation for the divergence of the trajectories in phase space is valid for time intervals over which the after-effect function differs significantly from zero.

Equation (7.6.12) is easily generalised to the case in which the external field also varies in space. If the unperturbed system is spatially uniform, the response is determined by an after-effect function $\Phi_{BA}(\mathbf{r}, t)$ through the relation

$$\langle \Delta B(\mathbf{r}, t) \rangle = \int_{-\infty}^{t} \mathrm{d}s \int \Phi_{BA}(\mathbf{r} - \mathbf{r}', t - s) \mathcal{F}(\mathbf{r}', s) \mathrm{d}\mathbf{r}' \tag{7.6.15}$$

or, in terms of Fourier components, by

$$\langle \Delta B_{\mathbf{k}}(t) \rangle = \int_{-\infty}^{t} \Phi_{BA}(\mathbf{k}, t - s) \mathcal{F}_{\mathbf{k}}(s) \mathrm{d}s \tag{7.6.16}$$

where

$$\Phi_{BA}(\mathbf{k}, t) = -\frac{\beta}{V} \langle \dot{B}_{\mathbf{k}}(t) A_{-\mathbf{k}} \rangle \tag{7.6.17}$$

Equation (7.6.16) shows that in the linear regime a perturbation of given wavevector induces a response only of the same wavevector; this is a consequence of the assumed uniformity of the unperturbed system and the property (7.4.11).

We now restrict the discussion to the case of isotropic fluids. If the external field has the monochromatic form of (7.6.5), the expression for the response becomes

$$
\begin{aligned}
\langle \Delta B_{\mathbf{k}}(t) \rangle &= \int_{-\infty}^{t} \Phi_{BA}(k, t - s) \mathcal{F}_{\mathbf{k}} \exp[-i(\omega + i\epsilon)s] ds \\
&= \mathcal{F}_{\mathbf{k}} \exp[-i(\omega + i\epsilon)t] \int_{-\infty}^{t} \Phi_{BA}(k, t - s) \exp[-i(\omega + i\epsilon)(s - t)] ds \\
&= \mathcal{F}_{\mathbf{k}} \exp[-i(\omega + i\epsilon)t] \int_{0}^{\infty} \Phi_{BA}(k, t) \exp[i(\omega + i\epsilon)t] dt \quad (7.6.18)
\end{aligned}
$$

or, taking the limit $\epsilon \to 0$:

$$
\langle \Delta B_{\mathbf{k}}(t) \rangle = \chi_{BA}(k, \omega) \mathcal{F}_{\mathbf{k}} \exp(-i\omega t) \quad (7.6.19)
$$

where $\chi_{BA}(k, \omega)$ is a complex *dynamic susceptibility* or *dynamic response function*:

$$
\begin{aligned}
\chi_{BA}(k, \omega) &= \chi'_{BA}(k, \omega) + i\chi''_{BA}(k, \omega) \\
&= \lim_{\epsilon \to 0+} \int_{0}^{\infty} \Phi_{BA}(k, t) \exp[i(\omega + i\epsilon)t] dt \quad (7.6.20)
\end{aligned}
$$

If we substitute for $\Phi_{BA}(k, t)$ from (7.6.17) and integrate by parts, we find that

$$
\chi_{BA}(k, \omega) = \frac{\beta}{V} [C_{BA}(k, t = 0) + i(\omega + i\epsilon)\tilde{C}_{BA}(k, \omega + i\epsilon)] \quad (7.6.21)
$$

When A and B are the same variable it follows from (7.1.19) that

$$
C_{AA}(k, \omega) = \frac{Vk_B T}{\pi \omega} \chi''_{AA}(k, \omega) \quad (7.6.22)
$$

The zero-frequency limit of $\chi_{AA}(k, \omega)$, i.e. the static susceptibility $\chi_{AA}(k)$, is obtained from (7.6.21) as

$$
\chi_{AA}(k) \equiv \chi_{AA}(k, \omega = 0) = \frac{\beta}{V} C_{AA}(k, t = 0) \quad (7.6.23)
$$

Thus the static version of (7.6.19) for the case when A and B are the same is

$$
\langle \Delta A_{\mathbf{k}} \rangle = \frac{\beta}{V} \langle A_{\mathbf{k}} A_{-\mathbf{k}} \rangle \mathcal{F}_{\mathbf{k}} \quad (7.6.24)
$$

Equation (7.6.22) is a particular form of the fluctuation-dissipation theorem. Indeed the name is often applied specifically to this relation between the power spectrum of the autocorrelation function of a dynamical variable and the imaginary part of the corresponding response function. Use of the term 'dissipation' is connected to the fact, well known in spectroscopy, that the energy absorbed from the external field and later dissipated as heat is proportional to $\omega \chi''_{AA}(k, \omega)$.

When A is the microscopic density some minor changes are needed to the formulae we have derived. Let $\phi_k \exp(-i\omega t)$ be a Fourier component of an external potential which couples to the component ρ_{-k} of the density. The term $\mathcal{H}'(t)$ in the hamiltonian (7.6.1) now has the form

$$\mathcal{H}'(t) = \frac{1}{V}\rho_{-k}\phi_k \exp(-i\omega t) \tag{7.6.25}$$

The resulting change in density is

$$\langle \Delta\rho_k(t) \rangle = \chi_{\rho\rho}(k, \omega)\phi_k \exp(-i\omega t) \tag{7.6.26}$$

which is a generalisation to non-zero frequencies of the static result (3.6.9). The after-effect function is

$$\Phi_{\rho\rho}(k, t) = \frac{\beta}{V} \langle \dot{\rho}_k(t)\rho_{-k} \rangle = \beta\rho\dot{F}(k, t) \tag{7.6.27}$$

and the imaginary part of the response function is related to the dynamic structure factor by

$$S(k, \omega) = -\frac{k_B T}{\pi\rho\omega}\chi''_{\rho\rho}(k, \omega) \tag{7.6.28}$$

The changes in sign relative to (7.6.17) and (7.6.22) arise from the difference in sign between the hamiltonian terms (7.6.4) and (7.6.25); the density response function is conventionally defined in terms of the response to an external potential rather than an external field. Similarly, the static susceptibility is now

$$\chi_{\rho\rho}(k) = -\frac{\beta}{V} \langle \rho_k\rho_{-k} \rangle = -\beta\rho S(k) \tag{7.6.29}$$

in agreement with (3.6.9).

The properties of the after-effect function $\Phi_{BA}(k, t)$ follow directly from its definition (7.6.17) and the general properties of time correlation functions. If A and B are different, we see from (7.1.9) and (7.6.17) that

$$\Phi_{BA}(k, t) = \epsilon_A\epsilon_B\Phi_{AB}(k, t) = -\epsilon_A\epsilon_B\Phi_{AB}(k, t) \tag{7.6.30}$$

Equation (7.6.30) is an expression of the Onsager *reciprocity relations*. If A and B are real quantities, $\Phi_{BA}(k, t)$ is also real, and from (7.6.20) we see that on the real axis

$$\chi_{BA}(k, -\omega) = \chi^*_{BA}(k, \omega) = \chi'_{BA}(k, \omega) - i\chi''_{BA}(k, \omega) \tag{7.6.31}$$

Thus the real and imaginary parts of χ_{BA} are, respectively, even and odd functions of frequency.

The response function $\chi_{BA}(k, \omega)$ can be interpreted as the limit of a Laplace transform $\chi(k, z)$ defined in the entire upper half of the complex plane (Im $z > 0$):

$$\chi_{BA}(k, z) = \int_0^\infty \Phi_{BA}(k, t) \exp(izt)dt \tag{7.6.32}$$

If we confine ourselves to the important special case when the variables A and B are the same we may discard the subscripts and consider the behaviour of the susceptibility $\chi(k, z) \equiv \chi_{AA}(k, z)$ as a function of the complex variable $z = \omega + i\epsilon$, with $\epsilon > 0$. By restricting ϵ to positive values we ensure that $\chi(k, z)$ is analytic in the upper half-plane, but the function is undefined in the lower half-plane because the integral in (7.6.32) diverges. Since (7.6.13) implies that the after-effect function (with $A = B$) is linear in t at short times, it follows that $\chi(k, z)$ behaves asymptotically as z^{-2} at large z.

Let the contour C in the complex plane be $C = C_1 + C_2$, where C_1 is the real axis and C_2 is the infinite semicircle in the upper half-plane. Application of Cauchy's integral formula shows that

$$\chi(k, z) = \frac{1}{2\pi i} \int_C \frac{\chi(k, z')}{z' - z} dz' \tag{7.6.33}$$

where z is any point inside C. On the other hand, because the conjugate variable z^* lies outside C, the function $\chi(k, z')/(z' - z^*)$ is analytic in and on the contour C. It follows from Cauchy's theorem that

$$\int_C \frac{\chi(k, z')}{z' - z^*} dz' = 0 \tag{7.6.34}$$

The contributions to the integrals (7.6.33) and (7.6.34) from the contour C_2 are both zero because $\chi(k, z)$ vanishes rapidly as $z \to \infty$. By adding quantities that are zero to the right-hand side of (7.6.33) and discarding the integral around C_2, $\chi(k, z)$ can be re-expressed either as

$$\chi(k, z) = \frac{1}{2\pi i} \int_{C_1} \chi(k, z') \left(\frac{1}{z' - z} + \frac{1}{z' - z^*} \right) dz' \tag{7.6.35}$$

or as

$$\chi(k, z) = \frac{1}{2\pi i} \int_{C_1} \chi(k, z') \left(\frac{1}{z' - z} - \frac{1}{z' - z^*} \right) dz' \tag{7.6.36}$$

Two further expressions for $\chi(k, z)$ are obtained by adding the real part of (7.6.35) to i times the imaginary part of (7.6.36) and vice versa:

$$\chi(k, z) = \frac{1}{\pi} \int_{-\infty}^\infty \frac{\chi''(k, \omega)}{\omega - z} d\omega \tag{7.6.37}$$

$$\chi(k, z) = \frac{1}{\pi i} \int_{-\infty}^\infty \frac{\chi'(k, \omega)}{\omega - z} d\omega \tag{7.6.38}$$

We now let $\epsilon \to 0$ in (7.6.37), so that $\chi(k, \omega + i\epsilon) \to \chi'(k, \omega) + i\chi''(k, \omega)$, and use the identity (7.1.18). In this way we find that

$$\chi'(k, \omega) = \mathcal{P}\frac{1}{\pi} \int_{-\infty}^{\infty} \frac{\chi''(k, \omega')}{\omega' - \omega} d\omega' \tag{7.6.39}$$

which is the Kramers–Kronig relation for $\chi'(k, \omega)$ in terms of $\chi''(k, \omega)$. The inverse relation, obtained by applying the rule (7.1.18) to (7.6.38), is

$$\chi''(k, \omega) = -\mathcal{P}\frac{1}{\pi} \int_{-\infty}^{\infty} \frac{\chi'(k, \omega')}{\omega' - \omega} \tag{7.6.40}$$

These results show that the real and imaginary parts of $\chi(k, \omega)$ are not independent of each other and a knowledge of one part is sufficient to determine the full response function.

7.7   APPLICATIONS OF THE LINEAR RESPONSE FORMALISM

The best known and most important of the applications of linear response theory is its use in the derivation of expressions for the transport coefficients of hydrodynamics, through which induced fluxes are related to certain gradients within the fluid. The simplest example concerns the mobility of a tagged particle under the action of a constant external force $\mathcal{F}$ that acts only on the tagged particles. We suppose that the force is applied along the x-direction from $t = 0$ onwards. Then the perturbation term in the hamiltonian is $\mathcal{H}'(t) = -\mathcal{F}x(t)\theta(t)$, where $x(t)$ is the x-coordinate of a tagged particle; if the fluid is isotropic, the drift velocity $\mathbf{u}$ of the particle will be in the same direction as the applied force. From (7.6.12) and (7.6.13) it follows that

$$\langle u_x(t) \rangle = \beta \int_{-\infty}^{t} \langle u_x(t')\dot{x} \rangle \mathcal{F}\theta(t')dt' = \beta\mathcal{F} \int_{0}^{t} \langle u_x(t')u_x \rangle dt' \tag{7.7.1}$$

This leads to the Einstein relation for the mobility ζ, defined as the ratio of the limiting drift velocity to the applied force:

$$\zeta = \lim_{t \to \infty} \frac{1}{k_B T} \int_{0}^{t} \langle u_x(t')u_x \rangle dt' = \frac{D}{k_B T} \tag{7.7.2}$$

where D is the diffusion coefficient. Equation (7.7.2) is a further example of the fluctuation-dissipation theorem: D is a quantity that characterises spontaneous fluctuations in the velocity of a tagged particle and ζ is a measure of the response of the tagged particle to an applied force.

It is instructive to consider an alternative derivation of (7.7.2). If the tagged particles are subjected to a weak, external force derived from a potential

$\exp{(\epsilon t)}\phi(\mathbf{r})$ $(\epsilon > 0)$, a concentration gradient is set up. The resulting induced current is

$$\left\langle \mathbf{j}^{(s)}(\mathbf{r}, t)\right\rangle = -\zeta \rho_s \exp{(\epsilon t)}\nabla\phi(\mathbf{r}) - D\nabla\left\langle \rho^{(s)}(\mathbf{r}, t)\right\rangle \qquad (7.7.3)$$

or, in terms of Fourier components:

$$\left\langle \mathbf{j}^{(s)}_{\mathbf{k}}(t)\right\rangle = -i\zeta \rho_s \exp{(\epsilon t)}\mathbf{k}\phi_{\mathbf{k}} - iD\mathbf{k}\left\langle \rho^{(s)}_{\mathbf{k}}(t)\right\rangle \qquad (7.7.4)$$

where ρ_s is the number of tagged particles per unit volume. The first term on the right-hand side of (7.7.3) represents the contribution to the current from the drift velocity of the tagged particles and the second term arises from Fick's law of diffusion (see below in Section 8.2). If the field is turned on sufficiently slowly, i.e. if $\epsilon \ll Dk^2$, the system will remain in a steady state. The two contributions to the current then cancel and (7.7.4) reduces to

$$\left\langle \rho^{(s)}_{\mathbf{k}}\right\rangle = -\frac{\zeta \rho_s}{D}\phi_{\mathbf{k}} \qquad (7.7.5)$$

If the concentration of tagged particles is sufficiently low for interactions between them to be negligible, it follows from (3.6.9) that $\left\langle \rho^{(s)}_{\mathbf{k}}\right\rangle$ and $\phi_{\mathbf{k}}$ are also related by[19]

$$\left\langle \rho^{(s)}_{\mathbf{k}}\right\rangle = -\beta\rho_s\phi_{\mathbf{k}} \qquad (7.7.6)$$

where $-\beta\rho_s$ is the static susceptibility of a non-interacting system of density ρ_s. Combination of (7.7.5) and (7.7.6) leads back to the Einstein expression (7.7.2).

The calculation of the electrical conductivity provides an example of a different type, in which a collective response of a system to an external field is involved. Suppose that a time-dependent electric field $\mathbf{E}(t)$ is applied to a system of charged particles. The field gives rise to a charge current, defined as

$$e\mathbf{j}^Z(t) = \sum_{i=1}^{N} z_i e\dot{\mathbf{r}}_i(t) = \dot{\mathbf{M}}(t) \qquad (7.7.7)$$

where $z_i e$ is the charge carried by the ith particle (e is the elementary charge) and $\mathbf{M}(t)$ is the total dipole moment of the sample. The interaction with the applied field is described by the hamiltonian

$$\mathcal{H}'(t) = -\sum_{i=1}^{N} \mathbf{M}(t) \cdot \mathbf{E}(t) \qquad (7.7.8)$$

If the system is isotropic and the field is applied, say, along the x-axis, then, in the statistical mean, only the x-component of the induced current will survive. The linear response to a real, periodic field can therefore be written as

$$e\left\langle j^Z_x(t)\right\rangle = \mathrm{Re}\,\sigma(\omega)E_0\exp{(-i\omega t)} \qquad (7.7.9)$$

where, according to the general formulae (7.6.13) and (7.6.20), the electrical conductivity per unit volume is given by

$$
\sigma(\omega) = \frac{\beta e}{V} \int_0^\infty \sum_{i=1}^N \left\langle j_x^Z(t) z_i e \dot{x}_i \right\rangle \exp{(i\omega t)} \mathrm{d}t
$$

$$
= \frac{\beta e^2}{V} \int_0^\infty \left\langle j_x^Z(t) j_x^Z \right\rangle \exp{(i\omega t)} \mathrm{d}t \tag{7.7.10}
$$

The usual static electrical conductivity σ is then identified as $\sigma = \lim_{\omega \to 0} \sigma(\omega)$. The statistical average in the second line of (7.7.10) is the autocorrelation function of the fluctuating charge current in the absence of the electric field. In deriving this result we have ignored any spatial variation of the field, thereby avoiding the difficulties that arise when taking the long-wavelength limit for coulombic systems; we shall return to a discussion of this problem in Chapter 10.

Correlation function formulae for transport coefficients have been obtained by many authors in a variety of ways. The derivation from linear response theory is not always as straightforward as it is in the case of electrical conductivity, the difficulty being that the dissipative behaviour described by hydrodynamics is generally induced not by external forces but by gradients of local thermodynamic variables, which cannot be represented by a perturbation term in the hamiltonian. The thermal conductivity provides an example; this is the transport coefficient that relates the induced heat flux to an imposed temperature gradient via Fourier's law. A temperature gradient is a manifestation of boundary conditions and cannot be formulated in mechanical terms because temperature is a statistical property of the system. However, a linear response argument can still be invoked by introducing an inhomogeneous field that couples to the energy density of the system and sets up a heat flow. Einstein's argument relating the diffusion coefficient to the mobility can then be extended to yield a correlation function expression for the thermal conductivity. We postpone a derivation of the microscopic expressions for thermal conductivity and shear and bulk viscosities to Chapter 8, where it is shown that these coefficients are related to the long-wavelength, low-frequency (or 'hydrodynamic') limit of certain space and time-dependent correlation functions.

The response to a weak, applied field can be measured directly in a molecular dynamics simulation in a way that allows the accurate calculation of transport coefficients with relatively modest computational effort.[20] To understand what is involved, we return to the problem of the electrical conductivity. Clearly we could hope to mimic a real experiment by adding to the equations of motion of the particles the force due to a steady electric field and computing the steady-state charge current to which the field gives rise. The practical value of such an approach is seriously limited by the fact that a very large field must be applied in order to produce a systematic response that is significantly greater than the natural fluctuations. Use of a large field leads to a rapid heating-up of the system, non-conservation of energy, and other undesirable effects.

The problems associated with the use of large fields can be overcome either by imposing constraints that maintain the system at constant kinetic energy or by a 'subtraction' technique closely related to linear response theory. In the subtraction method the response is computed as the difference in the property of interest along two phase space trajectories; both start from the same phase point at time $t = 0$ but in one case a very small perturbing force is applied. In the example of electrical conductivity the response is the difference in charge current after a time t, given by

$$\Delta j_x^Z(t) = \exp(i\mathcal{L}t)j_x^Z - \exp(i\mathcal{L}_0 t)j_x^Z \tag{7.7.11}$$

where $\mathcal{L}$ and $\mathcal{L}_0$ are the Liouville operators that determine the perturbed and unperturbed trajectories, respectively. The statistical response is obtained by averaging (7.7.11) over initial conditions:

$$\left\langle \Delta j_x^Z(t) \right\rangle = \iint f_0^{[N]}[\exp(i\mathcal{L}t) - \exp(i\mathcal{L}_0 t)]j_x^Z \, d\mathbf{r}^N d\mathbf{p}^N$$
$$= \left\langle j_x^Z(t) \right\rangle_{\mathcal{L}} - \left\langle j_x^Z(t) \right\rangle_{\mathcal{L}_0} \tag{7.7.12}$$

where the brackets denote averages over the unperturbed equilibrium distribution function and the nature of the mechanical evolution is indicated by the subscripts $\mathcal{L}$ and $\mathcal{L}_0$. The success of the method rests mostly on the fact that random fluctuations in the two terms in (7.7.12) are highly correlated and therefore largely cancel, leaving only the systematic part, i.e. the response to the perturbation. It is therefore possible to use a perturbing force that is very small. In principle, because the hamiltonian in the absence of the perturbation is symmetric under reflection $(x_i \rightarrow -x_i)$, the second term in (7.7.12) should vanish, but in practice this is not the case because the average is taken over a limited number of trajectories. The form of the statistical response depends on the time-dependence of the applied field. If a constant electric field is applied along the x-axis from $t = 0$ onwards, acting in opposite senses on charges of different sign, the mean response is proportional to the integral of the current autocorrelation function and therefore reaches a plateau value from which the conductivity can be calculated via (7.7.10); if a δ-function force is applied at $t = 0$, the response is proportional to the current autocorrelation function itself. The length of the trajectories must, of course, exceed the relevant relaxation time of the system, in this case the lifetime of spontaneous fluctuations in the electric current.

As a final example we show how the density response function of a non-interacting system can be calculated by a linear response argument. The time evolution of the single-particle phase space distribution function $f^{(1)}(\mathbf{r}, \mathbf{p}; t)$ of an ideal gas in an external potential $\phi(\mathbf{r}, t)$ is determined by the Boltzmann equation (2.1.24) with the collision term set equal to zero, i.e.

$$\left(\frac{\partial}{\partial t} + \frac{\mathbf{p}}{m} \cdot \frac{\partial}{\partial \mathbf{r}} - \frac{\partial \phi(\mathbf{r}, t)}{\partial \mathbf{r}} \cdot \frac{\partial}{\partial \mathbf{p}}\right) f^{(1)}(\mathbf{r}, \mathbf{p}; t) = 0 \qquad (7.7.13)$$

If we write the distribution function as

$$f^{(1)}(\mathbf{r}, \mathbf{p}; t) = \rho f_{\mathrm{M}}(\mathbf{p}) + \Delta f^{(1)}(\mathbf{r}, \mathbf{p}; t) \qquad (7.7.14)$$

where $f_{\mathrm{M}}(\mathbf{p})$ is the Maxwell distribution (2.1.26), the change $\Delta f^{(1)}$ induced by the external potential is linear in ϕ when the potential is weak. Substitution of (7.7.14) in (7.7.13) and linearisation with respect to small quantities yields an equation of motion for $\Delta f^{(1)}$:

$$\left(\frac{\partial}{\partial t} + \frac{\mathbf{p}}{m} \cdot \frac{\partial}{\partial \mathbf{r}}\right) \Delta f^{(1)}(\mathbf{r}, \mathbf{p}; t) - \rho \frac{\partial \phi(\mathbf{r}, t)}{\partial \mathbf{r}} \cdot \frac{\partial f_{\mathrm{M}}(\mathbf{p})}{\partial \mathbf{p}} = 0 \qquad (7.7.15)$$

and a double, Fourier–Laplace transform leads (in an obvious notation) to

$$\left(\omega + i\epsilon - \frac{\mathbf{p} \cdot \mathbf{k}}{m}\right) \Delta f^{(1)}(\mathbf{k}, \mathbf{p}; \omega + i\epsilon) + \rho \phi(\mathbf{k}, \omega + i\epsilon) \mathbf{k} \cdot \frac{\partial f_{\mathrm{M}}}{\partial \mathbf{p}} = 0 \quad (7.7.16)$$

The mean change in microscopic density due to the external potential is

$$\langle \Delta \rho(\mathbf{r}, t) \rangle = \int \Delta f^{(1)}(\mathbf{r}, \mathbf{p}; t) \mathrm{d}\mathbf{p} \qquad (7.7.17)$$

or, in terms of Fourier components:

$$\langle \rho_{\mathbf{k}}(\omega) \rangle = \int \Delta f^{(1)}(\mathbf{k}, \mathbf{p}; \omega) \mathrm{d}\mathbf{p} \qquad (7.7.18)$$

On dividing through (7.7.16) by $(\omega + i\epsilon - \mathbf{p} \cdot \mathbf{k}/m)$ and integrating over $\mathbf{p}$ we find that

$$\langle \rho_{\mathbf{k}}(\omega + i\epsilon) \rangle = -\rho \phi(\mathbf{k}, \omega + i\epsilon) \int \frac{\mathbf{k} \cdot (\partial f_{\mathrm{M}}/\partial \mathbf{p})}{\omega + i\epsilon - \mathbf{p} \cdot \mathbf{k}/m} \mathrm{d}\mathbf{p} \qquad (7.7.19)$$

Thus the density response function is

$$\chi_{\rho\rho}(k, \omega + i\epsilon) = -\rho \int \frac{\mathbf{k} \cdot (\partial f_{\mathrm{M}}/\partial \mathbf{p})}{\omega + i\epsilon - \mathbf{p} \cdot \mathbf{k}/m} \mathrm{d}\mathbf{p}$$

$$= \beta\rho \int \frac{(\mathbf{p} \cdot \mathbf{k}/m) f_{\mathrm{M}}(\mathbf{p})}{\omega + i\epsilon - \mathbf{p} \cdot \mathbf{k}/m} \mathrm{d}\mathbf{p} \qquad (7.7.20)$$

which, by adding and subtracting the quantity $(\omega + i\epsilon) f_{\mathrm{M}}(\mathbf{p})$ to and from the numerator of the integrand, can be rewritten as

$$\chi_{\rho\rho}(k, \omega + i\epsilon) = -\beta\rho + (\omega + i\epsilon)\beta\rho \int \frac{f_{\mathrm{M}}(\mathbf{p})}{\omega + i\epsilon - \mathbf{p} \cdot \mathbf{k}/m} \mathrm{d}\mathbf{p} \qquad (7.7.21)$$

In the limit $\epsilon \to 0$ the imaginary part of (7.7.21) is

$$\chi_{\rho\rho}''(k, \omega) = -\pi\beta\rho\omega \int f_M(\mathbf{p})\delta(\omega - \mathbf{p} \cdot \mathbf{k}/m)d\mathbf{p} \qquad (7.7.22)$$

This result follows immediately from the identity (7.1.18). On substituting for $f_M(\mathbf{p})$ and integrating over $\mathbf{p}$ we find that

$$\chi_{\rho\rho}''(k, \omega) = -\beta\rho\omega \left(\frac{\pi\beta m}{2k^2}\right)^{1/2} \exp\left(-\beta m\omega^2/2k^2\right) \qquad (7.7.23)$$

which, when combined with (7.6.28), is equivalent to the expression (7.5.17) derived earlier for the dynamic structure factor of an ideal gas.

Much of the early theoretical work on density fluctuations in liquids was based on attempts to modify the density response function of an ideal gas to allow for the effects of particle interactions through a variety of mean field or 'effective field' approximations. The problem with such approximations is that they account only for static and not for dynamic correlations between particles; they therefore fare badly at densities characteristic of the liquid state.

REFERENCES

[1] (a) Some of the material discussed in Chapters 7–9 is dealt with at greater length in a number of specialised texts. Berne, B.J. and Pecora, R. 'Dynamic Light Scattering'. John Wiley, New York, 1976. (b) Résibois, P. and DeLeener, M., 'Classical Kinetic Theory of Fluids'. John Wiley, New York, 1977. (c) Boon, J.P. and Yip, S., 'Molecular Hydrodynamics'. Dover Publications, New York, 1991. (d) Mazo, R.M., 'Brownian Motion'. Clarendon Press, Oxford, 2002.

[2] See, e.g. Ref. 1b, p. 240.

[3] Heyes, D.M., Powles, J.G. and Rickayzen, G., Mol. Phys. 100, 595 (2002).

[4] Lebowitz, J.L., Percus, J.K. and Sykes, J., Phys. Rev. 188, 487 (1969).

[5] Longuet-Higgins, H.C., and Pople, J.A., J. Chem. Phys. 25, 884 (1956).

[6] See, e.g. Ref. 1b, p. 324.

[7] When g(d) = 1, (7.2.2) reduces to the result obtained by solution of the Boltzmann equation in a first-order approximation; inclusion of the second-order correction has the effect of increasing the diffusion coefficient in (7.2.2) by approximately 2%. See Chapman, S. and Cowling, T.G., 'The Mathematical Theory of Non-Uniform Gases', 3rd edn. Cambridge University Press, Cambridge, 1970, pp. 258-60.

[8] (a) Alder, B.J., Gass, D.M. and Wainwright, T.E., Phys. Rev. A 1, 18 (1970). (b) Sigurgeirsson, H. and Heyes, D.M., Mol. Phys. 101, 469 (2003). (c) Heyes, D.M., Cass, M.J., Powles, J.G. and Evans, W.A.B., J. Phys. Chem. B 111, 1455 (2007).

[9] Heyes, D.M., J. Phys. Condens. Matter 19, 376106 (2007), Table 1.

[10] Comparison of molecular dynamics results for the Lennard-Jones fluid with an extended form of Enskog theory reveals a different behaviour at intermediate densities. See Miyazaki, K., Srinivas, G. and Bagchi, B., J. Chem. Phys. 114, 6276 (2001).

[11] van Hove, L., Phys. Rev. 95, 249 (1954).

[12] Berne, B.J., Boon, J.P. and Rice, S.A., J. Chem. Phys. 45, 1086 (1966).

[13] (a) Lovesey, S.W., 'Theory of Neutron Scattering from Condensed Matter', vol. 1. Clarendon Press, Oxford, 1984. (b) Squires, G.L., 'Introduction to the Theory of Thermal Neutron Scattering'. Dover Publications, New York, 1996.

[14] (a) See, in particular: Copley, J.R.D. and Rowe, J.M., *Phys. Rev. Lett.* **32**, 49 (1974); *Phys. Rev. A* **9**, 1656 (1974). (b) Bodensteiner, T., Morkel, C., Gläser, W. and Dorner, B., *Phys. Rev. A* **45**, 5709 (1992); erratum: *Phys. Rev. A* **46**, 3574 (1992).

[15] There is now a large literature on the subject but the classic papers remain those of Levesque, D., Verlet, L. and Kürkijarvi, J., *Phys. Rev. A* **7**, 1690 (1973) and Rahman, A., *Phys. Rev. Lett.* **32**, 52 (1974); *Phys. Rev. A,* **9**, 1667 (1974).

[16] For a discussion of the relative merits of the two techniques and a review of results obtained for a wide range of liquid metals, see Scopigno, T., Ruocco, G. and Sette, F., *Rev. Mod. Phys.* **77**, 881 (2005).

[17] See, e.g. Angelini, R., Giura, P., Monaco, G., Ruocco, G., Sette, F. and Verbeni, R., *Phys. Rev. Lett.* **88**, 255503 (2002) (hydrogen fluoride); Santucci, S.C., Fioretto, D., Comez, L., Gessini, A. and Masciovecchio, C., *Phys. Rev. Lett.* **97**, 225701 (2006) (water).

[18] See, e.g. Cunsolo, A., Leu, B.M., Said, A.H. and Cai, Y.Q., *J. Chem. Phys.* **134**, 184502 (2011) (glycerol).

[19] Equation (7.7.6) is just the ideal gas form of (3.6.9).

[20] Ciccotti, G., Jacucci, G. and McDonald, I.R., *J. Stat. Phys.* **21**, 1 (1979).

Chapter 8

Hydrodynamics and Transport Coefficients

Chapter 7 was concerned largely with the formal definition and general properties of time correlation functions and with the link that exists between spontaneous, time-dependent fluctuations and the response of a fluid to an external probe. The main objectives of the present chapter are, first, to show how the decay of fluctuations is described within the framework of linearised hydrodynamics and, secondly, to obtain explicit expressions for the macroscopic transport coefficients in terms of microscopic quantities. The hydrodynamic approach is valid only on scales of length and time much larger than those characteristic of the molecular level, but we show how the gap between the microscopic and macroscopic descriptions can be bridged by an essentially phenomenological extrapolation of the hydrodynamic results to shorter wavelengths and higher frequencies. The same problem is taken up in a more systematic way in Chapter 9.

8.1 THERMAL FLUCTUATIONS AT LONG WAVELENGTHS AND LOW FREQUENCIES

We have seen in Section 4.1 that the microscopic structure of a liquid is revealed experimentally by the scattering of radiation of wavelength comparable with the interparticle spacing. Examination of a typical pair distribution function, such as the one pictured in Figure 2.3, shows that positional correlations decay rapidly in space and are negligibly small at separations beyond a few molecular diameters. From a static point of view, therefore, a fluid behaves, for longer wavelengths, essentially as a continuum. When discussing the dynamics, however, it is necessary to consider simultaneously the scales of both length and time. In keeping with traditional kinetic theory it is conventional to compare wavelengths with the mean free path l_c and times with the mean collision time τ_c. The wavenumber-frequency plane may then be divided into three parts. The region in which $kl_c \ll 1$, $\omega\tau_c \ll 1$ corresponds to the *hydrodynamic* regime, in which the behaviour of the fluid is described by the phenomenological equations

Theory of Simple Liquids, Fourth Edition. http://dx.doi.org/10.1016/B978-0-12-387032-2.00008-8
311

of macroscopic fluid mechanics. The range of intermediate wavenumbers and frequencies ($kl_c \approx 1, \omega\tau_c \approx 1$) forms the *kinetic* regime, where allowance must be made for the molecular structure of the fluid and a treatment based on the microscopic equations of motion is required. Finally, the region where $kl_c \gg 1$, $\omega\tau_c \gg 1$ represents the *free-particle* or *ballistic* regime; here the distances and times involved are so short that the particles move almost independently of each other.

In this chapter we shall be concerned mostly with the hydrodynamic regime, where the local properties of the fluid vary slowly on microscopic scales of length and time. The set of *hydrodynamic variables* or *hydrodynamic fields* include the densities of mass (or particle number), energy and momentum; these are closely related to the conserved microscopic variables introduced in Section 7.4. Like their microscopic counterparts, the conserved hydrodynamic variables satisfy continuity equations of the form (7.4.3), which express the conservation of matter, energy and momentum. In addition, there exist certain *constitutive relations* between the fluxes (or currents) and gradients of the local variables, expressed in terms of phenomenological *transport coefficients*. Fick's law of diffusion and Fourier's law of heat transport are two of the more familiar examples of a constitutive relation.

One of the main tasks of the present chapter is to obtain microscopic expressions for the transport coefficients that are similar in structure to the formula (7.7.10) already derived for the electrical conductivity of an ionic fluid. This is achieved by calculating the *hydrodynamic limit* of the appropriate time correlation function. To understand what is involved in such a calculation it is first necessary to clarify the relationship between hydrodynamic and microscopic dynamical variables. As an example, consider the local density. The microscopic particle density $\rho(\mathbf{r}, t)$ is defined by (7.4.5); its integral over all volume is equal to N, the total number of particles in the system. The hydrodynamic local density $\bar{\rho}(\mathbf{r}, t)$ is obtained by averaging the microscopic density over a sub-volume v around the point $\mathbf{r}$ that is macroscopically small but still sufficiently large to ensure that the relative fluctuations in the number of particles inside v is negligible. Then

$$\bar{\rho}(\mathbf{r}, t) = \frac{1}{v} \int_v \rho(\mathbf{r}' - \mathbf{r}, t)\mathrm{d}\mathbf{r}' \qquad (8.1.1)$$

Strictly speaking, the definition of $\bar{\rho}(\mathbf{r}, t)$ also requires a smoothing or 'coarse graining' in time. This can be realised by averaging (8.1.1) over a time interval that is short on a macroscopic scale but long in comparison with the mean collision time. In practice, however, smoothing in time is already achieved by (8.1.1) if the sub-volume is sufficiently large. The Fourier components of the hydrodynamic density are defined as

$$\bar{\rho}_{\mathbf{k}}(t) = \int \bar{\rho}(\mathbf{r}, t) \exp(-i\mathbf{k} \cdot \mathbf{r})\mathrm{d}\mathbf{r} \qquad (8.1.2)$$

where the wavevector $\mathbf{k}$ must be such that k is less than about $2\pi/v^{1/3}$. The corresponding density autocorrelation function is then defined as in (7.4.20), except that the Fourier components of the microscopic density are replaced by $\bar{\rho}_{\mathbf{k}}$. Since we are now working at the macroscopic level, the average to be taken is not an ensemble average but an average over initial conditions, weighted by the probability density of thermodynamic fluctuation theory described in Appendix A. By forming such an average we are implicitly invoking the hypothesis of *local thermodynamic equilibrium*. In other words, we are assuming that although the hydrodynamic densities vary over macroscopic lengths and times, the fluid contained in each of the sub-volumes is in a state of thermodynamic equilibrium, and that the local density, pressure and temperature satisfy the usual relations of equilibrium thermodynamics. These assumptions are particularly plausible at high densities, since in that case local equilibrium is rapidly brought about by collisions between particles.

Once the calculation we have described in words has been carried out, the relations of interest are obtained by supposing that in the limit of long wavelengths ($\lambda \gg l_c$) and long times ($t \gg \tau_c$) or, equivalently, of small wavenumbers and low frequencies, correlation functions derived from the hydrodynamic equations are identical to the correlation functions of the corresponding microscopic variables. This intuitively appealing hypothesis, which is due to Onsager, can be justified on the basis of the fluctuation-dissipation theorem discussed in Section 7.6. In the example of the density autocorrelation function the assumption can be expressed by the statement that

$$\langle \rho_{\mathbf{k}}(t)\rho_{-\mathbf{k}} \rangle \sim \langle \bar{\rho}_{\mathbf{k}}(t)\bar{\rho}_{-\mathbf{k}} \rangle, \quad kl_c \ll 1, \ t/\tau_c \gg 1 \qquad (8.1.3)$$

with the qualification, explained above, that the meaning of the angular brackets is different for the two correlation functions. Since the sections that follow are concerned almost exclusively with the calculation of correlation functions of hydrodynamic variables, no ambiguity is introduced by dropping the bar we have used to distinguish the latter from the corresponding microscopic quantities.

One important implication of the assumption of local thermodynamic equilibrium is that the Maxwell distribution of velocities applies at the local level. The local velocity $\mathbf{u}(\mathbf{r}, t)$ is defined via the relation

$$\mathbf{p}(\mathbf{r}, t) = \rho_m(\mathbf{r}, t)\mathbf{u}(\mathbf{r}, t) \qquad (8.1.4)$$

where $\mathbf{p}(\mathbf{r}, t)$ is the momentum density and $\rho_m(\mathbf{r}, t) = m\rho(\mathbf{r}, t)$ is the mass density (we assume that the fluid consists of only one component). The single-particle distribution function is now a function of $\mathbf{r}$ and t and (2.1.26) is replaced by

$$f_{\text{l.e.}}(\mathbf{u}, \mathbf{r}; t) = \rho(\mathbf{r}, t)\left(\frac{m}{2\pi k_B T(\mathbf{r}, t)}\right)^{3/2} \exp\left(\frac{-m|\mathbf{u} - \mathbf{u}(\mathbf{r}, t)|^2}{2k_B T(\mathbf{r}, t)}\right) \qquad (8.1.5)$$

where $T(\mathbf{r}, t)$ is the local temperature. The function $f_{l.e.}(\mathbf{u}, \mathbf{r}; t)$ is called the 'local equilibrium' Maxwell distribution.

8.2 SPACE-DEPENDENT SELF MOTION

As an illustration of the general procedure described in the previous section we first consider the relatively simple problem of the diffusion of tagged particles. If the tagged particles are physically identical to the other particles in the fluid, and if their concentration is sufficiently low that their mutual interactions can be ignored, the problem is equivalent to that of single-particle motion as described by the self part of the van Hove correlation function $G_s(\mathbf{r}, t)$ (see Section 7.4). The macroscopic, tagged-particle density $\rho^{(s)}(\mathbf{r}, t)$ and current $\mathbf{j}^{(s)}(\mathbf{r}, t)$ satisfy a continuity equation of the form

$$\frac{\partial \rho^{(s)}(\mathbf{r}, t)}{\partial t} + \nabla \cdot \mathbf{j}^{(s)}(\mathbf{r}, t) = 0 \qquad (8.2.1)$$

and the corresponding constitutive equation is provided by Fick's law:

$$\mathbf{j}^{(s)}(\mathbf{r}, t) = -D\nabla\rho^{(s)}(\mathbf{r}, t) \qquad (8.2.2)$$

where the interdiffusion constant D is in this case the same as the self-diffusion constant. Combination of (8.2.1) and (8.2.2) yields the *diffusion equation*:

$$\frac{\partial \rho^{(s)}(\mathbf{r}, t)}{\partial t} = D\nabla^2\rho^{(s)}(\mathbf{r}, t) \qquad (8.2.3)$$

or, in reciprocal space:

$$\frac{\partial \rho_{\mathbf{k}}^{(s)}(t)}{\partial t} = -Dk^2\rho_{\mathbf{k}}^{(s)}(t) \qquad (8.2.4)$$

Equation (8.2.4) can be integrated immediately to give

$$\rho_{\mathbf{k}}^{(s)}(t) = \rho_{\mathbf{k}}^{(s)} \exp\left(-Dk^2 t\right) \qquad (8.2.5)$$

where $\rho_{\mathbf{k}}^{(s)}$ is a Fourier component of the tagged-particle density at $t = 0$. If we multiply both sides of (8.2.5) by $\rho_{-\mathbf{k}}^{(s)}$ and take the thermal average, we find that the normalised autocorrelation function is

$$\frac{1}{n}\left\langle \rho_{\mathbf{k}}^{(s)}(t)\rho_{-\mathbf{k}}^{(s)} \right\rangle = \frac{1}{n}\left\langle \rho_{\mathbf{k}}^{(s)}\rho_{-\mathbf{k}}^{(s)} \right\rangle \exp\left(-Dk^2 t\right) = \exp\left(-Dk^2 t\right) \qquad (8.2.6)$$

where n is the total number of tagged particles. Here we have used the fact that, because the concentration of tagged particles is low, their coordinates are mutually uncorrelated. It then follows from the general hypothesis discussed in Section 8.1 that in the hydrodynamic limit the self part of the density autocorrelation function (7.4.21), i.e. the self-intermediate scattering function defined by (7.5.12), behaves as

$$F_s(k,t) \sim \exp\left(-Dk^2 t\right), \quad kl_c \ll 1, \; t/\tau_c \gg 1 \tag{8.2.7}$$

The long-wavelength, low-frequency limit of the van Hove self correlation function is the spatial Fourier transform of (8.2.7):

$$G_s(r,t) = \frac{1}{(4\pi Dt)^{3/2}} \exp\left(-r^2/4Dt\right) \tag{8.2.8}$$

In the same limit the self dynamic structure factor is

$$S_s(k,\omega) = \frac{1}{\pi} \frac{Dk^2}{\omega^2 + (Dk^2)^2} \tag{8.2.9}$$

Equation (8.2.9) represents a single, Lorentzian curve centred at $\omega = 0$ with a width at half-height equal to $2Dk^2$. A spectrum of this type is typical of any diffusive process described by an equation similar to (8.2.3). Alternatively, the structure of the Laplace transform of (8.2.7), i.e.

$$\tilde{F}_s(k,z) = \frac{1}{-iz + Dk^2} \tag{8.2.10}$$

shows that a diffusive process is characterised by a purely imaginary pole at $z = -iDk^2$. It should be emphasised again that the simple result expressed by (8.2.9) is valid only for $kl_c \ll 1, \omega\tau_c \ll 1$. Its breakdown at high frequencies is reflected in the fact that the even frequency moments (beyond zeroth order) of $S_s(k,\omega)$ are all infinite. Note also that the transport coefficient D is related to the behaviour of $S_s(k,\omega)$ in the limit $k, \omega \to 0$. From (8.2.9) we see that

$$D = \lim_{\omega \to 0} \lim_{k \to 0} \frac{\omega^2}{k^2} \pi S_s(k,\omega) \tag{8.2.11}$$

where it is crucial that the limits are taken in the correct order, i.e. $k \to 0$ before $\omega \to 0$. In principle, (8.2.11) provides a means of determining D from the results of inelastic scattering experiments.

Equations (7.5.16) and (8.2.8) show that the van Hove self correlation function is a gaussian function of r both for $t \to 0$ (free-particle behaviour) and $t \to \infty$ (the hydrodynamic limit); it is therefore tempting to suppose that the function is gaussian at all times. To study this point in more detail we write $G_s(r,t)$ as a generalised gaussian function of r in the form

$$G_s(r,t) = \left(\frac{\alpha(t)}{\pi}\right)^{3/2} \exp\left[-\alpha(t)r^2\right] \tag{8.2.12}$$

where $\alpha(t)$ is a function of t but not of r; the hydrodynamic limit corresponds to taking $\alpha(t) = 1/4Dt$ and the ideal-gas model to $\alpha(t) = m/2k_B T t^2$. The mean-square displacement of tagged particles after a time t is the second moment of $G_s(r,t)$, i.e.

$$\left\langle r^2(t) \right\rangle \equiv \left\langle |\mathbf{r}(t) - \mathbf{r}(0)|^2 \right\rangle = \int r^2 G_s(r,t) d\mathbf{r} \tag{8.2.13}$$

and is therefore related to the unknown function $\alpha(t)$ by $\left\langle r^2(t) \right\rangle = 3/2\alpha(t)$. If we insert this result in (8.2.12) and take the Fourier transform, we find that in the gaussian approximation the self intermediate scattering function has the form

$$F_s(k,t) = \exp\left(-\frac{1}{6}k^2 \left\langle r^2(t) \right\rangle\right) \tag{8.2.14}$$

Systematic corrections to the gaussian approximation can be obtained from a cumulant expansion of $F_s(k,t)$ in powers of k^2. Comparison with molecular dynamics results for argon-like liquids shows that in the intermediate range of k between the free-particle and hydrodynamic regimes the first correction (of order k^4) to (8.2.14) is typically 10% or less and positive; corrections of higher order are even smaller.[1]

The Einstein expression for the long-time limit of the mean-square displacement of a tagged particle is a direct consequence of the hydrodynamic result for $G_s(r,t)$; substitution of (8.2.8) into the definition (8.2.13) leads immediately to (7.2.3). Since the mean-square displacement is also related to the velocity autocorrelation function through (7.2.6), there is a close connection between the functions $G_s(r,t)$ (or $F_s(k,t)$) and $Z(t)$. In fact, in the gaussian approximation represented by (8.2.14), $F_s(k,t)$ is entirely determined by $Z(t)$ and vice versa; more generally, only the second of these statements is true. To see the significance of this connection we return briefly to the description of the system in terms of microscopic variables. If we define the Fourier components of the microscopic current associated with a tagged particle i having velocity $\mathbf{u}_i$ as

$$\mathbf{j}_{\mathbf{k}i}(t) = \mathbf{u}_i(t) \exp\left[-i\mathbf{k} \cdot \mathbf{r}_i(t)\right] \tag{8.2.15}$$

and the self-current autocorrelation function as

$$C_s(k,t) = \left\langle \mathbf{k} \cdot \mathbf{j}_{\mathbf{k}i}(t) \mathbf{k} \cdot \mathbf{j}_{-\mathbf{k}i} \right\rangle \tag{8.2.16}$$

it is clear that

$$Z(t) = \langle u_{iz}(t) u_{iz} \rangle = \lim_{k \to 0} \frac{1}{k^2} C_s(k,t)$$

$$= -\lim_{k \to 0} \frac{1}{k^2} \frac{d^2}{dt^2} F_s(k,t) \tag{8.2.17}$$

where we have chosen $\mathbf{k}$ to lie along the z-axis and used the single-particle version of (7.4.26). The relation between the corresponding power spectra is

$$Z(\omega) = \frac{\omega^2}{2\pi} \lim_{k \to 0} \frac{1}{k^2} \int_{-\infty}^{\infty} F_s(k,t) \exp(i\omega t) \mathrm{d}t = \omega^2 \lim_{k \to 0} \frac{S_s(k,\omega)}{k^2} \quad (8.2.18)$$

Equation (8.2.18) may be regarded as a generalisation of (8.2.11) to non-zero frequencies in which $Z(\omega)$ appears as a frequency-dependent diffusion coefficient; it also provides a possible route to an experimental determination of the velocity autocorrelation function.

The relationship between $Z(t)$ and $F_s(k,t)$ (or $C_s(k,t)$) is further reflected in the short-time expansions of these functions. By analogy with (7.4.31) the expansion of $C_s(k,t)$ in powers of t can be written as

$$C_s(k,t) = \omega_0^2 \left(1 - \omega_{1s}^2 \frac{t^2}{2!} + \cdots \right) \quad (8.2.19)$$

From the general result (7.1.23) and the continuity equation (8.2.1) it follows that

$$\omega_0^2 \omega_{1s}^2 = -\langle \mathbf{k} \cdot \dot{\mathbf{j}}_{ki} \mathbf{k} \cdot \dot{\mathbf{j}}_{-ki} \rangle = \langle \ddot{\rho}_{ki} \ddot{\rho}_{-ki} \rangle$$

$$= k^4 \langle u_{iz}^4 \rangle + k^2 \langle \dot{u}_{iz}^2 \rangle = \omega_0^4 + (k^2/m^2) \langle F_{iz}^2 \rangle \quad (8.2.20)$$

and hence, from the definition (7.2.9), that

$$\omega_{1s}^2 = 3\omega_0^2 + \Omega_0^2 \quad (8.2.21)$$

The next term (of order t^4) in the Taylor expansion of $C_s(k,t)$ involves integrals over the triplet distribution function. Short-time expansions such as (8.2.19) are useful in extending the validity of hydrodynamic results to microscopic scales of length and time.

8.3 THE NAVIER–STOKES EQUATION AND HYDRODYNAMIC COLLECTIVE MODES

We turn now to the problem of describing the decay of long-wavelength fluctuations in the collective dynamical variables. For a one-component fluid the macroscopic local densities associated with the conserved variables are the number density $\rho(\mathbf{r},t)$, energy density $e(\mathbf{r},t)$ and momentum density $\mathbf{p}(\mathbf{r},t)$. The conservation laws for the local densities have the form

$$m\frac{\partial}{\partial t}\rho(\mathbf{r},t) + \nabla \cdot \mathbf{p}(\mathbf{r},t) = 0 \quad (8.3.1)$$

$$\frac{\partial}{\partial t}e(\mathbf{r},t) + \nabla \cdot \mathbf{J}^e(\mathbf{r},t) = 0 \quad (8.3.2)$$

$$\frac{\partial}{\partial t}\mathbf{p}(\mathbf{r}, t) + \nabla \cdot \mathbf{\Pi}(\mathbf{r}, t) = 0 \tag{8.3.3}$$

where $\mathbf{J}^e$ is the energy current and $\mathbf{\Pi}$ is the momentum current or *stress tensor*. These equations must be supplemented by two constitutive relations in which $\mathbf{J}^e$ and $\mathbf{\Pi}$ are expressed in terms of quantities representing dissipative processes in the fluid. We choose a frame of reference in which the mean velocity of the fluid is zero, i.e. $\langle \mathbf{u}(\mathbf{r}, t) \rangle = 0$, and assume that the local deviations of the hydrodynamic variables from their average values are small. The equations may then be linearised with respect to the deviations. We consider in turn each of the three conservation laws.

Conservation of particle number. Equation (8.3.1) is easily dealt with. The assumption that the local deviation in number density is small means that the momentum density can be written as

$$\mathbf{p}(\mathbf{r}, t) = m\left[\rho + \delta\rho(\mathbf{r}, t)\right]\mathbf{u}(\mathbf{r}, t) \approx m\rho\mathbf{u}(\mathbf{r}, t) \equiv m\mathbf{j}(\mathbf{r}, t) \tag{8.3.4}$$

which also serves as the definition of the local particle current $\mathbf{j}(\mathbf{r}, t)$. With this approximation, (8.3.1) becomes

$$\frac{\partial}{\partial t}\delta\rho(\mathbf{r}, t) + \nabla \cdot \mathbf{j}(\mathbf{r}, t) = 0 \tag{8.3.5}$$

Conservation of energy. The macroscopic energy current $\mathbf{J}^e$ is defined as

$$\mathbf{J}^e(\mathbf{r}, t) = (e + P)\mathbf{u}(\mathbf{r}, t) - \lambda\nabla T(\mathbf{r}, t) \tag{8.3.6}$$

where $e = U/V$ is the equilibrium energy density, P is the fixed, overall pressure, $e + P$ is the enthalpy density, λ is the thermal conductivity and $T(\mathbf{r}, t)$ is the local temperature already introduced in (8.1.5); terms corresponding to viscous heating have been omitted, since these are quadratic in the local velocity. Equations (8.3.2), (8.3.5) and (8.3.6) can now be combined to give the *energy equation*, i.e.

$$\frac{\partial}{\partial t}\delta q(\mathbf{r}, t) - \lambda\nabla^2\delta T(\mathbf{r}, t) = 0 \tag{8.3.7}$$

where $\delta q(\mathbf{r}, t)$ is the fluctuation in a quantity

$$q(\mathbf{r}, t) = e(\mathbf{r}, t) - \left(\frac{e + P}{\rho}\right)\rho(\mathbf{r}, t) \tag{8.3.8}$$

which can be interpreted as a density of heat energy. If the number of particles is held constant, the entropy change of the system in an infinitesimal process is $T\,\mathrm{d}S = \mathrm{d}U + P\,\mathrm{d}V$. Hence

$$T\,\mathrm{d}S = \mathrm{d}(eV) + P\,\mathrm{d}V = V\,\mathrm{d}e - \frac{eV}{\rho}\,\mathrm{d}\rho - \frac{PV}{\rho}\,\mathrm{d}\rho = V\,\mathrm{d}q \tag{8.3.9}$$

A change in q is therefore equal to the heat lost or gained by the system per unit volume when the change is carried out reversibly and $\delta q(\mathbf{r}, t)$ is related to the change in entropy density $s(\mathbf{r}, t)$ by

$$\delta q(\mathbf{r}, t) = T\delta s(\mathbf{r}, t) \tag{8.3.10}$$

If we invoke the hypothesis of local thermodynamic equilibrium the deviation of a local thermodynamic variable such as $s(\mathbf{r}, t)$ from its average value can be expressed in terms of a set of statistically independent quantities. We choose as independent variables the density and temperature (see Appendix A) and expand $q(\mathbf{r}, t)$ to first order in the deviations $\delta\rho(\mathbf{r}, t)$ and $\delta T(\mathbf{r}, t)$. Then, from (8.3.10), and remembering that N is fixed:

$$\delta q(\mathbf{r}, t) = \frac{T}{V} \left(\frac{\partial S}{\partial \rho} \right)_T \delta\rho(\mathbf{r}, t) + \frac{T}{V} \left(\frac{\partial S}{\partial T} \right)_\rho \delta T(\mathbf{r}, t)$$

$$= -\frac{T\beta_V}{\rho} \delta\rho(\mathbf{r}, t) + \rho c_V \delta T(\mathbf{r}, t) \tag{8.3.11}$$

where

$$\beta_V = \left(\frac{\partial P}{\partial T} \right)_\rho = -\rho \left(\frac{\partial (S/V)}{\partial \rho} \right)_T \tag{8.3.12}$$

is the thermal pressure coefficient, c_V is the heat capacity per particle at constant volume and use has been made of the Maxwell relation $(\partial S / \partial V)_T = (\partial P / \partial T)_V$ resulting from (2.3.8). If we now substitute (8.3.11) in (8.3.7), eliminate $(\partial / \partial t)\rho(\mathbf{r}, t)$ with the help of (8.3.5) and divide through by ρc_V, the energy equation becomes

$$\left(\frac{\partial}{\partial t} - a\nabla^2 \right) \delta T(\mathbf{r}, t) + \frac{T\beta_V}{\rho^2 c_V} \nabla \cdot \mathbf{j}(\mathbf{r}, t) = 0 \tag{8.3.13}$$

where

$$a = \frac{\lambda}{\rho c_V} \tag{8.3.14}$$

Conservation of momentum. The components of the stress tensor $\mathbf{\Pi}$ in (8.3.3) are given macroscopically by

$$\Pi^{\alpha\beta}(\mathbf{r}, t) = \delta_{\alpha\beta} P(\mathbf{r}, t) - \eta \left(\frac{\partial u_\alpha(\mathbf{r}, t)}{\partial r_\beta} + \frac{\partial u_\beta(\mathbf{r}, t)}{\partial r_\alpha} \right)$$

$$+ \delta_{\alpha\beta} \left(\frac{2}{3}\eta - \zeta \right) \nabla \cdot \mathbf{u}(\mathbf{r}, t) \tag{8.3.15}$$

where $P(\mathbf{r}, t)$ is the local pressure, η is the shear viscosity, ζ is the bulk viscosity and the bracketed quantity in the second term on the right-hand side is the rate-of-strain tensor.[2] Substitution of (8.3.15) in (8.3.3) and use of (8.3.5) leads to the Navier–Stokes equation in its linearised form:

$$\frac{\partial}{\partial t}\mathbf{j}(\mathbf{r}, t) + \frac{1}{m}\nabla P(\mathbf{r}, t) - \nu\nabla^2\mathbf{j}(\mathbf{r}, t) - \frac{\frac{1}{3}\eta + \zeta}{\rho m}\nabla\nabla \cdot \mathbf{j}(\mathbf{r}, t) = 0 \tag{8.3.16}$$

where

$$\nu = \frac{\eta}{\rho m} \tag{8.3.17}$$

is the kinematic shear viscosity. To first order in $\delta\rho(\mathbf{r},t)$ and $\delta T(\mathbf{r},t)$ the fluctuation in local pressure is

$$\delta P(\mathbf{r},t) = \left(\frac{\partial P}{\partial \rho}\right)_T \delta\rho(\mathbf{r},t) + \left(\frac{\partial P}{\partial T}\right)_\rho \delta T(\mathbf{r},t)$$

$$= \frac{1}{\rho \chi_T}\delta\rho(\mathbf{r},t) + \beta_V \delta T(\mathbf{r},t) \tag{8.3.18}$$

where χ_T is the isothermal compressibility (2.4.16). The Navier–Stokes equation can therefore be rewritten as

$$\frac{1}{\rho m \chi_T}\nabla\delta\rho(\mathbf{r},t) + \frac{\beta_V}{m}\nabla\delta T(\mathbf{r},t) + \left(\frac{\partial}{\partial t} - \nu\nabla^2 - \frac{\frac{1}{3}\eta + \zeta}{\rho m}\nabla\nabla\cdot\right)\mathbf{j}(\mathbf{r},t) = 0 \tag{8.3.19}$$

Equations (8.3.5), (8.3.13) and (8.3.19) form a closed set of linear equations for the variables $\delta\rho(\mathbf{r},t)$, $\delta T(\mathbf{r},t)$ and $\mathbf{j}(\mathbf{r},t)$. These are readily solved by taking the double transforms with respect to space (Fourier) and time (Laplace) to give

$$-iz\tilde{\rho}_\mathbf{k}(z) + i\mathbf{k}\cdot\tilde{\mathbf{j}}_\mathbf{k}(z) = \rho_\mathbf{k} \tag{8.3.20}$$

$$(-iz + ak^2)\tilde{T}_\mathbf{k}(z) + \frac{T\beta_V}{\rho^2 c_V}i\mathbf{k}\cdot\tilde{\mathbf{j}}_\mathbf{k}(z) = T_\mathbf{k} \tag{8.3.21}$$

$$\frac{1}{\rho m \chi_T}i\mathbf{k}\tilde{\rho}_\mathbf{k}(z) + \frac{\beta_V}{m}i\mathbf{k}\tilde{T}_\mathbf{k}(z) + \left(-iz + \nu k^2 + \frac{\frac{1}{3}\eta + \zeta}{\rho m}\mathbf{k}\mathbf{k}\cdot\right)\tilde{\mathbf{j}}_\mathbf{k}(z) = \mathbf{j}_\mathbf{k} \tag{8.3.22}$$

where, for example:

$$\tilde{\rho}_\mathbf{k}(z) = \int_0^\infty dt \exp(izt)\int \delta\rho(\mathbf{r},t)\exp(-i\mathbf{k}\cdot\mathbf{r})d\mathbf{r} \tag{8.3.23}$$

and $\rho_\mathbf{k}$, $T_\mathbf{k}$ and $\mathbf{j}_\mathbf{k}$ are the spatial Fourier components at $t = 0$. We now separate the components of the current $\mathbf{j}_\mathbf{k}$ into their longitudinal and transverse parts. Taking $\mathbf{k}$ along the z-axis, we rewrite (8.3.22) as

$$\frac{1}{\rho m \chi_T}ik\tilde{\rho}_\mathbf{k}(z) + \frac{\beta_V}{m}ik\tilde{T}_\mathbf{k}(z) + \left(-iz + bk^2\right)\tilde{j}^z_\mathbf{k}(z) = j^z_\mathbf{k}$$

$$\left(-iz + \nu k^2\right)\tilde{j}^\alpha_\mathbf{k} = j^\alpha_\mathbf{k}, \quad \alpha = x, y \tag{8.3.24}$$

where

$$b = \frac{\frac{4}{3}\eta + \zeta}{\rho m} \tag{8.3.25}$$

is the kinematic longitudinal viscosity.

Equations (8.3.20), (8.3.21) and (8.3.24) are conveniently summarised in matrix form:

$$
\begin{pmatrix}
-iz & 0 & ik & 0 & 0 \\
0 & -iz+ak^2 & \dfrac{T\beta_V ik}{\rho^2 c_V} & 0 & 0 \\
\dfrac{ik}{\rho m \chi_T} & \dfrac{\beta_V ik}{m} & -iz+bk^2 & 0 & 0 \\
0 & 0 & 0 & -iz+vk^2 & 0 \\
0 & 0 & 0 & 0 & -iz+vk^2
\end{pmatrix}
\begin{pmatrix}
\tilde{\rho}_\mathbf{k}(z) \\
\tilde{T}_\mathbf{k}(z) \\
\tilde{j}_\mathbf{k}^z(z) \\
\tilde{j}_\mathbf{k}^x(z) \\
\tilde{j}_\mathbf{k}^y(z)
\end{pmatrix}
=
\begin{pmatrix}
\rho_\mathbf{k} \\
T_\mathbf{k} \\
j_\mathbf{k}^z \\
j_\mathbf{k}^x \\
j_\mathbf{k}^y
\end{pmatrix}
$$

(8.3.26)

The matrix of coefficients in (8.3.26) is called the *hydrodynamic matrix*. Its block diagonal structure shows that the transverse current fluctuations are completely decoupled from fluctuations in the other, longitudinal variables. The determinant of the hydrodynamic matrix therefore factorises into the product of purely longitudinal (l) and purely transverse (t) parts, i.e.

$$D(k,z) = D_l(k,z)D_t(k,z) \tag{8.3.27}$$

with

$$
D_l(k,z) = -iz\left(-iz+ak^2\right)\left(-iz+bk^2\right)
$$
$$
+ \left(-iz+ak^2\right)\frac{k^2}{\rho m \chi_T} - iz\frac{T\beta_V^2 k^2}{\rho^2 m c_V}
\tag{8.3.28}
$$

and

$$D_t(k,z) = \left(-iz+vk^2\right)^2 \tag{8.3.29}$$

The dependence of frequency on wavenumber or *dispersion relation* for the collective modes is determined by the poles of the inverse of the hydrodynamic matrix and hence by the complex roots of the equation

$$D(k,z) = 0 \tag{8.3.30}$$

The factorisation in (8.3.27) shows that (8.3.30) has a double root associated with the two transverse modes, namely

$$z = -ivk^2 \tag{8.3.31}$$

while the complex frequencies corresponding to longitudinal modes are obtained as the solution to the cubic equation

$$iz^3 - z^2(a+b)k^2 - iz\left(abk^2 + c_s^2\right)k^2 + \frac{ac_s^2}{\gamma}k^4 = 0 \tag{8.3.32}$$

where $\gamma = c_P/c_V$ is the ratio of specific heats, c_s is the adiabatic speed of sound, given by

$$c_s^2 = \frac{\gamma}{\rho m \chi_T} \tag{8.3.33}$$

and use has been made of the thermodynamic relation[3]

$$c_P = c_V + \frac{T \chi_T \beta_V^2}{\rho} \qquad (8.3.34)$$

Since the hydrodynamic calculation is valid only in the long-wavelength limit, it is sufficient to calculate the complex frequencies to order k^2. The algebra is simplified by introducing the reduced variables $s = z/c_s k$; it is then straightforward to show[4] that the approximate solution to (8.3.32) is

$$z_0 = -i D_T k^2, \quad z_\pm = \pm c_s k - i \Gamma k^2 \qquad (8.3.35)$$

where

$$D_T = \frac{a}{\gamma} = \frac{\lambda}{\rho c_P} \qquad (8.3.36)$$

is the thermal diffusivity and

$$\Gamma = \frac{a(\gamma - 1)}{2\gamma} + \frac{1}{2} b \qquad (8.3.37)$$

is the sound attenuation coefficient. The imaginary roots in (8.3.31) and (8.3.35) represent diffusive processes of the type already discussed in the preceding section whereas the pair of complex roots in (8.3.35) correspond to propagating sound waves, as we shall see in Section 8.5.

8.4 TRANSVERSE CURRENT CORRELATIONS

The second relation in (8.3.24) shows that in the time domain the hydrodynamic behaviour of the transverse current fluctuations is described by a first-order differential equation of the form

$$\frac{\partial}{\partial t} j_{\mathbf{k}}^x(t) = -\nu k^2 j_{\mathbf{k}}^x(t) \qquad (8.4.1)$$

This result has precisely the same structure as the diffusion equation (8.2.4) and the kinematic shear viscosity has the same dimensions as the self-diffusion coefficient, but is typically two orders of magnitude larger than D for, say, an argon-like liquid near its triple point. If we multiply through (8.4.1) by $j_{-\mathbf{k}}^x$ and take the thermal average we find that the transverse current autocorrelation function satisfies the equation

$$\frac{\partial}{\partial t} C_t(k,t) + \nu k^2 C_t(k,t) = 0 \qquad (8.4.2)$$

the solution to which is

$$C_t(k,t) = C_t(k,0) \exp\left(-\nu k^2 t\right) = \omega_0^2 \exp\left(-\nu k^2 t\right) \qquad (8.4.3)$$

where ω_0 is the frequency defined by (7.4.29). The exponential decay in (8.4.3) is typical of a diffusive process (see Section 8.2).

The diffusive behaviour of the hydrodynamic 'shear' mode is also apparent in the fact that the Laplace transform of $C_t(k,t)$ has a purely imaginary pole corresponding to the root (8.3.31) of $D(k,z)$:

$$\tilde{C}_t(k,z) = \frac{\omega_0^2}{-iz + \nu k^2} \tag{8.4.4}$$

Let $z = \omega + i\epsilon$ approach the real axis from above ($\epsilon \to 0+$). Then $\tilde{C}_t(k,\omega)$ at small k is given approximately by

$$\tilde{C}_t(k,\omega) = \frac{\omega_0^2}{-i\omega}\left(1 - \frac{\nu k^2}{i\omega}\right)^{-1} \approx \frac{\omega_0^2}{-i\omega}\left(1 + \frac{\nu k^2}{i\omega}\right) \tag{8.4.5}$$

If we substitute for ω_0^2 and recall the definition (8.3.17) of ν we find that the shear viscosity, which must be real, is related to the long-wavelength, low-frequency behaviour of $\tilde{C}_t(k,\omega)$ by

$$\eta = \beta\rho m^2 \lim_{\omega\to 0}\lim_{k\to 0} \frac{\omega^2}{k^4}\mathrm{Re}\tilde{C}_t(k,\omega)$$

$$= \pi\beta\rho m^2 \lim_{\omega\to 0}\lim_{k\to 0} \frac{\omega^2}{k^4}C_t(k,\omega) \tag{8.4.6}$$

where $C_t(k,\omega)$ is the spectrum of transverse current fluctuations, i.e. the Fourier transform of $C_t(k,t)$; this result is the analogue of the expression for the self-diffusion coefficient given by (8.2.11). From the properties of the Laplace transform and the definition of $C_t(k,t)$ it follows that

$$\frac{k^2}{N}\int_0^\infty \langle j_\mathbf{k}^x(t) j_{-\mathbf{k}}^x\rangle \exp(i\omega t)\mathrm{d}t = -\int_0^\infty \frac{\mathrm{d}^2}{\mathrm{d}t^2}C_t(k,t)\exp(i\omega t)\mathrm{d}t$$

$$= \omega^2\tilde{C}_t(k,\omega) - i\omega\omega_0^2 \tag{8.4.7}$$

We may therefore rewrite (8.4.6) as

$$\eta = \frac{\beta m^2}{V}\lim_{\omega\to 0}\lim_{k\to 0}\mathrm{Re}\int_0^\infty \frac{1}{k^2}\langle j_\mathbf{k}^x(t) j_{-\mathbf{k}}^x\rangle\exp(i\omega t)\mathrm{d}t \tag{8.4.8}$$

The time derivative of the transverse current can be expressed in terms of the stress tensor via the conservation law (8.3.3). Taking the Fourier transform of (8.3.3), and remembering that $\mathbf{k}$ lies along the z-axis and $\mathbf{p}(\mathbf{r},t) = m\mathbf{j}(\mathbf{r},t)$, we find that

$$\frac{\partial}{\partial t}j_\mathbf{k}^x(t) + \frac{ik}{m}\Pi_\mathbf{k}^{xz}(t) = 0 \tag{8.4.9}$$

Combination of (8.4.8) and (8.4.9) shows that the shear viscosity is proportional to the time integral of the autocorrelation function of an off-diagonal element of the stress tensor taken in the limit $k \to 0$:

$$\eta = \frac{\beta}{V}\int_0^\infty \langle \Pi_0^{xz}(t)\Pi_0^{xz}\rangle \mathrm{d}t \equiv \int_0^\infty \eta(t)\mathrm{d}t \tag{8.4.10}$$

In order to relate the shear viscosity to the intermolecular forces it is necessary to have a microscopic expression for the stress tensor. It follows from the definition (7.4.7) of the microscopic particle current that

$$m\frac{\partial}{\partial t}j_{\mathbf{k}}^{\alpha} = m\sum_{i=1}^{N}\left(\dot{u}_{i\alpha} - \sum_{\beta}ik_{\beta}u_{i\alpha}u_{i\beta}\right)\exp\left(-i\mathbf{k}\cdot\mathbf{r}_i\right) \qquad (8.4.11)$$

where α, β denote any of x, y or z; the relation to the stress tensor is then established by use of (8.4.9), with $\alpha = x$ and $\beta = z$. To introduce the pair potential $v(r)$ we note that $\mathbf{r}_{ji} = -\mathbf{r}_{ij}$, and rewrite the first term on the right-hand side of (8.4.11) successively as

$$m\sum_{i=1}^{N}\dot{u}_{i\alpha}\exp\left(-i\mathbf{k}\cdot\mathbf{r}_i\right) = \sum_{i=1}^{N}\sum_{j\neq i}^{N}\frac{r_{ij\alpha}}{|\mathbf{r}_{ij}|}v'(r_{ij})\exp\left(-i\mathbf{k}\cdot\mathbf{r}_i\right)$$

$$= \frac{1}{2}\sum_{i=1}^{N}\sum_{j\neq i}^{N}\frac{r_{ij\alpha}}{|\mathbf{r}_{ij}|}v'(r_{ij})\left[\exp\left(-i\mathbf{k}\cdot\mathbf{r}_i\right) - \exp\left(-i\mathbf{k}\cdot\mathbf{r}_j\right)\right]$$

$$= \frac{1}{2}ik_{\beta}\sum_{i=1}^{N}\sum_{j\neq i}^{N}\frac{r_{ij\alpha}r_{ij\beta}}{ik_{\beta}r_{ij\beta}|\mathbf{r}_{ij}|}v'(r_{ij})\left[\exp\left(-i\mathbf{k}\cdot\mathbf{r}_i\right) - \exp\left(-i\mathbf{k}\cdot\mathbf{r}_j\right)\right]$$

$$(8.4.12)$$

where $v'(r) \equiv dv(r)/dr$; the second step is taken by writing each term in the double sum as half the sum of two equal terms. On introducing a quantity $\Phi_{\mathbf{k}}(\mathbf{r})$ defined as

$$\Phi_{\mathbf{k}}(\mathbf{r}) = rv'(r)\left(\frac{\exp\left(i\mathbf{k}\cdot\mathbf{r}\right) - 1}{i\mathbf{k}\cdot\mathbf{r}}\right) \qquad (8.4.13)$$

we finally obtain a microscopic expression for $\Pi_{\mathbf{k}}^{\alpha\beta}$ in the form

$$\Pi_{\mathbf{k}}^{\alpha\beta} = \sum_{i=1}^{N}\left(mu_{i\alpha}u_{i\beta} + \frac{1}{2}\sum_{j\neq i}^{N}\frac{r_{ij\alpha}r_{ij\beta}}{r_{ij}^2}\Phi_{\mathbf{k}}(\mathbf{r}_{ij})\right)\exp\left(-i\mathbf{k}\cdot\mathbf{r}_i\right) \qquad (8.4.14)$$

The Green–Kubo relation for the shear viscosity analogous to (7.2.7) is then obtained by inserting (8.4.14) (taken for $\mathbf{k} = 0$) in (8.4.10). Note that it follows from the virial theorem that

$$\langle\Pi_0^{\alpha\alpha}\rangle = PV \qquad (8.4.15)$$

whereas

$$\langle\Pi_0^{\alpha\beta}\rangle = 0, \quad \alpha \neq \beta \qquad (8.4.16)$$

Equation (8.4.10) is not directly applicable to the hard-sphere fluid because the pair potential has a singularity at $r = d$ (the hard-sphere diameter). However, the microscopic expression for the shear viscosity, together with formulae to be derived later for other transport coefficients, can be recast in a form that resembles the Einstein relation (7.2.3) for the self-diffusion coefficient and is valid even for hard spheres. A Green–Kubo formula for a transport coefficient K, including both (7.7.10) (taken for $\omega = 0$) and (8.4.10), can always be written as

$$K = \frac{\beta}{V} \int_0^\infty \langle \dot{A}(t) \dot{A} \rangle \, dt \qquad (8.4.17)$$

where A is some microscopic dynamical variable. The argument used to derive (7.2.7) from (7.2.3) can be extended to show that (8.4.17) is equivalent to writing

$$K = \frac{\beta}{V} \lim_{t \to \infty} \frac{1}{2t} \langle |A(t) - A(0)|^2 \rangle \qquad (8.4.18)$$

which may be regarded as a generalised form of the Einstein relation for D. In the case of the shear viscosity we see from (8.4.8) that the variable $A(t)$ is

$$A(t) = \lim_{k \to 0} \frac{im}{k} j_{\mathbf{k}}^x(t)$$

$$= \lim_{k \to 0} \frac{im}{k} \sum_{i=1}^N u_{ix}(t) \left[1 - ikr_{iz}(t) + \cdots \right] = m \sum_{i=1}^N u_{ix}(t) r_{iz}(t) \qquad (8.4.19)$$

where a frame of reference has been chosen in which the total momentum of the particles (a conserved quantity) is zero. Hence the generalised Einstein relation for the shear viscosity is

$$\eta = \frac{\beta m^2}{V} \lim_{t \to \infty} \frac{1}{2t} \left\langle \left| \sum_{i=1}^N \left[u_{ix}(t) r_{iz}(t) - u_{ix}(0) r_{iz}(0) \right] \right|^2 \right\rangle \qquad (8.4.20)$$

The quantity Π_0^{xz} in the Green–Kubo formula (8.4.10) is the sum of a kinetic and a potential term. There are consequently three distinct contributions to the shear viscosity: a purely kinetic term, corresponding to the transport of transverse momentum via the displacement of particles; a purely potential term, arising from the action of the interparticle forces ('collisional' transport); and a cross term. At liquid densities the potential term is much the largest of the three. In Enskog's theory (see Section 7.2) the shear viscosity of the hard-sphere fluid is

$$\frac{\eta_E}{\eta_0} = \frac{2\pi \rho d^3}{3} \left(\frac{1}{y} + 0.8 + 0.761y \right) \qquad (8.4.21)$$

where $y = \beta P / \rho - 1 = (2\pi \rho d^3 / 3) g(d)$ and $\eta_0 = (5/16d^2)(mk_B T/\pi)^{1/2}$ is the limiting, low-density result derived from the Boltzmann equation.[5] The three terms between brackets in (8.4.21) represent, successively, the kinetic, cross and potential contributions; the last of these is dominant close to the fluid-solid transition, where $g(d)$ (the pair distribution function at contact) ≈ 6 and $y \approx 11$. Note that the kinetic contribution scales with $g(d)$ in the same way as the diffusion constant (see Section 2.5); this is not surprising, since diffusion is a purely kinetic phenomenon.

Results obtained by molecular dynamics calculations for the self-diffusion constant and shear viscosity of the hard-sphere fluid are plotted as functions of density in Figure 8.1; the two transport coefficients are plotted in a reduced form that corresponds to setting $d = m = k_BT = 1$. A comparison between simulation and the predictions of Enskog theory was made earlier in Figure 7.3. In the case of the shear viscosity agreement is very good for densities up to $\rho d^3 \approx 0.7$. Near solidification, however, where η increases rapidly with density, the theory underestimates the shear viscosity by a factor of approximately two. The behaviour of the self-diffusion constant at high densities is the reverse of this. An inverse relationship between D and η is implicit in the Stokes–Einstein relation (7.3.19), which can be rewritten in reduced form as

$$\frac{1}{\pi D^* \eta^*} = \alpha \tag{8.4.22}$$

where $\alpha = 2$ or 3 for slip or stick boundary conditions, respectively. Figure 8.2 shows a plot of the quantity $1/\pi D^* \eta^*$ versus density in which the appearance of a plateau region extending from $\rho d^3 \approx 0.25$ up to $\rho d^3 \approx 0.9$ implies that the Stokes-Einstein relation is to a good approximation satisfied over a density range in which both D and η change by an order of magnitude but in opposite senses; the level of the plateau is close to that corresponding to slip boundary

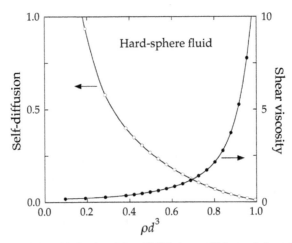

FIGURE 8.1   Variation with density of the self-diffusion coefficient and shear viscosity of the hard-sphere fluid, plotted in reduced units for which $d = m = k_BT = 1$. The points show the results of molecular dynamics calculations[6] and the curves are guides to the eye.

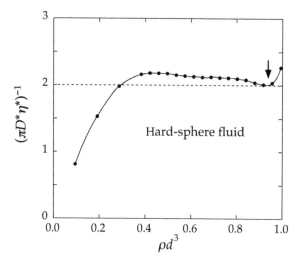

FIGURE 8.2 Test of the applicability of the Stokes–Einstein relation to the hard-sphere fluid. The points are calculated from the values of D and η shown in Figure 8.1 and the curve is a guide to the eye. The plateau in the results defines a region over which the relation is to a good approximation satisfied, with a value of α in (8.4.22) close to that corresponding to slip boundary conditions. The arrow marks the density at which the fluid–solid transition occurs. After Heyes et al.[7]

condition. No such plateau is seen in the curve computed from the Enskog values of D and η. Over the same range of density an even better fit[7] to the molecular dynamics data is provided by a 'fractional' Stokes–Einstein relation[8] of the form $D \propto \left(1/\eta\right)^{\gamma}$ with $\gamma \approx 0.975$. The same expression, with system-dependent values of $\gamma \approx 0.8$–1.0, has been used successfully in the correlation of transport data for a wide variety of liquids, but the justification for its use remains purely empirical.

The increase in shear viscosity at high densities is linked numerically to the appearance of a slowly decaying, quasi-exponential tail in the stress tensor autocorrelation function $\eta(t)$ defined by (8.4.10), colloquially called the 'molasses' tail.[9] The effect is not peculiar to hard spheres. For example, a persisting, positive tail is clearly present in the results shown in Figure 8.3 for a soft-sphere (r^{-12}) fluid at a high value of the coupling constant Γ, where $\eta(t)$ is well represented by the sum of two exponentials. At lower values of Γ, corresponding to lower densities or higher temperatures, the tail in $\eta(t)$ is barely perceptible even on a logarithmic scale.

8.5 LONGITUDINAL COLLECTIVE MODES

The longitudinal collective modes are those associated with fluctuations in density, temperature and the projection of the particle current along the direction of the wavevector $\mathbf{k}$. It is clear from the structure of the hydrodynamic matrix

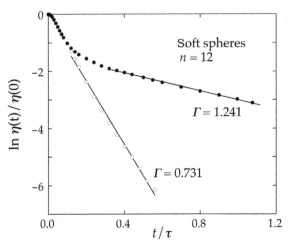

FIGURE 8.3 Normalised stress tensor autocorrelation function of a soft-sphere (r^{-12}) fluid at two values of the coupling parameter Γ defined by (5.4.13). The unit of time is $\tau = (\sigma^2/48\epsilon)^{1/2}$. Unpublished results of D.M. Heyes.

in (8.3.26) that the variables $\tilde{\rho}_{\mathbf{k}}(z)$, $\tilde{T}_{\mathbf{k}}(z)$ and $\tilde{j}_{\mathbf{k}}^z(z)$ are coupled to each other. The analysis is therefore more complicated than in the case of the transverse current fluctuations. There are three longitudinal modes, corresponding to the roots z_0, z_+ and z_- displayed in (8.3.35). The significance of the different roots is most easily grasped by solving the system of coupled, longitudinal equations represented by (8.3.26) to obtain the hydrodynamic limiting form of the dynamic structure factor $S(k,\omega)$. The solution for $\tilde{\rho}_{\mathbf{k}}(z)$ involves terms proportional to the initial values $\rho_{\mathbf{k}}$, $T_{\mathbf{k}}$ and $j_{\mathbf{k}}^z$. We may omit the term proportional to $j_{\mathbf{k}}^z$ because $\mathbf{k}$ can always be chosen to make $\mathbf{u}_{\mathbf{k}}$ (the Fourier transform of the initial local velocity $\mathbf{u}(\mathbf{r},0)$) perpendicular to $\mathbf{k}$, thereby ensuring that $j_{\mathbf{k}}^z = 0$. We can also ignore the term proportional to $T_{\mathbf{k}}$; this contributes nothing to the final expression for $S(k,\omega)$, since fluctuations in temperature and density are instantaneously uncorrelated, i.e. $\langle T_{\mathbf{k}}\rho_{-\mathbf{k}}\rangle = 0$ (see Appendix A). With these simplifications the solution for $\tilde{\rho}_{\mathbf{k}}(z)$ is

$$\frac{\tilde{\rho}_{\mathbf{k}}(z)}{\rho_{\mathbf{k}}} = \frac{(-iz + ak^2)(-iz + bk^2) + (\gamma - 1)c_s^2 k^2/\gamma}{D_l(k,z)} \qquad (8.5.1)$$

where all quantities are as defined in Section 8.3. Separation of the right-hand side of (8.5.1) into partial fractions shows that on the real axis $\tilde{\rho}_{\mathbf{k}}$ is given by

$$\frac{\tilde{\rho}_{\mathbf{k}}(\omega)}{\rho_{\mathbf{k}}} = \left(\frac{\gamma - 1}{\gamma}\right) \frac{1}{-i\omega + D_T k^2}$$
$$+ \frac{1}{2\gamma}\left(\frac{1}{-i\omega + \Gamma k^2 - ic_s k} + \frac{1}{-i\omega + \Gamma k^2 + ic_s k}\right) \qquad (8.5.2)$$

which, via an inverse transform, yields an expression for $\rho_{\mathbf{k}}(t)$ given by

$$\rho_{\mathbf{k}}(t) = \rho_{\mathbf{k}}\left[\left(\frac{\gamma - 1}{\gamma}\right)\exp\left(-D_T k^2 t\right) + \frac{1}{\gamma}\exp\left(-\Gamma k^2 t\right)\cos c_s k t\right] \quad (8.5.3)$$

The form of (8.5.3) shows that the purely imaginary root in (8.3.35) represents a fluctuation that decays without propagating, the lifetime of the fluctuation being determined by the thermal diffusivity defined by (8.3.36). By contrast, the complex roots correspond to a fluctuation that propagates through the fluid at the speed of sound, eventually decaying through the combined effects of viscosity and thermal conduction. The definition of Γ in (8.3.37) implies that the thermal damping of the sound mode is small when $\gamma \approx 1$, which is the case for many liquid metals. On multiplying through (8.5.3) by $\rho_{-\mathbf{k}}$, dividing by N and taking the thermal average, we obtain an expression for the density autocorrelation function $F(k, t)$; this is easily transformed to give

$$S(k, \omega) = \frac{S(k)}{2\pi}\left[\left(\frac{\gamma - 1}{\gamma}\right)\frac{2D_T k^2}{\omega^2 + (D_T k^2)^2}\right.$$
$$\left. + \frac{1}{\gamma}\left(\frac{\Gamma k^2}{(\omega + c_s k)^2 + (\Gamma k^2)^2} + \frac{\Gamma k^2}{(\omega - c_s k)^2 + (\Gamma k^2)^2}\right)\right](8.5.4)$$

The spectrum of density fluctuations therefore consists of three components: the *Rayleigh line*, centred at $\omega = 0$, and two *Brillouin lines* at $\omega = \pm c_s k$; a typical spectrum is plotted in Figure 8.4. The two shifted components correspond to propagating sound waves and are analogous to the longitudinal acoustic phonons of a solid, whereas the central line corresponds to the diffusive, thermal mode. The total integrated intensity of the Rayleigh line is

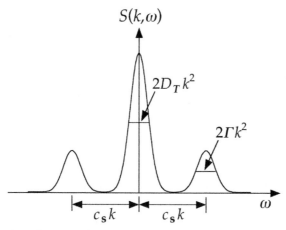

FIGURE 8.4 Dynamic structure factor in the hydrodynamic limit. D_T is the thermal diffusivity, Γ is the sound attenuation coefficient and c_s is the adiabatic speed of sound.

$$\mathcal{I}_R = \frac{\gamma - 1}{\gamma} S(k) \tag{8.5.5}$$

and that of each of the two Brillouin lines is

$$\mathcal{I}_B = \frac{1}{2\gamma} S(k) \tag{8.5.6}$$

Thus

$$\mathcal{I}_R + 2\mathcal{I}_B = S(k) \tag{8.5.7}$$

which is a particular case of the sum rule (7.4.23). The quantity

$$\frac{\mathcal{I}_R}{2\mathcal{I}_B} = \gamma - 1 \tag{8.5.8}$$

is called the Landau–Placzek ratio. As the values of c_P/c_V listed in Table 1.2 suggest, the Landau–Placzek ratio is typically an order of magnitude larger for the rare-gas liquids than for simple liquid metals. In passing from (8.5.1) to (8.5.2) we have, for sake of simplicity, omitted a non-Lorentzian term that in practice makes only a negligibly small, asymmetric correction to the Brillouin lines.[10]

We have chosen to discuss the behaviour of the longitudinal modes in terms of the local density and temperature, but it would have been equally appropriate to choose the pressure and entropy as variables, since these are also statistically independent (see Appendix A). The calculation is instructive because it shows that the first term in (8.5.2) can be identified with the decay of entropy fluctuations. It follows that the Brillouin doublet is associated with propagating pressure fluctuations at constant entropy (hence the appearance of the adiabatic speed of sound), while the Rayleigh line corresponds to non-propagating fluctuations in entropy at constant pressure.[4]

The wavelength of visible light is much greater than the nearest-neighbour spacing in liquids. Light scattering experiments are therefore ideally suited to measurements of the Rayleigh–Brillouin spectrum at long wavelengths and provide an accurate means of measurement of properties such as the thermal diffusivity, speed of sound and sound attenuation coefficient. However, the spectral lineshape is determined by a small number of macroscopic properties that are insensitive to details of either the interactions between particles or the molecular structure of the fluid. From the standpoint of microscopic theory the more interesting question is whether the propagating density fluctuations characteristic of the hydrodynamic regime can also be supported in simple liquids at wavelengths comparable with the spacing between particles. We have already seen in Section 7.5 that well-defined, collective excitations of the hydrodynamic type, manifesting themselves in a three-peak structure in $S(k, \omega)$, have been detected in neutron scattering experiments on liquid caesium, but comparable results have been obtained by neutron or X-ray scattering for the other alkali metals and for aluminium, gallium, magnesium and mercury.[11] Brillouin-type side-peaks have also been seen in molecular dynamics simulations of a variety of systems, including both the hard-sphere[12] and Lennard-Jones[13] fluids. The spectra are therefore qualitatively similar to those predicted by hydrodynamics, though there are some major differences in detail. Figure 8.5, for example, shows the dispersion of the sound-wave peak observed in neutron scattering experiments on liquid caesium. At the smallest wavenumbers the dispersion is approximately linear, in agreement with hydrodynamics, but corresponds to a speed of propagation significantly higher than the experimental speed of sound. At larger wavenumbers the dispersion is no longer linear and eventually becomes negative. The widths of the Rayleigh and Brillouin lines are also poorly described by the hydrodynamic result. As we shall see in Section 8.6 and again in Chapter 9, a description of the density fluctuations in the range of k explored in neutron or X-ray scattering experiments requires a generalisation of the hydrodynamic approach, the effect of which is to replace the transport coefficients and thermodynamic derivatives in (8.5.4) by quantities dependent on frequency and wavenumber.

For later purposes we also require an expression for the hydrodynamic limit of the longitudinal current autocorrelation function $C_l(k,t)$. We proceed, as before, by solving the system of equations (8.3.26) for the variable of interest, which in this case is the longitudinal particle current $\tilde{j}_{\mathbf{k}}^z(z)$. The terms in $\rho_{\mathbf{k}}$ and

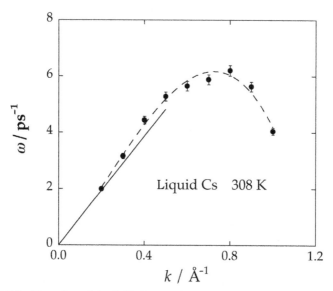

FIGURE 8.5 Dispersion of the Brillouin peak in liquid caesium near the normal melting temperature. The points are the results of inelastic neutron scattering experiments, the straight line shows the hydrodynamic dispersion corresponding to the experimental speed of sound, $c_s = 965$ m s^{-1}, and the dashed curve is a guide to the eye. After Bodensteiner et al.[14]

$T_{\mathbf{k}}$ may be omitted, since they are uncorrelated with $\tilde{j}^z_{-\mathbf{k}}$. For z on the real axis the result is

$$\tilde{j}^z_{\mathbf{k}}(\omega) = j^z_{\mathbf{k}} \frac{-i\omega(-i\omega + ak^2)}{D_l(k, \omega)} \tag{8.5.9}$$

Thus

$$\tilde{C}_l(k, \omega) = \frac{\omega_0^2}{-i\omega + bk^2 + c_s^2 k^2 \left(\frac{1}{-i\omega} + \frac{\gamma - 1}{-i\omega + ak^2}\right)} \tag{8.5.10}$$

Equation (8.5.10) shows that the spectrum of longitudinal current fluctuations at small k behaves as

$$C_l(k, \omega) = \frac{1}{\pi} \mathrm{Re}\, \tilde{C}_l(k, \omega) \approx \frac{\omega_0^2}{\pi \omega^2} \left(bk^2 + \frac{(\gamma - 1)ac_s^2 k^4}{\omega^2 + (ak^2)^2}\right) \tag{8.5.11}$$

Hence the longitudinal viscosity is given by a limiting operation analogous to (8.4.6) for the shear viscosity, i.e.

$$\frac{4}{3}\eta + \zeta = \rho m b = \lim_{\omega \to 0} \lim_{k \to 0} \frac{\omega^2}{k^4} C_l(k, \omega) \tag{8.5.12}$$

If we now follow steps similar to those that lead to the Green–Kubo formula (8.4.10), we find that the longitudinal viscosity can be expressed in terms of the autocorrelation function of a diagonal element of the microscopic stress tensor (8.4.14):

$$\frac{4}{3}\eta + \zeta = \lim_{\omega \to 0} \frac{\beta}{V} \int_0^\infty \left\langle \Pi_0^{zz}(t)\Pi_0^{zz}\right\rangle \exp(i\omega t)\,dt \tag{8.5.13}$$

In taking the limit $\omega = 0$ in (8.5.13) we find a discontinuity: the thermal average of Π_0^{zz} is non-zero (see (8.4.15)), so the integrand in (8.5.13) approaches a non-zero value as $t \to \infty$. The problem is overcome by subtracting the invariant part, the transport coefficient being linked only to fluctuations in the local variables. Thus

$$\frac{4}{3}\eta + \zeta = \frac{\beta}{V} \int_0^\infty \left\langle [\Pi_0^{zz}(t) - PV][\Pi_0^{zz} - PV]\right\rangle dt \tag{8.5.14}$$

To obtain the Green–Kubo relation for the thermal conductivity we require an expression for the rate of decay of a fluctuation in $q(\mathbf{r}, t)$, the macroscopic density of heat energy, which is related to the entropy density by (8.3.10). We first use (8.3.11) to eliminate the local temperature from the energy equation (8.3.13). The result is

$$\left(\frac{\partial}{\partial t} - a\nabla^2\right)\delta q(\mathbf{r}, t) - \frac{\lambda T \beta_V}{\rho^2 c_V}\nabla^2\delta\rho(\mathbf{r}, t) = 0 \tag{8.5.15}$$

which, after transformation to Fourier-Laplace variables and use of (8.3.12) together with the thermodynamic chain rule:

$$\left(\frac{\partial S}{\partial \rho}\right)_T = -\left(\frac{\partial S}{\partial T}\right)_\rho \left(\frac{\partial T}{\partial \rho}\right)_S = -\frac{Nc_V}{T}\left(\frac{\partial T}{\partial \rho}\right)_S \qquad (8.5.16)$$

gives

$$\left(-iz + ak^2\right)\tilde{q}_{\mathbf{k}}(z) + \lambda k^2 \left(\frac{\partial T}{\partial \rho}\right)_S \tilde{\rho}_{\mathbf{k}}(z) = q_{\mathbf{k}} \qquad (8.5.17)$$

Next, an equation relating $\tilde{\rho}_{\mathbf{k}}(z)$ to $\tilde{P}_{\mathbf{k}}(z)$ is obtained by taking the divergence of the Navier–Stokes equation (8.3.16) and transforming again to the variables k and z; the result in this case is

$$izm\left(-iz + bk^2\right)\tilde{\rho}_{\mathbf{k}}(z) - k^2\tilde{P}_{\mathbf{k}}(z) = -m\left(-iz + bk^2\right)\rho_{\mathbf{k}} \qquad (8.5.18)$$

where $\mathbf{k}$ has once more been chosen perpendicular to the initial particle current. Equation (8.5.18) can now be converted into a relation for $\tilde{q}_{\mathbf{k}}(z)$ by making the substitutions

$$\tilde{P}_{\mathbf{k}}(z) = \left(\frac{\partial P}{\partial \rho}\right)_S \tilde{\rho}_{\mathbf{k}}(z) + \frac{V}{T}\left(\frac{\partial P}{\partial S}\right)_\rho \tilde{q}_{\mathbf{k}}(z) \qquad (8.5.19)$$

and

$$\rho_{\mathbf{k}} = \left(\frac{\partial \rho}{\partial P}\right)_S P_{\mathbf{k}} + \frac{V}{T}\left(\frac{\partial \rho}{\partial S}\right)_\rho q_{\mathbf{k}} \qquad (8.5.20)$$

The final step is to eliminate $\tilde{\rho}_{\mathbf{k}}(z)$ between (8.5.17) and (8.5.18). The resulting expression for $\tilde{q}_{\mathbf{k}}(z)$ has some similarities with that obtained previously for $\tilde{\rho}_{\mathbf{k}}(z)$ in (8.5.1). In particular, there are two complex conjugate poles and a single imaginary pole. At small k the local pressure and entropy are uncorrelated (see Appendix A). The problem can therefore be simplified by discarding terms proportional to $P_{\mathbf{k}}$. The lowest-order solution for $\tilde{q}_{\mathbf{k}}(z)$ then reduces to

$$\tilde{q}_{\mathbf{k}}(z) = \frac{q_{\mathbf{k}}}{-iz + D_T k^2} \qquad (8.5.21)$$

Equation (8.5.21) describes a purely diffusive mode, thereby confirming the fact that the Rayleigh peak in $S(k, \omega)$ is associated with the decay of non-propagating entropy fluctuations.

Our main concern is with the behaviour at small k. Since $\lim_{k\to 0} q_{\mathbf{k}} = T\Delta S$, it follows from (A.8) of Appendix A that $\langle q_{\mathbf{k}} q_{-\mathbf{k}}\rangle$ can be replaced by

$$\left\langle q_0^2 \right\rangle = T^2 N k_{\mathrm{B}} c_P \qquad (8.5.22)$$

We now proceed as in the cases of the shear and longitudinal viscosities. On multiplying (8.5.21) through by $q_{-\mathbf{k}}$ and taking the thermal average we obtain an expression for the thermal conductivity of the form

$$\lambda = \rho c_P D_T = \frac{\beta}{VT} \lim_{\omega \to 0} \lim_{k \to 0} \frac{\omega^2}{k^2} \mathrm{Re} \, \langle \tilde{q}_\mathbf{k}(\omega) q_{-\mathbf{k}} \rangle \qquad (8.5.23)$$

If we introduce a fluctuating heat current $\mathbf{J}_\mathbf{k}^q(t)$ defined, by virtue of (8.3.8), as the Fourier transform of

$$\mathbf{J}^q(\mathbf{r}, t) = \mathbf{J}^e(\mathbf{r}, t) - \frac{e+P}{\rho} \mathbf{j}(\mathbf{r}, t) \qquad (8.5.24)$$

we see that the energy conservation equation (8.3.2) may be re-expressed as

$$\frac{\partial}{\partial t} q_\mathbf{k}(t) + i\mathbf{k} \cdot \mathbf{J}_\mathbf{k}^q(t) = 0 \qquad (8.5.25)$$

Hence, if the z-axis is taken parallel to $\mathbf{k}$, we can rewrite (8.5.23) in typical Green-Kubo form as

$$\lambda = \frac{\beta}{VT} \int_0^\infty \langle J_0^{qz}(t) J_0^{qz} \rangle \, dt \qquad (8.5.26)$$

For (8.5.26) to be useful we require a microscopic expression for the heat current. On taking the Fourier transform of (8.3.2) we find that the component of the microscopic energy current in the direction of $\mathbf{k}$ is

$$-ik J_\mathbf{k}^{ez} = \frac{\partial}{\partial t} e_\mathbf{k} = \frac{\partial}{\partial t} \sum_{i=1}^N \left(\frac{1}{2} m |\mathbf{u}_i|^2 + \frac{1}{2} \sum_{j \neq i}^N v(r_{ij}) \right) \exp(-i\mathbf{k} \cdot \mathbf{r}_i) \qquad (8.5.27)$$

where we have adopted the convention that the interaction energy of a pair of particles is shared equally between them. Differentiation of the quantity inside large brackets gives rise to a term that can be treated by the methods used in calculating the microscopic stress tensor; the final result for $\mathbf{k} = 0$ is

$$J_0^{ez} = \sum_{i=1}^N u_{iz} \left(\frac{1}{2} m |\mathbf{u}_i|^2 + \frac{1}{2} \sum_{j \neq i}^N v(r_{ij}) \right) - \frac{1}{2} \sum_{i=1}^N \sum_{j \neq i}^N \mathbf{u}_i \cdot \mathbf{r}_{ij} \frac{\partial v(r_{ij})}{\partial z_{ij}} \qquad (8.5.28)$$

The current J_0^{qz} is obtained from J_0^{ez} by subtracting the term $(e+P) \sum_i u_{iz}$; with a suitable choice of frame of reference this term will be zero. Thus we can equally well write the Green–Kubo formula for λ as

$$\lambda = \frac{\beta}{VT} \int_0^\infty \langle J_0^{ez}(t) J_0^{ez} \rangle \, dt \qquad (8.5.29)$$

The correlation function formulae (or the equivalent Einstein expressions) for D, η, ζ and λ have been used in simulations to determine the transport coefficients of a number of model systems. A particularly large body of results exists for the hard-sphere fluid, some of which have already been discussed in Section 8.4. As we saw there, the shear viscosity is in good agreement with the predictions of Enskog theory at densities up to about 80% of that corresponding to the fluid-solid transition, but close to the transition it is larger than the Enskog value by a factor of nearly two. The enhancement of the shear viscosity at high densities is linked numerically to the existence of a long-lived, positive tail in the corresponding autocorrelation function. The bulk viscosity is purely potential in origin and vanishes as $\rho \to 0$, but the Enskog result for the thermal conductivity has a structure similar to that displayed for η in (8.4.21), i.e.

$$\frac{\lambda_E}{\lambda_0} = \frac{2\pi\rho d^3}{3} \left(\frac{1}{y} + 1.2 + 0.757y\right) \tag{8.5.30}$$

where y has the same meaning as before and $\lambda_0 = (75k_B/64d^2)(k_B T/\pi m)^{1/2}$ is the conductivity in the low-density limit.[5] The potential term (the last term within brackets) again provides the dominant contribution at high densities, but good agreement with molecular dynamics results is now maintained up to the freezing transition, as Figure 8.6 reveals. The greater success of Enskog theory in the case of the thermal conductivity can plausibly be linked to the fact that there is no significant tail in the energy current autocorrelation function, which is a featureless curve that decays smoothly to zero.

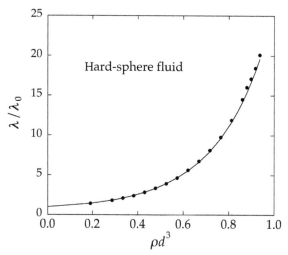

FIGURE 8.6 Thermal conductivity of the hard-sphere fluid as a function of density relative to its value in the low-density limit. The points are the results of molecular dynamics calculations[15] and the curve is the Enskog approximation (8.5.30).

8.6 GENERALISED HYDRODYNAMICS

In the earlier sections of this chapter we have shown in some detail how the equations of hydrodynamics can be used to calculate the time correlation functions of conserved variables in the long-wavelength, low-frequency limit. Two questions then arise. First, what are the scales of length and time over which it is possible to maintain the continuum description that underlies the hydrodynamic approach? Secondly, how may the hydrodynamic equations be modified to make their predictions applicable on the atomic scale, where lengths are typically of order a few ångström units and times are of order 10^{-13} s? We have seen in Chapter 7 that the behaviour of the correlation functions at short times is related to frequency sum rules involving static distribution functions descriptive of the molecular structure of the fluid. It is precisely these sum rules that are violated by hydrodynamic expressions such as (8.4.5) and (8.5.4), since the resulting frequency moments beyond zeroth order all diverge. In addition, an exponential decay, such as that in (8.4.3), cannot satisfy certain of the general properties of time correlation functions discussed in Section 7.1. The failure of the hydrodynamic approach at short times (or high frequencies) is linked to the presence of dissipative terms in the basic hydrodynamic equations; the latter, unlike the microscopic equations of motion, are not invariant under time reversal. In this section we describe some phenomenological generalisations of the hydrodynamic equations, based on the introduction of frequency and wavenumber-dependent transport coefficients, that have been developed in attempts to bridge the gap between the hydrodynamic (small k, ω) and kinetic (large k, ω) regimes. The use of non-local transport coefficients is closely related to the memory function approach of Section 7.3, which we develop in more systematic fashion in Chapter 9.

The ideas of *generalised hydrodynamics* are most easily illustrated by considering the example of the transverse current correlations. Equation (8.4.3) shows that in the hydrodynamic limit the correlation function $C_t(k, t)$ decays exponentially with a relaxation time equal to $1/\nu k^2$, where ν is the kinematic shear viscosity. The corresponding power spectrum is of Lorentzian form:

$$C_t(k, \omega) = \frac{1}{\pi} \operatorname{Re} \tilde{C}_t(k, \omega) = \frac{\omega_0^2}{\pi} \frac{\nu k^2}{\omega^2 + (\nu k^2)^2} \qquad (8.6.1)$$

The ω^{-2} behaviour at large ω is not compatible with the exact, high-frequency sum rules such as (7.4.38), nor does (8.6.1) yield the correct free-particle limit of $C_t(k, \omega)$ at large k; that limit is gaussian in form, similar to the longitudinal free-particle limit displayed in (7.5.17). Moreover, molecular dynamics calculations, which are the only source of 'experimental' information on transverse current fluctuations in atomic liquids, show that in an intermediate wavenumber range $C_t(k, t)$ decays in an oscillatory manner and its power spectrum has a peak at non-zero frequency, suggestive of the existence of a propagating shear mode. (Examples of the power spectra are shown later in Chapter 9, Figure 9.4.)

What this implies physically is that at high frequencies the fluid has insufficient time to flow in response to an applied strain rate, and instead reacts elastically in the manner of a solid. To account for the appearance of shear waves we need to extend the hydrodynamic description to include the effects of elasticity. Suppose that a shearing force is applied to a fluid. The strain at a point (x, y, z) is expressible in terms of the displacement $\mathbf{r}$ at that point and the rate of strain is expressible in terms of the velocity $\dot{\mathbf{r}}$. If the flow is purely viscous, the shearing stress (an off-diagonal component of the stress tensor $\mathbf{\Pi}$) is proportional to the rate-of-strain tensor and may be written as

$$\Pi^{xz} = -\eta \frac{\partial}{\partial t} \left(\frac{\partial r_x}{\partial z} + \frac{\partial r_z}{\partial x} \right) \tag{8.6.2}$$

which is the hydrodynamic form (see (8.3.15)). By contrast, if the force is applied suddenly, the instantaneous displacement is determined by the stress through a typical stress-strain relation, i.e.

$$\Pi^{xz} = -G_\infty \left(\frac{\partial r_x}{\partial z} + \frac{\partial r_z}{\partial x} \right) \tag{8.6.3}$$

where G_∞ is an instantaneous (high-frequency) modulus of rigidity. We can interpolate between these two extremes by making a *viscoelastic* approximation such that

$$\left(\frac{1}{\eta} + \frac{1}{G_\infty} \frac{\partial}{\partial t} \right) \Pi^{xz} = -\frac{\partial}{\partial t} \left(\frac{\partial r_x}{\partial z} + \frac{\partial r_z}{\partial x} \right) \tag{8.6.4}$$

By taking the Laplace transform of (8.6.4) it is easy to show that the viscoelastic approximation is equivalent to replacing η in (8.6.2) by a complex, frequency-dependent, shear viscosity given by

$$\tilde{\eta}(\omega) = \frac{G_\infty}{-i\omega + \tau_{\mathrm{M}}^{-1}} \tag{8.6.5}$$

The constant $\tau_M = \eta/G_\infty$ is called the Maxwell relaxation time. If $\omega\tau_M \ll 1$, $\tilde{\eta}(\omega) \approx \eta$, which corresponds to purely viscous flow, but if $\omega\tau_M \gg 1$, substitution of (8.6.5) in (8.6.4) yields a dispersion relation of the form $\omega^2 \approx (G_\infty/\rho m)\, k^2$, corresponding to elastic waves propagating at a speed

$$c_t = (G_\infty/\rho m)^{1/2} \tag{8.6.6}$$

Figure 8.7 shows the dispersion of the shear-wave peak observed in molecular dynamics simulations of liquid argon and liquid potassium at state conditions close to their respective triple points. Over the wavenumber range covered by the figure the dispersion is well described by a relation of the form $\omega = c_t(k - k_t)$, where k_t is the wavenumber below which the propagating mode vanishes. In the case of argon, for which a value of G_∞ is available from

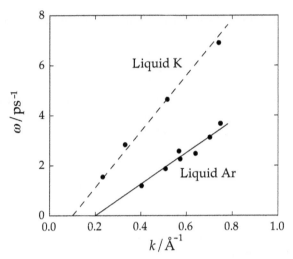

FIGURE 8.7 Dispersion of the shear-wave peak derived from molecular dynamics simulations of liquid argon[13,16] and liquid potassium[17] for state conditions close to the triple point. The dashed line through the data for potassium is a guide to the eye; the full line for argon is drawn with a slope given by the viscoelastic expression (8.6.6) for the speed of propagation (630 m s^{-1}). Results are shown only for the range of k in which the dispersion is approximately linear.

simulation, the slope of the dispersion curve is in surprisingly good agreement with that calculated from the viscoelastic approximation (8.6.6).

If account is to be taken of non-local effects in space the generalised shear viscosity must be a function of wavenumber as well as of frequency. The rigidity modulus is also dependent on k and related in a simple way to the second frequency moment ω_{1t}^2. These ideas can be formalised via a phenomenological generalisation of the hydrodynamic equation (8.4.2):

$$\frac{\partial}{\partial t}C_t(k,t) + k^2\int_0^\infty \nu(k,t-s)C_t(k,s)\mathrm{d}s = 0 \qquad (8.6.7)$$

The quantity $\nu(k,t)$ is a memory function; it describes a response that is non-local in both space and time and its Laplace transform $\tilde{\nu}(k,\omega)$ plays the role of a generalised kinematic shear viscosity. If we take the Laplace transform of (8.6.7) and compare the result with (8.4.4), we find that $\tilde{\nu}(k,\omega)$ must satisfy the constraint that

$$\lim_{\omega\to0}\lim_{k\to0}\tilde{\nu}(k,\omega) = \nu \qquad (8.6.8)$$

where ν is the macroscopic transport coefficient, given (apart from a factor ρm) by the Green–Kubo formula (8.4.10). If, on the other hand, we differentiate (8.6.7) with respect to t, set $t = 0$ and use (7.4.37), we find that

$$\nu(k,t=0) = \frac{\omega_{1t}^2}{k^2} \equiv \frac{G_\infty(k)}{\rho m} \qquad (8.6.9)$$

which acts as the definition of the k-dependent shear modulus $G_\infty(k)$. Equations (8.6.8) and (8.6.9) are useful in the construction of approximate forms of $\nu(k, t)$ that reduce to the hydrodynamic and viscoelastic expressions in the limits, respectively, $\omega \to 0$ and $\omega \to \infty$.

If molecular dynamics results for $C_t(k, t)$ are available, values of the generalised shear viscosity $\tilde{\eta}(k, \omega) = \rho m \tilde{\nu}(k, \omega)$ can be obtained by numerical inversion of (8.6.7) while its value at infinite wavelength, $\tilde{\eta}(k = 0, \omega) \equiv \tilde{\eta}(\omega)$, is given by the Laplace transform of the stress autocorrelation function $\eta(t)$ in (8.4.10). The generalised shear viscosity is believed to be a non-analytic function of both k and ω. For example, molecular dynamics calculations for hard spheres[18] have shown that $\eta(t)$ decays as $t^{-3/2}$ beyond about ten mean collision times, implying that $\tilde{\eta}(\omega)$ behaves as $\omega^{1/2}$ at low frequencies. If the zero-frequency shear viscosity $\eta(k) \equiv \tilde{\eta}(k, \omega = 0)$ could be expanded in a Taylor series in k about its macroscopic limit, $\eta \equiv \eta(0)$, the series would start as

$$\eta(k) = \eta + \eta_2 k^2 + \cdots \qquad (8.6.10)$$

since invariance under space inversion means that only even powers of k can appear. The quantity η_2 is called a Burnett coefficient. Burnett coefficients were introduced in an attempt to extend the range of validity of hydrodynamic equations through the addition of terms of higher order in the gradients of the hydrodynamic fields. However, the indications from mode coupling theories[19] of the type to be discussed in the section that follows are that the coefficients diverge, implying that the relation between the applied gradients and the induced hydrodynamic fluxes is non-analytic in character. This conclusion is supported by the results of computer simulations of a soft-sphere (r^{-12}) fluid[20], which are compatible with a small-k behaviour of the form

$$\eta(k) = \eta - \eta_{3/2} k^{3/2} + \cdots \qquad (8.6.11)$$

where $\eta_{3/2}$ is a positive quantity. These and related calculations[21] suggest that $\eta(k)$ and other generalised transport coefficients decrease smoothly with increasing wavenumber, becoming an order of magnitude smaller than their macroscopic ($k = 0$) values when the wavelength is comparable with the inter-particle spacing.

The longitudinal projections of the hydrodynamic equations may be treated in the same way through the introduction of wavenumber and frequency-dependent quantites that are generalisations of the coefficients a and b defined by (8.3.14) and (8.3.25). Similarly, the thermodynamic derivatives, which are related to static correlation functions, become functions of wavelength.[22] In particular, the macroscopic compressibility is replaced by its k-dependent generalisation, i.e. the structure factor $S(k)$ (see (3.6.11)), while the thermal pressure coefficient, which determines the coupling between momentum and energy, now contains a part that is explicitly dependent on frequency and vanishes in the limit $k \to 0$. A scheme in which the various thermodynamic and transport coefficients are assumed to be functions only of wavenumber and not of frequency has been found to reproduce satisfactorily a large part of the molecular dynamics results obtained for the dynamic structure factor of the hard-sphere fluid.[12] This approach breaks down, however, both for wavelengths shorter than the mean free path, corresponding to free-particle behaviour, and at densities close to crystallisation, where viscoelastic effects becomes important.

8.7   LONG-TIME TAILS IN TIME CORRELATION FUNCTIONS

Fluctuations in the conserved hydrodynamic variables decay infinitely slowly in the long-wavelength limit. The rates of relaxation are determined by the hydrodynamic eigenvalues (8.3.31) and (8.3.35) (multiplied by $-i$), all of which vanish with k. No such property holds for the non-conserved currents that enter the Green–Kubo integrands for the transport coefficients; if it did, the transport coefficients would not be well defined. Until the late 1960s it was generally believed that away from critical points the autocorrelation functions of non-conserved variables decay exponentially at long times. This, for example, is the behaviour predicted by the Boltzmann and Enskog equations. It therefore came as a surprise when analysis of the molecular dynamics results of Alder and Wainwright[23] on self diffusion in hard-disk ($\mathcal{D} = 2$) and hard-sphere ($\mathcal{D} = 3$) fluids showed that the velocity autocorrelation function apparently decays asymptotically as $t^{-\mathcal{D}/2}$, where $\mathcal{D}$ denotes the dimensionality of the system. Later simulations of hard-core fluids and other systems have also detected the presence of a long-time tail in the stress tensor autocorrelation function.

The presence of a slowly decaying tail in $Z(t)$ suggests that highly collective effects make a significant contribution to the process of self diffusion. The apparent involvement of large numbers of particles makes it natural to analyse the long-time behaviour in hydrodynamic terms, and Alder and Wainwright were led in this way to a simple but convincing explanation of their results. Underlying their argument is the idea that the initial motion of a tagged particle creates around that particle a vortex, which in turn causes a retarded current to develop in the direction of the initial velocity. At low densities, where the initial direction of motion is likely to persist, the effect of the current is to reduce the drag on the particle, thereby propelling it onwards in the forward direction. This results in a long-lasting, positive correlation between the initial velocity and its value at later times. At high densities, on the other hand, the initial direction of motion is on average soon reversed. In this case the retarded current gives rise to an extra drag at later times, causing $Z(t)$ to change sign; at very large times an enhancement of the forward motion can again be expected, but the effect is likely to be undetectable. That this physical picture is basically correct was confirmed in striking fashion by observation of the velocity field that forms around a moving particle in a fluid of hard disks. A vortex pattern quickly develops, which after a few mean collision times matches closely the pattern obtained by numerical solution of the Navier–Stokes equation. The persistence of the tail in $Z(t)$ is therefore associated with a coupling between the motion of the tagged particle and the hydrodynamic modes of the fluid. As we shall now show, this argument can be formalised in such a way as to predict the observed $t^{-\mathcal{D}/2}$ decay at long times.[24]

Suppose that at time $t = 0$ a particle i has a component of velocity $u_{ix}(0)$ in the x-direction. After a short time, τ say, collisions will have caused the initial momentum of particle i to be shared among the ρV_τ particles in a $\mathcal{D}$-dimensional volume V_τ centred on i. Local equilibrium now exists within the volume V_τ, and particle i will be moving with a velocity $u_{ix}(\tau) \approx u_{ix}(0)/\rho V_\tau$. (We have assumed, for simplicity, that the neighbours of i are initially at rest.) Further decay in the velocity $u_{ix}(t)$ for $t > \tau$ will occur as the result of enlargement of the volume V_τ, i.e. from the spread of the velocity field around particle i. At large times the dominant contribution to the growth of V_τ will come from diffusion of the transverse component of the velocity field and the radius of V_τ will therefore increase as $(\nu t)^{1/2}$. Thus $V_\tau \sim (\nu t)^{3/2}$ in the three-dimensional case, from which it follows that $Z(t) \sim (\nu t)^{-3/2}$. This argument assumes that particle i remains at the centre of V_τ; if the diffusive motion of i is taken into account it can be shown that

$$Z(t) \sim \left[(D + \nu)t\right]^{-3/2} \tag{8.7.1}$$

The analogous result in two dimensions implies that a self-diffusion coefficient does not exist, because the integral of $Z(t)$ diverges logarithmically.

The form of (8.7.1) has been confirmed by a number of more sophisticated calculations. In the case of hard-core fluids these include a microscopic treatment based on kinetic theory in which account is taken of the effect of correlated collision sequences (the ring collisions of Section 7.2) along with that of uncorrelated, binary collisions.[25] Though limited to low densities, the calculation shows that the velocity, stress tensor and energy current autocorrelation functions all decay as $t^{-\mathcal{D}/2}$; it also yields explicit expressions for the coefficients of the long-time tails. A more phenomenological approach has also been developed in which the existence of the long-time tails is explained by simple arguments concerning the decay of fluctuations into pairs of hydrodynamic modes. Since the physical content of this work is closely related to the mode coupling formalism to be discussed in Chapter 9, we give here a brief derivation of the result obtained in three dimensions for the velocity autocorrelation function.[26]

The definition (7.1.3) of a time correlation function involves an equilibrium ensemble average over the initial phase space coordinates of the system. This average can be replaced by a constrained ensemble average, characterised by an initial position $\mathbf{r}_0$ and initial velocity $\mathbf{u}_0$ of a tagged particle i, which is then integrated over all $\mathbf{r}_0$ and $\mathbf{u}_0$. The definition of $Z(t)$ is thereby reformulated as

$$Z(t) = \langle u_{ix}(t)u_{ix} \rangle = \int \mathrm{d}\mathbf{r}_0 \int \mathrm{d}\mathbf{u}_0\, u_{0x}\, \langle u_{ix}(t)\delta(\mathbf{u}_i - \mathbf{u}_0)\delta(\mathbf{r}_i - \mathbf{r}_0) \rangle \tag{8.7.2}$$

The constrained average in (8.7.2) can be written as a non-equilibrium ensemble average (subscript n.e.) defined through the relation

$$\langle u_{ix}(t)\delta(\mathbf{u}_i - \mathbf{u}_0)\delta(\mathbf{r}_i - \mathbf{r}_0) \rangle = \langle u_{ix}(t) \rangle_{\text{n.e.}}\, \langle \delta(\mathbf{u}_i - \mathbf{u}_0)\delta(\mathbf{r}_i - \mathbf{r}_0) \rangle \tag{8.7.3}$$

In the canonical ensemble the equilibrium average on the right-hand side of (8.7.3) is equal to $1/N$ times the single-particle distribution function defined by (2.1.15) (taken for $n = 1$) but with $\mathbf{p}$ replaced by $\mathbf{u}$ as independent variable. Equations (8.7.2) and (8.7.3) may therefore be combined to give

$$Z(t) = \frac{1}{V}\int \mathrm{d}\mathbf{r}_0 \int \mathrm{d}\mathbf{u}_0\, \phi_{\text{M}}(\mathbf{u}_0)u_{0x}\, \langle u_{ix}(t) \rangle_{\text{n.e.}} \tag{8.7.4}$$

where $\phi_{\text{M}}(\mathbf{u}_0)$ is the Maxwell distribution (2.1.28). By defining a tagged-particle distribution function in the non-equilibrium ensemble as

$$f^{(s)}(\mathbf{r}, \mathbf{u}; t) = \langle \delta\left[\mathbf{r}_i(t) - \mathbf{r}\right]\delta[\mathbf{u}_i(t) - \mathbf{u}] \rangle_{\text{n.e.}} \tag{8.7.5}$$

we can rewrite the non-equilibrium average in (8.7.4) as

$$\langle u_{ix}(t) \rangle_{\text{n.e.}} = \int \mathrm{d}\mathbf{r} \int \mathrm{d}\mathbf{u}\, u_x\, f^{(s)}(\mathbf{r}, \mathbf{u}; t) \tag{8.7.6}$$

The calculation thus far is exact. To make progress we assume that $f^{(s)}(\mathbf{r}, \mathbf{u}; t)$ relaxes towards the corresponding local equilibrium form on a time scale that is fast in comparison with the rate of decay of $Z(t)$. The long-time behaviour of the non-equilibrium average (8.7.6) is then obtained by replacing $f^{(s)}(\mathbf{r}, \mathbf{u}; t)$ by the tagged-particle analogue of (8.1.5) to give

$$\langle u_{ix}(t) \rangle_{\text{n.e.}} = \int \rho^{(s)}(\mathbf{r}, t)u_x(\mathbf{r}, t)\mathrm{d}\mathbf{r} \tag{8.7.7}$$

If this result is in turn substituted in (8.7.4), and the hydrodynamic variables $\mathbf{u}(\mathbf{r}, t)$ and $\rho^{(s)}(\mathbf{r}, t)$ are replaced by the sums of their Fourier components, we find that

$$Z(t) = \frac{1}{3V}\int \mathrm{d}\mathbf{r}_0 \int \mathrm{d}\mathbf{u}_0\, \phi_{\text{M}}(\mathbf{u}_0)\frac{1}{V^2}\sum_{\mathbf{k}}\sum_{\mathbf{k}'}\rho_{\mathbf{k}'}^{(s)}(t)$$

$$\times\, \mathbf{u}_{\mathbf{k}}(t) \cdot \mathbf{u}_0 \int \exp[-i(\mathbf{k} + \mathbf{k}') \cdot \mathbf{r}]\mathrm{d}\mathbf{r} \tag{8.7.8}$$

The integral over $\mathbf{r}$ is equal to $V\delta_{\mathbf{k}, -\mathbf{k}'}$ and (8.7.8) therefore reduces to

$$Z(t) = \frac{1}{3V}\int \mathrm{d}\mathbf{r}_0 \int \mathrm{d}\mathbf{u}_0\, \phi_{\text{M}}(\mathbf{u}_0)\frac{1}{V}\sum_{\mathbf{k}}\rho_{-\mathbf{k}}^{(s)}(t)\mathbf{u}_{\mathbf{k}}(t) \cdot \mathbf{u}_0 \tag{8.7.9}$$

Equation (8.7.9) is said to be of 'mode coupling' form because $Z(t)$ is expressed as a sum of products of pairs of hydrodynamic variables. We assume, in addition, that at times much longer than the mean collision time the decay of $Z(t)$ is dominated by the long-wavelength components of the hydrodynamic fields and that the time evolution of the latter is described by the equations of linearised hydrodynamics. The quantity $\rho_{-\mathbf{k}}^{(s)}(t)$ is then given by (8.2.5), while the hydrodynamic velocity field is conveniently divided into its longitudinal and transverse parts:

$$\mathbf{u_k}(t) = \mathbf{u}_{\mathbf{k}l}(t) + \mathbf{u}_{\mathbf{k}t}(t) \tag{8.7.10}$$

The term $\mathbf{u}_{\mathbf{k}t}(t)$ satisfies the transverse current diffusion equation (8.4.1) (with $\mathbf{j}_{\mathbf{k}t} = \rho\mathbf{u}_{\mathbf{k}t}$), the solution to which is

$$\mathbf{u}_{\mathbf{k}t}(t) = \mathbf{u}_{\mathbf{k}t} \exp\left(-\nu k^2 t\right) \tag{8.7.11}$$

The longitudinal velocity field may be treated in a similar way, but its contribution to $Z(t)$ turns out to decay exponentially, the physical reason for this being the fact that the momentum of the tagged particle is carried away by the propagating sound waves. Hence the long-time behaviour of $Z(t)$ is entirely determined by the transverse velocity field. Finally, the choice of initial conditions implies that

$$\rho_{-\mathbf{k}}^{(s)} = \exp\left(i\mathbf{k} \cdot \mathbf{r}_0\right) \tag{8.7.12}$$

and

$$\mathbf{j_k} = \rho\mathbf{u_k} = \mathbf{u}_0 \exp\left(-i\mathbf{k} \cdot \mathbf{r}_0\right) \tag{8.7.13}$$

An expression for $Z(t)$ is now obtained by substituting (8.7.11), (8.7.12) and the transverse projection of (8.7.13) into (8.7.9) (remembering that there are two transverse components), and integrating over $\mathbf{r}_0$ and $\mathbf{u}_0$. The result is

$$Z(t) = \frac{2k_{\mathrm{B}}T}{3\rho m V} \sum_{\mathbf{k}} \exp\left[-(D+\nu)k^2 t\right] \tag{8.7.14}$$

or, in the thermodynamic limit:

$$Z(t) = \frac{2k_{\mathrm{B}}T}{3\rho m}(2\pi)^{-3} \int \exp\left[-(D+\nu)k^2 t\right] d\mathbf{k} \tag{8.7.15}$$

Integration over all wavevectors is a questionable procedure, since the hydrodynamic equations on which (8.7.15) is based are not valid when k is large. However, we are interested only in the asymptotic form of $Z(t)$, and the main contribution to the integral comes from wavenumbers such that $k \approx [(D+\nu)t]^{-1/2}$; this is in the hydrodynamic range whenever t is much larger than typical microscopic times ($\sim 10^{-13}$) s. Alternatively, a natural upper limit on k can be introduced by a more careful choice of the initial spatial distribution of tagged particles. Use of such a cut-off has no effect on the predicted long-time behaviour that results from carrying out the integration in (8.7.15), namely

$$Z(t) \sim \frac{2k_B T}{3\rho m} \left[4\pi (D + v)t \right]^{-3/2}, \quad t \to \infty \tag{8.7.16}$$

This result has the same general form as (8.7.1) but it also provides an explicit expression for the coefficient of the long-time tail.

The result in (8.7.16) has been confirmed by molecular dynamics calculations for systems of hard discs and of particles interacting through a Lennard-Jones potential truncated at $r = 2^{1/6}\sigma$, the separation at which $v(r)$ has its minimum value; the simulations are difficult to carry out with the necessary precision because the long-time tail is very weak.[28] Results obtained for the truncated Lennard-Jones potential are shown in Figure 8.8, where $Z(t)$ is plotted versus t on a log-log scale. If (5.4.5) is used to define an effective hard-sphere diameter for the particles, the onset of the asymptotic behaviour is found to come after approximately 18 mean collision times. The predicted long-time behaviour of $Z(t)$ implies that at low frequencies its Fourier transform behaves as

$$Z(\omega) = \frac{D}{\pi} \left[1 - (\omega_0/\omega)^{1/2} + \cdots \right] \tag{8.7.17}$$

where ω_0 is related to the transport coefficients D and v. Experimentally, evidence for the presence of a long-time tail can be derived from neutron scattering measurements of the self dynamic structure factor, provided results are obtained at sufficiently small values of k to allow the extrapolation required

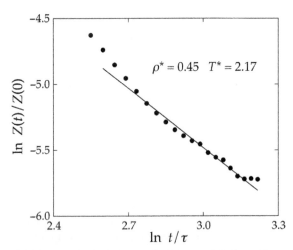

FIGURE 8.8 Log-log plot of the velocity autocorrelation function versus time for a system of particles interacting through a truncated Lennard-Jones potential. The points are molecular dynamics results and the line is drawn with a slope equal to $-3/2$. The unit of time is $\tau = (m\sigma^2/48\epsilon)^{1/2}$. *Redrawn with permission from Ref. 27 © 1974 American Physical Society.*

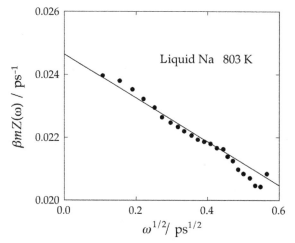

FIGURE 8.9 Power spectrum of the velocity autocorrelation function of liquid sodium as a function of $\omega^{1/2}$. The points are derived from inelastic neutron scattering measurements and the line is a least-squares fit to the data. *Redrawn with permission from Ref. 29 © 1987 American Physical Society.*

in (8.2.18) to be successfully carried through. Figure 8.9 shows some results obtained for liquid sodium at a temperature well above the melting point; at low temperatures the effect is too weak to be detectable. Not only is the square-root dependence on ω well reproduced, but the value obtained for ω_0 from a least-squares fit to the data lies within 2% of that predicted by mode coupling theory.

8.8 DYNAMICS OF SUPERCOOLED LIQUIDS

We know from Section 6.10 that when a liquid is quenched rapidly to temperatures below the freezing temperature T_f it may, rather than crystallising, undergo a transition to a glassy state at a glass transition temperature T_G. The freezing out of the translational and rotational degrees of freedom at the glass transition leads in many cases to anomalies in the temperature dependence of thermodynamic properties such as the specific heat. The change in behaviour at T_G is therefore described as a 'thermodynamic' or 'calorimetric' phase transition, though its nature is very different from that of an equilibrium phase transition. Section 6.10 was concerned with thermodynamic properties and the possible existence of an underlying ideal glass state; in this section we discuss the microscopic dynamics of liquids in the temperature range between T_f and T_G.

Relaxation times in the supercooled liquid measured, for example, in dielectric or shear stress relaxation experiments, increase dramatically with decreasing temperature; close to the glass transition they become comparable with macroscopic time scales. A rough but useful estimate of T_G is provided by the viscoelastic theory of Section 8.6, which shows that a crossover from viscous to elastic behaviour can be expected when the structural relaxation time of the system becomes comparable with the Maxwell relaxation time, defined as the ratio of shear viscosity to shear modulus, $\tau_M = \eta/G_\infty$. The shear modulus is of order 10^9 N m^{-2} for most materials and is only weakly dependent on temperature, but the shear viscosity rises by many orders of magnitude as the temperature approaches T_G. An implicit definition of T_G is obtained by identifying τ_M with some experimental time scale, τ_{exp}. A choice of 10^3 s for τ_{exp} leads to the conventional definition of T_G as the temperature at which the viscosity reaches a value of 10^{13} poise (1 P $\equiv 0.1$ N m^{-2} s).

Glass forming liquids fall into one of two broad classes: 'strong' and 'fragile'.[30] The difference between the two is particularly evident in the way in which the viscosity changes with temperature, as exemplified by the Arrhenius plots shown in Figure 8.10. Strong glass formers are covalently bonded, network forming substances such as silica; the network already exists in the high-temperature melt and gradually strengthens as the liquid is supercooled. The calorimetric anomalies near T_G are weak, or may be absent altogether, and the Arrhenius plots are essentially linear, implying that transport in the liquid is

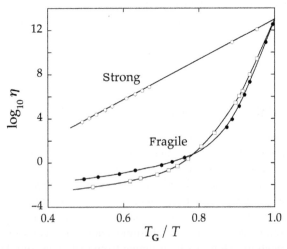

FIGURE 8.10 Arrhenius plots of the shear viscosities (in poise) of three glass-forming liquids, showing the difference in behaviour between strong and fragile glass formers. Open circles: silica; squares: o-terphenyl; filled circles: an ionic melt of composition [KNO_3]$_{0.6}$[$Ca(NO_3)_2$]$_{0.4}$. From C.A. Angell, 'Perspective on the glass transition', *J. Phys. Chem. Solids* **49**, 863–871 (1988), *with permission of Elsevier.*

largely governed by thermally activated processes or 'barrier hopping'. The anomalies are greater for the ionic and organic liquids that make up the class of fragile glass formers. Arrhenius plots for such materials show a marked change in curvature at a temperature T_C lying some 10–20% above T_G; this is suggestive of a qualitative change in character of the microscopic dynamics over a narrow temperature interval. When $T \approx T_C$, the Maxwell relaxation time is in the nanosecond range. This is a time scale well suited to studies of the dynamics by neutron and light scattering experiments and other experimental probes as well as by molecular dynamics simulation, and there is now ample evidence to show that as the temperature is lowered towards T_C there is a dramatic slowing down in the decay of time-dependent correlation functions. The crossover in behaviour near T_C seen, for example, in Figure 8.10, corresponds to what is called a *kinetic glass transition*. Experiment and simulation also show that structural and thermodynamic properties vary smoothly with temperature in the region of the transition. It is therefore reasonable to suppose that the supercooled liquid remains in a state of thermodynamic equilibrium and that equilibrium statistical mechanics applies once crystallisation has been by-passed. This is the key assumption underlying the mode coupling theory of the transition, described later in Section 9.6.

The nature of the changes that take place at the kinetic glass transition are well illustrated by the results shown in Figures 8.11 and 8.12. Those in Figure 8.11 are taken from a simulation of a binary,[31] soft-sphere (r^{-12}) fluid[32] and show the behaviour for one of the two species of the probability density

$$W(r,t) = 4\pi r^2 G_s(r,t) \tag{8.8.1}$$

where $G_s(r,t)$ is the self part of the van Hove function (7.4.19); the quantity $W(r,t)dr$ is the probability of finding a particle at time t at a distance in the range r to $r + dr$ from its position at $t = 0$. The thermodynamic state of the system is specified by a single coupling constant, Γ, defined in a manner similar to (5.4.13) but generalised to allow for the two-component nature of the system. A decrease in temperature is therefore strictly equivalent to an increase in density. The inset to the figure shows the results obtained for three different times at a value of Γ corresponding to a temperature above T_C. The curve has a single peak, which moves to larger r according to a $t^{1/2}$ law, in agreement with the result derived from Fick's law (see (8.2.8)). However, the qualitative behaviour changes dramatically above a threshold value of Γ, which can be identified with the crossover value Γ_C. The peak in $W(r,t)$ now appears frozen at a fixed value of r and its amplitude decreases only slowly with time as a secondary maximum builds up at a distance from the main peak roughly equal to the mean spacing between particles. The physical interpretation of this bimodal distribution is clear: most atoms vibrate around fixed, disordered positions, but some diffuse slowly by correlated hopping to neighbouring sites. The two values of Γ for which the results are shown correspond to temperatures differing by less

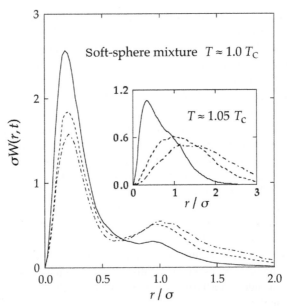

FIGURE 8.11 Molecular dynamics results for the probability density for diffusion of particles of one species in a two-component, soft-sphere fluid at temperatures in the supercooled region. Results are shown for three different values of the reduced time $t^* = t/\tau$. Full curves: $t^* = 100$; dashes: $t^* = 300$; chain curves: $t^* = 500$. For argon-like values of the potential parameters and particle masses, $\tau \approx 2\,\text{ps}$; σ is an averaged size parameter. From J.L. Barrat et al., 'Diffusion, viscosity and structural slowing down in soft-sphere alloys near the kinetic glass transition', *Chem. Phys.* **149**, 197–208 (1990), *with permission of Elsevier.*

than 6%. Thus the diffusion mechanism changes very rapidly from one that is hydrodynamic-like to one consisting of a succession of activated jumps. Further strong evidence of the change in diffusion mechanism is provided by increasing deviations from the Stokes-Einstein relation (7.3.19) as the temperature falls below T_C; the diffusion coefficient D is found to be substantially larger than predicted, by orders of magnitude at the lowest temperatures explored in simulations[32] or experiments.[33] This trend is indicative of an increased decoupling of the single-particle motion from collective, viscous flow.

The pronounced slowing down of single-particle motion as a threshold temperature is reached is also visible in the behaviour of the self intermediate scattering function $F_s(k,t)$ defined by (7.5.12) or, equivalently, by

$$F_s(k,t) = \frac{1}{N} \sum_{i=1}^{N} \left\langle \exp\left(i\mathbf{k}.[\mathbf{r}_i(t) - \mathbf{r}_i(0)]\right)\right\rangle \tag{8.8.2}$$

Some molecular dynamics results[34] obtained for a binary Lennard-Jones (the Kob–Andersen model) are shown in Figure 8.12. (Note that time is plotted on a logarithmic scale.) At high temperatures the correlation function relaxes

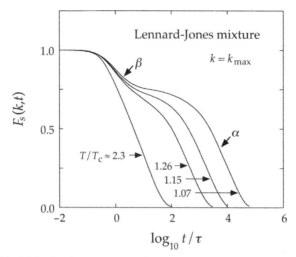

FIGURE 8.12 Molecular dynamics results for the self intermediate scattering function for particles of one species in a two-component Lennard-Jones fluid at temperatures in the supercooled region; k_{max} is the wavenumber corresponding to the main peak in the static structure factor. The labels α, β mark the two different relaxation regimes discussed in the text. For argon-like values of the potential parameters and particle masses, the unit of time is $\tau \approx 0.3$ ps. *Redrawn with permission from Ref. 34 © 1995 American Physical Society.*

to zero in near-exponential fashion. However, as the temperature is lowered into the supercooled region, the decay becomes very much slower and its exponential character is lost. As T approaches T_C, the relaxation proceeds in two, increasingly well-separated steps. After a fast initial decay on the time-scale of an inverse Einstein frequency, a first step (β-relaxation) leads to a plateau, where the function remains almost constant over two or more decades in time. The plateau is followed by a second step (α-relaxation) in which the correlation function finally decays to zero. The width of the plateau increases rapidly as the temperature is reduced. Eventually, when the temperature is sufficiently low, α-relaxation can be expected to set in only at times longer than those accessible in a simulation. The correlation function will then appear to level off at a non-zero value, signalling the onset of non-ergodic behaviour, at least on the nanosecond time scale of the simulation. The plateau value varies with k, but the general pattern seen in Figure 8.12 remains much the same over a wide range of molecular-scale wavenumber.

The decay of collective density fluctuations, as described by the full intermediate scattering function $F(k,t)$ (7.4.20) and measurable either experimentally or by simulation, shows a qualitatively similar behaviour to that of the single-particle function. The plateau value of $F(k,t)$ is analogous to the Debye–Waller factor of a solid; it provides a measure of the degree of structural arrest in the fluid, which persists for times that increase rapidly with decreasing temperature. Over a temperature range just above T_C, the decay of either function in the α-relaxation regime, normalised by its value at $t = 0$, is accurately represented by a function of the form

$$f(t) = f_k \Phi \left(t^* \right) \tag{8.8.3}$$

where f_k is the plateau value, $t^* \equiv t/\tau_k(T)$ and $\Phi(t^*)$ is a universal scaling function. The wavenumber and temperature dependence of the decay enter only through the relaxation time $\tau_k(T)$ and the correlation functions are said to satisfy a time-temperature superposition principle. The scaling function is distinctly non-exponential, but is generally well-approximated by a Kohlrausch, stretched-exponential function, i.e.

$$\Phi(t^*) \approx \exp\left[-(t^*)^\beta\right] \tag{8.8.4}$$

where the exponent[35] β (<1 for stretching) is material and wavenumber dependent but independent of temperature.[36] Stretched-exponential behaviour is typical of relaxation processes in which the experimentally observed rate is determined by a wide distribution of relaxation times.

At temperatures below that of the kinetic glass transition the viscosity of the liquid increases dramatically, particularly in the case of fragile glass formers, meaning that the time required for the system to come to equilibrium after some perturbation is applied becomes much longer as the temperature decreases towards T_G. On realistic time scales the system falls out of equilibrium and equilibrium statistical mechanics no longer applies, a situation accompanied by the appearance of phenomena such as ageing, deviations from the fluctuation–dissipation theorem and spatial inhomogeneities in the particle dynamics.

The ageing of glassy materials is seen most obviously in the fact that their macroscopic properties usually change with time; for example, the molar volume of many polymeric glasses slowly decreases from the time of their preparation. On a microscopic level the ageing of supercooled liquids is reflected in the loss of time translational invariance even at temperatures above T_G. Time dependent correlation functions, as defined by (7.1.1), are now functions of two time arguments, t' and t'', not merely of the time difference $t = t' - t''$. An out-of-equilibrium system retains a memory of its initial state at a time t'', and its subsequent time evolution depends explicitly on both t' and t''. Let us suppose that a system in thermal equilibrium at a temperature above T_C at time $t = 0$ is quenched to a temperature $T \approx T_G$. The subsequent relaxation of the quenched system can be probed, after some waiting period t_0, by measurement of a time autocorrelation function of the form

$$C_{AA}(t_0 + t, t_0) = \left\langle A(t_0 + t)A^*(t_0) \right\rangle \tag{8.8.5}$$

where the statistical average is taken over equilibrium configurations at the initial temperature. It is found that the decay of the correlation function

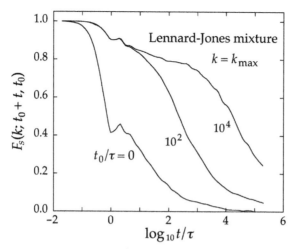

FIGURE 8.13 Decay of the self intermediate scattering function of a quenched Lennard-Jones mixture at a reduced temperature $T/T_C \approx 0.92$ for three values of the waiting time t_0 following the quench. For argon-like values of the potential parameters and particle masses the unit of time is $\tau \approx 0.3$ ps; k_{max} is the wavenumber corresponding to the main peak in the structure factor. *Redrawn from Ref. 37 with permission of European Physics Journal B © 2000 Springer.*

invariably slows down as the waiting time t_0 increases. An example taken from simulations of the Kob–Andersen model is shown in Figure 8.13, where the correlation function is again the self intermediate scattering function, $F_s(k; t_0+t, t_0)$, now defined in terms of the waiting time. Immediately following the quench, i.e. for $t_0 = 0$, the correlation function decays rapidly, but the rate of decay gradually reduces as t_0 increases. Data from both experiment and simulation, as well as theoretical considerations, suggest that at long times the autocorrelation functions of ageing systems satisfy a time/ageing-time superposition principle[37,38] of the form

$$C_{AA}(t_0 + t, t_0) \approx C_{AA}^s(t) + C_{AA}^a\left(\frac{h(t_0 + t)}{h(t_0)}\right) \tag{8.8.6}$$

The short-time contribution (superscript s) depends only on the time difference t, while the ageing part (superscript a) depends on the ratio $h(t_0 + t)/h(t_0)$, where $h(t)$ is a time-scaling function. In the case illustrated by Figure 8.13 the results at long times fall on a single curve if the scaling function is taken as $h(t) \propto t^\alpha$, with $\alpha \approx 0.88$.

Another feature of out-of-equilibrium systems, which is closely related to ageing, is the violation of the fluctuation-dissipation theorem. If the assumption of time translation invariance is dropped, the expression for the after-effect function in (7.6.13) becomes

$$\Phi_{BA}(t', t'') = \beta \langle B(t')\dot{A}(t'') \rangle = \beta \frac{\partial}{\partial t''} C_{BA}(t', t'') = \frac{\delta \langle \Delta B(t') \rangle}{\delta \mathcal{F}(t'')} \tag{8.8.7}$$

where the last equality follows from (7.6.12). The integrated response or susceptibility is then

$$
\chi_{BA}(t',t'') = \int_{t''}^{t'} \Phi_{BA}(t',s)ds = \beta \int_{t''}^{t'} \frac{\partial}{\partial s} C_{BA}(t',s)ds
$$
$$
= \beta \left[C_{BA}(t',t') - C_{BA}(t',t'') \right] \tag{8.8.8}
$$

For a system at equilibrium, (8.8.8) (with $t = t' - t''$) reduces to

$$
\chi_{BA}^{eq}(t) = \beta \left[C_{BA}(0) - C_{BA}(t) \right] \tag{8.8.9}
$$

Thus, within the linear response regime, a parametric plot of $\chi_{BA}^{eq}(t)$ versus $C_{AB}(t)$ yields a straight line of slope $-\beta$ irrespective of the choice of dynamical variables A and B. However, for systems far from equilibrium the fluctuation-dissipation theorem no longer holds. The non-equilibrium situation can be described by a generalisation of (7.6.13) having the form

$$
\Phi(t',t'') = \beta \, \Psi(t',t'') \frac{\partial}{\partial t''} C(t',t'') \tag{8.8.10}
$$

where we now restrict the discussion to autocorrelation functions but omit the subscript AA; in the case of time correlations described by (8.8.6), $t' \equiv t_0 + t$ and $t'' \equiv t_0$. The unknown function $\Psi(t',t'')$ is called the fluctuation-dissipation ratio. At equilibrium $\Psi(t',t'') = 1$ for all t',t''; away from equilibrium it is known[39] that within the mean field theory of spin glasses the fluctuation-dissipation ratio is a function only of the autocorrelation function to which it refers, i.e. $\Psi(t',t'') = \xi \left(C(t',t'') \right)$. It is generally assumed that the same simplifying property holds for structural glasses, for which the disorder is positional rather than magnetic. If that is so, then in its application to ageing (8.8.10) may be rewritten as

$$
\Phi(t_0 + t, t_0) = \beta \xi \left(C(t_0 + t, t_0) \right) \frac{\partial C(t_0 + t, t_0)}{\partial t_0} \tag{8.8.11}
$$

and the integrated response, as defined in the first line of (8.8.8) is now

$$
\chi(t_0 + t, t_0) = \beta \int_{t_0}^{t_0+t} \xi(C) \frac{\partial}{\partial s} C(t_0 + t, s)ds = -\beta \int_{C(t_0+t,t_0)}^{1} \xi(C)dC \tag{8.8.12}
$$

provided the correlation function is normalised by its value at $t = t_0$.

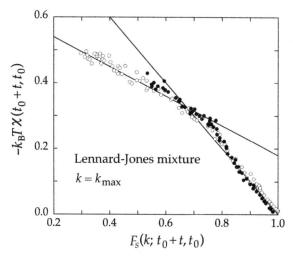

FIGURE 8.14 Parametric plot of susceptibility versus autocorrelation function for a supercooled Lennard-Jones mixture at two values of the waiting time t_0 and a reduced temperature $T^* = 0.3$. The dynamical variable involved is a Fourier component of the single-particle density. Open circles: $t_0/\tau = 1063$; closed circles; $t_0/\tau = 10^4$. For argon-like values of the potential parameters and particle masses the unit of time is $\tau \approx 0.3$ ps. See text for details. *Redrawn from Ref. 37 with permission of European Physics Journal B. © 2000 Springer.*

Examples of parametric plots of $\chi(C)$ versus C obtained by molecular dynamics calculations for the Kob–Andersen model at two values of the waiting time are shown in Figure 8.14. The correlation function in this case is again the self intermediate scattering function for a wavenumber corresponding to the main peak in the structure factor $S(k)$; the susceptibility measures the response of the system to a weak, external potential spatially modulated at the same wavenumber. The data are reasonably well fitted by two straight lines, with a clear break in slope when the correlation function has fallen to approximately 70% of its initial value. At short times the slope $p = -k_B T d\chi/dC \approx 1$, as it would be for a system at equilibrium, but at longer times $p \approx 0.45$; the fluctuation-dissipation theorem is satisfied at short but not at long times. The fact that the fluctuation-dissipation ratio at long times, ξ_∞, remains nearly constant as the correlation function decays to zero, i.e. as $t \to \infty$, suggests that relaxation in the long-time, ageing regime can be described in terms of an effective temperature

$$T_{\text{eff}} = \frac{T}{\xi_\infty} > T \qquad (8.8.13)$$

which depends on the temperature to which the system was quenched but is independent of t_0. Thus the rapidly relaxing modes of the glass, such as the 'rattling' motion of atoms in the cage formed by their nearest neighbours, respond to an external perturbation in a manner consistent with the equilibrium fluctuation-dissipation theorem for the physical temperature T, while the slowly relaxing modes, associated with collective structural relaxation, are characterised by a higher, effective temperature T_{eff}. In these calculations ξ_∞ was found to be independent of wavenumber but more generally it appears to be independent of the choice of dynamical variable that the autocorrelation function monitors. The existence of a two-valued fluctuation-dissipation ratio is also apparent in studies of glassy systems driven into a stationary non-equilibrium state by the application of a steady, external force field.[40,41] In that situation, ageing is halted and time translation invariance restored, while the role of the waiting time t_0 in ageing glasses is played by the strength of the driving force. A good example is that of a sheared, supercooled fluid, where the rate of shear γ defines a time scale γ^{-1}. Molecular dynamics calculations[41] have confirmed that under such conditions the effective temperature for a given system and physical temperature is again independent of the dynamical variable which is probed.

A last example of the way in which the dynamics of strongly supercooled liquids differ from those of fully equilibrated systems is provided by a number of effects that are grouped together under the heading of *dynamical heterogeneity*. Whereas in equilibrium systems single-particle and collective motions are spatially homogeneous in the sense that they relax on the same time scale throughout the volume of the system, there is considerable evidence from both experiment[42] and simulation[43] of a high degree of spatial-temporal heterogeneity in supercooled liquids. Over time scales shorter than the structural relaxation time the system is a patchwork of active and quiescent domains.[44] In one type of domain, particle motions are coordinated; in the other, motion is primarily one of small amplitude vibrations around frozen positions. The active and quiescent states are intermittent, with the boundaries between the two domains evolving slowly with time. It has been conjectured[45] that the coexistence of active and quiescent regions may explain the observed, stretched-exponential behaviour that defines the α-relaxation regime. It has also been suggested that a growing and possibly divergent length scale of the domains might be associated with the divergence of the structural relaxation time as the temperature is reduced towards T_G; in the active domains, an increasingly large number of particles must move cooperatively to allow structural relaxation. The earliest simulations of binary mixtures identified the presence of highly correlated, string-like motions of adjacent particles, involving nearly instantaneous jumps over distances comparable with the interparticle spacing.[46] These observations were later confirmed by more extensive simulations[43] and by optical imaging studies of jammed, colloidal dispersions.[47] The coexistence near the kinetic glass transition temperature of two populations of particles that differ in their local dynamics is already visible in the probability densities pictured in Figure 8.8, a conclusion reinforced by numerical studies of the self part of the van Hove function for a variety of supercooled liquids.[48] For liquids at temperatures above T_C, $G_s(\mathbf{r}, t)$ is given to a good approximation by the gaussian distribution (8.2.8). That result is a consequence of Fick's law (8.2.2), where the concept of a uniform diffusion coefficient is introduced. Below T_C a pronounced non-gaussian tail appears for displacements greater than roughly one particle diameter. The tail is well described by an exponential function of the form $G_s(r, t) \propto \exp[-r/\lambda(t)]$, where $\lambda(t)$ is a decay length that increases slowly with time. The picture that emerges is again one in which there are two populations of particles: those that are largely immobile and contribute to the central, gaussian part of the distribution; and those that are considerably more mobile and contribute to the exponential tail. Only at very long times is Fick's law behaviour recovered. A non-gaussian tail was also detected in the optical imaging experiments.

8.9 FLOW OF LIQUIDS AT THE INTERFACE WITH A SOLID

The static structure, interfacial thermodynamics and phase behaviour of confined, inhomogeneous fluids are by now well understood within the unifying framework of density functional theory, as described in Chapter 6. Significant progress has also been made in the experimental and theoretical investigation of dynamical processes at interfaces, stimulated in part by the increasing importance of nanofluidics and its technological applications.[49] In this section we show how the concepts and methods developed for the study of dynamics and transport in bulk liquids can be adapted to the flow of liquids close to a solid surface or confined to a narrow slit. On a macroscopic scale the flow pattern of a newtonian fluid, for which the local stress and strain rate are linearly related, is obtained by solution of the Navier–Stokes equation (8.3.16) subject to appropriate boundary conditions at any confining surfaces. An example is provided by Stokes's law (7.3.18). This gives an expression for the frictional force exerted by a flowing fluid on a suspended sphere in terms of a friction coefficient ξ, the value of which depends on the choice of boundary condition, stick or slip. The stick boundary condition is the one more appropriate for a sphere of diameter typical of large, colloidal particles. The situation on shorter length scales is more complex.

We take as an example the problem of fluid flow along a solid surface lying parallel to the xy-plane. Friction arises from the transfer of momentum from fluid to solid. If the surface is sufficiently rough, the assumption can be made that the tangential component of the local velocity field $\mathbf{u}(\mathbf{r}, t)$ vanishes at the interface. This corresponds to the stick or 'no slip' boundary condition, which is the one commonly assumed to apply at the interface of a fluid with a macroscopic surface. In the case of an ideal, atomistically smooth surface there can be no transfer of momentum parallel to the surface; hence there is no frictional force and the slip boundary condition applies. Between these two extremes there will be situations of 'partial' slip, where the tangential component of velocity at the interface is non-zero, though usually small. For the sake of simplicity we restrict the discussion to the case of laminar flow directed along the x-axis. Then the three possibilities we have listed can be accommodated within a single, phenomenological, boundary condition of the form[50]

$$\left(\frac{\partial u_x(z,t)}{\partial z}\right)_{z=z_{\mathrm{h}}} = \frac{1}{b}u_x(z_{\mathrm{h}},t) \qquad (8.9.1)$$

where b is the *slipping length*. Equation (8.9.1) applies at the 'hydrodynamic' boundary, positioned at $z = z_{\mathrm{h}}$, which will not in general coincide with the surface of the solid. The value of z_{h} appears instead as a parameter of the theory sketched below; typically the hydrodynamic boundary is found to lie above the physical interface by one to two particle diameters.[51]

The three different scenarios consistent with (8.9.1) are represented by the Couette flow patterns pictured schematically in Figure 8.15. Part (a) of the figure corresponds to the stick boundary condition; the fluid velocity vanishes at $z = z_{\mathrm{h}}$ and $b = 0$. For moderate degrees of roughness, but depending also on other features of the fluid-solid interaction, some 'velocity slip' may occur, as shown in part (b); this effect is measured by the *slip velocity*, $u_{\mathrm{S}} = u_x(z_{\mathrm{h}})$. We shall see later that b is the distance below the hydrodynamic boundary at which u_x vanishes when the flow profile is extrapolated linearly into the solid. Part (c) corresponds to the limiting case of a perfectly smooth surface, where $b \to \infty$.

The quantity $\partial u_x / \partial z$ is the xz-component of the rate-of-strain tensor $\dot{\gamma}$. Equation (8.9.1) may therefore be written as

$$\dot{\gamma}^{xz} = \frac{\partial u_x}{\partial z} \equiv \dot{\gamma} \qquad (8.9.2)$$

where $\dot{\gamma}$ is the shear rate, which in turn is related to the xz-component of the stress tensor and shear viscosity of the bulk fluid by the constitutive relation (8.3.15):

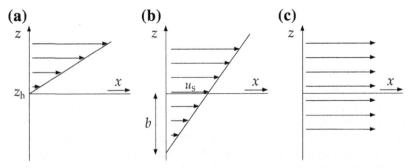

FIGURE 8.15 Schematic Couette flow patterns corresponding to (a) stick, (b) partial slip and (c) perfect slip boundary conditions, each applied at the hydrodynamic boundary $z = z_{\mathrm{h}}$. In part (b), b is the slipping length and u_{S} is the slip velocity, i.e. the x-component of the fluid velocity field at $z = z_{\mathrm{h}}$.

$$\Pi^{xz} = -\eta\dot{\gamma} \tag{8.9.3}$$

The stress tensor is also related to the tangential force F_λ exerted by the solid on the moving fluid:

$$\Pi^{xz} = -\lambda u_S = F_\lambda/\mathcal{A} \tag{8.9.4}$$

where $\mathcal{A}$ is the total surface area and λ is the fluid-solid friction coefficient. As $z \to z_h$, the internal friction (8.9.3) of the liquid must balance the friction (8.9.4) exerted by the wall, which implies that $u_S = (\eta/\lambda)\dot{\gamma}$. It then follows from (8.9.1) that the slipping length is given by the ratio of two transport coefficients:

$$b = \eta/\lambda \tag{8.9.5}$$

This result implies that the slipping length is an intrinsic property of a fluid at a given interface. It also shows that as the friction coefficient increases, b will decrease, and the boundary condition will become more and more stick-like. The two parameters that specify the boundary condition are b and z_h or, equivalently, for a fluid of given shear viscosity, λ and z_h. As we shall see, both λ and z_h are expressible in terms of microscopic dynamical variables[51], leading to Green–Kubo relations similar in nature to those that determine the transport coefficients of bulk fluids.

The hamiltonian of the fluid will contain the usual terms corresponding to interaction between particles of the fluid and that between fluid and solid, while the roughness of the surface can be represented by a periodic modulation of the fluid-solid potential. To complete a microscopic model of the system a stationary, Couette flow field $u_x(\mathbf{r})$ is imposed on the fluid by borrowing an idea used in non-equilibrium molecular dynamics calculations. A fictitious, non-newtonian, perturbation term $\mathcal{H}'$ is added to the unperturbed hamiltonian $\mathcal{H}_0$, where

$$\mathcal{H}' = \dot{\gamma} \sum_{i=1}^{N} (z_i - z_0)p_{xi} \equiv -\dot{\gamma}A \tag{8.9.6}$$

which also acts as the definition of a dynamical variable A. The x-component of the flow field induced by the perturbation is

$$\begin{aligned} \rho^{(1)}(\mathbf{r})u_x(\mathbf{r}) &= \left\langle \sum_{i=1}^{N} v_{xi}\delta(\mathbf{r}_i - \mathbf{r}) \right\rangle \\ &= \frac{\int \exp\left[-\beta\left(\mathcal{H}_0 + \mathcal{H}'\right)\right] \sum_{i=1}^{N} v_{xi}\delta(\mathbf{r}_i - \mathbf{r})d\mathbf{r}^N d\mathbf{p}^N}{\int \exp\left[-\beta\left(\mathcal{H}_0 + \mathcal{H}'\right)\right] d\mathbf{r}^N d\mathbf{p}^N} \end{aligned} \tag{8.9.7}$$

where $\rho^{(1)}(\mathbf{r})$ is the single-particle density (2.5.11). Linearisation of (8.9.7) with respect to $\mathcal{H}'$, and use of the fact that both $\mathcal{H}'$ and the dynamical variable to be averaged are odd functions of the particle momenta, shows that

$$\rho^{(1)}(\mathbf{r})u_x(\mathbf{r}) = \beta m \dot{\gamma} \left\langle \sum_i v_{xi}(z_i - z_0) \sum_j v_{xj}\delta(\mathbf{r}_j - \mathbf{r}) \right\rangle_0$$

$$= \dot{\gamma}(z - z_0)\left\langle \sum_i \delta(\mathbf{r}_i - \mathbf{r}) \right\rangle_0$$

$$= \dot{\gamma}(z - z_0)\rho^{(1)}(\mathbf{r}) \qquad (8.9.8)$$

In the planar geometry assumed here, both $\rho^{(1)}(\mathbf{r})$ and $u_x(\mathbf{r})$ depend only on the coordinate z. The perturbation therefore gives rise to a velocity field of the required form:

$$u_x(z) = \dot{\gamma}(z - z_0) \qquad (8.9.9)$$

which vanishes at $z = z_0$.

Let F_x^s be the instantaneous value of the force exerted by the solid on the moving fluid. Its mean value $\langle F_x^s(t)\rangle$ at a time t can be calculated by the linear response theory of Section 7.6. From the general relations (7.6.12) and (7.6.13) we find that

$$\langle F_x^s(t)\rangle = \beta\dot{\gamma}\int_0^t \langle F_x^s(t - t')\dot{A}(0)\rangle_0 \, dt' \qquad (8.9.10)$$

where the average is taken over the unperturbed system, the dynamical variable A is that defined in (8.9.6), and the shear rate $\dot{\gamma}$ acts only from $t = 0$ and is thereafter constant in time. The time derivative of A is

$$\dot{A} = \sum_{i=1}^N \left(\frac{p_{xi}\,p_{zi}}{m} + (z_i - z_0)F_{xi}\right) \qquad (8.9.11)$$

where $F_{xi} = F_{xi}^f + F_{xi}^s$ is the x-component of the total force acting on particle i; this is the sum of the microscopic forces due to other fluid particles (f) and the force arising from its interaction with the solid. From the definition (8.4.14) of the microscopic stress tensor in the $k \to 0$ limit, together with Newton's Third Law, which implies that $\sum_i F_{xi}^f = 0$, it follows that (8.9.11) may be rewritten as

$$\dot{A} = \Pi^{xz} - z_0 \sum_i F_{xi}^s = \Pi^{xz} - z_0 F_x^s \qquad (8.9.12)$$

By substituting (8.9.12) in (8.9.10) and taking the limit $t \to \infty$, we obtain an expression for the total frictional force F_λ:

$$F_\lambda = \lim_{t \to \infty} \langle F_x^s(t)\rangle$$

$$= -\beta\dot{\gamma}z_0\int_0^\infty \langle F_x^s(t)F_x^s(0)\rangle_0 \, dt + \beta\dot{\gamma}\int_0^\infty \langle F_x^s(t)\Pi^{xz}(0)\rangle_0 \, dt \qquad (8.9.13)$$

If we take as definitions of the two quantities λ and z_h:

$$\lambda = \frac{\beta}{\mathcal{A}} \int_0^\infty \langle F_x^s(t) F_x^s(0) \rangle_0 \, dt \qquad (8.9.14)$$

and

$$z_h = \frac{\int_0^\infty \langle F_x^s(t) \Pi^{xz}(0) \rangle_0 \, dt}{\int_0^\infty \langle F_x^s(t) F_x^s(0) \rangle_0 \, dt} \qquad (8.9.15)$$

the frictional force per unit area is

$$F_\lambda / \mathcal{A} = \dot{\gamma} \lambda (z_h - z_0) = -\lambda u_s \qquad (8.9.16)$$

Identification of (8.9.16) with the phenomenological relation (8.9.4) shows that (8.9.14) and (8.9.15) represent Green–Kubo formulae for the friction coefficient and location of the hydrodynamic boundary, respectively, while (8.9.1), (8.9.9) and (8.9.16) together show that the slipping length b is given by

$$b = |z_0 - z_h| \qquad (8.9.17)$$

where the boundary condition (8.9.1) is applied at $z = z_h$, with z_h determined by (8.9.15). Equivalently, given (8.9.5), b is determined by λ and η as evaluated from (8.9.14) and (8.4.20), and is therefore independent of the initial choice of z_0; a shift in z_0 leads only to a compensating shift in z_h. The expression for λ is analogous to that given for the friction coefficient ξ of Langevin theory by (7.3.8). Note, however, that whereas ξ is simply a frequency the dimensions of λ show that its meaning is that of momentum transfer per unit area per unit time.

Given a microscopic model of the fluid and surface, the correlation function expressions (8.9.14) and (8.9.15) can in principle be used to determine λ and z_h by equilibrium molecular dynamics simulation of a fluid confined between two parallel, planar surfaces. A less direct but more accurate method involves the computation of the transverse momentum density autocorrelation function $C_t(z, z'; t)$ of the confined fluid. The results of the simulation are then fitted to those obtained by analytical solution of the Navier–Stokes equation, similar to that of its **k**-space equivalent (8.4.2) in the bulk and based on the assumption that the boundary condition (8.9.1) applies at each surface; if the surfaces are identical, the fitting parameters are λ and b. The value thereby obtained for the slipping length is sensitive to the nature of the fluid-solid potential. If the interaction is strongly attractive, corresponding to a regime in which the liquid wets the surface, b is very small and the slip velocity $u_s \approx 0$. If the attraction is weak, b can become very large, of the order of tens of particle diameters, and 'near slip' boundary conditions apply.[52] Support for these conclusions has come experimentally from surface force measurements. In one investigation,[53] for example, a study was made of the flow behaviour of two, contrasting liquids, water and dodecane ($C_{12}H_{26}$), confined by hydrophilic or hydrophobic surfaces.

The stick boundary condition was found to apply for dodecane at both types of surface and for water at a hydrophilic surface. By contrast, in the hydrophobic case, where water does not wet the surface, a high degree of slip was observed, with $b \approx 19$ nm; this is some sixty times larger than the dimensions of a water molecule. The same experiments also showed that confinement had no measurable effect on the shear viscosity of either liquid down to film thicknesses of about 4 nm for dodecane and 10 nm for water.

REFERENCES

[1] Rahman, A., *Phys. Rev.* **136** A405 (1964).

[2] Landau, L.D. and Lifshitz, E.M.,'Fluid Mechanics', 2nd edn. Butterworth-Heinemann, Oxford, 1987, p. 44.

[3] Reif, F., 'Fundamentals of Statistical and Thermal Physics'. McGraw-Hill, New York, 1965, p. 168.

[4] (a) Mountain, R.D., *Rev. Mod. Phys.* **38**, 205 (1966). (b) Berne, B.J. and Pecora, R., 'Dynamic Light Scattering'. John Wiley, New York, 1976.

[5] Chapman, S. and Cowling, T.G., 'The Mathematical Theory of Non-Uniform Gases', 3rd edn. Cambridge University Press, Cambridge, 1970, p. 308. For a simplified discussion of the Boltzmann and Enskog equations see, e.g., Reed, T.M. and Gubbins, K.E., 'Applied Statistical Mechanics'. McGraw-Hill, New York, 1973.

[6] Heyes, D.M., *J. Phys. Condens. Matter* **19**, 376106 (2007), Table 1.

[7] Heyes, D.M., Cass, M.J., Powles, J.G. and Evans, W.A.B., *J. Phys. Chem. B* **111**, 1455 (2007).

[8] Harris, K.R., *J. Chem. Phys.* **131**, 054503 (2009).

[9] This is not the same effect as the $t^{-3/2}$ tail referred to in Sections 8.6 and 8.8.

[10] Boon, J.P. and Yip, S., 'Molecular Hydrodynamics'. McGraw-Hill, New York, 1980, pp. 249–50.

[11] For inelastic X-ray scattering and references to other neutron scattering experiments on liquid metals see, Scopigno, T., Ruocco, G. and Sette, F., *Re. Mod. Phys.* **77**, 031205 (2002).

[12] Alley, W.E., Alder, B.J. and Yip, S., *Phys. Rev. A* **27**, 3174 (1983).

[13] Levesque, D., Verlet, L. and Kürkijarvi, J., *Phys. Rev. A* **27**, 1690 (1973).

[14] Bodensteiner, T., Morkel, C., Gläser, W. and Dorner, B., *Phys. Rev. A* **45**, 5709 (1992); erratum: *Phys. Rev. A* **46**, 3574 (1992).

[15] Sigurgeirsson, H. and Heyes, D.M., *Mol. Phys.* **101**, 469 (2003).

[16] Rahman, A., *In* 'Neutron Inelastic Scattering', vol.1. IAEA, Vienna, 1968.

[17] Jacucci, G. and McDonald, I.R., *Mol. Phys.* **39**, 515 (1980).

[18] Erpenbeck, J.J. and Wood, W.W., *J. Stat. Phys.* **24**, 455 (1981).

[19] Keyes, T. and Oppenheim, I., *Physica* **70**, 100 (1973).

[20] Evans, D.J., *Mol. Phys.* **47**, 1165 (1982).

[21] (a) Alley, W.E. and Alder, B.J., *Phys. Rev. A* **27**, 3158 (1983). (b) Kambayashi, S. and Kahl, G., *Phys. Rev. A* **46**, 3255 (1992).

[22] Schofield, P., *Proc. Phys. Soc.* **88** 149 (1966).

[23] (a) Alder, B.J. and Wainwright, T.E., *Phys. Rev. Lett.* **18**, 988 (1967). (b) Alder, B.J. and Wainwright, T.E., *Phys. Rev. A* **1**, 18 (1970).

[24] Pomeau, Y. and Résibois, P., *Phys. Rep.* **19**, 63 (1975).

[25] (a) Dorfman, J.R. and Cohen, E.G.D., *Phys. Rev. A* **6**, 776 (1972). (b) Dorfman, J.R. and Cohen, E.G.D., *Phys. Rev. A* **12**, 292 (1975).

[26] (a) Ernst, M.H., Hauge, E.H. and van Leeuwen, J.M.J., *Phys. Rev. A* **4**, 2055 (1971). (b) Erpenbeck. J.J. and Wood, W.W., *Phys. Rev. A* **26**, 1648 (1982).

[27] Levesque, D. and Ashurst, W.T., *Phys. Rev. Lett.* **33**, 277 (1974).

[28] The conclusions of Ref. 27 were later confirmed, with much higher precision, by simulations of a lattice-gas model for which the equations of motion can be solved exactly: see van der Hoef, M.A. and Frenkel, D., *Phys. Rev. A* **41**, 4277 (1990).

[29] Morkel, C., Gronemeyer, C., Gläser, W. and Bosse, J., *Phys. Rev. Lett.* **58**, 1873 (1987).

[30] This classification is due to Angell, C.A., *J. Phys. Chem. Solids* **49**, 863 (1988).

[31] Simulations of the glass transition are commonly carried out on mixed systems as a device to inhibit crystallisation.

[32] Barrat, J.L., Roux, J.N. and Hansen, J.P., *Chem. Phys.* **149**, 198 (1990).

[33] Mapes, M.K., Swallen, S.F. and Ediger, M.D., *J. Phys. Chem. B* **110**, 507 (2006).

[34] Kob, W. and Andersen, H.C., *Phys. Rev. E* **52**, 4134 (1995).

[35] The conventional use of β for the Kohlrausch exponent is unfortunate, since it refers to the decay in the α-relaxation regime.

[36] For an example of a fit of the Kohlrausch law to experimental data, see Wuttke, J., Petry, W. and Pouget, S., *J. Chem. Phys.* **105**, 5177 (1996).

[37] Kob, W. and Barrat, J.L., *Eur. Phys. J. B* **13**, 319 (2000).

[38] Berthier, L. and Biroli, G., *Rev. Mod. Phys.* **83**, 587 (2011).

[39] Cugliandolo, L.F. and Kurchan, J., *J. Phys. A* **27**, 5749 (1994).

[40] Cugliandolo, L.F., Kurchan, J. and Peliti, L., *Phys. Rev. E* **55**, 3898 (1997).

[41] Berthier, L. and Barrat, J.L., *J. Chem. Phys.* **116**, 6228 (2002).

[42] Ediger, M.D., *Ann. Rev. Phys. Chem.* **51**, 99 (2000).

[43] Glotzer, S.C., *J. Non-Cryst. Solids* **274**, 342 (2000).

[44] Figure 5 of Perera, D.N. and Harrowell, P., *J. Chem. Phys.* **111**, 5441 (1999).

[45] Berthier, L., Biroli, G., Bouchaud, J.P., Cipilletti, L., El Masri, D., L'Hôte, D.L., Ladieu, F. and Pierno, M., *Science* **310**, 1797 (2005).

[46] Miyagawa, H., Hiwatari, Y., Bernu, B. and Hansen, J.P., *J. Chem. Phys.* **88**, 3879 (1988).

[47] Weeks, E.R., Crocker, J.C., Levitt, A.C., Schofield, A. and Weitz, D.A., *Science* **287**, 627 (2000).

[48] (a) Chaudhuri, P., Berthier, L. and Kob, W., *Phys. Rev. Lett.* **99**, 060604 (2007). (b) Chaudhuri, P., Sastry, S. and Kob, W., *Phys. Rev. Lett.* **101**, 190601 (2008).

[49] Bocquet, L. and Charlaix, E., *Chem. Soc. Rev.* **39**, 1073 (2010).

[50] Bocquet, L. and Barrat, J.L., *Soft Matter* **3**, 685 (2007).

[51] Bocquet, L. and Barrat, J.L., *Phys. Rev. E* **49**, 3079 (1994).

[52] Barrat, J.L. and Bocquet, L., *Faraday Disc.* **112**, 119 (1999).

[53] Cottin-Bizonne, C., Cross, B., Steinberger, A. and Charlaix, E., *Phys. Rev. Lett.* **94**, 056102 (2005).

Chapter 9

Theories of Time Correlation Functions

We turn now to the problem of devising a general theoretical scheme for the calculation of time correlation functions at wavelengths and frequencies on the molecular scale. Memory functions play a key role in the theoretical development and we begin by showing how the memory function approach can be formalised through use of the projection operator methods of Zwanzig[1] and Mori.[2] The calculation of the memory function in a specific problem is a separate task that can be tackled along two different lines. The first represents a systematic extension of the ideas of generalised hydrodynamics introduced in Section 8.6; the second is more microscopic in nature and based on the mode coupling approach already used in Section 8.7.

9.1   THE PROJECTION OPERATOR FORMALISM

Let A be some dynamical variable, dependent in general on the coordinates and momenta of all particles in the system of interest. The definition of A is assumed to be made in such a way that its mean value is zero, but this involves no loss of generality. We have seen in Section 7.1 that if the phase function A is represented by a vector in Liouville space the inner product $(B, A(t))$ of $A(t)$ with the vector representing a second variable B may be identified with the equilibrium time correlation function $C_{AB}(t)$. We can also use a vector in Liouville space to represent a set of dynamical variables of the system, but for the present we restrict ourselves to the single-variable case.

The time variation of the vector $A(t)$ is given by the exact equation of motion (2.1.14). Our aim is to find an alternative to (2.1.14) which is also exact but more easily usable. We proceed by considering the time evolution both of the projection of $A(t)$ onto A (the *projected* part), and of the component of $A(t)$ normal to A (the *orthogonal* part), which we denote by the symbol $A'(t)$. The projection of a second variable $B(t)$ onto A can be written in terms of a linear *projection operator* $\mathcal{P}$ as

$$\mathcal{P}B(t) = \left(A, B(t)\right)(A, A)^{-1}A \tag{9.1.1}$$

Theory of Simple Liquids, Fourth Edition. http://dx.doi.org/10.1016/B978-0-12-387032-2.00009-X

Thus

$$(\mathcal{P}B(t), A) = (A, B(t)) \equiv \langle B(t)A^* \rangle \tag{9.1.2}$$

The complementary operator $\mathcal{Q} = 1 - \mathcal{P}$ projects onto the subspace orthogonal to A. Hence the orthogonal part of $A(t)$ is

$$A'(t) = \mathcal{Q}A(t) \tag{9.1.3}$$

Both $\mathcal{P}$ and $\mathcal{Q}$ satisfy the fundamental properties of projection operators:

$$\mathcal{P}^2 = \mathcal{P}, \quad \mathcal{Q}^2 = \mathcal{Q}, \quad \mathcal{P}\mathcal{Q} = \mathcal{Q}\mathcal{P} = 0 \tag{9.1.4}$$

The projection of $A(t)$ along A is proportional to $Y(t)$, the normalised time autocorrelation function of the variable A, i.e.

$$\mathcal{P}A(t) = Y(t)A \tag{9.1.5}$$

with

$$Y(t) = (A, A(t))(A, A)^{-1} \equiv \langle A(t)A^* \rangle \langle AA^* \rangle^{-1} = C_{AA}(t)/C_{AA}(0) \tag{9.1.6}$$

The definitions (9.1.1)–(9.1.3) ensure that

$$(A, A'(t)) = 0 \tag{9.1.7}$$

The first step is to derive an equation for the time evolution of the projected part, $Y(t)$. The Laplace transform of the equation of motion (2.1.14) is

$$(z + \mathcal{L})\tilde{A}(z) \equiv (z + \mathcal{L})(\mathcal{P} + \mathcal{Q})\tilde{A}(z) = iA \tag{9.1.8}$$

Thus

$$\tilde{Y}(z) = \left(A, \int_0^\infty \exp(izt) \exp(i\mathcal{L}t) A \, dt\right)(A, A)^{-1}$$
$$= (A, i(z + \mathcal{L})^{-1}A)(A, A)^{-1} = (A, \tilde{A}(z))(A, A)^{-1} \tag{9.1.9}$$

where the 'resolvent' operator $i(z + \mathcal{L})^{-1}$ is the transform of the propagator $\exp(i\mathcal{L}t)$. We now project (9.1.8) parallel and perpendicular to A by application, respectively, of the operators $\mathcal{P}$ and $\mathcal{Q}$. Use of the properties (9.1.4) shows that

$$z\mathcal{P}\tilde{A}(z) + \mathcal{P}\mathcal{L}\mathcal{P}\tilde{A}(z) + \mathcal{P}\mathcal{L}\mathcal{Q}\tilde{A}(z) = iA \tag{9.1.10}$$

$$z\mathcal{Q}\tilde{A}(z) + \mathcal{Q}\mathcal{L}\mathcal{P}\tilde{A}(z) + \mathcal{Q}\mathcal{L}\mathcal{Q}\mathcal{Q}\tilde{A}(z) = 0 \tag{9.1.11}$$

and elimination of $\mathcal{Q}\tilde{A}(z)$ between (9.1.10) and (9.1.11) gives

$$z\mathcal{P}\tilde{A}(z) + \mathcal{P}\mathcal{L}\mathcal{P}\tilde{A}(z) - \mathcal{P}\mathcal{L}\mathcal{Q}(z + \mathcal{Q}\mathcal{L}\mathcal{Q})^{-1}\mathcal{Q}\mathcal{L}\mathcal{P}\tilde{A}(z) = iA \tag{9.1.12}$$

If we now take the inner product with A and multiply through by $-i(A, A)^{-1}$, (9.1.12) becomes

$$
\begin{aligned}
-iz\tilde{Y}(z) - i\big(A, \mathcal{LP}\tilde{A}(z)\big)(A, A)^{-1} \\
+ i\big(A, \mathcal{LQ}(z + \mathcal{QLQ})^{-1}\mathcal{QLP}\tilde{A}(z)\big)(A, A)^{-1} = 1
\end{aligned}
\tag{9.1.13}
$$

Since $i\mathcal{LP}\tilde{A}(z) = (A, A)^{-1}\big(A, \tilde{A}(z)\big)\dot{A}$, this expression can be rewritten as

$$
(-iz - i\Omega)\tilde{Y}(z) + \big(K, \tilde{R}(z)\big)(A, A)^{-1}\tilde{Y}(z) = 1
\tag{9.1.14}
$$

where

$$
K = Q\dot{A} = Q(i\mathcal{L})A
\tag{9.1.15}
$$

is the projection of $\dot{A}$ orthogonal to A and we have introduced the quantity

$$
\tilde{R}(z) = i(z + \mathcal{QLQ})^{-1}K
\tag{9.1.16}
$$

and defined a frequency Ω as

$$
i\Omega = (A, \dot{A})(A, A)^{-1} = \dot{Y}(0)
\tag{9.1.17}
$$

In the single-variable case the frequency Ω is identically zero for systems with continuous interactions, since all autocorrelation functions are even functions of time, but we retain the term in Ω here to facilitate the later generalisation to the multivariable description.

The projection K is conventionally termed a 'random force'. If A is the momentum of particle i, $\dot{A}$ is the total force acting on i and K is then the random force of the classic Langevin theory described in Section 7.3. In other cases, however, K is not a force in the mechanical sense. Instantaneously, K and $\dot{A}$ are the same, but the two quantities evolve differently in time. The time dependence of the random force is given by the inverse Laplace transform of $\tilde{R}(z)$:

$$
R(t) = \exp\left(i\mathcal{QLQ}t\right)K
\tag{9.1.18}
$$

with $R(0) = K$. The special form of its propagator means that $R(t)$ remains at all times in the subspace orthogonal to A, i.e.

$$
\big(A, R(t)\big) = 0 \quad \text{for all } t
\tag{9.1.19}
$$

This is easily proved by expanding the right-hand side of (9.1.18) in powers of t, since it is clear by inspection that every term in the series is orthogonal to A. The expansion also reveals that the propagator in (9.1.18) may equally well be written as $\exp\left(i\mathcal{QL}t\right)$ and both forms appear in the literature. The autocorrelation function of the random force defines the memory function $M(t)$ for the evolution of the dynamical variable A:

$$
M(t) = \big(R, R(t)\big)(A, A)^{-1}
\tag{9.1.20}
$$

or

$$\tilde{M}(z) = (R, \tilde{R}(z))(A, A)^{-1} \tag{9.1.21}$$

Equation (9.1.14) can be rewritten in terms of the memory function as

$$\tilde{Y}(z) = [-iz - i\Omega + \tilde{M}(z)]^{-1} \tag{9.1.22}$$

or, in the time domain, as

$$\dot{Y}(t) - i\Omega Y(t) + \int_0^t M(t-s)Y(s)\mathrm{d}s = 0 \tag{9.1.23}$$

The equation describing the time evolution of the orthogonal component $A'(t)$ is obtained along similar lines. From (9.1.11) we find that for $\tilde{A}'(z) = \mathcal{Q}\tilde{A}(z)$:

$$(z + \mathcal{Q}\mathcal{L}\mathcal{Q})\tilde{A}'(z) = -\mathcal{Q}\mathcal{L}\mathcal{P}\tilde{A}(z)$$
$$= -\mathcal{Q}\mathcal{L}\tilde{Y}(z)A = i\tilde{Y}(z)K \tag{9.1.24}$$

If we substitute for $\tilde{Y}(z)$ from (9.1.22) and use the definition of $\tilde{R}(z)$ in (9.1.16), (9.1.24) becomes

$$\tilde{R}(z) = [-iz - i\Omega + \tilde{M}(z)]\tilde{A}'(z) \tag{9.1.25}$$

or, in the time domain:

$$\dot{A}'(t) - i\Omega A'(t) + \int_0^t M(t-s)A'(s)\mathrm{d}s = R(t) \tag{9.1.26}$$

Equations (9.1.23) and (9.1.26) are the projections parallel and perpendicular to the variable A of a generalised Langevin equation for A:

$$\dot{A}(t) - i\Omega A(t) + \int_0^t M(t-s)A(s)\mathrm{d}s = R(t) \tag{9.1.27}$$

Apart from the introduction of the term in Ω, (9.1.27) has the same general form as the Langevin equation (7.3.21), but the random force $R(t)$ and memory function $M(t)$ now have the explicit definitions provided by (9.1.18) and (9.1.20).

There is a close connection between the behaviour of the functions $Y(t)$ and $M(t)$ at short times, a fact already exploited in Section 7.3. When differentiated with respect to time the memory function equation (9.1.23) becomes

$$\ddot{Y}(t) - i\Omega\dot{Y}(t) + M(0)Y(t) + \int_0^t \dot{M}(t-s)Y(s)\mathrm{d}s = 0 \tag{9.1.28}$$

Since $Y(0) = 1$ and $\dot{Y}(0) = i\Omega$, we see that

$$M(0) = -\ddot{Y}(0) - \Omega^2 = (\dot{A}, \dot{A})(A, A)^{-1} - \Omega^2 \tag{9.1.29}$$

Repeated differentiation leads to relations between the initial time derivatives of $Y(t)$ and $M(t)$ or, equivalently, given (7.1.24), between the frequency moments of the power spectra $Y(\omega)$ and $M(\omega)$. These relations are useful in constructing simple, approximate forms for $M(t)$ that satisfy the low-order sum rules on $Y(t)$. A link also exists between the autocorrelation function of the random force, i.e. the memory function, and that of the total force, $\dot{A}$. Let $\Phi(t)$ be the autocorrelation function of $\dot{A}$, defined as

$$\Phi(t) = \left(\dot{A}, \dot{A}(t)\right)(A, A)^{-1} = -\ddot{Y}(t) \qquad (9.1.30)$$

It follows from the properties of the Laplace transform that the functions $\tilde{\Phi}(z)$ and $\tilde{Y}(z)$ are related by

$$\tilde{\Phi}(z) = z^2 \tilde{Y}(z) - iz + i\Omega \qquad (9.1.31)$$

Since the term $i\Omega$ vanishes in the one-variable case, we may temporarily discard it. Then elimination of $\tilde{Y}(z)$ between (9.1.22) and (9.1.31) leads to the expression

$$\frac{1}{\tilde{M}(z)} = \frac{1}{\tilde{\Phi}(z)} + \frac{1}{iz} \qquad (9.1.32)$$

The two autocorrelation functions therefore vary with time in different ways except in the high-frequency (short-time) limit: the time dependence of $\Phi(t)$ is determined by the full Liouville operator $\mathcal{L}$ and that of $M(t)$ by the projected operator $\mathcal{QLQ}$.

There are two important ways in which the projection operator formalism can be extended. First, (9.1.23) may be regarded as the the leading member in a hierarchy of memory function equations. If we apply the methods already used to the case when R is treated as the dynamical variable, we obtain an equation similar to (9.1.23) for the time evolution of the projection of $R(t)$ along R. The kernel of the integral equation is now the autocorrelation function of a second-order random force which is orthogonal at all times to both R and A. As an obvious generalisation of this procedure we can write a memory function equation of the form

$$\dot{M}_n(t) - i\Omega_n M_n(t) + \int_0^t M_{n+1}(t-s)\Delta_{n+1}^2 M_n(s)\mathrm{d}s = 0 \qquad (9.1.33)$$

where

$$M_n(t) = \left(R_n, R_n(t)\right)(R_n, R_n)^{-1}$$
$$R_n(t) = \exp\left(i\mathcal{Q}_n\mathcal{L}\mathcal{Q}_n t\right)\mathcal{Q}_n \dot{R}_{n-1} \qquad (9.1.34)$$

and

$$\Delta_n^2 = (R_n, R_n)(R_{n-1}, R_{n-1})^{-1} \qquad (9.1.35)$$

The operator $\mathcal{P}_n$ projects a dynamical variable along R_{n-1} according to the rule (9.1.1). By construction, therefore, the complementary operator

$$\mathcal{Q}_n = 1 - \sum_{j=1}^{n} \mathcal{P}_j \qquad (9.1.36)$$

projects onto the subspace orthogonal to all R_j for $j < n$. Thus the nth-order random force $R_n(t)$ is uncorrelated at all times with random forces of lower order. Equation (9.1.23) is a special case of (9.1.33) with $Y \equiv M_0$. Repeated application of the Laplace transform to equations of the hierarchy leads to an expression for $\tilde{Y}(z)$ in the form of a continued fraction:

$$\tilde{Y}(z) = \cfrac{1}{-iz - i\Omega_0 + \cfrac{\Delta_1^2}{-iz - i\Omega_1 + \cfrac{\Delta_2^2}{-iz - i\Omega_2 + \cdots}}} \tag{9.1.37}$$

A second extension of the method, which has proved particularly useful for the description of collective modes in liquids, is one already mentioned. This is the generalisation to the case where the dynamical quantity of interest is not a single fluctuating property of the system but a set of n independent variables $A_1, A_2, \ldots, A_n$. We represent this set by a column vector $\mathbf{A}$ and its hermitian conjugate by the row vector $\mathbf{A}^*$. The derivation of the generalised Langevin equation for $\mathbf{A}$ follows the lines already laid down, due account being taken of the fact that the quantities involved are no longer scalars. The result may be written in matrix form as

$$\dot{\mathbf{A}}(t) - i\boldsymbol{\Omega} \cdot \mathbf{A}(t) + \int_0^t \mathbf{M}(t - s) \cdot \mathbf{A}(s)\mathrm{d}s = \mathbf{R}(t) \tag{9.1.38}$$

The definitions of the random force vector $\mathbf{R}(t)$, frequency matrix $\boldsymbol{\Omega}$ and memory function matrix $\mathbf{M}(t)$ are analogous to those of $R(t), \Omega$ and $M(t)$ in the single-variable case, the scalars A and A^* being replaced by the vectors $\mathbf{A}$ and $\mathbf{A}^*$. If we multiply (9.1.38) from the right by $\mathbf{A}^* \cdot (\mathbf{A}, \mathbf{A})^{-1}$ and take the thermal average we find that

$$\dot{\mathbf{Y}}(t) - i\boldsymbol{\Omega}\dot{\mathbf{Y}}(t) + \int_0^t \mathbf{M}(t - s) \cdot \mathbf{Y}(s)\mathrm{d}s = \mathbf{0} \tag{9.1.39}$$

where $\mathbf{Y}(t) = (\mathbf{A}, \mathbf{A}(t)) \cdot (\mathbf{A}, \mathbf{A})^{-1}$ is the correlation function matrix. Equation (9.1.39) is the multi-variable generalisation of (9.1.23); its solution in terms of Laplace transforms is

$$\tilde{\mathbf{Y}}(z) = [-iz\, \mathbf{I} - i\boldsymbol{\Omega} + \tilde{\mathbf{M}}(z)]^{-1} \tag{9.1.40}$$

where $\mathbf{I}$ is the identity matrix. Note that each diagonal element of $\mathbf{Y}(t)$ is an autocorrelation function, normalised by its value at $t = 0$, and the off-diagonal elements are cross-correlation functions.

The value of the memory function formalism is most easily appreciated by considering specific examples of its use. Before doing so, however, it is helpful to look at the problem from a wider point of view. Equation (9.1.38) represents an equation of motion for $\mathbf{A}(t)$ in which terms linear in $\mathbf{A}$ are displayed explicitly on the left-hand side while the random force vector describes the effects of non-linear terms, initial transient processes and the dependence of $\mathbf{A}(t)$ on variables not included in the set $\{A_i\}$. This separation of effects is most useful in cases where the random force fluctuates rapidly and the non-zero elements of the memory function matrix decay much faster than the correlation functions of interest. It is then not unreasonable to represent $\mathbf{M}(t)$ in some simple way, in particular by invoking a Markovian approximation whereby the non-zero elements are replaced by δ-functions in t. For this representation to be successful the vector $\mathbf{A}$ should contain as its components not only the variables of immediate interest but also those to which they are strongly coupled. If the set of variables is well chosen the effect of projecting $\mathbf{A}(t)$ onto the subspace spanned by $\mathbf{A}$ is to project out all the slowly varying properties of the system. The Markovian assumption can then be used with greater confidence in approximating the memory function matrix. By extending the dimensionality of $\mathbf{A}$ an increasingly detailed description can be obtained without departing from the Markovian hypothesis. In practice, as we shall see in later sections, this ideal state of affairs is often difficult to achieve, and some of the elements of $\mathbf{M}(t)$ may not be truly short ranged in time. The calculation of the frequency matrix $\mathbf{\Omega}$ is usually a straightforward problem, since it involves only static quantities; the same is true of the static correlation matrix $(\mathbf{A}, \mathbf{A})$.

As an alternative to the multi-dimensional description it is possible to work with a smaller set of variables and exploit the continued-fraction expansion, truncating the hierarchy at a suitable point in some simple, approximate way. This approach is particularly useful when insufficient is known about the dynamical behaviour of the system to permit an informed choice of a larger set of variables. Its main disadvantage is the fact that the physical significance of the memory function becomes increasingly obscure as the expansion is carried to higher orders.

9.2 SELF CORRELATION FUNCTIONS

As a simple example we consider first the application of projection operator methods to the calculation of the self-intermediate scattering function $F_s(k, t)$. This function is of interest because of its link to the velocity autocorrelation function via (8.2.17) and because its power spectrum, the self dynamic structure factor $S_s(k, \omega)$, is closely related to the cross-section for incoherent scattering of neutrons.

The most straightforward approach to the problem is to choose as the single variable A the fluctuating density $\rho_{\mathbf{k}i}$ of a tagged particle i and write a memory function equation for $\tilde{F}_s(k, z)$ in the form

$$\tilde{F}_s(k, z) = \frac{1}{-iz + \tilde{M}_s(k, z)} \tag{9.2.1}$$

Results given in Section 8.2 show that the short-time expansion of $F_s(k, t)$ starts as

$$F_s(k, t) = 1 - \omega_0^2 \frac{t^2}{2!} + \omega_0^2 (3\omega_0^2 + \Omega_0^2) \frac{t^4}{4!} + \cdots \tag{9.2.2}$$

where the coefficients of successive powers of t are related to the frequency moments of $S_s(k, \omega)$ via the general expression (7.1.24) and the quantities Ω_0 (the Einstein frequency) and ω_0 are defined by (7.2.9) and (7.4.29) respectively; it follows from (9.1.29) that the effect of setting $M_s(k, t = 0) = \omega_0^2 = k^2(k_B T / m)$ is to ensure that $S_s(k, \omega)$ has the correct second moment. We may also rewrite $\tilde{M}_s(k, z)$ as $k^2 \tilde{D}(k, z)$ where, by analogy with (8.2.10), $\tilde{D}(k, z)$ plays the role of a generalised self-diffusion coefficient such that $\lim_{z \to 0} \lim_{k \to 0} \tilde{D}(k, z) = D$. If the continued-fraction expansion is taken to second order we find that

$$\tilde{F}_s(k, z) = \cfrac{1}{-iz + \cfrac{\omega_0^2}{-iz + \tilde{N}_s(k, z)}} \tag{9.2.3}$$

By extension of the calculation that leads to (9.1.29) it is easy to show that the initial value of the second-order memory function $N_s(k, t)$ is related to the short-time behaviour of $M_s(k, t)$ by $N_s(k, 0) = -\ddot{M}_s(k, 0)/M_s(k, 0) = 2\omega_0^2 + \Omega_0^2$. Thus, if

$$\tilde{N}_s(k, z) = \left(2\omega_0^2 + \Omega_0^2\right)\tilde{n}_s(k, z) \tag{9.2.4}$$

where $n_s(k, t = 0) = 1$, the resulting expression for $S_s(k, \omega)$ also has the correct fourth moment regardless of the time dependence of $n_s(k, t)$.

As an alternative to making a continued-fraction expansion of $\tilde{F}_s(k, z)$ we can consider the multi-variable description of the problem that comes from the choice

$$\mathbf{A} = \begin{pmatrix} \rho_{ki} \\ \dot{\rho}_{ki} \\ \sigma_{ki} \end{pmatrix} \tag{9.2.5}$$

where the variable σ_{ki}, given by

$$\sigma_{ki} = \ddot{\rho}_{ki} - (\rho_{ki}, \ddot{\rho}_{ki})(\rho_{ki}, \rho_{ki})^{-1}\rho_{ki} \tag{9.2.6}$$

is orthogonal to both ρ_{ki} and $\dot{\rho}_{ki}$. From results derived in Sections 7.4 and 8.2 it is straightforward to show that the corresponding static correlation matrix is diagonal and given by

$$(\mathbf{A}, \mathbf{A}) = \begin{pmatrix} 1 & 0 & 0 \\ 0 & \omega_0^2 & 0 \\ 0 & 0 & \omega_0^2(2\omega_0^2 + \Omega_0^2) \end{pmatrix} \tag{9.2.7}$$

while the frequency matrix is purely off-diagonal:

$$i\Omega = (\mathbf{A}, \dot{\mathbf{A}}) \cdot (\mathbf{A}, \mathbf{A})^{-1} = \begin{pmatrix} 0 & 1 & 0 \\ -\omega_0^2 & 0 & 1 \\ 0 & -2\omega_0^2 - \Omega_0^2 & 0 \end{pmatrix} \tag{9.2.8}$$

Both $\dot{A}_1$ and $\dot{A}_2$ form part of the space spanned by the vector $\mathbf{A}$. In the case of $\dot{A}_1$ this is easy to see, since $\dot{A}_1 = A_2$. To understand why it is is also true for $\dot{A}_2$ it is sufficient to note that the projection of $\dot{A}_2$ along A_1 is obviously part of the space of $\mathbf{A}$, whereas the component orthogonal to A_1 is, according to the definition (9.2.6), the same as A_3. It follows that the random-force vector has only one non-zero component and the memory function matrix has only one non-zero entry:

$$
\mathbf{M}(k,t) = \begin{pmatrix} 0 & 0 & 0 \\ 0 & 0 & 0 \\ 0 & 0 & \mathcal{M}(k,t) \end{pmatrix} \tag{9.2.9}
$$

On collecting results and inserting them in (9.1.40), we find that the correlation function matrix has the form

$$
\tilde{\mathbf{Y}}(k,z) = \begin{pmatrix} -iz & -1 & 0 \\ \omega_0^2 & -iz & -1 \\ 0 & 2\omega_0^2 + \Omega_0^2 & -iz + \tilde{\mathcal{M}}(k,z) \end{pmatrix}^{-1} \tag{9.2.10}
$$

Inversion of (9.2.10) shows that $\tilde{F}_s(k,z)$ is given by

$$
\tilde{F}_s(k,z) = \tilde{Y}_{11}(k,z)
$$

$$
= \cfrac{1}{-iz + \cfrac{\omega_0^2}{-iz + \cfrac{2\omega_0^2 + \Omega_0^2}{-iz + \tilde{\mathcal{M}}(k,z)}}} \tag{9.2.11}
$$

and comparison with (9.2.3) and (9.2.4) makes it possible to identify $\mathcal{M}(k,t)$ as the memory function of $N_s(k,t)$. Similarly, the Laplace transform of the self current autocorrelation function $C_s(k,t)$ is

$$
\tilde{C}_s(k,z) = \omega_0^2 \tilde{Y}_{22}(k,z)
$$

$$
= \cfrac{\omega_0^2}{-iz + (2\omega_0^2 + \Omega_0^2)\tilde{n}_s(k,z) + \cfrac{\omega_0^2}{-iz}} \tag{9.2.12}
$$

The same result can be derived from (9.2.3) via the relation (8.2.17) between $C_s(k,t)$ and $F_s(k,t)$, which in turn implies that $\tilde{C}_s(k,z) = z^2 \tilde{F}_s(k,z) - iz$.

In the long-wavelength limit the memory function $n_s(k,t)$ is directly related to the memory function of the velocity autocorrelation function $Z(t)$. From (7.2.8), (8.2.17) and (9.2.12) we find that

$$
\tilde{Z}(z) = \frac{k_B T/m}{-iz + \Omega_0^2 \tilde{n}_s(0,z)} \tag{9.2.13}
$$

Thus

$$N_s(0,t) = \Omega_0^2 n_s(0,t) \equiv \xi(t) \tag{9.2.14}$$

where $\xi(t)$ is the memory function of $Z(t)$, introduced earlier in Section 7.3. Since $N_s(k,t)$ is also the memory function of $M_s(k,t)$ and $M_s(k,0) = k^2 Z(0)$, we see that $k^2 Z(t)$ becomes the memory function of $F_s(k,t)$ as $k \to 0$. For consistency with the hydrodynamic result (8.2.10) we also require that

$$\Omega_0^2 \tilde{n}_s(0,0) = \frac{k_B T}{mD} \tag{9.2.15}$$

A particularly simple (Markovian) approximation is to replace $N_s(k,t)$ by a quantity independent of t, $1/\tau_s(k)$ say, which is equivalent to assuming an exponential form for $M_s(k,t)$:

$$M_s(k,t) = \omega_0^2 \exp\left[-|t|/\tau_s(k)\right] \tag{9.2.16}$$

with the constraint, required to satisfy (9.2.15), that

$$\tau_s(0) = \frac{mD}{k_B T} \tag{9.2.17}$$

As we have seen in Section 7.3, this approximation leads to an exponential velocity autocorrelation function of the Langevin type, the quantity $1/\tau_s(0)$ appearing as a frequency independent friction coefficient. Better results are obtained by choosing an exponential form for $N_s(k,t)$, i.e.

$$N_s(k,t) = \left(2\omega_0^2 + \Omega_0^2\right) \exp\left[-|t|/\tau_s(k)\right] \tag{9.2.18}$$

with

$$\tau_s(0) = \frac{k_B T}{mD\Omega_0^2} \tag{9.2.19}$$

This second approximation is equivalent to neglecting the frequency dependence of $\tilde{\mathcal{M}}(k,z)$; it leads to an analytical form for $S_s(k,\omega)$ having the correct zeroth, second and fourth moments:

$$S_s(k,\omega) = \frac{1}{\pi} \frac{\tau_s(k)\omega_0^2(2\omega_0^2 + \Omega_0^2)}{\omega^2\tau_s^2(k)(\omega^2 - 3\omega_0^2 - \Omega_0^2)^2 + (\omega^2 - \omega_0^2)^2} \tag{9.2.20}$$

The corresponding expression for $\tilde{Z}(z)$ is that given in (7.3.26), with $\tau \equiv \tau_s(0)$.

In the absence of any well-based microscopic theory it is perhaps best to treat the relaxation time $\tau_s(k)$ as an adjustable parameter, but it is also tempting to look for some relatively simple prescription for this quantity. An argument based on a scaling of the memory function $M_s(k,t)$ has been used to derive the expression[3]

$$\tau_s^{-1}(k) = \gamma(2\omega_0^2 + \Omega_0^2)^{1/2} \tag{9.2.21}$$

where the parameter γ is taken to be independent of k, an assumption which is reasonably well-borne out in practice. If, in the limit $k \to 0$, we require (9.2.21) to yield the correct diffusion coefficient, it follows that $\gamma = mD\Omega_0/k_B T$; this leads to a value of γ of ≈ 0.9 at the triple point of liquid argon. On the other hand, for large wavenumbers, $S_s(k, 0)$ goes over correctly to the ideal gas result if $\gamma = 2/\pi^{1/2} \approx 1.13$.

Although the exponential approximation (9.2.18) has been used with some success in the interpretation of experimental neutron scattering data,[4] the true situation is known to be much less simple, at least at small wavenumbers. In particular, molecular dynamics calculations for a range of simple liquids have shown that the memory function of $Z(t)$, i.e. $N_s(0, t)$, cannot be adequately described by a model involving only a single relaxation time. Figure 9.1 shows the memory function obtained from a simulation of liquid sodium[5] in which a clear separation of time scales is apparent; the presence of the long-time tail in the memory function has the effect of reducing the self-diffusion coefficient by about 30%. In their analysis of the self-correlation functions of the Lennard-Jones fluid Levesque and Verlet[6] found it necessary to use a rather complicated expression for $N_s(k, t)$, which for $k = 0$ reduces to

$$\xi(t) = \Omega_0^2 \exp\left[-(t/\tau_1)^2\right] + At^4 \exp\left(-t/\tau_2\right) \qquad (9.2.22)$$

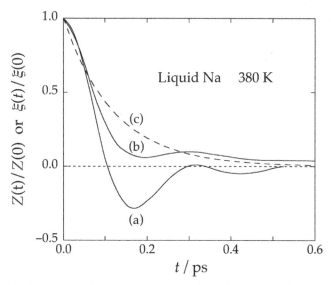

FIGURE 9.1 Velocity autocorrelation function, curve (a), and the associated memory function, curve (b), derived from molecular dynamics calculations for liquid sodium at state conditions close to the normal melting point. Curve (c) shows the exponential approximation (9.2.18) for the memory function, with $\tau_s(0)$ chosen to give the correct self-diffusion coefficient. From U. Balucani et al., *J. Non-Cryst. Solids* **205–207**, 299–303 (1996), *with permission of Elsevier.*

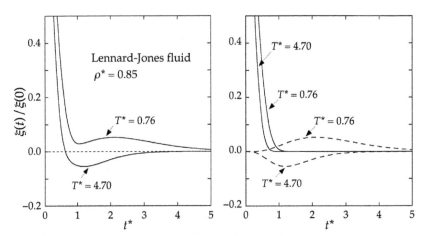

FIGURE 9.2 Normalised memory function associated with the velocity autocorrelation function of the Lennard-Jones fluid at a density $\rho^* = 0.85$ and two temperatures. Left-hand panel: the total memory function. Right-hand panel: the decomposition into short-lived (full curves) and long-lived (dashed curves) components, corresponding to the two terms in (9.2.22). After Levesque and Verlet.[6]

where A, τ_1 and τ_2 are adjustable parameters. A separation into a rapidly decaying part and a long-lived term that starts as t^4 is also an explicit ingredient of modern versions of kinetic theory, in which account is taken of correlated as well as uncorrelated collisions. The long-lived term represents collective effects and lends itself to calculation by mode coupling methods similar to that employed in Section 8.7, which we shall meet again later in this chapter. The relaxation time τ_1 decreases with temperature but is almost independent of density while τ_2 increases with density but is insensitive to changes in temperature. The behaviour of the parameter A is more complicated. Figure 9.2 shows the memory function at a high density and two temperatures. At the lower value of T^* the contribution from the long-lived term is positive and the memory function is very similar in form to that for liquid sodium pictured in Figure 9.1. As the temperature is raised – or the density lowered – A decreases in magnitude and eventually changes sign. This gives rise to a negative region in the memory function, which is the source of a persisting positive correlation of velocity of the type seen, for example, in Figure 7.1 for the case of the r^{-12}-fluid at a low value of the coupling parameter Γ.

The importance of including a long-lived component in the memory function $N_s(k,t)$ for $k > 0$ is illustrated for the case of the Lennard-Jones fluid close to the triple point in Figure 9.3. The quantity plotted there, as a function of k, is the width at half-height of $S_s(k, \omega)$ relative to its value in the hydrodynamic limit (where $\Delta \omega = 2Dk^2$). Comparison with results for $S_s(k\omega)$ itself is not very illuminating, since the spectrum is largely featureless, but the dependence of $\Delta \omega / Dk^2$ on k displays a structure that is very poorly described by the

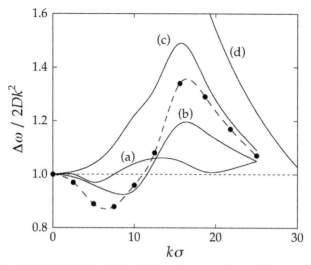

FIGURE 9.3 Width at half-height of the self dynamic structure factor relative to its value in the hydrodynamic limit. The points are molecular dynamics data for the Lennard-Jones fluid at a high density and low temperature ($\rho^* = 0.844, T^* = 0.722$) and the broken curve is drawn as an aid to the eye. The full curves show the results predicted (a) by the single-exponential approximation (9.2.18), (b) by the k-dependent generalisation of (9.2.22), (c) by the Gaussian approximation (8.2.14) and (d) in the ideal-gas ($k \to \infty$) limit. After Levesque and Verlet.[6]

single-exponential approximation (9.2.18); the same is true of the Gaussian approximation (8.2.14).

9.3 TRANSVERSE COLLECTIVE MODES

As we saw in Section 8.6, the appearance of propagating shear waves in dense fluids can be explained in qualitative or even semi-quantitative terms by a simple, viscoelastic model based on a generalisation of the hydrodynamic approach. In this section we show how such a theory can be developed in systematic fashion by use of the projection operator formalism.

Taking the viscoelastic relation (8.6.4) as a guide, we choose as components of the vector **A** the x-component of the mass current and the xz-component of the stress tensor (8.4.14), assuming as usual that the z-axis is parallel to **k**. Thus

$$\mathbf{A} = \begin{pmatrix} m j_{\mathbf{k}}^x \\ \Pi_{\mathbf{k}}^{xz} \end{pmatrix} \tag{9.3.1}$$

and

$$(\mathbf{A}, \mathbf{A}) = V k_B T \begin{pmatrix} \rho m & 0 \\ 0 & G_\infty(k) \end{pmatrix} \tag{9.3.2}$$

where $G_\infty(k)$ is the generalised elastic constant defined by (8.6.9). To calculate the frequency matrix we use the relations

$$(A_1, \dot{A}_1) = (A_2, \dot{A}_2) = 0 \tag{9.3.3}$$

$$(A_2, \dot{A}_1) = (A_1, \dot{A}_2) = -ikVk_BTG_\infty(k) \tag{9.3.4}$$

and find that

$$i\Omega = \begin{pmatrix} 0 & -ik \\ \frac{-ikG_\infty(k)}{\rho m} & 0 \end{pmatrix} \tag{9.3.5}$$

Because $\dot{A}_1$ is proportional to A_2 the projection of $\dot{A}_1$ orthogonal to $\mathbf{A}$ is identically zero. The memory function matrix therefore has only one non-zero element, which we denote by $M_t(k,t)$:

$$\mathbf{M}(k,t) = \begin{pmatrix} 0 & 0 \\ 0 & M_t(k,t) \end{pmatrix} \tag{9.3.6}$$

When these results are substituted in (9.1.40) we obtain an expression for the Laplace transform of the correlation function matrix in the form

$$\tilde{\mathbf{Y}}(k,z) = \begin{pmatrix} -iz & ik \\ \frac{ikG_\infty(k)}{\rho m} & -iz + \tilde{M}_t(k,z) \end{pmatrix}^{-1} \tag{9.3.7}$$

Thus the Laplace transform of the transverse current autocorrelation function is

$$\tilde{C}_t(k,z) = \omega_0^2 \tilde{Y}_{11}(k,z)$$

$$= \frac{\omega_0^2}{-iz + \dfrac{\omega_{1t}^2}{-iz + \tilde{M}_t(k,z)}} \tag{9.3.8}$$

where ω_{1t}^2, defined by (7.4.38), is related to $G_\infty(k)$ by (8.6.9). Consistency with the hydrodynamic result (8.4.4) in the long-wavelength, low-frequency limit is achieved by setting

$$\tilde{M}_t(0,0) = \frac{G_\infty(0)}{\eta} \tag{9.3.9}$$

The function $\tilde{M}_t(k,z)$ is the memory function of the generalised kinematic shear viscosity introduced in Section 8.6. This identification follows immediately from comparison of of (9.3.8) with the Laplace transform of (8.6.7), which shows that

$$\tilde{C}_t(k,z) = \frac{\omega_0^2}{-iz + k^2\tilde{v}(k,z)} \tag{9.3.10}$$

The viscoelastic approximation corresponds to ignoring the frequency dependence of $\tilde{M}_t(k,z)$ and replacing it by a constant, $1/\tau_t(k)$ say, implying that

$v(k, t)$ decays exponentially with a characteristic time $\tau_t(k)$ and hence, from (8.6.9), that

$$v(k, t) = \frac{G_\infty(k)}{\rho m} \exp[-|t|/\tau_t(k)] \tag{9.3.11}$$

Use of (9.3.11) ensures that the spectrum of transverse current fluctuations:

$$C_t(k, \omega) = \frac{1}{\pi} \operatorname{Re} \tilde{C}_t(k, \omega) = \frac{1}{\pi} \frac{\omega_0^2 \omega_{1t}^2 \tau_t(k)}{\omega^2 + \tau_t^2(k)(\omega_{1t}^2 - \omega^2)^2} \tag{9.3.12}$$

has the correct second moment irrespective of the choice of $\tau_t(k)$. If, as in Section 8.6, we define a wavenumber-dependent shear viscosity $\eta(k)$ as the zero-frequency limit of $\rho m \tilde{v}(k, \omega)$, we find in the approximation represented by (9.3.11) that

$$\eta(k) = \tau_t(k) G_\infty(k) \tag{9.3.13}$$

so that $\tau_t(k)$ appears as a wavenumber-dependent Maxwell relaxation time (see (8.6.5)). In particular:

$$\eta \equiv \eta(0) = \tau_t(0) G_\infty(0) \tag{9.3.14}$$

in agreement with (9.3.9).

It is easy to establish the criterion for the existence of propagating transverse modes within the context of the approximation represented by (9.3.12), characterised by a single relaxation time. The condition for $C_t(k, \omega)$ to have a peak at non-zero frequency at a given value of k is

$$\omega_{1t}^2 \tau_t^2(k) > \frac{1}{2} \tag{9.3.15}$$

and the peak, if it exists, is at a frequency ω such that $\omega^2 = \omega_{1t}^2 - \frac{1}{2}\tau_t^{-2}(k)$. It follows from the inequality (9.3.15) that shear waves will appear for values of k greater than k_c, where k_c is a critical wavevector given by

$$k_c^2 = \frac{\rho m}{2\tau_t^2(k) G_\infty(k)} \tag{9.3.16}$$

We can obtain an estimate for k_c by taking the $k \to 0$ limit of (9.3.16); this gives

$$k_c^2 \approx \frac{\rho m G_\infty(0)}{2\eta^2} \tag{9.3.17}$$

On inserting the values of η and $G_\infty(0)$ obtained by molecular dynamics calculations for the Lennard-Jones fluid close to its triple point we find that $k_c\sigma \approx 0.79$. This is apparently a rather good guide to what occurs in practice: the dispersion curve for liquid argon plotted in Figure 8.4 shows that shear waves first appear at $k_c \approx 2.0$ Å or, taking a value (3.4 Å) for σ appropriate to argon, $k_c\sigma \approx 0.7$. At sufficiently large values of k the shear-waves disappear again as the role of the interparticle forces becomes less important.

Given its simplicity, the viscoelastic approximation provides a satisfactory description of the transverse current fluctuations over a wide range of

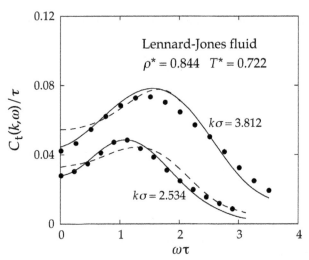

FIGURE 9.4 Spectrum of transverse-current fluctuations for the Lennard-Jones fluid near its triple point. The points are molecular dynamics results and the curves show results calculated from the viscoelastic approximation (9.3.11) (broken lines) and the two-exponential memory function (9.3.20) (full lines). The unit of time is $\tau = (m\sigma^2/48\epsilon)^{1/2}$. *Redrawn with permission from Ref. 7* © *1973 American Physical Society.*

wavelength. Careful study reveals, however, that there are some systematic discrepancies with the molecular dynamics data that persist even when the parameter $\tau_t(k)$ is chosen to fit the observed spectrum rather than calculated from some semi-empirical prescription. In particular, the shear-wave peaks at long wavelengths are significantly too broad and flat, as the results for the Lennard-Jones fluid shown in Figure 9.4 reveal. The structure of the correlation function matrix (9.3.7) provides a clue as to the origin of the deficiencies in the viscoelastic model. The element $\tilde{Y}_{22}(k, z)$ of the matrix is the Laplace transform of the normalised autocorrelation function of the xz-component of the stress tensor. Thus

$$\tilde{Y}_{22}(k, z) = \frac{\beta}{VG_\infty(k)} \int_0^\infty \left\langle \Pi_{\mathbf{k}}^{xz}(t)\Pi_{-\mathbf{k}}^{xz} \right\rangle \exp{(izt)}\mathrm{d}t$$

$$= \frac{1}{-iz + \tilde{M}_t(k, z) + \dfrac{\omega_{1t}^2}{-iz}} \qquad (9.3.18)$$

where the form of the normalisation factor follows from (8.4.10) and (8.6.9). If we again replace $\tilde{M}_t(k, z)$ by $1/\tau_t(k)$ and take the limit $k \to 0$, (9.3.18) can be inverted to give

$$\eta(t) = G_\infty(0)Y_{22}(0, t) = G_\infty(0) \exp[-G_\infty(0)|t|/\eta] \qquad (9.3.19)$$

which is consistent with (8.4.10). We saw in Section 8.6 that the memory function $\nu(k, t)$ and the stress autocorrelation function $\eta(t)$ become identical (apart from a multiplicative factor) as k tends to zero; within the viscoelastic approximation the identity is apparent immediately from intercomparison of (9.3.11), (9.3.14) and (9.3.19). At high densities, as Figure 8.3 illustrates, the correlation function $\eta(t)$ has a pronounced, slowly decaying tail and it is reasonable to suppose that the transverse current fluctuations at small wavevectors can be adequately described only if a comparably long-lived contribution is included in the memory function $\nu(k, t)$. In their classic analysis of the collective dynamical properties of the Lennard-Jones fluid, Levesque et al.[7] suggested the use of of a two-exponential memory function of the form

$$\nu(k,t)/\nu(k,0) = (1 - \alpha_k) \exp[-|t|/\tau_1(k)] + \alpha_k \exp[-|t|/\tau_2(k)] \quad (9.3.20)$$

which, as the discussion of the results in Figure 8.3 suggests, is also a useful approximation for other potential models. In practice, for the Lennard-Jones fluid, the slow relaxation time τ_2 turns out to be almost independent of k and some seven times larger than $\tau_1(0)$, while the parameter α_k, which at the smallest wavenumber studied has a value of approximately 0.1, decreases rapidly with increasing k. Thus, for large k, the single relaxation time approximation is recovered. At small k, however, inclusion of the long-lived tail in the memory function leads to a marked enhancement of the shear-wave peaks and significantly improved agreement with the molecular dynamics results, as illustrated in Figure 9.4; the price paid is the introduction of an additional two parameters. Broadly similar conclusions have emerged from calculations for liquid metals.[8]

9.4   DENSITY FLUCTUATIONS

The description of the longitudinal current fluctuations on the basis of the generalised Langevin equation is necessarily a more complicated task than in the case of the transverse modes. This is obvious from the much more complicated structure of the hydrodynamic formula (8.5.10) compared with (8.4.4). The problem of particular interest is to account for the dispersion and eventual disappearance of the collective mode associated with sound-wave propagation.

In discussion of the longitudinal modes a natural choice of components of the dynamical vector $\mathbf{A}$ is the set of conserved variables consisting of $\rho_\mathbf{k}$, $\mathbf{j_k}$ and the microscopic energy density $e_\mathbf{k}$ defined via (8.5.27). The variables $\rho_\mathbf{k}$ and $e_\mathbf{k}$ are both orthogonal to $\mathbf{j_k}$. In place of $e_\mathbf{k}$, however, it is more convenient to choose that part which is also orthogonal to $\rho_\mathbf{k}$ and plays the role of a microscopic temperature fluctuation; this we write as $T_\mathbf{k}$. Thus

$$T_\mathbf{k} = e_\mathbf{k} - (\rho_\mathbf{k}, e_\mathbf{k})(\rho_\mathbf{k}, \rho_\mathbf{k})^{-1} \rho_\mathbf{k} \quad (9.4.1)$$

The static correlation matrix is then diagonal. Since our attention is focused on the longitudinal fluctuations, we include only the projection of the current along $\mathbf{k}$, which we label $j_{\mathbf{k}}^z$. The vector $\mathbf{A}$ specified in this way, i.e.

$$\mathbf{A} = \begin{pmatrix} \rho_{\mathbf{k}} \\ j_{\mathbf{k}}^z \\ T_{\mathbf{k}} \end{pmatrix} \tag{9.4.2}$$

is only one of many possible choices; larger sets of variables that include both the stress tensor and heat current have also been considered. The static correlation matrix arising from (9.4.2) is

$$(\mathbf{A}, \mathbf{A}) = \begin{pmatrix} N S(k) & 0 & 0 \\ 0 & \dfrac{N k_{\mathrm{B}} T}{m} & 0 \\ 0 & 0 & \langle T_{\mathbf{k}} T_{-\mathbf{k}} \rangle \end{pmatrix} \tag{9.4.3}$$

and the corresponding frequency matrix is

$$-i\boldsymbol{\Omega} = \begin{pmatrix} 0 & -ik & 0 \\ \dfrac{-ik}{S(k)}\left(\dfrac{k_{\mathrm{B}}T}{m}\right) & 0 & \dfrac{\langle j_{\mathbf{k}}^z T_{-\mathbf{k}} \rangle}{\langle T_{\mathbf{k}} T_{-\mathbf{k}} \rangle} \\ 0 & -\dfrac{\langle T_{\mathbf{k}} j_{-\mathbf{k}}^z \rangle}{N k_{\mathrm{B}} T / m} & 0 \end{pmatrix} \tag{9.4.4}$$

It is unnecessary for our purposes to write more explicit expressions for the statistical averages appearing in (9.4.3) and (9.4.4).

Since $\dot{A}_1$ is proportional to A_2, it follows that the component R_1 of the random-force vector is zero and the memory function matrix reduces to

$$\mathbf{M}(k, t) = \begin{pmatrix} 0 & 0 & 0 \\ 0 & M_{22}(k, t) & M_{23}(k, t) \\ 0 & M_{32}(k, t) & M_{33}(k, t) \end{pmatrix} \tag{9.4.5}$$

The correlation function matrix is therefore given by

$$\tilde{\mathbf{Y}}(k, z) = \begin{pmatrix} -iz & ik & 0 \\ \dfrac{ik}{S(k)}\left(\dfrac{k_{\mathrm{B}}T}{m}\right) & -iz + \tilde{M}_{22}(k, z) & -i\Omega_{23} + \tilde{M}_{23}(k, z) \\ 0 & -i\Omega_{32} + \tilde{M}_{32}(k, z) & -iz + \tilde{M}_{33}(k, z) \end{pmatrix}^{-1} \tag{9.4.6}$$

and the Laplace transform of the longitudinal current autocorrelation function is

$$\tilde{C}_l(k, z) = \omega_0^2 \tilde{Y}_{22}(k, z) = \cfrac{\omega_0^2}{-iz + \cfrac{\omega_0^2}{-izS(k)} + \tilde{N}_l(k, z)} \tag{9.4.7}$$

where the memory function $N_l(k, t)$ is defined through its Laplace transform as

$$\tilde{N}_l(k, z) = \tilde{M}_{22}(k, z) - \frac{\Theta(k, z)}{-iz + \tilde{M}_{33}(k, z)} \tag{9.4.8}$$

with

$$\Theta(k, z) = \left(\tilde{M}_{23}(k, z) - \frac{\langle \dot{j}_{\mathbf{k}}^z T_{-\mathbf{k}} \rangle}{\langle T_{\mathbf{k}} T_{-\mathbf{k}} \rangle} \right) \left(\tilde{M}_{32}(k, z) + \frac{\langle T_{\mathbf{k}} \dot{j}_{-\mathbf{k}}^z \rangle}{N(k_{\mathrm{B}} T /m)} \right) \tag{9.4.9}$$

The physical significance of the four unknown memory functions in (9.4.5) can be inferred from their definitions in terms of the random forces $\mathcal{Q}\dot{j}_{\mathbf{k}}^z$ and $\mathcal{Q}\dot{T}_{\mathbf{k}}$. The functions M_{23} and M_{32} describe a coupling between the momentum current (the stress tensor) and heat flux whereas M_{22} and M_{33} represent, respectively, the relaxation processes associated with viscosity and thermal conduction. By comparison of (9.4.7)–(9.4.9) with the hydrodynamic result in (8.5.10) we can make the following identifications in the limit $k \to 0$:

$$\lim_{k \to 0} \tilde{M}_{22}(k, 0) = \frac{\left(\frac{4}{3}\eta + \zeta \right) k^2}{\rho m} = bk^2 \tag{9.4.10}$$

$$\lim_{k \to 0} \tilde{M}_{33}(k, 0) = \frac{\lambda k^2}{\rho c_V} = ak^2 \tag{9.4.11}$$

and

$$\lim_{k \to 0} \frac{|\langle \dot{j}_{\mathbf{k}}^z T_{-\mathbf{k}} \rangle|^2}{\langle T_{\mathbf{k}} T_{-\mathbf{k}} \rangle} = Nk^2 \left(\frac{k_{\mathrm{B}} T}{m} \right)^2 \frac{\gamma - 1}{S(k)} \tag{9.4.12}$$

Finally, by requiring that

$$N_l(k, t = 0) = \omega_{1l}^2 - \frac{\omega_0^2}{S(k)} \tag{9.4.13}$$

with ω_{1l}^2 given by (7.4.35), we guarantee that the first three non-zero moments of $S(k, \omega)$ are correct.

The derivation of (9.4.6) brings out clearly the advantage of working with a multi-variable description of a problem, such as that provided by (9.4.2). For example, we can immediately write down an expression for the fluctuations in temperature analogous to (9.4.7) for the current fluctuations. If we define a temperature autocorrelation function as

$$C_T(k, t) = \langle T_{\mathbf{k}}(t) T_{-\mathbf{k}} \rangle \tag{9.4.14}$$

we find from (9.4.6) that

$$\tilde{C}_T(k, z) = \langle T_{\mathbf{k}} T_{-\mathbf{k}} \rangle \, \tilde{Y}_{33}(k, z)$$

$$= \cfrac{\langle T_{\mathbf{k}} T_{-\mathbf{k}} \rangle}{-iz - \cfrac{\Theta(k, z)}{-iz + \cfrac{\omega_0^2}{-iz S(k)} + \tilde{M}_{22}(k, z)} + \tilde{M}_{33}(k, z)} \tag{9.4.15}$$

The key point to note is that $\tilde{C}_T(k, z)$ can be expressed in terms of the same memory functions used to describe $\tilde{C}_l(k, z)$. Similarly, by solving for $\tilde{Y}_{11}(k, z)$, we obtain an expression for the density autocorrelation function:

$$\tilde{F}(k, z) = S(k)\tilde{Y}_{11}(k, z) = \cfrac{S(k)}{-iz + \cfrac{1}{S(k)}\left(\cfrac{\omega_0^2}{-iz + \tilde{N}_l(k, z)}\right)} \tag{9.4.16}$$

This is a less interesting result than that obtained for $\tilde{C}_T(k, z)$ because $F(k, t)$ and $C_l(k, t)$ are in any case related by (7.4.26). It nevertheless brings out a second important feature of the multi-variable approach. An expression for $\tilde{F}(k, z)$ having the same form as (9.4.16) can more easily be obtained by setting $A = \rho_{\mathbf{k}}$ and making a continued-fraction expansion of $\tilde{F}(k, z)$ truncated at second order. What the more elaborate calculation yields is detailed information on the structure of the memory function $N_l(k, t)$, enabling contact to be made with the hydrodynamic result and allowing approximations to be introduced in a controlled way.

If we write the complex function $\tilde{N}_l(k, z)$ on the real axis ($z = \omega + i\epsilon, \epsilon \to 0+$) as the sum of its real and imaginary parts, i.e.

$$\tilde{N}_l(k, \omega) = N_l'(k, \omega) + iN_l''(k, \omega) \tag{9.4.17}$$

we find from (9.4.7) that the spectrum of longitudinal current fluctuations is given by

$$C_l(k, \omega) = \frac{1}{\pi} \frac{\omega^2 \omega_0^2 N_l'(k, \omega)}{[\omega^2 - \omega_0^2/S(k) - \omega N_l''(k, \omega)]^2 + [\omega N_l'(k, \omega)]^2} \tag{9.4.18}$$

If the memory function were small, there would be a resonance at a frequency determined by the static structure of the fluid, i.e. at $\omega^2 \approx \omega_0^2/S(k)$. The physical role of the memory function – the generalised 'friction' – is therefore to shift and damp the resonance.

The task of calculating the function $N_l(k, t)$ remains a formidable one, even with the restrictions we have discussed. Some recourse to modelling is therefore needed if tractable expressions for $C_l(k, \omega)$ and $S(k, \omega)$ are to be obtained. The limiting form of $\tilde{N}_l(k, \omega)$ when $k, \omega \to 0$ (the hydrodynamic limit) follows from (9.4.8)–(9.4.12):

$$\lim_{\omega \to 0} \lim_{k \to 0} \tilde{N}_l(k, \omega) = bk^2 + \frac{\omega_0^2}{S(k)} \frac{\gamma - 1}{-i\omega + ak^2} \tag{9.4.19}$$

The first term on the right-hand side of this expression describes viscous relaxation and corresponds to $\tilde{M}_{22}(k, \omega)$ in (9.4.8), while the second term arises from temperature fluctuations. We now require a generalisation of (9.4.19) that is valid for microscopic wavelengths and frequencies. An obvious first approximation is to assume that the coupling between the momentum and heat currents, represented by the memory functions $M_{23}(k, t)$ and $M_{32}(k, t)$, makes no contribution to the density fluctuations. This is true in the hydrodynamic limit and it is true instantaneously at finite wavelengths because the random forces $Q j_{\mathbf{k}}^{jz}$ and $Q \dot{T}_{\mathbf{k}}$ are instantaneously uncorrelated; the two memory functions therefore vanish at $t = 0$. If we also assume that the effect of thermal fluctuations is negligible, an approximation that can be justified at large wavenumbers, we are left only with the problem of representing the generalised longitudinal viscosity $\tilde{M}_{22}(k, \omega)$. Since the viscoelastic model (9.3.11) works moderately well in the case of the transverse currents, it is natural to make a similar approximation here by writing

$$N_l(k,t) = \left(\omega_{1l}^2 - \frac{\omega_0^2}{S(k)} \right) \exp[-|t|/\tau_l(k)] \qquad (9.4.20)$$

which is compatible with the constraint (9.4.13). The resulting expression for the dynamic structure factor is

$$S(k,\omega) = \frac{1}{\pi} \frac{\tau_l(k)\omega_0^2[\omega_{1l}^2 - \omega_0^2/S(k)]}{\omega^2\tau_l^2(k)(\omega^2 - \omega_{1l}^2)^2 + [\omega^2 - \omega_0^2/S(k)]^2} \qquad (9.4.21)$$

A variety of proposals have been made for the calculation of the relaxation time $\tau_l(k)$. For example, arguments similar to those used in the derivation of (9.2.21) lead in this case to the expression[9]

$$\tau_l^{-1}(k) = \frac{2}{\pi^{1/2}} \left(\omega_{1l}^2 - \frac{\omega_0^2}{S(k)} \right)^{1/2} \qquad (9.4.22)$$

The usefulness of this approach is illustrated[11] in Figure 9.5, which shows the dispersion of the sound wave peak obtained from molecular dynamics calculations for liquid rubidium and compares the results with those predicted by the viscoelastic approximation (9.4.21) in conjunction with (9.4.22). The agreement is good but the detailed shape of $S(k, \omega)$ is less well reproduced, particularly at small k. As the example shown in Figure 9.6 reveals, the discrepancies occur mostly at low frequencies. This is not surprising, since the low-frequency region of the spectrum is dominated by temperature fluctuations, which the viscoelastic model ignores.

The type of scheme outlined above is clearly an oversimplification and one that fails badly in the case of the Lennard-Jones fluid, where it cannot explain the appearance of the Brillouin peak seen in molecular dynamics calculations, an example of which is pictured in Figure 9.7. It can be shown from (9.4.22) that the viscoelastic model predicts the existence of a propagating mode at values of k such that

$$\omega_{1l}^2 < \frac{3\omega_0^2}{S(k)} \qquad (9.4.23)$$

If k is small this inequality can be rewritten as

$$\chi_T \left[\frac{4}{3}G_\infty(0) + K_\infty(0) \right] < 3 \qquad (9.4.24)$$

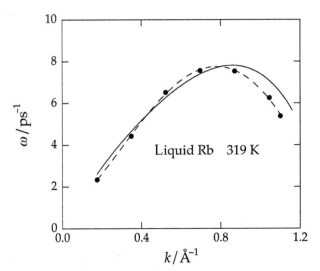

FIGURE 9.5 Sound-wave dispersion curve for a model of liquid rubidium near the normal melting temperature. The points are molecular dynamics results[10] and the broken curve is drawn as a guide to the eye. The full curve is calculated from the viscoelastic approximation (9.4.21) in conjunction with (9.4.22). From J.R.D. Copley and S.W. Lovesey, 'The dynamic properties of monatomic liquids', *Rep. Prog. Phys.* **38**, 461–563 (1975). © *IOP Publishing 1975. Reproduced by permission of IOP Publishing. All rights reserved.*

when ω_{1l}^2 is expressed in terms of the long-wavelength limits of the instantaneous shear modulus (8.6.9) and the instantaneous bulk modulus $K_\infty(k)$ defined by the relation

$$\frac{4}{3}G_\infty(k) + K_\infty(k) = \frac{\rho m \omega_{1l}^2}{k^2} \tag{9.4.25}$$

In the case of the alkali metals the inequality (9.4.24) is easily satisfied, but for the Lennard-Jones fluid under triple-point conditions the left-hand side of (9.4.24) has a value of ≈ 4.9. Given the structure of (9.4.24), it seems plausible to conclude that the fact that the sound-wave peak in liquid metals is found to persist to larger wavenumbers than in rare-gas liquids is associated with the lower compressibility of the metals (see Table 1.2). This difference in behaviour can in turn be correlated with the softer nature of the interatomic potentials in metals compared with those in the rare gases.

In order to describe the small-k behaviour of the Lennard-Jones system it is necessary to go beyond the viscoelastic approximation (9.4.20) by including the effect of temperature fluctuations. A generalisation of the hydrodynamic result (9.4.19) that satisfies the short-time constraint (9.4.13) is obtained by setting

$$\tilde{N}_l(k,\omega) = \left(\omega_{1l}^2 - \frac{\omega_0^2 \gamma(k)}{S(k)}\right)\tilde{n}_{1l}(k,\omega) + \frac{\omega_0^2}{S(k)}\frac{\gamma(k)-1}{-i\omega + a(k)k^2} \tag{9.4.26}$$

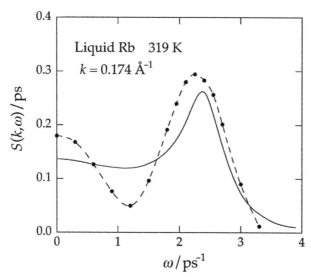

FIGURE 9.6 Dynamic structure factor at a wavenumber $k = 0.174$ Å^{-1} for a model of liquid rubidium near the normal melting temperature. The points are molecular dynamics results[10] and the broken curve is drawn as a guide to the eye. The full curve is calculated from the viscoelastic approximation (9.4.21) in conjunction with (9.4.22). From J.R.D. Copley and S.W. Lovesey, 'The dynamic properties of monatomic liquids', *Rep. Prog. Phys.* **38**, 461–563 (1975). © *IOP Publishing 1975. Reproduced by permission of IOP Publishing. All rights reserved.*

with $n_{1l}(k, t = 0) = 1$; this ignores any frequency dependence of the generalised thermal diffusivity $a(k)$ (the quantity $a(0)$ is defined by (8.3.14)). If, in addition, $\gamma(k)$ (a k-dependent ratio of specific heats) is set equal to one, the term representing temperature fluctuations disappears and (9.4.26) reduces to the viscoelastic approximation; the latter, as we have seen, works reasonably well for liquid metals, for which $\gamma(0) \approx 1$ (see Table 1.2). The first term on the right-hand side of (9.4.26) can be identified as $\tilde{M}_{22}(k, \omega)$. Then, if we assume a simple, exponential form for $n_{1l}(k, t)$, i.e.

$$n_{1l}(k, t) = \exp\left[-|t|/\tau_l(k) \right] \tag{9.4.27}$$

we find that in the hydrodynamic limit $\tilde{M}_{22}(k, 0)$ approaches the value

$$\lim_{k \to 0} \frac{\tilde{M}_{22}(k, 0)}{k^2} = \frac{\tau_l(0)}{\rho m}\left[\frac{4}{3}G_\infty(0) + K_\infty(0) - \gamma/\chi_T \right] \tag{9.4.28}$$

Comparison of (9.4.28) with (9.4.10) shows that $\tau_l(0)$ is given by

$$\tau_l(0) = \frac{\frac{4}{3}\eta + \zeta}{\frac{4}{3}G_\infty(0) + K_\infty(0) - \gamma/\chi_T} \tag{9.4.29}$$

Equations (9.4.26)–(9.4.29) make up the set of generalised hydrodynamic equations used by Levesque et al.[7] in their study of the Lennard-Jones fluid; together they yield a good fit to the dynamic structure factor over a wide range of k. Among the satisfying features of the analysis is the fact that at long wavelengths $\tau_l(k)$, as determined by a least-squares fitting procedure, tends correctly to its limiting value (9.4.29) as $k \to 0$. Moreover, $\gamma(k) \approx 1$ beyond $k\sigma \approx 2$. The large-k behaviour of $\gamma(k)$ implies that the viscoelastic model is a good approximation at short wavelengths because the coupling with the thermal mode becomes negligible. On the other hand, at small k, $\gamma(k)$ tends to a value that is greater by a factor of ≈ 2 than the thermodynamic value derived from the simulation, $\gamma = 1.86 \pm 0.01$. This fault can be eliminated by inclusion of a slowly relaxing part in the generalised longitudinal viscosity $\tilde{M}_{22}(k, \omega)$. If a two-exponential form is used for $n_{1l}(k, t)$, with the two decay times $\tau_1(k)$ (fast) and $\tau_2(k)$ (slow) given the same values[12] and the two terms in the memory function given the same relative weight as in the transverse current memory function (9.3.20), a very good fit is obtained, as Figure 9.7 shows, for which $\gamma(k)$ tends to its thermodynamic value as $k \to 0$. The agreement obtained with a single exponential is to some extent fortuitous. At small k the relaxation time $1/a(k)k^2$ associated with the thermal term in (9.4.26) is similar in value to the slow relaxation time τ_2. The omission of the long-time part of the viscous

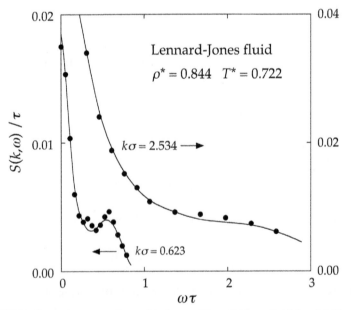

FIGURE 9.7 Dynamic structure factor of the Lennard-Jones fluid near its triple point. The points are molecular dynamics results and the curves show results calculated from (9.4.26) with a two-exponential approximation to $n_{1l}(k, t)$. The unit of time is $\tau = (m\sigma^2/48\epsilon)^{1/2}$. *Redrawn with permission from Ref. 7 © 1973 American Physical Society.*

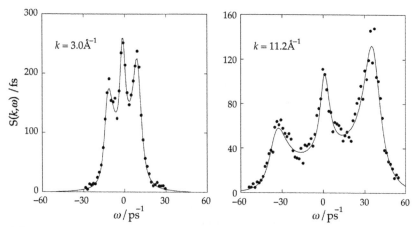

FIGURE 9.8 Dynamic structure factor of liquid lithium at 475 K for two values of k. The points are the results of inelastic X-ray scattering experiments and the curves show results calculated from (9.4.26) with a two-exponential approximation to $n_{1l}(k,t)$ and the experimental values of γ and a. *Redrawn with permission from Ref. 13 © 2000 American Physical Society.*

contribution to the memory function can therefore be offset, at least in part, by an increase in the size of the thermal contribution.

The slowly decaying contribution to the memory function is most important at those values of k for which a propagating mode is seen; at shorter wavelengths the dynamic structure factor is well described by a simpler, rapidly decaying, memory function. Calculations for the alkali metals therefore provide a more severe test of the need to include the contributions from both fast and slow processes, since the wavenumber range over which the Brillouin peaks are detectable is considerably wider than it is for argon-like liquids. The high quality data provided by inelastic X-ray scattering experiments make such a test possible. Figure 9.8 shows some results of X-ray measurements[13] on liquid lithium, together with those derived from a memory function having the same form as that used in the work on the Lennard-Jones fluid, again with an approximation for $n_l(k,t)$ involving two significantly different time scales; to allow direct comparison with the experimental data the theoretical results have been modified to allow for detailed balance (see (7.5.15)) and instrumental resolution. One important difference compared with the earlier work is the fact that the very high thermal conductivity of a liquid metal means that at wavenumbers relevant to X-ray scattering experiments the thermal contribution to the memory function is essentially a delta-function in time. A different fitting procedure was also used, in which the relaxation times and the relative weight attached to the fast and slow contributions, all taken to be functions of k, were treated as free parameters. The agreement between theory and experiment is very good and significantly better than that obtained with a single relaxation time. Inclusion of the long-lived term is necessary in order to reproduce the high degree of structure seen in the experimental results. At small wavenumbers the two relaxation times were found to differ by roughly an order of magnitude, with the weight of the slow process being 5–10 times smaller than that of the faster one, results which are broadly consistent with those obtained for the Lennard-Jones fluid. Similar conclusions have emerged from the analysis of X-ray scattering experiments for a number of other liquid metals.[14]

9.5 MODE COUPLING THEORY I. THE VELOCITY AUTOCORRELATION FUNCTION

The applications of the projection operator formalism studied thus far are largely phenomenological in character in the sense that a simple functional form has generally been assumed to describe the decay of the various memory functions. Such descriptions may be looked upon as interpolation schemes between the short-time behaviour of correlation functions, which is introduced via frequency sum rules, and the hydrodynamic regime, which governs the choice of dynamical variables to be included in the vector $\mathbf{A}$. A more ambitious programme would be to derive expressions for the memory functions from first principles, starting from the formally exact definitions of Section 9.1. A possible route towards such a microscopic theory is provided by the mode coupling approach, which we have already used in Section 8.7 to investigate the slow decay of the velocity autocorrelation function at long times. In this section we show how mode coupling concepts can be applied to the calculation of time correlation functions and their associated memory functions within the framework of the projection operator approach. The basic idea behind mode coupling theory is that a fluctuation (or 'excitation') in a given dynamical variable decays predominantly into pairs of hydrodynamic modes associated with conserved single-particle or collective dynamical variables. The possible 'decay channels' of a fluctuation are determined by 'selection rules' based, for example, on time reversal symmetry or on physical considerations. If a further, decoupling approximation is made, time correlation functions become expressible as sums of products of the correlation functions of conserved variables.

To illustrate the method we shall first rederive the asymptotic form (8.7.15) of the velocity autocorrelation function. Let u_{ix} be the x-component of the velocity of a tagged particle i. In the notation of Section 9.1 the velocity autocorrelation function has the from

$$Z(t) = \left(u_{ix}, \exp\left(i\mathcal{L}t\right)u_{ix}\right) \tag{9.5.1}$$

From the discussion in Section 8.7 we can expect the tagged-particle velocity to be strongly coupled to the longitudinal and transverse components of the collective particle current, while the form of (8.7.8) suggests that we take the tagged-particle density $\rho_{\mathbf{k}'i}$ and the current $\mathbf{j}_{-\mathbf{k}''}$ to be the modes into which fluctuations in u_{ix} decay. Translational invariance implies that the only products of Fourier components whose inner product with the tagged-particle velocity are non-zero are those for which $\mathbf{k}' = \mathbf{k}''$. The first approximation of the mode coupling treatment therefore consists in replacing the full evolution operator $\exp(i\mathcal{L}t)$ by its projection onto the subspace of the product variables $\rho_{\mathbf{k}i}\mathbf{j}_{-\mathbf{k}}$, i.e.

$$\exp(i\mathcal{L}t) \approx \mathcal{P}\exp(i\mathcal{L}t)\mathcal{P} \tag{9.5.2}$$

The projection operator $\mathcal{P}$ is defined, as in (9.1.1), by its action on a dynamical variable B:

$$\mathcal{P}B = \sum_{\mathbf{k}}\sum_{\alpha}\left(\rho_{\mathbf{k}i}j^{\alpha}_{-\mathbf{k}}, B\right)\left(\rho_{\mathbf{k}i}j^{\alpha}_{-\mathbf{k}}, \rho_{\mathbf{k}i}j^{\alpha}_{-\mathbf{k}}\right)^{-1}\rho_{\mathbf{k}i}j^{\alpha}_{-\mathbf{k}} \tag{9.5.3}$$

where the sum on α runs over all Cartesian components. Thus

$$\exp(i\mathcal{L}t)\mathcal{P}u_{ix} = \sum_{\mathbf{k}'}\sum_{\beta}\left(\rho_{\mathbf{k}'i}j^{\beta}_{-\mathbf{k}'}, u_{ix}\right)\left(\rho_{\mathbf{k}'i}j^{\beta}_{-\mathbf{k}'}, \rho_{\mathbf{k}'i}j^{\beta}_{-\mathbf{k}'}\right)^{-1}$$
$$\times \exp(i\mathcal{L}t)\rho_{\mathbf{k}'i}j^{\beta}_{-\mathbf{k}'} \tag{9.5.4}$$

and

$$Z(t) \approx \left(u_{ix}, \mathcal{P}\exp(i\mathcal{L}t)\mathcal{P}u_{ix}\right)$$
$$= \sum_{\mathbf{k},\mathbf{k}'}\sum_{\alpha}\sum_{\beta}\left(\rho_{\mathbf{k}'i}j^{\beta}_{-\mathbf{k}'}, u_{ix}\right)\left(\rho_{\mathbf{k}'i}j^{\beta}_{-\mathbf{k}'}, \rho_{\mathbf{k}'i}j^{\beta}_{-\mathbf{k}'}\right)^{-1}$$
$$\times \left(\rho_{\mathbf{k}i}j^{\alpha}_{-\mathbf{k}}, \exp(i\mathcal{L}t)\rho_{\mathbf{k}'i}j^{\beta}_{-\mathbf{k}'}\right)$$
$$\times \left(\rho_{\mathbf{k}i}j^{\alpha}_{-\mathbf{k}}, \rho_{\mathbf{k}i}j^{\alpha}_{-\mathbf{k}}\right)^{-1}\left(u_{ix}, \rho_{\mathbf{k}i}j^{\alpha}_{-\mathbf{k}}\right) \tag{9.5.5}$$

In this expression the time correlation functions of the product variables are bracketed by two, time-independent 'vertices', each of which has the same value. For example, since $\left\langle \rho_{\mathbf{k}i}j^{\alpha}_{-\mathbf{k}}\rho_{-\mathbf{k}i}j^{\alpha}_{\mathbf{k}}\right\rangle = N(k_{\mathrm{B}}T/m)$ and $\left\langle u_{ix}\rho_{-\mathbf{k}i}j^{\alpha}_{\mathbf{k}}\right\rangle = (k_{\mathrm{B}}T/m)\delta_{\alpha x}$, it follows that

$$\left(\rho_{\mathbf{k}i}j^{\alpha}_{-\mathbf{k}}, \rho_{\mathbf{k}i}j^{\alpha}_{-\mathbf{k}}\right)^{-1}\left(u_{ix}, \rho_{\mathbf{k}i}j^{\alpha}_{-\mathbf{k}}\right) = \frac{1}{N}\delta_{\alpha x} \tag{9.5.6}$$

The time correlation functions appearing on the right-hand side of (9.5.5) are of an unusual type, since they involve four, rather than two, dynamical variables. A second approximation usually made is to assume that the two modes appearing in the product variables propagate independently of each other. This means that the four-variable functions factorise into products of two-variable functions. In the present case:

$$\left(\rho_{\mathbf{k}i}j^{\alpha}_{-\mathbf{k}}, \exp(i\mathcal{L}t)\rho_{\mathbf{k}'i}j^{\beta}_{-\mathbf{k}'}\right) \approx \left(\rho_{\mathbf{k}i}, \exp(i\mathcal{L}t)\rho_{\mathbf{k}'i}\right)\left(j^{\alpha}_{-\mathbf{k}}, \exp(i\mathcal{L}t)j^{\beta}_{-\mathbf{k}'}\right)\delta_{\mathbf{k}\mathbf{k}'}$$
$$\equiv \langle\rho_{\mathbf{k}i}(t)\rho_{-\mathbf{k}i}\rangle\left\langle j^{\beta}_{-\mathbf{k}}(t)j^{\alpha}_{\mathbf{k}}\right\rangle \tag{9.5.7}$$

and use of (9.5.6) and (9.5.7) reduces (9.5.5) to the simpler form given by

$$Z(t) = \frac{1}{N^2} \sum_{\mathbf{k}} \langle \rho_{\mathbf{k}i}(t) \rho_{-\mathbf{k}i} \rangle \langle j_{\mathbf{k}}^x(t) j_{-\mathbf{k}}^x \rangle \tag{9.5.8}$$

The first factor in the sum over wavevectors is the self intermediate scattering function $F_s(k, t)$ and the second is a current correlation function; the latter can be decomposed into its longitudinal and transverse parts in the manner of (7.4.24). On switching from a sum to an integral and replacing the current correlation function by its average over a sphere, (9.5.8) becomes

$$Z(t) = \frac{1}{3\rho}(2\pi)^{-3} \int F_s(k, t) \frac{1}{k^2} \big[C_l(k, t) + 2C_t(k, t) \big] \mathrm{d}\mathbf{k} \tag{9.5.9}$$

If the time correlation functions on the right-hand side of (9.5.9) are replaced by the corresponding hydrodynamic expressions, (9.5.9) leads back to (8.7.15), which is valid for long times. At short times, however, (9.5.9) breaks down: as $t \to 0$, $Z(t)$ diverges, since $F_s(k, t = 0) = 1$ and $C_l(k, t = 0) = C_t(k, t = 0) = k^2(k_\mathrm{B}T/m)$. To overcome this difficulty a cut-off at large wavenumbers must be introduced in the integration over $\mathbf{k}$. Such a cut-off occurs naturally in the so-called velocity-field approach,[15] in which a result very similar to (9.5.9) is obtained on the basis of a microscopic expression for the local velocity of the tagged particle. This expression involves a 'form factor' $f(r)$, which in the simplest model used is represented by a unit step function that vanishes for distances greater than the particle 'radius' a and has the effect of making the velocity field constant over the range $r \le a$. Replacement of the Fourier components of the velocity field by their projections along the particle current leads to an expression of the form

$$Z(t) = \frac{1}{3}(2\pi)^{-3} \int \hat{f}(k) F_s(k, t) \frac{1}{k^2} [C_l(k, t) + 2C_t(k, t)] \, \mathrm{d}\mathbf{k} \tag{9.5.10}$$

where $\hat{f}(k = 0) = 1/\rho$ and $\lim_{k \to \infty} \hat{f}(k) = 0$. This result reduces to that obtained by the mode coupling approach in the long-wavelength limit but the behaviour at short times is much improved compared with (9.5.9). In particular, the zero-time value is now correct:

$$Z(0) = (2\pi)^{-3} \int \hat{f}(k) \frac{k_\mathrm{B}T}{m} \, \mathrm{d}\mathbf{k} = \frac{k_\mathrm{B}T}{m} f(r = 0) = \frac{k_\mathrm{B}T}{m} \tag{9.5.11}$$

Equation (9.5.10) does not represent a complete theory, since its evaluation requires a knowledge of the intermediate scattering function and the two current correlation functions. For numerical purposes, however, use can be made of the viscoelastic approximations for $C_l(k, t)$ and $C_t(k, t)$ and the Gaussian approximation (8.2.14) for $F_s(k, t)$. As Figures 9.9 and 9.10 show, results obtained in this way for the velocity autocorrelation function and corresponding

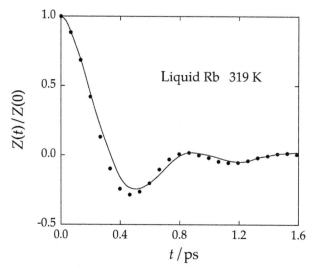

FIGURE 9.9 Normalised velocity autocorrelation function for a model of liquid rubidium. The points are molecular dynamics results and the curve is calculated from the velocity field approximation (9.5.10). From T. Gaskell and S. Miller, 'Longitudinal modes, transverse modes and velocity correlations in liquids: I', *J. Phys. C* **11**, 3749–3761 (1978). © *IOP Publishing 1978. Reproduced by permission of IOP Publishing. All rights reserved.*

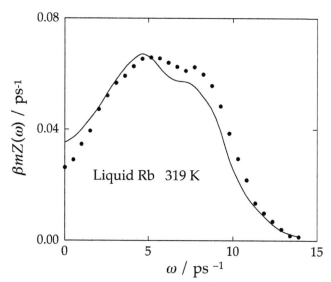

FIGURE 9.10 The power spectrum corresponding to the autocorrelation function plotted in Figure 9.9. The points are molecular dynamics results and the curve is calculated from the velocity field approximation (9.5.10). From T. Gaskell and S. Miller, 'Longitudinal modes, transverse modes and velocity correlations in liquids: I', *J. Phys. C* **11**, 3749–3761 (1978). © *IOP Publishing 1978. Reproduced by permission of IOP Publishing. All rights reserved.*

power spectrum of liquid rubidium are in good agreement with those obtained by molecular dynamics. The pronounced, low-frequency peak in the power spectrum arises from the coupling to the transverse current and the shoulder at higher frequencies comes from the coupling to the longitudinal current.

Another method whereby the short-time behaviour of the mode coupling approximation can be improved is to include the exact, low-order frequency moments of $Z(\omega)$ in a systematic way by working in the continued-fraction representation.[16] Truncation of (9.1.37) at second order gives

$$\tilde{Z}(z) = \cfrac{1}{-iz + \cfrac{\Omega_0^2}{-iz + \tilde{N}_2(z)}} \tag{9.5.12}$$

where Ω_0 is the Einstein frequency (7.2.13) and $\tilde{N}_2(z) \equiv \Delta_2^2 \tilde{M}_2(z)$. The Laplace transform of $\tilde{N}_2(z)$ is related to the autocorrelation function of the second-order random force $R_2 = \mathcal{Q}_2(i\mathcal{L})^2 u_{ix} = \mathcal{Q}_1(i\mathcal{L})^2 u_{ix}$ by

$$
\begin{aligned}
N_2(t) &= \left(R_2, \exp\left(i\mathcal{Q}_2\mathcal{L}\mathcal{Q}_2 t\right) R_2 \right)(R_1, R_1)^{-1} \\
&= \frac{m}{\Omega_0^2 k_B T} \left(\mathcal{Q}_1 \mathcal{L}^2 u_{ix}, \exp\left(i\mathcal{Q}_2\mathcal{L}\mathcal{Q}_2 t\right) \mathcal{Q}_1 \mathcal{L}^2 u_{ix} \right)
\end{aligned} \tag{9.5.13}
$$

The operator $\mathcal{Q}_1 = 1 - \mathcal{P}_1$ projects onto the subspace orthogonal to u_{ix} while $\mathcal{Q}_2 = \mathcal{Q}_1 - \mathcal{P}_2$ projects onto the subspace orthogonal to both u_{ix} and the acceleration $\dot{u}_{ix} = i\mathcal{L}u_{ix}$. The fact that $(i\mathcal{L})^2 u_{ix}$ is automatically orthogonal to $(i\mathcal{L})u_{ix}$ makes it possible to replace $\mathcal{Q}_2$ by $\mathcal{Q}_1$ in the definition of R_2.

If the product variables $\rho_{ki}\mathbf{j}_{-k}$ are again chosen as the basis set, use of the approximation (9.5.2) allows (9.5.13) to be rewritten as

$$
\begin{aligned}
N_2(t) \approx \frac{m}{\Omega_0^2 k_B T} \sum_{k,k'} \sum_{\alpha} \sum_{\beta} & \left(\rho_{k'i} j^{\beta}_{-k'}, \mathcal{Q}_1 \mathcal{L}^2 u_{ix} \right) \left(\rho_{k'i} j^{\beta}_{-k'}, \rho_{k'i} j^{\beta}_{-k'} \right)^{-1} \\
& \times \left(\rho_{ki} j^{\alpha}_{-k}, \exp\left(i\mathcal{Q}_2\mathcal{L}\mathcal{Q}_2 t\right) \rho_{k'i} j^{\beta}_{-k'} \right) \\
& \times \left(\rho_{ki} j^{\alpha}_{-k}, \rho_{ki} j^{\alpha}_{-k} \right)^{-1} \left(\mathcal{Q}_1 \mathcal{L}^2 u_{ix}, \rho_{ki} j^{\alpha}_{-k} \right)
\end{aligned} \tag{9.5.14}
$$

If we again assume that the variables ρ_{ki} and $\mathbf{j}_k^{\alpha}$ evolve in time independently of each other, and make the further approximation of replacing the projected operator $\mathcal{Q}_2\mathcal{L}\mathcal{Q}_2$ by the full Liouville operator $\mathcal{L}$ in the propagator governing the time evolution of the factorised correlation functions, (9.5.14) becomes

$$
\begin{aligned}
N_2(t) \approx \frac{m}{\Omega_0^2 k_B T} \sum_{k,k'} \sum_{\alpha} \sum_{\beta} & \left(\rho_{k'i} j^{\beta}_{-k'}, \mathcal{Q}_1 \mathcal{L}^2 u_{ix} \right) \left(\rho_{k'i} j^{\beta}_{-k'}, \rho_{k'i} j^{\beta}_{-k'} \right)^{-1} \\
& \times \left(\rho_{ki}, \rho_{k'i}(t) \right) \left(j^{\alpha}_{-k}, j^{\beta}_{-k'}(t) \right) \delta_{kk'} \\
& \times \left(\rho_{ki} j^{\alpha}_{-k}, \rho_{ki} j^{\alpha}_{-k} \right)^{-1} \left(\mathcal{Q}_1 \mathcal{L}^2 u_{ix}, \rho_{ki} j^{\alpha}_{-k} \right)
\end{aligned} \tag{9.5.15}
$$

The time correlation functions appearing here are the same as in (9.5.7), but the time-independent vertices have a more complicated form; a detailed calculation shows that

$$(\mathcal{Q}_1 \mathcal{L}^2 u_{i\alpha}, \rho_{\mathbf{k}i} j^\beta_{-\mathbf{k}}) = -\frac{\Omega_0^2 k_B T}{m} \mathcal{V}_{\alpha\beta}(k) \tag{9.5.16}$$

where

$$\mathcal{V}_{\alpha\beta}(k) = \frac{\rho}{\Omega_0^2 m} \int \exp(-i\mathbf{k}\cdot\mathbf{r}) g(r) \boldsymbol{\nabla}_\alpha \boldsymbol{\nabla}_\beta \, v(r) \, d\mathbf{r} \tag{9.5.17}$$

is a normalised 'vertex function'. Then, proceeding as before by switching from a sum over wavevectors to an integral, we find that

$$N_2(t) = N_{2l}(t) + 2N_{2t}(t) \tag{9.5.18}$$

with

$$N_{2l,2t}(t) = \frac{\Omega_0^2 m}{3\rho k_B T} (2\pi)^{-3} \int \mathcal{V}_{l,t}^2(k) F_s(k,t) \frac{1}{k^2} C_{l,t}(k,t) \, d\mathbf{k} \tag{9.5.19}$$

where $\mathcal{V}_{l,t}$ are the longitudinal and transverse components of the vertex tensor, defined in a manner analogous to (7.4.24).

There is a striking similarity between the structure of (9.5.19) and that of the mode coupling expression (9.5.9) obtained earlier for $Z(t)$ except that (9.5.19) contains the vertex factors $\mathcal{V}_{l,t}$. Inclusion of these factors ensures that the integral over wavevectors converges for all t; they therefore play a similar role to that of the form factor $\hat{f}(k)$ in the velocity-field approach, but have the advantage of being defined unambiguously through (9.5.17). The theory is also self-consistent, since the correlation functions required as input may be obtained by a mode coupling calculation of the same type. Numerically, however, the results are less satisfactory than those pictured in Figures 9.9 and 9.10.

9.6 MODE COUPLING THEORY II. THE KINETIC GLASS TRANSITION

The mode coupling ideas introduced in Section 9.5 were first used by Kawasaki[17] to study the 'critical slowing down' of density fluctuations near the liquid-gas critical point. Here we describe the application of the same general approach[18] to the not dissimilar phenomena associated with the kinetic glass transition of a fragile glass former already discussed in a qualitative way in Section 8.8. The theory shows that the structural arrest and associated dynamical anomalies that appear in the supercooled liquid at a well-defined temperature (on cooling) or density (on compression) are a direct consequence of a non-linear, feedback mechanism, the source of which is the fact that the memory function of the density autocorrelation function $F(k,t)$ may be expressed, at least approximately, in terms of $F(k,t)$ itself. Although real glass-forming liquids are usually multi-component in nature, we limit the discussion to one-component systems; the generalisation to mixtures is straightforward.

We saw in Section 9.4 that the decay of density fluctuations in a simple liquid above its triple point is well described within the memory function formalism by choosing as components of the dynamical vector $\mathbf{A}$ the three variables $\rho_\mathbf{k}$ (particle density), $j_\mathbf{k} \equiv \mathbf{k} \cdot \mathbf{j_k}/k$ (longitudinal particle current) and $T_\mathbf{k}$ (a microscopic temperature variable). It turns out, however, that temperature fluctuations are not important for the description of structural arrest and for present purposes the variable $T_\mathbf{k}$ can therefore be omitted. To simplify the resulting equations we first introduce a normalised density autocorrelation function

$$\phi(k,t) = F(k,t)/S(k) \tag{9.6.1}$$

with $\phi(k, t = 0) = 1$. Then, by following steps similar to those used to derive the memory function equation (9.4.16), we arrive at an expression for the Laplace transform of $\phi(k, t)$ in the form

$$\tilde{\phi}(k,z) = \cfrac{1}{-iz + \cfrac{\Omega_k^2}{-iz + \tilde{M}(k,z)}} \tag{9.6.2}$$

where $\Omega_k^2 = v_T^2 k^2 / S(k)$ and $v_T = (k_B T/m)^{1/2}$ is the thermal velocity. The structure of this result is identical with that in (9.4.16) and the function $\tilde{M}(k, z)$, like $\tilde{N}_l(k, z)$ in (9.4.16), is again the memory function of the longitudinal current, but the choices made for the vector $\mathbf{A}$ means that the explicit form of the memory function is different in the two cases. In the two-variable description the random-force vector has only one component, given by

$$K_\mathbf{k} = \mathcal{Q}(i\mathcal{L} j_\mathbf{k}) \tag{9.6.3}$$

and the corresponding memory function is

$$M(k,t) = \frac{1}{N v_T^2} \left(K_\mathbf{k}, R_\mathbf{k}(t) \right) \tag{9.6.4}$$

with $R_\mathbf{k}(t) = \exp{(i\mathcal{Q}\mathcal{L}\mathcal{Q}t)}K_\mathbf{k}$, where the operator $\mathcal{Q} = 1 - \mathcal{P}$ projects an arbitrary dynamical variable onto the subspace orthogonal to the variables $\rho_\mathbf{k}$ and $j_\mathbf{k}$. The time dependence of $\phi(k,t)$ is obtained from (9.6.2) via an inverse Laplace transform:

$$\ddot{\phi}(k,t) + \Omega_k^2 \phi(k,t) + \int_0^t M(k, t-t')\dot{\phi}(k,t')\,dt' \tag{9.6.5}$$

which can be recognised as the equation of motion of a harmonic oscillator of frequency Ω_k, damped by a time-retarded, frictional force.

The theoretical task is to derive an expression for the memory function that accounts for the structural slowing down near the transition temperature T_C; to achieve this, we follow the original arguments of Götze and collaborators.[19]

The random force $K_{\mathbf{k}}$ is by construction orthogonal to the slow variable $\rho_{\mathbf{k}}$ and the simplest slow variables having a non-zero correlation with $K_{\mathbf{k}}$ are the pair products

$$A_{\mathbf{p},\mathbf{q}} = \rho_{\mathbf{p}}\rho_{\mathbf{q}} \tag{9.6.6}$$

Hence the first approximation, one of typical mode coupling type, is to replace the random force $K_{\mathbf{k}}$ in (9.6.3) by its projection onto the subspace spanned by all pair products, i.e.

$$K_{\mathbf{k}} \approx \sum_{\mathbf{p},\mathbf{q}} \sum_{\mathbf{p}',\mathbf{q}'} (A_{\mathbf{p}',\mathbf{q}'}, K_{\mathbf{k}})(A_{\mathbf{p},\mathbf{q}}, A_{\mathbf{p}',\mathbf{q}'})^{-1} A_{\mathbf{p},\mathbf{q}} \tag{9.6.7}$$

Substitution of (9.6.7) and the corresponding expression for $R_{\mathbf{k}}(t)$ in (9.6.4) gives

$$
\begin{aligned}
M(k,t) = \frac{1}{Nv_T^2} &\sum_{\mathbf{p},\mathbf{q}} \sum_{\mathbf{p}',\mathbf{q}'} (A_{\mathbf{p}',\mathbf{q}'}, K_{\mathbf{k}})(A_{\mathbf{p},\mathbf{q}}, A_{\mathbf{p}',\mathbf{q}'})^{-1} \\
&\times \sum_{\mathbf{p}'',\mathbf{q}''} \sum_{\mathbf{p}''',\mathbf{q}'''} (A_{\mathbf{p}''',\mathbf{q}'''}, K_{\mathbf{k}})(A_{\mathbf{p}'',\mathbf{q}''}, A_{\mathbf{p}''',\mathbf{q}'''})^{-1} \\
&\times (A_{\mathbf{p},\mathbf{q}}, \exp(i\mathcal{QLQ}t) A_{\mathbf{p}'',\mathbf{q}''})
\end{aligned} \tag{9.6.8}
$$

The next step is to factorise the static and dynamic four-point correlation functions in (9.6.8) into products of two-point functions, and simultaneously to replace the propagator of the projected dynamics by the full propagator. Thus

$$
\begin{aligned}
(A_{\mathbf{p},\mathbf{q}}, \exp(i\mathcal{QLQ}t) A_{\mathbf{p}'',\mathbf{q}''}) &= (\rho_{\mathbf{p}}\rho_{\mathbf{q}}, \exp(i\mathcal{QLQ}t)\rho_{\mathbf{p}''}\rho_{\mathbf{q}''}) \\
&\approx (\rho_{\mathbf{p}}, \exp(i\mathcal{L}t)\rho_{\mathbf{p}''})(\rho_{\mathbf{q}}, \exp(i\mathcal{L}t)\rho_{\mathbf{q}''}) \\
&= \delta_{\mathbf{p},\mathbf{p}''}\delta_{\mathbf{q},\mathbf{q}''} N^2 S(p)S(q)\phi(p,t)\phi(q,t)
\end{aligned} \tag{9.6.9}
$$

while for $t = 0$:

$$(A_{\mathbf{p},\mathbf{q}}, A_{\mathbf{p}',\mathbf{q}'})^{-1} = \frac{\delta_{\mathbf{p},\mathbf{p}'}\delta_{\mathbf{q},\mathbf{q}'}}{N^2 S(p)S(q)} \tag{9.6.10}$$

The three-point static correlation functions that appear in the terms involving $K_{\mathbf{k}}$ in (9.6.8) can be eliminated with a help of a generalisation of the Yvon equality (7.2.11), i.e.

$$\langle \dot{A}B^* \rangle = \langle (i\mathcal{L}A)B^* \rangle \equiv -\langle \{\mathcal{H}, A\}B^* \rangle = k_B T \langle \{A, B^*\} \rangle \tag{9.6.11}$$

the proof of which now requires a double integration by parts. We also make use of the Ornstein–Zernike relation in the form $S(k) = 1/(1 - \rho\hat{c}(k))$ and the convolution approximation (4.2.10). Then, for example:

$$(\rho_{\mathbf{p}',\mathbf{q}'}, i\mathcal{L}j_{\mathbf{k}}) = -\frac{iv_T^2 N\delta_{\mathbf{k},\mathbf{p}'+\mathbf{q}'}}{k}[\mathbf{k}\cdot\mathbf{p}'S(q') + \mathbf{k}\cdot\mathbf{q}'S(p')] \tag{9.6.12}$$

The final result of these manipulations is

$$M(k,t) = \frac{v_T^2 \rho^2}{2Nk^2} \sum_{\mathbf{p},\mathbf{q}} \delta_{\mathbf{k},\mathbf{p}+\mathbf{q}} S(p)S(q)\big[\hat{c}(p)\mathbf{k}\cdot\mathbf{p} + \hat{c}(q)\mathbf{k}\cdot\mathbf{q}\big]^2 \phi(p,t)\phi(q,t)$$

(9.6.13)

The factor $\frac{1}{2}$ on the right-hand side arises from the fact that all double sums over pairs of wavevectors must be ordered in such a way that each product variable $A_{\mathbf{p},\mathbf{q}}$ appears only once.

The appearance of the product $\phi(p,t)\phi(q,t)$ in (9.6.13) means that the memory function decays on the same timescale as the correlation function. This represents only the long-time contribution to the total memory function and cannot describe the behaviour at short times, which is dominated by nearly instantaneous, binary collisions. To describe the effect of collisions it is assumed that the short-time contribution $M^{(0)}(k,t)$ can be represented by a δ-function, i.e.

$$M^{(0)}(k,t) = v(k)\delta(t)$$

(9.6.14)

The complete memory function is therefore written as

$$M(k,t) = v(k)\delta(t) + \Omega_k^2 m(k,t)$$

(9.6.15)

Comparison with (9.6.13) shows that

$$m(k,t) = \frac{1}{2V} \sum_{\mathbf{p},\mathbf{q}} \delta_{\mathbf{k},\mathbf{p}+\mathbf{q}} V(\mathbf{k},\mathbf{p},\mathbf{q})\phi(p,t)\phi(q,t)$$

(9.6.16)

where the vertex function V is

$$V(\mathbf{k},\mathbf{p},\mathbf{q}) = \frac{\rho}{k^4} S(k)S(p)S(q)\big[\hat{c}(p)\mathbf{k}\cdot\mathbf{p} + \hat{c}(q)\mathbf{k}\cdot\mathbf{q}\big]^2$$

(9.6.17)

The non-linear, integro-differential equation (9.6.5) may then be rewritten as

$$\ddot{\phi}(k,t) + \Omega_k^2\phi(k,t) + v(k)\dot{\phi}(k,t) + \Omega_k^2\int_0^\infty m(k,t-t')\dot{\phi}(k,t')\mathrm{d}t' = 0$$

(9.6.18)

The coupled equations (9.6.16) and (9.6.18) form a closed, self-consistent set; the only input required for their solution is the static structure factor of the supercooled liquid, which determines the value of the vertex function via (9.6.17). The feedback mechanism is provided by the quadratic dependence of the memory function on $\phi(k,t)$, with the density and temperature dependence of the effect coming from the vertex function. Numerical solution of the coupled equations reveals the existence of a sharp crossover from ergodic to non-ergodic behaviour of $\phi(k,t)$ at a well-defined temperature (at constant density) or density (at constant temperature). The predicted correlation function can also be used as input to a similar set of equations for the self-correlation function $F_s(k,t)$, where the memory function now involves the product $\phi(k,t)F_s(k,t)$.

In the case of hard spheres the theory outlined above predicts a kinetic glass transition at a packing fraction $\eta_C \approx 0.516$ when the Percus–Yevick approximation for the structure factor is used. At the critical packing fraction the order parameter[20] $f_k = \lim_{t \to \infty} \phi(k, t)$ changes discontinuously from zero to a wavenumber-dependent value $0 < f_k \leq 1$. That this transition is a direct consequence of the non-linearity of the equation of motion (9.6.18) can be demonstrated with the help of some further approximations.[19,21] The largest contribution to the vertex function comes from the region $k \approx k_{max}$ of the main peak in the structure factor. It is therefore not unreasonable to ignore the sum over wavevectors by putting $S(k) \approx 1 + a\delta(k - k_{max})$, where a is the area under the main peak. With this assumption, (9.6.18) becomes an equation for the single correlation function $\phi(k_{max}, t) \equiv \phi(t)$, which we write as

$$\ddot{\phi}(t) + \Omega^2 \phi(t) + \nu \dot{\phi}(t) + \lambda \Omega^2 \int_0^\infty [\phi(t - t')]^2 \dot{\phi}(t') \, dt' = 0 \qquad (9.6.19)$$

where $\Omega \equiv \Omega_{k_{max}}$, ν can be interpreted as a collision frequency and λ, which replaces the complicated vertex function, acts as a 'control parameter', a role played by inverse temperature or density in the more complete theory. By taking the Laplace transform of (9.6.19) we recover (9.6.2) in the form

$$\tilde{\phi}(z) = \cfrac{1}{-iz + \cfrac{\Omega^2}{-iz + \nu + \Omega^2 \tilde{m}(z)}} \qquad (9.6.20)$$

with

$$\tilde{m}(z) = \lambda \int_0^\infty [\phi(t)]^2 \exp(izt) dt \qquad (9.6.21)$$

Equation (9.6.20) can be rearranged to give

$$\frac{\tilde{\phi}(z)}{1 + iz\tilde{\phi}(z)} = \frac{1}{\Omega^2}\left[-iz + \nu + \Omega^2 \tilde{m}(z)\right] \qquad (9.6.22)$$

Let $\lim_{t \to \infty} \phi(t) = f$, where the order parameter f is now independent of k. Then

$$\lim_{z \to 0} \tilde{\phi}(z) = \frac{f}{-iz} \qquad (9.6.23)$$

and hence, from substitution in (9.6.22):

$$\lim_{z \to 0} \tilde{m}(z) = \frac{f}{-iz(1 - f)} \qquad (9.6.24)$$

In the non-ergodic or structurally arrested phase, where $f > 0$, the power spectrum $\phi(\omega)$ will contain a fully elastic component, $f\delta(\omega)$; experimentally this would correspond to scattering from the frozen structure.

Equation (9.6.21) shows that

$$\lim_{z \to 0} \tilde{m}(z) = \frac{\lambda f^2}{-iz} \qquad (9.6.25)$$

Identification of (9.6.25) with (9.6.24) leads to a simple equation for the order parameter:

$$\frac{f}{1-f} = \lambda f^2 \tag{9.6.26}$$

the solutions to which are

$$f = 0, \quad f = \frac{1}{2}[1 \pm (1 - 4/\lambda)^{1/2}] \tag{9.6.27}$$

Since f must be real, the only acceptable solution for $\lambda < 4$ is $f = 0$, corresponding to the ergodic phase. This remains a solution at larger values of λ, but at the critical value, $\lambda_C = 4$, there is a bifurcation to the non-ergodic solution, $f = \frac{1}{2}[1+(1-4/\lambda)^{1/2}]$; for $\lambda = 4$, $f = \frac{1}{2}$. The root $f = \frac{1}{2}[1-(1-4/\lambda)^{1/2}]$ is not acceptable, since it implies that the system would revert to ergodic behaviour in the limit $\lambda \to \infty$. Let $\lambda = 4(1 + \sigma\epsilon)$, where $\sigma = -1$ and $+1$ in the ergodic and arrested phases, respectively. The quantity $\epsilon = (\lambda - \lambda_C)/\sigma\lambda_C$ is a positive number which measures the distance from the transition. Substitution in (9.6.27) shows that for $\sigma = +1$, f has a square-root cusp as $\epsilon \to 0$:

$$\lim_{\epsilon \to 0} f = \frac{1}{2}\left(1 + \epsilon^{1/2}\right) \tag{9.6.28}$$

The dependence of f on λ calculated from (9.6.27) and (9.6.28) is sketched in Figure 9.11.

Equation (9.6.28) describes the infinite-time behaviour of the correlation function in the arrested phase for $\lambda \approx \lambda_C$. To extend this result to finite times,

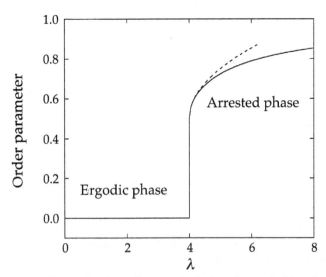

FIGURE 9.11 Predictions of mode coupling theory for the dependence on λ of the order parameter f. The full curve is the result obtained from the equation of motion (9.6.19) and the dashes show the approximate solution (9.6.27).

we look for a solution to (9.6.19) of the form

$$\phi(t) = \frac{1}{2} + \epsilon^{1/2} g_\epsilon(\tau) \tag{9.6.29}$$

where $\tau = \epsilon^s t$ is a scaled time and $g_\epsilon(\tau)$ is a scaling function. The quantity $s(> 0)$ is a scaling exponent, which is determined later by requiring $\phi(t)$ to be independent of ϵ in the short-time limit. This restriction on ϕ follows from the fact that the short-time behaviour is controlled by the collision frequency ν, not by the mode coupling contribution to the memory function. The Laplace transform of (9.6.29) is

$$\tilde{\phi}(z) = \epsilon^{-s} \left(\frac{1}{-2i\zeta} + \epsilon^{1/2} \tilde{g}_\epsilon(\zeta) \right) \tag{9.6.30}$$

where $\zeta \equiv \epsilon^{-s} z$. If we substitute (9.6.29) in (9.6.21) (with $\lambda = 4 + 4\sigma\epsilon$) and (9.6.30) in (9.6.22), combine the two results and let $\epsilon \to 0$, we obtain an equation for the scaling function at the critical point ($\epsilon = 0$):

$$-8i\zeta[\tilde{g}_0(\zeta)]^2 - 4 \int_0^\infty [g_0(\tau)]^2 \exp(i\zeta\tau)\mathrm{d}\tau = \frac{\sigma}{-i\zeta} \tag{9.6.31}$$

To derive this result it must be assumed that $\epsilon^{1/2}\tilde{g}_0(\zeta)$ vanishes with ϵ; the solution obtained below is consistent with that assumption.

The β-relaxation regime corresponds to scaled times $\tau \ll 1$ (or $\zeta \gg 1$). We look for a power-law solution for $g_0(\tau)$ such that

$$g_0(\tau) = a_0\tau^{-a}, \quad \tau \to 0 \tag{9.6.32}$$

with a Laplace transform given by

$$\tilde{g}_0(\zeta) = a_0\Gamma(1-a)(-i\zeta)^{a-1}, \quad \zeta \to \infty \tag{9.6.33}$$

where $\Gamma(x)$ is the gamma function. Substitution in (9.6.31) gives

$$(-i\zeta)^{2a-1}4a_0^2[2\Gamma^2(1-a) - \Gamma(1-2a)] = 0 \tag{9.6.34}$$

i.e. $2\Gamma^2(1-a) = \Gamma(1-2a)$, the positive solution to which is $a \approx 0.395$. When written in terms of the original time variable t, combination of (9.6.29) and (9.6.32) shows that

$$\phi(t) = \frac{1}{2} + a_0\epsilon^{-as+1/2}t^{-a} \tag{9.6.35}$$

Since $\phi(t)$ must be independent of ϵ in the limit $t \to 0$, it follows that $s = 1/2a \approx 1.265$. Thus the correlation function decays as

$$\phi(t) = \frac{1}{2} + a_0 t^{-a} \tag{9.6.36}$$

The result expressed by (9.6.36) is independent of σ. It therefore describes both the decay of $\phi(t)$ towards its non-zero, asymptotic value in the arrested phase and the first relaxation process in the ergodic phase, where the power-law behaviour will persist so long as $\tau \ll 1$. Times $\tau \gg 1$ correspond to α-relaxation in the ergodic phase. A scaling analysis similar to the previous one starts from the ansatz

$$g_0(\tau) = -b_0 \tau^b, \quad \tau \to \infty \tag{9.6.37}$$

and leads to the exponent relation

$$2\Gamma^2(1+b) - \Gamma(1+2b) = 0 \tag{9.6.38}$$

The only acceptable solution to this equation is $b = 1$. Thus

$$\phi(t) = \frac{1}{2} - b_0 \epsilon^{1/2} \tau, \quad 1 \ll \tau \ll 1/b_0 \epsilon^{1/2} \tag{9.6.39}$$

The upper limit on τ in (9.6.39) appears because $|\epsilon^{1/2} g_0(\tau)|$ must be less than unity for the asymptotic analysis to be valid. At yet longer times a purely exponential decay is predicted, in contrast to the stretched exponential decay seen both experimentally and in simulations (see Section 8.8). To reproduce the observed behaviour the simplified model represented by (9.6.19), in which $m(t)$ behaves as $[\phi(t)]^2$, must be generalised[18] to include more control parameters and other powers of $\phi(t)$.

The scaling predictions of mode coupling theory have been tested against experimental data and the results of simulations, and generally good agreement is found at temperatures just above T_C. However, the distinction between ergodic and strictly non-ergodic phases that appears in the original version of the theory is unrealistic. At sufficiently long times thermally activated processes of the type evident, for example, in Figure 8.11 will eventually cause ergodicity to be restored. Such effects can be accommodated within the theory by inclusion of the coupling of fluctuations in the microscopic density with those in particle current.[22] The 'ideal' transition is then suppressed and the correlation function is found to decay to zero even below T_C, though only after a period of near-complete structural arrest that rapidly lengthens as the temperature is lowered.

Mode coupling theory has also been extended to the case of concentrated, colloidal solutions. As we shall see in Chapter 12, these are systems that are well described by potentials consisting of a hard-core repulsion and a very short-ranged attractive interaction which favours aggregation of particles. Higher-order glass transition singularities are predicted, resulting in a re-entrant, liquid–glass transition line in the temperature–density plane and a transition line between two glass states.[23] Another generalisation of the theory deals with inhomogeneous, supercooled liquids,[24] and is therefore relevant to the phenomenon of dynamical heterogeneity discussed in Section 8.8.[25]

REFERENCES

[1] Zwanzig, R., *In* 'Lectures in Theoretical Physics', vol. III (W.E. Britton, B.W. Downs and J. Downs, eds). Wiley Interscience, New York, 1961.

[2] (a) Mori, H., *Prog. Theor. Phys.* **33**, 423 (1965). (b) Mori, H., *Prog. Theor. Phys.* **34**, 399 (1965).

[3] Lovesey, S.W., *J. Phys. C* **6**, 1856 (1973).

[4] See, e.g., de Jong, P.H.K., Verkerk, P. and de Graaf, L.A., *J. Phys.: Cond. Matter* **6**, 8391 (1994).

[5] Balucani, U., Torcini, A., Stangl, A. and Morkel, C., *J. Non-Cryst. Solids* **205–207**, 299 (1996).

[6] Levesque, D. and Verlet, L., *Phys. Rev. A* **2**, 2514 (1970).

[7] Levesque, D., Verlet, L. and Kürkijarvi, J., *Phys. Rev. A* **7**, 1690 (1973).

[8] See, e.g., Kambayashi, S. and Kahl, G., *Phys. Rev. A* **46**, 3255 (1992).

[9] Lovesey, S.W., *J. Phys. C* **4**, 3057 (1971).

[10] Rahman, A., *In* 'Statistical Mechanics: New Concepts, New Problems, New Applications' (S.A. Rice, K.F. Freed and J.C. Light, eds). University of Chicago Press, Chicago, 1972.

[11] Copley, J.R.D. and Lovesey, S.W., *Rep. Prog. Phys.* **38**, 461 (1975).

[12] Values of $\tau_1(k)$ obtained from separate fits of $C_t(k, \omega)$ and $S(k, \omega)$ are nearly the same; the small differences that do occur can be ascribed to statistical uncertainties in the molecular dynamics data.

[13] (a) Scopigno, T., Balucani, U., Ruocco, G. and Sette, F., *Phys. Rev. Lett.* **85**, 4076 (2000). (b) Scopigno, T., Balucani, U., Ruocco, G. and Sette, F., *J. Non-Cryst. Solids* **312–314**, 121 (2002).

[14] See Monaco, A., Scopigno, T., Benassi, P., Giugni, A., Monaco, G., Nardone, M., Ruocco, G. and Sampoli, M., *J. Chem. Phys.* **120**, 8089 (2004) (for K) and references therein to work on Na, Al and Ga.

[15] Gaskell, T. and Miller, S., *J. Phys. C* **11**, 3749 (1978).

[16] Bosse, J., Götze, W. and Zippelius, A., *Phys. Rev. A* **18**, 1214 (1978).

[17] Kawasaki, K., *Ann. Phys. NY* **61**, 1 (1970).

[18] (a) Götze, W. and Sjögren, L., *Rep. Prog. Phys.* **55**, 241 (1992). (b) Götze, W., *In* 'Liquids, Freezing and the Glass Transition: Les Houches, Session LI' (J.P. Hansen, D. Levesque and J. Zinn-Justin, eds). Elsevier, Amsterdam, 1991. (c) Das, S.P., *Rev. Mod. Phys.* **76**, 785 (2004).

[19] Bengtzelius, U., Götze, W. and Sjölander, A., *J. Phys. C* **17**, 5915 (1984).

[20] The terms non-ergodicity parameter or Edwards–Anderson parameter are also used, the latter by analogy with a related problem in the theory of spin glasses: Edwards, S.F. and Anderson P.W., *J. Phys. F* **5**, 965 (1975).

[21] Leutheusser, E., *Phys. Rev. A* **29**, 2765 (1984).

[22] (a) Das, S.P. and Mazenko, G.F., *Phys. Rev. A* **34**, 2265 (1986). (b) Götze, W. and Sjögren, L., *Z. Phys. B* **65**, 415 (1987).

[23] Dawson, K., Foffi, G., Fuchs, M., Götze, W., Sciortino, F., Sperl, M., Tartaglia, P., Voigtmann, Th. and Zaccarelli, E., *Phys. Rev. E* **63**, 011401 (2000).

[24] Biroli, G., Bouchaud, J.P., Miyazaki, K. and Reichman, D.R., *Phys. Rev. Lett.* **97**, 195701 (2006).

[25] For a full review of all aspects of dynamical slowing down, see Götze, W., 'Complex Dynamics of Glass-Forming Liquids: A Mode-Coupling Theory'. Oxford University Press, New York, 2009.

Chapter 10

Ionic Liquids

10.1 CLASSES AND MODELS OF IONIC LIQUIDS

We have been concerned thus far primarily with fluids in which the range of the interparticle forces is of the order of a few atomic radii. This chapter is devoted to systems in which the particles carry an electric charge. Ionic liquids have certain properties that are absent in fluids composed of neutral particles and many of their distinguishing features are associated in some way with the slow decay of the Coulomb potential. Our attention will be focused on three types of system: molten salts, ionic solutions and liquid metals. *Molten salts* are in many respects the simplest class of ionic liquids. We shall concentrate mostly on the case in which there is a single cation and a single anion species, of which the alkali halides are the best understood examples. Molten salts are characterised by large cohesive energies and high temperatures, and by ionic conductivities of the order of $1\,\Omega^{-1}\,\mathrm{cm}^{-1}$. There exist also certain crystalline salts that have conductivities comparable with those of the molten phase. These are the so-called 'fast-ion' conductors, or 'solid electrolytes', in which one of the ionic species becomes liquid-like in behaviour above a certain temperature. *Ionic solutions* are liquids consisting of a solvent formed from neutral, polar molecules and a solute that dissociates into positive and negative ions. They vary widely in complexity. In the classic *electrolyte solutions* the cations and anions are of comparable size and absolute charge, whereas macromolecular ionic solutions contain both macroions (charged polymer chains, micelles, charged colloidal particles, etc.) and microscopic counterions. Despite their complexity some systems of the latter type, including charged colloidal suspensions, can be treated quantitatively by standard methods of liquid state theory, as we shall see in Chapter 12. Finally, *liquid metals* are similar in composition to molten salts, the anion of the salt being replaced by electrons from the valence or conduction bands of the metal. The analogy is a superficial one, however, because the small mass of the electron leads to a pronounced asymmetry between the two charge-carrying species. Whereas the behaviour of the ions can be discussed within the framework of classical statistical mechanics, the electrons form a degenerate Fermi gas for which a quantum mechanical treatment is required. The presence

Theory of Simple Liquids, Fourth Edition. http://dx.doi.org/10.1016/B978-0-12-387032-2.00010-6
403

of 'free' electrons is also the origin of the very high electrical conductivities of liquid metals, which are typically three to four orders of magnitude larger than those of molten salts. 'Simple' metals are those in which the electronic valence states are well separated in energy from the tightly bound, core states; they include the alkali metals, magnesium, zinc, mercury, gallium and aluminium.

The systems we have listed vary widely in character but they have two important features in common: first, that of overall, macroscopic charge neutrality and, secondly, the presence of mobile charge carriers. The condition of overall charge neutrality imposes a constraint on the relative concentrations of the ions. If the fluid contains $\rho_\nu = N_\nu/V$ ions per unit volume of species ν and if the charge carried by ions of that species is $q_\nu = z_\nu e$, where e is the elementary charge, overall charge neutrality requires that

$$\sum_\nu z_\nu \rho_\nu = 0 \qquad (10.1.1)$$

We shall see in the next section that a tendency towards charge neutrality exists even at the local, microscopic level. This effect gives rise in turn to the phenomenon of *screening*. Introduction of an external charge into an ionic fluid causes a rearrangement, or polarisation, of the surrounding charge density with the result that the net electrostatic potential due to the external charge and the 'polarisation cloud' decays much faster than the bare Coulomb potential. In fact, as we shall show later, the potential decays exponentially. Since it is permissible to regard any ion in the fluid as an 'external' charge, it follows that the screening mechanism determines the long-range behaviour of the ionic distribution functions. Screening also requires that the distribution functions satisfy a number of important sum rules. In ionic liquids of high density, such as molten salts, there is a competition between packing effects and screening; this leads to a *charge ordering* of the ions, which manifests itself as an alternation in sign of the charge carried by successive coordination shells around a central ion.

The presence of mobile charge carriers plays an important role in determining the dynamical properties of ionic liquids. It leads in particular to new forms of transport, of which electrical conduction is the most familiar example. In addition, the interplay between Maxwell's equations and the equations of hydrodynamics causes the long-wavelength charge fluctuations to relax in a manner qualitatively different from that of concentration fluctuations in mixtures of uncharged particles. Under conditions achievable, in particular, in molten salts, fluctuations in charge may give rise to propagating, high-frequency, collective modes. These excitations are similar in character to the optic modes of ionic crystals and are also closely related to the charge oscillations found in plasmas.

Theories of ionic liquids rely heavily on the use of simple hamiltonian models that retain only the essential features of the ionic interactions. One simplifying approximation commonly made is to ignore the polarisability of the ions and represent the interactions by a *rigid-ion model*. The total potential energy is then assumed to be pairwise additive and written as the sum of short-range (S) and coulombic (C) terms in the form

$$V_N(\mathbf{r}^N) = V_N^S(\mathbf{r}^N) + V_N^C(\mathbf{r}^N) = V_N^S(\mathbf{r}^N) + \sum_{i=1}^{N} \sum_{j>i}^{N} \frac{z_i z_j e^2}{\epsilon |\mathbf{r}_j - \mathbf{r}_i|} \qquad (10.1.2)$$

where N is the total number of ions and ϵ is the dielectric constant of the medium in which the ions are immersed. It is often convenient to replace the Coulomb term in (10.1.2) by a sum in reciprocal space. Let $\rho_{\mathbf{k}}^Z$ be a Fourier component of the microscopic charge density, given by

$$\rho_{\mathbf{k}}^Z = \sum_{\nu} z_\nu \rho_{\mathbf{k}}^\nu \qquad (10.1.3)$$

where $\rho_{\mathbf{k}}^\nu$ is a Fourier component of the microscopic number density of species ν. Then the total Coulomb energy of a periodic system of volume V is

$$V_N^C(\mathbf{r}^N) = \frac{1}{2V} \sum_{\mathbf{k}} \hat{v}(k) \left(\rho_{\mathbf{k}}^Z \rho_{-\mathbf{k}}^Z - \sum_{i=1}^{N} z_i^2 \right) \qquad (10.1.4)$$

where the sum on $\mathbf{k}$ runs over wavevectors compatible with the assumed periodic boundary conditions and the (negative) second term inside brackets cancels the infinite self-energy of the ions. The function $\hat{v}(k)$ is the Fourier transform of the Coulomb potential between two elementary charges, i.e.

$$\hat{v}(k) = \frac{4\pi e^2}{k^2} \qquad (10.1.5)$$

The same expression was used earlier in the derivation of the Debye–Hückel result (4.6.26); the k^{-2} singularity in the limit $k \to 0$ is a key characteristic of Coulomb systems. In the thermodynamic limit the sum over wavevectors in (10.1.4) becomes an integral over $\mathbf{k}$ divided by $(2\pi)^3$; the equivalence of the two expressions for V_N^C in (10.1.2) and (10.1.4) is then an immediate consequence of elementary properties of the Fourier transform.

If electrical neutrality is to be achieved, an ionic fluid must contain at least two species of opposite charge. The simplest representation of such a system is obtained by replacing one of the species by a uniformly smeared-out, structureless background, the total charge of which must cancel that of the discrete ions. When the ions are identical point charges the resulting model (already discussed in Section 4.6) is called the *one-component plasma* or OCP.[1] The total potential energy of an OCP in which the ions carry a charge ze is given by the sum over $\mathbf{k}$ in (10.1.4), with $\rho_{\mathbf{k}}^Z = z\rho_{\mathbf{k}}$, except that the presence of the neutralising background means that the term for $\mathbf{k} = 0$ must be omitted. The OCP has certain unphysical features. For example, mass and charge fluctuations are proportional to each other and the system therefore has zero resistivity, because conservation of total momentum is equivalent to conservation of the microscopic electric current. Nevertheless, as the prototypical ionic fluid, the OCP plays a conceptual role similar to that filled by the hard-sphere model in the theory of simple, insulating liquids. It provides, in particular, a useful starting point for the study of liquid metals, where the mobile species corresponds to the metal ions and the background represents the conduction electrons.

To illustrate the usefulness of the OCP in the qualitative discussion of the properties of ionic liquids we return briefly to the question of the high-frequency, charge-fluctuation modes mentioned earlier. The characteristic frequency of the longitudinal mode is the plasma frequency, ω_p. In the case of the OCP an expression for ω_p can be obtained by a simple argument based on a δ-function representation of the dynamic structure factor. Use of such a model is justified by the fact that conservation of momentum of the ions means that there is no damping of charge fluctuations in the long-wavelength limit. We therefore assume that $S(k, \omega)$ consists of a pair of δ-functions located at frequencies $\pm\omega_k$, and identify the plasma frequency as $\omega_p = \lim_{k\to 0} \omega_k$. If the spectrum is to satisfy the sum rules (7.4.23) and (7.4.30), ω_k must be such that

$$\omega_k^2 = \frac{\omega_0^2}{S(k)} = \frac{k_B T}{m S(k)} k^2 \tag{10.1.6}$$

The long-wavelength limit of $S(k)$ can be estimated within the random phase approximation of Section 5.6. If we choose the ideal gas as reference system and make the substitution $\hat{c}(k) = -\beta z^2 \hat{v}(k)$, (5.6.25) becomes

$$S(k) = \frac{1}{1 + \beta\rho z^2 \hat{v}(k)} = \frac{1}{1 + 4\pi\beta\rho z^2 e^2/k^2} \sim \frac{k^2}{k_D^2}, \quad k \to 0 \tag{10.1.7}$$

where k_D is the Debye wavenumber defined by (4.6.25); as we shall see later, (10.1.7) is exact for the OCP. If we now substitute for $S(k)$ in (10.1.6), we find that

$$\omega_p^2 = \lim_{k\to 0} \omega_k^2 = \frac{4\pi\rho z^2 e^2}{m} \tag{10.1.8}$$

The frequency of the propagating mode therefore remains non-zero even in the long-wavelength limit. Such behaviour is characteristic of an optic-type excitation and is a direct consequence of the k^{-2} singularity in $\hat{v}(k)$, since it is this singularity that determines the small-k behaviour of $S(k)$. Note also that the plasma frequency is independent of temperature.

If the fluid is genuinely two component in character, a short-range repulsion is essential if the system is to be stable against the collapse of oppositely charged pairs. Within a model, stability is most easily achieved by imposing a hard-sphere repulsion between ions, a choice of interaction that defines the *primitive model* of electrolytes and molten salts. The primitive model has been widely adopted in studies of the osmotic properties of ionic solutions, the solvent being replaced by a continuum of dielectric constant ϵ which acts to reduce the Coulomb interaction between ions; the *restricted* version of the model is one in which all ions have the same diameter, d, and the same absolute valency, z.

The restricted primitive model with $\epsilon = 1$ provides the simplest example of a rigid-ion model of a molten salt. Alternatively, the short-range interactions in the salt can be modelled by soft-core repulsions characterised by a single length parameter σ. For example, the short-range contribution to the pair potential can be written as

$$v^S_{\nu\mu}(r) = \frac{z^2 e^2}{n\sigma} \left(\frac{\sigma}{r}\right)^n \tag{10.1.9}$$

for all pairs ν, μ; the parameter σ is the separation at which the total cation–anion potential has its minimum value. Equation (10.1.9), together with the coulombic term, defines what we shall call the 'simple molten salt'. This provides a fair representation of the ionic interactions in the molten alkali halides, particularly of salts in which the positive and negative ions are of approximately equal size. The values of n appropriate to the alkali halides are in the range $n = 8$–10; in the limit $n \to \infty$ the simple molten salt reduces to the restricted primitive model. If the two ionic species have equal masses, the hamiltonian of the system is fully symmetric under charge conjugation, meaning that cations and anions play identical roles.

The examples given in later sections of this chapter draw heavily on calculations for the restricted primitive model and the simple molten salt, but a number of more realistic models appropriate to molten salts have also been extensively studied both theoretically and by simulation. The best known of these are the rigid-ion potentials derived by Fumi and Tosi[2] for salts of the alkali halide family in which the short-range interaction between a given ion pair is written as the sum of an exponential repulsion and attractive terms arising from dipole–dipole and dipole–quadrupole dispersion forces. However, if the ions are highly polarisable, as is usually the case for the anion, the effect of induction forces cannot be ignored. A variety of schemes have therefore been devised that allow the incorporation of ionic polarisation into molecular-dynamics simulations of molten salts. Much of the early work on polarisable systems was based on the 'shell model' of lattice dynamics, in which the total charge of the ion is divided between a core and a massless shell.[3] The shell is bound to the core by a harmonic potential and polarisation of the ion corresponds to a bodily shift of the shell relative to the core; the shells, being of zero mass, are assumed to adjust themselves instantaneously in such a way as to minimise the total potential energy. Though the shell model provides a good description of the properties of ionic crystals, its use in studies of the molten phase has proved less successful. From simulations it is now known, for example, that the shell model exaggerates the changes in structure relative to those found for rigid-ion models. In later years, starting with a paper by Wilson and Madden,[4] a different approach has gradually evolved in which emphasis is placed on the development of potential models which include, in addition to pair interactions, many-body terms that are sensitive to fluctuations in the local environment of an ion.[5] The parameters of the model are evaluated by matching the forces to those obtained from high level *ab initio* calculations on multiple configurations of the system of interest. The many-body terms are chosen to represent not only the polarisation of the ion, which is sufficient for the alkali halides, but also the compression or distortion of its electronic structure, which is necessary when dealing with more complex ionic materials. The time evolution of the components of the induced dipole moment, and of the variables which describe other many-body interactions that may be present, can be treated either within an extended lagrangian scheme,[6] along the lines described in Section 2.7, or by minimisation of the potential energy with respect to these quantities at each time step in the simulation. We shall see examples of the results of this approach in later sections.

10.2 SCREENING AND CHARGE ORDERING

The microscopic structure of an n-component ionic fluid can be discussed in terms of $\frac{1}{2}n(n+1)$ partial structure factors $S_{\nu\mu}(k)$ with ν, $\mu = 1$ to n, but it is certain linear combinations of these functions that are of most physical relevance. If

$$\rho_{\mathbf{k}}^N = \sum_\nu \rho_{\mathbf{k}}^\nu \qquad (10.2.1)$$

is a Fourier component of the microscopic number density, and if the components of the charge density are defined as in (10.1.3), fluctuations in the densities are described by three static structure factors of the form

$$S_{NN}(k) = \frac{1}{N}\left\langle \rho_{\mathbf{k}}^N \rho_{-\mathbf{k}}^N \right\rangle = \sum_\nu \sum_\mu S_{\nu\mu}(k)$$

$$S_{NZ}(k) = \frac{1}{N}\left\langle \rho_{\mathbf{k}}^N \rho_{-\mathbf{k}}^Z \right\rangle = \sum_\nu \sum_\mu z_\mu S_{\nu\mu}(k) \qquad (10.2.2)$$

$$S_{ZZ}(k) = \frac{1}{N}\left\langle \rho_{\mathbf{k}}^Z \rho_{-\mathbf{k}}^Z \right\rangle = \sum_\nu \sum_\mu z_\nu z_\mu S_{\nu\mu}(k)$$

Of these three functions the number-number structure factor $S_{NN}(k)$ is the closest in significance to the single structure factor of a one-component fluid.

Let $\delta\phi_\mu(\mathbf{r})$ be a weak, external potential that couples to the microscopic number density of species μ. We saw in Section 3.6 that the change induced in a Fourier component of the single-particle density of species ν is

$$\delta\hat{\rho}_\nu^{(1)}(\mathbf{k}) = \chi_{\nu\mu}(k)\delta\hat{\phi}_\mu(\mathbf{k}) \qquad (10.2.3)$$

where the static response function $\chi_{\nu\mu}(k)$ is related to the corresponding partial structure factor by

$$\chi_{\nu\mu}(k) = -\beta\rho S_{\nu\mu}(k) \qquad (10.2.4)$$

where ρ is the total number density. The problem of greatest interest here concerns the response of the fluid to a weak field produced by an external charge density with Fourier components $e\delta\hat{\rho}^{\text{ext}}(\mathbf{k})$; to simplify the discussion we consider a system of rigid ions *in vacuo* ($\epsilon = 1$). The electric potential due to the external charge density is obtained from the **k**-space version of Poisson's equation, i.e.

$$\delta\hat{\phi}^{\text{ext}}(\mathbf{k}) = \frac{4\pi e}{k^2}\delta\hat{\rho}^{\text{ext}}(\mathbf{k}) \tag{10.2.5}$$

The electric potential couples directly to the microscopic charge density of the fluid, giving rise to a mean induced charge density $\delta\hat{\rho}_Z(\mathbf{k})$. The latter is proportional to $e\delta\hat{\phi}^{\text{ext}}(\mathbf{k})$, the constant of proportionality being, by definition, the charge density response function, $\chi_{ZZ}(k)$. Thus

$$\delta\hat{\rho}_Z(\mathbf{k}) = \sum_{\nu} z_\nu \delta\hat{\rho}_\nu^{(1)}(\mathbf{k})$$

$$= \chi_{ZZ}(k)e\delta\hat{\phi}^{\text{ext}}(\mathbf{k}) \tag{10.2.6}$$

If we put $\delta\hat{\phi}_\mu(\mathbf{k}) = z_\mu e\delta\hat{\phi}^{\text{ext}}(\mathbf{k})$ in (10.2.3) and then substitute for $\delta\hat{\rho}_\nu^{(1)}(\mathbf{k})$ in (10.2.6), we find that the response function can be identified as

$$\chi_{ZZ}(k) = \sum_{\nu}\sum_{\mu} z_\nu z_\mu \chi_{\nu\mu}(k) \tag{10.2.7}$$

and combination of (10.2.4) and (10.2.7) with the definition of the charge–charge structure factor in (10.2.2) leads to the charge response version of the fluctuation-dissipation theorem:

$$\chi_{ZZ}(k) = -\beta\rho S_{ZZ}(k) \tag{10.2.8}$$

The electrostrictive behaviour of the fluid, i.e. the number-density response to an external electric potential, is characterised by a cross response function $\chi_{NZ}(k)$ through an expression analogous to (10.2.6):

$$\delta\hat{\rho}_N(\mathbf{k}) = \sum_{\nu}\delta\hat{\rho}_\nu^{(1)}(\mathbf{k})$$

$$= \chi_{NZ}(k)e\delta\hat{\phi}^{\text{ext}}(\mathbf{k}) \tag{10.2.9}$$

The charge response to the external potential may equally well be described in terms of a longitudinal dielectric function $\epsilon(k)$; this is a wavenumber-dependent generalisation of the macroscopic dielectric constant of elementary electrostatics. If **E** is the electric field and **D** the electric displacement, $\epsilon(k)$ is given by

$$\frac{1}{\epsilon(k)} = \frac{\mathbf{k}\cdot\hat{\mathbf{E}}(k)}{\mathbf{k}\cdot\hat{\mathbf{D}}(k)} = 1 + \frac{\delta\hat{\rho}_Z(k)}{\delta\hat{\rho}^{\text{ext}}(\mathbf{k})} \tag{10.2.10}$$

where Maxwell's equations have been used to relate **E** and **D**, respectively, to the total and external charge densities. Equations (10.2.5), (10.2.6) and (10.2.10) can now be combined to yield the fundamental relation between the dielectric and charge response functions:

$$\frac{1}{\epsilon(k)} = 1 + \frac{4\pi e^2}{k^2}\chi_{ZZ}(k) \tag{10.2.11}$$

The definition in (10.2.2) shows that $S_{ZZ}(k)$ can never be negative. Equations (10.2.8) and (10.2.11) therefore imply that $1/\epsilon(k) \leq 1$ for all k.

It is known experimentally that an external charge distribution is completely screened by a conducting fluid. In other words, the total charge density vanishes in the long-wavelength limit, or

$$\lim_{k\to 0} \left[\delta\hat{\rho}^{\text{ext}}(\mathbf{k}) + \delta\hat{\rho}_Z(\mathbf{k}) \right] = 0 \tag{10.2.12}$$

If this result is to be consistent with (10.2.10), it follows that

$$\lim_{k\to 0} \epsilon(k) = \infty \tag{10.2.13}$$

In combination with (10.2.8) and (10.2.11) the assumption of *perfect screening* contained in (10.2.13) implies that the charge structure factor at long wavelengths behaves as

$$\lim_{k\to 0} \frac{k_D^2}{k^2} S_{ZZ}(k) = \sum_{\nu} x_\nu z_\nu^2 \tag{10.2.14}$$

where $x_\nu = \rho_\nu/\rho$ and k_D, the Debye wavenumber, is given by a generalisation of (4.6.25):

$$k_D^2 = \frac{4\pi\beta\rho e^2}{\epsilon} \sum_{\nu} x_\nu z_\nu^2 \tag{10.2.15}$$

The quantity $\Lambda_D = 1/k_D$ is the Debye screening length, familiar from ionic solution theory; in a dilute electrolyte it is the distance beyond which the electric potential due to an ion is completely screened by the local, induced charge distribution. From comparison of (10.2.14) with the compressibility equation (3.6.11) we see that large scale (long-wavelength) charge fluctuations are strongly inhibited in comparison with the fluctuations in number density of a fluid of uncharged particles. It has been proved rigorously[7] that the fluctuations in the total charge Q_V contained in a volume V, i.e. ($\langle Q_V^2 \rangle - \langle Q_V \rangle^2$), is proportional only to the surface area bounding the volume. By contrast, (2.4.23) shows that the fluctuations in the number of particles within V is proportional to V itself.

Equation (10.2.14) leads directly to two important relations between the partial pair distribution functions of an ionic fluid, known as the Stillinger–Lovett sum rules.[8] We see from (3.6.15) and (10.2.2) that the charge–charge structure factor is related to the partial pair correlation functions $h_{\nu\mu}(r)$ by

$$S_{ZZ}(k) = \sum_{\nu}\sum_{\mu} z_\nu z_\mu \left(x_\nu \delta_{\nu\mu} + 4\pi\rho x_\nu x_\mu \int_0^\infty \frac{\sin kr}{kr} h_{\nu\mu}(r) r^2 \, dr \right) \tag{10.2.16}$$

If the functions $h_{\nu\mu}(r)$ decay sufficiently rapidly at large r, the Fourier integrals in (10.2.16) can be expanded to order k^2. The two sum rules are then obtained by equating terms of zeroth and second order in k in (10.2.14) and (10.2.16) and exploiting the condition of overall charge neutrality expressed by (10.1.1). The results derived in this way are

$$\rho \sum_{\nu} x_\nu z_\nu \sum_{\mu} \int x_\mu z_\mu g_{\nu\mu}(r) d\mathbf{r} = -\sum_{\nu} x_\nu z_\nu^2$$

$$\rho \sum_{\nu} x_\nu z_\nu \sum_{\mu} \int x_\mu z_\mu g_{\nu\mu}(r) r^2 \, d\mathbf{r} = -6\Lambda_D^2 \sum_{\nu} x_\nu z_\nu^2 \qquad (10.2.17)$$

The assumption concerning the large-r behaviour of the correlation functions is equivalent to a 'clustering' hypothesis for the particle densities. An n-particle density $\rho^{(n)}(\mathbf{r}^n)$ is said to have a clustering property if, for all $m < n$, it reduces to the product $\rho^{(m)}(\mathbf{r}^m)\rho^{(n-m)}(\mathbf{r}^{(n-m)})$ faster than a prescribed inverse power of the distance between the centres of mass of the clusters $(\mathbf{r}_1, \ldots, \mathbf{r}_m)$ and $(\mathbf{r}_{m+1}, \ldots, \mathbf{r}_n)$ as the clusters become infinitely separated. If the clustering hypothesis is used, the Stillinger–Lovett sum rules can be derived from the YBG hierarchy of Section 4.2 without making any assumption about the behaviour of $S_{ZZ}(k)$ at small k. The derivation is therefore not dependent on the perfect-screening condition (10.2.13); perfect screening appears instead as a consequence of the sum rules.

The first of the Stillinger–Lovett rules is just a linear combination of local electroneutrality conditions of the form

$$\rho \sum_{\mu} \int x_\mu z_\mu g_{\nu\mu}(r) d\mathbf{r} = -z_\nu \qquad (10.2.18)$$

The physical meaning of (10.2.18) is that the total charge around a given ion must exactly cancel the charge of the ion. This is the first of a series of sum rules satisfied by the multipole moments of the charge distribution in the vicinity of a given number of fixed ions.[9] The sum rules can again be derived from the YBG hierarchy if appropriate clustering assumptions are made. In particular, if correlations are assumed to decay exponentially, it can be shown that the charge distribution around any number of fixed ions has no multipole moment of any order. The local electroneutrality condition may be re-expressed in terms of the long-wavelength limits of the partial structure factors. In the case of a two-component system, (10.2.18) becomes $z_1^2 S_{11}(0) = -z_1 z_2 S_{12}(0) = z_2^2 S_{22}(0)$ or, because the fluid is electrically neutral overall:

$$x_2^2 S_{11}(0) = x_1 x_2 S_{12}(0) = x_1^2 S_{22}(0) \qquad (10.2.19)$$

No such relation holds for the partial structure factors of a mixture of neutral fluids.

The $k \to 0$ limits of the partial structure factors of a binary ionic fluid are related to the isothermal compressibility via the Kirkwood–Buff formula

(3.6.17). The conditions imposed by charge neutrality mean, however, that direct substitution of (10.2.19) in (3.6.17) leads to an indeterminate result. This problem can be avoided by inverting the system of linear equations represented by (10.2.2) and rewriting (3.6.17) in terms of $S_{NN}(k)$, $S_{NZ}(k)$ and $S_{ZZ}(k)$ in the form

$$\rho k_B T \chi_T = \lim_{k \to 0} \frac{S_{NN}(k) S_{ZZ}(k) - S_{NZ}^2(k)}{S_{ZZ}(k)} \tag{10.2.20}$$

The small-k limits of the three structure factors in (10.2.20) can be deduced from the asymptotic behaviour of the partial direct correlation functions $c_{\nu\mu}(r)$. At large r we may expect these functions to decay as $c_{\nu\mu}(r) \sim -\beta z_\nu z_\mu e^2 / r$. It is therefore natural to separate $c_{\nu\mu}(r)$ into short-range and coulombic parts; in **k**-space $\hat{c}_{\nu\mu}(k)$ becomes

$$\hat{c}_{\nu\mu}(k) = \hat{c}_{\nu\mu}^S(k) - \frac{4\pi \beta z_\nu z_\mu e^2}{k^2} \tag{10.2.21}$$

where $\hat{c}_{\nu\mu}^S(k)$ is a regular function in the limit $k \to 0$. Substitution of (10.2.21) in the Ornstein–Zernike relation (3.6.12) leads, after some straightforward algebra and use of (10.1.1) and (10.2.2), to the required results: at small k, $S_{NN}(k) \sim k^0$, $S_{NZ}(k) \sim k^2$ and $S_{ZZ}(k) \sim k^2$; the last result agrees with (10.2.14). Thus (10.2.20) reduces to the simpler expression

$$\rho k_B T \chi_T = \lim_{k \to 0} S_{NN}(k) \tag{10.2.22}$$

while (3.6.16) becomes

$$\frac{1}{\rho k_B T \chi_T} = 1 - \rho \lim_{k \to 0} \sum_\nu \sum_\mu x_\nu x_\mu \hat{c}_{\nu\mu}^S(k) \tag{10.2.23}$$

Because fluctuations in concentration correspond to fluctuations in charge density and such fluctuations are suppressed at long wavelengths, all reference to the two-component nature of the fluid has vanished from (10.2.22), which therefore resembles the corresponding result for a one-component system of uncharged particles.

The coefficients of terms of order k^4 in the small-k expansions of $S_{ZZ}(k)$ and $S_{NZ}(k)$ and those of order k^2 in the expansion of $S_{NN}(k)$ can be determined by macroscopic arguments based on linearised hydrodynamics or thermodynamic fluctuation theory. We give here the corresponding calculation for the OCP, where the problem is simplified by the fact that fluctuations in particle number are equivalent to fluctuations in charge. In the absence of any flow the force per unit volume due to the electric field must exactly balance the force due to the pressure gradient. Thus

$$z e \rho \mathbf{E}(\mathbf{r}) = \nabla P(\mathbf{r}) \tag{10.2.24}$$

where $ze\rho$ is the mean charge density of the mobile ions and the electric field $\mathbf{E}(\mathbf{r})$ is related to the sum of external and induced charge densities by Poisson's equation:

$$\nabla \cdot \mathbf{E}(\mathbf{r}) = 4\pi e\left[\delta\rho^{\mathrm{ext}}(\mathbf{r}) + \delta\rho_Z(\mathbf{r})\right] \tag{10.2.25}$$

If the system is in local thermodynamic equilibrium the pressure change in an isothermal process is

$$\delta P(\mathbf{r}) \equiv P(\mathbf{r}) - P = \left(\frac{\partial P}{\partial \rho}\right)_T \delta\rho(\mathbf{r})$$

$$= \frac{1}{z\rho\chi_T}\delta\rho_Z(\mathbf{r}) \tag{10.2.26}$$

Equations (10.2.24)–(10.2.26) may now be combined to give a differential equation for $\delta\rho_Z(\mathbf{r})$ of the form

$$\frac{1}{k_s^2}\nabla^2\delta\rho_Z(\mathbf{r}) - \delta\rho_Z(\mathbf{r}) = \delta\rho^{\mathrm{ext}}(\mathbf{r}) \tag{10.2.27}$$

where

$$k_s^2 = 4\pi z^2 e^2 \rho^2 \chi_T = k_D^2 \frac{\chi_T}{\chi_T^{\mathrm{id}}} \tag{10.2.28}$$

The solution to (10.2.27), obtained by taking Fourier transforms, is

$$\delta\hat{\rho}_Z(\mathbf{k}) = -\frac{\delta\hat{\rho}^{\mathrm{ext}}(\mathbf{k})}{1 + k^2/k_s^2} \tag{10.2.29}$$

Comparison of (10.2.29) with (10.2.10) shows that the long-wavelength limit of $\epsilon(k)$ is

$$\lim_{k\to 0}\epsilon(k) = 1 + k_s^2/k^2 \tag{10.2.30}$$

which clearly satisfies the perfect-screening condition (10.2.13). The corresponding long-wavelength expression for $S_{ZZ}(k)$ $(= z^2 S(k))$, derived from (10.2.8) and (10.2.11), is

$$\lim_{k\to 0} S_{ZZ}(k) = \frac{z^2 k^2/k_D^2}{1 + k^2/k_s^2} \tag{10.2.31}$$

in agreement, to leading order, with (10.1.7). Equations (10.2.30) and (10.2.31) also apply to mixtures of oppositely charged ions with $z_1 = -z_2 = z$, except that k_s is differently defined.[10]

The Fourier components of the total electrostatic potential $\delta\phi(\mathbf{r})$ are related to the components of the total charge density by the analogue of (10.2.5). In the long-wavelength limit it follows from (10.2.10) and (10.2.30) that

$$\delta\hat{\phi}(\mathbf{k}) = \frac{4\pi e}{k^2}\left[\delta\hat{\rho}^{\text{ext}}(\mathbf{k}) + \delta\hat{\rho}_Z(\mathbf{k})\right]$$

$$= \frac{4\pi e}{k^2\epsilon(k)}\delta\hat{\rho}^{\text{ext}}(\mathbf{k})$$

$$= \frac{4\pi e}{k^2 + k_{\text{s}}^2}\delta\hat{\rho}^{\text{ext}}(\mathbf{k}) \tag{10.2.32}$$

If an ion of species ν in the fluid is regarded as an 'external' charge placed at the origin, the 'external' charge density is $e\delta\rho^{\text{ext}}(\mathbf{r}) = z_\nu e\delta(\mathbf{r})$, and (10.2.32) shows that the effective potential due to the ion decays as

$$\phi_\nu(r) = \frac{z_\nu e}{r}\exp\left(-k_{\text{s}}r\right) \tag{10.2.33}$$

The quantity $\phi_\nu(r)$ $(= \delta\phi(r))$ is the potential of mean force for ions of species ν. In the case of the OCP, k_{s} is given by (10.2.28); this becomes equal to the Debye wavenumber in the weak-coupling limit ($\rho \to 0$ or $T \to \infty$), where χ_T may be replaced by its ideal gas value, $\chi_T^{\text{id}} = \beta\rho$. With these simplifications, (10.2.33) reduces to the Debye–Hückel result (4.6.27). In the strong-coupling regime the compressibility of the OCP becomes negative, k_{s} takes on imaginary values, and the potential of mean force develops the oscillations characteristic of systems with short-range order.

Oscillations of the charge density around a given ion are also a feature of two-component ionic fluids, where they arise as a result of competition between hard-core packing and local charge neutrality. In the case of the restricted primitive model a simple argument[8a] based on the sum rules (10.2.17) shows that the radial charge distribution function $[g_{11}(r) - g_{12}(r)]$ (or $[g_{22}(r) - g_{12}(r)]$) must change sign as a function of r if $k_D d \geq \sqrt{6}$. Charge ordering of this type is a very strong effect in molten salts and oscillations in the charge density around a central ion extend over many ionic radii. Some molecular dynamics results for the simple molten salt defined by (10.1.19) (with $n = 9$) are shown in Figure 10.1 for a thermodynamic state corresponding roughly to that of sodium chloride at $T \approx$ 1270 K. The regular alternation of concentric shells of oppositely charged ions is clearly visible in the pair distribution functions plotted in the upper panel of the figure. In $\mathbf{k}$-space the effects of charge ordering reflect themselves in the very sharp main peak in the charge–charge structure factor $S_{ZZ}(k)$, shown in the lower panel; by contrast, $S_{NN}(k)$ is a relatively structureless function. The symmetry of the model means that charge and number fluctuations are completely decoupled; thus $S_{NZ}(k)$ is zero at all k. In the general case, the fluctuations are strictly independent only in the long-wavelength limit, since $S_{NZ}(k) \sim k^2$ as $k \to 0$.

Computer simulations of a variety of monovalent molten salts show that the pair distribution functions for ions of like sign, $g_{11}(r)$ and $g_{22}(r)$, are similar in form even when the difference in ionic radius of the two species is large, and that

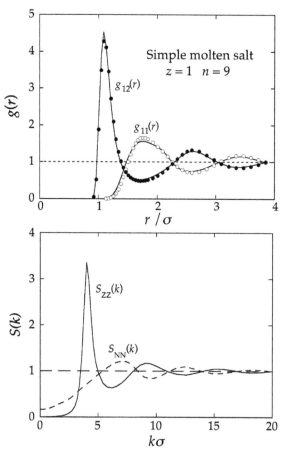

FIGURE 10.1 Charge ordering in the simple molten salt. Upper panel: the partial pair distribution functions; the points show the results of molecular dynamics calculations and the curves are calculated from the HNC approximation. Lower panel: molecular dynamics results for the static number-number and charge-charge structure factors. *Redrawn with permission from Ref. 11 © 1975 American Physical Society.*

the oscillations in these two functions are almost exactly out of phase with those in the much more sharply peaked, cation–anion distribution function $g_{12}(r)$. Thus the charge distribution around both cation and anion is essentially the same and strongly oscillatory. The same behaviour is seen in results derived from neutron scattering experiments, which rely on the use of isotopic substitution to separate the contributions of the partial structure factors $S_{\nu\mu}(k)$ to the measured cross-section.[12] The similarity between $g_{11}(r)$ and $g_{22}(r)$ gives support to the use of the simple molten salt as a model of the alkali halides. The left-hand panel of Figure 10.2 shows the partial distribution functions obtained by molecular dynamics calculations for sodium chloride based on a polarisable-ion model.[5b]

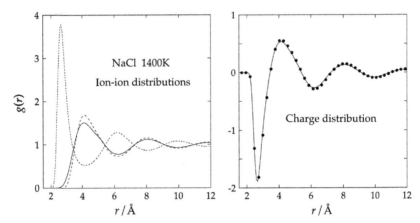

FIGURE 10.2 Results of molecular dynamics calculations for the radial distribution functions of molten sodium chloride at 1400 K. Left-hand panel: partial distribution functions of a polarisable-ion model derived by the force matching method of Section 10.1; full curve, Na^+–Na^+; short dashes, Na^+–Cl^-; long dashes, Cl^-–Cl^-. Right-hand panel: charge distribution function (10.2.34) with (curve) and without (circles) the inclusion of polarisation terms in the potential. Unpublished results of M. Salanne.

The effect of polarisation is small, as is evident in the right-hand panel, which shows the results obtained for the charge distribution function

$$g_{ZZ}(r) = \frac{1}{4}\left(g_{11}(r) - 2g_{+-}(r) + g_{22}(r)\right) \qquad (10.2.34)$$

with and without inclusion of the polarisation terms in the potential. The minor differences that do exist between the results for the two models can be ascribed almost exclusively to changes in the form of the sodium–sodium distribution function; inclusion of polarisation leads to a reduction in height and an inward shift of the first peak and a weak damping of the later oscillations. The importance of polarisation is significantly greater for salts containing divalent cations. It has, for example, proved impossible to devise a rigid-ion model that accounts satisfactorily for the structural properties of either the crystalline or molten phases of zinc chloride.[13]

10.3   INTEGRAL EQUATION THEORIES

The techniques introduced in Chapters 3–5 provide a number of possible routes to the calculation of thermodynamic and structural properties of simple ionic liquids. Versions of the optimised cluster theory of Section 5.5 and other, closely related methods have proved particularly successful in describing the thermodynamic behaviour of dilute systems. In this section, however, we focus on the integral equation approach, in which the emphasis is placed on the calculation of the pair distribution functions. Much of the published work in this field is concerned with the alkali halides, either in the molten phase or in solution, though there has also been considerable interest in the properties of 2:1 and 2:2 electrolyte solutions, the structure of which is characterised by a high degree of ionic association. The physical conditions are, of course, very different in the molten salt and electrolyte regimes. If we adopt the primitive model of Section 10.1, the thermodynamic state is conveniently characterised by the reduced density $\rho^* = Nd^3/V$, where N is the total number of ions and $d = \frac{1}{2}(d_{11} + d_{22})$ is the mean ionic diameter, and a reduced Coulomb coupling parameter, or inverse temperature, defined as

$$\beta^* = \frac{|z_1 z_2| e^2}{\epsilon k_B T d} \tag{10.3.1}$$

Near the melting point of an alkali halide, $\rho^* \approx 0.4$ and $\beta^* \approx 65$, while for a 1 M aqueous solution of the same salt at room temperature, $\rho^* \approx 0.01$ and $\beta^* \approx 3$. We must therefore expect the nature of the interionic correlations to be very different in the two cases. The liquid–vapour phase diagram of a molten alkali halide is qualitatively similar to that, say, of a rare gas, but the reduced critical densities of the salts are only about one-third of those of typical insulating liquids.

The value of different theoretical approaches can be illustrated by limiting attention initially to systems of charged hard spheres and, in particular, to the restricted primitive model, with $z_1 = -z_2 = 1$. A convenient starting point for the discussion is the mean spherical approximation (MSA) introduced in Section 4.5, since in this case the MSA has a completely analytical solution.[14] The MSA for equisized hard spheres of diameter d is

$$g_{\nu\mu} = 0, \quad r < d; \qquad c_{\nu\mu}(r) = -\frac{\beta z_\nu z_\mu e^2}{\epsilon r}, \quad r > d \tag{10.3.2}$$

which must be used in conjunction with the Ornstein–Zernike relation for equimolar binary mixtures; this is obtained as a special case of (3.6.12), with $x_1 = x_2 = \frac{1}{2}$. The symmetry of the restricted primitive model allows the Ornstein–Zernike relation to be rewritten as two independent equations for the linear combinations

$$h_S(r) = \frac{1}{2}[h_{11}(r) + h_{12}(r)], \quad h_D(r) = h_{11}(r) - h_{12}(r) \tag{10.3.3}$$

and the corresponding direct correlation functions $c_S(r)$ and $c_D(r)$; $h_S(r)$ is a number density correlation function and $h_D(r)$ describes the correlation in charge density. When written in terms of the new functions the MSA becomes

$$h_S(r) = -1, \quad r < d; \qquad c_S(r) = 0, \quad r > d \tag{10.3.4}$$

$$h_D(r) = 0, \quad r < d; \qquad c_D(r) = -\frac{2\beta e^2}{\epsilon r}, \quad r > d \tag{10.3.5}$$

The closure relation (10.3.4) is just the Percus–Yevick approximation for hard spheres, for which the solution is known (see Section 4.4). The solution to (10.3.5) and the associated Ornstein–Zernike relation between $h_D(r)$ and $c_D(r)$ can also be obtained in closed form by incorporating the sum rules (10.2.17) into generalised versions of the methods used to solve the PY equation for hard spheres. The result for $c_D(r)$ inside the hard core is

$$c_D(r) = -\frac{\beta e^2}{\epsilon d}\left(2 - \frac{Br}{d}\right)B, \quad r < d \qquad (10.3.6)$$

with $B = [\xi + 1 - (1 + 2\xi)^{1/2}]/\xi$, where $\xi^2 = k_D^2 d^2 = 4\pi\rho^*\beta^*$ and k_D is the Debye wavenumber defined by (10.2.15). The excess internal energy has a very simple form:

$$\frac{U^{\mathrm{ex}}}{N} = -\frac{e^2}{\epsilon d}B \qquad (10.3.7)$$

and is a function of the single coupling constant ξ and not separately of ρ^* and β^*. In the high temperature or low density (or low concentration) limit, i.e. for $\xi \ll 1$, (10.3.7) reduces to the Debye–Hückel result:

$$\frac{U_{\mathrm{DH}}^{\mathrm{ex}}}{N} = -\frac{e^2}{2\epsilon d}\xi = -\frac{k_B T}{8\pi\rho}k_D^3 \qquad (10.3.8)$$

The limiting law (10.3.8) is valid when ion size effects are negligible; it corresponds to the case when the direct correlation functions $c_{\nu\mu}(r)$ are replaced by their asymptotic forms (10.3.2) for all r. The virial pressure in the MSA is the sum of a hard-sphere contact term and the contribution of the Coulomb forces, i.e.

$$\frac{\beta P^v}{\rho} = 1 + \frac{2\pi\rho^*}{3}g_S(d) + \frac{\beta U^{\mathrm{ex}}}{3N} \qquad (10.3.9)$$

Alternatively, the pressure can be calculated by first integrating (10.3.7) to give the free energy and then differentiating with respect to density. The comparison made in Figure 10.3 for the case of a 1:1 electrolyte shows that the results for the excess internal energy are in good agreement with those of Monte Carlo calculations, but there is a serious inconsistency between the pressures calculated by the two different routes. In the molten salt regime the results, not surprisingly, are much less satisfactory.[16]

Although the MSA provides a good starting point for the calculation of thermodynamic properties of the restricted primitive model it is less reliable in predicting the correlation functions. If the density and temperature are such that $\xi \gg 1$, use of the MSA leads to distribution functions $g_{11}(r)$ and $g_{22}(r)$ that become negative at separations close to contact. The situation is improved if, at small r, the direct correlation functions $c_S(r)$ and $c_D(r)$ are allowed to deviate from their asymptotic forms.[17] In the 'generalised' mean spherical approximation or GMSA the deviations are expressed in terms of Yukawa functions and the closure relations for $c_S(r)$ and $c_D(r)$ in (10.3.4) and (10.3.5)

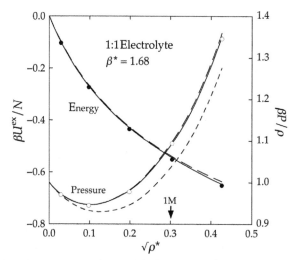

FIGURE 10.3 Thermodynamic properties of the restricted primitive model of a 1:1 electrolyte. The points show the results of Monte Carlo simulations and the curves are calculated from the MSA and the HNC approximation. Energy: dashes, MSA; full curve, HNC. Pressure: long and short dashes, MSA via the energy and virial equations, respectively; full curve, HNC via the virial (or energy) equation. The value of β^* corresponds to an aqueous solution of ions of diameter 4.25 Å at $T = 298$ K; the arrow marks the value of $\sqrt{\rho^*}$ corresponding to a 1 M solution. After Rasaiah et al.[15]

are replaced by

$$c_S(r) = \frac{A_1}{r} \exp\left[-t_1(r-d)\right], \quad r > d$$

$$c_D(r) = -\frac{2\beta e^2}{\epsilon r} - \frac{A_2}{r} \exp\left[-t_2(r-d)\right], \quad r > d \tag{10.3.10}$$

The parameters A_1, t_1, A_2 and t_2 are related via a set of algebraic equations to the internal energy, compressibility, virial pressure and contact value of $g_D(r)$, and can be fitted to those quantities if the necessary data are available from an independent source. Where that is possible, the resulting pair distribution functions represent a significant improvement over the MSA, but in this form the theory is not self-contained.

The main appeal of theories such as the MSA or GMSA in the calculation of the pair distribution functions is the fact that they can be solved analytically in closed or nearly closed form, but their applicability is limited, at least in their conventional forms, to systems of charged hard spheres. These 'primitive' models display certain structural features that are artefacts of the hard-sphere interaction. In particular, for values of ρ^* and β^* appropriate to molten salts, the main peak in the distribution functions for ions of like charge, i.e. $g_{++}(r)$ or $g_{--}(r)$, shows a marked splitting that is not seen experimentally. The splitting disappears when the short-range repulsion is softened but different theoretical methods are then required. Of the integral equation theories described in Chapter 4 the HNC approximation is far better suited to ionic systems than its PY counterpart. Equation (4.4.3) shows that the PY approximation cannot account for the exponential screening of the pair correlations at large separations, since within that approximation the pair distribution function decays as the pair potential. The HNC equation does describe the long-range correlations correctly and there is also a close connection between HNC theory and the traditional form of the Debye–Hückel approach. When generalised to a system of more than one component the exact relation (4.6.13) becomes

$$\ln\left[h_{\nu\mu}(r) + 1\right] = -\beta v_{\nu\mu}(r) + b_{\nu\mu}(r) + h_{\nu\mu}(r) - c_{\nu\mu}(r) \qquad (10.3.11)$$

and the HNC approximation corresponds to setting the bridge function $b_{\nu\mu}(r) = 0$ for all pairs ν, μ. As Figure 10.3 shows, the thermodynamic results obtained in this way for a 1:1 electrolyte agree very well with those calculated by the Monte Carlo method. The degree of thermodynamic consistency in the theory is high; even at the highest concentration studied, pressures calculated via the compressibility and virial (or energy) routes differ by less than 1%. Good results are also obtained for the thermodynamic properties of the restricted primitive model of a 2:2 electrolyte, where the superiority of the HNC approximation over the MSA becomes more obvious.[18] On the other hand, over a range of low to moderate concentrations the calculated like-ion distribution function of the 2:2 system has a pronounced peak at $r \approx 2d$, a feature that persists even when the hard-sphere term in the pair potential is replaced by an inverse-power repulsion.[19] No similar peak is seen in simulations of the same potential models, as the examples shown in Figure 10.4 illustrate; instead, the distribution function increases monotonically towards its limiting value at large r. Conversely, the HNC calculations significantly underestimate the height (of order 100) of the peak in the unlike-ion distribution function, the strength of which provides a measure of the degree of ion pairing in the system. These defects in the theory are linked to the difference in form of the bridge functions for like and unlike pairs. The results of simulations[19,20] show that the function $b_{++}(r)$ (or $b_{--}(r)$) is negative at all separations, and therefore resembles the bridge function of the Lennard-Jones fluid (see Figure 4.6), but $b_{+-}(r)$ is everywhere positive. Thus the HNC approximation acts in such a way as to weaken both the effective repulsion between ions of like charge and the effective attraction between those of unlike charge, with differing consequences for the calculated distribution functions.[21] At the concentration increases the bridge functions maintain their difference in sign but their magnitude is greatly reduced. The error involved in neglecting them is therefore small and the spurious peak in the like-ion distribution function becomes progressively less pronounced.

The HNC approximation is also successful in reproducing the pair structure under state conditions typical of molten salts, as shown by the results for the simple molten salt plotted in Figure 10.1. The deficiencies in the approximation are evident only in the small-k region of $S_{NN}(k)$; the error there means that the calculated compressibility is about twice as large as that obtained by simulation.

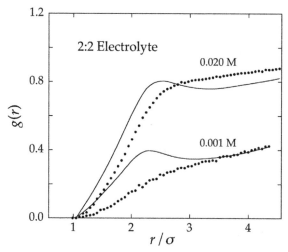

FIGURE 10.4 Pair distribution function for like-charged ions in a 2:2 electrolyte solution under state conditions corresponding to an aqueous solution at $T = 298$ K. Short-range repulsions are represented by a soft-sphere (r^{-9}) potential and σ is the separation at which the cation–anion potential is a minimum. The points show the results of molecular-dynamics simulations and the curves are calculated from the HNC approximation. The spurious peak in the HNC results is less pronounced at the higher concentration and disappears for concentrations greater than about 0.06 M. *Redrawn with permission from Ref. 19 © American Institute of Physics.*

A systematic study of the alkali halides has confirmed that HNC theory is able to reproduce quantitatively all the main features of the pair distribution functions of more realistic potential models; still better results are obtained by including the contributions from the bridge diagrams in a semi-empirical way[22] or by enforcing thermodynamic self-consistency through the hybrid, HMSA scheme[23] mentioned in Section 4.7.

The pole analysis of the asymptotic decay of pair correlation functions introduced in Section 4.8 can be extended to ionic fluids,[24] which are binary 'mixtures' with the constraint on composition provided by (10.1.1). The discussion that follows is restricted to fully symmetric systems, such as the restricted primitive model of electrolyte solutions and molten salts. The linear combinations $h_S(r)$ and $h_D(r)$ of the partial pair correlation functions, defined by (10.3.3), describe, respectively, the correlations in number density and charge density. These functions and the corresponding combinations of direct correlation functions, $c_S(r)$ and $c_D(r)$, satisfy two independent Ornstein–Zernike relations which in k-space take the form

$$\hat{h}_S(k) = \frac{\hat{c}_S(k)}{1 - \rho\hat{c}_S(k)}$$
$$\hat{h}_D(k) = \frac{\hat{c}_D(k)}{1 - \frac{1}{2}\rho\hat{c}_D(k)}$$

$$(10.3.12)$$

The partial functions $\hat{c}_{\nu\mu}(k)$ and hence their combinations $\hat{c}_S(k)$ and $\hat{c}_D(k)$ can be split in the manner of (10.2.21) into short-range and coulombic parts:

$$\hat{c}_S(k) = \hat{c}_S^{sr}(k)$$
$$\hat{c}_D(k) = \hat{c}_D^{sr}(k) - \frac{8\pi\beta z^2 e^2}{\epsilon k^2} \qquad (10.3.13)$$

where $\hat{c}_S^{sr}(k)$ is assumed to be short-ranged in the sense of Section 4.8 and can therefore be expanded in even powers of k. It follows that $h_S(r)$, as calculated by contour integration of (4.8.2), should behave asymptotically in exactly the same way as the pair correlation function of a one-component fluid of particles interacting via a short-range pair potential. Hence, as the density increases, the decay of $h_S(r)$ at large r will show a cross-over from exponential to damped oscillatory form along a Fisher-Widom line.

Combination of the second relations in (10.3.12) and (10.3.13), together with the definition (10.2.15) of the Debye wavenumber, yields an expression for the charge density function:

$$\rho\hat{h}_D(k) = \frac{k^2\rho\hat{c}_D^{sr}(k) - 2k_D^2}{k_D^2 + k^2\left[1 - \frac{1}{2}\rho\hat{c}_D^{sr}(k)\right]} \qquad (10.3.14)$$

This is an even function of k, so $h_D(r)$ can again be calculated by contour integration of (4.8.2). The poles of $\hat{h}_D(k)$ in the complex upper-half plane are determined by the zeros of the denominator and can be calculated for the restricted primitive model from either the MSA or the more accurate GMSA equation. What emerges is a pole structure which differs significantly from that shown schematically in Figure 4.9. At low densities or, equivalently, for small values of the parameter $x = k_D d$, there are only two purely imaginary poles. The pole closest to the real axis, which determines the dominant exponential decay at large r, remains very close to k_D for $x \leq 1$, as expected from Debye–Hückel theory. As x becomes larger, the two poles draw closer to each other, eventually coalescing at a critical value $x_c \approx 1.2$. For $x > x_c$ the imaginary pole is replaced by two complex-conjugate poles with real parts that grow as x increases, giving rise to a damped, oscillatory decay of the form (4.8.10). The cross-over in the qualitative nature of the decay at a given temperature defines a point on a 'Kirkwood line' in the density–temperature plane,[24,25] which would lie far to the left of the Fisher–Widom line in a plot of the type pictured in Figure 4.12. Damped oscillations therefore appear in the charge density correlation function at densities much lower than in the case of the number density function $h_S(r)$, an effect which is linked to the strong charge ordering characteristic of ionic fluids. Near the liquid–vapour critical point the dominant imaginary pole of $h_S(r)$ approaches the real axis, meaning that the correlation function becomes infinitely long ranged, but h_D remains of finite range, showing that charge correlations are unaffected by the divergence of number density fluctuations.

In the more general, asymmetric case, account must be taken of cross-correlations between fluctuations in number density and charge density. However, the analysis sketched above remains approximately correct, at least for moderate asymmetry.[24]

10.4 FREQUENCY-DEPENDENT ELECTRIC RESPONSE

We have seen in earlier sections of this chapter that the static properties of ionic liquids are strongly affected by the long-range nature of the Coulomb potential or, equivalently, the k^{-2} singularity in its Fourier transform. We now turn to the question of how the same factors influence the dynamical correlations. The discussion here is limited to two-component systems of ions *in vacuo*, the case of liquid metals being deferred until Section 10.9. The phenomena of greatest interest are those linked to charge fluctuations; these generate a local electric field that acts as a restoring force on the local charge density. At low frequencies the charge density responds in a diffusive manner, but at high frequencies there is a reactive behaviour, which gives rise to a propagating mode of the type briefly discussed in Section 10.1.

The linear combinations of microscopic partial densities that arise naturally in a discussion of the collective dynamics are the mass (M) and charge (Z) densities, defined in terms of Fourier components as

$$\rho_{\mathbf{k}}^{M}(t) = \sum_{\nu} m_{\nu} \rho_{\mathbf{k}}^{\nu}(t), \quad \rho_{\mathbf{k}}^{Z}(t) = \sum_{\nu} z_{\nu} \rho_{\mathbf{k}}^{\nu}(t) \qquad (10.4.1)$$

where m_{ν} is the mass of an ion of species ν. With each fluctuating density we may associate a current. Thus

$$\mathbf{j}_{\mathbf{k}}^{M}(t) = \sum_{\nu} m_{\nu} \mathbf{j}_{\mathbf{k}}^{\nu}(t), \quad \mathbf{j}_{\mathbf{k}}^{Z}(t) = \sum_{\nu} z_{\nu} \mathbf{j}_{\mathbf{k}}^{\nu}(t) \qquad (10.4.2)$$

where the partial currents $\mathbf{j}_{\mathbf{k}}^{\nu}$ are given by an expression identical to (7.4.7) except that the sum on i is now restricted to ions of a given species. Each current can be divided into longitudinal (l) and transverse (t) parts in the manner of (7.4.25); the longitudinal currents satisfy equations of continuity analogous to (7.4.4). The mass current is related to the stress tensor $\mathbf{\Pi}_{\mathbf{k}}$ by

$$\frac{\partial}{\partial t} \mathbf{j}_{\mathbf{k}}^{M}(t) + i\mathbf{k} \cdot \mathbf{\Pi}_{\mathbf{k}} = 0 \qquad (10.4.3)$$

where the components of $\mathbf{\Pi}_{\mathbf{k}}$ are given by a two-component generalisation of (8.4.14). Equation (10.4.3) shows that the time derivative of $\mathbf{j}_{\mathbf{k}}^{M}(t)$ vanishes as $k \to 0$. The mass current is therefore a conserved variable in the sense of Section 7.4, but the charge current is not. Although the total momentum of the ions is conserved, there is a continuous exchange of momentum between the two species; this momentum exchange is the source of the electrical resistivity of the fluid.

The mass and charge densities can be used to construct three, independent, time correlation functions $F_{AB}(k,t)$ (with $A, B = M$ or Z), the definitions of which are similar to that of the intermediate scattering function (7.4.20). The initial values of the correlation functions are equal to the static structure factors in (10.2.2), but with number N replaced by mass M, and their Fourier transforms with respect to t are the corresponding dynamic structure factors. A function of particular interest for our purposes is the charge–charge dynamic structure factor, defined as

$$S_{ZZ}(k,\omega) = \frac{1}{2\pi N} \int_{-\infty}^{\infty} \left\langle \rho_{\mathbf{k}}^{Z}(t)\rho_{-\mathbf{k}}^{Z}\right\rangle \exp{(i\omega t)}dt \qquad (10.4.4)$$

Finally, three longitudinal and three transverse current correlation functions can be defined through straightforward generalisations of (7.4.25):

$$C_{AB,l}(k,t) = \frac{k^2}{N}\left\langle j_{\mathbf{k}}^{Az}(t)j_{-\mathbf{k}}^{Bz}\right\rangle$$
$$C_{AB,t}(k,t) = \frac{k^2}{N}\left\langle j_{\mathbf{k}}^{Ax}(t)j_{-\mathbf{k}}^{Bx}\right\rangle \qquad (10.4.5)$$

where, as usual, the z-axis is chosen parallel to $\mathbf{k}$. Each $C_{AB,l}(k,t)$ is related to the corresponding $F_{AB}(k,t)$ by an analogue of (7.4.26).

We next consider how the response of the system to an external electric field can be described in terms of the correlation functions introduced above. This requires a generalisation to frequency-dependent perturbations of the result in (10.2.6). As an extension of the linear response relation (7.6.26), we find that the mean induced charge density is

$$\delta\hat{\rho}_Z(\mathbf{k},t) = \left\langle \rho_{\mathbf{k}}^{Z}(t)\right\rangle = \chi_{ZZ}(k,\omega)e\phi_{\mathbf{k}}^{\text{ext}}\exp{(-i\omega t)} \qquad (10.4.6)$$

The imaginary part of the complex dynamic susceptibility $\chi_{ZZ}(k,\omega)$ is related to the dynamic structure factor $S_{ZZ}(k,\omega)$ through a trivial modification of the fluctuation-dissipation theorem (7.6.28), i.e.

$$S_{ZZ}(k,\omega) = -\frac{k_B T}{\pi\rho\omega}\chi_{ZZ}''(k,\omega) \qquad (10.4.7)$$

and the susceptibility can also be expressed in terms of the complex dielectric function $\epsilon(k,\omega)$ by a frequency-dependent generalisation of (10.2.11):

$$\frac{1}{\epsilon(k,\omega)} = 1 + \frac{4\pi e^2}{k^2}\chi_{ZZ}(k,\omega) \qquad (10.4.8)$$

The functions $\chi_{ZZ}(k,\omega)$ and $1/\epsilon(k,\omega)$ measure the linear response of a fluid of charged particles to an external electric field. The external field polarises the fluid and the local, internal field (the Maxwell field) is the sum of the field due to the external charge distribution and that due to the induced charge density. The local field is, of course, the field experienced by the ions. The response of the system to the local electric potential is described by a screened response function $\chi_{ZZ}^{sc}(k, \omega)$, defined through the expression

$$\delta\hat{\rho}_Z(\mathbf{k}, t) = \chi_{ZZ}^{sc}(k, \omega)e\left[\phi_{\mathbf{k}}^{ext}\exp(-i\omega t) + \delta\hat{\phi}^{ind}(\mathbf{k}, \omega)\right] \tag{10.4.9}$$

where the induced electric potential $\delta\hat{\phi}^{ind}(\mathbf{k}, \omega)$ is related to the induced charge density by Poisson's equation (cf. (10.2.5)):

$$\delta\hat{\phi}^{ind}(\mathbf{k}, t) = \frac{4\pi e}{k^2}\delta\hat{\rho}_Z(\mathbf{k}, t) \tag{10.4.10}$$

Comparison of (10.4.9) with (10.4.6) shows that the relation between the external and screened susceptibilities is

$$\chi_{ZZ}(k, \omega) = \frac{\chi_{ZZ}^{sc}(k, \omega)}{1 - \frac{4\pi e^2}{k^2}\chi_{ZZ}^{sc}(k, \omega)} \tag{10.4.11}$$

and hence, from (10.4.8), that

$$\epsilon(k, \omega) = 1 - \frac{4\pi e^2}{k^2}\chi_{ZZ}^{sc}(k, \omega) \tag{10.4.12}$$

The response function $\chi_{ZZ}(k, \omega)$ satisfies the Kramers–Kronig relations (7.6.39) and (7.6.40), which are merely consequences of causality. The same is not necessarily true of the screened function $\chi_{ZZ}^{sc}(k, \omega)$, which determines the response of the system to the local field. Since the local field depends on the material properties of the system, it cannot be controlled at will in an experiment.

The electric response of an ionic fluid can also be discussed in terms of the induced electric current. Let $\mathbf{E}(\mathbf{k}, \omega)$ be a Fourier component of the local electric field. Ohm's Law in its most general form states that the induced electric current $\mathbf{J}^Z$ is linearly related to the field, i.e.

$$\mathbf{J}^Z(\mathbf{k}, \omega) = \sigma(\mathbf{k}, \omega) \cdot \mathbf{E}(\mathbf{k}, \omega) \tag{10.4.13}$$

The quantity σ is the conductivity tensor, which can be divided into longitudinal and transverse parts in the form

$$\sigma(\mathbf{k}, \omega) = \frac{\mathbf{kk}}{k^2}\sigma_l(\mathbf{k}, \omega) + \left(1 - \frac{\mathbf{kk}}{k^2}\right)\sigma_t(\mathbf{k}, \omega) \tag{10.4.14}$$

where σ_l and σ_t are scalars. The longitudinal and transverse projections of the current are related, respectively, to the longitudinal (or irrotational) and transverse (or divergence free) components of the local electric field. Thus

$$\mathbf{J}_l^Z(\mathbf{k}, \omega) = \sigma_l(\mathbf{k}, \omega)\mathbf{E}_l(\mathbf{k}, \omega), \quad \mathbf{J}_t^Z(\mathbf{k}, \omega) = \sigma_t(\mathbf{k}, \omega)\mathbf{E}_t(\mathbf{k}, \omega) \tag{10.4.15}$$

Since $\mathbf{E} = -\nabla\delta\phi$, it follows that the longitudinal component of the local field is related to the total electric potential by the expression

$$\mathbf{E}_l(\mathbf{k},\omega) = -i\mathbf{k}\delta\hat{\phi}(\mathbf{k},\omega) = -i\mathbf{k}\left[\phi_{\mathbf{k}}^{\text{ext}}\exp\left(-i\omega t\right) + \delta\hat{\phi}^{\text{ind}}(\mathbf{k},\omega)\right] \quad (10.4.16)$$

Equations (7.4.4), (10.4.9), (10.4.12), (10.4.15) and (10.4.16) can now be combined to yield the fundamental relation between the dielectric function and the conductivity tensor:

$$\epsilon(k,\omega) = 1 + \frac{4\pi i}{\omega}\sigma_l(k,\omega) \quad (10.4.17)$$

Note that $\sigma_l(k,\omega)$ is a screened response function in the same sense as $\chi_{ZZ}^{\text{sc}}(k,\omega)$, since it measures a response to the internal field.

Linear response theory was used in Section 7.7 to derive a microscopic expression for the frequency-dependent electrical conductivity; this 'external' conductivity measures the response of a fluid to a uniform ($\mathbf{k} = 0$) applied electric field. A uniform field corresponds to a situation in which the boundaries of the system are removed to infinity, thereby avoiding the appearance of a surface polarisation. The electric response to an inhomogeneous ($\mathbf{k}$-dependent) applied field is measured by a wavenumber-dependent external conductivity that can be related to the time autocorrelation function of the fluctuating charge current $\mathbf{j}_{\mathbf{k}}^Z(t)$. In the case of the longitudinal component the required generalisation of (7.7.10) is simply

$$\sigma_l^{\text{ext}}(k,\omega) = \frac{\beta e^2}{V}\int_0^\infty \left\langle j_{\mathbf{k}}^{Zz}(t) j_{-\mathbf{k}}^{Zz}\right\rangle \exp\left(i\omega t\right)\mathrm{d}t \quad (10.4.18)$$

However, the macroscopic electrical conductivity σ given by the low-frequency limit of (7.7.10) is not the same as the $k,\omega \to 0$ limit of $\sigma_l^{\text{ext}}(k,\omega)$. Indeed it follows from the continuity equation (see (7.4.4)) that the integral in (10.4.18) can be re-expressed as

$$\int_0^\infty \left\langle j_{\mathbf{k}}^{Zz}(t) j_{-\mathbf{k}}^{Zz}\right\rangle \exp\left(i\omega t\right)\mathrm{d}t = \frac{1}{k^2}\int_0^\infty \left\langle \dot{\rho}_{\mathbf{k}}^Z(t)\dot{\rho}_{-\mathbf{k}}^Z\right\rangle \exp\left(i\omega t\right)\mathrm{d}t$$

$$= \frac{-i\omega N S_{ZZ}(k) + \omega^2 N \tilde{F}_{ZZ}(k,\omega)}{k^2} \quad (10.4.19)$$

Written in this form it is easy to see that the integral vanishes as $k,\omega \to 0$, since $S_{ZZ}(k) \sim k^2$ for small k. (Note that $\tilde{F}_{ZZ}(k,\omega)$ is the Laplace transform of $F_{ZZ}(k,t)$, which is bounded above by $S_{ZZ}(k)$: see (7.1.14).) On the other hand, the rotational invariance of an isotropic fluid implies that the macroscopic longitudinal and transverse conductivities must be the same, i.e. $\sigma_l^{\text{ext}}(0,\omega) = \sigma_t^{\text{ext}}(0,\omega) = \sigma(\omega)$. Hence σ may be defined in terms of the transverse charge current autocorrelation function; the transverse current is not related to the charge density by a continuity equation and is therefore unaffected by the small-k divergence of the longitudinal electric field. Thus

$$\sigma = \lim_{\omega \to 0} \lim_{k \to 0} \frac{\beta e^2}{V} \int_0^\infty \left\langle j_{\mathbf{k}}^{Zx}(t) j_{-\mathbf{k}}^{Zx} \right\rangle \exp{(i\omega t)}\mathrm{d}t$$

$$= \lim_{\omega \to 0} \lim_{k \to 0} \frac{\beta \rho e^2}{k^2} \tilde{C}_{ZZ,t}(k,\omega) \tag{10.4.20}$$

The differing behaviour of the longitudinal and transverse charge current autocorrelation functions is also evident from the sum rules for the corresponding spectra. The short-time expansions of $C_{ZZ,l}(k,t)$ and $C_{ZZ,t}(k,t)$ can be written in a form similar to (7.4.31) and (7.4.36), namely

$$C_{ZZ,l}(k,t) = \omega_0^2 \left(1 - \omega_{1l}^2 \frac{t^2}{2!} + \cdots \right)$$

$$C_{ZZ,t}(k,t) = \omega_0^2 \left(1 - \omega_{1t}^2 \frac{t^2}{2!} + \cdots \right) \tag{10.4.21}$$

where, in the case when $z_1 = -z_2 = z$:

$$\omega_0^2 = z^2 k^2 \left(\frac{k_\mathrm{B} T}{2M} \right) \tag{10.4.22}$$

with $M = m_1 m_2/(m_1 + m_2)$. The frequency moments ω_{1l}^2 and ω_{1t}^2 are the charge current analogues of the quantities defined in Section 7.4. If the interionic potentials are separated into their coulombic and short-range parts, the derivation of (7.4.35) and (7.4.38) can be suitably generalised.[26] The resulting expressions are lengthy but reduce in the limit $k \to 0$ to the simpler forms given by

$$\lim_{k \to 0} \omega_{1l}^2(k) = \frac{2}{3}\omega_\mathrm{p}^2 + \frac{\rho}{6M} \int \nabla^2 v_{12}^\mathrm{S}(r) g_{12}(r)\mathrm{d}\mathbf{r}$$

$$\lim_{k \to 0} \omega_{1t}^2(k) = -\frac{1}{3}\omega_\mathrm{p}^2 + \frac{\rho}{6M} \int \nabla^2 v_{12}^\mathrm{S}(r) g_{12}(r)\mathrm{d}\mathbf{r} \tag{10.4.23}$$

where $v_{12}^\mathrm{S}(r)$ is the short-range part of the cation–anion potential and ω_p is the plasma frequency (10.1.8), generalised to the two-component case:

$$\omega_\mathrm{p}^2 = \sum_\nu \frac{4\pi \rho_\nu z_\nu^2 e^2}{m_\nu} \tag{10.4.24}$$

Thus, in contrast to the results obtained in Section 7.4, the characteristic frequencies of the charge current fluctuations remain non-zero as $k \to 0$. In addition, the longitudinal and transverse frequencies at $\mathbf{k} = 0$ are split according to the rule

$$\omega_{1l}^2(0) - \omega_{1t}^2(0) = \omega_\mathrm{p}^2 \tag{10.4.25}$$

This result has the same form as the well-known relation between the longitudinal and transverse optic frequencies of ionic crystals. The behaviour of $\omega_{1l}(k)$ and $\omega_{1t}(k)$ at finite wavelengths is also similar to that of the corresponding phonon dispersion curves for the crystal: initially, $\omega_{1l}(k)$ falls rapidly with increasing k, but the curve of $\omega_{1t}(k)$ is almost flat. In the case of the alkali halides, $\omega_{1l}(0)$ is typically 20–30% larger than ω_p.

The nature of the collective modes associated with fluctuations in mass, charge and temperature in a molten salt can be analysed by methods described in Chapters 8 and 9. By analogy with the spectra of ionic crystals, the collective modes are expected to be of acoustic and optic character, corresponding respectively to low-frequency sound waves and high-frequency 'plasma' oscillations. The different fluctuations are, in general, strongly coupled and the associated memory functions have a complicated structure. A considerable simplification occurs when the anions and cations differ only in the sign of their electrical charge. Under such conditions, charge fluctuations are completely decoupled from fluctuations in mass and temperature at all frequencies and all wavenumbers. The same is true for any molten salt in the long-wavelength limit, thereby making it possible to calculate the spectrum of charge fluctuations at long wavelengths by the following, simple, macroscopic argument.[10] The Laplace transform of the continuity equation for the induced charge density is

$$- i\omega \delta \tilde{\rho}_Z(\mathbf{k}, \omega) = \delta \hat{\rho}_Z(\mathbf{k}, t = 0) + i\mathbf{k} \cdot \mathbf{J}^Z(\mathbf{k}, \omega) \qquad (10.4.26)$$

while Poisson's equation may be written as

$$- i\mathbf{k} \cdot \mathbf{E}(\mathbf{k}, \omega) = 4\pi \delta \tilde{\rho}_Z(\mathbf{k}, \omega) \qquad (10.4.27)$$

These two expressions can be combined with the longitudinal projection of Ohm's Law to give

$$\delta \tilde{\rho}_Z(\mathbf{k}, \omega) = \frac{\delta \hat{\rho}_Z(\mathbf{k}, t = 0)}{-i\omega + 4\pi \sigma_l(k, \omega)} \qquad (10.4.28)$$

If we multiply (10.4.28) through by $\delta \hat{\rho}_Z(-\mathbf{k}, t = 0)$ and take the thermal average we find that

$$\tilde{F}_{ZZ}(k, \omega) = \frac{S_{ZZ}(k)}{-i\omega + 4\pi \sigma_l(k, \omega)} \qquad (10.4.29)$$

In the limit $k \to 0$, $\sigma_l(k, \omega)$ can be replaced by $\sigma(\omega)$. This gives an important result:

$$\lim_{k \to 0} \frac{\tilde{F}_{ZZ}(k, \omega)}{S_{ZZ}(k)} = \frac{1}{-i\omega + 4\pi \sigma(\omega)} \qquad (10.4.30)$$

Comparison with (7.3.23) shows that the frequency-dependent, complex conductivity is the memory function for the long-wavelength limit of the charge density autocorrelation function. The spectrum of charge density fluctuations may therefore be expressed in terms of the real (σ') and imaginary (σ'') parts of $\sigma(\omega)$ in the form

$$\lim_{k \to 0} \frac{S_{ZZ}(k,\omega)}{S_{ZZ}(k)} = \frac{1}{\pi} \frac{4\pi\sigma'(\omega)}{[\omega - 4\pi\sigma''(\omega)]^2 + [4\pi\sigma'(\omega)]^2} \tag{10.4.31}$$

In the low-frequency limit, $\sigma'(\omega) \to \sigma, \sigma''(\omega) \to 0$ and (10.4.31) reduces to

$$S_{ZZ}(k,\omega) \sim \frac{1}{\pi} \frac{4\pi\sigma(k/k_D)^2}{\omega^2 + (4\pi\sigma)^2}, \qquad k, \omega \to 0 \tag{10.4.32}$$

Charge fluctuations in the low-frequency, long-wavelength regime are therefore of a non-propagating type. The same is true of concentration fluctuations in a mixture of uncharged particles but the two cases differ in a significant way. If the coupling to other hydrodynamic variables is weak, a Fourier component of a fluctuation in the local concentration $c(\mathbf{r}, t)$ in a non-ionic, binary mixture decays in approximately the same way as a component of the density of tagged particles in a one-component system (see (8.2.5)), i.e.

$$c_k(t) \approx c_k \exp(-Dk^2 t) \tag{10.4.33}$$

where D is the interdiffusion coefficient.[27] The functional form of the spectrum of concentration fluctuations is therefore similar to that the self dynamic structure factor (8.2.9):

$$\begin{aligned} S_{cc}(k,\omega) &= \frac{1}{2\pi} \int_{-\infty}^{\infty} \langle c_k(t)c_{-k} \rangle \exp(i\omega t)\mathrm{d}t \\ &\approx \frac{\langle |c_k|^2 \rangle}{\pi} \frac{Dk^2}{\omega^2 + (Dk^2)^2} \end{aligned} \tag{10.4.34}$$

Equation (10.4.34) represents a lorentzian curve centred at $\omega = 0$ and having a width that varies as k^2, whereas the width of the charge fluctuation spectrum (10.4.32) remains non-zero even in the long-wavelength limit. The source of this difference in behaviour is the fact that in the coulombic case the 'restoring force' is proportional to the charge density fluctuation, while in the neutral system it is proportional to the laplacian of the concentration fluctuation.

Although the hydrodynamic analysis yields the correct low-frequency behaviour the possibility that a propagating charge density oscillation could occur at higher frequencies has to be investigated within the framework of either generalised hydrodynamics or the memory function formalism. The memory function approach in particular lends itself easily to a unified treatment of transverse and longitudinal charge fluctuations. Here, however, we consider only the more interesting question of the nature of the longitudinal fluctuations. We also restrict the discussion to long wavelengths and to the case when $z_1 = -z_2 = z$, and use the fact that

$$\lim_{k \to 0} \frac{\omega_0^2}{S_{ZZ}(k)} = \omega_p^2 \tag{10.4.35}$$

which follows from the long-wavelength relation (10.2.14) and the definitions (10.4.22) and (10.4.24). When adapted to the problem of the longitudinal charge current, the memory function equation (9.4.7) becomes

$$\tilde{C}_{ZZ,l}(k,\omega) = \cfrac{\omega_0^2}{-i\omega + \cfrac{\omega_p^2}{-i\omega} + \tilde{N}_l(k,\omega)} \tag{10.4.36}$$

Use of (10.4.19) shows that the corresponding expression for the charge density autocorrelation function is given in terms of Laplace transforms by

$$\tilde{F}_{ZZ}(k,\omega) = \cfrac{S_{ZZ}(k)}{-i\omega + \cfrac{\omega_p^2}{-i\omega + \tilde{N}_l(k,\omega)}} \tag{10.4.37}$$

The high-frequency behaviour may now be studied in an approximate way by assuming that the memory function $N_l(k,t)$ decays exponentially with a relaxation time τ_l. This is the characteristic approximation of the viscoelastic model introduced in Chapter 9, and leads, for small k, to

$$\tilde{N}_l(k,\omega) = \frac{\omega_{1l}^2 - \omega_p^2}{-i\omega + 1/\tau_l} \tag{10.4.38}$$

A simple calculation then shows that if $\omega\tau_l \gg 1$, the charge–charge dynamic structure factor (proportional to Re $\tilde{F}_{ZZ}(k,\omega)$) has peaks at $\omega = 0$ and $\omega = \pm\omega_{1l}$; those at $\pm\omega_{1l}$ correspond to charge fluctuations that propagate at a frequency comparable with the plasma frequency but modified by the short-range interactions between ions. The calculation is a crude one, limited as it is to high frequencies and long wavelengths, but it provides a fair description of the dispersion of the propagating mode observed in simulations (see below in Figure 10.7).

10.5 MICROSCOPIC DYNAMICS IN MOLTEN SALTS

Much of our current understanding of the microscopic dynamics in strongly coupled ionic systems comes from molecular dynamics simulations. In this section we give some examples, taken from studies of monovalent molten salts, that illustrate the richness of the observed single-particle and collective behaviour.

Single-particle motion is conveniently discussed in terms of the velocity autocorrelation functions $Z_\nu(t)$ and self-diffusion coefficients D_ν of the two ionic species; D_ν is related to $Z_\nu(t)$ in the manner of (7.2.7). For mixtures of neutral particles in which cross-correlations of velocity of the type $\langle \mathbf{u}_i(t) \cdot \mathbf{u}_j \rangle$ ($i \neq j$) are negligible, the two self-diffusion coefficients are related to the interdiffusion coefficient $\mathcal{D}$ by the expression

$$\mathcal{D} \approx \mathcal{F}\frac{x_1 x_2}{N k_B T}(x_2 D_1 + x_1 D_2) \tag{10.5.1}$$

where $\mathcal{F} = (\partial^2 G/\partial x_1^2)_{P,T}$ is a purely thermodynamic quantity.[27] If, in addition, the mixture is nearly ideal, which is a good approximation for mixtures of simple liquids, $\mathcal{F} \approx N k_B T / x_1 x_2$, and (10.5.1) becomes

$$\mathcal{D} \approx x_2 D_1 + x_1 D_2 \tag{10.5.2}$$

In an ionic liquid interdiffusion is equivalent to electrical conduction. We have shown in Section 7.7 that the static electrical conductivity σ is proportional to the time integral of the electric current autocorrelation function $J(t)$, defined as

$$J(t) = \left\langle \mathbf{j}^Z(t) \cdot \mathbf{j}^Z \right\rangle = \sum_{i=1}^{N} \sum_{j=1}^{N} \left\langle z_i \mathbf{u}_i(t) \cdot z_j \mathbf{u}_j \right\rangle \tag{10.5.3}$$

If the self-correlation terms ($i = j$) in (10.5.3) are separated from the cross terms ($i \neq j$), integration over time and use of (7.7.10) shows that

$$\sigma = \beta e^2 \rho \left(x_1 z_1^2 D_1 + x_2 z_2^2 D_2 \right) (1 - \Delta) \tag{10.5.4}$$

or, in the case of a monovalent salt:

$$\sigma = \frac{1}{2} \beta e^2 \rho (D_1 + D_2)(1 - \Delta) \tag{10.5.5}$$

Equation (10.5.4), with $\Delta = 0$, is called the Nernst–Einstein relation; the value of the deviation factor Δ is a measure of the importance of cross-correlations. If $\Delta = 0$, (10.5.4) becomes the ionic equivalent of the approximate relation (10.5.2). In practice, at least for the alkali halides, Δ is significantly different from zero and always positive. The importance of cross-correlations in monovalent salts is illustrated in Figure 10.5, where molecular dynamics

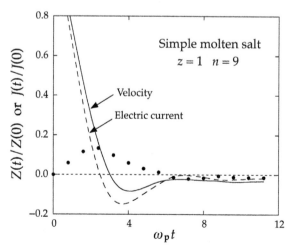

FIGURE 10.5 Normalised velocity and electric current autocorrelation functions of the simple molten salt under the state conditions described in the caption to Figure 10.1. Full curve: $Z(t)/Z(0)$; dashes: $J(t)/J(0)$. The points show the difference between the two functions. *Redrawn with permission from Ref. 11* © *1975 American Physical Society.*

results for the velocity and electric current autocorrelation functions of the simple molten salt are plotted. The symmetry of the model means that the velocity autocorrelation functions of cations and anions are identical; if cross-correlations of velocities were negligible, the normalised curves of $Z(t)$ and $J(t)$ would also be the same. At short times, however, there are substantial differences between the two functions, and the calculated Nernst–Einstein deviation factor for the case shown is $\Delta = 0.19$, a value which is typical of those measured experimentally for the alkali halides. A positive value of Δ corresponds physically to the fact that motion in the same direction by a pair of oppositely charged ions contributes to self-diffusion but not to electrical conduction. The observed deviations from the Nernst–Einstein relation therefore have a natural explanation in terms of positive correlations between the velocities of nearest-neighbour ions that persist for times which, for a real molten salt, would be of order 10^{-13} s. Such correlations are of a nature that physical intuition would lead one to expect but it is not necessary to assume the existence of well-defined ion pairs.[28] The velocity autocorrelation function shown in Figure 10.5 has a negative plateau similar to that seen in argon-like liquids. Both the shape of $Z(t)$ and the value of the diffusion coefficient are reasonably well reproduced by a mode coupling calculation[29] of the type discussed in Section 9.5. The mode coupling results for the electric current autocorrelation function are much less satisfactory and the theoretical value for the case illustrated in the figure is about 30% too small, a discrepancy which has been attributed to the neglect of temperature fluctuations.

Other molecular dynamics studies of self-diffusion have been made for models of the alkali halides in which allowance is made for the differences in mass and size of the two ions. Where the mass difference is large, the velocity autocorrelation function of the lighter ion is strongly oscillatory. This effect is the result of a 'rattling' motion of the ion in the relatively long-lived cage formed by its heavier neighbours and is particularly marked in the case of the very light Li^+ ion. An example of such behaviour is pictured in Figure 10.6, which shows the results obtained for the velocity autocorrelations of both ion species in a simulation of lithium chloride based on the rigid-ion pair potentials of Fumi and Tosi.[2] By contrast, the autocorrelation function for the Cl^- ion decays to zero in near-monotonic fashion, but when a polarisable-ion model is used, a negative region appears, thereby reducing the diffusion coefficient to a value closer to that measured experimentally. It has long been known[30] that use of the Fumi–Tosi potentials leads to results for the transport coefficients of the alkali halides that are in generally fair agreement with experiment. In general, inclusion of polarisation in the potential model is found to improve that agreement, as in the example just described, but intercomparison of the adequacy of the two classes of model has mostly been made in only piecemeal fashion A systematic study of the electrical and thermal conductivities, self-diffusion coefficients and shear viscosities has, however, been carried out for the chloride salts of lithium, sodium and potassium.[31] This has shown that overall

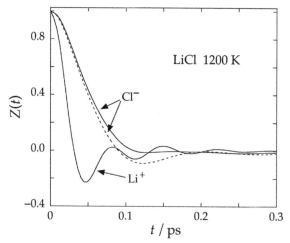

FIGURE 10.6 Normalised velocity autocorrelation functions of the Li^+ and Cl^- ions obtained by molecular dynamics calculations for models of molten lithium chloride. Full curves: based on the Fumi–Tosi potentials; dashes: based on a model that allows for polarisation of the Cl^- ion. Unpublished results of M. Salanne.

the introduction of polarisation via the force matching method of Section 10.1 reduces the discrepancies seen in calculations for the Fumi–Tosi potentials by roughly a factor of 2. The most striking difference between the two sets of results lies in the values of thermal conductivity, which are greatly overestimated by the rigid-ion model; otherwise the improvement achieved by including polarisation is only marginal. The discrepancies that remain are broadly comparable with the combined computational and experimental uncertainties.

The wavenumber-dependent collective motions in molten salts have also been studied by molecular dynamics. The simple molten salt is particularly well-suited to theoretical investigation of the collective modes[11] because the fluctuations in mass and charge densities are strictly independent at all wavelengths (see Section 10.4). The main objects of interest are the optic-type modes associated with charge fluctuations, since these are specific to ionic fluids. The results of the simulations show that the charge density autocorrelation function $F_{ZZ}(k, t)$ is strongly oscillatory at wavelengths up to about twice the mean interionic spacing. These oscillations give rise to a 'plasmon' peak in the dynamic structure factor $S_{ZZ}(k, \omega)$, as shown in Figure 10.7. The frequency ω_k at which the optic peak is seen is in the region of the plasma frequency ω_p, but its dispersion is strongly negative and described reasonably well by the relation $\omega_k \approx \omega_{1l}(k)$, as suggested by the rough calculation made in the previous section. The peak eventually disappears at a value of k close to the position of the main peak in the charge–charge structure factor $S_{ZZ}(k)$. More surprising is the fact that at small wavenumbers the optic peak initially sharpens as k increases, i.e. the damping of the plasmon mode becomes weaker. This behaviour is in

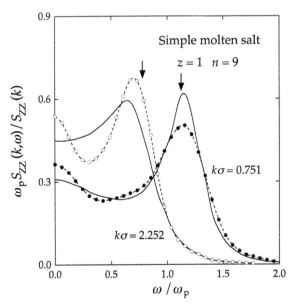

FIGURE 10.7 Charge–charge dynamic structure factor of the simple molten salt at two values of k under the state conditions described in the caption to Figure 10.1. The points show the results of molecular dynamics calculations, the dashes are guides to the eye and the curves are calculated from a single relaxation-time approximation for the memory function $N_l(k, t)$ with the relaxation time determined by a least-squares fit to the simulated spectra. The arrows mark the value of $\omega_{1l}(k)$. *Redrawn with permission from Ref. 11 © 1975 American Physical Society.*

striking contrast to that of the sound-wave mode; in molten salts, as in systems of neutral particles, the sound-wave damping increases rapidly with k.

The main features of the charge fluctuation spectrum of the simple molten salt are also seen in simulations of more realistic rigid-ion models; the effect of including polarisation is to broaden the optic peak and shift it to lower frequencies. It can be seen from Figure 10.7 that in the case of the simple molten salt the single relaxation time, viscoelastic approximation cannot account for the detailed shape of the spectrum. At least two relaxation times are required, and other calculations have confirmed that the memory function for the longitudinal charge current correlation function consists of a rapidly decaying term and a long-time, quasi-exponential tail; it therefore has a structure similar to that required to describe the density fluctuations in argon-like liquids (see Section 9.4). A fair description of the spectra of mass and charge fluctuations in the simple molten salt has been obtained by mode coupling methods[32] along the general lines of Section 9.5. In particular, a mode coupling calculation has shown that the width of the plasmon peak should decrease with increasing k in a certain wavenumber range, in qualitative agreement with the unexpected behaviour observed in the simulations.

Several attempts have been made to detect a collective, plasmon-like excitation in molten salts by inelastic neutron scattering. If b_1 and b_2 are the coherent neutron scattering lengths of the two ionic species, and if $z_1 = -z_2$, a straightforward extension of the derivation given for a one-component fluid in Section 7.5 shows that the coherent, inelastic cross-section for a monovalent salt can be written in the form

$$\frac{d^2\sigma}{d\Omega d\omega} \propto (b_1 + b_2)^2 S_{NN}(k,\omega) + 2(b_1^2 - b_2^2) S_{NZ}(k,\omega)$$
$$+ (b_1 - b_2)^2 S_{ZZ}(k,\omega) \tag{10.5.6}$$

Thus a single experiment yields only a linear combination of the three dynamic structure factors (number–number, number–charge and charge–charge). Moreover, the contribution made by the charge fluctuation component is very low at small wavenumbers, since $S_{ZZ}(k)$ (the integral of $S_{ZZ}(k,\omega)$) is proportional to k^2 in the limit $k \to 0$. Only when the scattering lengths are such that $b_1 \approx -b_2$ does the component $S_{ZZ}(k,\omega)$ dominate, which is a situation not easily achievable with readily available isotopes. The weakness of the contribution to the total scattered intensity also makes it unlikely that a propagating optic mode could be detected by inelastic X-ray scattering.

The autocorrelation functions of the transverse components of the mass and charge currents have been calculated in molecular dynamics simulations of a number of model systems. The frequency of the transverse optic mode lies roughly an amount ω_p below that of its longitudinal counterpart, as suggested by the sum rule (10.4.25), and is relatively insensitive to wavenumber. As in the case of the longitudinal modes, an accurate memory function representation of the transverse current spectra requires the introduction of two relaxation times that are very different in value.[33]

10.6 THE ELECTRIC DOUBLE LAYER

So far in this chapter the emphasis has been placed on the bulk properties of ionic liquids. We turn now to a discussion of some of the new phenomena that arise in the vicinity of a charged surface and show how the resulting inhomogeneities can be described within the framework of the density functional theory developed in Chapters 3 and 6.

When colloidal particles or macromolecules are dissolved in a highly polar solvent such as water, they will normally release counterions into the solvent, leaving behind a 'polyion' carrying a surface charge of opposite sign. The solvent will in general be an electrolyte solution and is therefore itself a source of both counterions and coions, coions being those of like charge to that of the polyion. Counterions are attracted by the surface charge, but this is counterbalanced by the tendency for ions to spread into the bulk solution in order to maximise the entropy. These competing effects lead to the formation of an electric double layer at the charged surface, to which both coions and counterions contribute. In the discussion that follows we restrict ourselves to the situation in which only two ionic species are present, with charges $z_\nu e$, $\nu = +$ or $-$. The inhomogeneous solution in the vicinity of the surface is assumed to be in chemical equilibrium with a bulk solution (or reservoir) of the same ions at chemical potentials μ_ν. The surface charge is the source of an external field acting on the ions and the solution of the electrostatic problem involves boundary conditions on the local electrostatic field.

Within a confined dielectric medium of permittivity ϵ the electrostatic potential at $\mathbf{r}'$ due to a unit point charge at $\mathbf{r}$ is given by the Green's function $\mathcal{G}(\mathbf{r}, \mathbf{r}')$ that satisfies Poisson's equation:

$$\nabla^2 \mathcal{G}(\mathbf{r}, \mathbf{r}') = -\frac{4\pi}{\epsilon}\delta(\mathbf{r}' - \mathbf{r}) \qquad (10.6.1)$$

for given boundary conditions at any interfaces.[34] If there are no boundaries, the Green's function is the usual Coulomb potential, $\mathcal{G}(\mathbf{r}, \mathbf{r}') = \mathcal{G}(\mathbf{r}' - \mathbf{r}) = 1/\epsilon|\mathbf{r}' - \mathbf{r}|$; when boundaries are present, the solution can be obtained by the method of images, at least for sufficiently simple geometries.[35] Let $\rho_Z(\mathbf{r})$ be the local charge density of the fluid, defined as[36]

$$\rho_Z(\mathbf{r}) = \sum_\nu z_\nu \rho_\nu^{(1)}(\mathbf{r}) \qquad (10.6.2)$$

where $\rho_\nu^{(1)}(\mathbf{r})$ is the single-particle density of species ν. The local electrostatic potential $\Phi^C(\mathbf{r})$ that satisfies Poisson's equation:

$$\nabla^2 \Phi^C(\mathbf{r}) = -\frac{4\pi e}{\epsilon}\rho_Z(\mathbf{r}) \qquad (10.6.3)$$

subject to any boundary conditions, is

$$\Phi^C(\mathbf{r}) = \int \mathcal{G}(\mathbf{r}, \mathbf{r}') e \rho_Z(\mathbf{r}') d\mathbf{r}' \qquad (10.6.4)$$

The electrostatic energy of the system is then given by

$$U^C = \frac{1}{2}e \int \Phi^C(\mathbf{r})\rho_Z(\mathbf{r})d\mathbf{r} = \frac{1}{2}e^2 \iint \rho_Z(\mathbf{r})\mathcal{G}(\mathbf{r}, \mathbf{r}')\rho_Z(\mathbf{r}')d\mathbf{r}\, d\mathbf{r}' \quad (10.6.5)$$

where the integral extends over the region occupied by the fluid. From now on, however, we shall restrict ourselves to the situation in which there are no dielectric discontinuities and the permittivity is the same throughout space. Lifting this restriction introduces only technical complications.

The grand potential functional of the fluid is

$$\Omega[\rho_+^{(1)}, \rho_-^{(1)}] = \mathcal{F}[\rho_+^{(1)}, \rho_-^{(1)}] - \sum_\nu \int [\mu_\nu - \phi_\nu(\mathbf{r})]\rho_\nu^{(1)}(\mathbf{r})d\mathbf{r} \qquad (10.6.6)$$

where $\phi_\nu(\mathbf{r})$ is the total external potential acting on ions of species ν, which may contain both coulombic and non-coulombic components. The intrinsic free energy functional $\mathcal{F}$ can be split, as usual, into ideal and excess parts:

$$\mathcal{F}[\rho_+^{(1)}, \rho_-^{(1)}] = \sum_\nu \mathcal{F}_\nu^{\mathrm{id}}[\rho_\nu^{(1)}] + \mathcal{F}^{\mathrm{ex}}[\rho_+^{(1)}, \rho_-^{(1)}] \qquad (10.6.7)$$

where the ideal contributions are defined as in (3.1.22) and the excess contribution is given by a two-component generalisation of (3.5.23). If the reference state, corresponding to $\lambda = 0$ in (3.5.18), is taken as one in which the chemical potentials are the same as those of the bulk solution, then

$$\mathcal{F}^{\mathrm{ex}}[\rho_+^{(1)}, \rho_-^{(1)}] = F^{\mathrm{ex}}(n_+, n_-) + \sum_\nu \mu_\nu^{\mathrm{ex}} \int \Delta\rho_\nu^{(1)}(\mathbf{r}) d\mathbf{r}$$

$$- k_{\mathrm{B}}T \sum_\nu \sum_\mu \int_0^1 d\lambda(1 - \lambda) \iint \Delta\rho_\nu^{(1)}(\mathbf{r}) c_{\nu\mu}(\mathbf{r}, \mathbf{r}'; \lambda)$$

$$\times \Delta\rho_\mu^{(1)}(\mathbf{r}') d\mathbf{r} \, d\mathbf{r}' \qquad (10.6.8)$$

where n_+, n_- are the number densities in the bulk.

The direct correlation functions in (10.6.8) may be decomposed in the form

$$c_{\nu\mu}(\mathbf{r}, \mathbf{r}') = c_{\nu\mu}^{\mathrm{S}}(\mathbf{r}, \mathbf{r}') - z_\nu z_\mu l_{\mathrm{B}}/|\mathbf{r}' - \mathbf{r}| \qquad (10.6.9)$$

where the quantity $l_{\mathrm{B}} = e^2/\epsilon k_{\mathrm{B}}T$ is called the Bjerrum length. The second term on the right-hand side is the asymptotic value of the function; the first term therefore represents the short-range correlations. If we now substitute for $c_{\nu\mu}(\mathbf{r}, \mathbf{r}')$ in (10.6.8), the excess free energy functional separates into a mean field, purely coulombic part, $\mathcal{F}^{\mathrm{C}}$, and a correlation term, $\mathcal{F}^{\mathrm{corr}}$. The mean field part is given by (10.6.5), with $\mathcal{G}(\mathbf{r}, \mathbf{r}')$ taking its coulombic form, and the correlation term is formally identical to (10.6.8) but with the direct correlation functions replaced by their short-range parts. Thus

$$\mathcal{F}^{\mathrm{ex}} = \mathcal{F}^{\mathrm{C}} + \mathcal{F}^{\mathrm{corr}}, \quad \mathcal{F}^{\mathrm{C}} = \frac{1}{2}e^2 \iint \frac{\rho_Z(\mathbf{r})\rho_Z(\mathbf{r}')}{\epsilon|\mathbf{r}' - \mathbf{r}|} d\mathbf{r} \, d\mathbf{r}' \qquad (10.6.10)$$

A particularly simple approximation is to set $\mathcal{F}^{\mathrm{corr}} = 0$, implying that the fluid behaves as an ideal gas in which each ion experiences only the average electrostatic potential due to other ions and the charges at any interfaces. The density profile $\rho_\nu^{(1)}(\mathbf{r})$ derived from the variational principle (3.4.3) is then

$$\rho_\nu^{(1)}(\mathbf{r}) = \xi_\nu \exp\left(-\beta[\phi_\nu(\mathbf{r}) + z_\nu e\Phi^{\mathrm{C}}(\mathbf{r})]\right) \qquad (10.6.11)$$

where the electrostatic potential $\Phi^{\mathrm{C}}(\mathbf{r})$ is given by (10.6.4) and $\xi_\nu = \exp(\beta\mu_\nu)/\Lambda_\nu^3$ is the activity of species ν, which in the mean field approximation is equal to the bulk density n_ν.

If the external potentials have a coulombic component arising from an external charge density $\rho_Z^{\text{ext}}(\mathbf{r})$, (10.6.11) can be rewritten as

$$\rho_\nu^{(1)}(\mathbf{r}) = n_\nu \exp\left(-\beta[\phi_\nu^S(\mathbf{r}) + z_\nu e \Phi(\mathbf{r})]\right) \tag{10.6.12}$$

where $\phi_\nu^S(\mathbf{r})$ is the short-range, non-coulombic contribution to $\phi_\nu(\mathbf{r})$ and $\Phi(\mathbf{r})$ is the total electrostatic potential, which is related to the total charge density by

$$\nabla^2 \Phi(\mathbf{r}) = -\frac{4\pi e}{\epsilon}\left[\rho_Z^{\text{ext}}(\mathbf{r}) + \rho_Z(\mathbf{r})\right] \tag{10.6.13}$$

The coupled equations (10.6.11) (or (10.6.12)) and (10.6.3) (or (10.6.13)) are the equations of Poisson–Boltzmann theory.

As a first application of the theory we take the case of an electric double layer near an impenetrable, planar wall located at $z = 0$. The wall separates the ionic solution for $z > 0$ from a dielectric medium of the same permittivity for $z < 0$; the density profiles now depend only on z. The wall carries a surface charge density σ and overall charge neutrality requires that

$$\int_0^\infty e\rho_Z(z)\mathrm{d}z = -\sigma \tag{10.6.14}$$

If we assume that the absolute charges of the two ionic species are equal, it follows that $n_+ = n_- = \frac{1}{2}n_0$, and combination of (10.6.2), (10.6.12) and (10.6.13) gives

$$\frac{\mathrm{d}^2\Phi(z)}{\mathrm{d}z^2} = \frac{4\pi e n_0}{\epsilon}\sinh\left[\beta e \Phi(z)\right], \quad z > 0 \tag{10.6.15}$$

with the constraint, valid for point ions, that $\rho_\nu^{(1)}(z) = 0$ for $z < 0$. Equation (10.6.15) is the Poisson–Boltzmann equation; this must be solved subject to two boundary conditions:

$$\lim_{z\to\infty}\frac{\mathrm{d}\Phi(z)}{\mathrm{d}z} = 0, \qquad \frac{\mathrm{d}\Phi(z)}{\mathrm{d}z}\bigg|_{z=0} = -\frac{4\pi\sigma}{\epsilon} \tag{10.6.16}$$

The local number density of microions is $\rho_N(z) = \rho_+^{(1)}(z) + \rho_-^{(1)}(z)$, the gradient of which is easily obtained from (10.6.2), (10.6.12) and (10.6.13):

$$\frac{\mathrm{d}\rho_N(z)}{\mathrm{d}z} = -\beta\frac{\mathrm{d}\Phi(z)}{\mathrm{d}z}e\rho_Z(z) = \frac{\beta\epsilon}{4\pi}\frac{\mathrm{d}\Phi(z)}{\mathrm{d}z}\frac{\mathrm{d}^2\Phi(z)}{\mathrm{d}z^2} = \frac{\beta\epsilon}{8\pi}\frac{\mathrm{d}}{\mathrm{d}z}\left(\frac{\mathrm{d}\Phi(z)}{\mathrm{d}z}\right)^2 \tag{10.6.17}$$

Integration of both sides of (10.6.17) from z to infinity yields a relation between the local number density and the local electric field $E(z) = -\mathrm{d}\Phi(z)/\mathrm{d}z$:

$$k_B T\left[\rho_N(z) - n_0\right] = \frac{\epsilon}{8\pi}\left[E(z)\right]^2 \tag{10.6.18}$$

Since the microions behave as an ideal gas, the left-hand side of (10.6.18) is the difference in local osmotic pressure $P(z) = k_B T \rho_N(z)$ between a point z and a point in the bulk, where $\rho_N(z) = n_0$; the right-hand side is the electrostatic pressure,[34] which vanishes in the bulk. Differentiation of (10.6.18) with respect to z and use of Poisson's equation leads to the condition necessary for hydrostatic equilibrium, i.e.

$$\frac{dP(z)}{dz} = eE(z)\rho_Z(z) = f(z) \tag{10.6.19}$$

where $f(z)$ is the local force per unit volume acting on the solution. By evaluating (10.6.18) at $z = 0$ and making use of the second of the boundary conditions (10.6.16), we obtain an expression for the enhancement of the microion density at contact over its bulk value:

$$k_B T \rho_N(0) = k_B T n_0 + \frac{\epsilon [E(0)]^2}{8\pi} = k_B T n_0 + \frac{2\pi \sigma^2}{\epsilon} \tag{10.6.20}$$

This result is a special case of the contact theorem for ionic systems[37]:

$$k_B T \rho_N(0) = P + \frac{2\pi \sigma^2}{\epsilon} \tag{10.6.21}$$

where P is the bulk osmotic pressure, which for an ideal solution is equal to $k_B T n_0$. Equation (10.6.21) is the generalisation of (6.6.3) that applies to uncharged systems. As the surface charge increases, the contact density will eventually become sufficiently large that the role of ion–ion correlations can no longer be ignored. The correlation term in the free energy functional (10.6.10) must then be included in some approximate form,[38] such as a weighted density approximation of the type discussed in Section 6.2.

Equation (10.6.15) can be solved analytically. The dimensionless potential $\Phi^*(z) = \beta e \Phi(z)$ satisfies the equation

$$\frac{d^2 \Phi^*(z)}{dz^2} = k_D^2 \sinh \Phi^*(z) \tag{10.6.22}$$

where k_D is the Debye wavenumber (10.2.15). The solution to (10.6.22) is

$$\Phi^*(z) = 4 \tanh^{-1} \left[g \exp(-k_D z) \right] \tag{10.6.23}$$

where g is related to the dimensionless surface potential $\Phi^*(0)$ by

$$g = \tanh \frac{1}{4} \Phi^*(0) \tag{10.6.24}$$

The density profiles follow from (10.6.11):

$$\rho_\pm^{(1)}(z) = \frac{1}{2} n_0 \left(\frac{1 \mp g \exp(-k_D z)}{1 \pm g \exp(-k_D z)} \right)^2 \tag{10.6.25}$$

At distances $z \approx k_D^{-1}$ or larger, the density profiles approach their bulk values exponentially, so the thickness of the double layer is of the order of Λ_D, the Debye screening length.

We next consider the question of what the effective interaction is between charged surfaces separated by an inhomogeneous, ionic solution. The simplest geometry is that of two infinite, parallel, uniformly charged planes placed at $z = \pm\frac{1}{2}L$. If the two surface charge densities are the same, there is a plane of symmetry at $z = 0$ where the local electric field must vanish. The ionic fluid is assumed to be in chemical equilibrium with a reservoir of non-interacting, monovalent microions, which fixes the chemical potentials of the two species at their ideal values, $\mu_\nu = k_B T \ln (\Lambda_\nu^3 n_\nu)$. The mirror symmetry means that it is necessary to solve the Poisson–Boltzmann equation only in the interval $-\frac{1}{2}L \leq z \leq 0$, with the boundary conditions

$$\frac{d\Phi(z)}{dz}\bigg|_{z=-L/2} = -\frac{4\pi\sigma}{\epsilon}, \quad \frac{d\Phi(z)}{dz}\bigg|_{z=0} = 0 \qquad (10.6.26)$$

For this problem, apart from the somewhat academic case when the solution contains only counterions, the solution to the non-linear differential equation (10.6.15) must be obtained numerically. If the surface charge σ is sufficiently low, however, it is justifiable to linearise (10.6.22) by setting $\sinh \Phi^*(z) \approx \Phi^*(z)$. The resulting linear equation can then be solved to give

$$\Phi(z) = \frac{\Phi_0}{\sinh (k_D L/2)} \cosh (k_D z) \qquad (10.6.27)$$

with $\Phi_0 = 4\pi\sigma/\epsilon k_D$.

The normal component $P_N(z)$ of the pressure tensor determines the force per unit area on a test surface placed at z within the fluid. In mechanical equilibrium, P_N must be constant throughout the interval between the planes, i.e.

$$\frac{d P_N(z)}{dz} = 0, \quad -\frac{1}{2}L < z < \frac{1}{2}L \qquad (10.6.28)$$

The quantity $P_N(z)$ is the sum of the osmotic pressure of the ions, $P(z) = k_B T \rho_N(z)$, and an electrostatic contribution, which is related to Maxwell's electrostatic stress tensor[34]:

$$P_N = P(z) - \frac{\epsilon}{8\pi}\left(\frac{d\Phi(z)}{dz}\right)^2 = k_B T \rho_N(z) - \frac{\epsilon}{8\pi}\left[E(z)\right]^2 \qquad (10.6.29)$$

Taken together, (10.6.28) and (10.6.29) lead back to the equilibrium condition (10.6.19). The pressure difference

$$\Delta P = P_N(L) - P_N(\infty) \qquad (10.6.30)$$

is the force per unit area that must be applied to the charged planes in order to maintain them at a separation L; it can therefore be identified with the solvation force f_S introduced in Section 6.1. Since the local electrical field is zero at $z = 0$, it follows from (10.6.29) that

$$f_S \equiv \Delta P = k_B T [\rho_N(0) - n_0] \qquad (10.6.31)$$

Combination of (10.6.27) and (10.6.31) and the linearised version of (10.6.12) shows that to lowest, non-vanishing order in $\Phi(z = 0)$:

$$f_S(L) = \frac{1}{2} k_B T n_0 [\beta e \Phi(0)]^2 = \frac{2\pi\sigma^2}{\epsilon} \frac{1}{\sinh^2 (k_D L/2)}$$
$$\approx \frac{8\pi\sigma^2}{\epsilon} \exp(-k_D L) \qquad (10.6.32)$$

Thus the effective interaction between the charged plates is always repulsive; the same conclusion is reached within non-linear Poisson–Boltzmann theory. However, when correlations between ions are taken into account, the force between the planes may become attractive at small separations.[39] Such correlations are particularly strong in the case of divalent (or polyvalent) counterions, as illustrated by the results of Monte Carlo calculations shown in Figure 10.8.

Attraction between two like-charged surfaces can be accounted for within density functional theory only if the correlation term in the excess free energy functional is adequately approximated. If the ions are modelled as charged hard spheres, the correlations between ions arise both from hard-core effects and from short-range, coulombic interactions. This suggests that $\mathcal{F}^{ex}$ can be

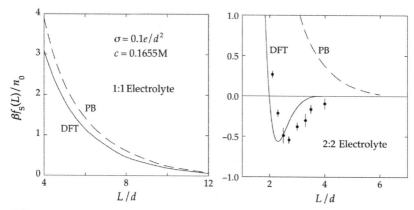

FIGURE 10.8 Electric double-layer force between charged plates in restricted primitive models of 1:1 and 2:2 electrolyte solutions as a function of the plate separation L. The state conditions correspond in each case to an aqueous solution of ions of diameter $d = 4.2\,\text{Å}$ at 298 K. The curves are calculated from the Poisson–Boltzmann approximation (PB) or from density-functional theory (DFT) and the points are the results of Monte Carlo simulations.[39b] See text for details. *Redrawn with permission from Ref. 40 © American Institute of Physics.*

usefully rewritten as

$$\mathcal{F}^{ex}[\rho_+^{(1)}, \rho_-^{(1)}] = \frac{1}{2} \int e\rho_Z(\mathbf{r})\Phi(\mathbf{r})d\mathbf{r} + \mathcal{F}^{HS}[\rho_+^{(1)}, \rho_-^{(1)}]$$

$$- k_B T \sum_{\nu} \sum_{\mu} \int_0^1 d\lambda (1-\lambda) \iint \Delta\rho_{\nu}^{(1)}(\mathbf{r})$$

$$\times \Delta c_{\nu\mu}(\mathbf{r}, \mathbf{r}'; \lambda)\Delta\rho_{\mu}^{(1)}(\mathbf{r}')d\mathbf{r}\,d\mathbf{r}' \qquad (10.6.33)$$

The first term on the right-hand side of (10.6.33) is the mean field, purely coulombic contribution; the second is the excess free energy functional for a binary hard-sphere mixture, corresponding to uncharged ions; and the last term contains the 'residual' direct correlation functions, defined as

$$\Delta c_{\nu\mu}(\mathbf{r}, \mathbf{r}'; \lambda) = c_{\nu\mu}(\mathbf{r}, \mathbf{r}'; \lambda) + \frac{z_{\nu} z_{\mu} \lambda l_B}{|\mathbf{r} - \mathbf{r}'|} - c_{\nu\mu}^{HS}(\mathbf{r}, \mathbf{r}'; \lambda) \qquad (10.6.34)$$

which represents the remaining correlations.[40] The hard-sphere direct correlation functions $c_{\nu\mu}^{HS}(\mathbf{r}'\mathbf{r}')$ are those compatible with the assumed form of the functional $\mathcal{F}^{HS}$, for which a weighted density approximation can be used, and the residual direct correlation functions can be replaced by those of the bulk solution obtained, for example, from the solution of the MSA given in Section 10.3. Figure 10.8 makes a comparison between the results obtained in this way and those of Poisson–Boltzmann theory for restricted primitive models of both 1:1 and 2:2 electrolyte solutions. In the case of the 1:1 solution, where the force is everywhere repulsive, the two theories give similar results. In the divalent system, however, the inclusion of correlations gives rise to a strongly attractive force at small separations with a minimum at $L \approx 2d$; the results are in good agreement with those obtained by simulation for the same system. Poisson–Boltzmann theory, by contrast, again predicts that the force should be repulsive for all L. Ion correlations may also lead to charge inversion or 'overscreening' of the surface charge, meaning that the total charge of the double layer, integrated over a few ionic diameters, can be of opposite sign to that of the planes. Similar results have been reached on the basis of numerical solution of the so-called anisotropic HNC equation, which represents an extension of bulk HNC theory to inhomogeneous situations.[41]

Both here and in the discussion of the properties of electrolyte solutions in Section 10.3 the molecular nature of the solvent has been ignored, since in the primitive model the solvent is treated as a dielectric continuum. In real electrolyte solutions the solvent is invariably highly polar and the electrostatic coupling between ions and molecules leads to ion solvation; each ion is surrounded by solvent molecules, distributed in such a way as to minimise the solvation free energy. In the 'civilised' model of ionic solutions[42] allowance is made for solvation by treating the solvent molecules as dipolar hard spheres. Mixtures of charged and dipolar hard spheres have been used in studies of a range of problems: bulk properties of electrolyte solutions,[43] the solvent structure around an ion,[44] the formation of electrical double layers[45] and the wetting of charged substrates by ionic solutions.[46] More complete descriptions of, say, an aqueous ionic solution would require a realistic potential model for water and the ion–water interaction, and possibly also the inclusion of terms arising from the polarisation both of the water molecule and the ion.

10.7 LIQUID METALS: ELECTRONS AND IONS

Pure liquid metals are two-component fluids consisting of N_i positive ions and $N_e = zN_i$ conduction electrons, where z is the ionic valency. The ionic core radius is usually only a small fraction of the mean interionic spacing, with the result that the ion cores occupy less than 10% of the total volume of the metal. In the 'nearly free electron' picture the conduction electrons are assumed to move more or less freely through the liquid, interacting only rarely with the ions; the mean-free path of the electrons is typically ten to a hundred times larger than the separation of neighbouring ions. In the crudest approximation, interactions are neglected altogether, and the electrons are treated as an ideal Fermi gas characterised by the energy ϵ_F of the highest occupied (Fermi) level, i.e. $\epsilon_F = \hbar^2 k_F^2/2m_e = \hbar^2 \left(3\pi^2 \rho_e\right)^{2/3}/2m_e$, where k_F is the Fermi wavenumber, ρ_e is the number density of conduction electrons and m_e is the electron mass. The Fermi temperature, $T_F = \epsilon_F/k_B$, is always some two orders of magnitude higher than the melting temperature. It is therefore a good approximation to assume that the electron gas is completely degenerate under normal liquid metal conditions.

The simplest model that takes account of electron–ion interactions is the 'jellium' model of Wigner. This is the quantum mechanical analogue of the classical one-component plasma (OCP) discussed in Section 10.1. It treats the conduction electrons as an interacting Coulomb gas moving in the uniform background provided by the positively charged ions, with a hamiltonian $\mathcal{H} = K_{N_e} + V_{N_e}$, where K_{N_e} is the kinetic energy operator and the potential energy V_{N_e} is the sum of electron–electron, electron–background and background–background terms. In a **k**-space representation, V_{N_e} is given as a special case of (10.1.4) by

$$V_{N_e} = \frac{1}{2V} {\sum_{\mathbf{k}}}' \hat{v}_{ee}(k) \left(\rho_{\mathbf{k}}^e \rho_{-\mathbf{k}}^e - N_e\right) \qquad (10.7.1)$$

where $\rho_{\mathbf{k}}^e$ is a Fourier component of the microscopic electron density, $\hat{v}_{ee}(k) = 4\pi e^2/k^2$ is the Fourier transform of the electron–electron potential and the prime on the summation means that the contribution for $\mathbf{k} = 0$ is omitted because of cancellation by the background. The ground-state energy has been calculated by methods of quantum mechanical many-body theory[47]; it is the sum of kinetic, exchange and correlation terms and is expressible as a function of the single, dimensionless parameter $r_S = a_e/a_0$, where $a_e = \left(3/4\pi\rho_e\right)^{1/3}$ is the 'electron-sphere' radius and a_0 is the Bohr radius. The minimum energy, corresponding to zero pressure, occurs at $r_S \approx 4.2$. This figure is independent of the chemical nature of the system, but is in fair agreement with experimental results for the alkali metals, which range from 3.30 (for Li) to 5.78 (for Cs).

In a more realistic model the hamiltonian of a liquid metal is written as the sum of a purely electronic term $\mathcal{H}_e$, a purely ionic term $\mathcal{H}_i$ and an electron–ion interaction V_{ei}. The Coulomb repulsion between ions is in general sufficiently strong to prevent any short-range forces coming into play, while dispersion forces are weak because the ion cores are only weakly polarisable. It is therefore a good approximation to set the ion–ion interaction $v_{ii}(R)$ equal to $z^2 e^2/R$ for all R. The electron–electron interaction $v_{ee}(r)$ is purely coulombic and the electron–ion potential v_{ei} is also coulombic outside the ion core; we shall see below that inside the core, v_{ei} can be replaced by a weak 'pseudopotential'. We proceed[48] by adding to and subtracting from the hamiltonian the two contributions that would arise if the electrons were replaced by a uniform background of charge density $-e\rho_e$. The terms involved are the ion–background interaction V_{ib} and the background self-energy V_{bb}, given by

$$V_{ib} = -\rho_e \sum_{j=1}^{N_i} \int \frac{ze^2}{|\mathbf{R}_j - \mathbf{r}|} d\mathbf{r}, \quad V_{bb} = \frac{1}{2}\rho_e^2 \iint \frac{e^2}{|\mathbf{r} - \mathbf{r}'|} d\mathbf{r}\, d\mathbf{r}' \quad (10.7.2)$$

where $\mathbf{R}_j$ denotes the coordinates of ion j. The hamiltonian can then be written as

$$\mathcal{H} = \mathcal{H}_e' + \mathcal{H}_i' + V_{ei}' \quad (10.7.3)$$

with

$$\mathcal{H}_e' = \mathcal{H}_e - V_{bb}, \quad \mathcal{H}_i' = \mathcal{H}_i + V_{ib} + V_{bb}, \quad V_{ei}' = V_{ei} - V_{ib} \quad (10.7.4)$$

In $\mathbf{k}$-space:

$$\mathcal{H}_e' = K_e + \frac{1}{2V}\sum_{\mathbf{k}}{}' \frac{4\pi e^2}{k^2}\left(\rho_{\mathbf{k}}^e \rho_{-\mathbf{k}}^e - N_e\right)$$

$$\mathcal{H}_i' = K_i + \frac{1}{2V}\sum_{\mathbf{k}}{}' \frac{4\pi z^2 e^2}{k^2}\left(\rho_{\mathbf{k}}^i \rho_{-\mathbf{k}}^i - N_i\right) \quad (10.7.5)$$

$$V_{ei}' = U_0 + \frac{1}{V}\sum_{\mathbf{k}}{}' \hat{v}_{ei}(k)\rho_{\mathbf{k}}^i \rho_{-\mathbf{k}}^e$$

where K_i is the kinetic energy of the ions and

$$U_0 = \frac{1}{V}\lim_{k\to 0}\left(\hat{v}_{ei}(k) + \frac{4\pi z e^2}{k^2}\right)\rho_{\mathbf{k}}^i \rho_{-\mathbf{k}}^e = N_i \rho_e \int\left(v_{ei}(r) + \frac{ze^2}{r}\right) d\mathbf{r} \quad (10.7.6)$$

The term $\mathcal{H}_e'$ is the jellium hamiltonian and $\mathcal{H}_i'$ is the hamiltonian of an OCP of positive ions in a uniform background. In this formulation of the problem a liquid metal emerges as a 'mixture' of a classical OCP and a quantum mechanical jellium, the two components being coupled together through the term V'_{ei}.

Inside the ion core the interaction of the conduction electrons with the ion is determined by details of the charge distribution of the core electrons. The true electron–ion interaction is therefore a complicated, non-local function for $r < r_C$, where r_C is the ion-core radius. In addition, the potential has a singularity at $r = 0$. Despite these difficulties it is possible to treat the electron–ion coupling by perturbation theory if the interaction is recast in pseudopotential form. The procedure adopted in practice is to parameterise an assumed functional form for the pseudopotential by fitting to experimental results for quantities that are sensitive to electron–ion collisions. A particularly simple and widely adopted pseudopotential $v^*_{ei}(r)$ consists in taking

$$v^*_{ei}(r) = 0, \qquad r < r_C$$
$$= -ze^2/r, \quad r > r_C \qquad (10.7.7)$$

This is called the 'empty-core' pseudopotential[49]; values of the parameter r_C can be derived from transport and Fermi-surface data and lie close to generally accepted values for the ionic radii of simple metals.

If the pseudopotential is weak, the electron–ion term in (10.7.5) can be treated as a perturbation, the reference system being a superposition of a classical OCP and a degenerate, interacting electron gas. To lowest order in perturbation theory, the structure of each component of the reference system is unaffected by the presence of the other. In this approximation, assuming the two fluids to be homogeneous:

$$\left\langle \rho^i_k \rho^e_{-k} \right\rangle = \left\langle \rho^i_k \right\rangle \left\langle \rho^e_{-k} \right\rangle = 0, \quad k \neq 0 \qquad (10.7.8)$$

Hence, on averaging the perturbation V'_{ei}, only the structure-independent term survives. The internal energy of the metal is then the sum of three terms: the energy of the degenerate electron gas, given by the jellium model; the internal energy of the classical OCP, which is known from Monte Carlo calculations[50] as a function of the dimensionless coupling constant $\Gamma = z^2e^2/a_ik_BT$, where $a_i = (3/4\pi\rho_i)^{1/3}$; and the quantity U_0, which can be calculated from (10.7.6) and (10.7.7). When combined, these results allow the calculation of the internal energy and equation of state as functions of the density parameter r_S for values of Γ and r_C appropriate to a particular metal. Figure 10.9 shows the equation of state obtained in this way for four alkali metals along isotherms corresponding to the experimental melting temperatures. Given the crudeness of the model, the agreement between theory and experiment for the zero-pressure value of r_S is surprisingly good.

A more accurate calculation has to take account of the influence of the ionic component on the structure of the electron gas and vice versa. To do so we must go to second order in perturbation theory. We also use an adiabatic approximation in which the electrons are assumed to adjust themselves instantaneously

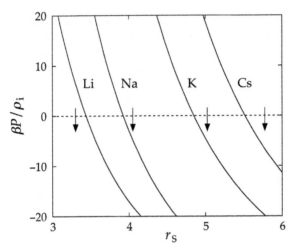

FIGURE 10.9 Equation of state of four alkali metals along isotherms corresponding to the experimental melting temperatures (Li, 452 K; Na, 371 K; K, 337 K; Cs, 303 K). The curves are calculated from the first-order perturbation theory described in the text and the arrows mark the experimental values of r_S at atmospheric pressure.

to the much slower changes in the ionic coordinates. Thus the problem to be considered is that of an inhomogeneous, interacting electron gas in the external field produced by a given ionic charge distribution; because the electron–ion pseudopotential is assumed to be weak, the influence of the external field can be treated by linear response theory. The polarisation of the electron gas by the ionic charge distribution leads to a screening of the external field and hence, as we shall see, to a new, effective interaction between the ions.

The partition function corresponding to the hamiltonian (10.7.3) is

$$Q_{N_i N_e} = \frac{1}{N_i! h^{3N_i}} \iint \exp\left(-\beta \mathcal{H}'_i\right) \mathrm{Tr}_e \exp\left[-\beta(\mathcal{H}'_e + V'_{ei})\right] \mathrm{d}\mathbf{R}^{N_i} \, \mathrm{d}\mathbf{P}^{N_i}$$

$$(10.7.9)$$

where $\mathbf{P}^{N_i} \equiv \{\mathbf{P}_j\}$ represents the momenta of the ions. The trace is taken over a complete set of quantum states of the electron gas in the field due to a fixed ionic configuration; the free energy F'_e of the inhomogeneous electron gas is a function of the ionic coordinates $\{\mathbf{R}_j\}$ and given by

$$F'_e(\{\mathbf{R}_j\}) = -k_B T \ln \left(\mathrm{Tr}_e \exp\left[-\beta(\mathcal{H}'_e + V'_{ei})\right]\right) \qquad (10.7.10)$$

If the homogeneous electron gas is taken as the reference system and V'_{ei} is again treated as a perturbation, F'_e is obtained from the coupling-parameter formula (5.2.5) as[51]

$$F'_e = F_e + \int_0^1 \langle V'_{ei} \rangle_\lambda \, \mathrm{d}\lambda \qquad (10.7.11)$$

where F_e is the free energy of the reference system and the subscript λ shows that the average is taken over an ensemble characterised by the hamiltonian $\mathcal{H}'_e + \lambda V'_{ei}$. From (10.7.5), with $\hat{v}_{ei}(k)$ replaced by $\hat{v}^*_{ei}(k)$, we find that for a fixed ionic configuration:

$$\langle V'_{ei} \rangle_\lambda = U_0 + \frac{1}{V} {\sum_{\mathbf{k}}}' \hat{v}^*_{ei}(k) \rho^i_{\mathbf{k}} \langle \rho^e_{-\mathbf{k}} \rangle_\lambda \qquad (10.7.12)$$

The result of first-order perturbation theory corresponds to setting $\lambda = 0$. But $\langle \rho^e_{-\mathbf{k}} \rangle_0$ is zero because the reference system is homogeneous; the second term on the right-hand side of (10.7.12) therefore disappears and we are led back to our earlier result. To obtain the second-order term it is sufficient to calculate the components of the induced electron density to first order in $\lambda V'_{ei}$. If $\chi_e(k)$ is the static electron-density response function, the induced density is

$$\langle \rho^e_{-\mathbf{k}} \rangle_\lambda = \chi_e(k) \lambda \hat{v}^*_{ei}(k) \rho^i_{-\mathbf{k}} \qquad (10.7.13)$$

If we now substitute for $\langle V'_{ei} \rangle$ in (10.7.11) and integrate over λ, we find that the free energy of the electron gas is given to second order in the electron–ion coupling by

$$F'_e = F_e + U_0 + \frac{1}{2V} {\sum_{\mathbf{k}}}' \chi_e(k) \left[\hat{v}^*_{ei}(k) \right]^2 \rho^i_{\mathbf{k}} \rho^i_{-\mathbf{k}} \qquad (10.7.14)$$

On comparing this result with (10.7.9) and (10.7.10) we see that the system can be regarded as a one-component fluid for which the total interaction energy is

$$V_{N_i}(\{\mathbf{R}_j\}) = V_0 + \frac{1}{2V} {\sum_{\mathbf{k}}}' \left(\hat{v}_{ii}(k) + \chi_e(k) \left[\hat{v}^*_{ei}(k) \right]^2 \right) \left(\rho^i_{\mathbf{k}} \rho^i_{-\mathbf{k}} - N_i \right) \qquad (10.7.15)$$

where

$$V_0 = U_0 + F_e + \frac{1}{2} \rho_i {\sum_{\mathbf{k}}}' \chi_e(k) \left[\hat{v}^*_{ei}(k) \right]^2 \qquad (10.7.16)$$

is independent of the structure of the liquid. Since T is normally much less than T_F, F_e can be replaced by the ground-state energy of the interacting electron gas (the jellium model).

The total interaction energy may now be rewritten in a form that involves a sum of pair interactions:

$$V_{N_i} = V_0 + \sum_{j=1}^{N_i} \sum_{j'>j}^{N_i} v^{\text{eff}}_{ii}(|\mathbf{R}_{j'} - \mathbf{R}_j|) \qquad (10.7.17)$$

The effective ion–ion potential $v^{\text{eff}}_{ii}(R)$ is the Fourier transform of the sum of the bare ion–ion interaction $v_{ii}(R)$ and an electron-induced term $v'_{ii}(R)$ or, in

k-space:

$$\hat{v}_{ii}^{\text{eff}}(k) = \hat{v}_{ii}(k) + \hat{v}_{ii}'(k) = \frac{4\pi z^2 e^2}{k^2} + \left[\hat{v}_{ei}^*(k)\right]^2 \chi_e(k)$$

$$= \frac{4\pi z^2 e^2}{k^2} + \frac{[\hat{v}_{ei}^*(k)]^2}{4\pi e^2/k^2}\left(\frac{1}{\epsilon_e(k)} - 1\right) \tag{10.7.18}$$

where $\epsilon_e(k)$, the dielectric function of the electron gas, is related to the susceptibility $\chi_e(k)$ in the manner of (10.2.11).[47] In the long-wavelength limit $\epsilon_e(k)$ behaves as

$$\lim_{k \to 0} \epsilon_e(k) = 1 + k_e^2/k^2 \tag{10.7.19}$$

with

$$k_e^2 = k_{\text{TF}}^2 \frac{\chi_{Te}}{\chi_{Te}^{\text{id}}} \tag{10.7.20}$$

where χ_{Te} and χ_{Te}^{id} are the isothermal compressibilities, respectively, of the interacting and non-interacting electron gas, and $k_{\text{TF}} = 2(k_F/\pi a_0)^{1/2}$ is the Thomas–Fermi wavenumber. Equation (10.7.19) is the electronic counterpart of the relation (10.2.30) satisfied by the classical OCP and k_e is the analogue of the ionic screening wavenumber k_s. In the same limit, $\hat{v}_{ei}^*(k) \to 4\pi z e^2/k^2$. It follows that the effective interaction $\hat{v}_{ii}^{\text{eff}}(k)$ is a regular function in the limit $k \to 0$, the k^{-2} singularity in the bare potential $\hat{v}_{ii}(k)$ being cancelled by the same singularity in $\hat{v}_{ii}'(k)$. In physical terms this means that the bare ion–ion potential $v_{ii}(R)$ is completely screened by the polarisation of the electron gas; the effective potential $v_{ii}^{\text{eff}}(R)$ is therefore a short-range function. A typical effective potential for an alkali metal has a soft repulsive core, an attractive well with a depth (in temperature units) of a few hundred kelvin and a weakly oscillatory tail.[52] An example of a calculated effective potential for liquid potassium has been shown earlier in Figure 1.4.

The results of the second-order calculation may be summarised by saying that we have reduced the liquid metal problem to one of calculating the classical partition function of a fluid of N_i pseudoatoms in which the particles interact through a short-range effective potential $v_{ii}^{\text{eff}}(R)$. After integration over momenta the partition function becomes

$$Q_{N_i} = \frac{\exp(-\beta V_0)}{N_i! \Lambda_i^{3N_i}} \int \exp\left(-\beta V_{N_i}^{\text{eff}}\right) d\mathbf{R}^{N_i} \tag{10.7.21}$$

where $V_{N_i}^{\text{eff}}$ is the sum of the pairwise-additive, effective interactions for a given ionic configuration and Λ_i is the de Broglie thermal wavelength of the ions. Equation (10.7.21) differs from the usual partition function of a monatomic fluid in two important ways: first, in the appearance of a large, structure-independent energy V_0; and, secondly, in the fact that both V_0 and the pair potential from which $V_{N_i}^{\text{eff}}$ is derived are functions of density by virtue of the density dependence

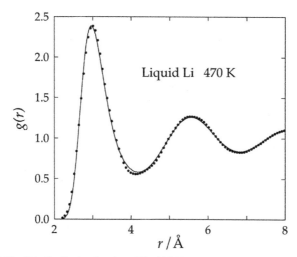

FIGURE 10.10 Pair distribution function of liquid lithium near the normal melting temperature. The curve shows results obtained by molecular dynamics calculations for an effective ion–ion potential and the points are the results of neutron scattering measurements. From P.S. Salmon et al., "Structure of liquid lithium", *J. Phys. Condens. Matter* **16** 195–222 (2004). © *IOP Publishing (2004). Reproduced by permission of IOP Publishing. All rights reserved.*

of the properties of the electron gas. The reduction of the problem to the form described by (10.7.21) means that the theoretical methods developed for the calculation of static properties of simple classical liquids can also be applied to liquid metals. Special care is needed only when evaluating volume derivatives of the free energy, notably the pressure, because the density dependence of the effective interaction gives rise to extra terms. Computer simulations have shown that effective ion–ion potentials can account quantitatively for many of the observed properties of simple liquid metals. From Figure 10.10, for example, we see that the pair distribution function obtained in this way for liquid lithium[53] is in excellent agreement with that derived from neutron diffraction data.[54]

10.8 IONIC DYNAMICS IN LIQUID METALS

The microscopic dynamics of the ions in liquid metals do not differ in any fundamental way from the corresponding motions in simple, insulating liquids such as the rare gases. This is not surprising, since the pair potentials for metallic pseudoatoms and rare-gas atoms are qualitatively similar. For the same reason, experimental and theoretical methods that have been used successfully to study and describe the dynamics of argon-like liquids have, for the most part, met with comparable success in their application to simple liquid metals. However, as comparison of Figures 1.3 and 1.4 shows, the interactions in, say, potassium and argon do differ considerably in detail, and this gives rise to quantitative differences in the dynamical behaviour of the two types of system. For example, as we have seen in earlier chapters, experiments and simulations have combined to show that propagating collective modes of both transverse and longitudinal character persist over ranges of wavelength relative to the particle diameters that are considerably wider in liquid metals than in argon-like liquids.[55]

A different insight into the dynamics can be obtained through the representation of a liquid metal as an ion–electron plasma along the lines followed for static properties in Section 10.7. In this picture of the liquid, the ionic and electronic components are only weakly coupled through the electron–ion pseudopotential, so that each component may be regarded as an external perturbation on the other. Let $\phi_v(\mathbf{k}, \omega)$ be an external potential that acts on component v, where $v = 1$ for the ions and 2 for the electrons. Within linear response theory the Fourier components of the induced densities are related to the external potentials by a matrix of density response functions:

$$\langle \rho_{\mathbf{k}}^v(\omega) \rangle = \sum_\mu \chi_{v\mu}(k, \omega) \phi_\mu(\mathbf{k}, \omega) \tag{10.8.1}$$

The response to the internal field is described by a similar matrix of screened response functions, $\chi_{v\mu}^{sc}$. Written in matrix form the response is

$$\langle \rho_{\mathbf{k}}(\omega) \rangle = \chi^{sc}(k, \omega) \cdot \left[\phi(\mathbf{k}, \omega) + \hat{\mathbf{v}}(k) \cdot \langle \rho_{\mathbf{k}}(\omega) \rangle \right] \tag{10.8.2}$$

where $\hat{\mathbf{v}}(k)$ is the matrix of bare potentials $\hat{v}_{v\mu}(k)$ and the second term in square brackets is the 'polarisation potential'. Elimination of $\langle \rho_{\mathbf{k}}^v(\omega) \rangle$ between (10.8.1) and (10.8.2) leads to a matrix generalisation of the relation (10.4.11) between the external and screened response functions:

$$\chi(k, \omega) = \chi^{sc}(k, \omega) + \chi^{sc}(k, \omega) \cdot \hat{\mathbf{v}}(k) \cdot \chi(k, \omega) \tag{10.8.3}$$

or, in terms of elements of the inverse matrices:

$$\left[\chi(k, \omega) \right]_{v\mu}^{-1} = \left[\chi^{sc}(k, \omega) \right]_{v\mu}^{-1} - \hat{v}_{v\mu}(k) \tag{10.8.4}$$

To lowest order in the ion–electron coupling the two species respond to the internal field as two, independent, one-component plasmas. The off-diagonal elements of χ^{sc} are then zero, and the diagonal elements $\chi_{vv}^{sc}(k, \omega)$ are the screened response functions of the classical OCP (for $v = 1$) and the degenerate electron gas in a uniform background (jellium) (for $v = 2$). It follows, given (10.4.11) and (10.4.12), that

$$\begin{aligned}
\left[\chi^{sc}(k, \omega) \right]_{11}^{-1} &= \frac{1}{\chi_{OCP}(k, \omega)} + \hat{v}_{11}(k) \\
\left[\chi^{sc}(k, \omega) \right]_{22}^{-1} &= \frac{\hat{v}_{22}(k)}{1 - \epsilon_e(k, \omega)}
\end{aligned} \tag{10.8.5}$$

and the external susceptibility of the ions is obtained from (10.8.4) as

$$\chi_{11}(k,\omega) = \frac{\chi_{OCP}(k,\omega)}{1 - \hat{v}(k,\omega)\chi_{OCP}(k,\omega)} \tag{10.8.6}$$

where $\hat{v}(k,\omega)$ describes the dynamical screening of the ion–ion interaction by the electrons:

$$\hat{v}(k,\omega) = \frac{k^2[\hat{v}_{12}(k)]^2}{4\pi e^2}\left(\frac{1}{\epsilon_e(k,\omega)} - 1\right) \tag{10.8.7}$$

The frequency scale of the electronic motion is much higher than any frequency associated with the ions. It is therefore reasonable to make an adiabatic approximation in which $\hat{v}(k,\omega)$ is replaced by $\hat{v}(k,0)$. The characteristic frequencies of the longitudinal modes of the screened ionic plasma are given by the roots of the denominator in (10.8.6) or, in the adiabatic approximation, by the solution to the equation

$$1 - \hat{v}(k,0)\chi_{OCP}(k,\omega) = 0 \tag{10.8.8}$$

In the limit $k \to 0$, the ratio $\tilde{F}_{OCP}(k,\omega)/S_{OCP}(k)$ is related to the frequency-dependent electrical conductivity by (10.4.30). Thus, from (7.6.21)[56]:

$$\lim_{k\to 0}\chi_{OCP}(k,\omega) = -\beta\rho_i\lim_{k\to 0}\lim_{\epsilon\to 0}\left[S_{OCP}(k) + i(\omega + i\epsilon)\tilde{F}_{OCP}(k,\omega + i\epsilon)\right]$$

$$= -\beta\rho_i\lim_{k\to 0}S_{OCP}(k)\lim_{\epsilon\to 0}\frac{4\pi\sigma(\omega + i\epsilon)}{-i(\omega + i\epsilon) + 4\pi\sigma(\omega + i\epsilon)} \tag{10.8.9}$$

The long-wavelength limit of $S_{OCP}(k)$ is given by (10.2.31) and the complex conductivity $\sigma(\omega + i\epsilon)$ can be expressed, via (7.7.10), in the form

$$\sigma(\omega + i\epsilon) = \frac{\beta}{V}\int_0^\infty J(t)\exp[i(\omega + i\epsilon)t]dt \tag{10.8.10}$$

where $J(t)$ is the charge current autocorrelation function. In the OCP the proportionality of mass and charge means that the conservation of total linear momentum is equivalent to the conservation of charge current, i.e. the resistivity is zero. Hence

$$J(t) = J(0) = \frac{N_iz^2e^2k_BT}{m_i} \tag{10.8.11}$$

and, from (10.8.10):

$$\sigma(\omega + i\epsilon) = \frac{i\omega_{pi}^2}{4\pi(\omega + i\epsilon)} \tag{10.8.12}$$

where $\omega_{pi}^2 = 4\pi\rho_iz^2e^2/m_i$ is the square of the ionic plasma frequency. Substitution of (10.8.12) in (10.8.9) shows that

$$\lim_{k\to 0}\frac{\chi_{OCP}(k,\omega)}{S_{OCP}(k)} = \frac{\beta\rho_i\omega_{pi}^2}{\omega^2 - \omega_{pi}^2} \tag{10.8.13}$$

At small k, $\epsilon_e(k, 0) \approx k_e^2/k^2$, from (10.7.19), and $\hat{v}_{12}(k)$, in the empty-core model, behaves as

$$\hat{v}_{12}(k) = \frac{4\pi z^2 e^2 \cos k r_c}{k^2} \approx \frac{4\pi z e^2}{k^2} \left(1 - \frac{1}{2}k^2 r_c^2 \right) \tag{10.8.14}$$

so that

$$\lim_{k \to 0} \hat{v}(k, 0) = \frac{4\pi z^2 e^2}{k^2}(1 - k^2 r_c^2)\left(\frac{k^2}{k_e^2} - 1\right) \tag{10.8.15}$$

When results are brought together we find that to order k^2 the solution to (10.8.8) leads to a dispersion relation characteristic of a propagating sound wave, i.e.

$$\omega = \omega_{pi}\left(k_e^{-2} + k_s^{-2} + r_c^2\right)^{1/2} k = ck \tag{10.8.16}$$

where k_s is the ionic screening wavenumber defined by (10.2.28) and c is the speed of sound. Thus the effect of electron screening is to convert the plasmon mode at frequency ω_{pi} into a sound wave of a frequency that vanishes linearly with k. A more detailed analysis shows that c can be identified with the isothermal speed of sound, but this differs little from the adiabatic value, since the ratio of specific heats is close to unity for liquid metals.

REFERENCES

[1] For a review of properties of the OCP, see Baus, M. and Hansen, J.P., *Phys. Rep.* **59**, 1 (1980).

[2] (a) Fumi, F.G. and Tosi, M.P., *J. Phys. Chem. Solids* **25**, 31 (1964). (b) Tosi, M.P. and Fumi, F.G., *J. Phys. Chem. Solids* **25**, 45 (1964).

[3] Sangster, M.J.L. and Dixon, M., *Adv. Phys.* **25**, 247 (1976).

[4] Wilson, M. and Madden, P.A., *J. Phys. Condens. Matter* **5**, 2687 (1993).

[5] (a) Madden, P.A., Heaton, R., Aguado, A. and Jahn, S., *J. Mol. Struct.*: THEOCHEM **771**, 9 (2006). (b) Salanne, M. and Madden, P.A., *Mol. Phys.* **109**, 2299 (2011).

[6] This method resembles closely one devised earlier for the treatment of polarisation in polar fluids. See Sprik, M. and Klein, M.L., *J. Chem. Phys.* **89**, 7558 (1988).

[7] Martin, P.A. and Yalcin, T., *J. Stat. Phys.* **31**, 691 (1983).

[8] (a) Stillinger, F.H. and Lovett, R., *J. Chem. Phys.* **48**, 3858 (1968). (b) Stillinger, F.H. and Lovett, R., *J. Chem. Phys.* **49**, 1991 (1968).

[9] Blum, L., Gruber, C., Lebowitz, J.L. and Martin, P.A., *Phys. Rev. Lett.* **48**, 1769 (1982).

[10] Giaquinta, P.V., Parrinello, M. and Tosi, M.P., *Phys. Chem. Liq.* **5**, 305 (1976).

[11] Hansen, J.P. and McDonald, I.R., *Phys. Rev. A* **11**, 2111 (1975).

[12] See Rollet, A. and Salanne, M., *Annu. Rep. Prog. Chem, Sect. C* **107**, 88 (2011) and references therein.

[13] (a) Wilson, M. and Madden, P.A., *J. Phys: Condens. Matter* **5**, 6833 (1993). (b) Madden, P.A. and Wilson, M., *Chem. Soc. Rev.* **25**, 339 (1996).

[14] Waisman, E. and Lebowitz, J.L., *J. Chem. Phys.* **56**, 3086 and 3093 (1972).

[15] Rasaiah, J.C., Card, D.N. and Valleau, J.P., *J. Chem. Phys.* **56**, 248 (1972).

[16] Larsen, B., *J. Chem. Phys.* **65**, 3431 (1976).

[17] (a) Høye, J.S., Stell, G. and Waisman, E., *Mol. Phys.*. **32**, 209 (1976). (b) Høye, J.S. and Stell, G., *J. Chem. Phys.* **67**, 524 (1977).

[18] Valleau, J.P., Cohen, L.K. and Card, D.N., *J. Chem. Phys.* **72**, 5942 (1980).

[19] See, e.g., Duh, D.-M. and Haymet, A.J., *J. Chem. Phys.* **97**, 7716 (1992).

[20] Bresme, F., Lomba, E., Weis, J.J. and Abascal, J.L.F., *Phys. Rev. E* **51**, 289 (1995).

[21] See the discussion surrounding (4.7.2).

[22] Ballone, P., Pastore, G. and Tosi, M.P., *J. Chem. Phys.* **81**, 3174 (1984).

[23] Zerah, G. and Hansen, J.P., *J. Chem. Phys.* **84**, 2336 (1986).

[24] Leote de Carvalho, R.J.F. and Evans, R., *Mol. Phys.* **83**, 619 (1994).

[25] Kirkwood, J.G., *Chem. Rev.* **19**, 275 (1936).

[26] Abramo, M.C., Parrinello, M. and Tosi, M.P., *J. Non-Metals* **2**, 67 (1973).

[27] Cohen, C., Sutherland, J.W.H. and Deutch, J.M., *Phys. Chem. Liq.* **2**, 213 (1971).

[28] For a review of the issues involved, see Harris, K.R., *J. Phys. Chem. B* **114**, 9572 (2010).

[29] (a) Sjögren, L. and Yoshida, F., *J. Chem. Phys.* **77**, 3703 (1982). (b) Munakata, T. and Bosse, J., *Phys. Rev. A* **27**, 455 (1983).

[30] See, e.g., Ciccotti, G., Jacucci, G. and McDonald, I.R., *Phys. Rev. A* **13**, 426 (1976).

[31] Ohtori, N., Salanne, M. and Madden, P.A., *J. Chem. Phys.* **130**, 104507 (2009).

[32] Bosse, J. and Munakata, T., *Phys. Rev. A* **25**, 2763 (1982).

[33] See, e.g., Madden, P.A. and O'Sullivan, K.F., *J. Chem. Phys.* **95**, 1980 (1991).

[34] Jackson, J.D., "Classical Electrodynamics", 3rd edn. John Wiley, New York, 1999.

[35] Stillinger, F.H., *J. Chem. Phys.* **35**, 1584 (1961). See also Allen, R.J., Hansen, J.P. and Melchionna, S., *Phys. Chem. Chem. Phys.* **3**, 4177 (2001).

[36] This definition is already implicit in (10.2.6).

[37] Henderson, D. and Blum, L., *J. Chem. Phys.* **69**, 5441 (1978).

[38] For a review of approximations beyond Poisson–Boltzmann theory, see Hansen, J.P. and Löwen, H., *Ann. Rev. Phys. Chem.* **51**, 209 (2000).

[39] (a) Guldbrand, L., Jönsson, B., Wennerström, H. and Linse, P., *J. Chem. Phys.* **80**, 2221 (1984). (b) Valleau, J.P., Ivkov, R. and Torrie, G.M., *J. Chem. Phys.* **95**, 520 (1991).

[40] Tang, Z., Scriven, L.E. and Davis, H.T., *J. Chem. Phys.* **97**, 9258 (1992).

[41] (a) Kjellander, R., Åkesson, T., Jönsson, B. and Marčelia, S., *J. Chem. Phys.* **97**, 1424 (1992). (b) Greberg, H. and Kjellander, R., *J. Chem. Phys.* **108**, 2940 (1998).

[42] Augousti, A.T. and Rickayzen, G., *J. Chem. Soc. Faraday Trans. 2* **80**, 141 (1984).

[43] (a) Carnie, S.L. and Chan, D.Y.C., *J. Chem. Phys.* **73**, 2949 (1980). (b) Blum, L. and Henderson, D., *J. Chem. Phys.* **74**, 1902 (1981).

[44] Chan, D.Y.C., Mitchell, D.J. and Ninham, B.W., *J. Chem. Phys.* **70**, 2946 (1979).

[45] Biben, T., Hansen, J.P. and Rosenfeld, Y., *Phys. Rev. E* **57**, R3727 (1998).

[46] Oleksy, A. and Hansen, J.P., *J. Chem. Phys.* **132**, 204702 (2010).

[47] Pines, D. and Nozières, P., "The Theory of Quantum Liquids". W.A. Benjamin, New York, 1966.

[48] Ashcroft, N.W. and Stroud, D., *Sol. State Phys.* **33**, 1 (1978).

[49] Ashcroft, N.W., *J. Phys. C* **1**, 232 (1966).

[50] (a) Hansen, J.P., *Phys. Rev. A* **8**, 3096 (1973). (b) Slattery, W.L., Doolen, G.D. and DeWitt, H.E., *Phys. Rev. A* **21**, 2087 (1980).

[51] Note that the prime has different meanings in (5.2.5) and (10.7.11).

[52] The oscillatory behaviour at large separations (called Friedel oscillations) is linked to a logarithmic singularity in the dielectric function. See Faber, T.E., "An Introduction to the Theory of Liquid Metals". Cambridge University Press, Cambridge, 1972, p. 28.

[53] The potential used in the simulations was derived from an empty-core pseudo-potential with a core radius obtained by fitting to the height of the first peak in the experimental structure factor: Anento, N., Canale, M. and Gonzàlez, L.E., unpublished results. See also Canales, M., Gonzàlez, L.E. and Pàdro, J.A., *Phys. Rev. E* **50**, 3656 (1994).

[54] Salmon, P.S., Petri, I., de Jong, P.H.K., Verkerk, P., Fischer, H.E. and Howells, W.S., *J. Phys. Condens. Matter* **16**, 195 (2004).

[55] For a review of experimental results on liquid metals, see Scopigno, T., Ruocco, G. and Sette, F., *Rev. Mod. Phys.* **77**, 881 (2005).

[56] The origin of the minus sign in (10.8.9) is explained in Section 7.6.

Chapter 11

Molecular Liquids

The earlier parts of the book have dealt almost exclusively with atomic systems. In this chapter we consider some of the new problems that arise when the theory is extended to include molecular fluids.

11.1 THE MOLECULAR PAIR DISTRIBUTION FUNCTION

The description of the structure of a molecular fluid in terms of particle densities and distribution functions can be developed along lines similar to those followed in the atomic case. The main added complication is the fact that the phase space probability density for particles with rotational degrees of freedom is not immediately factorisable into kinetic and configurational parts. This problem is very well treated in the book by Gray and Gubbins[1] and we shall not dwell on it here. The final expressions for the molecular distribution functions resemble closely those obtained for atomic fluids, except that all quantities are now functions of the molecular orientations.

We take as our starting point a suitably generalised form of the definition (2.5.13) of the pair density of a uniform fluid. Let $\mathbf{R}_i$ be the translational coordinates of molecule i and let $\boldsymbol{\Omega}_i$ be the orientation of i in a laboratory-fixed frame of reference. If the molecule is linear, $\boldsymbol{\Omega}_i \equiv (\theta_i, \phi_i)$, where θ_i, ϕ_i are the usual polar angles; if it is non-linear, $\boldsymbol{\Omega}_i \equiv (\theta_i, \phi_i, \chi_i)$, where θ_i, ϕ_i, χ_i are the Euler angles. Then the molecular pair density is defined as

$$\rho^{(2)}(\mathbf{R}, \mathbf{R}', \boldsymbol{\Omega}, \boldsymbol{\Omega}') = \left\langle \sum_{i=1}^{N} \sum_{j \neq i}^{N} \delta(\mathbf{R} - \mathbf{R}_i)\delta(\mathbf{R}' - \mathbf{R}_j)\delta(\boldsymbol{\Omega} - \boldsymbol{\Omega}_i)\delta(\boldsymbol{\Omega}' - \boldsymbol{\Omega}_j) \right\rangle$$

(11.1.1)

and the molecular pair distribution function as

$$g(\mathbf{R}_{12}, \boldsymbol{\Omega}_1, \boldsymbol{\Omega}_2) = (\Omega/\rho)^2 \rho^{(2)}(\mathbf{R}_{12}, \boldsymbol{\Omega}_1, \boldsymbol{\Omega}_2)$$

(11.1.2)

where $\Omega \equiv \int d\boldsymbol{\Omega}_i$. The definition of Ω means that

Theory of Simple Liquids, Fourth Edition. http://dx.doi.org/10.1016/B978-0-12-387032-2.00011-8

$$\Omega = \iint d(\cos\theta_i)d\phi_i = 4\pi, \quad \text{if linear}$$

$$= \iiint d(\cos\theta_i)d\phi_i\,d\chi_i = 8\pi^2, \quad \text{if non-linear} \tag{11.1.3}$$

The coordinates $\mathbf{R}_i$ are often taken to be those of the molecular centre of mass or some other point of high symmetry in the molecule, but the choice of molecular 'centre' is entirely arbitrary. To simplify the notation it is convenient to use the symbol $i \equiv (\mathbf{R}_i, \mathbf{\Omega}_i)$ to denote both the coordinates of the molecular centre and the orientation of molecule i. Thus the molecular pair distribution function will often be written simply as $g(1,2)$ and the molecular pair correlation function as $h(1,2) = g(1,2) - 1$. The quantities $e(1,2) = \exp[-\beta v(1,2)]$, $f(1,2) = e(1,2) - 1$ and $y(1,2) = g(1,2)/e(1,2)$ have the same significance as in the atomic case, but are now functions of the orientations $\mathbf{\Omega}_1, \mathbf{\Omega}_2$. Finally, the molecular direct correlation function $c(1,2)$ is related to $h(1,2)$ by a generalisation of the Ornstein–Zernike relation (3.5.12):

$$h(1,2) = c(1,2) + \frac{\rho}{\Omega}\int c(1,3)h(3,2)d3 \tag{11.1.4}$$

Integration of $g(1,2)$ over the variables $\mathbf{\Omega}_1, \mathbf{\Omega}_2$ yields a function $g_c(R)$ (with $R \equiv |\mathbf{R}_{12}|$) which describes the radial distribution of molecular centres:

$$g_c(R) = \frac{1}{\Omega^2}\iint g(\mathbf{R}, \mathbf{\Omega}_1, \mathbf{\Omega}_2)d\mathbf{\Omega}_1\,d\mathbf{\Omega}_2 \equiv \langle g(1,2)\rangle_{\mathbf{\Omega}_1\mathbf{\Omega}_2} \tag{11.1.5}$$

Here and elsewhere in this chapter we use angular brackets with one or more orientations $\mathbf{\Omega}_i$ as subscripts to denote an unweighted average over the orientations of the molecules involved. Thus

$$\langle\cdots\rangle_{\mathbf{\Omega}_1} \equiv \frac{1}{\Omega}\int \cdots d\mathbf{\Omega}_1, \quad \langle\cdots\rangle_{\mathbf{\Omega}_1\mathbf{\Omega}_1} \equiv \frac{1}{\Omega^2}\int \cdots d\mathbf{\Omega}_1\mathbf{\Omega}_2 \tag{11.1.6}$$

With this convention the Ornstein–Zernike relation (11.1.4) may be re-expressed as

$$h(1,2) = c(1,2) + \rho\int \langle c(1,3)h(3,2)\rangle_{\mathbf{\Omega}_3}\,d\mathbf{R}_3 \tag{11.1.7}$$

If $g(1,2)$ is multiplied by some function of the orientations $\mathbf{\Omega}_1, \mathbf{\Omega}_2$ and then integrated over all coordinates of the pair 1 and 2, the result is a quantity that measures the importance of angular correlations of a specific type. Let us suppose that molecule i has an axis of symmetry and let $\mathbf{u}_i$ be a unit vector along that axis. A set of angular order parameters that are of interest both theoretically and experimentally are those defined as

$$G_l = \rho\int \langle P_l(\mathbf{u}_1 \cdot \mathbf{u}_2)g(\mathbf{R}_{12}, \mathbf{\Omega}_1, \mathbf{\Omega}_2)\rangle_{\mathbf{\Omega}_1\mathbf{\Omega}_2}\,d\mathbf{R}_{12}$$

$$= \langle(N-1)P_l(\mathbf{u}_1 \cdot \mathbf{u}_2)\rangle \tag{11.1.8}$$

where $P_l(\cdots)$ denotes a Legendre polynomial. The value of the first-rank order parameter G_1 determines the dielectric constant of a polar fluid, as we show in Section 11.5, while G_2 is related to a number of measurable quantities, including the integrated intensity of the spectrum observed in depolarised light scattering experiments.

When the total potential energy of the fluid is a sum of pair terms the internal energy and equation of state can both be written as integrals over $g(1,2)$. The excess internal energy, for example, is given by

$$
\begin{aligned}
\frac{U^{\text{ex}}}{N} &= \frac{\rho}{2\Omega^2} \iiint v(1,2)g(1,2)\mathrm{d}\mathbf{R}_{12}\,\mathrm{d}\mathbf{\Omega}_1\,\mathrm{d}\mathbf{\Omega}_2 \\
&= 2\pi\rho \int_0^\infty \langle v(1,2)g(1,2)\rangle_{\mathbf{\Omega}_1\mathbf{\Omega}_2}\,R_{12}^2\,\mathrm{d}R_{12}
\end{aligned}
\tag{11.1.9}
$$

which is the molecular analogue of (2.5.20). The corresponding result for the pressure is a generalisation of (2.5.22):

$$
\frac{\beta P}{\rho} = 1 - \frac{2\pi\beta\rho}{3} \int_0^\infty \langle v'(1,2)g(1,2)\rangle_{\mathbf{\Omega}_1\mathbf{\Omega}_2}\,R_{12}^3\,\mathrm{d}R_{12}
\tag{11.1.10}
$$

where the prime denotes differentiation with respect to R_{12} with $\mathbf{\Omega}_1, \mathbf{\Omega}_2$ held constant. Irrespective of whether or not the potential energy is pairwise additive, an argument similar to that leading to (2.6.12) shows that the isothermal compressibility is given by

$$
\rho k_{\mathrm{B}} T \chi_T = 1 + \rho \int \langle g(1,2) - 1\rangle_{\mathbf{\Omega}_1\mathbf{\Omega}_2}\,\mathrm{d}\mathbf{R}_{12} = 1 + \rho \int \big[g_{\mathrm{c}}(R) - 1\big]\mathrm{d}\mathbf{R}
\tag{11.1.11}
$$

This result is of particular interest insofar as all reference to angular coordinates has disappeared.

Equations (11.1.9)–(11.1.11) are identical to their atomic counterparts apart from the fact that the pair functions (or products of pair functions) in the integrands are replaced by their unweighted angular averages. Their significance, however, is largely formal. The many-dimensional character of the molecular pair distribution function means that, in general, these results do not represent practical routes to the calculation of thermodynamic properties. The shape of $g(1,2)$ is difficult even to visualise and if progress is to be made the basic problem must be cast in simpler form. Two different approaches have been widely used. In one, which we review in the next section, $g(1,2)$ (or $h(1,2)$) is expanded in a series of suitably chosen, angle-dependent basis functions; in the other, which we discuss in Section 11.3, the fluid structure is described in terms of *site–site distribution functions*. Use of site–site distribution functions is particularly appropriate when the intermolecular potential is cast in site–site form, as in (1.2.6).

11.2 EXPANSIONS OF THE PAIR DISTRIBUTION FUNCTION

The pair distribution function for molecules of arbitrary symmetry can be expanded in terms of the Wigner rotation matrices or generalised spherical harmonics.[2] That formalism has not been widely used, however, and the discussion that follows is limited to linear molecules. The natural expansion functions are then the usual spherical harmonics, which we denote by $Y_{lm}(\theta, \phi)$.[3] Let $\mathbf{\Omega}_1, \mathbf{\Omega}_2$ be the orientations of molecules 1, 2 in a system of polar coordinates in which the z-axis lies along the vector $\mathbf{R}_{12} = \mathbf{R}_2 - \mathbf{R}_1$ (the 'intermolecular' frame). Then $g(1, 2)$ may be written as

$$g(1, 2) = 4\pi \sum_{l_1} \sum_{l_2} \sum_m g_{l_1 l_2 m}(R) Y_{l_1 m}(\mathbf{\Omega}_1) Y_{l_2 \bar{m}}(\mathbf{\Omega}_2) \qquad (11.2.1)$$

where $R \equiv |\mathbf{R}_{12}|$ and $\bar{m} \equiv -m$. The sum on m runs from $-l$ to l, where l is the lesser of l_1 and l_2; the indices m of the two harmonics are equal (apart from sign) by virtue of the cylindrical symmetry with respect to the axis $\mathbf{R}_{12}$. Important properties of the spherical harmonics include the fact that they are normalised and orthogonal:

$$\int Y_{lm}^*(\mathbf{\Omega}) Y_{l'm'}(\mathbf{\Omega}) d\mathbf{\Omega} = \delta_{ll'} \delta_{mm'} \qquad (11.2.2)$$

and that $Y_{l\bar{m}}(\mathbf{\Omega}) = (-1)^m Y_{lm}^*(\mathbf{\Omega})$.

If (11.2.1) is multiplied through by $Y_{l_1 \bar{m}}^*(\mathbf{\Omega}_1) Y_{l_2 m}^*(\mathbf{\Omega}_2)$ and integrated over angles, it follows from the properties just quoted that

$$\begin{aligned} g_{l_1 l_2 m}(R) &= \frac{1}{4\pi} \iint Y_{l_1 \bar{m}}(\mathbf{\Omega}_1) Y_{l_2 m}(\mathbf{\Omega}_2) g(1, 2) d\mathbf{\Omega}_1 d\mathbf{\Omega}_2 \\ &= 4\pi \langle Y_{l_1 \bar{m}}(\mathbf{\Omega}_1) Y_{l_2 m}(\mathbf{\Omega}_2) g(1, 2) \rangle_{\mathbf{\Omega}_1 \mathbf{\Omega}_2} \end{aligned} \qquad (11.2.3)$$

The expansion coefficients $g_{l_1 l_2 m}(R)$ are called the 'projections' of $g(1, 2)$ onto the corresponding angular functions and are easily calculated by computer simulation. Certain projections of $g(1, 2)$ are closely related to quantities introduced in Section 11.1. Given that $Y_{00}(\mathbf{\Omega}) = (1/4\pi)^{1/2}$, we see that $g_{000}(R)$ is identical to the centres distribution function $g_c(R)$; this is the reason for the inclusion of the factor 4π in (11.2.1). Moreover, the order parameters defined by (11.1.8) can be re-expressed as

$$G_l = \frac{\rho}{2l + 1} \sum_m (-1)^m \int g_{llm}(R) \, d\mathbf{R} \qquad (11.2.4)$$

This result is a consequence of the addition theorem for spherical harmonics, i.e.

$$P_l(\cos \gamma_{12}) = \frac{4\pi}{2l + 1} \sum_m Y_{lm}^*(\mathbf{\Omega}_1) Y_{lm}(\mathbf{\Omega}_2) \qquad (11.2.5)$$

where γ_{12} is the angle between two vectors with orientations $\mathbf{\Omega}_1$ and $\mathbf{\Omega}_2$.

An expansion similar to (11.2.1) can be made of any scalar function of the variables $\mathbf{R}_{12}$, $\boldsymbol{\Omega}_1$ and $\boldsymbol{\Omega}_2$, including both the intermolecular potential $v(1,2)$ and its derivative $v'(1,2)$ with respect to R_{12}. The corresponding expansion coefficients $v_{l_1 l_2 m}(R)$ and $v'_{l_1 l_2 m}(R)$ can be calculated numerically for any pair potential and in some cases are expressible in analytical form. If we introduce the expansions of $g(1,2)$ and $v(1,2)$ into (11.1.9) and integrate over angles, the energy equation becomes

$$\frac{U^{\mathrm{ex}}}{N} = 2\pi\rho \sum_{l_1} \sum_{l_2} \sum_{m} \int_0^\infty v_{l_1 l_2 m}(R) g_{l_1 l_2 m}(R) R^2 \, \mathrm{d}R \qquad (11.2.6)$$

The pressure equation (11.1.10) can be similarly rewritten in terms of the coefficients $v'_{l_1 l_2 m}(R)$ and $g_{l_1 l_2 m}(R)$. The multi-dimensional integrals appearing on the right-hand sides of (11.1.9) and (11.1.10) are thereby transformed into infinite sums of one-dimensional integrals. In general, however, the new expressions do not represent an improvement in the computational sense. The evidence from Monte Carlo calculations for systems of diatomic molecules is that on the whole the rate of convergence of the sums is poor and becomes rapidly worse as the elongation of the molecule increases.[4]

A different expansion of $g(1,2)$ is obtained if the orientations $\boldsymbol{\Omega}_1, \boldsymbol{\Omega}_2$ are referred to a laboratory-fixed frame of reference (the 'laboratory' frame). Let $\boldsymbol{\Omega}_R$ be the orientation of the vector $\mathbf{R}_{12}$ in the laboratory frame. Then $g(1,2)$ may be expanded in the form

$$g(1,2) = \sum_{l_1} \sum_{l_2} \sum_{l} g(l_1 l_2 l; R) \sum_{m_1} \sum_{m_2} \sum_{m} C(l_1 l_2 l; m_1 m_2 m)$$
$$\times Y_{l_1 m_1}(\boldsymbol{\Omega}_1) Y_{l_2 m_2}(\boldsymbol{\Omega}_2) Y_{lm}^*(\boldsymbol{\Omega}_R) \qquad (11.2.7)$$

where $C(\cdots)$ is a Clebsch–Gordan coefficient. The coefficients $g(l_1 l_2 l; R)$ are linear combinations of the coefficients in (11.2.1) and the two expansions are equivalent if the z-axis of the laboratory frame is taken parallel to $\mathbf{R}_{12}$. The relation between the two sets of coefficients is then

$$g(l_1 l_2 l; R) = \left(\frac{64\pi^3}{2l+1}\right)^{1/2} \sum_m C(l_1 l_2 l; m\bar{m}0) g_{l_1 l_2 m}(R) \qquad (11.2.8)$$

with, as a special case, $g(000; R) = (4\pi)^{3/2} g_{000}(R)$. Equation (11.2.7) is often written in the abbreviated form

$$g(1,2) = \sum_{l_1} \sum_{l_2} \sum_{l} g(l_1 l_2 l; R) \Phi^{l_1 l_2 l}(\boldsymbol{\Omega}_1, \boldsymbol{\Omega}_2, \boldsymbol{\Omega}_R) \qquad (11.2.9)$$

where $\Phi^{l_1 l_2 l}$ is a 'rotational invariant'.

Use of (11.2.7) in preference to (11.2.1) does not help in resolving the problem of slow convergence in expansions such as (11.2.6), but it does have some advantages, particularly in the manipulation of Fourier transforms. We shall use the notation $\hat{g}(1,2) \equiv \hat{g}(\mathbf{k}, \mathbf{\Omega}_1, \mathbf{\Omega}_2)$ to denote a Fourier transform with respect to $\mathbf{R}_{12}$, i.e.

$$\hat{g}(\mathbf{k}, \mathbf{\Omega}_1, \mathbf{\Omega}_2) = \int g(\mathbf{R}_{12}, \mathbf{\Omega}_1, \mathbf{\Omega}_2) \exp(-i\mathbf{k} \cdot \mathbf{R}_{12}) \, d\mathbf{R}_{12} \qquad (11.2.10)$$

Then $\hat{g}(1,2)$ can be written in terms of laboratory-frame harmonics as

$$\hat{g}(1,2) = \sum_{l_1} \sum_{l_2} \sum_{l} g(l_1 l_2 l; k) \sum_{m_1} \sum_{m_2} \sum_{m} C(l_1 l_2 l; m_1 m_2 m)$$
$$\times Y_{l_1 m_1}(\mathbf{\Omega}_1) Y_{l_2 m_2}(\mathbf{\Omega}_2) Y_{lm}^*(\mathbf{\Omega}_k) \qquad (11.2.11)$$

where $\mathbf{\Omega}_k$ is the orientation of $\mathbf{k}$ in the laboratory frame. The reason that this expansion and the corresponding expansions of $\hat{h}(1,2)$ and $\hat{c}(1,2)$ are so useful is the fact that the coefficients $g(l_1 l_2 l; k)$ and $g(l_1 l_2 l; R)$ are related by a generalised Fourier or Hankel transform, i.e.

$$g(l_1 l_2 l; k) = 4\pi i^l \int_0^\infty j_l(kR) g(l_1 l_2 l; R) R^2 \, dR \qquad (11.2.12)$$

where $j_l(\cdots)$ is the spherical Bessel function of order l. No equivalent simplification is found in the case of the intermolecular-frame expansion. We shall not give a general proof of (11.2.12), since in this book we are concerned only with $l = 0$ and $l = 2$. The case when $l = 0$ corresponds to the usual Fourier transform of a spherically symmetric function; the case when $l = 2$ is considered in detail in Section 11.4.

Expansions of $g(1,2)$ and other pair functions along the lines of (11.2.1) and (11.2.7) have been applied most successfully in the theory of polar fluids, as we shall see in Sections 11.5 and 11.6.

11.3   SITE–SITE DISTRIBUTION FUNCTIONS

When an interaction-site model is used to represent the intermolecular potential the natural way to describe the structure of the fluid is in terms of site–site distribution functions. If the coordinates of site α on molecule i are denoted by $\mathbf{r}_{i\alpha}$ and those of site β on molecule j ($j \neq i$) by $\mathbf{r}_{j\beta}$, then the site–site pair distribution function $g_{\alpha\beta}(r)$ is defined in a manner similar to (2.5.15):

$$\rho g_{\alpha\beta}(r) = \left\langle \frac{1}{N} \sum_{i=1}^{N} \sum_{j \neq i}^{N} \delta(\mathbf{r} + \mathbf{r}_{2\beta} - \mathbf{r}_{1\alpha}) \right\rangle$$
$$= \langle (N-1)\delta(\mathbf{r} + \mathbf{r}_{2\beta} - \mathbf{r}_{1\alpha}) \rangle \qquad (11.3.1)$$

where N is the number of molecules; the corresponding site–site pair correlation function is defined as $h_{\alpha\beta}(r) = g_{\alpha\beta}(r) - 1$. The site–site distribution functions are, of course, of interest in a wider context than that of interaction–site models. For any real molecular fluid the most important site–site distribution functions are those that describe the distribution of atomic sites.

The definition (11.3.1) can be used to relate the site–site distribution functions to the molecular pair distribution function $g(1, 2)$. Let $\boldsymbol{\ell}_{i\alpha}$ be the vector displacement of site α in molecule i from the molecular centre $\mathbf{R}_i$, i.e.

$$\boldsymbol{\ell}_{i\alpha} = \mathbf{r}_{i\alpha} - \mathbf{R}_i \tag{11.3.2}$$

Then $g_{\alpha\beta}(r)$ is given by the integral of $g(1, 2)$ over all coordinates subject to the constraint that the vector separation of sites α, β is equal to $\mathbf{r}$:

$$
\begin{aligned}
g_{\alpha\beta}(\mathbf{r}) &= \frac{1}{\Omega^2} \iiiint d\mathbf{R}_1\, d\mathbf{R}_2\, d\boldsymbol{\Omega}_1\, d\boldsymbol{\Omega}_2\, g(1, 2) \\
&\quad \times \delta\big[\mathbf{R}_1 + \boldsymbol{\ell}_{1\alpha}(\boldsymbol{\Omega}_1)\big]\delta\big[\mathbf{R}_2 + \boldsymbol{\ell}_{2\beta}(\boldsymbol{\Omega}_2) - \mathbf{r}\big] \\
&= \frac{1}{\Omega^2} \iiint d\mathbf{R}_{12}\, d\boldsymbol{\Omega}_1\, d\boldsymbol{\Omega}_2\, g(1, 2) \\
&\quad \times \delta\big[\mathbf{R}_{12} + \boldsymbol{\ell}_{2\beta}(\boldsymbol{\Omega}_2) - \boldsymbol{\ell}_{1\alpha}(\boldsymbol{\Omega}_1) - \mathbf{r}\big]
\end{aligned}
\tag{11.3.3}
$$

It follows from (11.3.3) that the Fourier transform of $g_{\alpha\beta}(\mathbf{r})$ with respect to $\mathbf{r}$ is

$$
\begin{aligned}
\hat{g}_{\alpha\beta}(\mathbf{k}) &= \frac{1}{\Omega^2} \iiiint d\mathbf{R}_{12}\, d\boldsymbol{\Omega}_1\, d\boldsymbol{\Omega}_2\, g(1, 2) \\
&\quad \times \delta\big[\mathbf{R}_{12} + \boldsymbol{\ell}_{2\beta}(\boldsymbol{\Omega}_2) - \boldsymbol{\ell}_{1\alpha}(\boldsymbol{\Omega}_1) - \mathbf{r}\big] \exp\left(-i\mathbf{k} \cdot \mathbf{r}\right) d\mathbf{r} \\
&= \frac{1}{\Omega^2} \iiint d\mathbf{R}_{12}\, d\boldsymbol{\Omega}_1\, d\boldsymbol{\Omega}_2\, g(1, 2) \exp\left(-i\mathbf{k} \cdot \mathbf{R}_{12}\right) \\
&\quad \times \exp\left[-i\mathbf{k} \cdot \boldsymbol{\ell}_{2\beta}(\boldsymbol{\Omega}_2)\right] \exp\left[i\mathbf{k} \cdot \boldsymbol{\ell}_{1\alpha}(\boldsymbol{\Omega}_1)\right] \\
&= \big\langle \hat{g}(1, 2) \exp\left[-i\mathbf{k} \cdot \boldsymbol{\ell}_{2\beta}(\boldsymbol{\Omega}_2)\right] \exp\left[i\mathbf{k} \cdot \boldsymbol{\ell}_{1\alpha}(\boldsymbol{\Omega}_1)\right] \big\rangle_{\boldsymbol{\Omega}_1\boldsymbol{\Omega}_2}
\end{aligned}
\tag{11.3.4}
$$

where $\hat{g}(1, 2)$ is defined by (11.2.10). There is an analogous expression for $\hat{h}_{\alpha\beta}(\mathbf{k})$ in terms of $h(1, 2)$.

The site–site distribution functions have a simple physical interpretation. They are also directly related to the structure factors measured in X-ray and neutron-scattering experiments. On the other hand, the integrations in (11.3.3) involve an irretrievable loss of information, and $g(1, 2)$ cannot be reconstructed exactly from any finite set of site–site distribution functions.

Many of the quantities that are expressible as integrals over $g(1, 2)$ may also be written in terms of site–site distribution functions. For example, if the intermolecular potential is of the interaction–site form and the site–site potentials are spherically symmetric, the excess internal energy is given by

$$\frac{U^{\mathrm{ex}}}{N} = 2\pi\rho \sum_{\alpha} \sum_{\beta} \int_0^{\infty} v_{\alpha\beta}(r) g_{\alpha\beta}(r) r^2\, dr \tag{11.3.5}$$

Equation (11.3.5) is a straightforward generalisation of (2.5.20) and can be derived by the same, intuitive approach discussed in connection with the earlier result. The generalisation of the virial equation (2.5.22) is more complicated and knowledge of all site-site distribution functions and pair potentials is not sufficient to determine the pressure. The equation of state can, however, be determined by integration of the compressibility equation (11.1.11). Because the choice of molecular centre is arbitrary, and need not be the same for each molecule, (11.1.11) can be written as

$$\rho k_B T \chi_T = 1 + \rho \int [g_{\alpha\beta}(r) - 1]d\mathbf{r} = 1 + \rho \hat{h}_{\alpha\beta}(0) \tag{11.3.6}$$

where α, β refer to any pair of sites. Finally, the angular correlation parameters G_l defined by (11.1.8) can be expressed[5] as integrals over combinations of the functions $h_{\alpha\beta}(r)$. In the case of a heteronuclear but non-polar molecule with atomic sites α and β the result for G_1 is[6]

$$G_1 = -\frac{\rho}{2L^2} \int_0^\infty r^2 \Delta h(r) d\mathbf{r} \tag{11.3.7}$$

where L is the bond length and

$$\Delta h(r) = h_{\alpha\alpha}(r) + h_{\beta\beta}(r) - 2h_{\alpha\beta}(r) \tag{11.3.8}$$

If $\hat{h}_{\alpha\beta}(k)$ is expanded in powers of k in the form

$$\hat{h}_{\alpha\beta}(k) = 4\pi \int_0^\infty h_{\alpha\beta}(r)\frac{\sin kr}{kr} r^2\, dr = \hat{h}_{\alpha\beta}(0) + h_{\alpha\beta}^{(2)} k^2 + \cdots \tag{11.3.9}$$

we find that G_1 is proportional to the coefficient of k^2 in the small-k expansion of $\Delta \hat{h}(k)$:

$$G_1 = \frac{3\rho}{L^2} \Delta h^{(2)} \tag{11.3.10}$$

Similarly, G_l for $l > 1$ can be written in terms of the higher-order coefficients $\Delta h^{(n)}$. The example given is somewhat artificial, since any real heteronuclear molecule will have a dipole moment; in that case (11.3.7) is no longer correct. Nonetheless, it serves to illustrate the general form of the results, and we shall see in Section 11.5 how (11.3.7) can be recovered from the expression appropriate to polar molecules. If the molecule is homonuclear, all site–site distribution functions are the same and G_1 vanishes, as it must do on grounds of symmetry.

Information on the atom–atom distribution functions of real molecules is gained experimentally from the analysis of radiation scattering experiments. Consider first the case of a homonuclear diatomic. Let $\mathbf{u}_i$ be a unit vector along the internuclear axis of molecule i. Then the coordinates of atoms α, β relative to the centre of mass $\mathbf{R}_i$ are

$$\mathbf{r}_{i\alpha} = \mathbf{R}_i + \frac{1}{2}\mathbf{u}_i L, \quad \mathbf{r}_{i\beta} = \mathbf{R}_i - \frac{1}{2}\mathbf{u}_i L \tag{11.3.11}$$

We define the Fourier components of the atomic density as

$$\rho_{\mathbf{k}} = \sum_{i=1}^{N} \left[\exp\left(-i\mathbf{k} \cdot \mathbf{r}_{i\alpha}\right) + \exp\left(-i\mathbf{k} \cdot \mathbf{r}_{i\beta}\right) \right] \tag{11.3.12}$$

and the molecular structure factor as

$$S(k) = \left\langle \frac{1}{4N} \rho_{\mathbf{k}} \rho_{-\mathbf{k}} \right\rangle \tag{11.3.13}$$

The factor $\frac{1}{4}$ is included in order to make the definition of $S(k)$ reduce to that of an atomic fluid in the limit $L \to 0$. The statistical average in (11.3.13) may be rewritten in terms of either the atomic or molecular pair distribution functions. In the first case, by exploiting the fact that atoms α, β in each molecule play equivalent roles, we can write

$$\left\langle \frac{1}{4N} \rho_{\mathbf{k}} \rho_{-\mathbf{k}} \right\rangle = \frac{1}{2} + \frac{1}{2N} \sum_{i=1}^{N} \langle \cos\left(\mathbf{k} \cdot \mathbf{u}_i L\right) \rangle_{\Omega_i}$$

$$+ \left\langle \frac{1}{4N} \sum_{i=1}^{N} \sum_{j \neq i}^{N} \exp[-i\mathbf{k} \cdot (\mathbf{r}_{j\beta} - \mathbf{r}_{i\alpha})] \right\rangle \tag{11.3.14}$$

The second term on the right-hand side involves an average over angles alone and the third term can be related to any of the four identical distribution functions $g_{\alpha\beta}(r)$ via the definition (11.3.1). Thus

$$S(k) = S_{\text{intra}}(k) + S_{\text{inter}}(k) \tag{11.3.15}$$

The first term on the right-hand side of (11.3.15) is the intramolecular contribution:

$$S_{\text{intra}}(k) = \frac{1}{2}\left(1 + \langle \cos \mathbf{k} \cdot \mathbf{u}_i L \rangle_{\Omega_i}\right) = \frac{1}{2}\left[1 + j_0(kL)\right] \tag{11.3.16}$$

where $j_0(x) = x^{-1} \sin x$. The intermolecular part is given by

$$S_{\text{inter}}(k) = \rho \int h_{\alpha\beta}(r) \exp\left(-i\mathbf{k} \cdot \mathbf{r}\right) d\mathbf{r} = S_{\alpha\beta}(k) - 1 \tag{11.3.17}$$

where $S_{\alpha\beta}(k)$ is the atomic structure factor and a physically unimportant term in $\delta(\mathbf{k})$ has been omitted. The total intensity of scattered radiation at a given value of k is proportional to the structure factor (11.3.15); this can be inverted to yield the atomic pair distribution function if the intramolecular part is first removed.[7]

In order to relate $S(k)$ to the molecular pair distribution function we start from the definition (11.3.13) and proceed as follows:

$$S(k) = \left\langle \frac{1}{4N} \rho_{\mathbf{k}} \rho_{-\mathbf{k}} \right\rangle$$

$$= \left\langle \frac{1}{N} \sum_{i=1}^{N} \sum_{j=1}^{N} \exp\left(-i\mathbf{k} \cdot \mathbf{R}_{ij}\right) \cos\left(\frac{1}{2}\mathbf{k} \cdot \mathbf{u}_i L\right) \cos\left(\frac{1}{2}\mathbf{k} \cdot \mathbf{u}_j L\right) \right\rangle$$

$$= \frac{1}{2}\left[1 + j_0(kL)\right]$$

$$+ \left\langle \frac{1}{N} \sum_{i=1}^{N} \sum_{j \neq i}^{N} \exp\left(-i\mathbf{k} \cdot \mathbf{R}_{ij}\right) \cos\left(\frac{1}{2}\mathbf{k} \cdot \mathbf{u}_i L\right) \cos\left(\frac{1}{2}\mathbf{k} \cdot \mathbf{u}_j L\right) \right\rangle$$

$$= S_{\text{intra}}(k) + \frac{\rho}{\Omega^2} \iiint [g(1,2) - 1] \exp\left(-i\mathbf{k} \cdot \mathbf{R}_{12}\right)$$

$$\times \cos\left(\frac{1}{2}\mathbf{k} \cdot \mathbf{u}_1 L\right) \cos\left(\frac{1}{2}\mathbf{k} \cdot \mathbf{u}_2 L\right) d\mathbf{R}_{12} \, d\Omega_1 \, d\Omega_2 \qquad (11.3.18)$$

Equation (11.3.18) is an exact relation between $S(k)$ and $g(1, 2)$. Comparison with (11.3.4) shows that the second term on the right-hand side is $\hat{h}_{\alpha\beta}(k)$; this can also be deduced from inspection of Eqs. (11.3.15)–(11.3.17). A more tractable expression is obtained by replacing $g(1, 2)$ by its spherical harmonic expansion (11.2.1). The structure factor can then be written as

$$S(k) = S_{\text{intra}}(k) + f(k)\left[S_c(k) - 1\right] + S_{\text{aniso}}(k) \qquad (11.3.19)$$

where

$$f(k) = \left\langle \cos\left(\frac{1}{2}\mathbf{k} \cdot \mathbf{u}_1 L\right) \cos\left(\frac{1}{2}\mathbf{k} \cdot \mathbf{u}_2 L\right) \right\rangle_{\Omega_1 \Omega_2} = \left[j_0\left(\frac{1}{2}kL\right)\right]^2$$
$$(11.3.20)$$

and $S_c(k)$ is the Fourier transform of the centres distribution function $g_c(r)$. The term $S_{\text{aniso}}(k)$ in (11.3.19) represents the contribution to $S(k)$ from the angle-dependent terms in $g(1, 2)$, i.e. from all spherical harmonics beyond $(l_1, l_2, m) = (0, 0, 0)$. If the intermolecular potential is only weakly anisotropic, $S_{\text{aniso}}(k)$ will be small. It then follows from (11.3.15), (11.3.17) and (11.3.19) that

$$S_{\alpha\beta}(k) \approx 1 + f(k)\left[S_c(k) - 1\right] \qquad (11.3.21)$$

Equation (11.3.21) represents the 'free rotation' approximation. This can be expected to work well only when the intermolecular potential is very weakly anisotropic, as in the case of liquid nitrogen, for example. At the same time, even in the absence of strong orientational correlations, the modulating role of the function $f(k)$ means that the intermolecular contribution to $S(k)$ will differ from the structure factor of an atomic fluid. This is evident in Figure 11.1, which shows the results of X-ray scattering experiments on liquid nitrogen. Although the function $S_c(k)$ cannot usually be determined experimentally,[9] the evidence from computer simulations[10] is that for small molecules it has a strongly oscillatory character and can be well fitted by the structure factor

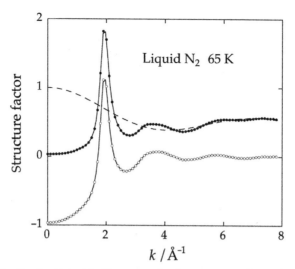

FIGURE 11.1 Results obtained by X-ray scattering for the structure factor of liquid nitrogen near its triple point. Filled circles: $S(k)$; open circles: $S_{inter}(k)$; dashes: $S_{intra}(k)$. *Redrawn with permission from Ref. 8 © 1980 American Institute of Physics.*

of an atomic system. By contrast, as comparison of Figures 3.2 (or 5.1) and 11.1 reveals, the first peak in the molecular structure factor is significantly weaker and the later oscillations are more strongly damped than in the case of a typical atomic fluid. Note also that beyond the first peak the behaviour of the molecular structure factor is dominated by the intramolecular term. The free rotation approximation becomes exact in the limit $k \to 0$ because the cosine terms in (11.3.18) all approach unity. Thus

$$\rho k_B T \chi_T = \lim_{k \to 0} S(k) = 1 + \rho \int [g_c(R) - 1]\, d\mathbf{R} \tag{11.3.22}$$

which is the same result as in (11.1.11).

For heteronuclear molecules there is normally little value in defining a structure factor through a formula analogous to (11.3.13). It is more useful instead to focus attention on those combinations of atomic structure factors that are experimentally accessible. In the case of neutron scattering the measured structure factor can again be written in the form of (11.3.15), but now

$$\left(\sum_\alpha b_\alpha\right)^2 S_{intra}^N(k) = \sum_\alpha b_\alpha^2 + \sum_\alpha \sum_{\beta \neq \alpha} b_\alpha b_\beta\, j_0(kL_{\alpha\beta}) \tag{11.3.23}$$

$$\left(\sum_\alpha b_\alpha\right)^2 S_{inter}^N(k) = \rho \sum_\alpha \sum_\beta b_\alpha b_\beta \int [g_{\alpha\beta}(r) - 1]\exp(-i\mathbf{k}\cdot\mathbf{r})d\mathbf{r}$$
$$= \sum_\alpha \sum_\beta b_\alpha b_\beta [S_{\alpha\beta}(k) - 1] \tag{11.3.24}$$

where the sums run over all nuclei in the molecule, b_α is the coherent neutron-scattering length of nucleus α and $L_{\alpha\beta}$ is the separation of nuclei α, β. These expressions reduce to (11.3.16) and (11.3.17) for a diatomic molecule with $b_\alpha = b_\beta$. After removal of the intramolecular term, Fourier transformation yields a weighted sum of atomic pair distribution functions of the form

$$g^N(r) = \left(\sum_\alpha b_\alpha\right)^{-2} \sum_\alpha \sum_\beta b_\alpha b_\beta g_{\alpha\beta}(r) \qquad (11.3.25)$$

Isotopic substitution makes it possible to vary the weights with which the different $g_{\alpha\beta}(r)$ contribute to $g^N(r)$ and hence, in favourable cases, to determine some or all of the individual atom–atom distribution functions.

Formulae similar to (11.3.23) apply also to X-ray scattering, the only difference being that the nuclear scattering lengths are replaced by the atomic form factors (see Section 4.1). Since the form factors are functions of k, the weighted distribution function $g^X(r)$ obtained by Fourier transformation of the measured structure factor $S^X(k)$ is not a linear combination of the functions $g_{\alpha\beta}(r)$, but for large atoms the error introduced by ignoring this fact is small.

In Figure 11.2 we show some results obtained by X-ray scattering for the carbon–carbon distribution function $g_{CC}(r)$ in liquid ethylene near its triple point. Although ethylene is a polyatomic molecule, $g_{CC}(r)$ resembles the pair distribution function for diatomics, as seen in both simulations and experiments. However, the main peak is appreciably weaker than in argon-like liquids and there is a pronounced shoulder on the large-r side. Both these features are consequences of the interference between inter and intramolecular correlations.

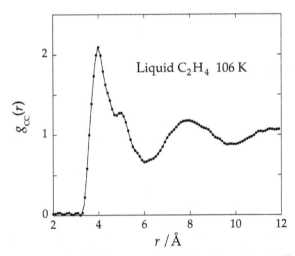

FIGURE 11.2 Results obtained by X-ray scattering for the carbon–carbon distribution function in liquid ethylene. *Redrawn with permission from Ref. 11 © 1981 American Institute of Physics.*

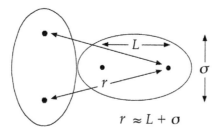

FIGURE 11.3 The T-shaped configuration for a pair of homonuclear diatomics.

Simple geometry suggests that shoulders might be seen at combinations of distances such that $r_{\alpha\gamma} \approx |\sigma_{\alpha\beta} \pm L_{\beta\gamma}|$, where σ is an atomic diameter and $L_{\beta\gamma}$ is a bond length, but they are often so smooth as to be undetectable. In the case of fused hard-sphere models of the intermolecular potential the shoulders appear as cusps in the site–site distribution functions, i.e. as discontinuities in the derivative of $g_{\alpha\beta}(r)$ with respect to r. The shoulder seen in Figure 11.2 is associated with 'T-shaped' configurations of the type pictured in Figure 11.3. This particular feature is enhanced for molecules having a large quadrupole moment, such as bromine,[12] since the quadrupolar interaction strongly favours the T-configuration.

11.4 CORRELATION FUNCTION EXPANSIONS FOR SIMPLE POLAR FLUIDS

In the simplest models of a polar fluid the intermolecular potential can be written as the sum of a small number of spherical harmonic components. The prospects for success of theories are therefore greater than in situations where the potential contains an infinite number of harmonics and the series expansions are only slowly convergent, as is true, for example, in the case of Lennard-Jones diatomics.[4] In this section we discuss some of the general questions that arise in attempts to treat polar fluids in this way.

Consider a polar fluid for which the intermolecular potential is the same as in (1.2.4), but which we rewrite here as

$$v(1,2) = v_0(R) - \frac{\mu^2}{R^3} D(1,2) \qquad (11.4.1)$$

with

$$D(1,2) = 3(\mathbf{u}_1 \cdot \mathbf{s})(\mathbf{u}_2 \cdot \mathbf{s}) - \mathbf{u}_1 \cdot \mathbf{u}_2 \qquad (11.4.2)$$

where $R \equiv |\mathbf{R}_{12}|$, $\mathbf{s}$ is a unit vector in the direction of $\mathbf{R}_{12}$, $\mathbf{u}_i$ is a unit vector parallel to the dipole moment of molecule i, $v_0(R)$ is assumed to be spherically symmetric and the angle-dependent terms represent the ideal dipole–dipole interaction. It was first shown by Wertheim[13] and subsequently elaborated by others[14] that an adequate description of the static properties of such a fluid can be obtained by working with a basis set consisting of only three functions: $S(1,2) = 1, \Delta(1,2) = \mathbf{u}_1 \cdot \mathbf{u}_2$ and $D(1,2)$, defined above. The solution for $h(1,2)$ is therefore assumed to be of the form

$$h(1,2) = h_S(R) + h_\Delta(R)\Delta(1,2) + h_D(R)D(1,2) \qquad (11.4.3)$$

On multiplying through (11.4.3) successively by S, Δ and D and integrating over angles we find that the projections $h_S(R), h_\Delta(R)$ and $h_D(R)$ are given by

$$h_S(R) = \langle h(1,2)\rangle_{\Omega_1\Omega_2}$$
$$h_\Delta(R) = 3\langle h(1,2)\Delta(1,2)\rangle_{\Omega_1\Omega_2} \qquad (11.4.4)$$
$$h_D(R) = \frac{3}{2}\langle h(1,2)D(1,2)\rangle_{\Omega_1\Omega_2}$$

Equation (11.4.3) is equivalent to an expansion in laboratory-frame harmonics, since the functions Δ and D are the same, respectively, as the rotational invariants Φ^{110} and Φ^{112} introduced in (11.2.9).

The direct correlation function $c(1,2)$ can be treated in similar fashion to $h(1,2)$. We therefore write

$$c(1,2) = c_S(R) + c_\Delta(R)\Delta(1,2) + c_D(R)D(1,2) \qquad (11.4.5)$$

and introduce both (11.4.3) and (11.4.5) into the molecular Ornstein–Zernike relation (11.1.4). After taking Fourier transforms we find that

$$\hat{h}(1,2) = \hat{c}(1,2) + \rho\langle \hat{c}(1,3)\hat{h}(3,2)\rangle_{\Omega_3} \qquad (11.4.6)$$

where, for example:

$$\hat{h}(1,2) = \hat{h}_S(k) + \hat{h}_\Delta(k)\Delta(1,2) + \int h_D(R)D(1,2)\exp\left(-i\mathbf{k}\cdot\mathbf{R}\right)d\mathbf{R} \qquad (11.4.7)$$

The term in D can be transformed by taking the direction of $\mathbf{k}$ as the z-axis and making the substitution $\mathbf{s} = (\sin\theta\cos\phi, \sin\theta\sin\phi, \cos\theta)$. Two integrations by parts show that

$$\int_{-1}^{1}\int_0^{2\pi}(\mathbf{u}_1\cdot\mathbf{s})(\mathbf{u}_2\cdot\mathbf{s})\exp\left(-ikR\cos\theta\right)d\phi\,d(\cos\theta)$$
$$= -4\pi R^2\left(3u_{1z}u_{2z}j_2(kR) - \mathbf{u}_1\cdot\mathbf{u}_2[j_0(kR) + j_2(kR)]\right) \qquad (11.4.8)$$

where $j_2(x) = 3x^{-3}\sin x - 3x^{-2}\cos x - x^{-1}\sin x$. Thus

$$\int h_D(R)D(1,2)\exp\left(-i\mathbf{k}\cdot\mathbf{R}\right)d\mathbf{R} = D_k(1,2)\bar{h}_D(k) \qquad (11.4.9)$$

with

$$D_k(1,2) = 3u_{1z}u_{2z} - \mathbf{u}_1\cdot\mathbf{u}_2 = \frac{3(\mathbf{u}_1\cdot\mathbf{k})(\mathbf{u}_2\cdot\mathbf{k})}{k^2} - \mathbf{u}_1\cdot\mathbf{u}_2 \qquad (11.4.10)$$

and the Hankel transform $\bar{h}_D(k)$ is

$$\bar{h}_D(k) = -4\pi \int_0^\infty j_2(kR)h_D(R)R^2\,dR \qquad (11.4.11)$$

Equation (11.4.9) is a particular case of the general result contained in (11.2.12); the transform of $c_D(R)D(1,2)$ is handled in the same way.

In order to summarise the effect of the integration over angles in (11.4.6) we define the angular convolution of two functions A, B as

$$A*B = B*A = \frac{1}{\Omega}\int A(1,3)B(3,2)d\mathbf{\Omega}_3 \equiv \langle A(1,3)B(3,2)\rangle_{\mathbf{\Omega}_3} \qquad (11.4.12)$$

For the functions of interest here the 'multiplication' rules shown in Table 11.1 are easily established. We see from the table that the functions S, Δ and D_k form a closed set under the operation (11.4.12) in the sense that convolution of any two functions yields only a function in the same set (or zero). The practical significance of this result is the fact that if $h(1,2)$ is assumed to be of the form (11.4.3), then $c(1,2)$ is necessarily given by (11.4.5), and vice versa. A closure of the Ornstein–Zernike relation is still required. However, if this does not generate any new harmonics, (11.4.3) and (11.4.5) together form a self-consistent approximation, to which a solution can be found either analytically (as in the MSA, discussed in Section 11.6) or numerically.

At large R, $c(1,2)$ behaves as $-\beta v(1,2)$. Hence $c_D(R)$ must be long ranged, decaying asymptotically as R^{-3}. It turns out, as we shall see in Section 11.5, that $h_D(R)$ also decays as R^{-3}, the strength of the long-range part being related to the dielectric constant of the fluid, but the other projections of $h(1,2)$ and $c(1,2)$ are all short ranged. The slow decay of $h_D(R)$ and $c_D(R)$ creates difficulties in numerical calculations. It is therefore convenient to introduce two short-range, auxiliary functions $h_D^0(R)$ and $c_D^0(R)$. These are defined in terms, respectively, of $h_D(R)$ and $c_D(R)$ in such a way as to remove the long-range parts. Thus

$$h_D^0(R) = h_D(R) - 3\int_R^\infty \frac{h_D(R')}{R'}\,dR' \qquad (11.4.13)$$

with an analogous definition of $c_D^0(R)$; we see from (11.4.13) that $h_D^0(R)$ vanishes for R in the range where $h_D(R)$ has reached its asymptotic value.

TABLE 11.1 Rules for the evaluation of angular convolutions of the functions S, D and D_k.

	S	Δ	D_k
S	S	0	0
Δ	0	$\Delta/3$	$D_k/3$
D_k	0	$D_k/3$	$(D_k + 2\Delta)/3$

The inverse of (11.4.13) is

$$h_D(R) = h_D^0(R) - \frac{3}{R^3} \int_0^R h_D^0(R') R'^2 \, dR' \tag{11.4.14}$$

which can be checked by first differentiating (11.4.14) with respect to R and then integrating from R to $R = \infty$ (where both $h_D(R)$ and $h_D^0(R)$ are zero); this leads back to (11.4.13). Equation (11.4.14) shows that $h_D(R)$ behaves asymptotically as

$$\lim_{R \to \infty} h_D(R) = -\frac{3}{4\pi R^3} \lim_{k \to 0} \hat{h}_D^0(k) \tag{11.4.15}$$

The short-range functions $h_D^0(R)$ and $c_D^0(R)$ play an important part in the analytical solution of the MSA for dipolar hard spheres.

We have seen that use of the approximation (11.4.3) has some attractive mathematical features. The solution is of physical interest, however, only because the projections $h_S(R)$, $h_A(R)$ and $h_D(R)$ contain between them all the information needed to calculate both the thermodynamic and static dielectric properties of the fluid. We postpone discussion of the difficult problem of dielectric behaviour until the next section, but expressions for thermodynamic properties are easily derived. If $v_0(R)$ in (11.4.1) is the hard-sphere potential, the excess internal energy is determined solely by the dipole–dipole interaction and (11.1.9) becomes

$$\frac{U^{ex}}{N} = -2\pi\rho \int_0^\infty \frac{\mu^2}{R_{12}} \langle D(1,2) g(1,2) \rangle_{\Omega_1 \Omega_2} \, dR_{12}$$

$$= -\frac{4\pi\mu^2\rho}{3} \int_0^\infty \frac{h_D(R)}{R} \, dR \tag{11.4.16}$$

where we have used the definition of $h_D(R)$ in (11.4.4) and the fact that the angle average of $D(1,2)$ is zero. If $v_0(R)$ is the Lennard-Jones potential or some other spherically symmetric but continuous interaction, there will be a further contribution to U^{ex} that can be expressed as an integral over $h_S(R)$. Similarly, (11.1.10) can be used to relate the equation of state to the projections $h_S(R)$ and $h_D(R)$. Thermodynamic properties are therefore not explicitly dependent on $h_A(R)$.

Examples of $h_A(R)$ and $h_D(R)$ for the dipolar hard-sphere fluid are shown in Figure 11.4. For the state point concerned, corresponding to a static dielectric constant $\epsilon \approx 30$, the curves retain a pronounced oscillatory character over a range of 3–4 molecular diameters. The structure in $h_A(R)$ and $h_D(R)$ disappears as the dipole moment is reduced, but $h_S(R)$ (not shown) is much less sensitive to the value of ϵ and bears a strong resemblance to the pair correlation function of a fluid of non-polar hard spheres. The structure seen in the Δ and D projections is also depressed by addition of a quadrupole moment, as we discuss again in Section 11.6.

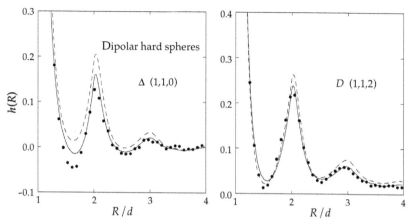

FIGURE 11.4 Projections $h_\Delta(R)$ (left) and $h_D(R)$ (right) of $h(1,2)$ for a fluid of dipolar hard spheres at $\rho d^3 = 0.80, \beta\mu^2/d^3 = 2$. The points are Monte Carlo results and the curves are calculated from the LHNC (dashes) and RHNC (continuous lines) approximations discussed in Section 11.6. *Redrawn with permission from Ref. 15 © American Institute of Physics.*

11.5 THE STATIC DIELECTRIC CONSTANT

Our goal in this section is to obtain molecular expressions for the static dielectric constant. We show, in particular, that ϵ is related to the long-wavelength behaviour of each of the functions $\hat{h}_\Delta(k)$ and $\bar{h}_D(k)$ introduced in the previous section.[16] By suitably combining the two results it is also possible to express ϵ in terms of site–site distribution functions.[17]

Consider a sample of dielectric material (a polar fluid) placed in an external electric field. Let $\mathbf{E}(\mathbf{R},t)$ be the field at time t at a point $\mathbf{R}$ inside the sample (the Maxwell field), let $\mathbf{P}(\mathbf{R},t)$ be the polarisation induced in the sample and let $\mathbf{E}^0(\mathbf{R},t)$ be the field that would exist at the same point if the sample were removed (the external field). The polarisation is related to the Maxwell field by

$$\mathbf{P}(\mathbf{R},t) = \int d\mathbf{R}' \int_{-\infty}^{t} \chi(\mathbf{R}-\mathbf{R}',t-t') \cdot \mathbf{E}(\mathbf{R}',t')dt' \qquad (11.5.1)$$

where the tensor $\chi(\mathbf{R},t)$ is an after-effect function of the type introduced in Section 7.6. A Fourier–Laplace transform of (11.5.1) (with z on the real axis) gives

$$\hat{\mathbf{P}}(\mathbf{k},\omega) = \chi(\mathbf{k},\omega) \cdot \hat{\mathbf{E}}(\mathbf{k},\omega) \qquad (11.5.2)$$

where the susceptibility $\chi(\mathbf{k},\omega)$ is related to the dielectric permittivity $\epsilon(\mathbf{k},\omega)$ by

$$\chi(\mathbf{k},\omega) = \frac{1}{4\pi}\big[\epsilon(\mathbf{k},\omega) - \mathbf{I}\big] \qquad (11.5.3)$$

The polarisation is also related to the external field via a second susceptibility, $\chi^0(\mathbf{k}, \omega)$:

$$\hat{\mathbf{P}}(\mathbf{k}, \omega) = \chi^0(\mathbf{k}, \omega) \cdot \hat{\mathbf{E}}^0(\mathbf{k}, \omega) \qquad (11.5.4)$$

The external field and Maxwell field will not, in general, be the same because the polarisation of the sample makes a contribution to the Maxwell field. The relation between the two fields, and hence also that between χ and χ^0, is dependent on sample geometry. We shall assume that the system is infinite, in which case the relation between $\mathbf{E}$ and $\mathbf{E}^0$ is

$$\mathbf{E}(\mathbf{R}, t) = \mathbf{E}^0(\mathbf{R}, t) + \int \mathbf{T}(\mathbf{R} - \mathbf{R}') \cdot \mathbf{P}(\mathbf{R}', t) \mathrm{d}\mathbf{R}' \qquad (11.5.5)$$

where $\mathbf{T}(\mathbf{R})$ is the dipole–dipole interaction tensor defined by (1.2.5). Integrals involving the dipole–dipole tensor must be handled with care, since $\mathbf{T}(\mathbf{R})$ has a singularity at the origin; the usual procedure is to cut-off the integrand inside a sphere of radius σ centred on the origin and take the limit $\sigma \to 0$ after integration.[18] The transform of (11.5.5) is then given by

$$\hat{\mathbf{E}}(\mathbf{k}, \omega) = \hat{\mathbf{E}}^0(\mathbf{k}, \omega) - \frac{4\pi}{k^2}\mathbf{k}\mathbf{k} \cdot \hat{\mathbf{P}}(\mathbf{k}, \omega) \qquad (11.5.6)$$

The relationship between the two susceptibilities follows immediately from consideration of (11.5.2), (11.5.4) and (11.5.6)[19]:

$$\chi^0(\mathbf{k}, \omega) = \left[\mathbf{I} + (4\pi/k^2)\mathbf{k}\mathbf{k} \cdot \chi(\mathbf{k}, \omega)\right]^{-1} \cdot \chi(\mathbf{k}, \omega) \qquad (11.5.7)$$

It is an experimental fact that the dielectric permittivity is an intensive property of the fluid, having a value that for given $\mathbf{k}$ and ω is independent of sample size and shape. The same is therefore true of the susceptibility $\chi(\mathbf{k}, \omega)$, since the two quantities are trivially linked by (11.5.3). It follows, provided the system is isotropic, that both ϵ and χ must be independent of the direction of $\mathbf{k}$. Thus, in the limit $\mathbf{k} \to 0$:

$$\lim_{\mathbf{k}\to 0} \epsilon(\mathbf{k}, \omega) = \epsilon(\omega)\mathbf{I}, \quad \lim_{\mathbf{k}\to 0} \chi(\mathbf{k}, \omega) = \chi(\omega)\mathbf{I} \qquad (11.5.8)$$

where $\epsilon(\omega)$ and $\chi(\omega)$ are scalars. On the other hand, the longitudinal (parallel to $\mathbf{k}$) and transverse (perpendicular to $\mathbf{k}$) components of $\chi^0(\mathbf{k}, \omega)$ must behave differently in the long-wavelength limit; this is inevitable, given that the relation between χ and ϵ is shape dependent. Taking the z-axis along the direction of $\mathbf{k}$, we find from (11.5.3), (11.5.7) and (11.5.8) that

$$4\pi \lim_{\mathbf{k}\to 0} \chi^0_{\alpha\alpha}(\mathbf{k}, \omega) = \epsilon(\omega) - 1, \quad \alpha = x, y$$

$$4\pi \lim_{\mathbf{k}\to 0} \chi^0_{zz}(\mathbf{k}, \omega) = \frac{\epsilon(\omega) - 1}{\epsilon(\omega)} \qquad (11.5.9)$$

and

$$4\pi \lim_{\mathbf{k} \to 0} \mathrm{Tr}\, \boldsymbol{\chi}^0(\mathbf{k}, \omega) = \frac{[\epsilon(\omega) - 1][2\epsilon(\omega) + 1]}{\epsilon(\omega)} \tag{11.5.10}$$

The statistical mechanical problem is to obtain expressions for the components of $\boldsymbol{\chi}^0$ in terms of microscopic variables. The microscopic expression for the polarisation induced by the external field is

$$\mathbf{P}(\mathbf{R}, t) = \langle \mathbf{M}(\mathbf{R}, t) \rangle_{\mathbf{E}^0} = \left\langle \mu \sum_{i=1}^{N} \mathbf{u}_i(t) \delta[\mathbf{R} - \mathbf{R}_i(t)] \right\rangle_{\mathbf{E}^0} \tag{11.5.11}$$

where $\mathbf{M}(\mathbf{R}, t)$ is the dipole-moment density, $\langle \cdots \rangle_{\mathbf{E}^0}$ denotes a statistical average in the presence of the external field and the other symbols have the same meaning as in earlier sections of this chapter. The susceptibility $\boldsymbol{\chi}^0$ can now be calculated by the methods of linear response theory described in Section 7.6. (Note that $\boldsymbol{\chi}$ cannot be treated in the same way as $\boldsymbol{\chi}^0$, because the Maxwell field is not an 'external' field in the required sense.) As an application of the general result given by (7.6.21) we find that

$$\boldsymbol{\chi}^0(\mathbf{k}, \omega) = \frac{\beta}{V} \left(\langle \mathbf{M}_{\mathbf{k}} \mathbf{M}_{-\mathbf{k}} \rangle + i\omega \int_0^\infty \langle \mathbf{M}_{\mathbf{k}}(t) \mathbf{M}_{-\mathbf{k}} \rangle \exp(i\omega t) \mathrm{d}t \right) \tag{11.5.12}$$

where the statistical averages are now computed in the absence of the field, $\mathbf{M}_{\mathbf{k}} \equiv \mathbf{M}_{\mathbf{k}}(t = 0)$ and

$$\mathbf{M}_{\mathbf{k}}(t) = \mu \sum_{i=1}^{N} \mathbf{u}_i(t) \exp\left[-i\mathbf{k} \cdot \mathbf{R}_i(t) \right] \tag{11.5.13}$$

On taking the limit $\omega \to 0$, (11.5.12) reduces to

$$\chi_{\alpha\alpha}^0(\mathbf{k}, 0) = \frac{\beta}{V} \langle M_{\mathbf{k}}^\alpha M_{-\mathbf{k}}^\alpha \rangle, \quad \alpha = x, y, z \tag{11.5.14}$$

By combining this result with (11.5.10) we find that

$$\frac{(\epsilon - 1)(2\epsilon + 1)}{9\epsilon} = g_K y \tag{11.5.15}$$

where $\epsilon \equiv \epsilon(0)$ is the static dielectric constant, y is a molecular parameter defined as

$$y = \frac{4\pi \mu^2 \rho}{9 k_B T} \tag{11.5.16}$$

and g_K, the Kirkwood 'g-factor', is given by

$$g_K = \langle |\mathbf{M}|^2 \rangle / N\mu^2 \tag{11.5.17}$$

where $\mathbf{M} \equiv \mathbf{M}_{\mathbf{k}=0}$ is the total dipole moment of the sample. Equation (11.5.17) can be rewritten, with the help of (11.4.4), as

$$
\begin{aligned}
g_K &= 1 + \langle (N-1)\mathbf{u}_1 \cdot \mathbf{u}_2 \rangle \\
&= 1 + \frac{4\pi\rho}{3} \int_0^\infty h_\Delta(R) R^2 \, \mathrm{d}R = 1 + \frac{1}{3}\rho \hat{h}_\Delta(0) \qquad (11.5.18)
\end{aligned}
$$

where $h_\Delta(R)$ is the function appropriate to an infinite system.

Equation (11.5.15) is the first of two key results of this section. It was originally derived by Kirkwood[20] via a calculation of the fluctuations in total dipole moment of a spherical region surrounded by a dielectric continuum and is commonly referred to as the Kirkwood formula. By setting $g_K = 1$ we obtain the result known as the Onsager equation; this amounts to ignoring the short-range angular correlations represented by the function $h_\Delta(R)$. The Kirkwood formula could have been obtained by working throughout in the $\omega = 0$ limit, but the frequency-dependent results are needed for the discussion of dielectric relaxation in Section 11.11.

The next task is to relate ϵ to the function $h_D(R)$. To do this we must consider separately the longitudinal and transverse components of χ^0. For the longitudinal component we find from (11.5.13) and (11.5.14) that

$$
\begin{aligned}
\chi^0_{zz}(\mathbf{k},0) &= \frac{\beta}{V} \langle M^z_{\mathbf{k}} M^z_{-\mathbf{k}} \rangle \\
&= \frac{1}{3}\mu^2 \rho\beta + \mu^2 \rho\beta \langle (N-1) u_{1z} u_{2z} \exp(-i\mathbf{k}\cdot\mathbf{R}_{12}) \rangle \\
&= \frac{1}{3}\mu^2 \rho\beta \\
&\quad + \frac{\mu^2 \rho^2 \beta}{\Omega^2} \iiint \frac{(\mathbf{k}\cdot\mathbf{u}_1)(\mathbf{k}\cdot\mathbf{u}_2)}{k^2} h(1,2) \\
&\quad \times \exp(-i\mathbf{k}\cdot\mathbf{R}_{12}) \mathrm{d}\mathbf{R}_{12} \, \mathrm{d}\Omega_1 \, \mathrm{d}\Omega_2 \\
&= \frac{1}{3}\mu^2 \rho\beta + \mu^2 \rho^2 \beta \left\langle k^{-2}(\mathbf{k}\cdot\mathbf{u}_1)(\mathbf{k}\cdot\mathbf{u}_2)\hat{h}(1,2) \right\rangle_{\Omega_1 \Omega_2} \quad (11.5.19)
\end{aligned}
$$

We now substitute for $\hat{h}(1,2)$ from (11.4.7) and evaluate the angular averages with the help of the equalities

$$
\langle (\mathbf{n}\cdot\mathbf{u}_1)(\mathbf{n}\cdot\mathbf{u}_2)(\mathbf{u}_1 \cdot \mathbf{u}_2) \rangle_{\Omega_1 \Omega_2} = \left\langle (\mathbf{n}\cdot\mathbf{u}_1)^2 (\mathbf{n}\cdot\mathbf{u}_2)^2 \right\rangle_{\Omega_1 \Omega_2} = \frac{1}{9} \quad (11.5.20)
$$

where $\mathbf{n}$ is a unit vector of fixed orientation. A simple calculation shows that

$$
\lim_{\mathbf{k}\to 0} \chi^0_{zz}(\mathbf{k},0) = \frac{1}{3}\mu^2 \rho\beta \left[1 + \frac{1}{3}\rho\hat{h}_\Delta(0) + \frac{2}{3}\rho\bar{h}_D(0) \right] \quad (11.5.21)
$$

Although we have used the approximation (11.4.9), (11.5.21) is an exact result, since the terms ignored in (11.4.3) make no contribution to the angular average in (11.5.19).

The transverse component could be treated in a similar way. It is possible, however, to take a short-cut, since we are interested only in the $\mathbf{k} \to 0$ limit. Equations (11.5.10), (11.5.15) and (11.5.18), taken together, show that the trace of the tensor $\chi^0(\mathbf{k}, \omega)$ in the long-wavelength, low-frequency limit is

$$\lim_{\mathbf{k} \to 0} \mathrm{Tr}\, \chi^0(\mathbf{k}, 0) = \mu^2 \rho \beta \left[1 + \frac{1}{3} \hat{h}_\Delta(0) \right] \tag{11.5.22}$$

As the two transverse components are equivalent, we find from (11.5.21) and (11.5.22) that

$$\lim_{\mathbf{k} \to 0} \chi^0_{xx}(\mathbf{k}, 0) = \frac{1}{3} \mu^2 \rho \beta \left[1 + \frac{1}{3} \rho \hat{h}_\Delta(0) - \frac{1}{3} \rho \bar{h}_D(0) \right] \tag{11.5.23}$$

Use of (11.5.9) leads to the second main result:

$$\frac{(\epsilon - 1)^2}{\epsilon} = 4\pi \lim_{\mathbf{k} \to 0} \left[\chi^0_{xx}(\mathbf{k}, 0) - \chi^0_{zz}(\mathbf{k}, 0) \right] = -3y\rho \bar{h}_D(0) \tag{11.5.24}$$

It can be shown[21] that the Hankel transform in (11.5.24) is also the Fourier transform of the short-range function $h^0_D(R)$ defined by (11.4.13), i.e. $\bar{h}_D(k) = \hat{h}^0_D(k)$. Equations (11.4.15) and (11.5.24) may therefore be combined to give

$$\lim_{R \to \infty} h_D(R) = \frac{(\epsilon - 1)^2}{4\pi y \rho \epsilon} \frac{1}{R^3} \tag{11.5.25}$$

This calculation demonstrates that $h(1, 2)$ is long ranged and that the long range of the correlations is responsible for the difference in behaviour of the longitudinal and transverse components of the susceptibility $\chi^{(0)}(\mathbf{k}, 0)$.

The expansion of $h(1, 2)$ in terms of the functions S, $\Delta(1, 2)$ and $D(1, 2)$ is particularly well suited to treating the type of potential model described by (11.4.1), but its range of applicability is wider than this. It can be used, in particular, to discuss the dielectric properties of linear, interaction–site molecules. Consider a diatomic molecule of bond length L with charges $\pm q$ located on atoms α, β and a dipole moment $\mu = qL$. If ℓ_α is the distance of atom α from the molecular centre, (11.3.4) shows that the Fourier transform of any of the atomic pair correlation functions may be written as

$$\hat{h}_{\alpha\beta}(k) = \left\langle \hat{h}(1, 2) \exp\left(-i\mathbf{k} \cdot \mathbf{u}_1 \ell_\alpha\right) \exp\left(i\mathbf{k} \cdot \mathbf{u}_2 \ell_\beta\right) \right\rangle_{\Omega_1 \Omega_2} \tag{11.5.26}$$

with $\ell_\alpha + \ell_\beta = L$. The plane-wave functions can be replaced by their Rayleigh expansions[22]:

$$\exp\left(-i\mathbf{k} \cdot \mathbf{r}\right) = \sum_{n=0}^{\infty} (2n + 1) i^n j_n(kr) P_n(\mathbf{k} \cdot \mathbf{r}/kr) \tag{11.5.27}$$

but since our concern is with the behaviour of $\hat{h}_{\alpha\beta}(k)$ to order k^2 it is sufficient to retain only the contributions from $n = 0$ and $n = 1$. If, in addition, we substitute for $\hat{h}(1,2)$ from (11.4.7), (11.5.26) becomes

$$
\begin{aligned}
\hat{h}_{\alpha\beta}(k) = \Big\langle \Big(& \hat{h}_S(k) + \hat{h}_\Delta(k)\mathbf{u}_1 \cdot \mathbf{u}_2 + \bar{h}_D(k)\big[3k^{-2}(\mathbf{k} \cdot \mathbf{u}_1)(\mathbf{k} \cdot \mathbf{u}_2) - \mathbf{u}_1 \cdot \mathbf{u}_2\big] \Big) \\
& \times \big[j_0(-k\ell_\alpha) + 3i j_1(-k\ell_\alpha)\mathbf{k} \cdot \mathbf{u}_1/k \big] \\
& \times \big[j_0(k\ell_\beta) + 3i j_1(k\ell_\beta)\mathbf{k} \cdot \mathbf{u}_2/k \big] \Big\rangle_{\Omega_1\Omega_2}
\end{aligned}
\tag{11.5.28}
$$

where $j_1(x) = x^{-2}\sin x - x^{-1}\cos x$.

The terms in (11.5.28) that survive the integration over angles are those of the type shown in (11.5.20). On multiplying out, integrating with the help of (11.5.20) and collecting terms we find that

$$
\hat{h}_{\alpha\beta}(k) = \hat{h}_S(k) j_0(-k\ell_\alpha) j_0(k\ell_\beta) - \big[\hat{h}_\Delta(k) + 2\bar{h}_D(k)\big] j_1(-k\ell_\alpha) j_1(k\ell_\beta)
\tag{11.5.29}
$$

The functions $\hat{h}_{\alpha\alpha}(k)$, $\hat{h}_{\beta\alpha}(k)$ and $\hat{h}_{\beta\beta}(k)$ can be expressed in a similar way. If we now expand the Bessel functions to order k^2, the result obtained for the Fourier transform of the function $\Delta h(r)$ in (11.3.8) is

$$
\Delta\hat{h}(k) = \frac{k^2 L^2}{9}\big[\hat{h}_\Delta(0) + 2\bar{h}_D(0)\big] + \mathcal{O}(k^4)
\tag{11.5.30}
$$

or, from the second relation in (11.5.9) and (11.5.21):

$$
\Delta h^{(2)} = \frac{L^2}{9\rho}\left(\frac{\epsilon - 1}{y\epsilon} - 3\right)
\tag{11.5.31}
$$

where $\Delta h^{(2)}$ is the coefficient introduced in (11.3.10). Equation (11.5.31) expresses the dielectric constant as a combination of integrals involving only the site–site distribution functions and may be rewritten as

$$
\sum_\alpha \sum_\beta q_\alpha q_\beta h^{(2)}_{\alpha\beta} = \frac{\mu^2}{9\rho}\left(\frac{\epsilon - 1}{y\epsilon} - 3\right)
\tag{11.5.32}
$$

where q_α is the charge on site α. The result in this form is not limited to diatomics: it applies to any interaction–site molecule.[23]

It is clear from (11.5.18) that $\hat{h}_\Delta(0)$ is related to the angular correlation parameter (11.3.10) by $G_1 = \frac{1}{3}\rho\hat{h}_\Delta(0)$. This is true whether or not the molecule has a dipole moment but the analysis that leads to (11.3.10) is valid only in the non-polar case. The difference between polar and non-polar molecules lies in the long-range function $h_D(R)$. The significance of $h_D(R)$ can be seen in the fact that whereas $\hat{h}_\Delta(0)$ contributes equally to the longitudinal and transverse components of the long-wavelength susceptibility $\chi^0(\mathbf{k}, 0)$, $\bar{h}_D(0)$ does not. The effect of long-range correlations is therefore suppressed by setting $\bar{h}_D(0) = 0$ in (11.5.30), which then reduces to (11.3.10).

11.6 INTEGRAL EQUATION APPROXIMATIONS FOR DIPOLAR HARD SPHERES

The expansion of $h(1,2)$ or $c(1,2)$ in terms of S, $\Delta(1,2)$ and $D(1,2)$ was first exploited by Wertheim[13] in obtaining the analytic solution to the MSA (mean spherical approximation) for dipolar hard spheres. Although the MSA is not a quantitatively satisfactory theory, Wertheim's methods have had a considerable influence on later work on simple models of polar fluids.

The groundwork for the solution has already been laid in Section 11.4. The next stage in the calculation consists in substituting for $\hat{h}(1,2)$ and $\hat{c}(1,2)$ in the Ornstein–Zernike relation (11.4.6), integrating over angles with the help of Table 11.1, and equating coefficients of S, Δ and D_k on the two sides of the equation. The terms in S separate from those in Δ and D_k to give

$$\hat{h}_S(k) = \hat{c}_S(k) + \rho\hat{c}_S(k)\hat{h}_S(k)$$

$$\hat{h}_\Delta(k) = \hat{c}_\Delta(k) + \frac{1}{3}\rho\big[\hat{c}_\Delta(k)\hat{h}_\Delta(k) + 2\bar{c}_D(k)\bar{h}_D(k)\big]$$

$$\bar{h}_D(k) = \bar{c}_D(k) + \frac{1}{3}\rho\big[\bar{c}_D(k)\bar{h}_D(k) + \bar{c}_D(k)\hat{h}_\Delta(k) + \hat{c}_\Delta(k)\bar{h}_D(k)\big]$$

$$(11.6.1)$$

The Hankel transforms in these equations are the Fourier transforms of the short-range functions $h_D^0(R)$ and $c_D^0(R)$; this fact has already been used in the derivation of (11.5.25). The inverse Fourier transforms of $\hat{h}_S(k), \hat{h}_\Delta(k)$ and $\bar{h}_D(k)$ can therefore all be written in terms of spatial convolution integrals (denoted by the symbol $\otimes$):

$$h_S(R) = c_S(R) + \rho c_S \otimes h_S$$

$$h_\Delta(R) = c_\Delta(R) + \frac{1}{3}\rho(c_\Delta \otimes h_\Delta + 2c_D^0 \otimes h_D^0)$$

$$(11.6.2)$$

$$h_D(R) = c_D(R) + \frac{1}{3}\rho(c_D^0 \otimes h_D^0 + c_D^0 \otimes h_\Delta + c_\Delta \otimes h_D^0)$$

These equations are to be solved subject to the MSA closure relations (4.5.2). For dipolar hard spheres (4.5.2) becomes

$$h(1,2) = -1, \quad R < d; \qquad c(1,2) = \frac{\beta\mu^2 D(1,2)}{R^3}, \quad R > d \qquad (11.6.3)$$

or, equivalently:

$$h_S(R) = -1, \quad R < d; \qquad h_\Delta(R) = h_D(R) = 0, \quad R < d$$

$$c_D(R) = \frac{\beta\mu^2}{R^3}, \quad R > d; \qquad c_S(R) = c_\Delta(R) = 0, \quad R > d \qquad (11.6.4)$$

It is clear from (11.6.2) and (11.6.4) that within the MSA the functions $h_S(R)$ and $c_S(R)$ are simply the solution to the PY equation for non-polar hard spheres:

the dipolar interaction has no effect on the distribution of molecular centres. The closure relations involving the projections $h_D(R)$ and $c_D(R)$ can also be written as

$$h_D^0(R) = -3K, \quad R < d; \qquad c_D^0(R) = 0, \quad R > d \tag{11.6.5}$$

where K is the dimensionless parameter defined as

$$K = \int_d^\infty \frac{h_D(R)}{R}\, dR \tag{11.6.6}$$

We now look for a linear combination of functions that causes the expressions for the Δ and D projections in (10.6.2) to become decoupled. Direct substitution shows that this is achieved by taking

$$h_+(R) = \frac{1}{3K}\left[h_D^0(R) + \frac{1}{2}h_\Delta(R) \right]$$
$$h_-(R) = \frac{1}{3K}\left[h_D^0(R) - h_\Delta(R) \right] \tag{11.6.7}$$

with analogous expressions for $c_+(R)$ and $c_-(R)$. The new functions satisfy the equations

$$h_+(R) = c_+(R) + 2K\rho c_+ \otimes h_+$$
$$h_-(R) = c_-(R) - K\rho c_- \otimes h_- \tag{11.6.8}$$

Equations (11.6.8) are to be solved subject to the closure relations $h_+(R) = h_-(R) = -1, R < d$ (this is why the factor $1/3K$ is included in (11.6.7)) and $c_+(R) = c_-(R) = 0, R > d$.

The original problem has now been greatly simplified. The effect of decoupling the different projections, first in (11.6.2) and then in (11.6.8), means that the Ornstein–Zernike relation has been reduced to three independent equations: the first of those in (11.6.2) and the two in (11.6.8). These equations, with their corresponding closure relations, are just the Percus–Yevick approximation for hard spheres at densities equal, respectively, to $\rho, 2K\rho$ and $-K\rho$. The fact that one solution is required at a negative density poses no special difficulty.

To complete the analytical solution it is necessary to relate the quantity K to hard-sphere properties. Given the analogue of (11.4.14) for $c_D(R)$, the closure relation (11.6.5) requires that

$$c_D(R) = -\frac{3}{R^3}\int_0^d c_D^0(R')R'^2\, dR', \quad R > d \tag{11.6.9}$$

Because $c_D^0(R)$ vanishes for $R > d$, comparison of (11.6.4) with (11.6.9) shows that

$$\beta\mu^2 = -3\int_0^d c_D^0(R)R^2\, dR = -\frac{3}{4\pi}\bar{c}_D(0) \tag{11.6.10}$$

The function $c_D^0(R)$ may be written as

$$c_D^0(R) = K[c_+(R) + c_-(R)] = K[2c_{PY}(R; 2K\rho) + c_{PY}(R; -K\rho)] \qquad (11.6.11)$$

where $c_{PY}(R; \rho)$ is the PY hard-sphere direct correlation function at a density ρ. Let $Q(\eta) = \beta(\partial P/\partial \rho)_T$ be the PY approximation to the inverse compressibility of the hard-sphere fluid at a packing fraction η. Integrals over $c_{PY}(R; \rho)$ can be related to $Q(\eta)$ via the general expression (3.8.8) and the approximate result (4.4.12). A short calculation shows that

$$Q(\eta) = 1 - 4\pi\rho \int_0^d c_{PY}(R; \rho) R^2 \, dR = \frac{(1 + 2\eta)^2}{(1 - \eta)^4} \qquad (11.6.12)$$

On combining the last three equations we find that

$$\begin{aligned} \beta\mu^2 &= -3K \int_0^d [2c_{PY}(R; 2K\rho) + c_{PY}(R; -K\rho)] R^2 \, dR \\ &= \frac{3}{4\pi\rho}[Q(2K\eta) - Q(-K\eta)] \end{aligned} \qquad (11.6.13)$$

or

$$3y = Q(2K\eta) - Q(-K\eta) \qquad (11.6.14)$$

where y is the parameter defined by (11.5.16). Equations (11.6.12) and (11.6.14) determine K implicitly for given choices of y and η; as y varies from 0 to ∞, $K\eta$ varies from 0 to $\frac{1}{2}$.

As an alternative to (11.6.12) we can write

$$\frac{1}{Q(\eta)} = 1 + 4\pi\rho \int_0^\infty h_{PY}(R; \rho) R^2 \, dR \qquad (11.6.15)$$

whence, from (11.6.7):

$$\begin{aligned} \rho\hat{h}_\Delta(0) &= 8\pi\rho K \int_0^\infty [h_+(R) - h_-(R)] R^2 \, dR \\ &= 8\pi\rho K \int_0^\infty [h_{PY}(R; 2K\rho) - h_{PY}(R; -K\rho)] R^2 \, dR \\ &= \frac{1}{Q(2K\eta)} + \frac{2}{Q(-K\eta)} - 3 \end{aligned} \qquad (11.6.16)$$

Combination of (11.5.15), (11.6.14) and (11.6.16) leads to a remarkably simple expression for the dielectric constant:

$$\epsilon = \frac{Q(2K\eta)}{Q(-K\eta)} \qquad (11.6.17)$$

The same result is obtained if (11.5.24) is used instead of (11.5.15).

Although the method of solution is very elegant, comparison with the results of Monte Carlo calculations shows that the MSA does not provide a quantitatively acceptable description of the properties of the dipolar hard-sphere fluid. As is evident from comparison of (11.6.14) with (11.6.17), the dielectric constant in the MSA is dependent only on the parameter y and not separately on the two independent parameters $\rho^* = \rho d^3$ and $\mu^{*2} = \beta \mu^2 / d^3$ required to specify the thermodynamic state of the system. When both these variables are large (for liquid water, $\mu^{*2} \approx 3$), use of the MSA gives values of ϵ that are much too small, as shown by the results in Figure 11.5.

The analytical solution to the MSA has also been found for dipolar hard-sphere mixtures[25] and for dipolar hard spheres with a Yukawa tail.[26] The numerical results obtained for dipolar mixtures show again that the MSA seriously underestimates the dielectric constant. Addition of a Yukawa term to the pair potential leads to changes in thermodynamic properties, but the dielectric constant remains the same.

Of the developments inspired by Wertheim's work on the MSA the simplest to implement is the 'linearised HNC' or LHNC approximation of Patey and coworkers.[27] The LHNC approximation is equivalent to one proposed earlier by Wertheim himself and called by him the 'single-superchain' approximation.[28] In the case of dipolar hard spheres the LHNC approximation resembles the MSA in basing itself on expansions of $h(1, 2)$ and $c(1, 2)$ limited to the terms in S, $\Delta(1, 2)$ and $D(1, 2)$, but improves on it by employing a closure relation that is applicable to other simple models of polar liquids, such as the Stockmayer fluid. As the name implies, the LHNC closure corresponds to a linearisation of

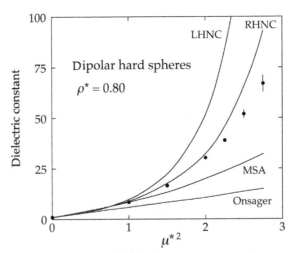

FIGURE 11.5   Dielectric constant of the dipolar hard-sphere fluid at $\rho d^3 = 0.80$ as a function of $\mu^{*2} = \beta \mu^2 / d^3$, showing a comparison between Monte Carlo results[24] (points) and the predictions of theories discussed in the text (curves). After Fries and Patey.[15]

the HNC approximation, which in its general form is

$$c(1,2) = h(1,2) - \ln g(1,2) - \beta v(1,2) \qquad (11.6.18)$$

The LHNC closure is obtained by substituting for $h(1,2)$ and $c(1,2)$ from (11.4.3) and (11.4.5) and linearising with respect to the functions Δ and D. The result is

$$c(1,2) = h_S(R) - \ln g_S(R) - \beta v_0(R) + h_\Delta(R)\left[1 - 1/g_S(R)\right]\Delta(1,2)$$
$$+ \left(h_D(R)\left[1 - 1/g_S(R)\right] + \beta\mu^2/R^3\right) D(1,2) \qquad (11.6.19)$$

where $g_S(R) = h_S(R) + 1$. When $v_0(R)$ is the hard-sphere potential, (11.6.19) reduces to the MSA closure if the substitution $g_S(R) = 1$ for $R > d$ is made; the MSA may therefore be regarded as the low-density limit of the LHNC approximation.

The linearisation involved in (11.6.19) means that the closure relation involves only the harmonics S, Δ and D. This is consistent with the assumed form of $h(1,2)$ and $c(1,2)$ and the results in (11.6.2) remain valid. In other words, the relation between $h_S(R)$ and $c_S(R)$ remains independent of the other projections and the results for these two functions are just the solutions to the HNC equation for the potential $v_0(R)$. In contrast to the MSA, however, the projections on Δ and D are influenced by the projections on S through the appearance of $g_S(R)$ in the closure relations for $c_\Delta(R)$ and $c_D(R)$.

The method of solution of the LHNC equations for the problem of dipolar hard spheres parallels that used for the MSA up to the point at which the linear combinations (11.6.7) are introduced. In the LHNC approximation the functions $h_+(R), c_+(R)$ remain coupled to $h_-(R), c_-(R)$ through the closure relations; the solution must therefore be completed numerically. Some results for the projections $h_\Delta(R)$ and $h_D(R)$ are compared with those obtained by the Monte Carlo method in Figure 11.4. The general agreement between theory and simulation is fair and improves markedly as the value of the parameter μ^* is reduced. However, in contrast to the MSA, the calculated values of the dielectric constant are now everywhere too large, as is evident from Figure 11.5, and the discrepancy between theory and simulation increases rapidly with μ^*. The LHNC approximation has also been applied to systems of quadrupolar hard spheres and to fluids of hard spheres carrying both dipoles and quadrupoles.[27b,d] The calculations are more complicated than in the purely dipolar case because the pair potentials contain additional harmonics and still more are generated by the angular convolutions in the Ornstein–Zernike relation. The results for the mixed, dipolar–quadrupolar system are of particular interest for the light they throw on the way in which the quadrupolar interaction modifies the dipolar correlations in the fluid. The effect on the projection $h_\Delta(R)$ is particularly striking. In the purely dipolar case, when both ρ^* and μ^* are large, $h_\Delta(R)$ is positive nearly everywhere and significantly different from zero out to values of R corresponding to 10 or more molecular diameters. Since ϵ is determined by the integral of $R^2 h_\Delta(R)$ over all R (see (11.5.18)) these effects combine to give very large values for the dielectric constant. The addition of even a small quadrupole moment leads to a marked falling off in both the magnitude and range of $h_\Delta(R)$; ϵ therefore decreases rapidly as the quadrupole moment is increased. This could have been anticipated from the discussion in Section 11.3, since $\Delta = 0$ for the ideal T-shaped configurations favoured by the quadrupolar interaction.

The LHNC approximation for dipolar hard spheres resembles the MSA to the extent that the function $g_S(R)$ is the pair distribution function of the underlying hard-sphere system, and is therefore independent of the strength of the dipole–dipole interaction. This unrealistic feature disappears when the expansion of the HNC closure relation is taken to second order, since $h_S(R)$ and $c_S(R)$ can no longer be decoupled from the other projections. In other respects, the results are not always an improvement on those of the the linearised version, and the theory becomes computationally more awkward to implement. Rather than pursuing the expansion to higher orders it seems preferable to return to the full HNC closure (11.6.18) or its 'reference' (RHNC) modification.[15,29] The molecular generalisation of the RHNC closure (4.7.1) is

$$\ln g(1,2) = -\beta\left[v(1,2) - k_\mathrm{B}T b_0(1,2)\right] + h(1,2) - c(1,2) \qquad (11.6.20)$$

where $b_0(1,2)$ is the bridge function of some anisotropic reference system. In the case of dipolar hard spheres, however, hard spheres are the obvious choice of reference system. Because the closure relation couples together all harmonic components of $h(1,2)$ and $c(1,2)$, the results obtained depend on the number of harmonics retained when expanding the pair functions, but essentially complete convergence is achieved with a basis set of easily manageable size. Some results for $h_\Delta(R)$ and $h_D(R)$ are shown in Figure 11.4. The theoretical curves lie systematically below those given by the LHNC approximation; the dielectric constant is therefore much reduced, and the agreement with simulations correspondingly improved, as Figure 11.5 confirms.

It is known experimentally that the Kirkwood g-factors of many non-associating, polar liquids are close to unity, and the very large discrepancies seen in Figure 11.5 between the Monte Carlo results and the predictions of the Onsager approximation (for which $g_K = 1$) show that the dipolar hard-sphere model gives dielectric constants that are unrealistically large.[30] The role played by quadrupolar forces provides a possible explanation of the experimental facts, but a more realistic model of a polar fluid must also make allowance for the inevitable anisotropy in the short-range, repulsive forces. The simplest such model consists of a hard, homonuclear diatomic with a dipole moment superimposed at the mid-point between the two spheres. Within the RHNC approximation the natural choice of reference system is now the underlying hard-dumbell fluid, the bridge function of which can be calculated from the molecular version of the PY approximation.[31] The same general approach can be used for heteronuclear molecules having either soft or hard cores. Some good results have been achieved in this way, though a strong empirical element is often involved both in the choice of reference system and in the form of closure relation used to calculate the corresponding bridge function.[32]

11.7 INTERACTION–SITE DIAGRAMS

The diagrammatic expansions of $c(1,2)$, $h(1,2)$ and $y(1,2)$ given in Chapters 3 and 4 are also applicable to molecular fluids if some minor changes in interpretation are made. First, the circles in a 'molecular' diagram are associated with both the translational and orientational coordinates of a molecule and the black circles imply integration over both sets of coordinates. Secondly, black circles carry a weight factor equal to $1/\Omega$, where Ω is defined by (11.1.3). As an illustration of these rules, the diagram

which appears at order ρ in the ρ-circle, f-bond expansion of $h(1,2)$ (see (4.6.2)) now represents the integral

$$\frac{\rho}{\Omega} \iint f(\mathbf{R}_{13}, \mathbf{\Omega}_1, \mathbf{\Omega}_3) f(\mathbf{R}_{23}, \mathbf{\Omega}_2, \mathbf{\Omega}_3) \mathrm{d}\mathbf{R}_3 \, \mathrm{d}\mathbf{\Omega}_3$$

and is therefore much more complicated to evaluate than in the atomic case.

The diagrammatic expansion of $h(1,2)$ is not immediately useful in cases where the focus of interest is the set of site–site distribution functions $h_{\alpha\beta}(r)$ rather than $h(1,2)$ itself. Ladanyi and Chandler[33] have shown how the diagrammatic approach can be adapted to the needs of such a situation and this section is devoted to a brief review of their results. We give only a simplified treatment, restricting the detailed discussion to the case of rigid, diatomic (or two-site) molecules. The generalisation to larger numbers of interaction sites is straightforward but requires a more complex notation.

The first step is to rewrite the molecular Mayer function $f(1,2)$ as a product of interaction–site Mayer functions $f_{\alpha\beta}(r)$:

$$f(1,2) = \exp\left[-\beta v(1,2)\right] - 1 = \exp\left(-\beta \sum_\alpha \sum_\beta v_{\alpha\beta}\left(|\mathbf{r}_{2\beta} - \mathbf{r}_{1\alpha}|\right)\right) - 1$$

$$= -1 + \prod_{\alpha,\beta}\left[f_{\alpha\beta}\left(|\mathbf{r}_{2\beta} - \mathbf{r}_{1\alpha}|\right) + 1\right] \tag{11.7.1}$$

The subscripts α, β run over all interaction sites in the molecule; if there are two sites per molecule, the right-hand side of (11.7.1) consists of 15 separate terms.

Equation (11.7.1) can be used to rewrite the integrals occurring in the density expansion of $h(1, 2)$. As the simplest possible example, consider the low-density limit of $h(1, 2)$, namely $\lim_{\rho \to 0} h(1, 2) = f(1, 2)$. The corresponding approximation to, say, $h_{\alpha\alpha}(\mathbf{r}, \mathbf{r}')$ is

$$\lim_{\rho \to 0} h_{\alpha\alpha}(\mathbf{r}, \mathbf{r}') = \iint f(1, 2)\delta(\mathbf{r}_{1\alpha} - \mathbf{r})\delta(\mathbf{r}_{2\alpha} - \mathbf{r}')\mathrm{d}1 \, \mathrm{d}2 \qquad (11.7.2)$$

When $f(1, 2)$ is replaced by (11.7.1), (11.7.2) becomes

$$\lim_{\rho \to 0} h_{\alpha\alpha}(\mathbf{r}, \mathbf{r}') = f_{\alpha\alpha}(|\mathbf{r}' - \mathbf{r}|) + [1 + f_{\alpha\alpha}(|\mathbf{r}' - \mathbf{r}|)]$$

$$\times \iint [f_{\alpha\beta}(|\mathbf{r}_{2\beta} - \mathbf{r}_{1\alpha}|) + \text{six other terms}]$$

$$\times \delta(\mathbf{r}_{1\alpha} - \mathbf{r})\delta(\mathbf{r}_{2\alpha} - \mathbf{r}')\mathrm{d}1 \, \mathrm{d}2 \qquad (11.7.3)$$

The integrals appearing on the right-hand side of (11.7.3) can be re-expressed in terms of an intramolecular site–site distribution function $s_{\alpha\beta}(\mathbf{x} - \mathbf{y})$ defined as

$$s_{\alpha\beta}(\mathbf{x} - \mathbf{y}) = (1 - \delta_{\alpha\beta}) \int \delta(\mathbf{R}_1 + \mathbf{u}_1 \ell_\alpha - \mathbf{x})\delta(\mathbf{R}_1 - \mathbf{u}_1 \ell_\beta - \mathbf{y})\mathrm{d}1$$

$$= (1 - \delta_{\alpha\beta}) \langle \delta(\mathbf{x} - \mathbf{y} - \mathbf{u}_1 L) \rangle_{\Omega_1}$$

$$= \frac{(1 - \delta_{\alpha\beta})}{4\pi L^2} \delta(|\mathbf{x} - \mathbf{y}| - L) \qquad (11.7.4)$$

where ℓ_α, ℓ_β and $\mathbf{u}_1$ have the same meaning as in (11.5.26) and $L = \ell_\alpha + \ell_\beta$. The function $s_{\alpha\beta}(\mathbf{r})$ is the probability density for finding site β of a molecule at a position $\mathbf{r}$, given that site α of the same molecule is at the origin. The definition (11.7.4) satisfies the obvious conditions that the interpretation as an intramolecular distribution function requires, i.e. $s_{\alpha\beta}(\mathbf{r}) = s_{\beta\alpha}(\mathbf{r}), s_{\alpha\alpha}(\mathbf{r}) = 0$ and $\int s_{\alpha\beta}(\mathbf{r})\mathrm{d}\mathbf{r} = 1, \alpha \neq \beta$. The integral shown explicitly in (11.7.3) can now be transformed as follows:

$$\iint \delta(\mathbf{r}_{1\alpha} - \mathbf{r})\delta(\mathbf{r}_{2\alpha} - \mathbf{r}') f_{\alpha\beta}(|\mathbf{r}_{2\beta} - \mathbf{r}_{1\alpha}|)\mathrm{d}1 \, \mathrm{d}2$$

$$= \int \mathrm{d}\mathbf{x} \iint \delta(\mathbf{r}_{1\alpha} - \mathbf{r})\delta(\mathbf{r}_{2\alpha} - \mathbf{r}') f_{\alpha\beta}(\mathbf{r}_{1\alpha} - \mathbf{x})\delta(\mathbf{r}_{2\beta} - \mathbf{x})\mathrm{d}1 \, \mathrm{d}2$$

$$= \int \mathrm{d}1 \, \delta(\mathbf{r}_{1\alpha} - \mathbf{r}) \int \mathrm{d}\mathbf{x} \, f_{\alpha\beta}(\mathbf{r}_{1\alpha} - \mathbf{x}) \int \mathrm{d}2 \, \delta(\mathbf{r}_{2\alpha} - \mathbf{r}')\delta(\mathbf{r}_{2\beta} - \mathbf{x})$$

$$= \int f_{\alpha\beta}(|\mathbf{r} - \mathbf{x}|)s_{\alpha\beta}(\mathbf{x} - \mathbf{r}')\mathrm{d}\mathbf{x} \qquad (11.7.5)$$

All other integrals in (11.7.3) may be treated in the same way and each can then be represented by an *interaction–site diagram*. The circles (white or black) of an interaction–site diagram are associated with the coordinates of interaction sites and the bonds, in the two-site case, represent components of the 2×2 matrices $\mathbf{f}$ and $\mathbf{s}$ formed by the functions $\{f_{\alpha\beta}\}$ and $\{s_{\alpha\beta}\}$, respectively. The symmetry number and value of an interaction–site diagram are defined as in the atomic case (see Section 3.7), except that black circles imply a summation over all sites in the molecule in addition to integration over site coordinates. For example, if we denote an f-bond by a solid line and an s-bond by a dotted line, the diagrammatic representation of the sum of integrals in (11.7.3) is

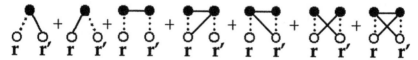

The diagrams shown all have a symmetry number of one. They are of zeroth order in density, since they arise from a molecular diagram – their 'molecular origin' – which represents the low-density limit of $h(1,2)$. Thus all circles, white or black, are 1-circles. The order in density of any interaction–site diagram in the expansion of a site–site pair correlation function is equal to the number of black circles in its molecular origin, which in turn is equal to the number of black circles in the interaction–site diagram minus the number of s-bonds.

The procedure outlined above can be applied to each of the integrals appearing in the density expansion of $h(1,2)$. This yields an expansion of any of the functions $h_{\alpha\beta}(\mathbf{r})$ in terms of interaction–site diagrams. As the example (11.7.3) demonstrates, each molecular diagram is replaced by a large number of interaction–site diagrams, but the interaction–site diagrams are mathematically simpler objects because all reference to orientational coordinates has disappeared. (Note that the black circles no longer carry the weight factor Ω^{-1} associated with the black circles of a molecular diagram.)

The topology of allowed interaction–site diagrams is restricted in certain ways. Diagrams must be simple and connected; white circles must not be connected by an s-bond (because different white circles always refer to different molecules); all black circles must be intersected by at least one f-bond (otherwise they contribute nothing to the intermolecular correlations); no circle may be intersected by more than one s-bond (for reasons to be explained below); and diagrams must be free of articulation circles and articulation s-bonds, i.e. s-bonds whose removal causes the diagram to separate into two or more components of which at least one contains no white circle. The last restriction is imposed because any such diagram would have as its molecular origin a diagram containing one or more articulation circles; as we showed in Chapters 3 and 4, the expansions of the pair functions of interest here consist entirely of irreducible diagrams.

Given the restrictions listed above, the site–site pair correlation functions may be characterised as follows:

$$h_{\alpha\beta}(\mathbf{r}_{1\alpha}, \mathbf{r}_{2\beta}) = \big[\text{all interaction–site diagrams consisting of two white}$$

1-circles labelled 1α and 2β, black 1-circles, f-bonds and s-bonds, each diagram to be multiplied by ρ^n, where n is the number of black circles minus the number of s-bonds$\big]$

$$(11.7.6)$$

The generalisation of this result to molecules with more than two interaction sites requires the introduction of three-body and higher-order intramolecular distribution functions. It remains true, however, that no circle may be intersected by more than one s-bond or, indeed, by more than one intramolecular bond of any order. Consider the diagram shown in (a) below. For a two-site molecule such a diagram is physically meaningless because one site is bonded to two others. But it is also not an allowed diagram even for a three-site (or larger) molecule, because the three black circles would then be linked, as in (b), by a single bond or 'face', representing a three-body intramolecular distribution function.

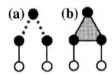

The diagrammatic formalism can be extended to flexible molecules, but in that case the intramolecular distribution functions become statistically averaged quantities.

11.8   INTERACTION-SITE MODELS: THE RISM EQUATIONS

We saw in Section 11.3 that the static structure factors measured in neutron and X-ray scattering experiments on molecular liquids are weighted sums of atomic pair distribution functions. In this section we describe an integral equation theory that has been widely used in the interpretation of diffraction experiments and, more generally, in the calculation of site–site distribution functions for interaction-site potential models: this is the 'reference interaction-site model' or RISM approximation of Andersen and Chandler.[34] The theory has been applied with particular success in calculations for model fluids composed of hard molecules. From experience with atomic systems we can expect the structure of simple molecular liquids to be dominated by the strongly repulsive part of the pair potential, and an obvious way represent to the short-range repulsions is through an interaction–site model consisting of fused hard spheres.

The key ingredient of the RISM approximation is an Ornstein–Zernike-like relation between the site–site pair correlation functions $h_{\alpha\beta}(r)$ and a set of direct correlation functions $c_{\alpha\beta}(r)$. In the atomic case the meaning of the Ornstein–Zernike relation is that the total correlation between particles 1 and 2 is the sum of all possible paths of direct correlations that propagate via intermediate particles 3, 4, . . . The same, intuitive idea can be applied to site–site correlations, but allowance must now be made for the fact that correlations also propagate via the intramolecular distribution functions. Hence, whereas in an atomic fluid $h(1, 2)$ is given diagrammatically by the sum of all simple chains consisting of c-bonds, $h_{\alpha\beta}(r)$ consists of all simple chains formed from c-bonds and s-bonds. We make this idea precise by writing $h_{\alpha\beta}(r)$ as a sum of interaction–site diagrams in the form

$$h_{\alpha\beta}(\mathbf{r}_{1\alpha}, \mathbf{r}_{2\beta})$$

= [all interaction-site chain diagrams consisting of two white
terminal 1-circles labelled 1α and 2β, black 1-circles,
at least one c-bond, and s-bonds, each diagram to be
multiplied by ρ^{n-1}, where n is the number of c-bonds]

$$= \underset{1\alpha \quad 2\beta}{\circ\!\!-\!\!\circ} + \underset{1\alpha \qquad 2\beta}{\circ\!\!-\!\!\bullet\!\cdots\!\circ} + \underset{1\alpha \qquad 2\beta}{\circ\!\cdots\!\bullet\!\!-\!\!\circ}$$

$$+ \rho \left(\underset{1\alpha \qquad 2\beta}{\circ\!\!-\!\!\bullet\!\!-\!\!\circ} + \underset{1\alpha \qquad 2\beta}{\circ\!\!-\!\!\bullet\!\cdots\!\bullet\!\!-\!\!\circ} + \cdots \right) + \cdots$$

$$(11.8.1)$$

where a full line denotes a c-bond and a broken line denotes an s-bond. We recall that within the diagrammatic formalism a black circle implies a summation over all sites in the molecule. Thus, for example, the value of the third diagram on the right-hand side of (11.8.1) is

$$\sum_{\gamma} \int s_{\alpha\gamma}(\mathbf{r}_{1\gamma} - \mathbf{r}_{1\alpha}) c_{\gamma\beta}(|\mathbf{r}_{2\beta} - \mathbf{r}_{1\gamma}|) \mathrm{d}\mathbf{r}_{1\gamma}$$

The term for which $\alpha = \gamma$ contributes nothing to the sum because the intramolecular distribution function is zero when the two sites are the same.

We now have to sum the chain diagrams in (11.8.1). To do so, we use the same techniques as in Section 5.6, because the diagrams have the same topology as those in the diagrammatic expansion (5.6.16) of the renormalised potential $C(1, 2)$. We define a matrix of functions $\mathbf{w}(r)$ by

$$w_{\alpha\beta}(|\mathbf{r}_{1\beta} - \mathbf{r}_{1\alpha}|) = \delta_{\alpha\beta}\delta(\mathbf{r}) + s_{\alpha\beta}(\mathbf{r}) \qquad (11.8.2)$$

and represent $w_{\alpha\beta}(|\mathbf{r}_{1\beta} - \mathbf{r}_{1\alpha}|)$ by the hypervertex

Then the sum of all chain diagrams with n c-bonds becomes a single diagram consisting of $(n + 1)$ hypervertices and n c-bonds. For example, when $n = 1$:

$$= \sum_{\gamma} \sum_{\delta} \iint w_{\alpha\gamma}(|\mathbf{r}_{1\alpha} - \mathbf{r}_{1\gamma}|) c_{\gamma\delta}(|\mathbf{r}_{1\gamma} - \mathbf{r}_{2\delta}|)$$

$$\times w_{\delta\beta}(|\mathbf{r}_{2\delta} - \mathbf{r}_{2\beta}|) \, d\mathbf{r}_{1\gamma} \, d\mathbf{r}_{2\delta} \tag{11.8.3}$$

A hypervertex corresponds to a single molecule and incorporates all the intramolecular constraints represented by the s-bonds. The Fourier transform of (11.8.3) is the $\alpha\beta$-component of the matrix $\hat{\mathbf{w}}(k) \cdot \hat{\mathbf{c}}(k) \cdot \hat{\mathbf{w}}(k)$, i.e.

$$\sum_{\gamma} \sum_{\delta} \hat{w}_{\alpha\gamma}(k) \hat{c}_{\gamma\delta}(k) \hat{w}_{\delta\beta}(k) \equiv \left(\hat{\mathbf{w}}\hat{\mathbf{c}}\hat{\mathbf{w}}\right)_{\alpha\beta} \tag{11.8.4}$$

The components of the matrix $\hat{\mathbf{w}}(k)$ are

$$\hat{w}_{\alpha\beta}(k) = \delta_{\alpha\beta} + (1 - \delta_{\alpha\beta}) j_0(k L_{\alpha\beta}) \tag{11.8.5}$$

where $L_{\alpha\beta}$ is the intramolecular separation of sites α, β. Similarly, the Fourier transform of the sum of all chain diagrams containing precisely n c-bonds is $\rho^{n-1}((\hat{\mathbf{w}}\hat{\mathbf{c}})^n \hat{\mathbf{w}})_{\alpha\beta}$ (cf. (5.6.22)), and $\hat{h}_{\alpha\beta}(k)$ is the sum of a geometric series (cf. (5.6.23)). The matrix $\hat{\mathbf{h}}(k)$ is therefore given by

$$\hat{\mathbf{h}}(k) = \hat{\mathbf{w}}(k) \cdot \hat{\mathbf{c}}(k) \cdot \left[\mathbf{1} - \rho\hat{\mathbf{w}}(k) \cdot \hat{\mathbf{c}}(k)\right]^{-1} \cdot \hat{\mathbf{w}}(k)$$

$$= \hat{\mathbf{w}}(k) \cdot \hat{\mathbf{c}}(k) \cdot \hat{\mathbf{w}}(k) + \rho\hat{\mathbf{w}}(k) \cdot \hat{\mathbf{c}}(k) \cdot \hat{\mathbf{h}}(k) \tag{11.8.6}$$

Equation (11.8.6) is the Ornstein–Zernike-like relation. It can be derived in ways other than the one we have described, but the the diagrammatic method[35] has a strong intuitive appeal. We shall refer to it as the RISM-OZ relation, though we shall see later that its status differs from that of the molecular Ornstein–Zernike relation (11.1.4). If $\hat{\mathbf{w}}$ is the identity matrix and ρ is appropriately reinterpreted, it reduces to the Ornstein–Zernike relation for a mixture of atomic fluids.

If the RISM-OZ relation is to be useful it must be combined with some approximate closure relation. For systems of fused hard spheres the obvious choice is a generalisation of the PY approximation for atomic hard spheres, i.e.

$$h_{\alpha\beta}(r) = -1, \quad r < d_{\alpha\beta}; \qquad c_{\alpha\beta}(r) = 0, \quad r > d_{\alpha\beta} \tag{11.8.7}$$

where $d_{\alpha\beta}$ is the $\alpha - \beta$ hard-sphere diameter. When the site–site potentials are continuous, generalisations of either the PY or HNC approximations may

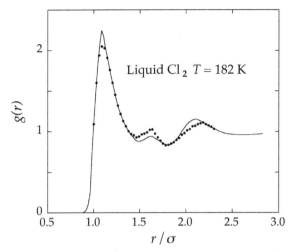

FIGURE 11.6 Atom–atom distribution function for a Lennard-Jones diatomic model of liquid chlorine. The points show the results of a molecular dynamics simulation and the curve is calculated from the RISM approximation with PY closure. *Redrawn with permission from Ref. 36 © Taylor & Francis Limited.*

be used. A number of schemes have been devised for numerical solution of the resulting system of equations and calculations have been made for a wide variety of molecular liquids. Figure 11.6 shows the results of RISM calculations based on the PY closure relation for the atomic pair distribution function of a two-site Lennard-Jones model of liquid chlorine. There are some differences in detail, but all the main features seen in molecular dynamics calculations for the same potential model are well reproduced. Note that the shoulder in the ethylene results of Figure 11.2 appears here as a clearly defined, subsidiary peak.

The agreement between theory and simulation seen in Figure 11.6 is typical of that achieved for other small, rigid molecules.[37]

11.9 ANGULAR CORRELATIONS AND THE RISM FORMALISM

Although successful in many applications, the RISM formalism suffers from a number of defects. First, it does not lend itself readily to a calculation of the equation of state and the results obtained are thermodynamically inconsistent in the sense of Section 4.4. Secondly, calculated structural properties show an unphysical dependence on the presence of 'auxiliary' sites, which are sites that label a point in the molecule but contribute nothing to the intermolecular potential. Thirdly, and most unexpectedly, trivial and incorrect results are obtained for certain quantities descriptive of angular correlations in the fluid.[38]

As an example, we show below that the order parameter G_1 defined by (11.1.8) is identically zero for any asymmetric but non-polar diatomic. The only assumption made is that the site–site potentials are short ranged.

We note first that all elements of the matrix $\hat{\mathbf{w}}(k)$ defined by (11.8.5) are unity when $k = 0$. If we define a matrix $\mathbf{Q}$ as

$$\mathbf{Q} = \mathbf{I} - n^{-1}\hat{\mathbf{w}}(0) = \begin{pmatrix} \frac{1}{2} & -\frac{1}{2} \\ -\frac{1}{2} & \frac{1}{2} \end{pmatrix} \tag{11.9.1}$$

where n is the number of sites (here equal to two), then

$$\mathbf{Q} \cdot \hat{\mathbf{w}}(0) = \hat{\mathbf{w}}(0) \cdot \mathbf{Q} = 0 \tag{11.9.2}$$

Next we write the RISM-OZ relation (11.8.6) in the form

$$\hat{\mathbf{h}}(k) = \hat{\mathbf{w}}(k) \cdot \mathbf{R}(k) \tag{11.9.3}$$

where

$$\mathbf{R}(k) \equiv \hat{\mathbf{c}}(k) \cdot \left[\hat{\mathbf{w}}(k) + \rho\hat{\mathbf{h}}(k)\right] \tag{11.9.4}$$

On expanding $\hat{\mathbf{w}}(k)$ in powers of k we find that to order k^2:

$$\hat{\mathbf{h}}(k) = \left[\hat{\mathbf{w}}(0) + k^2\mathbf{w}^{(2)} + \cdots\right] \cdot \mathbf{R}(k) \tag{11.9.5}$$

If multiplied on the left by $\mathbf{Q}$, (11.9.5) reduces, by virtue of the property (11.9.2), to

$$\mathbf{Q} \cdot \hat{\mathbf{h}}(k) = \mathbf{Q} \cdot \left[k^2\mathbf{w}^{(2)} + \cdots\right] \cdot \mathbf{R}(k) \tag{11.9.6}$$

We now suppose that $\hat{\mathbf{h}}(k)$ and $\hat{\mathbf{c}}(k)$ (and hence also $\mathbf{R}(k)$) have small-k expansions at least up to order k^2. This is plausible, since the site–site potentials are assumed to be short ranged. Then

$$\mathbf{Q} \cdot \left[\hat{\mathbf{h}}(0) + k^2\mathbf{h}^{(2)} + \cdots\right] = \mathbf{Q} \cdot \left[k^2\mathbf{w}^{(2)} + \cdots\right] \cdot \left[\mathbf{R}(0) + k^2\mathbf{R}^{(2)} + \cdots\right] \tag{11.9.7}$$

and by equating coefficients of k^2 we find that

$$\mathbf{Q} \cdot \mathbf{h}^{(2)} = \mathbf{Q} \cdot \mathbf{w}^{(2)} \cdot \mathbf{R}(0) \tag{11.9.8}$$

We have seen in Section 11.3 that all elements of $\hat{\mathbf{h}}(0)$ are the same and related to the compressibility by (11.3.6). Thus $\mathbf{R}(0)$ may be written as

$$\mathbf{R}(0) = \left[1 + \rho\hat{h}_0(0)\right]\hat{\mathbf{c}}(0) \cdot \hat{\mathbf{w}}(0) \tag{11.9.9}$$

where $\hat{h}_0(0)$ (a scalar) is any element of $\hat{\mathbf{h}}(0)$. Inserting (11.9.9) in (11.9.8), multiplying on the right by $\mathbf{Q}$ and using again the property (11.9.2), we find that

$$\mathbf{Q} \cdot \mathbf{h}^{(2)} \cdot \mathbf{Q} = 0 \tag{11.9.10}$$

But every element of the matrix $\mathbf{Q} \cdot \mathbf{h}^{(2)} \cdot \mathbf{Q}$ is proportional to $\Delta h^{(2)}$, where $\Delta h(r)$ is defined by (11.3.8). Thus $\Delta h^{(2)} = 0$ and hence, from (11.3.10), $G_1 = 0$. As we pointed out in Section 11.3, this result is obvious on symmetry grounds for a homonuclear molecule, but in the general case it will be true (unless accidentally) only in the ideal-gas limit. Similarly, by considering terms of order k^4 in (11.9.7), it can be shown that $G_2 = 0$ for an asymmetric, linear, triatomic molecule.

If the molecule is polar, with the interaction sites carrying point charges, the problem becomes more complicated. When expanding $\hat{c}(k)$, allowance must be made for a term in k^{-2}, corresponding to an r^{-1} decay of the site–site potential. This term must be treated separately, but it is then possible to show that for any interaction–site molecule

$$\rho \sum_{\alpha} \sum_{\beta} q_\alpha q_\beta h^{(2)}_{\alpha\beta} = -\frac{y\mu^2}{1 + 3y} \tag{11.9.11}$$

where q_α is the charge carried by site α. Comparison of (11.9.11) with the exact result (11.5.32) shows that within the RISM approximation

$$\epsilon = 1 + 3y \tag{11.9.12}$$

which is a well-known result for the dielectric constant of an ideal gas of polar molecules.

The results in (11.9.10) and (11.9.12) are consequences solely of the RISM-OZ relation (11.8.6). They are independent of the choice of closure relation except insofar as the latter must be consistent with the assumed small-k behaviour of $\hat{\mathbf{h}}(k)$ and $\hat{c}(k)$. It follows that the RISM-OZ relation, when combined with a conventional closure approximation, cannot describe correctly certain long-wavelength properties of molecular systems, of which G_1, G_2 and ϵ are important examples.

Attempts to develop a more satisfactory theory while retaining the essential features of the RISM approach have developed along two different lines.[39] The first relies on treating the RISM-OZ relation as providing the definition of the site–site direct correlation functions. So far as the calculation of angular order parameters is concerned, it then appears necessary to abandon the assumption that $c_{\alpha\beta}(r)$ is a short-range function, even when the corresponding site–site potential is short ranged. For example,[40] a non-zero value of G_1 for a symmetric diatomic is obtained if $c_{\alpha\beta}(r)$ is assumed to decay as r^{-1}. In such circumstances the concept of 'direct correlation' no longer has a clear physical meaning. In the alternative approach the view is taken that the RISM-OZ relation, though plausible, does not provide an adequate basis for a complete theory of interaction–site models of molecular liquids. Accordingly, it is there rather than in the closure relation that improvement must be sought.[41] The argument is based on the difference in diagrammatic structure between the RISM-OZ relation and the molecular Ornstein–Zernike relation (11.1.4). In the latter case the indirect correlation function $h(1, 2) - c(1, 2)$ is given, as in (4.6.1), by the sum of all simple chain diagrams containing two or more c-bonds. In any such diagram, every black circle is a nodal circle, and $c(1, 2)$ consists of the subset of diagrams in the ρ-circle, f-bond expansion of $h(1, 2)$ that are free of nodal circles. By analogy, it might be supposed that the c-bonds in (11.8.1) represent the subset of diagrams in the expansion of $h_{\alpha\beta}(r)$ that are free of nodal circles. This is not the case. For example, the diagram

which appears at zeroth order in the density expansion of $h_{\alpha\beta}(r)$ is a diagram without nodal circles. If this is substituted into the second and third diagrams on the right-hand side of (11.8.1), it yields, respectively, diagrams (a) and (b) below:

(a)   **(b)**

Diagram (a) is a diagram in the exact expansion of $h_{\alpha\beta}(r)$, but (b) is not. In fact (b) is not even an allowed diagram, because two s-bonds intersect the same black circle. The problem can be overcome by the introduction of another Ornstein–Zernike-like relation that reduces to (11.8.6) in the limit $\rho \to 0$ but in which the direct correlation functions correspond to well-defined subsets of diagrams that contribute to the pair correlation functions. Calculations based on the new relation lead to non-trivial results for the angular order parameters and dielectric constant when approximate closures of conventional type are employed. However, the method has not been widely applied, and so far as the description of short-range order is concerned it is not clear that it represents an improvement on the original formulation of the theory.[42]

11.10 ASSOCIATING LIQUIDS

Although hydrogen-bonded liquids and other associating fluids are not normally classed as 'simple', our understanding of the link between the macroscopic properties of such systems and their behaviour at the microscopic level has improved greatly in recent years. This is a development to which both experiment and simulation have made major contributions. For understandable reasons, much of the effort has been focused on studies of water. The particular geometry of the water molecule gives rise to structural features in the liquid that are not seen for other small, hydrogen-bonded species, and the macroscopic properties of water display a number of anomalies that are directly attributable to hydrogen bonding, of which the best known is the fact that the density at atmospheric pressures passes through a maximum at a temperature of

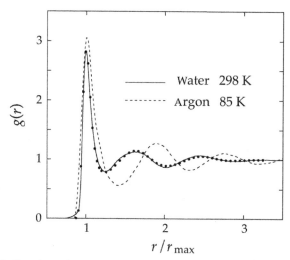

FIGURE 11.7 Experimental results for the pair distribution function for oxygen atoms in water at room temperature (from X-ray scattering[43]) and for liquid argon near its triple point (from neutron scattering[45]). The points are the results for $g_{OO}(r)$ obtained by Monte Carlo calculations[46] for an interaction–site model of water. The quantity r_{max} is the separation at which the corresponding experimental curve has its main peak: 2.74 Å for water and 3.68 Å for argon.

approximately 4 °C. One of the most important advances on the experimental front has been the resolution of significant differences that had previously existed between the results of X-ray and neutron-scattering measurements of the structure of liquid water. X-ray scattering is sensitive primarily to the oxygen–oxygen correlations,[43] while neutron scattering is the main source of information on correlations involving hydrogen atoms.[44] Figure 11.7 shows the results obtained by X-ray scattering for the distribution function of oxygen atoms in water at room temperature, contrasting these with the results for liquid argon already shown in Chapter 2. To assist comparison, the horizontal axis is scaled so as to bring the two main peaks into coincidence. Clearly the structure is very different in the two liquids. Two features in particular stand out. First, the area under the main peak is significantly smaller for water than it is for argon, leading to a large reduction in the nearest-neighbour coordination number defined in Section 2.5 from ≈ 12 for argon to about four for water. Secondly, the oscillations in the two curves are out of phase. The second peak for water is displaced inwards with respect to that for argon and appears at a distance $r/r_{max} = 1.61 \pm 0.01$, which is very close to the value found for the ratio of the second-neighbour separation to that of first neighbours in the ideal ice structure, i.e. $2\sqrt{(2/3)} \approx 1.63$. Both the value of the coordination number and the position of the second peak in the oxygen–oxygen distribution function provide clear evidence that the molecules in liquid water form a hydrogen-bonded network that represents a strained version of the tetrahedral ordering found in ice.

A similar picture to that outlined above emerges from the many simulations of water that have been carried out.[47] Such calculations provide a level of detail that cannot be matched experimentally concerning the number, energies and lifetimes of the hydrogen bonds formed by individual molecules. A very large number of empirical intermolecular potentials have been designed for use in simulations, which differ from each other mainly in the way in which the electrostatic interaction between molecules is described. The majority are based on rigid charge distributions represented by three or more point charges, though a number of polarisable models have also been developed, and the best of these empirical potentials give results in impressive agreement with experiment for a wide range of properties. An example of what can be achieved is illustrated by the Monte Carlo results shown in Figure 11.7. These were obtained with a model[46] (TIP5P) consisting of a Lennard-Jones interaction centred on the oxygen atom and four rigid charges, one on each hydrogen site and two at sites chosen to represent the lone-pair electrons. The same model is also successful in reproducing the density anomaly at 4 °C, while a predecessor (TIP4P) has been shown to capture all the main features of the experimental phase diagram.[48]

A variety of RISM-based, integral equation approximations have been used in calculations for models of specific hydrogen-bonded liquids, including water,[49] but progress has also been achieved in the development of a general approach to the theory of associating fluids.[50–52] Of these theories the best known and most successful is that of Wertheim, which in its commonly used form has the character of a thermodynamic perturbation theory.[52b] The theory is designed for application to a class of highly simplified models in which the associating species are treated as particles with repulsive cores in which a number of attractive interaction sites are embedded; it is at these sites that association occurs. In the examples discussed below the particles are taken to be hard spheres of diameter d and the association sites are represented by off-centre, square-well potentials with a well depth ϵ_A. Because hard spheres cannot overlap, the square-well potential can always be made sufficiently short ranged that the formation of more than one bond at any given site is forbidden, as in the example shown in Figure 11.8. A model with one association site describes a dimerising fluid; with two sites, illustrated in the figure, the spheres can form chains and rings; with three sites, chain branching and network formation become possible; and a sphere with four, tetrahedrally disposed sites serves as a crude model of a water molecule.

If the attractive interaction between particles is sufficiently strong to promote association we cannot expect a conventional perturbation calculation to succeed. In Wertheim's approach this difficulty is circumvented by treating different association aggregates as distinct species, each described by a separate single-particle density within a 'multi-density' formalism. The theory leads ultimately to an expression for the free energy in terms of the densities of particles in different bonding states. As a specific example, consider the case of a system of hard spheres with a single association site. Since only dimer formation is

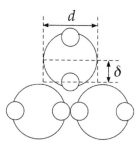

FIGURE 11.8 A simple interaction–site model of an associating liquid. The large circles represent hard spheres and the small circles represent square-well interaction sites displaced from the centre of the hard sphere by a distance δ. The range of the square-well potential is sufficiently short to ensure that multiple bonding at any association site is forbidden, since that would require hard spheres to overlap.

allowed, the total number density of spheres can be written as

$$\rho = \rho_M + 2\rho_D \qquad (11.10.1)$$

where ρ_M and ρ_D are the number densities of monomers and dimers, respectively. Diagrammatic arguments along the general lines of those pursued in Section 3.8 can then be used to show that

$$\rho = \rho_M + \rho_M^2 \int g_{MM}(1,2) f_A(1,2) d2 \qquad (11.10.2)$$

where $g_{MM}(1,2)$ is the pair distribution function of the free monomers and $f_A(1,2)$ is the Mayer function for the association potential. Equation (11.10.2) can be recognised as a statement of the law of mass action when applied to the 'reaction' $M + M \rightleftharpoons D$, for which the equilibrium constant K is defined as

$$K = \frac{\rho_D}{\rho_M^2} \qquad (11.10.3)$$

This result can be rearranged in the form

$$\rho = \rho_M + 2K\rho_M^2 \qquad (11.10.4)$$

thereby identifying the integral in (11.10.2) with the quantity $2K$.

The starting point in the derivation of (11.10.2) is the activity expansion of $\ln \Xi$ provided by (3.8.3). By decomposing the Mayer function for the full pair potential in the form

$$f(1,2) = f_0(1,2) + \Phi(1,2), \quad \Phi(1,2) = e_0(1,2) f_A(1,2) \qquad (11.10.5)$$

where $e_0(1,2)$ and $f_0(1,2)$ are, respectively, the Boltzmann factor and Mayer function for the hard-sphere potential, the right-hand side of (3.8.3) can be written a sum of diagrams consisting of z^*-circles, f_0-bond and Φ-bonds. The assumption that multiple bonding at a single site is blocked by steric effects means that many of the diagrams either vanish or are cancelled by other diagrams; this greatly simplifies the subsequent analysis. The diagrammatic representation of the total single-particle density $\rho^{(1)}(1)$ is again obtained from the prescription given by (3.8.4), i.e. as the sum of all topologically distinct diagrams obtained from $\ln \Xi$ by whitening a black circle and labelling it 1. Now, however, the diagrams that contribute to $\rho^{(1)}(1)$ can be divided into two classes:

$$\rho^{(1)}(1) = \rho_M^{(1)}(1) + \rho_A^{(1)}(1) \tag{11.10.6}$$

where $\rho_M^{(1)}(1)$ is the density of unassociated spheres (monomers) and $\rho_A^{(1)}(1)$ is the density of spheres that form part of an associated aggregate, which in the present case can only be a dimer; the class of monomer diagrams consists of those diagrams in which the white circle is not intersected by a Φ-bond. The last step in the derivation involves a topological reduction in which the z^*-circles in the z^*-expansion of $\ln \Xi$ are replaced by $\rho^{(1)}$ or $\rho_M^{(1)}$-circles, which in turn leads to expressions for the free energy and pressure as functionals of the two densities. The monomer density is not a free parameter; it is determined self-consistently by a relation between $\rho^{(1)}(1)$ and $\rho_M^{(1)}(1)$, which in the homogeneous limit reduces to (11.10.2). The full calculation is too lengthy to reproduce here, but the brief sketch we have given is enough to show that the derivation of the expression that relates ρ and ρ_M does not rely on the assumption that the association potential is in some sense weak. It is, however, straightforward to show, starting from (11.10.2), that a first-order perturbation treatment based on a generalisation to the molecular case of the f-expansion[53] of Section 5.3 leads to an expression for the free energy of the associated system.

The first step is to replace the unknown function $g_{MM}(1, 2)$ in (11.10.2) by the pair distribution function of the underlying hard-sphere reference system, $g_0(1, 2)$:

$$\rho \approx \rho_M + \rho_M^2 \int g_0(1, 2) \langle f_A(1, 2) \rangle_{\Omega_1 \Omega_2} \, d\mathbf{r}_{12}$$
$$\equiv \rho_M + \rho_M^2 D(\rho, T) \tag{11.10.7}$$

Within perturbation theory the total free energy of the mixture of monomers and dimers that constitutes the system of interest is the sum of the free energy of the reference system and the free energy due to association, F^A. The quantity F^A is itself the sum of two terms:

$$F^A = F_1^A + F_2^A \tag{11.10.8}$$

where F_1^A is the contribution from the interaction between monomers, given in terms of the integral $D(\rho, T)$ by the standard, first-order expression[54]

$$\frac{\beta F_1^A}{V} = -\frac{1}{2}\rho_M^2 D(\rho, T) \tag{11.10.9}$$

The second term in (11.10.8) represents the contribution to the free energy that arises from dimerisation. Its form must be such that the equilibrium composition obtained by minimisation of the resulting expression for F^A with respect to ρ_M satisfies (11.10.7). A simple calculation confirms that this requirement is met by taking

$$\frac{\beta F_2^A}{V} = \rho \ln \rho_M \Lambda^3 - \rho_M \tag{11.10.10}$$

Equation (11.10.7) can be rewritten as

$$x^2 \rho D(\rho, T) + x - 1 = 0 \tag{11.10.11}$$

where $x = \rho_M/\rho$, given by the positive root of this equation, is the fraction of monomers that remain unassociated. Combination of the results in (11.10.9) and (11.10.11) shows that the contribution from monomer–monomer interactions is equal, apart from a sign, to the dimer density:

$$\frac{\beta F_1^A}{V} = -\frac{1}{2}(\rho - \rho_M) = -\rho_D \tag{11.10.12}$$

The total free energy of association is therefore

$$\frac{\beta F^A}{V} = \rho \ln \rho_M \Lambda^3 - (\rho_M + \rho_D) \tag{11.10.13}$$

On recalling the chemical potential of a sphere must be the same in its monomer and dimer states it is easy to see that this is the free energy density of a dimerising ideal gas. Thus the total free energy density due to association relative to that of a system of non-interacting monomers, $F_0^A = \rho \ln \rho \Lambda^3 - \rho$, is given by a very simple expression:

$$\frac{\beta(F^A - F_0^A)}{V} = \rho \ln x + \frac{1}{2}\rho(1 - x) \tag{11.10.14}$$

which, when x is obtained by solution of (11.10.11), is the result derived by Wertheim. The equivalence that emerges in Wertheim's theory between the equilibrium, associated system and a mixture of non-interacting monomers and dimers extends in a generalised form[55] to systems of particles with multiple bonding sites. To be useful, however, that equivalence must be supplemented by a prescription for determining the equilibrium composition.

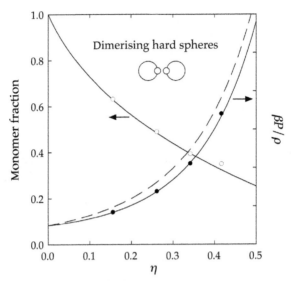

FIGURE 11.9 Equilibrium composition and equation of state of a dimerising hard-sphere fluid at a reduced inverse temperature $\epsilon_A/k_B T = 7$, where ϵ_A is the depth of the square-well potential. The points are the results of Monte Carlo calculations and the full curves are obtained by perturbation theory. The broken curve shows the Carnahan–Starling equation of state for the hard-sphere reference system. After Jackson et al.[56]

Because the association potential is very short ranged, the integral $D(\rho, T)$ can be adequately approximated in the form

$$D(\rho, T) \propto g_0(d)d^3 \langle f_A(1, 2)\rangle_{\Omega_1 \Omega_2} \equiv g_0(d)D'(T) \qquad (11.10.15)$$

where $g_0(d)$ is the value of the hard-sphere distribution function at contact and $D'(T)$ is a microscopic volume,[56] the value of which is dependent on the depth and range of the square-well potential and the displacement δ pictured in Figure 11.8. Once the free energy (including the hard-sphere term) is known, other thermodynamic properties can be obtained by differentiation. Figure 11.9 shows results obtained for the equilibrium composition and equation of state as a function of the hard-sphere packing fraction at a temperature such that $\epsilon_A/k_B T = 7$. The agreement between theory and simulation is very good.

An important feature of Wertheim's approach is the fact that it leads naturally to a theory of polymerisation.[57] This is easily illustrated, again in a non-rigorous way, for the case of dimer formation. The degree of dimerisation approaches unity as the depth of the square-well potential is increased to values appropriate to a covalent bond and the Mayer function $f_A(1, 2)$ becomes correspondingly large. Equation (11.10.11) implies that as the limit of complete dimerisation is approached the monomer fraction must vanish as $x \to 1/[\rho D(\rho, T)]^{1/2}$. If the approximation (11.10.15) is used the free energy of association in the limit

$\rho \to 2\rho_D$ is

$$\frac{\beta F^A}{N} \approx -\frac{1}{2} \ln \rho D'(T) + \frac{1}{2} \qquad (11.10.16)$$

Hence the equation of state of the fully dimerised system, i.e. a fluid of hard diatomics of bond length d, or 'tangent' hard spheres, is

$$\beta(P - P_0) = -\frac{\partial \beta F^A}{\partial V} = -\frac{1}{2}\rho \left(1 + \rho g_0(d)\frac{\partial \ln g_0(d)}{\partial \rho}\right) \qquad (11.10.17)$$

where P_0 is the pressure of the hard-sphere fluid. If the Carnahan–Starling equation of state is used for P_0 the contact value $g_0(d)$ is given by

$$g_0(d) = \frac{(1 - \frac{1}{2}\eta)}{(1 - \eta)^3} \qquad (11.10.18)$$

and (11.10.17) (with $\rho = 2\rho_D$) becomes

$$\frac{\beta P}{\rho_D} = \frac{2(1 + \eta + \eta^2 - \eta^3)}{(1 - \eta)^3} - \frac{\left(1 + \eta - \frac{1}{2}\eta^2\right)}{(1 - \eta)\left(1 - \frac{1}{2}\eta\right)} \qquad (11.10.19)$$

Equation (11.10.19) proves to be remarkably accurate. It yields results that agree with those of simulations of systems of tangent hard spheres to within 0.2% over the full density range.[57a]

As an alternative to (11.10.14) the free energy density due to association can be written as

$$\frac{\beta(F^A - F_0^A)}{V} = \rho \ln(1 - p_B) + \frac{1}{2}\rho p_B \qquad (11.10.20)$$

where $p_B = 1 - x$ is the probability that a bond is formed.[55] The generalisation to systems of particles with multiple but equal numbers of bonding sites is straightforward if it is assumed that bonds are formed independently with equal probabilities and that closed loops are not permitted. If those conditions are met the free energy is directly proportional to f, the number of sites or 'functionality', and consequently

$$\frac{\beta(F^A - F_0^A)}{V} = \rho \ln(1 - p_B)^f + \frac{1}{2}\rho f p_B \qquad (11.10.21)$$

where p_B is given by the mass action equation

$$\frac{p_B}{(1 - p_B)^2} = \rho f D(\rho, T) \qquad (11.10.22)$$

The restriction to open structures, thereby excluding rings, is also present in Wertheim's theory if confined to the lowest level of approximation. This does not, in general, pose a major problem, but there are cases, notably that of hydrogen fluoride, where ring formation cannot be ignored. An extension of the theory described thus far is required to deal with such situations.[58]

Wertheim's theory forms the basis for the general approach called 'statistical associating fluid theory' or SAFT, within which equations of state have been developed for complex fluids of importance in chemical engineering.[59] In its simplest version SAFT is based on a molecular model in which a number of hard-sphere 'segments' are covalently bonded to form a chain-like structure. The segments may represent either single atoms or, more commonly, functional groups, and attractive, dispersion forces act between segments in different chains. Certain segments may also carry a small number of association sites, which can give rise to chain dimerisation or aggregation in larger clusters. The excess free energy of the fluid is therefore given by the sum of four different terms: a hard-sphere term, for which the Carnahan–Starling approximation is used; the contributions from chain formation and association, expressions for which are provided by Wertheim's theory; and a dispersion term, which is treated in a mean field, van der Waals manner. Empirical parameterisations of these and other, more elaborate models have made implementation of SAFT possible for a very wide range of mostly organic fluid and their mixtures, in which particular emphasis has been laid on the study of phase equilibria.[60]

11.11 REORIENTATIONAL TIME-CORRELATION FUNCTIONS

The description of the dynamical properties of molecular liquids differs most obviously from that appropriate to atomic systems through the appearance of a class of reorientational time-correlation functions. We end this chapter by briefly considering some of the properties of these functions, limiting ourselves mainly to the case of linear molecules. We consider first the simpler problem of the single-molecule functions, leaving until later the question of collective reorientational properties.

Reorientation of a linear molecule can be described in a compact way through the introduction of a family of time-correlation functions defined as

$$C^{(l)}(t) = \langle P_l[\mathbf{u}_i(t) \cdot \mathbf{u}_i] \rangle \qquad (11.11.1)$$

where, as before, $\mathbf{u}_i$ is a unit vector parallel to the internuclear axis of molecule i and $P_l(\cdots)$ is again a Legendre polynomial. The functions $C^{(l)}(t)$ are time-dependent generalisations of the angular order parameters G_l of Section 11.1. Apart from their application to linear molecules they are also the most important functions for the description of the reorientational motion of spherical-top molecules, i.e. those in which all three principal moments of inertia are the same (CCl_4, SF_6, etc.), and of the reorientation of the main symmetry axis of symmetric-top molecules, i.e. those in which two of the principal moments of inertia are equal (NH_3, CH_3I, etc.). The $l = 1$ and $l = 2$ functions are related to the spectral bandshapes measured in infrared absorption ($l = 1$) and Raman or depolarised light scattering ($l = 2$) experiments. Information on the correlation functions can be obtained by Fourier transformation of the experimental spectra, but the interpretation of the results is complicated by a number of factors, including uncertainty about the contributions to the spectra from vibrational relaxation and collision-induced effects or, in the case of depolarised light scattering, the importance of angular correlations of the type described by the order parameter G_2.

Figure 11.10 shows some typical results for the $l = 2$ function derived from spectroscopic measurements on carbon dioxide[61] in two very different thermodynamic states and, in the inset, liquid methyl cyanide[62] (or acetonitrile, CH_3CN, a symmetric-top molecule). Under liquid state conditions the function is approximately exponential in form, except at short times, but at low densities oscillations appear; infrared-absorption experiments on polar molecules give qualitiatively similar results for the $l = 1$ function. The oscillations seen at low densities can be understood by considering the behaviour of the correlation functions in the ideal-gas limit. Let $\boldsymbol{\omega} = \mathbf{u} \times \dot{\mathbf{u}}$ be the angular velocity of a linear molecule of moment of inertia I. In the absence of any interactions the angular velocity is a constant of the motion, and in a time t the molecule will rotate through an angle $\omega t = \cos^{-1} \mathbf{u}(t) \cdot \mathbf{u}(0)$, where $\omega \equiv |\boldsymbol{\omega}|$. The probability that a molecule will rotate through such an angle is therefore determined by the probability that $|\boldsymbol{\omega}|$ lies in the range $\omega \to \omega + d\omega$. Thus the correlation function $C^{(l)}(t)$ is the value of $P_l(\cos \omega t)$ averaged over a Maxwell distribution of angular velocities and appropriately normalised, i.e.

$$C^{(l)}(t) = \frac{I}{k_B T} \int_0^\infty P_l(\cos \omega t) \exp\left(-\frac{1}{2}\beta I \omega^2\right) \omega \, d\omega \qquad (11.11.2)$$

These functions are oscillatory and tend to zero as $t \to \infty$ only for odd l. They are commonly called the 'free-rotor' correlation functions and the oscillations seen in gas phase experimental results are the remnants of free-rotor behaviour. Similar results are obtained for the free-rotor functions of non-linear molecules; the principle of the calculation is the same but the final expressions have a more complicated form.[63]

The short-time expansion of the Legendre polynomial in (11.11.2) starts as

$$P_l(\cos \omega t) = 1 - \frac{1}{4}l(l+1)\omega^2 t^2 + \cdots \qquad (11.11.3)$$

If we expand the correlation function in powers of t:

$$C^{(l)}(t) = 1 - M_2^{(l)} \frac{t^2}{2!} + \cdots \qquad (11.11.4)$$

a simple integration shows that

$$M_2^{(l)} = l(l+1)\frac{k_B T}{I} \qquad (11.11.5)$$

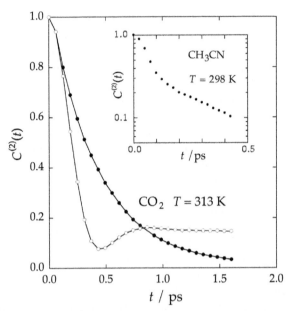

FIGURE 11.10 Main figure: the $l = 2$ reorientational correlation function derived from experiments on liquid and gaseous carbon dioxide. Open and closed circles show the results for $\rho/\rho_c = 0.09$ and $\rho/\rho_c = 2.35$, respectively, where ρ_c is the critical density. *Redrawn with permission from Ref. 61* © *Taylor & Francis Limited.* Inset: the $l = 2$ function for liquid methyl cyanide plotted on a logarithmic scale. From T. Bien et al. "Studies of molecular motions and vibrational relaxation in acetonitrile. VII. *Chem. Phys.* **56**, 203–211 (1981), *with permission of Elsevier.*

At sufficiently short times a molecule rotates freely. Hence, although (11.11.5) has been derived only in the free-rotor limit, it is also valid for interacting molecules; there is an analogy here with the short-time behaviour of the mean-square translational displacement. From the general properties of time-correlation functions discussed in Section 7.1 it follows that the coefficient $M_2^{(l)}$ is the second moment of the power spectrum of $C^{(l)}(t)$. The mean-square width of the experimental bandshape is therefore independent of the molecular interactions. The fourth moment, however, contains a contribution proportional to the mean-square torque acting on the molecule.

The quasi-exponential behaviour of the correlation functions at high densities can be rationalised by invoking an approximation similar in spirit to the Langevin equation (7.3.21). We begin by writing a memory function equation for $C^{(l)}(t)$ and taking the Laplace transform to give

$$\tilde{C}^{(l)}(\omega) = \frac{1}{-i\omega + \tilde{N}^{(l)}(\omega)} \tag{11.11.6}$$

From (11.11.5) it follows that the memory function $N^{(l)}(t)$ behaves as

$$N^{(l)}(t) = l(l+1)\left(\frac{k_{\mathrm{B}}T}{I}\right)n^{(l)}(t) \qquad (11.11.7)$$

with $n^{(l)}(0) = 1$. We now suppose that reorientation occurs as the result of a succession of small, uncorrelated steps. This is the Debye approximation or 'small-step-diffusion' model. In memory function terms the Debye approximation is equivalent to the assumption that $\tilde{N}^{(l)}(\omega)$ is independent of frequency. To preserve the l-dependence contained in the exact result (11.11.7) we approximate the memory function in the form $\tilde{N}^{(l)} \approx l(l+1)D_{\mathrm{R}}$, where D_{R} (a frequency) is a 'rotational diffusion coefficient'. Then

$$C^{(l)}(t) = \exp\left[-l(l+1)D_{\mathrm{R}}t\right] \qquad (11.11.8)$$

In this approximation the correlation functions decay exponentially at all times and for all values of l, and the entire family of functions is characterised by the single parameter D_{R}; for small molecules under triple-point conditions, D_{R} is typically of order 10^{11} s^{-1}. The characteristic decay times for different values of l are related by the simple rule that

$$\frac{\tau_l}{\tau_{l+1}} = \frac{l+2}{l} \qquad (11.11.9)$$

The correlation times derived from infrared and Raman measurements should therefore be in the ratio $\tau_1/\tau_2 = 3$. This is approximately true of many liquids and also of correlation times obtained by simulation.

A weakness of the Debye approximation is its neglect of the fact that molecules rotate freely at short times. It therefore cannot account for the quadratic time dependence of the reorientational correlation functions at small t. A more complete theory must also describe correctly the details of the transition to the long-time, quasi-exponential behaviour. In the case of methyl cyanide, for example, Figure 11.10 shows that the transition region is characterised by a marked change in slope of the curve of $\ln C^{(2)}(t)$ versus t. The behaviour in the different time regimes can be described in a unified way[64] by relating the form of the reorientational correlation functions to that of the angular velocity autocorrelation function $C_\omega(t)$. By analogy with (7.2.6) and (7.2.7) the rotational diffusion coefficient of a linear molecule can be defined as

$$D_{\mathrm{R}} = \frac{k_{\mathrm{B}}T}{I}\lim_{t\to\infty}\int_0^t\left(1-\frac{s}{t}\right)C_\omega(s)\,\mathrm{d}s \qquad (11.11.10)$$

where

$$C_\omega(t) = \frac{\langle\boldsymbol{\omega}_i(t)\cdot\boldsymbol{\omega}_i\rangle}{\langle|\boldsymbol{\omega}_i|^2\rangle} = \frac{I}{2k_{\mathrm{B}}T}\langle\boldsymbol{\omega}_i(t)\cdot\boldsymbol{\omega}_i\rangle \qquad (11.11.11)$$

Then substitution of (11.11.10) in (11.11.8) gives an expression for $C^{(l)}(t)$ in terms of $C_\omega(t)$:

$$\ln C^{(l)}(t) = -l(l+1)\left(\frac{k_B T}{I}\right)\int_0^t (t-s)C_\omega(s)\mathrm{d}s \qquad (11.11.12)$$

The main merit of this approximation is the fact that it contains the correct short-time behaviour yet goes over to the Debye model at long times. Let τ_ω be the integral correlation time for the angular velocity, i.e.

$$\tau_\omega = \int_0^\infty C_\omega(t)\mathrm{d}t \qquad (11.11.13)$$

At times $t \ll \tau_\omega$, $C_\omega(t) \approx 1$ and (11.11.12) becomes

$$\ln C^{(l)}(t) \approx -l(l+1)\left(\frac{k_B T}{I}\right)\frac{t^2}{2} \qquad (11.11.14)$$

in agreement with the exact result (11.11.5). In the opposite limit, $t \gg \tau_\omega$, (11.11.12) becomes

$$\ln C^{(l)}(t) \approx -l(l+1)\left(\frac{k_B T}{I}\right)\tau_\omega t \qquad (11.11.15)$$

which is equivalent to the Debye approximation (11.11.8) with the identification $D_R = (k_B T/I)\tau_\omega$. Finally, the behaviour at intermediate times is related to the shape of the function $C_\omega(t)$. Differentiating (11.11.12) twice with respect to t we find that

$$\frac{\mathrm{d}^2 \ln C^{(l)}(t)}{\mathrm{d}t^2} = -l(l+1)\left(\frac{k_B T}{I}\right)C_\omega(t) \qquad (11.11.16)$$

The angular velocity autocorrelation function is not measurable experimentally, but molecular dynamics calculations show that for liquids such as methyl cyanide, in which the intermolecular torques are strong, it decays rapidly at short times and then becomes negative. The change in sign occurs because the direction of the angular velocity vector is on average soon reversed; the behaviour is similar to that seen in the linear velocity autocorrelation function at high densities and low temperatures (see Figure 7.1). Equation (11.11.16) shows that a change in sign of $C_\omega(t)$ corresponds to a point of inflection in $\ln C^{(l)}(t)$ of the type visible in Figure 11.10, which in turn is a common feature of the reorientational correlation functions of high-torque fluids.

A variety of theoretical schemes have been devised to treat those cases in which the Debye model is inadequate. Many of these are expressible in terms of simple approximations for the relevant memory functions, but none has proved to be satisfactory either for any large group of molecules or for any particular molecule over a wide range of density and temperature. The failure to develop a quantitatively reliable theory is striking in view of the apparent simplicity in structure of the correlation functions themselves.

We have focused until now on the reorientational motion of single molecules. There are, in addition, a number of collective reorientational correlation functions of experimental significance that are many-particle generalisations of single-particle functions. It is therefore of interest to establish an approximate relation between corresponding collective and single-particle quantities and, in particular, between the two correlation times, since this allows a connection to be made between the results of very different experiments. We take as an example the collective motions that determine the frequency-dependent, dielectric behaviour of a polar fluid,[16] as described by the complex dielectric permittivity $\epsilon(\omega)$ introduced in Section 11.5. The quantities of interest in the study of dielectric relaxation are the correlation functions and associated power spectra of the longitudinal (l) and transverse (t) components of the dipole-moment density (11.5.13), i.e.

$$C_l(k,t) = \frac{\langle M_{\mathbf{k}}^z(t) M_{-\mathbf{k}}^z \rangle}{\langle |M_{\mathbf{k}}^z|^2 \rangle}, \quad C_t(k,t) = \frac{\langle M_{\mathbf{k}}^x(t) M_{-\mathbf{k}}^x \rangle}{\langle |M_{\mathbf{k}}^x|^2 \rangle} \tag{11.11.17}$$

where we have followed the usual convention that $\mathbf{k}$ is parallel to the z-axis. These two functions are collective analogues, generalised to finite wavelengths, of the single-molecule function $C^{(1)}(t)$. It follows from (11.5.9) and (11.5.12) that the long-wavelength limits of the Laplace transforms $\tilde{C}_l(k,\omega)$ and $\tilde{C}_t(k,\omega)$ are related to $\epsilon(\omega)$ by

$$\frac{4\pi\beta}{V} \lim_{\mathbf{k}\to 0} \langle |M_{\mathbf{k}}^z|^2 \rangle [1 + i\omega\tilde{C}_l(k,\omega)] = \frac{\epsilon(\omega) - 1}{\epsilon(\omega)}$$
$$\frac{4\pi\beta}{V} \lim_{\mathbf{k}\to 0} \langle |M_{\mathbf{k}}^x|^2 \rangle [1 + i\omega\tilde{C}_t(k,\omega)] = \epsilon(\omega) - 1 \tag{11.11.18}$$

We begin by writing memory function equations for $C_l(k,t)$ and $C_t(k,t)$ in the form

$$\tilde{C}_l(k,\omega) = \frac{1}{-i\omega + \tilde{N}_l(k,\omega)}, \quad \tilde{C}_t(k,\omega) = \frac{1}{-i\omega + \tilde{N}_t(k,\omega)} \tag{11.11.19}$$

The initial values of the memory functions $N_l(k,t)$ and $N_t(k,t)$ in the limit $k \to 0$ can be deduced from the general property (9.1.29) and the limiting behaviour described by (11.11.18), taken for $\omega = 0$:

$$\lim_{\mathbf{k}\to 0} N_l(k, t=0) = \lim_{\mathbf{k}\to 0} \frac{\langle |\dot{M}_{\mathbf{k}}^z|^2 \rangle}{\langle |M_{\mathbf{k}}^z|^2 \rangle} = \frac{4\pi\beta\epsilon}{3V(\epsilon - 1)} \langle |\dot{\mathbf{M}}|^2 \rangle$$

$$\lim_{\mathbf{k}\to 0} N_t(k, t=0) = \lim_{\mathbf{k}\to 0} \frac{\langle |\dot{M}_{\mathbf{k}}^x|^2 \rangle}{\langle |M_{\mathbf{k}}^x|^2 \rangle} = \frac{4\pi\beta}{3V(\epsilon - 1)} \langle |\dot{\mathbf{M}}|^2 \rangle \tag{11.11.20}$$

where $\epsilon \equiv \epsilon(0)$ and $\dot{\mathbf{M}} \equiv \dot{\mathbf{M}}_{\mathbf{k}\to 0}$. In deriving these results we have exploited the fact that the different components of $\langle |\dot{\mathbf{M}}_{\mathbf{k}}|^2\rangle$ (unlike those of $\langle |\mathbf{M}_{\mathbf{k}}|^2\rangle$) are equivalent and, in particular, that $\lim_{\mathbf{k}\to 0}\langle |\dot{M}_{\mathbf{k}}^\alpha|^2\rangle = \frac{1}{3}\langle |\dot{\mathbf{M}}|^2\rangle$, where $\alpha = x, y$ or z.

The form of (11.11.20) makes it convenient to write the memory functions at long wavelengths as

$$\lim_{\mathbf{k}\to 0} \tilde{N}_l(k,\omega) = \frac{\epsilon \tilde{R}_l(\omega)}{\epsilon - 1}, \quad \lim_{\mathbf{k}\to 0} \tilde{N}_t(k,\omega) = \frac{\tilde{R}_t(\omega)}{\epsilon - 1} \tag{11.11.21}$$

It is clear from comparison of (11.11.20) with (11.11.21) that $R_l(t = 0) = R_t(t = 0) = (4\pi\beta/3V)\langle |\dot{\mathbf{M}}|^2\rangle$. More generally, if the two parts of (11.11.18) are to be consistent with each other in the sense of giving the same result for $\epsilon(\omega)$, some straightforward algebra shows that $R_l(t)$ and $R_t(t)$ must be the same for all t. Thus

$$R_l(t) = R_t(t) = R(t), \text{ say} \tag{11.11.22}$$

This has the immediate consequence that in the long-wavelength limit the correlation times for the longitudinal and transverse functions differ by a factor ϵ, i.e.

$$\lim_{\mathbf{k}\to 0} \tilde{C}_l(k,0) = \epsilon^{-1} \lim_{\mathbf{k}\to 0} \tilde{C}_t(k,0) \tag{11.11.23}$$

or

$$\lim_{\mathbf{k}\to 0} \int_0^\infty C_l(k,t)\mathrm{d}t = \epsilon^{-1} \lim_{\mathbf{k}\to 0} \int_0^\infty C_t(k,t)\mathrm{d}t \tag{11.11.24}$$

The diffusion approximation analogous to (11.11.8) now corresponds to setting

$$R(t) = R(0)\delta(t) = \frac{4\pi\beta}{3V}\langle |\dot{\mathbf{M}}|^2\rangle\delta(t) \tag{11.11.25}$$

so that both $\tilde{N}_l(k,\omega)$ and $\tilde{N}_t(k,\omega)$ are assumed to be independent of frequency in the limit $\mathbf{k}\to 0$. If we define a characteristic time τ_{D} as

$$\tau_{\mathrm{D}} = \frac{3V}{4\pi\beta}\frac{\epsilon - 1}{\langle |\dot{\mathbf{M}}|^2\rangle} \tag{11.11.26}$$

it follows from (11.11.19) and (11.11.21) that

$$\lim_{\mathbf{k}\to 0} C_l(k,t) = \exp(-\epsilon t/\tau_{\mathrm{D}}), \quad \lim_{\mathbf{k}\to 0} C_t(k,t) = \exp(-t/\tau_{\mathrm{D}}) \tag{11.11.27}$$

which represents a special case of the general result in (11.11.24). Simulations of strongly polar fluids confirm that the longitudinal and transverse correlation functions at small $\mathbf{k}$ do decay on very different timescales and that the ratio of correlation times is approximately equal to the value of ϵ derived from fluctuations in the mean-square dipole moment of the sample.[65] The transverse function is also approximately exponential in form with a decay time only weakly dependent on $\mathbf{k}$, in qualitative agreement with (11.11.27), but the longitudinal function has oscillations at large t; to describe these oscillations it is necessary to allow for some frequency dependence of the memory function. The approximation for $\epsilon(\omega)$ corresponding to (11.11.27) is

$$\frac{\epsilon(\omega) - 1}{\epsilon(0) - 1} = \frac{1}{1 - i\omega\tau_D} \qquad (11.11.28)$$

This is an expression much used in the analysis of experimental data on $\epsilon(\omega)$, in which context τ_D is invariably called the Debye relaxation time. A feature of the approximation is the fact that the curve, or Cole–Cole plot, of the real versus imaginary part of $\epsilon(\omega)$ is a semicircle with a maximum at a frequency such that $\omega\tau_D = 1$. Many real liquids have Cole–Cole plots that are approximately semicircular. Because of its neglect of short-time, inertial effects, the diffusion approximation is least satisfactory at high frequencies, where the deviations from (11.11.28) are mostly to be found. However, as in the case of the single-molecule problem, it has proved difficult to develop an alternative theory having a wide range of applicability.

One goal of dielectric relaxation theory is to relate the decay times that characterise the collective functions (11.11.17) and the single-molecule correlation function $C^{(1)}(t)$. The necessary link can be established by postulating some relationship between the memory functions $R(t)$ and $N^{(1)}(t)$. A simple but useful result is obtained by supposing that the two memory functions have the same time dependence, but also have their correct initial values. It follows from (9.1.29) that

$$N^{(1)}(0) = \left\langle |\dot{\mathbf{u}}_i|^2 \right\rangle = \frac{\left\langle |\dot{\mathbf{M}}|^2 \right\rangle}{N\mu^2} = \frac{R(0)}{3y} \qquad (11.11.29)$$

where y is the molecular parameter defined by (11.5.16). If, for simplicity, we adopt the diffusion model, we find immediately from the definition (11.11.26) that

$$\tau_D = \left(\frac{\epsilon - 1}{3y} \right) \tau_1 \qquad (11.11.30)$$

or, after substitution from the Kirkwood formula (11.5.15):

$$\tau_D = \left(\frac{3\epsilon g_K}{2\epsilon + 1} \right) \tau_1 \qquad (11.11.31)$$

This expression relates the dielectric relaxation time to the correlation time measured by infrared spectroscopy in a form determined solely by static dielectric properties of the fluid.

REFERENCES

[1] Gray, C.G. and Gubbins, K.E., 'Theory of Molecular Fluids', vol. 1. Clarendon Press, Oxford, 1984, p. 151.
[2] Steele, W.A., J. Chem. Phys. **39**, 3197 (1963).

[3] For a discussion of the properties of the spherical harmonics and their use in liquid state theory, see Ref. 1, particularly Appendix A.

[4] See, e.g., Streett, W.B. and Tildesley, D.J., *Proc. Roy. Soc.* **A355**, 239 (1977).

[5] The general procedure is described by Høye, J.S. and Stell, G., *J. Chem. Phys.* **66**, 795 (1977).

[6] Sullivan, D.E. and Gray, C.G., *Mol. Phys.* **42**, 443 (1981).

[7] A correction for the effects of nuclear vibration is also usually made.

[8] Narten, A.H., Johnson, E. and Habenschuss A., *J. Chem. Phys.* **73**, 1248 (1980).

[9] X-ray scattering from methane is a special case. See Habenschuss, A., Johnson, E. and Narten, A.H., *J. Chem. Phys.* **74**, 5234 (1981).

[10] Weis, J.J. and Levesque, D., *Phys. Rev. A* **13**, 450 (1976).

[11] Narten, A.H. and Habenschuss, A., *J. Chem. Phys.* **75**, 3073 (1981).

[12] (a) Narten, A.H., Agrawal, R. and Sandler, S.I., *Mol. Phys.* **35**, 1077 (1978). (b) Agrawal, R., Sandler, S.I. and Narten, A.H., *Mol. Phys.* **35**, 1087 (1978).

[13] Wertheim, M.S., *J. Chem. Phys.* **55**, 4291 (1971).

[14] See, e.g., Stell, G., Patey, G.N. and Høye, J.S., *Adv. Chem. Phys.* **48**, 183 (1981).

[15] Fries, P.H. and Patey, G.N., *J. Chem. Phys.* **82**, 429 (1985).

[16] We follow in outline the arguments of Madden, P.A. and Kivelson, D., *Adv. Chem. Phys.* **56**, 467 (1984).

[17] (a) Høye, J.S. and Stell, G., *J. Chem. Phys.* **65**, 18 (1976). (b) Chandler, D. *J. Chem. Phys.* **67**, 1113 (1977).

[18] Ramshaw, J.D., *J. Chem. Phys.* **70**, 1577 (1979).

[19] Note the similarity with the relation (10.4.11) between the external and screened susceptibilities of charged fluids.

[20] Kirkwood, J.G., *J. Chem. Phys.* **7**, 911 (1939).

[21] Høye, J.S., Lebowitz, J.L. and Stell, G., *J. Chem. Phys.* **61**, 3253 (1974).

[22] See Ref. 1, p. 527.

[23] The same result follows directly from the Stillinger–Lovett sum rules (10.2.17). See Martin, P.A., *Rev. Mod. Phys.* **60**, 1075 (1988).

[24] Weis, J.J., unpublished results.

[25] Cummings, P. and Blum, L., *J. Chem. Phys.* **85**, 6658 (1986).

[26] Henderson, D., Boda, D., Szala, I. and Chan, K.Y., *J. Chem. Phys.* **110**, 7348 (1999).

[27] (a) Patey, G.N., *Mol. Phys.* **34**, 427 (1977). (b) Patey, G.N., *Mol. Phys.* **35**, 1413 (1978). (c) Patey, G.N., Levesque, D. and Weis, J.J., *Mol. Phys.* **38**, 219 (1979). (d) Patey, G.N., Levesque, D. and Weis, J.J., *Mol. Phys.* **38**, 1635 (1979).

[28] Wertheim, M.S., *Mol. Phys.* **26**, 1425 (1973).

[29] Lee, L.Y., Fries, P.H. and Patey, G.N., *Mol. Phys.* **55**, 751 (1985).

[30] The dipolar hard-sphere model has other unsatisfactory features. See Teixeira, P.I.C., Tavares, J.M. and Telo da Gama, M.M. *J. Phys. Condens. Matter* **12**, R4111 (2000).

[31] (a) Lado, F., *Mol. Phys.* **47**, 283 (1982). (b) Lado, F., Lombardero, M., Enciso, E., Abascal, J.L.F. and Lago, S., *J. Chem. Phys.* **85**, 2916 (1986). (c) Lomba, E., Lombardero, M. and Abascal, J.L.F., *J. Chem. Phys.* **91**, 2581 (1989).

[32] See, e.g., Lombardero, M., Martin, C., Lomba, E. and Lado, F., *J. Chem. Phys.* **104**, 6710 (1996).

[33] Ladanyi, B.M. and Chandler, D., *J. Chem. Phys.* **62**, 4308 (1975).

[34] Andersen, H.C. and Chandler, D., *J. Chem. Phys.* **53**, 547 (1970).

[35] Chandler, D., *Mol. Phys.* **31**, 1213 (1976).

[36] Monson, P.A., *Mol. Phys.* **47**, 435 (1982).

[37] See, e.g., Munaò, G., Costa, D., Saija, F. and Caccamo, C., *J. Chem. Phys.* **132**, 084506 (2010).

[38] (a) This remarkable fact was first pointed out by Chandler, D., *Faraday Disc. Chem. Soc.* **66**, 74 (1978) and later expanded upon by Sullivan and Gray, Ref. 6. (b) The derivation given in the text follows the method of Sullivan and Gray. See also Morriss, G.P. and Perram, J.W., *Mol. Phys.* **43**, 669 (1981).

[39] A third, more ambitious approach is based on the recognition that a 'diagrammatically proper' theory for interaction site models should be consistent with the full molecular description provided by 11.2.9. See Dyer, K.P., Perkyns, J.S. and Pettitt, B.M., *J. Chem. Phys.* **127**, 194506 (2007).

[40] (a) Cummings, P.T. and Stell, G., *Mol. Phys.* **46**, 383 (1982). (b) For later work on closure of the RISM-OZ relation, see, e.g., Raineri, F.O. and Stell, G., *J. Phys. Chem. B* **105**, 11880 (2001).

[41] (a) Chandler, D., Silbey, R. and Ladanyi, B.M., *Mol. Phys.* **46**, 1335 (1982). (b) Chandler, D., Joslin, C.G. and Deutch, J.M., *Mol. Phys.* **47**, 871 (1982). (c) Chandler, D. and Richardson, D.M., *J. Phys. Chem.* **87**, 2060 (1983). (d) Rossky, P.J. and Chiles, R.A., *Mol. Phys.* **51**, 661 (1984).

[42] See, e.g., Reddy, G., Lawrence, C.P., Skinner, J.L. and Yethiraj, A., *J. Chem. Phys.* **119**, 13012 (2003).

[43] Sorensen, J.M., Hura, G., Glaeser, R.M. and Head-Gordon, T., *J. Chem. Phys.* **113**, 9149 (2000).

[44] Soper, A.K., *Chem. Phys.* **258**, 121 (2000).

[45] Yarnell, J.L., Katz, M.J., Wenzel, R.G. and Koenig, S.H., *Phys. Rev. A* **7**, 2130 (1973).

[46] Mahoney, M.W. and Jorgensen, W.L., *J. Chem. Phys.* **112**, 8910 (2000). TIP5P is one of a series of 'transferable intermolecular potentials' devised for the simulation of water in different environments.

[47] For a review of early work, see Guillot, B., *J. Mol. Liq.* **101**, 219 (2002).

[48] Sanz, E., Vega, C., Abascal, J.L.F. and MacDowell, L.G., *Phys. Rev. Lett.* **92**, 255701 (2004).

[49] Examples include: (a) Lue, L. and Blankschtein, D., *J. Chem. Phys.* **102**, 4203 (1995). (b) Lombardero, M., Martin, C., Jorge, S., Lado, F. and Lomba, E., *J. Chem. Phys.* **110**, 1148 (1999). (c) Dyer, K.P., Perkyns, J.S., Stell, G. and Pettitt, B.M., *J. Chem. Phys.* **129**, 104512 (2008). (d) Kežić, B. and Perera, A., *J. Chem. Phys.* **135**, 234104 (2011).

[50] (a) Andersen, H.C., *J. Chem. Phys.* **59**, 4714 (1973). (b) Andersen, H.C., *J. Chem. Phys.* **61**, 4985 (1974).

[51] Cummings, P.T. and Stell, G., *Mol. Phys.* **51**, 253 (1984).

[52] (a) Wertheim, M.S., *J. Stat. Phys.* **35**, 19 (1984). (b) Wertheim, M.S., *J. Stat. Phys.* **35**, 35 (1984).

[53] The argument which follows is a simplified version of that given by Joslin, C.G., Gray, C.G., Chapman, W.G. and Gubbins, K.E., *Mol. Phys.* **62**, 843 (1987).

[54] See (5.3.6). Use of a first-order approximation is justified by the fact that when $D(\rho, T)$ is very large, as is true for strong association, ρ_M^2 will be very small.

[55] Bianchi, E., Tartaglia, P., Zaccarelli, E. and Sciortino, F., *J. Chem. Phys.* **128**, 144504 (2008).

[56] Jackson, G., Chapman, W.G. and Gubbins, K.E., *Mol. Phys.* **65**, 1 (1988).

[57] (a) Wertheim, M.S., *J. Chem. Phys.* **85**, 2929 (1986). (b) Wertheim, M.S., *J. Chem. Phys.* **87**, 7323 (1987). (c) Chapman, W.G., Jackson, G. and Gubbins, K.E., *Mol. Phys.* **65**, 1057 (1988).

[58] Galindo, A., Burton, S.J., Jackson, G., Visco, D.P. and Kofke, D.A., *Mol. Phys.* **100**, 2241 (2002).

[59] (a) Chapman, W.G., Gubbins, K.E., Jackson, G. and Radosz, M., *Fluid Phase Equil.* **52**, 31 (1989). (b) Chapman, W.G., Gubbins, K.E., Jackson, G. and Radosz, M., *Ind. Eng. Chem. Res.* **29**, 1709 (1990).

[60] For a review, see McCabe, C. and Galindo, A., *In* 'Applied Thermodynamics of Fluids' (A.R.H. Goodwin, J.V. Sengers and C.J. Peters, eds). RSC Publishing, Cambridge, 2010.

[61] Versmold, H., *Mol. Phys.* **43**, 383 (1981).

[62] Bien, T., Possiel, M., Döge, G., Yarwood, J. and Arnold, K.E., *Chem. Phys.* **56**, 203 (1981).

[63] St. Pierre, A.G. and Steele, W.A., *Mol. Phys.* **43**, 123 (1981).

[64] Lynden-Bell, R.M. and McDonald, I.R., *Mol. Phys.* **43**, 1429 (1981).

[65] Pollock, E.L. and Alder, B.J., *Phys. Rev. Lett.* **46**, 950 (1981).

Chapter 12

Applications to Soft Matter

In this chapter we describe some of the ways in which methods taken from the theory of simple liquids have been adapted to the study of much more complicated, molecular or macromolecular assemblies, generically referred to as *complex fluids* or *soft matter*.[1] Those labels encompass a wide variety of systems including, among others, polymer solutions and polymer melts, dispersions of colloidal particles of various sizes and shapes, thermotropic and lyotropic liquid crystals, and micelles, membranes or vesicles formed by ampiphilic surfactant molecules such as lipids. A theme common to all the example we discuss is the abandonment of the microscopic picture employed for simple systems in favour of a coarse-grained representation from which unnecessary detail has been eliminated.

12.1 COARSE GRAINING AND EFFECTIVE INTERACTIONS

The main barrier to the development of a fully microscopic description of complex fluids lies in the coexistence of widely different scales of length and time within the same system. Consider, for example, the case of an aqueous dispersion of colloidal particles that carry a high surface charge. The dimensions of the dispersed particles are typically of the order of tens or hundreds of nanometres and they move on time scales of the order of a nanosecond or microsecond, whereas the water molecules and microscopic counterions are much smaller (a fraction of a nanometre) and move much faster (on a picosecond time scale). A treatment of such highly asymmetric systems by the theoretical methods of earlier chapters would therefore be impractical, as would the use of numerical methods such as molecular dynamics simulation. However, the focus of interest often lies in the mesoscopic structure and dynamics of the colloidal particles rather than in the microscopic behaviour of the much smaller and lighter molecules and ions, which provide a thermal bath through which the large particles move. In that situation the problem can be greatly simplified by adoption of a coarse-graining strategy in which statistical averages are taken over the microscopic degrees of freedom for given configurations of the large

Theory of Simple Liquids, Fourth Edition. http://dx.doi.org/10.1016/B978-0-12-387032-2.00012-X

particles. This approach leads to the definition of effective interactions between 'dressed' colloidal particles which no longer depend on the microscopic details of the bath. Dynamical coarse graining switches from the newtonian equations of motion of both small and large particles to Langevin or brownian dynamics of the large particles alone; this involves a more elaborate procedure. We shall see examples of both forms of coarse graining in later sections.

We show first how a formally exact expression for the effective interaction between large particles can be derived, taking as a simple example that of a system consisting of N_1 large, spherical particles with coordinates $\{\mathbf{R}_i\}$ and $N_2 \gg N_1$ small particles with coordinates $\{\mathbf{r}_j\}$, all contained in a volume Ω. The total potential energy of the system is the sum of three terms:

$$V_{N_1 N_2}(\{\mathbf{R}_i\}, \{\mathbf{r}_j\}) = V_{11}(\{\mathbf{R}_i\}) + V_{22}(\{\mathbf{r}_j\}) + V_{12}(\{\mathbf{R}_i\}, \{\mathbf{r}_j\}) \quad (12.1.1)$$

corresponding to the interactions between large particles, that between small particles and the cross-interaction between them. The configuration integral of the mixture is then

$$
\begin{aligned}
Z_{N_1 N_2} &= \int \exp\left(-\beta V_{N_1 N_2}\right) \mathrm{d}\mathbf{R}^{N_1} \, \mathrm{d}\mathbf{r}^{N_2} \\
&= \int \mathrm{d}\mathbf{R}^{N_1} \exp\left(-\beta V_{11}\right) \int \mathrm{d}\mathbf{r}^{N_2} \exp[-\beta(V_{22} + V_{12})] \\
&= \Omega^{N_2} \int \mathrm{d}\mathbf{R}^{N_1} \exp[-\beta V_{11}(\{\mathbf{R}_i\}) - \beta F_2^{\mathrm{ex}}(\{\mathbf{R}_i\})] \quad (12.1.2)
\end{aligned}
$$

where the definition (2.3.20) has been used. The quantity $F_2^{\mathrm{ex}}(\{\mathbf{R}_i\})$ is the excess free energy of the system of small particles in the external field V_{12} due to large particles fixed at positions $\{\mathbf{R}_i\}$. The dimensionless configuration integral $Z_{N_1 N_2}/\Omega^N$, with $N = N_1 + N_2$, may therefore be written as

$$\frac{Z_{N_1 N_2}}{\Omega^N} = \frac{Z_{N_1}}{\Omega^{N_1}} = \frac{1}{\Omega^{N_1}} \int \exp[-\beta V_{11}^{\mathrm{eff}}(\{\mathbf{R}_i\})] \mathrm{d}\mathbf{R}^{N_1} \quad (12.1.3)$$

where the total, effective interaction between the large particles is

$$V_{11}^{\mathrm{eff}}(\{\mathbf{R}_i\}) = V_{11}(\{\mathbf{R}_i\}) + F_2^{\mathrm{ex}}(\{\mathbf{R}_i\}) \quad (12.1.4)$$

The first term on the right-hand side corresponds to the direct interactions between large particles; the second term represents the interactions induced by the small ones. The induced term arises from integration over the coordinates of the small particles. It is therefore a free energy that depends parametrically on the positions of the large particles, meaning that the effective interaction is a state-dependent quantity to which there is an entropic contribution given by $-T S_2^{\mathrm{ex}}(\{\mathbf{R}_i\})$. Because the induced term is many-body in nature, the effective interaction between the large particles will not be pairwise additive, even when that is true of all three contributions to $V_{N_1 N_2}$.

Explicit calculation of the effective potential is, in general, a difficult problem. Its solution relies on the use of approximations, often implemented within the framework of the density functional theory of an inhomogeneous fluid of small particles in the spatially varying potential $V_{12}(\{\mathbf{R}_i\}, \{\mathbf{r}_j\})$ due to the large particles. Two classic examples are the screened interaction between electric double layers, already discussed in Section 10.6, and the depletion interaction between colloidal particles induced by non-adsorbing polymers. Similar methods have been used in coarse-grained descriptions of interacting polymers in solution. We shall say more about all three problems in later sections. Two other, important examples of effective interaction should also be mentioned. One is the hydrophobic attraction between nano-sized or larger particles referred to at the end of Section 6.7. This plays a key role in molecular biology in relation to the stability of protein solutions. The second example is that of the fluctuation-induced Casimir force between surfaces, an effect that results from spatial confinement of critical fluctuations near a second-order phase transition. Let us take the case of a binary, liquid mixture near its critical consolute point, which is analogous to the liquid–gas critical point. The correlation length ξ, which measures the distance over which local concentrations are correlated, diverges when the critical temperature is approached at the critical concentration. If a near-critical mixture is confined to the region between the surfaces of two large, colloidal particles, critical fluctuations cannot develop beyond the surface-to-surface distance L. If $\sigma \ll L < \xi$, where σ measures the size of the microscopic fluid particles, confinement leads to a long-range attraction between the surfaces. The resulting force has been determined very accurately by use of a sophisticated dynanometer in conjunction with total internal reflection microscopy,[2] with results that are in excellent agreement with theoretical predictions. The strength of the interaction is comparable with the thermal energy and can strongly affect the stability of colloidal dispersions in a near-critical solvent.

The coarse-graining approach described here is applicable to complex fluids involving length scales that may differ by several orders of magnitude. Related approaches have been applied to assemblies of large molecules, such as alkanes or phospho-lipids, where small groups of atoms or functional groups are replaced by single interaction sites within a 'united atom' description, thereby reducing the resolution achievable by roughly an order of magnitude. Force fields of this type are used to speed up molecular dynamics simulations of complex processes such as aggregation and self-assembly. The same method has been used, mostly in *ad hoc* form, since the early days of computer simulation, but a more refined approach has subsequently evolved in which emphasis is placed on matching the results obtained from coarse-grained and atomistic descriptions.[3] At a more fundamental level a variational principle has been derived that provides a recipe for construction of a coarse-grained potential model for which the interaction-site distribution functions are consistent with the atomic distributions in the underlying, atomically detailed model.[4]

12.2 POLYMER SOLUTIONS

Polymer is a generic name for elongated, linear or branched molecules of high molecular weight. *Homopolymers* consist of long sequences of a single unit or *monomer*. Their dimensionless, reduced properties obey certain scaling laws in the limit in which the number of monomers becomes very large; these laws are independent of the chemical nature of the monomer involved. *Heteropolymers* consist of sequences of different monomeric species, which may be distributed along the molecule in either regular or random fashion. Homopolymers and regular, block copolymers are the systems which, by virtue of their chemical homogeneity, are best adapted to a coarse-grained description of their properties. Much of the material covered in this section and the one that follows is concerned specifically with linear (or 'chain') polymers, but some reference is also made to star polymers and block copolymers. For readers unfamiliar with the field, Appendix G provides a brief summary of the basic properties of polymers and establishes the notation used in this and later sections. Excellent introductory texts[5] also exist, together with several classic books[6] devoted to theory at a more advanced level.

The properties of polymer solutions are largely controlled by two factors: polymer density and solvent quality. The effect of changes in density can be discussed in terms of the *overlap density*

$$\rho^{\circ} = \frac{3}{4\pi R_{\mathrm{g}}^{3}} \tag{12.2.1}$$

where R_{g} is the radius of gyration defined by (G.1). The overlap density is the number density beyond which polymer coils will, on average, overlap, since the mean volume of a single polymer, equal to $1/\rho^{\circ}$, will then exceed the volume per polymer, $V/N = 1/\rho$. The range of monomer density is conventionally divided into the three regions pictured in Figure 12.1, corresponding to a dilute solution ($\rho < \rho^{\circ}$), the onset of chain overlap ($\rho \sim \rho^{\circ}$) and a semi-dilute solution

(a)   **(b)**   **(c)**

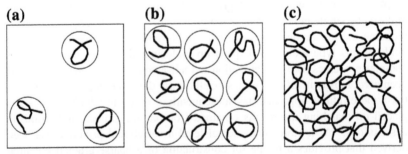

FIGURE 12.1   The three ranges of density described in the text for a system of chain polymers: (a) $\rho < \rho^{\circ}$, (b) $\rho \sim \rho^{\circ}$ and (c) $\rho > \rho^{\circ}$.

($\rho > \rho^0$). Still higher densities are found in polymer melts, discussion of which is postponed until Section 12.3. The quality of a solvent is described as either 'good' or 'poor'. Good solvent conditions correspond to high temperatures, at which the effect of the short-range, repulsive interaction between monomers outweighs that of the attractive interaction; this leads to a swelling of the polymer coils. At low temperatures the reverse is true. The attractive forces are now dominant and the coils contract as the temperature is reduced until, in dilute solutions, the polymer becomes unstable against *coil-globule collapse*. The temperature at which the interactions balance each other is called the θ-temperature, T_θ. In 'θ-solvent' conditions the polymer behaves like an ideal chain of non-interacting monomers in a manner similar to that of an imperfect gas at the Boyle temperature.

In dilute solution polymer coils rarely overlap and in the limit $\rho \to 0$ the properties of individual polymers depend only on temperature. In particular, the osmotic pressure Π of an ultra-dilute solution is given by van't Hoff's law, $\beta\Pi = \rho$. As the density increases the probability of binary overlap also increases. This leads to deviations from ideal behaviour of a magnitude determined by the osmotic second virial coefficient:

$$\beta\Pi = \rho + B_2(L,T)\rho^2 + \mathcal{O}(\rho^3) \tag{12.2.2}$$

The virial coefficient, a function of polymer length and temperature, is defined by a generalisation of (3.9.7):

$$B_2(L,T) = \frac{1}{2}\int \langle 1 - \exp[-\beta V_2(1,2)]\rangle_{0,\mathbf{R}} \, d\mathbf{R} \tag{12.2.3}$$

where the centre of mass or some other chosen centre of coil 1 is placed at the origin, that of coil 2 at $\mathbf{R}$, and $V_2(1,2)$ is the total energy of interaction between monomers on different coils. The statistical average in (12.2.3) must be taken over all chain conformations of the two coils for a given microscopic model of the polymer; its evaluation usually relies on data obtained by computer simulation. The dimensionless ratio

$$A_2(L,T) = \frac{B_2(L,T)}{R_g^3(L,T)} = A_2^* + \frac{a_2(T)}{L^\delta} + \cdots \tag{12.2.4}$$

involves a quantity A_2^* which has a universal value in the scaling limit, where the polymer length $L \to \infty$, and a universal exponent $\delta \approx 0.517$, whereas the coefficient $a_2(T)$ of the first finite-size correction is model dependent.[7] The value of A_2^* has been calculated by extrapolation to infinite L of the results of Monte Carlo calculations of $A_2(L,T)$ for two self-avoiding walk (SAW) polymers, giving $A_2^* = 5.494 \pm 0.005$. A numerical estimate of the third virial coefficient has also been obtained but the calculation of higher-order coefficients poses severe computational problems.

A different description of dilute solutions is that based on a coarse-graining approach, first proposed by Flory and Krigbaum,[8] which involves integration over monomer degrees of freedom for fixed coordinates of the polymer centres of mass. Let $\mathbf{R}_i$ be the centre-of-mass coordinates and let $\mathbf{r}_{i\alpha}$ be the coordinates of a monomer α of a chain labelled i. Then the probability distribution of centres of mass is

$$P_N(\{\mathbf{R}_i\}) = \frac{1}{Z_N} \int \exp\left[-\beta V_N(\mathbf{r}_{i\alpha})\right] \prod_{i=1}^{N} \delta\left(\mathbf{R}_i - \frac{1}{M}\sum_{\alpha=1}^{M}\mathbf{r}_{i\alpha}\right) \prod_{i=1}^{N}\prod_{\alpha=1}^{M} d\mathbf{r}_{i\alpha}$$

(12.2.5)

where M is the number of monomers per chain, V_N is the total interaction energy and Z_N is the configuration integral. The effective interaction energy between the centres of mass is related to the probability distribution by

$$V_N^{\text{eff}}(\{\mathbf{R}_i\}) = -k_B T \ln\left[\mathcal{C} P_N(\{\mathbf{R}_i\})\right] \qquad (12.2.6)$$

where $\mathcal{C}$ is an irrelevant constant which fixes the zero of energy. As discussed in Section 12.1, the effective energy is a free energy, and will therefore, in general, be many-body in nature. In the low-density limit, however, it will be pairwise additive, with an effective pair potential between centres of mass given by

$$\beta v_2^{\text{eff}}(R) = -\ln\left[\mathcal{C} P_2(|\mathbf{R}_2 - \mathbf{R}_1| = R)\right]$$
$$= -\ln\left\langle\exp\left[-\beta V_2(\{\mathbf{r}_{1\alpha}\}, \{\mathbf{r}_{2\beta}\})\right]\right\rangle_{|\mathbf{R}_2 - \mathbf{R}_1| = R} \qquad (12.2.7)$$

The statistical average is taken over conformations of the two chains for a fixed value of the separation R of their centres of mass. In common with the second virial coefficient, the pair potential is dependent on both L and T. Once the effective pair potential has been determined, typically from a Monte Carlo simulation of a system of two interacting polymers, the second virial coefficient is obtained by integration:

$$B_2(L, T) = \frac{1}{2} \int \left(1 - \exp[-\beta v_2(R)]\right) d\mathbf{R} \qquad (12.2.8)$$

where the superscript 'eff' has now been dropped. A knowledge of the virial coefficient allows the calculation of the lowest-order correction to van't Hoff's law, but the potential may also be used to determine other properties of the solution.

The effective pair potential is expected to be soft even if the monomer–monomer potential contains a strongly repulsive component. We show in Appendix G that in good solvent the radius of gyration of a chain behaves as bL^{ν}, where b is the monomer diameter and $\nu \approx \frac{3}{5}$. This implies that above the θ-temperature the monomer volume fraction ϕ_m within a single coil behaves as $L/R_g^3 \sim L^{1-3\nu} \sim L^{-4/5}$. That fraction will be very small when $L \gg 1$, in which case two polymer coils can easily interpenetrate. At full overlap, where the two centres of mass coincide, the potential has been shown[9] to scale as L^0, meaning that it is finite and independent of L in the scaling limit. Its range is expected to be of the same order as the coil size, R_g. These findings have been confirmed by Monte Carlo simulations[10] of pairs of hard-sphere chains and SAW polymers with results which show that $\beta v_2(R)$ is reasonably well-represented by a gaussian function of the form

$$\beta v_2(R) \approx A \exp\left[-\alpha(R/R_g)^2\right] \qquad (12.2.9)$$

where A and α are constants. For these athermal models the effective potentials are of purely entropic origin and therefore proportional to $k_B T$, but their application is limited to high temperatures where the interaction between monomers is dominated by excluded volume effects. As the temperature is reduced towards T_θ, the models must be augmented by terms that takes account of solvent induced, attractive interactions. When this is done, the effective potential acquires an additional temperature dependence; the amplitude A in (12.2.9) decreases with T and an attractive tail appears[11] for distances beyond R_g. It is also found that below a well-defined 'stability' temperature T_S the potential violates the condition required[12] for the existence of a thermodynamic limit, namely

$$I_2 = \int v_2(R)d\mathbf{R} > 0 \qquad (12.2.10)$$

The critical temperature T_S may be identified as the temperature below which coil-globule collapse occurs, given implicitly for a chain of length L by the relation

$$I_2(L, T = T_S) = 0 \qquad (12.2.11)$$

On the other hand the θ-temperature is the temperature at which attractive and repulsive forces balance and the second virial coefficient vanishes. It is therefore determined by the condition

$$\lim_{L \to \infty} B_2(L, T = T_\theta) = 0 \qquad (12.2.12)$$

Equation (12.2.12) has been used to estimate the θ-temperature of the SAW model with results that are consistent with those obtained in other ways.[13] The relationship between T_S and T_θ is less clear.

The low density, effective pair potential (12.2.9) is a special case of the purely repulsive, gaussian-core model,[14] for which

$$v_2(R) = \epsilon \exp\left[-(R/R_0)^2\right] \qquad (12.2.13)$$

where ϵ and R_0 measure, respectively, the strength and range of the interaction. This model is of interest in its own right as one that exhibits a re-entrant fluid–solid phase diagram in the density–temperature plane; freezing is followed by a structural phase transition and then by remelting.[15] At high densities, where each particle is overlapped by many others, the behaviour is of mean field type[16] and accurately described by the random phase approximation of

(3.5.17), which becomes exact in the limit $\rho R_0^3 \to \infty$. Since there is no hard core, the reference part of the potential is zero, $w(r) = v_2(r)$ and (within the RPA) $c(r) = -\beta v_2(r)$. The compressibility equation (3.5.15) then shows that in its application to polymer solutions the osmotic pressure is given by

$$\beta \Pi = \rho + \frac{1}{2}\hat{v}_2(k=0)\rho^2 = \rho + \frac{\pi^{3/2}}{T^*}R_0^3\rho^2 \qquad (12.2.14)$$

This result applies to the high-density limit and the coefficient of the quadratic term inevitably differs from the second virial coefficient, which determines the leading correction to the van't Hoff equation of state.

For polymer solutions in good solvent the semi-dilute regime corresponds to polymer densities greater than the overlap density. Overlap of polymers is now greatly increased, as pictured in Figure 12.1, the identity of individual coils is lost, and the polymer chains form a network characterised by a spatially homogeneous distribution of monomers and a mesh size, or correlation length, ζ. The dependence of ζ on density can be derived by a simple scaling argument. Let $R_{g0} \sim bL^\nu$ be the radius of gyration of an isolated polymer, which we assume to be related to ζ in typical scaling form:

$$\zeta = R_{g0} f\left(\frac{\rho_m}{\rho_m^o}\right) \qquad (12.2.15)$$

where $f(x)$ is a dimensionless scaling function, $\rho_m = M\rho \sim L\rho$ is the monomer density and $\rho_m^o = M\rho^o \sim b^3 L^{1-3\nu}$. When polymers overlap, ζ will be independent of L, implying that the scaling function represents a power law, $f(x) \sim x^\gamma$, say. Thus

$$\zeta \sim bL^\nu (\rho_m b^3 L^{3\nu}/L)^\gamma \sim b^{1+3\gamma}\rho_m^\gamma L^{\nu+3\nu\gamma-\gamma} \qquad (12.2.16)$$

which shows that γ must be equal to $-\nu/(3\nu - 1)$ and hence that

$$\zeta = R_{g0}\left(\frac{\rho_m}{\rho_m^o}\right)^{-\nu/(3\nu-1)} \sim b(\rho_m b^3)^{-3/4} \qquad (12.2.17)$$

for $\nu = \frac{3}{5}$.

A similar argument can be used to determine the asymptotic variation of osmotic pressure with monomer concentration. For obvious dimensional reasons $\beta \Pi$ may be written in the form

$$\beta \Pi = \rho f\left(\frac{\rho_m}{\rho_m^o}\right) = \frac{\rho_m}{L} f\left(\frac{\rho_m}{\rho_m^o}\right) \qquad (12.2.18)$$

where $f(x)$ is another scaling function and $\rho_m^o \sim L\rho^o \sim b^{-3}L^{1-3\nu}$. In the regime of high polymer overlap the length of individual polymers becomes irrelevant and $\beta \Pi$ should therefore be independent of L. This implies that $f(x)$ is again a simple power law, $f(x) \sim x^\alpha$, where $\alpha = 1/(3\nu - 1)$, leading in turn to the expression for the osmotic pressure due to des Cloizeaux[17] in which Π

behaves asymptotically as $\rho_m^{9/4}$:

$$\beta \Pi b^3 \sim (\rho_m b^3)^{3\nu/(3\nu-1)} \sim (\rho_m b^3)^{9/4} \sim b^3/\zeta^3 \qquad (12.2.19)$$

The scaling argument provides a value for the exponent but not the prefactor. The des Cloizeaux exponent is larger than the quadratic value predicted by the RPA, which is based on the use of an effective, gaussian core interaction between polymer coils. Since the RPA is asymptotically exact for the gaussian potential, the breakdown of (12.2.14) in the semi-dilute regime must be ascribed to the inadequacy of the effective pair potential for densities above ρ^0. In the semi-dilute regime the role of many-body forces become significant, but their effect can be allowed for in an approximate way by introduction of a state-dependent effective pair potential. The density-dependent potential for the athermal SAW model has been calculated by inversion of the pair distribution function of polymer centres of mass obtained by Monte Carlo simulations.[18] Inversion is achieved by use of the HNC closure relation (4.3.19), which is known to be very accurate for soft, penetrable-core models. In this approximation the pair potential is given by

$$\beta v_2(R) = h(R) - c(R) - \ln g(R) \qquad (12.2.20)$$

where all quantities are functions of density; $g(R)$ (and hence $h(R)$) are given by simulation and the direct correlation function is derived from the Ornstein–Zernike relation. The distribution function that serves as input depends weakly on L and the results must be extrapolated to the scaling limit, $L \to \infty$. Corrections to the scaling limit of the potential have also been obtained[7]; the leading correction is proportional to $1/L^\delta$, with $\delta \approx 0.517$, as in (12.2.4).

Examples of the pair distribution function and the resulting pair potential are shown in Figure 12.2. As the density increases, the size of the 'correlation hole' in $g(R)$ at small R decreases and $g(R = 0)$ increases towards unity. Such behaviour is indicative of the fact that short-range correlations become weaker with increasing density. This is the reverse of what is found in the case of simple liquids, but is consistent with Flory's conjecture, discussed in the section which follows, that polymers behave like ideal chains at high densities. The pair potential has a gaussian-like form with a range that increases with density. The only unexpected feature is that the value at full overlap, $v_2(R = 0)$, varies non-monotonically with density, passing through a maximum at $\rho \approx 2\rho^0$. The state dependence of the potential precludes the use of the energy and virial equations, (2.5.20) and (2.5.22), for the calculation of thermodynamic properties, but the compressibility equation (3.5.15) remains applicable if the direct correlation function is calculated from the pair potential via the HNC equation. The osmotic pressure is then obtained by integration over density:

$$\beta \Pi = \int_0^\rho \rho'[1 - \rho'\hat{c}(k = 0; \rho')]\mathrm{d}\rho' \qquad (12.2.21)$$

and is found to follow the des Cloizeaux scaling law for densities greater than about $2\rho^0$.

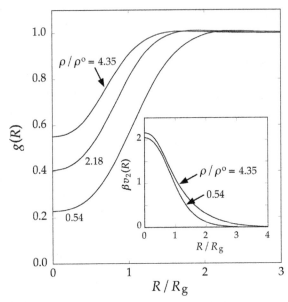

FIGURE 12.2 Pair distribution functions and effective pair potentials for polymers in a semi-dilute solution. *Redrawn with permission from Ref. 18 © 2001 American Institute of Physics.*

The same strategy can be pursued when the temperature is reduced from good solvent conditions to the θ-temperature and below,[11] but the potential is now dependent on both density and temperature. The collapse of isolated polymer coils into globules seen in dilute solutions at temperatures below T_θ is replaced by a polymer–solvent phase separation into a low-concentration phase of collapsed globules and a concentrated phase of stable polymer coils, with a concomitant lowering of the solvent volume fraction. This 'restabilisation' of concentrated polymer solutions is reflected in a strong density dependence of the pair potential, which satisfies the stability criterion (12.2.10) at sufficiently high densities. A disadvantage of the inversion procedure is the fact that its implementation relies on simulations of a microscopic model to extract the distribution function of centres of mass, which is a computationally costly procedure at high densities. The problem could be overcome by development of a theory that relates the mesoscopic distribution of centres of mass to the microscopic, monomer–monomer distribution. This can be achieved, at least approximately, by an extension (discussed below) of the RISM formalism of Section 11.8.

Soft-potential coarse graining applies to other polymer topologies and in particular to star polymers. These are made up of $f = 3$ or more linear branches, as shown in Figure G.1. The connection point of the branches is the natural choice of centre for a star polymer rather than the centre of mass. Scaling arguments[19] have shown that the effective interaction between two star polymers with SAW branches and centres separated by R behaves logarithmically when

R is less than R_g, i.e.

$$\beta v_2(R) = -\alpha f^{3/2} \ln (R/R_g), \quad R < R_g \qquad (12.2.22)$$

where the prefactor α is a positive quantity. The potential therefore has a very soft but ultimately impenetrable core, since it diverges as $r \to 0$, in contrast to the effective potentials between the centres of mass of linear molecules. It becomes more stiffly repulsive as the number of branches increases and interpenetration of two star polymers becomes entropically more costly. The form of (12.2.22) has been confirmed by Monte Carlo simulations, which also provide both the value of α and the variation of $v_2(R)$ with separation for $R > R_g$; the potential is everywhere repulsive.[20] Integral equations, thermodynamic perturbation theory and simulations, all based on an effective pair potential of that type, have been used to determine the phase diagram of star polymer solutions[19b] over a range of values of f.

For linear polymers the complications associated with use of a state-dependent effective pair potential in the semi-dilute regime can be by-passed[21] by representing each chain not by a single, penetrable sphere located at the centre of mass but by a chain of n tethered 'blobs', each representing a sequence of $m = M/n$ monomers, as pictured in Figure 12.3. The average size of a blob is given by its radius of gyration, $r_g \sim b\ell^\nu$, with $\ell = m - 1$ and $\nu \approx \frac{3}{5}$. It is natural to choose n such that r_g is equal to the correlation length ζ given by (12.2.16), i.e.

$$b \left(\frac{L}{n} \right)^\nu = \zeta \approx bL^\nu \left(\frac{\rho_m}{\rho_m^0} \right)^{-\nu/(3\nu-1)} \qquad (12.2.23)$$

which shows that

$$n \approx \left(\frac{\rho_m}{\rho_m^0} \right)^{1/(3\nu-1)} \approx \left(\frac{\rho}{\rho^0} \right)^{5/4} \qquad (12.2.24)$$

If n is chosen to be at least as large as that given by this expression, the blob density $\rho_b = n\rho$ remains at or below the blob overlap density $\rho_b^0 = 3/4\pi r_g^3$, corresponding to the dilute blob regime, even though the polymer density may be greater than ρ^0. In that situation we may assume that the effective interaction between blobs, $v_{bb}(r)$, is well approximated by its low-density limit. It is also

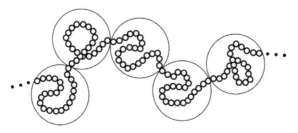

FIGURE 12.3 A multi-blob representation of part of a linear polymer.

reasonable to assume that the potential between non-bonded blobs is identical for all pairs, whether on the same or different polymers, and given by the gaussian form (12.2.9) but with R_g replaced by r_g, i.e.

$$v_{bb}(r) = A_{bb} \exp\left[-\beta \alpha_{bb}(r/r_g)^2\right] \tag{12.2.25}$$

where A_{bb} and α_{bb} each has approximately the same value as the corresponding quantity in (12.2.9). The effective potential between bonded blobs can then be written as the superposition of $v_{bb}(r)$ and a tethering, harmonic spring potential $\phi(r)$ that acts between blobs, similar to the entropic spring of an ideal, gaussian chain, described in Appendix G:

$$\phi(r) = \frac{1}{2}k(r - r_0)^2 \tag{12.2.26}$$

The potential parameters k and r_0 are determined by measurement of the intramolecular distribution function $P(r)$ of the distance r between the centres of mass of the two blobs in a simulation of a single dimer of SAW polymers, from which the potential is obtained via the relation

$$\phi(r) = -k_B T \ln P(r) \tag{12.2.27}$$

The total potential energy of the two polymers is then

$$V_2(\{\mathbf{r}_{1\alpha}\}, \{\mathbf{r}_{2\beta}\}) = \sum_{i=1}^{2}\sum_{\alpha=1}^{n-1}\phi(|\mathbf{r}_{i,\alpha+1} - \mathbf{r}_{i,\alpha}|) + \sum_{i=1}^{2}\sum_{\alpha=1}^{n-1}\sum_{\beta=\alpha+1}^{n} v_{bb}(|\mathbf{r}_{i\beta} - \mathbf{r}_{i\alpha}|)$$
$$+ \sum_{\alpha=1}^{n}\sum_{\beta=1}^{n} v_{bb}(|\mathbf{r}_{2\beta} - \mathbf{r}_{1\alpha}|) \tag{12.2.28}$$

where $\mathbf{r}_{i\alpha}$ now represents the coordinates of the centre of mass of blob α on polymer i. The three terms on the right-hand side represent, successively, the potential energies due to tethering and those due to interaction between blobs on the same or different polymers.

As examples of the use of the multi-blob representation we show first how a tractable expression can be obtained for the static structure factor.[21] The structure factor of a system of N chains, each composed of n blobs, is

$$S(k) = \frac{1}{Nn}\langle \rho_{\mathbf{k}}\rho_{-\mathbf{k}}\rangle = \hat{\omega}(k) + \frac{1}{n}\sum_{\alpha}\sum_{\beta} s_{\alpha\beta}(k) \tag{12.2.29}$$

where $\rho_{\mathbf{k}}$ is a Fourier component of the monomer density, $\hat{\omega}(k)$ is the form factor (G.6) of a single polymer and $s_{\alpha\beta}(k)$ is the partial, intermolecular structure factor for blobs α and β. By anticipating the results of PRISM theory, described in Section 12.3, we may assume that all $s_{\alpha\beta}(k)$ are identical and equal to a unique blob–blob structure factor $s(k)$, which in turn is related to a blob–blob direct correlation function $\hat{c}(k)$ and the form factor by a generalised Ornstein–Zernike relation. This leads to the PRISM expression for $S(k)$:

$$S(k) = \frac{\hat{\omega}(k)}{1 - \rho_b \hat{\omega}(k)\hat{c}(k)} \tag{12.2.30}$$

Within the RPA (3.5.17), $\hat{c}(k) = -\beta \hat{v}_{bb}(k)$ and substitution of the Fourier transform of (12.2.25) in (12.2.30) shows that

$$S_{RPA}(k) = \frac{\hat{\omega}(k)}{1 + \rho_b r_g^3 \hat{\omega}(k) A_{bb}(\pi/\alpha_{bb})^{3/2} \exp\left(-k^2 r_g^2/4\alpha_{bb}\right)} \tag{12.2.31}$$

which is a generalisation to the multi-blob model of the classic RPA expression for the structure factor of a polymer solution.[22] Use of (3.6.11) and the fact that $\hat{\omega}(k = 0) = n$, together with substitution for n from (12.2.24), leads to an expression for the osmotic compressibility:

$$\rho k_B T \chi_T = \frac{1}{1 + c(\rho/\rho^\circ)^{1/(3\nu-1)}} \tag{12.2.32}$$

with $c = 3A_{bb}(\pi/\alpha_{bb})^{3/2}/4\pi \approx 3.23$. For $\rho \gg \rho^\circ$, (12.2.32) leads back to the des Cloizeaux scaling law (12.2.19) in the form

$$\frac{\beta \Pi^{ex}}{\rho^\circ} = \frac{3\nu - 1}{3\nu} c \left(\frac{\rho}{\rho^\circ}\right)^{3\nu/(3\nu-1)} \tag{12.2.33}$$

The RPA therefore yields an expression for the prefactor, which the scaling argument does not.

As a second example, advantage can be taken of the weakness of the effective blob–blob potential and its gaussian form in a way that allows the use of perturbation theory to calculate the free energy F_N of a system of N interacting n-blob polymers.[21] The reference system is chosen to be the solution of N non-interacting gaussian chains of length n and spring constant $k_0 = 3k_B T/b'^2$, where the bond length b' is used as as a variational parameter in minimising the right-hand side of the Gibbs–Bogoliubov inequality (5.2.27):

$$F_N \leq F_N^{(0)} + \left\langle V_N - V_N^{(0)} \right\rangle_0 \tag{12.2.34}$$

where V_N is the total potential energy of the N interacting polymers, $V_N^{(0)}$ is the corresponding energy of N gaussian coils, obtained from (G.3), and the average is taken with gaussian statistics. The two terms on the right-hand side can be calculated analytically as functions of b'; their sum is then minimised with respect to the reduced bond length $b_0 = b'/r_g$ to provide the best variational estimate of the radius of gyration of the interacting polymers relative to that of the reference, gaussian chain, given by (G.5). The result is

$$R_g = \frac{n^{1/2} b_0 r_g}{\sqrt{6}} \approx \frac{n^{1/2} b_0}{\sqrt{6}} b \left(\frac{L}{n}\right)^\nu \tag{12.2.35}$$

or, after substitution for n from (12.2.24) and setting $\nu = \frac{3}{5}$:

$$R_g \approx bL^\nu \left(\frac{\rho}{\rho^0}\right)^{-(2\nu-1)/2(3\nu-1)} \approx R_{g0}\left(\frac{\rho}{\rho^0}\right)^{-1/8} \quad (12.2.36)$$

where $R_{g0} \approx bL^\nu$ is the radius of gyration of an isolated SAW polymer. The optimum value of the dimensionless variational parameter b_0 is found to increase with the number of blobs and saturates at a value $b_0 \approx 2.24$. The result in (12.2.36) agrees with the prediction of a scaling argument similar to the one that leads to (12.2.17); it shows that the radius of gyration of interacting polymers in a semi-dilute solution slowly contracts as ρ/ρ^0 increases.

Use of the multi-blob representation in the examples discussed hinges on the assumption of transferability implied in the use for any value of n of effective pair potentials derived in the low-density limit. Since intramolecular interactions beyond the two-body level are neglected, the effective pair potentials cannot account quantitatively for intramolecular correlations, nor for thermodynamic properties at finite concentrations. Improvements have been proposed that include the use of effective bending and torsion angle potentials[23] or blobs of fluctuating size.[24] Multi-blob calculations have also been made for AB diblock copolymers in which different effective and tethering potentials are used for AA, AB and BB pairs. The differences in solvent-induced potentials lead to self-assembly and phase separation into ordered or disordered micellar, cylindrical, lamellar or bicontinuous phases, reminiscent of the phase behaviour of ampiphilic molecules in water or oil.[25a] This complex phase behaviour has been reproduced, at least qualitatively, by Monte Carlo simulations.[25b]

12.3 POLYMER MELTS

Polymer melts are solvent-free, viscoelastic liquids consisting of entangled macromolecules with a monomer volume fraction $\eta_m = \pi \rho_m b^3/6$ comparable with that of simple liquids. The large volume fraction means that monomer concentration fluctuations are strongly suppressed and the liquid is highly incompressible. The identity of the polymer to which a given monomer belongs is therefore largely irrelevant; the spatial constraints associated with high packing fraction dominate those due to connectivity. In other words, a polymer behaves, in a first approximation, like a liquid of disconnected monomers, at least as regards its structure and thermodynamics. Entanglement of the chains does play a major role in determining the dynamics of the melt,[6c] but we confine ourselves here to the discussion of static properties.

In the multi-blob representation of semi-dilute solutions described in the previous section we see from (12.2.24) that the number of blobs required to ensure the no-overlap condition increases with monomer concentration. This trend carries over to the melt, where $n \to M$. Each blob now corresponds to a single monomer, and the effective interaction between blobs is the microscopic, excluded volume interaction that acts between monomers under good solvent conditions. On the other hand, if $n \approx M$, if follows from (12.2.35) that in the melt the radius of gyration of a linear polymer should scale as $bL^{1/2}$, as it does for an ideal chain. This surprising result was first noted by Flory, who conjectured that the conformations of individual polymer chains in a melt should follow gaussian statistics, as do ideal chains or interacting polymers in θ-solvent. No proof has been given of Flory's hypothesis but theoretical arguments have been forward that support it. At the simplest level, for example, the local monomer density $\rho_m(\mathbf{r})$ will be constant throughout the melt if spatial fluctuations can be ignored. Hence the local potential energy $u(\mathbf{r})$ will also be constant and the force $-\nabla u(\mathbf{r})$ acting on a monomer located at $\mathbf{r}$ therefore vanishes. Thus the polymer behaves like an ideal chain of non-interacting monomers. The repulsive interactions between monomers along a given chain, which would lead to a swelling of the polymer if it were isolated, are now screened by the monomers of neighbouring chains.[26]

The screening mechanism can be quantified by an RPA calculation[27] in which the excluded volume interaction between monomers is modelled by a contact potential:

$$v(\mathbf{r}_{j\beta} - \mathbf{r}_{i\alpha}) = v_x k_B T \delta(\mathbf{r}_{j\beta} - \mathbf{r}_{i\alpha}) \qquad (12.3.1)$$

where the excluded volume parameter $v_x \approx b^3$. If the reference system is taken as one consisting of non-interacting gaussian chains, the RPA expression (5.6.24) for the structure factor may be written as

$$S(k) = \frac{S_0(k)}{1 + \rho_m \beta \hat{v}(k) S_0(k)} = \frac{S_0(k)}{1 + \rho_m v_x S_0(k)} \qquad (12.3.2)$$

where $S_0(k) = \hat{\omega}(k)$, defined by (G.7), is the structure factor of the reference system. If the Lorentzian approximation in (G.8) is used for $\hat{\omega}(k)$, with $R_g^2 = Mb^2/6$, we find that

$$S(k) = \frac{12}{b^2} \frac{1}{k^2 + \zeta^{-2}} \qquad (12.3.3)$$

The quantity ζ is a correlation length defined as

$$\zeta^2 = \frac{Mb^2}{12(1 + M\rho_m v_x)} \approx \frac{b^2}{12\rho_m v_x} \qquad (12.3.4)$$

where the approximation used is justified when $M \gg 1$. Equation (12.3.3) resembles the expression (5.7.22) for the structure factor of a fluid close to the critical point. The corresponding monomer–monomer pair correlation function is therefore of the same form as (5.7.23):

$$h(r) = \frac{3}{\pi \rho_m b^2} \frac{\exp(-r/\zeta)}{r} \qquad (12.3.5)$$

We now show that ζ plays the role of a screening length associated with the screening mechanism invoked earlier. Linear response theory can be used to determine the induced modulation $\delta\rho_m(r)$ of the monomer concentration and the corresponding potential energy profile $u(r)$ at a distance r from a monomer placed at the origin. According to (3.6.9):

$$\delta\hat{\rho}_m(k) = -\beta\rho_m\hat{v}(k)S(k) = -\rho_m v_x S(k) \qquad (12.3.6)$$

On combining this result with (3.6.10) and (12.3.5) we find that in real space

$$\delta\rho_m(r) = \rho_m(r) - \rho_m = -\frac{3\rho_m v_x}{\pi b^2}\frac{\exp(-r/\zeta)}{r} \qquad (12.3.7)$$

and hence that

$$u(r) = v(r) + \int v(\mathbf{r}' - \mathbf{r})\delta\rho_m(r')\mathrm{d}\mathbf{r}'$$

$$= k_B T v_x \left(\delta(\mathbf{r}) - \frac{\exp(-r/\zeta)}{4\pi\zeta^2 r}\right) \qquad (12.3.8)$$

where the first term on the right-hand side is the contribution to the local potential energy from the monomer at the origin and the second is that generated by modulation of the surrounding monomer density. The quantity $u(r)$ may be treated as the effective pair potential between the central monomer and a neighbouring monomer at a distance r from the origin; this accounts for the presence of other monomers located between the two, which reduces the bare potential. In fact integration of $u(r)$ over all space yields zero, meaning that the attraction induced by intermediate monomers exactly cancels the direct interaction for distances larger than ζ; the situation is identical to that of polymer coils in solution under θ-point conditions. Thus the individual coils in melts behave like non-interacting polymers, obeying gaussian statistics with $R_g \sim L^{1/2}$ on length scales greater than ζ.

Equation (12.3.8) is reminiscent of the screened Coulomb interaction in ionic systems, as represented by (4.6.27), with the correlation length playing the role of the Debye screening length; both quantities decrease as the number concentration increases. In a dense melt, $\rho_m v_x \approx 1$, and polymers behave like ideal chains on all length scales. The derivation that leads to (12.3.8) also applies to concentrated solutions of overlapping polymers, but in that case $\rho_m v_x \ll 1$ and hence $\zeta \gg b$. Non-ideal behaviour, characterised by a swollen radius of gyration, $R_g \sim L^\nu$ with $\nu \approx \frac{3}{5}$, prevails over length scales r such that $b < r < \zeta$.

It is possible is to formulate an accurate theory of the monomer–monomer pair structure that exploits Flory's hypothesis but goes beyond the RPA result given by (12.2.31); the RPA does not account properly for short-range correlations and (12.2.31) is therefore valid only for small wavenumbers. Each monomer may instead be regarded as an interaction site on a polymer chain so that a system of polymers can be treated within the RISM formalism of Section 11.8. The difference here is the fact that when $M \gg 1$, the matrices $\hat{\mathbf{h}}, \hat{\mathbf{c}}$ and $\hat{\mathbf{w}}$ are very large. The application to polymers, known as PRISM,[28] is based on the assumption that all monomers are geometrically equivalent and that the pair functions $h_{\alpha\beta}(r)$ and $c_{\alpha\beta}(r)$ are therefore the same for all α, β. The $M \times M$ RISM-OZ relation (11.8.6) then reduces to a single scalar equation

$$\hat{h}(k) = \hat{\omega}(k)\hat{c}(k)\hat{\omega}(k) + \rho_m\hat{\omega}(k)\hat{c}(k)\hat{h}(k) \qquad (12.3.9)$$

where

$$\hat{\omega}(k) = \frac{1}{M}\sum_{\alpha}\sum_{\beta}\langle\hat{\omega}_{\alpha\beta}(k)\rangle \qquad (12.3.10)$$

is the form factor (G.7) of the single polymer. For the rigid molecules discussed in Section 11.8 the quantities $\hat{\omega}_{\alpha\beta}$ depend only on fixed intramolecular bond lengths, but here they must be averaged over macromolecular conformations, as indicated in (12.3.10). The basic assumption underlying the PRISM equation (12.3.9), namely the equivalence of all monomers, is true for ring polymers and for linear polymers if end effects can be ignored. Equation (12.3.9) is easily generalised, in matrix form, to multi-component melts, such as binary mixtures of homopolymers or block copolymer systems, in which monomers of two or more different chemical species are present.[29]

When combined with a suitable closure relation, such as PY, HNC or a molecular closure better adapted to multi-site systems,[30] the PRISM-OZ relation (12.3.9) can be solved numerically to yield the monomer–monomer pair correlation function $h_{mm}(r)$. All that is required is an expression for the form factor $\hat{\omega}(k)$, one possible choice of which is the Debye function (G.7), corresponding to a gaussian chain. The static structure factor of the melt, as measured by diffraction experiments, is given by a generalisation of (3.6.10) in which the monomers now take the role of atoms:

$$S(k) = \hat{\omega}(k) + \rho_m\hat{h}(k) = \frac{\hat{\omega}(k)}{1 - \rho_m\hat{\omega}(k)\hat{c}(k)} \qquad (12.3.11)$$

The isothermal compressibility follows from (11.3.6):

$$\rho k_B T \chi_T = 1 + \rho\hat{h}(k = 0) = \frac{S(k = 0)}{M} \qquad (12.3.12)$$

where $\rho = \rho_m/M$ is the polymer density. An example of the agreement achievable between theory and experiment is illustrated by Figure 12.4, which shows a comparison between the results of molecular dynamics calculations and the predictions of PRISM for a system of Lennard-Jones chains consisting of 200 monomers. The form factor used is the one computed in the simulation, a procedure designed to test the internal consistency of the theory.[28b]

A different application of PRISM leads to an accurate relation between the monomer and centre-of-mass pair correlation functions in both polymer solutions and melts,[31] as already referred to in Section 12.2. The key idea is to consider the centre of mass of each polymer i as an additional, auxiliary site $\mathbf{R}_i$, linked to the monomer coordinates $\mathbf{r}_{i\alpha}$ by the constraint that

$$\mathbf{R}_i = \frac{1}{M}\sum_{\alpha=1}^{M}\mathbf{r}_{i\alpha} \qquad (12.3.13)$$

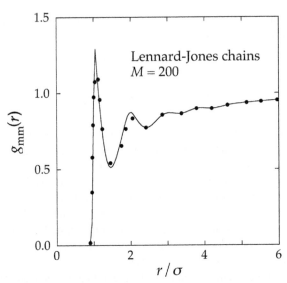

FIGURE 12.4 Monomer–monomer radial distribution function for a system of Lennard-Jones chains with $M = 200$ at a reduced temperature $T^* = 1.0$ and $\rho_m \sigma^3 = 0.85$. The points are the results of molecular dynamics calculations and the curve is calculated from the PRISM equation with a form factor computed in the simulation. *Redrawn with permission from Ref. 28(b) © 1989 American Institute of Physics.*

Since the auxiliary site does not interact with the monomers, it must be treated separately. Thus each polymer contains two types of site, the single, auxiliary one, labelled x, and the M equivalent, interaction sites associated with the monomers, labelled m. The three relevant form factors are then $\hat{\omega}_{mm}(k)$, $\hat{\omega}_{mx}(k)$ and $\hat{\omega}_{xx}(k)$, of which $\hat{\omega}_{xx} = 1$ and

$$
\begin{aligned}
\hat{\omega}_{mx}(k) &= \left\langle \sum_\alpha \exp\left[i\mathbf{k}\cdot(\mathbf{r}_{i\alpha} - \mathbf{R}_i)\right] \right\rangle \\
&\approx \frac{\sqrt{\pi}M}{kR_g} \exp\left(-k^2 R_g^2\right) \mathrm{erf}\left(\frac{1}{2}kR_g\right)
\end{aligned}
\tag{12.3.14}
$$

where the second line holds for gaussian statistics.

The single PRISM-OZ relation (12.3.9) is now replaced by a 2×2 matrix of relations, one of which is identical to (12.3.9) but with an mm subscript for all pair functions. The other three can be simplified by assuming that the direct correlation functions between auxiliary sites and between the auxiliary site in one polymer and the monomers of another are all identically zero since there is no physical interaction involved, i.e.

$$
\hat{c}_{xx}(k) \equiv 0; \quad \hat{c}_{mx}(k) = \hat{c}_{xm}(k) \equiv 0
\tag{12.3.15}
$$

What remains is

$$\hat{h}_{mx}(k) = \hat{\omega}_{mx}(k)\hat{c}_{mm}(k)\left[\hat{\omega}_{mm}(k) + \rho_m\hat{h}_{mm}(k)\right]$$
$$\hat{h}_{xx}(k) = \hat{\omega}_{mx}(k)\hat{c}_{mm}(k)\left[\hat{\omega}_{mx}(k) + \rho_m\hat{h}_{mx}(k)\right] \qquad (12.3.16)$$

Combination of (12.3.9) and (12.3.16) leads immediately to the desired result:

$$\hat{h}_{xx}(k) = \frac{\hat{\omega}_{mx}^2(k)}{\hat{\omega}_{mm}^2(k)}\hat{h}_{mm}(k) \qquad (12.3.17)$$

which, apart from the assumption (12.3.15), is independent of the choice of closure relation. To extract $\hat{h}_{xx}(k)$ from $\hat{h}_{mm}(k)$ requires a knowledge of the two form factors involved, which are available analytically for gaussian chains. Equation (12.3.17) is therefore directly applicable in the melt regime. It has also been successfully applied to coarse-grained multi-blob representations of polymer melts[32] similar to that adopted for semi-dilute solutions in Section 12.2. In that case the greater importance of concentration fluctuations and the consequent swelling of individual coils means that the ratio $\hat{\omega}_{mx}^2(k)/\hat{\omega}_{mm}^2(k)$ deviates significantly from its ideal form.

12.4 COLLOIDAL DISPERSIONS

Colloidal dispersions are highly asymmetric systems of mesoscopic particles suspended in a molecular solvent, with particle sizes typically in the range from 10 to 10^3 nm. They include polymer latex dispersions (in paints), casein micelles (in dairy products), oil-in-water emulsions and clays. Colloidal systems have an enormous range of applications, from the oil industry (drilling fluids) to the manufacture of cosmetics, food and pharmaceutical products.[33] This section is concerned with the relatively simple case of spherical, solid particles such as mineral silica or synthetic PMMA spheres. These are systems that have been much studied experimentally, in particular by photon correlation spectroscopy; visible light is well adapted to probe the structure and dynamics of micron-sized colloids, since the wavelength of the radiation is comparable with the dimensions of the particles. One complication that arises with colloidal systems, but not with atomic or molecular liquids, is that of size polydispersity. Even for carefully prepared samples the diameters of the particles may have a spread of 5–10%, but in theoretical work it is usually assumed that the system is monodisperse.

Colloidal particles are composed of very large numbers of atoms, roughly of order 10^6–10^{12}. A straightforward calculation[34] shows that summing the dispersion interactions ($\sim -C/r^6$) between pairs of atoms contained in two spheres of diameter d and centre-to-centre separation R leads to an attractive potential between spheres of the form

$$w(R) = -\frac{A}{12}h(x) \qquad (12.4.1)$$

where $A = \pi^2 C \rho^2$ is the Hamaker constant; ρ is the number density of atoms within each sphere, $x = R/d$ and

$$h(x) = \frac{1}{x^2 - 1} + \frac{1}{x^2} + 2\ln\left(1 - \frac{1}{x^2}\right) \qquad (12.4.2)$$

with

$$h(x) \approx \frac{1}{3x^6}, \qquad x \gg 1$$
$$\approx \frac{1}{2(x - 1)}, \qquad x \approx 1$$

The Hamaker constant is typically of order $10 k_B T$ at room temperature and the potential diverges as contact is approached ($x \to 1^+$). Such a strong, attractive interaction would lead to irreversible aggregation or *flocculation* unless balanced by a strong repulsive force. A dispersion can be stabilised against flocculation either by grafting polymers onto the surface of the colloidal particles or by formation of electric double layers around the particles, which acquire a surface charge in strongly polar solvents. *Charge stabilisation* is discussed in the following section; here we consider only *steric stabilisation* by grafted polymers.

Consider first a polymer 'brush' of identical polymer chains end-grafted onto a planar substrate, as pictured in Figure 12.5a. If σ is the surface grafting density, the mean distance between adjacent grafting sites will be $D \approx \sigma^{-1/2}$. If D is smaller than the radius of gyration of the polymer, then in good solvent the repulsion between monomers will cause the polymer to stretch in the direction normal to the substrate. The mean height h of the brush can be estimated by a simple scaling argument[35] based on the blob picture of Figure 12.5a. Let us suppose that the blob size is set equal to the correlation length ζ appropriate

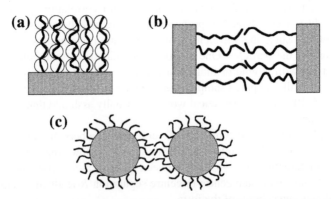

FIGURE 12.5 Steric stabilisation of colloid dispersions by grafted polymers. (a) A single polymer brush and its multi-blob representation. (b) Two polymer brushes repel each other when brought into contact. (c) Two colloidal particles stabilised by repulsion between their polymer layers.

to the semi-dilute regime, introduced in (12.2.17). Then, by taking $\zeta = D$, we ensure that blobs will not, on average, overlap. The number of monomers per blob is $\ell \approx (\zeta/b)^{1/\nu}$, where $\nu \approx \frac{3}{5}$. If Lb is the length of a chain, it follows that

$$h = \frac{L\zeta}{\ell} = \frac{L\zeta}{(\zeta/b)^{1/\nu}} = Lb(\sigma b^2)^{(1-\nu)/2\nu} \approx Lb(\sigma b^2)^{1/3} \qquad (12.4.3)$$

Thus the stretching of the polymers increases with grafting density, as one would expect, and maximum stretching is achieved in the limit $\sigma = b^{-1/2}$, when the monomers of neighbouring chains come into contact; the same result can be derived by a free energy minimisation of the type that leads to Flory's estimate of the exponent ν (see Appendix G). These simple arguments imply that the monomer density profile is a step function:

$$\rho_m(z) = \frac{\sigma L}{h}, \quad z < h$$
$$= 0, \qquad z > h \qquad (12.4.4)$$

which is strictly true only at $D = b$. For moderate grafting densities the monomer density as a function of distance z above the surface has been determined by a self-consistent calculation which relates the density to a mean field expression for the local chemical potential of the monomers[36]; the resulting profile turns out to be parabolic. The same result can be obtained by the simpler, density functional argument sketched in Appendix H. Monte Carlo calculations[37] of SAW polymer brushes show that a rapid transition from a partly stretched, parabolic profile to a fully stretched, near-rectangular profile similar to (12.4.4) occurs at a reduced grafting density $\sigma R_g^2 \approx 25$.

Now consider the more complicated case of two parallel polymer brushes facing each other, illustrated in Figure 12.5b. As the distance z between the two substrates is reduced, the polymer brushes are compressed when $z < 2h$; the resulting decrease in entropy gives rise to an effective repulsion between the brushes. The force per unit area acting on each substrate is the *disjoining pressure*, i.e. the osmotic pressure of the grafted polymers, given by (12.2.19).[38] The correlation length ζ is given by (12.2.16), while for $z < 2h$ the mean monomer density between substrates is $\rho_m = \sigma L/z$. Thus $\zeta \approx b(\sigma Lb^3/z)^{-\nu(3\nu-1)}$ and

$$\beta \Pi b^3 \approx \left(\frac{\sigma Lb^3}{z}\right)^{3\nu/(3\nu-1)} \approx \left(\frac{\sigma Lb^3}{z}\right)^{9/4} \qquad (12.4.5)$$

The disjoining pressure increases with grafting density and polymer length and diverges as $z \to 0$; it is this strong repulsion which provides the mechanism for steric stabilisation.

Colloidal particles carrying a dense, grafted layer of short polymers, such as those pictured schematically in Figure 12.5c, are essentially hard-sphere-like in their behaviour. By observing a dense suspension of slowly sedimenting, sterically stabilised PMMA particles in a mixture of decalin and carbon disulphide, Pusey and van Megen[39] were able to locate the freezing transition from a dense colloidal fluid to an opalescent crystal that diffracts visible light; the measured coexistence densities lay within roughly 1% of the values provided by simulations of hard-sphere systems. Also observed was a transition to an amorphous solid phase at $\eta \approx 0.59$, a density lying well below that of random close packing. The existence of such a phase could be explained by a small degree of size polydispersity in the sample and it is now recognised that polydispersity is not merely an unwanted complication but a variable of interest in its own right. An extension to a polydisperse, hard-sphere system of the density functional theory of freezing of Section 6.6 has shown that beyond a critical polydispersity crystallisation is inhibited in favour of transition to a high density, disordered state,[40] while specialised Monte Carlo simulations[41] have identified qualitative differences between the phase behaviour of monodisperse and polydisperse systems.

Studies of sedimentation equilibria of colloidal dispersions provide a direct, experimental route to the determination of the osmotic equation of state over a wide range of density. Optical techniques may be used to measure the density profile $\rho(z)$ of an equilibrated suspension of charged colloidal particles in a gravitational field or an ultracentrifuge, from which the osmotic pressure can be derived as a function of density. In the first case the external potential is

$$\beta\phi(z) = \frac{mg}{k_B T}z = \alpha z \tag{12.4.6}$$

where g is the acceleration due to gravity and $m = M - 4\pi R^3 d_m/3$ is the buoyant mass, with M and R being the mass and radius of the particles and d_m the mass density of the suspending fluid; α is the inverse gravitational length. By adjusting d_m to be close to the mass density of the colloidal particles, α can be made sufficiently small to ensure that the potential varies very slowly with z. Under those conditions the free energy functional $\mathcal{F}[\rho(z)]$ may be replaced by its local density approximation (6.2.5). This leads to the condition for mechanical equilibrium given by (6.2.7), which here takes the form

$$\frac{d\Pi(z)}{dz} = -mg\rho(z) \tag{12.4.7}$$

This is easily integrated to give

$$\beta\Pi(z) - \beta\Pi(z=0) = \beta\Pi(z) - \alpha n_s - \alpha \int_0^z \rho(z')dz' \tag{12.4.8}$$

where

$$n_s = \int_0^\infty \rho(z)dz \tag{12.4.9}$$

is the number of colloids per unit area at $z = 0$. A single measurement of the equilibrium density profile is therefore sufficient[42] to determine the equation of state as a function of the overall number density by elimination of z between

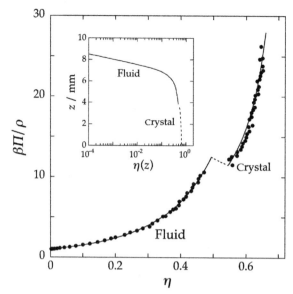

FIGURE 12.6 The main figure shows the osmotic equation of state as a function of the hard-sphere packing fraction for a sterically stabilised colloidal dispersion. The points are experimental results and the curves are calculated from the Carnahan–Starling equation (fluid branch) or an empirical equation of state[44] (solid branch). The broken line links the values obtained by simulation for the packing fractions of the two phases at coexistence. The inset shows the packing fraction profile from which the equation of state is deduced; see text for details. *Redrawn with permission from Ref. 43 © 2007 American Physical Society.*

$\rho(z)$ and $\Pi(z)$. Figure 12.6 shows the results of such an experiment[43] on an aqueous suspension of spherical, colloidal particles (a fluorinated polymer), sterically stabilised by addition of surfactant. The particles again behave, to a very good approximation, as hard spheres. There is excellent agreement with both the Carnahan–Starling equation in the fluid phase and an empirical equation of state[44] in the solid phase, with the fluid–solid transition occurring very close to that expected from computer simulations of hard-sphere systems.

The argument leading to (12.4.8) holds for colloidal particles much larger than those of the solvent; in that case it is justified to employ an expression for the buoyant mass which derives from Archimedes' principle. When that condition is not met, subtle effects due to the depletion forces discussed in the next section come into play, the solvent can no longer be treated as a continuum, and a generalised form of Archimedes' principle must be employed.[45]

12.5 COLLOID–POLYMER MIXTURES

In the previous section it was shown how grafted polymers can stabilise colloidal dispersions against flocculation. By contrast, free, non-adsorbing polymers may destabilise homogeneous colloidal suspensions by driving a separation into

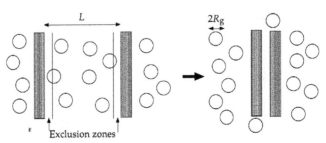

FIGURE 12.7 Exclusion zones at the surfaces of two plates immersed in a a polymer solution. When the distance between the plates is reduced to a value at which the exclusion zones overlap, the polymers initially found between the plates are expelled, leading to an imbalance in osmotic pressure.

concentrated (or 'liquid') and dilute (or 'gas') colloid phases. Phase separation arises from a polymer *depletion* effect, which is of essentially entropic origin.[33] The physical basis of the effect is easily understood by considering the very simple case of two parallel plates immersed in a solution of non-interacting polymers, as shown in Figure 12.7. The polymers are free to interpenetrate but their centres cannot approach the plates within a distance equal to their radius of gyration; this leads to the polymer exclusion zones represented in the figure. When the separation L of the plates is less than $2R_g$ the two exclusion zones overlap and the polymer coils are expelled. This creates an osmotic pressure imbalance which represents an attractive force per unit area between the plates equal to the decrease in pressure, given by van't Hoff's equation:

$$\Delta \Pi = -\rho k_B T \Theta (L - 2R_g) \tag{12.5.1}$$

where ρ is the bulk polymer density and $\Theta(x)$ is the Heaviside step function. The force is the (negative) derivative of an effective depletion potential given by

$$V^{\text{eff}}(L) = -\rho k_B T (2R_g - L)\Theta(L - 2R_g) \tag{12.5.2}$$

Both force and potential vanish when $L > 2R_g$. The depletion attraction can also be understood in terms of a gain in total volume accessible to the polymers, and hence an increase in their entropy, when the exclusion zones around the plates overlap. The increase in entropy, and hence the decrease in free energy or grand potential when $L < 2R_g$ lead to the effective attraction between the two plates.

These simple arguments can be formalised within the general framework of the coarse-graining strategy described in Section 12.1. Consider a dispersion of N_C spherical, colloidal particles of radius R_0 in a solution of polymer coils of fluctuating number N_P in osmotic equilibrium with a pure solution of polymers which serves as a reservoir and fixes the chemical potential μ of the polymers. The system can be described within the semi-grand canonical ensemble (rather than the canonical ensemble for a binary mixture considered in Section 12.1), characterised by the variables N_C, V (the total volume of the suspension), T, and μ. Let $V_{CC}(\{\mathbf{R}_i\})$, $V_{PP}(\{\mathbf{r}_j\})$ and $V_{CP}(\{\mathbf{R}_i\}, \{\mathbf{r}_j\})$ be the contributions to the total potential energy from colloid–colloid, polymer–polymer and colloid–polymer interactions. Then the semi-grand partition function may be written as a generalisation of (2.4.6):

$$\Xi(V, T, N_C, \mu) = \frac{1}{N_C! \Lambda_C^{3N_C}} \sum_{N_P=0}^{\infty} \frac{z^{N_P}}{N_P!} Z(V, T, N_C, N_P) \qquad (12.5.3)$$

where Λ_C is the de Broglie thermal wavelength of the colloidal particles, $z = \exp(\beta\mu/\Lambda_P^3)$ is the activity of the polymers (Λ_P being the thermal wavelength associated with their centre-of-mass motion) and Z is the configuration integral:

$$Z(V, T, N_C, \mu) = \int \exp\Big[-\beta\big(V_{CC}(\{\mathbf{R}_i\}) + V_{PP}(\{\mathbf{r}_j\})$$
$$+ V_{CP}(\{\mathbf{R}_i\}, \{\mathbf{r}_j\})\big]d\mathbf{R}^{N_C}\,d\mathbf{r}^{N_P} \quad (12.5.4)$$

The grand partition function of the system of polymers for a given configuration $\{\mathbf{R}_i\}$ of the colloids is

$$\Xi_P(V, T, \mu; \{\mathbf{R}_i\})$$
$$= \sum_{N_P=0}^{\infty} \frac{z^{N_P}}{N_P!} \int \exp\big(-\beta\big[V_{PP}(\{\mathbf{r}_j\}) + V_{CP}(\{\mathbf{R}_i\}, \{\mathbf{r}_j\})\big]\big)d\mathbf{r}^{N_P} \quad (12.5.5)$$

and the polymer grand potential is

$$\Omega_P(V, T, \mu; \{\mathbf{R}_i\}) = -k_B T \ln \Xi_P(V, T, \mu; \{\mathbf{R}_i\}) \qquad (12.5.6)$$

Combination of (12.5.3)–(12.5.6) yields an expression for the semi-grand partition function:

$$\Xi(V, T, N_C, \mu)$$
$$= \frac{1}{N_C! \Lambda^{3N_C}} \int \exp\big(-\beta\big[V_{CC}(\{\mathbf{R}_i\}) + \Omega_P(V, T, \mu; \{\mathbf{R}_i\})\big]\big)d\mathbf{R}^{N_C} \quad (12.5.7)$$

This result shows that the initial colloid–polymer mixture has been mapped onto a system of N_C colloidal particles for which the total interaction energy is

$$V^{\mathrm{eff}}(\{\mathbf{R}_i\}) = V_{CC}(\{\mathbf{R}_i\}) + \Omega_P(V, T, \mu; \{\mathbf{R}_i\}) \qquad (12.5.8)$$

which depends on the thermodynamic state variables V, T and μ. The semi-grand partition function (12.5.7) is therefore equivalent to the canonical partition function of a system of N_C interacting colloidal particles for which the total potential energy is given by (12.5.8). The derivation of (12.5.8) takes no account the internal partition function of the polymer coil; it implicitly assumes that a coarse-grained representation of the polymer coils has been used. The internal partition function would contribute a temperature-dependent term to the effective potential energy but is independent of the colloid configuration. This contribution is conventionally referred to as a 'volume' term; it may affect the thermodynamic properties of the suspension but not its structure.

As an illustration of (12.5.8) consider a model of hard sphere colloids and non-interacting polymers which are excluded from a sphere of radius $R_x = R_0 + R_g$ around each of the colloids. This is an extreme example of a non-additive hard-sphere mixture with diameters (in the notation of Section 3.10) given by $d_{11} = 2R_0, d_{22} = 0$ and $d_{12} = \frac{1}{2}(d_{11} + d_{22})(1 + \Delta)$, where $\Delta = 1 + R_g/R_0$; it is equivalent to an ideal gas confined to an accessible volume $\mathcal{V}(\{\mathbf{R}_i\})$, the magnitude of which depends on the instantaneous colloid configuration. The accessible volume is the volume of the suspension less the volume occupied by colloidal particles and their exclusion shells. Since exclusion spheres may overlap, $\mathcal{V}(\{\mathbf{R}_i\})$ will in general be a highly complicated function of the colloid coordinates. Formally, according to (2.4.11), which is valid for ideal particles:

$$\Xi_P^{id} = \exp[z\mathcal{V}(\{\mathbf{R}_i\})] \tag{12.5.9}$$

where z is equal to ρ_R, the polymer reservoir density, not the polymer density in the mixture. Thus

$$\Omega_P(V, T, \mu; \{\mathbf{R}_i\}) = -\rho_R k_B T \mathcal{V}(\{\mathbf{R}_i\}) = -\Pi\mathcal{V}(\{\mathbf{R}_i\}) \tag{12.5.10}$$

where Π is the osmotic pressure of the polymers, which is assumed to take its ideal value.

Consider first the case of a pair of colloidal particles, as pictured in Figure 12.8; $\mathcal{V}$ is now dependent only on the distance R between the centres of the spheres and the effective, depletion-induced, pair potential is

$$v_2(R) = \Omega_P(T, \rho_R; R) - \Omega_P(T, \rho_R; R \to \infty)$$
$$= 0, \quad R > 2R_x \tag{12.5.11}$$
$$= -\rho_R k_B T \frac{4\pi}{3} R_x^3 \left[1 - \frac{3R}{4R_x} + \frac{1}{16}\left(\frac{R}{R_x}\right)^3\right], \quad 2R_0 < R < 2R_x$$

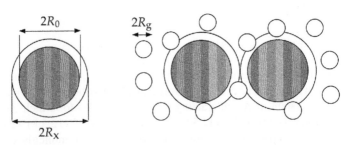

FIGURE 12.8 Exclusion zones around two hard-sphere colloids in a polymer solution.

where $R_x = R_0 + R_g$ is the radius of the exclusion sphere shown in Figure 12.8 and the last line involves the volume of intersection of two exclusion spheres separated by a distance R, which vanishes when $R > 2R_x$. Equation (12.5.11) is a famous result, first obtained by Asakura and Oosawa.[46]

The range of interaction increases with polymer size and its amplitude increases with polymer concentration. The effective interaction between colloids can therefore be tuned by changes in the two physical parameters. As the colloid concentration increases, overlap of the exclusion spheres of more than two colloids becomes increasingly probable, leading to effective, many-body interactions. All pair and higher-order interactions can be taken into account within thermodynamic perturbation theory,[47] whereby the system of unperturbed hard-sphere colloids is chosen as a reference system and the accessible volume $\mathcal{V}(\{\mathbf{R}_i\})$ in (12.5.10) is replaced by its mean value, $\bar{\mathcal{V}} = \alpha V$, obtained by averaging over all configurations of the reference system. The accessible volume fraction α depends only on the colloid packing fraction $\eta_C = 4\pi R_0^3 N_C/3V$ and the polymer–colloid size ratio $q = R_g/R_0$. The free energy of the colloid–polymer system therefore splits into two parts:

$$F = F_C(V, T, N_C) + F_P(\alpha V, T, N_P) \qquad (12.5.12)$$

The excess free energy of the hard-sphere system is given by the Carnahan–Starling result (3.9.21) for the fluid phase or by an empirical equation of state for the face-centred cubic, solid phase,[44] while the free energy of the non-interacting polymers is, from (2.3.16):

$$F_P(\alpha V, T, N_P) = V\rho k_B T \ln(\rho \Lambda_P^3/\alpha - 1) \qquad (12.5.13)$$

The remaining task is to estimate the accessible volume fraction $\alpha(\eta_C)$. This is achieved by a straightforward generalisation of Widom's particle insertion formula (2.4.33) in which the test particle in Figure 2.2 is of radius R_x and the probability p_0 is equal to α. Thus

$$\alpha = \exp\left(-\beta\mu^{ex}\right) \qquad (12.5.14)$$

where μ^{ex} is the excess chemical potential of a particle of radius R_x in a binary mixture with hard spheres of smaller radius R_0 in the limit of vanishing concentration of the test particle species. If the Percus–Yevick approximation is used for μ^{ex}, we find that α is given by a simple formula:

$$\alpha = (1 - \eta_C)\exp\left(-A\gamma - B\gamma^2 - C\gamma^3\right) \qquad (12.5.15)$$

where $\gamma = \eta_C/(1 - \eta_C)$, $A = 3q + 3q^2 + q^3$, $B = 9q^2/2 + 3q^3$ and $C = 3q^3$. By equating the chemical potentials and osmotic pressures of the two species in coexisting colloid gas, liquid or crystal phases it is possible to deduce the phase diagram of the colloid–polymer mixture for any value of q in the η_R versus η_C plane, where $\eta_R = 4\pi\rho_R R_g^3/3$ is the polymer packing fraction in the reservoir. The polymer-induced, effective attraction between the colloidal

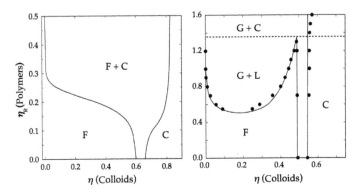

FIGURE 12.9 Phase diagrams of the Asakura–Oosawa model of a colloid–polymer mixture for two values of the size ratio q. Left-hand panel: $q = 0.1$. From H.N.W. Lekkerkerker et al., 'Phase behaviour of colloid + polymer mixtures', *Europhys. Lett.* **20**, 559–564 (1992). Right-hand panel: $q = 0.6$; the points show the results of Monte Carlo calculations. *Redrawn with permission from Ref. 47 © 2006 American Physical Society.*

particles drives a gas–liquid phase separation, while the hard-sphere repulsion leads to a liquid–crystal phase transition at sufficiently high values of η_C; in the limit of vanishing polymer density ($\mu_P \to -\infty$), colloid crystallisation reduces to that of hard spheres.

Figure 12.9 shows the phase diagrams[47,48] of the Asakura–Oosawa model, augmented by the correction for many-body interactions provided by (12.5.13), for two values of the size ratio q. For $q = 0.1$ the gas–liquid transition is pre-empted by a fluid–crystal transition; hence there is no colloid liquid phase. For q greater than approximately 0.32 a gas–liquid critical point emerges and the phase diagram resembles that of a simple liquid, such as that shown in Figure 1.1, with η_R playing the role of an inverse temperature. As q increases further, the critical packing fractions of polymers and colloids both decrease, while the ratio of critical to triple point packing fraction for the polymers increases. The phase diagram is therefore very sensitive to the range of the depletion-induced attraction relative to the size of the colloidal particles. If the range is too short, the liquid phase disappears altogether, or is at best metastable to freezing, as seen in experiments[49] on PMMA–polystyrene mixtures, where three-phase coexistence occurs only for q greater than approximately 0.25.

The phase diagrams of Figure 12.9 should provide a fair description of the phase behaviour of real colloidal systems under θ-solvent conditions but in good solvent the interactions between polymer coils cannot be ignored. More realistic models[50] that include effective polymer–polymer and colloid–polymer interactions, constructed in ways similar to those described in Section 12.2, have been used in Gibbs ensemble Monte Carlo calculations[51] for a range of values of q. The calculated phase diagrams are topologically similar to those in Figure 12.9, but there are major quantitative and even qualitative differences. For $q = 0.34$ there is reasonable agreement between the results for interacting and non-interacting polymers; in each case the gas–liquid transition is metastable relative to freezing. As q increases,[52] the differences become greater. These are particularly pronounced in the location of the triple point, for which the inclusion of interactions leads to very good agreement with experimental results,[49] whereas in the non-interacting case the triple point moves to polymer reservoir packing fractions that are much too high.

In binary mixtures of large and small hard-sphere colloids, or of colloids and nanoparticles, the small hard spheres can act as depletants. The interaction thereby induced between two large spheres is attractive at short range and damped oscillatory at larger distances; the oscillations arise from a layering of the small spheres around the large ones.[53] Monte Carlo simulations[54] based on an effective pair potential lead, for q less than about 0.2, to a fluid–solid coexistence curve that widens considerably as the small sphere packing fraction increases; this has the effect of pre-empting a fluid–fluid transition. This behaviour is qualitatively similar to that observed for a colloid–polymer mixture and confirmed by simulations of the underlying hard-sphere mixture. It implies that the fluid–fluid phase separation predicted for a highly asymmetric binary mixture by thermodynamically self-consistent integral equations[55] is only metastable, but may nonetheless be observable, given the slow rates of crystal nucleation expected for such a system.

As we have seen, a feature of colloid–polymer mixtures is that no fluid–fluid critical point exists if the range of the effective, attractive forces between colloidal particles is too short. Similar considerations apply to simpler systems. Analysis of results obtained for a number of spherically symmetric pair potentials has shown, as would be expected, that the reduced critical temperature $T_c^* = k_B T / \epsilon$, where ϵ is the depth of the potential well, decreases rapidly with a reduction in range of the attractive interaction. What is surprising is the fact that the value of the reduced second virial coefficient $B_2^*(T_c)$ is insensitive to variation in range or choice of potential and given[56] to a good approximation by

$$B_2^*(T_c) = \frac{B_2(T_c)}{v_0} \approx -6 \qquad (12.5.16)$$

where $v_0 = \pi d^3 / 6$ is the volume of a particle of effective diameter d. A striking illustration of this apparent 'universality' is the fact that for the mean field equation of state (5.7.4), which is most accurate when the range of attraction is large, $B_2^* \approx -6.66$, while for the Baxter sticky-sphere model (4.4.18), representative of very short-ranged attractions, $B_2^* \approx -6.2$. Use of (12.5.16) therefore allows a rough estimate of the critical temperature to be made from a simple integral involving only the pair potential. This semi-empirical line of reasoning has been carried further by the suggestion that an extended law of corresponding states[57] applies to non-conformal potentials, the difference compared with conformal potentials (3.10.1) being that the compressibility factor $\beta P / \rho$ is now a function of three, rather than two reduced variables: $T^* = k_B T / \epsilon$, $\rho^* = \rho d^3$ and $B_2^* = B_2 / v_0$, where d is related to the repulsive part of the potential by the

Barker–Henderson prescription (5.4.11). The effective range q of an arbitrary, attractive potential is then taken to be the range of attraction, $q = \gamma - 1$ (see Figure 1.2), for a square-well potential having the same, reduced second virial coefficient as the potential of interest. With this definition of q, the boundary of stability of the fluid–fluid transition for a wide variety of non-conformal potentials is found to lie within a narrow range extending from about 0.13 to 0.15.

12.6 CHARGE-STABILISED COLLOIDS

We now show how the methods developed in earlier sections can be adapted to the calculation of the effective interaction between large polyions in solution. As our main example we take the case of a dispersion of hard sphere, colloidal particles of radius R_0 in a polar solvent of dielectric constant ϵ. We assume that each particle carries a uniform charge density σ, corresponding to a total charge equal to Ze ($|Z| \gg 1$), which gives rise an electric double layer at the surface similar to that formed near a charged, planar surface, as described in Section 10.6. We again adopt a primitive-model description of the solvent, with both coions and counterions represented as charged hard spheres of diameter $d(\ll R_0)$, and assume that the dispersion is in equilibrium with a salt reservoir that fixes the chemical potentials μ_+, μ_- of the microions. The three-component system can be described within the semi-grand canonical ensemble introduced in the previous section, characterised here by the variables V, T, N_0, μ_+ and μ_-. Thus the number of polyions, N_0, is fixed but the numbers of microions, N_+ and N_-, are allowed to fluctuate. If, as before, we denote the coordinates of the large particles (polyions, subscript 0) by $\{\mathbf{R}_i\}$ and those of the small particles (microions, subscript M) by $\{\mathbf{r}_j\}$, the total potential energy of the system may be written in short-hand form as

$$V_{\{N\}}\left(\{\mathbf{R}_i\}, \{\mathbf{r}_j\}\right) = V_{00}\left(\{\mathbf{R}_i\}\right) + V_{MM}\left(\{\mathbf{r}_j\}\right) + V_{0M}\left(\{\mathbf{R}_i\}, \{\mathbf{r}_j\}\right) \quad (12.6.1)$$

where V_{00}, V_{MM} and V_{0M} are all sums of pair potentials of primitive-model type and $\{N\} \equiv N_0, N_+, N_-$.

The coarse-graining approach that we adopt has a now familiar pattern in which the degrees of freedom of the microions are averaged out, thereby reducing the problem to that of an effective, one-component system of polyions dressed by their electric double layers. This reduction is accomplished by writing the semi-grand partition function of the semi-grand canonical system in a form similar to (12.5.7):

$$\Xi(V, T, N_0, \mu_+, \mu_-) = \frac{1}{N_0! \Lambda_0^{3N_0}}$$

$$\times \int \exp[-\beta V_{00}(\{\mathbf{R}_i\})] \Xi_M\left(V, T, \mu_+, \mu_-; \{\mathbf{R}_i\}\right) d\mathbf{R}^{N_0} \quad (12.6.2)$$

where

$$
\Xi_M = \sum_{N_+=0}^{\infty} \sum_{N_-=0}^{\infty} \frac{z_+^{N_+} z_-^{N_-}}{N_+! N_-!}
$$

$$
\iint \exp\left[-\beta(V_{MM}(\{\mathbf{r}_j\}) + V_{0M}(\{\mathbf{R}_i\}, \{\mathbf{r}_j\}))\right] d\mathbf{r}^{N_+} d\mathbf{r}^{N_-} \quad (12.6.3)
$$

is the grand partition function of the microions in the external potential $\phi_\nu(\mathbf{r})$ of the polyions in a configuration $\{\mathbf{R}_i\}$:

$$
\phi_\nu(\mathbf{r}) = \sum_{i=1}^{N_0} v_{0\nu}(|\mathbf{r} - \mathbf{R}_i|), \quad \nu = +, - \quad (12.6.4)
$$

and z_+, z_- are the activities of the microions. Equation (12.6.2) may be re-expressed as

$$
\Xi(V, T, N_0, \mu_+, \mu_-) = \frac{1}{N_0! \Lambda_0^{3N_0}} \int \exp\left(-\beta V^{\mathrm{eff}}(\{\mathbf{R}_i\})\right) d\mathbf{R}^{N_0} \quad (12.6.5)
$$

in which the effective interaction between the dressed polyions is

$$
V^{\mathrm{eff}}(\{\mathbf{R}_i\}) = V_{00}(\{\mathbf{R}_i\}) + \Omega_M(V, T, \mu_+, \mu_-; \{\mathbf{R}_i\}) \quad (12.6.6)
$$

where $\Omega_M = -k_B T \ln \Xi_M$ is the grand potential of the microions. The first term on the right-hand side of (12.6.6) arises from the direct interaction between polyions, while the second is a state-dependent, microion-induced interaction, which depends parametrically on the coordinates $\{\mathbf{R}_i\}$. The direct interaction is pairwise additive but the effective interaction is not; the effective interaction also includes a volume term which is independent of the polyion coordinates.[58]

The grand potential Ω_M can be evaluated by the methods of density functional theory. If we limit ourselves to a mean field approach we may take over the grand potential functional defined by (10.6.6), (10.6.7) and (10.6.10) (with $\mathcal{F}^{\mathrm{corr}} = 0$). The solution of the resulting Euler–Lagrange equations for the local densities $\rho_\nu^{(1)}(\mathbf{r})$ in the multi-centre, external potential (12.6.4) poses a formidable task. Numerical results can be obtained through a molecular dynamics scheme in which the Fourier components of the local densities are treated as dynamical variables,[59] a scheme inspired by the Car–Parrinello method for simulating systems of classical ions and quantum mechanical, valence electrons.[60] Further progress can be made analytically if the inhomogeneities induced by the polyions are assumed to be weak. In that case it is justifiable to expand the ideal free energy functional (3.1.22) to second order in the deviation $\Delta\rho_\nu^{(1)}(\mathbf{r})$ of the local density from its bulk value, i.e.

$$
\Delta\rho_\nu^{(1)}(\mathbf{r}) = \rho_\nu^{(1)}(\mathbf{r}) - n_\nu \quad (12.6.7)
$$

The intrinsic free energy functional of the microions is then

$$\mathcal{F}[\rho_+^{(1)}, \rho_-^{(1)}] = \sum_\nu \left(F^{\mathrm{id}}(n_\nu) + k_B T \ln (n_\nu \Lambda_\nu^3) \int \Delta \rho_\nu^{(1)}(\mathbf{r}) d\mathbf{r} \right.$$
$$\left. + \frac{k_B T}{2n_\nu} \int [\Delta \rho_\nu^{(1)} \mathbf{r}]^2 d\mathbf{r} \right) + \frac{1}{2} \int e\rho_Z(\mathbf{r}) \Phi^C(\mathbf{r}) d\mathbf{r}$$

$$(12.6.8)$$

where the electrostatic potential $\Phi^C(\mathbf{r})$ satisfies Poisson's equation (10.6.3). Substitution of (12.6.8) in (10.6.6), replacement of the chemical potentials μ_ν by their ideal values, and use of the variational principle (3.4.3) gives

$$\frac{\Delta \rho_\nu^{(1)}(\mathbf{r})}{n_\nu} + z_\nu \Phi^C(\mathbf{r}) = -\beta \phi_\nu(\mathbf{r}), \quad \nu = +, - \qquad (12.6.9)$$

The last two equations are coupled through the terms in $\Phi^C(\mathbf{r})$. If the polyions were point particles, the coulombic contribution to $\phi_\nu(\mathbf{r})$ would be everywhere equal to $z_\nu e \Phi^{\mathrm{ext}}(\mathbf{r})$, where $\Phi^{\mathrm{ext}}(\mathbf{r})$ is the 'external' electrostatic potential acting on the microions.[61] If there were no boundaries, the total electrostatic potential within the fluid would then be

$$\Phi(\mathbf{r}) = \Phi^C(\mathbf{r}) + \Phi^{\mathrm{ext}}(\mathbf{r}) = e \int \frac{\rho_Z(\mathbf{r}') + Z\rho^{\mathrm{ext}}(\mathbf{r}')}{\epsilon |\mathbf{r} - \mathbf{r}'|} d\mathbf{r}' \qquad (12.6.10)$$

where $\rho^{\mathrm{ext}}(\mathbf{r}) = \sum_i \delta(\mathbf{r} - \mathbf{R}_i)$ is the microscopic density of the polyions. Equation (12.6.9) now becomes

$$\Delta \rho_\nu^{(1)}(\mathbf{r}) = -\frac{n_\nu z_\nu e^2}{k_B T} \int \frac{\rho_Z(\mathbf{r}') + Z\rho^{\mathrm{ext}}(\mathbf{r}')}{\epsilon |\mathbf{r} - \mathbf{r}'|} d\mathbf{r}' \qquad (12.6.11)$$

To simplify the problem, we consider only the salt-free case, where all microions are counterions. The coupled equations (12.6.11) then reduce to a single integral equation from which the subscript ν can be dropped and the charge density $\rho_Z(\mathbf{r})$ replaced by $z\rho^{(1)}(\mathbf{r})$. On taking Fourier transforms of both sides of (12.6.11), applying the convolution theorem and incorporating the result in (10.1.5), we find that the Fourier transform of $\Delta \rho^{(1)}(\mathbf{r})$ is

$$\hat{\rho}^{(1)}(\mathbf{k}) = \frac{Zk_D^2}{k^2 + k_D^2} \sum_{i=1}^{N_0} \exp(-i\mathbf{k} \cdot \mathbf{R}_i) \qquad (12.6.12)$$

where $k_D^2 = 4\pi n z^2 e^2 / \epsilon k_B T$ is the square of the Debye wavenumber associated with the counterions. Inverse Fourier transformation of (12.6.12) leads to a counterion density profile given by

$$\rho^{(1)}(\mathbf{r}) = \sum_{i=1}^{N_0} \frac{Zk_D^2}{4\pi} \frac{\exp(-k_D |\mathbf{r} - \mathbf{R}_i|)}{|\mathbf{r} - \mathbf{R}_i|} \equiv \sum_{i=1}^{N_0} \rho_i^{(1)}(\mathbf{r}) \qquad (12.6.13)$$

The total profile is therefore a superposition of profiles associated with each of the polyions. The radius of the polyions is now reintroduced by imposing the constraint that $\rho_i^{(1)}(\mathbf{r})$ must be zero whenever $|\mathbf{r} - \mathbf{R}_i| < R$. Charge neutrality means that $\rho_i^{(1)}(\mathbf{r})$ must be normalised such that

$$\int_{|\mathbf{r}-\mathbf{R}_i|>R} \rho_i^{(1)}(\mathbf{r})\mathrm{d}\mathbf{r} = |Z/z| \tag{12.6.14}$$

For the profile defined by (12.6.13) this requirement would be met if the polyion charge Ze were replaced by an apparent charge $Z'e$, where

$$Z' = Z\frac{\exp(k_\mathrm{D}R)}{1 + k_\mathrm{D}R} \tag{12.6.15}$$

The normalisation in (12.6.14) implicitly assumes that the colloid concentration is low and hence that the electric double layers associated with neighbouring polyions have, on average, little overlap. From Poisson's equation it is evident that the total electrostatic potential may similarly be written as a superposition of N_0 screened potentials:

$$\Phi(\mathbf{r}) = \sum_{i=1}^{N_0} \frac{Z'e}{\epsilon}\frac{\exp(-k_\mathrm{D}|\mathbf{r} - \mathbf{R}_i|)}{|\mathbf{r} - \mathbf{R}_i|} \equiv \sum_{i=1}^{N_0} \Phi_i(\mathbf{r}) \tag{12.6.16}$$

If the density profile (12.6.13) and the potential (12.6.16) are substituted in the free energy functional (12.6.8), we find that the effective interaction energy (12.6.6) takes the form

$$V^{\mathrm{eff}}(\{\mathbf{R}_i\}) = V_0 + \sum_{i}\sum_{j>i} v_2(|\mathbf{R}_j - \mathbf{R}_i|) \tag{12.6.17}$$

where the effective pair potential $v_2(R)$ provides the electrostatic contribution to the Derjaguin–Landau–Verwey–Overbeek (DLVO) potential[62]:

$$\begin{aligned} v_2(|\mathbf{R}_j - \mathbf{R}_i|) &= \int \Phi_i(\mathbf{r})\rho_j^{(1)}(\mathbf{r})\mathrm{d}\mathbf{r} \\ &= \frac{Z'^2 e^2}{\epsilon}\frac{\exp(-k_\mathrm{D}|\mathbf{R}_j - \mathbf{R}_i|)}{|\mathbf{R}_j - \mathbf{R}_i|} \end{aligned} \tag{12.6.18}$$

The pairwise additivity is a consequence of the quadratic form of the approximate functional (12.6.8).

The effective interaction energy (12.6.17) contains a structure-independent term, V_0. This term has no effect on the forces acting between the polyions, but it has a significant influence on the phase diagram.[58] It includes, among other contributions, the self-energy of the double layers associated with individual polyions. The DLVO potential is a function of density and temperature through its dependence on the Debye wavenumber; its form remains the same even in the presence of coions provided the contributions of all microions are included in the definition of k_D and in V_0. It is strictly repulsive, which would stabilise the system against flocculation induced by the van der Waals forces that are present in any real dispersion. On the other hand, if the salt concentration is sufficiently low, the structure-independent term can drive a phase transition into colloid-rich and colloid-poor dispersions even in the absence of attractive forces.

A quadratic functional would seem unlikely to be adequate for use in calculations for systems of highly charged particles. In practice the strong electrostatic attraction exerted by the polyions on the counterions leads to a substantial fraction of the latter becoming tightly bound to the colloid surface; this 'counterion condensation' reduces[63] the magnitude of the bare polyion charge to an effective value $|Z_{eff}| < |Z|$. The remaining counterions therefore experience a much weaker external potential, so the diffuse part of the double layer may still be described within the quadratic approximation. The functional can also be extended in a way that allows the determination of effective, three-body interactions.[64] For triplet configurations close to contact the three-body interaction provides a substantial, attractive correction to the pairwise-additive, repulsive interaction described by (12.6.18). Nonetheless, direct measurement of the effective pair potential between charged colloidal particles shows that (12.6.18) provides a good representation of the data when Z_{eff} is suitably chosen, as the results shown in Figure 12.10 illustrate.

An explicit relation between $|Z|$ and $|Z_{eff}| < |Z|$ can be derived within Poisson–Boltzmann theory.[66] Consider the case of an isolated planar surface

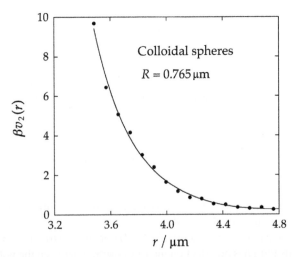

FIGURE 12.10 Effective pair potential between polystyrene sulphate spheres of radius $0.765 \pm 0.01\,\mu m$ dispersed in water. The points are experimental results and the curve is calculated from (12.6.18) for an effective charge $|Z_{eff}| = 22,793$. *Redrawn with permission from Ref. 65 © 1994 American Physical Society.*

immersed in a symmetric electrolyte of bulk concentration n_0, already discussed in Section 10.6. Use of the identity $\tanh^{-1} x = \frac{1}{2} \ln[(1+x)/(1-x)]$ allows the solution (10.6.23) of the Poisson–Boltzmann equation for the dimensionless electrostatic potential to be written in the form

$$\Phi^*(z) = 2 \ln \left(\frac{1 + g \exp(-k_D z)}{1 - g \exp(-k_D z)} \right) \tag{12.6.19}$$

while the boundary condition (10.6.16) yields an expression for g:

$$g = (1 + x^2)^{1/2} - x \tag{12.6.20}$$

where $x = k_D e / 2\pi l_B \sigma$. Far from the charged plane, $k_D z \gg 1$, and to linear order $\Phi^*(z)$ reduces to

$$\Phi^*(z) \approx \Phi_S^* \exp(-k_D z), \quad k_D z \gg 1 \tag{12.6.21}$$

where $\Phi_S^* = 4g$. Thus for $z = 0$:

$$\left. \frac{\partial \Phi^*(z)}{\partial z} \right|_{z=0} = -4 k_D g \tag{12.6.22}$$

The linearised Poisson–Boltzmann equation

$$\frac{d^2}{dz^2} \Phi^*(z) = k_D^2 \Phi^*(z) \tag{12.6.23}$$

has a solution of the same form as (12.6.21) but if that solution is to be consistent with (12.6.22) the bare charge density in the boundary condition (10.6.16) must be replaced by an effective charge

$$\sigma_{\text{eff}} = \frac{e k_D g}{\pi l_B} \tag{12.6.24}$$

with g given by (12.6.20). On defining what turns out to be a saturation charge density $\sigma_{\text{sat}} = k_D e / \pi l_B$, we see that $x = \sigma_{\text{sat}} / 2\sigma$. If the surface charge density is very low, i.e. if $x \gg 1$, then $g \approx (1 - 1/4x^2)/2x$ and

$$\sigma_{\text{eff}} \approx \sigma \left[1 - (\sigma / \sigma_{\text{sat}})^2 \right], \quad \sigma \ll \sigma_{\text{sat}} \tag{12.6.25}$$

while in the opposite limit of very high charge density, $x \ll 1$, $g \approx 1 - x$ and

$$\sigma_{\text{eff}} \approx \sigma_{\text{sat}} \left[1 - (\sigma_{\text{sat}} / 2\sigma) \right], \quad \sigma \gg \sigma_{\text{sat}} \tag{12.6.26}$$

As expected, the effective surface charge reduces to its bare value when that is low; counterion condensation is negligible and linear Poisson–Boltzmann theory is adequate. For high surface charge, the effective value reaches its maximum value σ_{sat}, which is independent of σ. In that case counterion condensation strongly reduces the absolute value of the bare charge and the linear theory remains applicable to the diffuse part of the double layer.

A similar argument to that just outlined applies in the case of an isolated colloidal sphere of radius R_0 and bare charge Ze in a symmetric electrolyte. The radial Poisson–Boltzmann equation for this problem has no analytical solution, but in the limit $R_0 \gg k_D^{-1}$, where planar geometry is recovered, the calculations become simpler and show that the effective charge at saturation is[66]

$$Z_{\text{sat}} = \frac{4R_0}{l_B}(1 + k_D R_0) \qquad (12.6.27)$$

An improved estimate of Z_{sat}, valid down to $k_D R_0 \approx 1$ has also been obtained, together with an analytical approximation for its dependence on the bare charge.[67]

Thus far we have considered only the case of a single polyion in a bulk electrolyte, corresponding to the limit of infinite dilution of the polyions. At high concentrations the polyions form a dense, colloidal fluid in which each particle is confined to a cage formed by its nearest neighbours. This situation is well described by a Wigner–Seitz cell model in which each polyion is placed at the centre of a cell of volume $v = V/N_0$ together with monovalent coions and counterions in osmotic equilibrium with a reservoir of overall concentration n_0; the total charge within the cell is assumed to vanish.[68] The geometry of the cell mimics the shape of the polyion; thus for spherical ions the cell is a sphere of radius $a = (3v/4\pi)^{1/3}$. Since the cell is overall neutral, it does not interact with the cells associated with neighbouring polyions, so calculations are needed for only a single sphere. The radial Poisson–Boltzmann equation is now

$$\left(\frac{d^2}{dr^2} + \frac{2}{r}\frac{d}{dr} \right) \Phi^*(r) = k_{D0}^2 \sinh \Phi^*(r), \quad R_0 < r < a \qquad (12.6.28)$$

where $k_{D0}^2 = 4\pi n_0 l_B$. Equation (12.6.28) must be solved subject to boundary conditions at $r = R_0$ (surface of the polyion) and $r = a$ (surface of the Wigner–Seitz cell):

$$\frac{d\Phi^*(r)}{dr}\bigg|_{r=R} = -\frac{Z l_B}{R^2}, \quad \frac{d\Phi^*(r)}{dr}\bigg|_{r=a} = 0 \qquad (12.6.29)$$

The microion charge density within the Wigner–Seitz sphere is

$$\rho_Z(r) = \rho_+^{(1)}(r) - \rho_-^{(1)}(r) = -\frac{k_{D0}^2}{4\pi l_B} \sinh \Phi^*(r) \qquad (12.6.30)$$

and its integral over the available volume, for a polyion of charge $-Ze$, is equal to Z:

$$4\pi \int_{R_0}^{a} \rho_Z(r) r^2 \, dr = Z \qquad (12.6.31)$$

Equation (12.6.28) can be solved numerically for the electrostatic potential, in particular for its value at $r = a$; this is the only numerical input needed to determine the effective charge. The analytical solution involves expansion of

$\Phi^*(r)$ and $\rho_Z(r)$ to first order around $\Phi^*(a)$; the resulting, linear equation for $\delta\Phi^*(r) = \Phi^*(r) - \Phi^*(a)$ is

$$\left(\frac{d^2}{dr^2} + \frac{2}{r}\right)\delta\Phi^*(r) = k_{D0}^2\left(\gamma_a + \delta\Phi^*\right), \quad R < r < a \tag{12.6.32}$$

where $\gamma_a = \tanh\Phi^*(a)$ and the boundary conditions are now

$$\delta\Phi^*(a) = 0, \quad \left.\frac{d}{dr}\delta\Phi^*(r)\right|_{r=a} = 0 \tag{12.6.33}$$

The solution to this equation determines the charge density via the linearised form of (12.6.30), i.e.

$$\rho_Z(r) \approx -\frac{k_D^2}{4\pi l_B}\left[\gamma_a + \delta\Phi^*(r)\right] \tag{12.6.34}$$

where $k_D^2 = k_{D0}^2 \cosh\Phi^*(a)$, and the effective charge of the polyion is obtained by substitution in (12.6.31). The value of Z_{eff} is given by a lengthy expression:

$$Z_{\text{eff}} = \frac{\gamma_0}{k_D l_B}\left[(k_D^2 Ra - 1)\sinh\left(k_D(a - R)\right) + k_D(a - R)\right.$$
$$\left. \times \cosh\left(k_D(a - R)\right)\right] \tag{12.6.35}$$

but its key feature is that Z_{eff} depends on the bare charge only through the term $\tanh\Phi^*(a)$, which is calculated from (12.6.28). The effective charge is therefore that charge which, within linearised Poisson–Boltzmann theory, leads at $r = a$ to the same electrostatic potential and its gradient as the full theory does for the bare charge Z. The fact that $|Z_{\text{eff}}| < |Z|$ accounts implicitly for the effects of counterion condensation, thereby justifying the use of the linear theory at small values of r. This implies that the quantity Z' in the DLVO potential should be replaced by $Z'_{\text{eff}} = Z_{\text{eff}}\exp(k_D R)/(1 + k_D R)$.

While accounting approximately for the non-linearity of mean field Poisson–Boltzmann theory, charge renormalisation does not allow for microion correlations, which can no longer be ignored when multivalent microions are present. As shown in Section 10.6 for the case of parallel, charged plates, microion correlations may lead to charge inversion and hence to an effective attraction between like-charged polyions.[69] A simple model of charge inversion in the presence of multivalent counterions is based on the idea that highly correlated ions tightly bound to a planar, charged surface will form a two-dimensional, hexagonal Wigner crystal.[70] These periodic charge patterns lead to an attraction between two plates carrying surface charges of the same sign when the patterns on opposite plates are favourably staggered.[71] This is an appealing picture, but not one that is wholly consistent; it ignores the discreteness of the distribution of ionised sites on the colloid surface, which is replaced by a uniform charge density.

Application of the methods discussed thus far can be extended to systems of non-spherical particles. For example, much theoretical as well as experimental work has been devoted to aqueous suspensions of laponite particles. Laponite is a synthetic clay consisting of disc-like particles which have a thickness to diameter ratio of approximately 0.03 and carry a substantial surface charge. In the semi-dilute regime, suspensions are found experimentally to undergo a transition from a liquid-like sol to a network-forming gel[72] or a colloidal glass,[73] depending on experimental conditions. At high concentration, the suspensions, like certain natural clays of geophysical importance, form lamellar stacks that swell as the ionic strength is increased. The swelling is linked to the effective interaction between electric double layers. It can be described by non-linear Poisson–Boltzmann theory applied to the problem of two parallel plates confined, together with coions and counterions, within a cylindrical Wigner–Seitz cell.[74] The DLVO potential (12.6.18) can also be generalised to the case of charged, anisometric particles by introduction of anisotropy factors that are dependent on the orientations of the two particles involved[75] and an explicit expression for the anisotropy factor of rod-like particles has been derived.

12.7 COLLOIDAL LIQUID CRYSTALS

The discussion of colloidal systems up till now has been focused on the properties of spherical, hard-core particles, but rod-like and plate-like, mesoscopic particles are also very common both in nature and in the laboratory. Examples include elongated particles, such as the tobacco mosaic virus, which has a length to thickness ratio of ≈ 15, and thin clay platelets such as gibbsite, a form of aluminium hydroxide. The orientational degrees of freedom confer on dispersions of elongated or flat hard bodies a very rich phase behaviour, with partially ordered *mesophases* appearing between the fully isotropic, fluid phase, and the three-dimensionally ordered, crystal phase. Dispersions of highly anisometric colloidal particles are referred to as *lyotropic* liquid crystals; their thermodynamic and structural properties are largely controlled by excluded volume effects. *Thermotropic* liquid crystals, by contrast, are dense assemblies of smaller molecules for which long-range, attractive interactions play an important role; temperature is therefore the key control parameter. Although the length scales involved may be very different, lyotropic and thermotropic liquid crystals are structurally similar, but are generated by varying the packing fraction at constant temperature in one case and by changes in temperature at constant pressure in the other. The properties of thermotropic liquid crystals and their theoretical description are dealt with at length in classic texts[76]; this section is devoted exclusively to the lyotropic case.[77]

The equilibrium, single-particle density of a system of rigid, non-spherical particles is a function of the centre-of-mass coordinates $\mathbf{R}$ of the particle and its orientation $\mathbf{\Omega}$ relative to a laboratory-fixed frame:

$$\rho^{(1)}(\mathbf{R}, \mathbf{\Omega}) = \left\langle \sum_{i=1}^{N} \delta(\mathbf{R}_i - \mathbf{R})\delta(\mathbf{\Omega}_i - \mathbf{\Omega}) \right\rangle \qquad (12.7.1)$$

As we shall limit the discussion to the case of uniaxial particles, $\mathbf{\Omega} = (\theta, \phi)$, where θ and ϕ are the usual polar angles. In the homogeneous, isotropic phase, $\rho^{(1)}$ is just the mean number density ρ, while the crystal phase has full translational periodicity and orientational order. In the *nematic* liquid crystal phase the particles are preferentially ordered along the *director*, a unit vector $\hat{\mathbf{n}}$ parallel to the polar axis, but the translational invariance characteristic of the liquid state persists. Thus

$$\rho^{(1)}(\mathbf{R}, \mathbf{\Omega}) = \rho\Psi(\mathbf{\Omega}) \quad \text{(nematic)} \tag{12.7.2}$$

where $\Psi(\mathbf{\Omega})$ is an orientational distribution function normalised such that $\int \Psi(\mathbf{\Omega})d\mathbf{\Omega} = 1$. As there is axial symmetry around the director, $\Psi(\mathbf{\Omega})$ is a function only of $\cos\theta = \hat{\mathbf{n}} \cdot \hat{\mathbf{\Omega}}$, where $\hat{\mathbf{\Omega}}$ is another unit vector. The degree of alignment of the particles is measured by a nematic order parameter $\mathcal{Q}$ defined as

$$\mathcal{Q} = \langle P_2(\cos\theta) \rangle_{\mathbf{\Omega}} = 2\pi \int_0^\pi \Psi(\cos\theta) P_2(\cos\theta) \sin\theta \, d\theta \tag{12.7.3}$$

where $P_2(x)$ is the second-order Legendre polynomial. For perfectly aligned particles, $\mathcal{Q} = 1$, while in the isotropic phase $\mathcal{Q} = 0$.

The *smectic-A* phase has an orientationally ordered, lamellar structure which is also translationally ordered along the director. The single-particle density is now a function of the vertical coordinate z and orientation $\mathbf{\Omega}$:

$$\rho^{(1)}(\mathbf{R}, \mathbf{\Omega}) = \rho^{(1)}(z, \mathbf{\Omega}) = \rho^{(1)}(z, \hat{\mathbf{n}} \cdot \hat{\mathbf{\Omega}}) \quad \text{(smectic-A)} \tag{12.7.4}$$

This is a periodic function of z, meaning that $\rho^{(1)}(z + \ell h, \hat{\mathbf{n}} \cdot \hat{\mathbf{\Omega}}) = \rho^{(1)}(z, \hat{\mathbf{n}} \cdot \hat{\mathbf{\Omega}})$ for any integer ℓ, where h is the smectic layer thickness. The smectic-C phase also has a lamellar structure, but one in which the orientations of the particles are tilted relative to $\hat{\mathbf{n}}$. Plots of configurations of the isotropic, nematic, smectic-A and crystal phases of a system of elongated particles are shown in Figure 12.11.

Three hard-core models of lyotropic liquid crystals have been widely studied both theoretically and by simulation:

- Ellipsoids of revolution with major and minor axes of length, respectively, L and d, are prolate (elongated) if $L > d$ or oblate (flat) for $L < d$; the quantity $\kappa = L/d$ is called the *aspect ratio*.
- Cylinders of length L and diameter d are rod-like if $\kappa > 1$, and plate-like if $\kappa < 1$. Their behaviour in the so-called needle limit, where $\kappa \to \infty$, was studied by Onsager in a paper[78] that set the standard for theoretical work on lyotropic liquid crystals.
- Spherocylinders, pictured in Figure 12.11, are cylinders that are capped at both ends by hemispheres of the same diameter as the cylinder.

Transitions involving mesophases are accompanied by the appearance of orientational and translational inhomogeneities. The natural choice for their description within statistical mechanics is therefore density functional theory,

FIGURE 12.11 Plots, reading clockwise from upper left, of configurations of the isotropic, nematic, smectic-A and ordered crystal phases of a system of hard spherocylinders.
Picture by courtesy of P. Bolhuis.

as described in Chapters 3 and 6; see, in particular, Section 6.8 for the application to freezing. The free energy is now a functional of $\rho^{(1)}(\mathbf{R}, \boldsymbol{\Omega})$ and can be written, as a generalisation of (3.3.2), in the form

$$\mathcal{F}[\rho^{(1)}] = \mathcal{F}^{\text{id}}[\rho^{(1)}(\mathbf{R}, \boldsymbol{\Omega})] + \mathcal{F}^{\text{ex}}[\rho^{(1)}(\mathbf{R}, \boldsymbol{\Omega})] \qquad (12.7.5)$$

where the ideal contribution, which for spheres is given by (3.1.22), must now be integrated over all orientations. In the case of the nematic phase, $\rho^{(1)}$ is given by (12.7.2), and

$$\mathcal{F}^{\text{id}}[\rho^{(1)}(\mathbf{R}, \boldsymbol{\Omega})] = Nk_{\text{B}}T\left(\ln\rho\Lambda^3 - 1 + \int \Psi(\boldsymbol{\Omega})\ln\left[4\pi\Psi(\boldsymbol{\Omega})\right]\mathrm{d}\boldsymbol{\Omega}\right) \quad (12.7.6)$$

In the isotropic phase, where $\Psi(\boldsymbol{\Omega}) = 1/4\pi$, the orientational contribution to $\mathcal{F}^{\text{id}}$ vanishes.

We focus first on the isotropic–nematic transition. An approximate expression for the excess part of the free energy functional is required which, following Onsager,[78] can be based on the virial expansion of the free energy. To lowest order in density (6.2.21), which applies to hard spheres, may be generalised to the case of non-spherical particles in the form

$$\beta \mathcal{F}^{ex}[\rho^{(1)}(\mathbf{R}, \mathbf{\Omega})] = -\frac{1}{2} \int d\mathbf{\Omega} \int d\mathbf{\Omega}' \int d\mathbf{R} \int d\mathbf{R}' \rho^{(1)}(\mathbf{R}, \mathbf{\Omega})$$
$$f(\mathbf{R}' - \mathbf{R}, \mathbf{\Omega}, \mathbf{\Omega}') \rho^{(1)}(\mathbf{R}', \mathbf{\Omega}') \qquad (12.7.7)$$

where $f(x)$ is the Mayer function, which, for hard particles, is equal to -1 when the particles overlap but is otherwise zero. Substitution of (12.7.2) in (12.7.7), followed by integration over $\mathbf{R}'$ gives

$$f^{ex} = \frac{\beta \mathcal{F}^{ex}}{N} = \frac{1}{2} \rho \int d\mathbf{\Omega} \int d\mathbf{\Omega}' \, \Psi(\mathbf{\Omega}) v_x(\mathbf{\Omega}, \mathbf{\Omega}') \Psi(\mathbf{\Omega}') \qquad (12.7.8)$$

where

$$v_x(\mathbf{\Omega}, \mathbf{\Omega}') = -\int f(\mathbf{r}, \mathbf{\Omega}, \mathbf{\Omega}') d\mathbf{r} \qquad (12.7.9)$$

is the excluded volume for two particles of orientations $\mathbf{\Omega}, \mathbf{\Omega}'$, calculated by integration over relative coordinates $\mathbf{r} = \mathbf{R}' - \mathbf{R}$. The excluded volume for two hard cylinders of aspect ratio sufficiently large that end effects can be neglected is related to the angle $\gamma = \cos^{-1}(\hat{\mathbf{\Omega}} \cdot \hat{\mathbf{\Omega}}')$ between the long axes by[77b]

$$v_x = 2L^2 d |\sin \gamma(\mathbf{\Omega}, \mathbf{\Omega}')| \qquad (12.7.10)$$

In the isotropic phase, $\Psi(\mathbf{\Omega}) = 1/4\pi$, and (12.7.10) reduces to the second virial coefficient approximation for f^{ex}:

$$f^{ex} = \frac{1}{2} \rho \frac{1}{(4\pi)^2} \int d\mathbf{\Omega} \int d\mathbf{\Omega}' v_x(\mathbf{\Omega}, \mathbf{\Omega}') = B_2 \rho = B_2^* \eta \qquad (12.7.11)$$

where $B_2 = \pi L^2 d/4$ is the second virial coefficient in the limit $\kappa \gg 1$, $B_2^* = B_2/v_0$ and $\eta = \rho v_0$, $v_0 = \frac{1}{4} \pi L d^2$ being the volume of a rod.

If the rods are sufficiently long, higher order terms in the virial expansion of the excess free energy can be neglected. The term of order ρ^2 would be

$$\frac{1}{2} B_3 \rho^2 = \frac{1}{2} (B_3/B_2^2)(\rho B_2)^2$$

Onsager showed by use of geometric arguments that $B_3/B_2^2 \sim \kappa^{-1} \ln \kappa \to 0$ as $\kappa \to \infty$; the same is true for ratios of all higher-order coefficients, B_n/B_2^{n-1}. This behaviour has been confirmed by explicit, numerical calculation[79] of B_n/B_2^{n-1} as functions of κ for n in the range $3 \le n \le 5$, but convergence of the expansion is slow when the aspect ratio falls below $\kappa \approx 100$.

Combination of (12.7.6)–(12.7.10) leads to Onsager's free energy functional for the nematic phase:

$$f[\Psi(\mathbf{\Omega})] = \ln \rho \Lambda^3 - 1 + \int \Psi(\mathbf{\Omega}) \ln \left[4\pi \Psi(\mathbf{\Omega})\right] d\mathbf{\Omega}$$
$$+ \rho L^2 d \int d\mathbf{\Omega} \int d\mathbf{\Omega}' \, \Psi(\mathbf{\Omega}) |\sin \gamma(\mathbf{\Omega}, \mathbf{\Omega}')| \Psi(\mathbf{\Omega}') + \mathcal{O}(\rho^2)$$
$$(12.7.12)$$

which must be minimised with respect to the orientational distribution function $\Psi(\mathbf{\Omega})$. The minimum arises from the competition between the orientational entropy, which favours the isotropic phase where all orientations are equally probable, and the excess term, which favours the nematic phase. The extremum condition analogous to (3.3.1), according to which

$$\frac{\delta f[\Psi(\mathbf{\Omega})]}{\delta \Psi(\mathbf{\Omega})} = \beta \mu_{or} \qquad (12.7.13)$$

where μ_{or} is the orientational contribution to the chemical potential, leads to a non-linear equation for $\Psi(\mathbf{\Omega}) \equiv \Psi(\theta)$ in the form:

$$\Psi(\theta) = (4\pi)^{-1} \exp\left(\beta\mu_{or} - 2\rho L^2 d \int \Psi(\theta)|\sin\gamma(\mathbf{\Omega}, \mathbf{\Omega}')| d\mathbf{\Omega}'\right) \qquad (12.7.14)$$

Equation (12.7.14) always has the isotropic solution, for which $\Psi(\theta) = 1/4\pi$ and $\beta\mu_{or} = 2B_2\rho$. As the density increases, an anisotropic solution appears. Detailed analysis[77b,80] based on an expansion of $\Psi(\theta)$ around its isotropic form, i.e. $\Psi(\theta) = [1+\epsilon\Delta\Psi(\theta)]/4\pi$, shows that the isotropic phase becomes unstable beyond a dimensionless concentration $c = \eta\kappa = 4$.

A more straightforward but approximate method,[77b] similar to that used in Section 6.8, is to use a trial function $\Psi_\alpha(\theta)$, dependent on a variational parameter α, and to minimise the free energy resulting from (12.7.12) with respect to α. A convenient choice for $\Psi_\alpha(\theta)$ is the properly normalised gaussian function:

$$\Psi_\alpha(\theta) = A\exp\left(-\frac{1}{2}\alpha\theta^2\right), \qquad 0 \leq \theta \leq \frac{1}{2}\pi$$

$$= A\exp\left(-\frac{1}{2}\alpha(\pi - \theta)^2\right), \qquad \frac{1}{2}\pi < \theta < \pi \qquad (12.7.15)$$

where the two ranges of θ correspond, respectively, to orientations parallel or anti-parallel to the director. For large α, $\Psi_\alpha(\theta)$ is sharply peaked for orientations close to $\hat{n}$. With this simplification the prefactor is determined by the condition

$$\int_0^{\pi/2} \exp\left(-\frac{1}{2}\alpha\theta^2\right)\sin\theta\, d\theta \approx \int_0^\infty \exp\left(-\frac{1}{2}\alpha\theta^2\right)\left[\theta - \frac{1}{6}\theta^3 + \mathcal{O}(\theta^5)\right] d\theta$$

$$= 1/4\pi A(\alpha) \qquad (12.7.16)$$

from which it follows that

$$A(\alpha) \approx \frac{\alpha}{4\pi}\left(1 + \frac{1}{3\alpha} + \cdots\right) \qquad (12.7.17)$$

Substitution of (12.7.15) and (12.7.17) in (12.7.12) gives

$$f(\alpha) \approx c + \ln\alpha + \frac{4c}{(\pi\alpha)^{1/2}} \qquad (12.7.18)$$

and minimisation of $f(\alpha)$ shows that to leading order

$$\alpha \approx 4c^2/\pi \tag{12.7.19}$$

The corresponding value of the nematic order parameter is

$$\mathcal{Q} \approx 1 - \frac{3}{\alpha} \approx 1 - \frac{3\pi}{4c^2} \tag{12.7.20}$$

Coexistence between the isotropic (I) and nematic (N) phases is determined by equating the osmotic pressures and chemical potentials derived by the usual thermodynamic relations from the free energies of the two phases, (12.7.11) and (12.7.18). This leads to values of the packing fractions $\eta = c/\kappa = cd/L$ at coexistence given by

$$\eta_{\mathrm{I}} \approx 3.45d/L, \quad \eta_{\mathrm{N}} \approx 5.12d/L$$

and to an order parameter $\mathcal{Q} \approx 0.910$. The transition is strongly first order; not only are the rods highly aligned in the nematic phase but the increase in packing fraction is large ($\Delta\eta/\eta_{\mathrm{N}} \approx 0.33$). The values of the order parameter and of the two packing fractions are sensitive to the choice of trial function. For the function used by Onsager, i.e.

$$\Psi_\alpha(\theta) = \frac{\alpha \cosh(\alpha \cos\theta)}{4\pi \sinh\alpha} \tag{12.7.21}$$

the packing fractions are[78]

$$\eta_{\mathrm{I}} \approx 3.34d/L, \quad \eta_{\mathrm{N}} \approx 4.48d/L \quad \text{(Onsager)}$$

and $\mathcal{Q} \approx 0.848$. The important point to bear in mind, however, is that whatever the choice of trial function the theory is valid only for $L/d \gg 1$, a regime in which the packing fractions are sufficiently small for use of a second virial approximation to be justified. When the aspect ratio is less than about 100, the contributions from higher order terms are no longer negligible.

Several efforts have been made to extend Onsager's theory to physically relevant values of κ. A simple, phenomenological approach[81] uses a hardsphere fluid as a reference system, the packing fraction of which is set equal to that of the system of interest. In the case of hard cylinders, the excess part of the free energy functional in (12.7.12) is replaced by

$$f^{\mathrm{ex}} = f_d^{\mathrm{ex}} \int d\mathbf{\Omega} \int d\mathbf{\Omega}' \Psi(\mathbf{\Omega}) \frac{v_{\mathrm{x}}(\mathbf{\Omega}, \mathbf{\Omega}')}{8v_0} \Psi(\mathbf{\Omega}') \tag{12.7.22}$$

where v_{x} is given by (12.7.10), $f_d^{\mathrm{ex}} = \eta(4 - 3\eta)/(1 - \eta)^2$ is the Carnahan–Starling expression for the excess free energy of the reference system, v_0 is the hard-sphere volume and the quantity $8v_0$ is the excluded volume around a hard sphere. At low densities, $f_d^{\mathrm{ex}} \approx 4\eta$ and Onsager's expression for $f^{\mathrm{ex}}[\Psi(\mathbf{\Omega})]$ in (12.7.12) is recovered. Though simple, the results obtained represent a significant improvement over Onsager's theory. In particular, for $\kappa = 5$, the predicted increase in density at the isotropic–nematic transition is reduced by an order of magnitude to a value close to that obtained by Monte Carlo calculations for hard spherocylinders.[79]

At a more fundamental level density functional methods have been employed that go beyond the second virial approximation. The case of hard ellipsoids, for example, has been treated[82] by factorisation of the direct correlation function into translational and orientational parts in the form

$$c(\mathbf{R} - \mathbf{R'}, \mathbf{\Omega}, \mathbf{\Omega'}) \approx \frac{v_x(\mathbf{\Omega}, \mathbf{\Omega'})}{v_0} c_0(|\mathbf{R} - \mathbf{R'}|/d_0; \eta) \qquad (12.7.23)$$

where $v_x(\mathbf{\Omega}, \mathbf{\Omega'})$ is the excluded volume of two ellipsoids of volume $v_0 = \pi L d^2/6$, $c_0(r)$ is the direct correlation function of a fluid of hard spheres of diameter d_0, chosen such that its volume matches that of the ellipsoid, and $\eta = \rho v_0$. The Percus–Yevick expression (4.4.10) is used for $c_0(r)$ and the free energies of the isotropic and nematic phases are calculated relative to that of the hard-sphere fluid by an angle-dependent generalisation of the exact expression (3.5.23); the free energy of the nematic phase is again obtained by minimisation with respect to a trial orientational distribution function. An interesting feature of the theory is that thermodynamic properties display a prolate–oblate symmetry; thermodynamic properties at a given packing fraction are the same for aspect ratios κ and κ^{-1}, a finding confirmed to a good approximation by Monte Carlo simulations.[83]

The transition between nematic and smectic-A phases of rod-like particles has been studied within density functional theory by imposing a one-dimensional, periodic, modulation of the single-particle density such that $\rho^{(1)}(\mathbf{r}, \mathbf{\Omega}) = \rho^{(1)}(z, \hat{\mathbf{n}} \cdot \hat{\mathbf{\Omega}})$, where $\hat{\mathbf{n}}$ is the common director of the two phases.[84] Calculations are simplified for systems of parallel rods aligned along the director, which show[85] that the transition for this restricted model is continuous for all values of κ. The importance of particle shape is well illustrated by the fact that, unlike the case of parallel spherocylinders, no smectic phase exists for parallel ellipsoids. By scaling all z-coordinates by $1/\kappa$, the spherocylinders can be mapped onto hard spheres, leaving the partition function invariant,[86] and hard spheres cannot form a stable, lamellar phase. The case of freely rotating spherocylinders is much more complicated but density functional methods have been devised to study the stability of the nematic phase with respect to a smectic-A perturbation of the form

$$\delta\rho^{(1)}(\mathbf{R}, \mathbf{\Omega}) = \Psi(\mathbf{\Omega}) \sum_{n=1}^{\infty} c_n \cos(2\pi n z/a) \qquad (12.7.24)$$

where $a \approx L + d$ is the periodicity of the density wave.[87] The calculations show that an isotropic–nematic–smectic-A triple point should exist at $\kappa \approx 3.2$, $\eta \approx 0.46$, in fair agreement with Monte Carlo simulations in which the complete phase diagram of hard spherocylinders in the $\kappa - \eta$ plane was mapped out.[88] Part of the phase diagram is pictured in Figure 12.12, from which it can be seen that

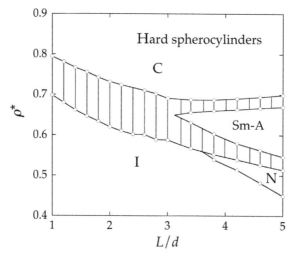

FIGURE 12.12 Part of the phase diagram of hard spherocylinders in the aspect ratio – packing fraction plane, showing the boundaries between isotropic, nematic, smectic-A and ordered crystal phases; $\rho^* = \rho/\rho_{cp}$, where ρ_{cp} is the close-packed density. A plastic crystal phase appears at lower values of κ (not shown). The shaded area is the region of two-phase coexistence. *Redrawn with permission from Ref. 88* © *American Institute of Physics.*

a direct transition between isotropic and smectic-A phases is possible for aspect ratios in the range from approximately 3.1 to 3.7. All transitions are first order.[89]

The limit in which $\kappa \rightarrow 0$ has received less attention. Early Monte Carlo studies[90] of infinitesimally thin ($\kappa = 0$) discs of diameter d detected a weakly first order, isotropic to nematic transition at a reduced density $\rho^* = Nd^3/V \approx 4.0$. Onsager's theory can be adapted to this case, with

$$v_x(\mathbf{\Omega}, \mathbf{\Omega}') = \frac{1}{2}\pi d^3 |\sin\gamma(\mathbf{\Omega}, \mathbf{\Omega}')| \qquad (12.7.25)$$

replacing the corresponding term in the functional (12.7.12). Now, however, the virial series is not rapidly convergent and the results of the theory are poor; calculations based on the trial function (12.7.21) lead to a transition that is strongly first order. A more successful approach to the problem of thin, hard platelets is based on the PRISM formalism[91] originally developed for polymer melts (see Section 12.3). Each platelet is assumed to carry a regular, rigid array of v interaction sites, and the total interaction between two platelets labelled 1 and 2 is a sum of v^2 site–site pair potentials. If it is assumed that all v sites are equivalent, which here amounts to neglecting edge effects, the PRISM-OZ relation (12.3.9) between site–site total and direct correlation functions carries over to the case of platelets, with the interaction sites playing the role of monomers and the monomer density replaced by the quantity $v\rho$. A change is also required in the definition of the form factor (12.3.10). In the isotropic phase, the conformational average required for flexible polymers is replaced by an orientational average, i.e.

$$\langle \hat{\omega}_{\alpha\beta} \rangle = \frac{\sin(kr_{\alpha\beta})}{r_{\alpha\beta}} \qquad (12.7.26)$$

where $r_{\alpha\beta}$ is the separation of sites α, β on a given platelet. If, in addition, the sites are uniformly distributed over a disc of radius $R_0 = \frac{1}{2}d$, the form factor is

$$\hat{\omega}(k) = \frac{2v}{(kR_0)^2} \left(1 - \frac{\mathcal{J}_1(2kR_0)}{kR_0} \right) \qquad (12.7.27)$$

where $\mathcal{J}_1(x)$ is the first-order, cylindrical Bessel function which reduces to the Lorentzian function

$$\hat{\omega}(k) \approx \frac{2v}{1 + k^2 R_0^2} \qquad (12.7.28)$$

at both small and large k. For an anisotropic phase with preferential orientation along the z-axis, an approximate, orientation-dependent form factor has been proposed[92] as a generalisation of (12.7.28):

$$\hat{\omega}(k) = \frac{2v}{2 + (k_z R_0)^2 (1 + 2\mathcal{Q}) + (k_\perp R_0)^2 (1 - \mathcal{Q})} \qquad (12.7.29)$$

where the subscript $\perp$ refers to the plane orthogonal to the z-axis, and $\mathcal{Q}$ is the nematic order parameter.

With this approximation, an analytic solution of the PRISM-OZ relation (12.3.9) can be derived for infinitesimally thin discs if the Percus–Yevick closure $c(r) = c_0 \delta(\mathbf{r})$ is adopted, the parameter $c_0 = \hat{c}(k)$ being determined by the core condition $h(r = 0) = -1$. The compressibility can then be determined from the resulting structure factor via (3.6.11) and the osmotic pressure follows by thermodynamic integration:

$$\frac{\beta P}{\rho} = 1 + 4\pi\rho \left(\frac{R_0'}{\sqrt{2}} \right)^3 + \frac{16}{3}\pi^2 \rho^2 \left(\frac{R_0'}{\sqrt{2}} \right)^6 \qquad (12.7.30)$$

where $R_0' = R_0[(1 - \mathcal{Q})(1 + \mathcal{Q})^{1/2}]^{1/3}$ is an effective platelet radius; R_0' is equal to R_0 in the isotropic phase and vanishes in the fully aligned nematic phase. Despite appearances, (12.7.30) is not a truncated virial expansion. In the isotropic phase the coefficients of the terms of order ρ and ρ^2 are, respectively, $B_2' = \sqrt{2}\pi R_0^3$ and $B_3' = 2\pi^2 R_0^6/3 \approx 6.58 R_0^6$, while the corresponding virial coefficients are $B_2 = \pi^2 R_0^3/2$ and $B_3 \approx 10.83 R_0^6$.

The free energy for a given density must be minimised with respect to a trial orientational distribution function; the ideal contribution, given by (12.7.6), may be expressed in terms of $\mathcal{Q}$, while the excess part is obtained from the equation of state (12.7.30). Minimisation with respect to $\mathcal{Q}$ shows that the isotropic phase is stable for a reduced density $\rho^* = 8R_0^3\rho \leq 3.12$, while for higher densities a non-zero value of $\mathcal{Q}$ is obtained, corresponding to the onset of nematic

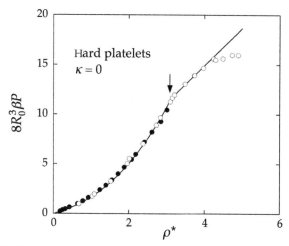

FIGURE 12.13 Osmotic equation of state of a suspension of infinitesimally thin, hard platelets of radius R_0. The points show the results of Monte Carlo calculations[90,93] and the curve is calculated from the PRISM equation (12.7.30); the arrow marks the density at which the isotropic–nematic transition is predicted to occur. From L. Harnau et al., 'A solvable interaction site model for lamellar colloids', *Europhys. Lett.* **53**, 729–734 (2001).

ordering. Figure 12.13 shows a comparison between results of the theory[91] and those obtained by Monte Carlo calculations; the agreement is excellent up to $\rho^* \approx 4$, but the isotropic–nematic transition is predicted to occur at $\rho^* \approx 3.2$ rather than $\rho^* \approx 4.1$.

Although infinitesimally thin platelets can form both isotropic and nematic phases the fact that they occupy zero volume means that no other phases exist. For $\kappa > 0$ we may expect to see the emergence of other phases, including a columnar phase in which the platelets form a hexagonal array of parallel, cylindrical stacks. Monte Carlo calculations[94] for oblate, hard spherocylinders have revealed the existence of isotropic, nematic, columnar and both tilted and aligned crystal phases, with an isotropic–columnar–crystal triple point for κ in the range 0.12–0.13. Overall the phase diagram is similar to that calculated earlier for the related system of cut, hard spheres.[95] The isotropic–nematic transition has been detected experimentally[96] for gibbsite platelets with $\kappa \approx 0.08$.

12.8 CLUSTERING AND GELATION

We have seen in earlier sections that stabilised suspensions of colloidal particles behave much like simple liquids, with the added freedom that the amplitude and range of the effective interactions can be tuned by changes, for example, in the concentration of non-adsorbing polymers or of added salt. Tunability of the interaction leads to a greater variety of phase behaviour than is seen in the phase diagram of simple liquids, as Figure 12.9 illustrates. Over the last decade much progress has been made in the design and synthesis of colloidal particles involving competing interactions or specific surface patterns; these systems display even richer behaviour, such as clustering, microphase separation and gelation. Gelation corresponds to the formation of a space-spanning, fractal network of particles that represents a low-density, disordered phase with solid-like elastic properties. Unlike the gelation of polymers, which is an irreversible process characterised by the formation of cross-linking chemical bonds, colloidal systems can form physical, reversible gels because the bonds due to colloid–colloid interactions are typically of the order of a few $k_B T$. Here we describe briefly two examples that have been studied by a combination of experiment, simulation and theory.[97,98]

We first consider colloids with competing, spherically symmetric interactions, involving a hard core, a short-range attraction and a long-range repulsion. The attraction may be induced by non-adsorbing polymers acting as depletant, while the repulsion arises from a residual surface charge screened by microions. If the solvent is weakly polar, the concentration of disassociated ions is low, and the screening length $\lambda_D = k_D^{-1}$ (see (10.2.15)) may be much larger than the range of the depletion potential, which is roughly equal to the radius of gyration of the depletant. Fluorescence microscopy has shown that at low colloid volume fractions ($\eta < 0.1$) the particles form an equilibrium cluster phase; as the density is increased, the clusters are seen to grow and become increasingly anisotropic, tending ultimately to aggregate and thereby to form a percolating, dynamically arrested network, in other words a gel.[100] Reversible cluster formation has also been observed in small-angle X-ray and neutron scattering experiments on colloid–polymer mixtures and globular protein solutions, in which a well-defined pre-peak appears in the static structure factor at a wavenumber k_c lying well below that of the main peak at $k \approx 2\pi/d$, where d is the particle diameter; the pre-peak arises from cluster–cluster correlations.[99] A similar pre-peak, the amplitude of which grows rapidly as the temperature is lowered, has been seen in molecular dynamics simulations of a simple model which combines a generalised, $2\ell - \ell$, Lennard-Jones potential, with $\ell = 100$, and a long-range repulsion of Yukawa form. The high value chosen for ℓ ensures the attraction is very short ranged[101]; the resulting interaction between clusters is therefore both long ranged and repulsive. At very low densities this leads to the formation of a cluster glass as the temperature is lowered. If η is greater than about 0.12, the effect of reducing the temperature is different; the clusters first form a percolating network, which then evolves into a reversible gel. Dynamical arrest in the gel is characterised by an intermediate scattering function which resembles that observed at the kinetic glass transition of supercooled liquids, shown in Figure 8.12, with non-ergodic behaviour appearing at sufficiently low temperatures.

The balance between short-range attraction and long-range repulsion, which depends sensitively on their relative range, means that formation of the cluster phase preempts condensation into a colloidal liquid phase. Simple calculations show that the internal energy of a spherical cluster goes through a minimum for a finite aggregation number.[101] The cluster phase is therefore stable, since the entropic contribution to the free energy will always favour small clusters.

Most theoretical calculations are based on the two-Yukawa model, consisting of hard spheres of diameter d with short-range attractive and long-range repulsive terms, both of Yukawa form, i.e.

$$v(x) = \infty, \quad x < 1$$
$$= -\frac{\epsilon}{x}\left(\exp[-z_1(x-1)] - A\exp[-z_2(x-1)]\right), \quad x > 1 \quad (12.8.1)$$

where $x = r/d$, ϵ (a positive quantity) is a characteristic energy, z_1 and z_2 are the dimensionless range parameters of the attractive and repulsive contributions (with $z_2 < z_1$) and $0 < A < 1$; the depth of the potential well at contact is $-\epsilon(1-A)$. Physically, $z_1 \approx d/R_g$ and $z_2 \approx d/\lambda_D$. Figure 12.14 shows the form of the potential for typical choices of the parameters A, z_1 and z_2. The relative importance of the attractive and repulsive components of the potential (12.8.1) may be quantified by the integral

$$I = 4\pi \int_1^\infty v(x)x^2\, dx = -4\pi\epsilon\left(\frac{z_1+1}{z_1^2} - A\frac{z_2+1}{z_2^2}\right) \quad (12.8.2)$$

For $A = 0$, the attraction leads to a first-order, gas–liquid transition. Within the mean field theory of Section 5.7 the critical temperature is largely determined by the value of the quantity $a = -\frac{1}{2}I$. As the long-range repulsion is switched

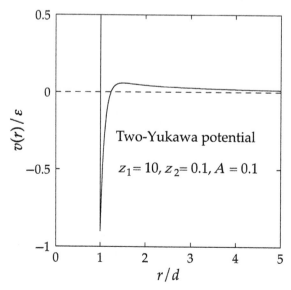

FIGURE 12.14 A two-Yukawa potential for typical values of the parameters A, z_1 and z_2.

on, a decreases and T_c is expected to fall.[102] The position of the prepeak in the structure factor, which signals the aggregation of particles into clusters, can be calculated within the random phase approximation (3.5.17) for the direct correlation function[103,104]:

$$\hat{c}(q) = \hat{c}_d(q) - \beta\hat{v}(q) \tag{12.8.3}$$

Here $q = kd$, $\hat{c}_d(q)$ is the Fourier transform of the hard-sphere direct correlation function, which is well approximated by the Percus–Yevick expression (4.4.10), and $\hat{v}(q)$ is the Fourier transform of the pair potential, with the continuous part extrapolated inside the hard core. Accordingly, $S(q) = 1/D(q)$, where, from (3.6.10) and (12.8.3):

$$D(q) = 1 - \rho^*\left[\hat{c}_d(q) - \beta\hat{v}(q)\right]$$
$$= 1 - \rho^*\left(\hat{c}_d(q) + \frac{4\pi\beta\epsilon_1}{z_1^2 + q^2} - \frac{4\pi\beta\epsilon_2}{z_2^2 + q^2}\right) \tag{12.8.4}$$

where $\rho^* = \rho d^3$, $\epsilon_1 = \epsilon\exp(z_1)$ and $\epsilon_2 = A\epsilon\exp(z_2)$; a peak in $S(q)$ corresponds to a minimum in $D(q)$. We are interested in the low-q region, $q \ll 2\pi$ (2π is roughly the position of the main peak); in that range the hard-sphere structure factor is very flat. Thus $\hat{c}_d(q)$ is almost independent of q and therefore contributes little to the extremum condition, $\delta D(q)/\delta q = 0$, from which it follows that

$$\frac{8\pi\beta\epsilon_1 q}{\left(z_1^2 + q^2\right)^2} - \frac{8\pi\beta\epsilon_2 q}{\left(z_2^2 + q^2\right)^2} = 0 \tag{12.8.5}$$

Equation (12.8.5) has one root, $q = 0$, corresponding to a minimum in $S(q)$, and another given by

$$q_c = \frac{\left(z_1^2 - z_2^2\right)^{1/2}}{\alpha - 1} \tag{12.8.6}$$

where $\alpha = \left(\epsilon_1/\epsilon_2\right)^{1/2} > 1$, at which the cluster prepeak appears. Note that q_c is independent of density, but the amplitude of the peak depends on both density and temperature. The value of $S(q_c)$ will diverge as the temperature is reduced at constant density and the locus of points in the density–temperature plane at which this occurs is called the λ-line. The λ-line plays a role analogous to that of the spinodal line of the gas–liquid transition, along which $S(q)$ diverges at $q = 0$. Its relation to the gas–liquid coexistence curve and spinodal for a two-Yukawa fluid is illustrated in Figure 12.15, where it encloses a large portion of the density–temperature plane above the critical temperature.[104] Along the spinodal line the fluid becomes unstable against macroscopic phase separation; along the λ-line it becomes unstable against density modulations of mesoscopic wavelength $\lambda = 2\pi/q_c$, leading to a microphase transition into a spatially ordered, cluster phase. A one-dimensional modulation, for example, leads to

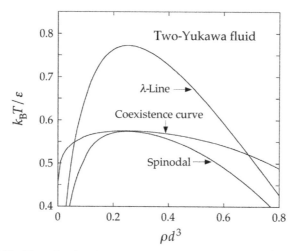

FIGURE 12.15 The λ-line, liquid–gas coexistence curve and spinodal line for a typical two-Yukawa fluid. From A.J. Archer et al., 'Theory for the phase behaviour of a colloidal fluid with competing interactions', *J. Phys. Condens. Matter* **20**, 41506 (2008). © *IOP Publishing 2008. Reproduced by permission of IOP Publishing. All rights reserved.*

a lamellar phase. The response to a modulation of that form has been studied in detail by mean field, density functional theory based on the free energy functional (3.4.10), but without the correlation term, and the local density approximation (6.2.5) for the hard-sphere contribution.[104]

Examples of the structure factor obtained by solution of a thermodynamically consistent integral equation[105] are compared with Monte Carlo data in Figure 12.16 for the cases when $z_1 = 10$, $z_2 = 0.10$ and $A = 0.01$ or 0.10. At the larger value of A the intensity of the pre-peak increases rapidly as the temperature is lowered, while for $A = 0.01$ the pre-peak appears at a significantly smaller wavenumber, indicative of the formation of increasingly larger clusters as the long-range repulsion is weakened.

A second type of model system suitable for the study of gelation is that of patchy particles of functionality f, already described in Section 11.10, in which a hard-sphere core has a pattern of f interaction sites on its surface.[98,106] Sites on different particles interact via a short-range, attractive potential, which in practice is taken to be of square-well form. If f is not too large and the sites are regularly distributed, it is possible to choose the range of the square-well interaction to be sufficiently short that at most only one bond can form between any two particles and no more than one particle can bond to a given site; this implies that f is the maximum number of bonds that a particle can form. Figure 11.8 shows how this comes about in the simplest case, when $f = 2$. Regular patterns of surface sites for which calculations have been made[107] include two diametrically opposite sites ($f = 2$) (as in the Figure), an equilateral, triangular distribution ($f = 3$), a tetrahedral distribution ($f = 4$) and the vertices

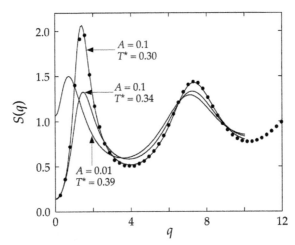

FIGURE 12.16 Structure factor of the two-Yukawa model at different reduced temperatures $T = k_B T/\epsilon$ and different values of the parameter A. The points show the results of Monte Carlo calculations and the curves are calculated from a thermodynamically consistent integral equation. The prepeak at $q < 2$ is a signal of cluster formation. *Redrawn with permission from Ref. 105 © American Institute of Physics.*

of two face-sharing tetrahedra ($f = 5$). Larger patterns can be used, such as an octahedral distribution for $f = 6$, but as f increases a regular distribution of sites leads to less directional, quasi-isotropic interactions between particles. Mean, non-integral functionalities can be studied by taking binary mixtures of components having different values of f, for which $\bar{f} = x f_1 + (1 - x) f_2$. The case $f_1 = 2, f_2 = 3$ is of particular interest[108] as it allows $\bar{f}$ to vary continuously towards 2^+. The $f = 2$ model, where each particle interacts with at most two neighbours, can lead only to the formation of independent, linear chains. Bulk condensation into a liquid phase is inhibited, while gelation into a volume-spanning, cross-linked network is impossible, since branching would require at least some of the particles to be of higher functionality.

Figure 12.17 shows the phase diagrams of monodisperse systems of patchy particles for $f = 3, 4$ and 5, obtained by Monte Carlo calculations, together with the results[107] of Wertheim's thermodynamic perturbation theory, also described in Section 11.10. The agreement is fair for $f = 3$ but rapidly deteriorates as the functionality increases. What both sets of results show, however, is that the critical temperature and density decrease rapidly with f and the liquid–gas coexistence curve gradually shrinks towards the lower, left-hand corner of the density–temperature plane. This is confirmed by the spinodal curves calculated[108] from Wertheim's theory for a binary mixture of mean valence $2 < \bar{f} < 3$. As surmised earlier, gas–liquid phase separation disappears at $f = 2$.

The coexistence curves for $\bar{f} \leq 3$ show that the liquid state occupies a much wider range of density than is the case for simple liquids. These low density,

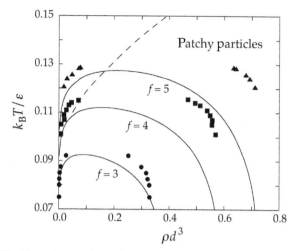

FIGURE 12.17 Phase diagrams for patchy particles of different functionalities. The curves show the predictions of Wertheim's theory, described in Section 11.10, and the points are the results of Monte Carlo calculations. The broken curve is the percolation line for $f = 4$. *Redrawn with permission from Ref. 107 © American Institute of Physics.*

equilibrium states are referred to as 'empty liquids', since the particles occupy only a very small fraction of the available volume. For example, the critical packing fraction is approximately 0.070 for $f = 3$ and 0.046 for $\bar{f} = 2.5$. Thus, when a suspension of low-functionality, patchy particles is cooled along an isochore to the right of the coexistence curve, particles begin to aggregate into mostly linear clusters. Since phase separation is avoided, the liquid-like suspension remains in equilibrium throughout and the thermodynamic path is reversible down to low temperatures. Beyond a percolation threshold the system forms a transient, space-filling network. Figure 12.17 includes the percolation line predicted[109] by Wertheim's theory for $f = 4$. This merges into the coexistence curve at low density; the behaviour for $f = 3$ and $f = 4$ is qualitatively similar.

Thermal fluctuations in fluids consisting of patchy particles will lead to frequent breakage and recombination of the bonds between particles, so that bonded networks continually switch from percolating to disconnected configurations, thereby ensuring that ergodicity is maintained. The bond lifetime τ_B can be expected to satisfy an Arrhenius law, $\tau_B \sim \exp{(\beta \epsilon)}$, where ϵ is a measure of the strength of the bonding potential. When τ_B exceeds the experimental time scale, reversible gelation occurs, an effect similar to the glass transition in denser fluids. The difference is that at high densities arrested dynamics are a consequence of the caging of tightly packed particles, whereas in the low-density liquids formed by patchy particles dynamical arrest arises from the long-lived bonding.[110]

Reversible network formation and gelation at low packing fractions are features that are not peculiar to low-functionality, patchy particles. They may also result from the long-range, dipolar interactions typical of ferrofluids. Spherical particles with embedded point particles, of which the simplest example is that of dipolar hard spheres, favour head-to-tail configurations. This leads to the formation of linear chains at low temperatures,[111] as in the case of patchy particles with $f = 2$. However, slightly elongated particles that carry extended rather than point dipoles favour branching of the chains, which can then interconnect to form a three-dimensional network. This has been demonstrated in the case of dipolar dumbbells, similar to the model pictured in Figure 1.5 except that the charges are now q and $-q$ and the atoms are represented by repulsive, soft spheres. The point dipole limit is recovered in the limit in which the bondlength $L \to 0$ and $q \to \infty$ at a fixed value of the dipole moment $\mu = qL$. Molecular dynamics calculations[112] have found clear evidence of clustering, branching, network formation, percolation, gel formation and dynamical arrest at packing fractions as low as 0.02, behaviour similar to that observed for patchy particles of low functionality. Other simulations have revealed no sign of a gas–liquid transition in point dipolar systems[113]; though contrary to earlier speculation, this is consistent with results obtained for patchy particles with $f = 2$.

12.9   THE FOKKER–PLANCK AND SMOLUCHOWSKI EQUATIONS

Earlier parts of this chapter were largely concerned with the calculation of static properties based on coarse-graining strategies that lead to effective interactions between the mesoscopic particles that make up the systems of interest. We turn now to the problem of describing the dynamics of colloidal particles on the mesoscopic time scale by a coarse graining in time. The traditional approach to the problem is a stochastic one,[114,115] exploited more than a century ago by Einstein, Langevin and Smoluchowski. The underlying assumption is that there exists a separation of time scales which allows the fast processes to be treated in an averaged way. A classic example is provided by the Langevin equation introduced in Section 7.3, which yields an expression for the velocity autocorrelation function of a brownian particle. In this section we show how, starting from first principles, an expression can be obtained for the time evolution of the single-particle distribution function $f^{(1)}(\mathbf{R}, \mathbf{P}; t)$ of a brownian particle, of mass M and diameter Σ, suspended in a fluid consisting of particles of much smaller mass m and diameter σ. The result is called the Fokker–Planck or Kramers equation.

On the microscopic level the shortest time scale is the Enskog mean collision time $\tau_E = 1/\Gamma_E$, defined by (7.2.19), while a typical collective time scale is the time $\tau_s = \Sigma/c_s$ required for a sound wave of velocity c_s to propagate over a distance of order Σ. For a solvent such as water and a colloidal particle of diameter of order 10^2 nm, $\tau_E \approx 10^{-13}$ s, while $\tau_s \approx 10^{-11}$ s. Relevant mesoscopic time scales are the time $\tau_B = 1/\xi$ over which the velocity of a brownian particle relaxes, where ξ is the friction coefficient of Langevin theory (see (7.3.6)), and a configurational relaxation time τ_c, which is the time required for an isolated brownian particle to diffuse over a distance equal to its diameter. An estimate of τ_c in terms of ξ is given by the Einstein relation (7.3.17):

$$\tau_c = \frac{\Sigma^2}{D} = \frac{\xi M \Sigma^2}{k_B T} \tag{12.9.1}$$

At room temperature, for typical values of M, Σ and ξ, we find that $\tau_B \approx 10^{-9}$ s and $\tau_c \approx 10^{-3}$ s. Thus the assumption of a clear separation of microscopic (τ_E, τ_s) and mesoscopic (τ_B, τ_c) time scales appears to be well justified.

The derivation of the Fokker–Planck equation starts from the Liouville equation (2.1.9) for the phase space probability density of a system of N bath particles and a single brownian particle. It relies on an expansion of the distribution function in powers of the natural small parameter $\epsilon = (m/M)^{1/2}$, which is the ratio of the thermal velocities of the brownian and bath particles. The task is nor a straightforward one, since a conventional perturbation expansion leads to secular divergence of the solution at sufficiently long times, irrespective of how small ϵ may be. A similar difficulty is encountered even in the simple case of a weakly damped harmonic oscillator when expansion is made in powers of the damping coefficient.[116] In each case, however, the problem can be overcome by use of a 'multiple time scale' method, first applied to the problem of brownian motion by Cukier and Deutch.[117]

The Hamiltonian of the $(N + 1)$-particle system is

$$\mathcal{H} = \frac{P^2}{2M} + \sum_{i=1}^{N} \frac{p_i^2}{2m} + V_N(\mathbf{r}^N) + \Phi(\mathbf{R}, \mathbf{r}^N) \tag{12.9.2}$$

where V_N is the total interaction energy of the N bath particles and Φ is the potential energy of the bath particles in the field of a brownian particle placed at $\mathbf{R}$. The Liouville operator splits naturally into 'bath' and 'brownian' terms:

$$\mathcal{L} = \mathcal{L}_b + \mathcal{L}_B \tag{12.9.3}$$

with

$$\mathcal{L}_b = -i \sum_{i=1}^{N} \left(\frac{\mathbf{p}_i}{m} \cdot \frac{\partial}{\partial \mathbf{r}_i} + \mathbf{f}_i \cdot \frac{\partial}{\partial \mathbf{p}_i} \right)$$
$$\mathcal{L}_B = -i \left(\frac{\mathbf{P}}{M} \cdot \frac{\partial}{\partial \mathbf{R}} + \mathbf{F}_B \cdot \frac{\partial}{\partial \mathbf{P}} \right) \tag{12.9.4}$$

Here $\mathbf{f}_i = -\partial(V_N + \Phi)/\partial \mathbf{r}_i$ is the force acting on bath particle i and $\mathbf{F}_B = -\partial \Phi/\partial \mathbf{R}$ is the force exerted on the brownian particle by particles of the bath. The Liouville equation for the phase space probability density of the system of $(N + 1)$ particles is therefore

$$\frac{\partial}{\partial t} f^{[N+1]}(\mathbf{B}, \mathbf{b}^N; t) = -i\left(\mathcal{L}_b + \mathcal{L}_B\right) f^{[N+1]}(\mathbf{B}, \mathbf{b}^N; t) \qquad (12.9.5)$$

where we have used the short-hand notation $\mathbf{B} \equiv \{\mathbf{R}, \mathbf{P}\}$ and $\mathbf{b}^N \equiv \{\mathbf{r}^N, \mathbf{p}^N\}$ to represent, respectively, the phase space coordinates of the brownian and bath particles. Since ϵ is treated as a perturbation, it proves useful to express the momentum of the brownian particle in scaled form as $\mathbf{P}' = \epsilon \mathbf{P}$, in which case the kinetic energy $P^2/2M$ becomes $P'^2/2m$ and the brownian term in the Liouville operator scales as

$$\mathcal{L}_B = -i\epsilon \left(\frac{\mathbf{P}'}{m} \cdot \frac{\partial}{\partial \mathbf{R}} + \mathbf{F}_B \cdot \frac{\partial}{\partial \mathbf{P}'}\right) \equiv \epsilon \mathcal{L}'_B \qquad (12.9.6)$$

The single-particle distribution function $f_B(\mathbf{B}, t) \equiv f^{[1]}(\mathbf{R}, \mathbf{P}', t)$ is the integral of $f^{[N+1]}$ over the coordinates and momenta of the bath particles and its time evolution is similarly obtained by integration of the Liouville equation (12.9.5):

$$\begin{aligned}
\frac{\partial f_B(\mathbf{B}, t)}{\partial t} &= \int \left(-i\mathcal{L}_b\right) f^{[N+1]}(\mathbf{B}, \mathbf{b}^N; t) \, d\mathbf{b}^N \\
&+ \epsilon \int \left(-i\mathcal{L}'_B\right) f^{[N+1]}(\mathbf{B}, \mathbf{b}^N; t) \, d\mathbf{b}^N \\
&= -\epsilon \frac{\mathbf{P}'}{m} \cdot \frac{\partial}{\partial \mathbf{R}} f_B(\mathbf{B}, t) - \epsilon \int \mathbf{F}_B \cdot \frac{\partial}{\partial \mathbf{P}'} f^{[N+1]}(\mathbf{B}, \mathbf{b}^N; t) \, d\mathbf{b}^N
\end{aligned}$$
$$(12.9.7)$$

The term involving the operator $\mathcal{L}_b$ in the first equality vanishes for the same reasons as those invoked in passing from (2.1.17) to (2.1.18) in the derivation of the BBGKY hierarchy.

The multiple time scale method[116,118] involves the introduction of an auxiliary distribution function, $f_\epsilon^{[N+1]}(\mathbf{B}, \mathbf{b}^N; t_0, t_1, t_2, \dots)$, which is a function of multiple time variables $t_0, t_1, t_2, \dots$, corresponding to increasingly long time scales. The equation that describes the time evolution of the auxiliary function is then similar to (12.9.5) except that the single time derivative is replaced by a sum of derivatives with respect to $t_0, t_1, t_2, \dots$:

$$\left(\frac{\partial}{\partial t_0} + \epsilon \frac{\partial}{\partial t_1} + \epsilon^2 \frac{\partial}{\partial t_2} + \cdots\right) f_\epsilon^{[N+1]} = -i\left(\mathcal{L}_b + \epsilon \mathcal{L}'_B\right) f_\epsilon^{[N+1]} \qquad (12.9.8)$$

This equation can be solved perturbatively by expansion of $f_\epsilon^{[N+1]}$ in powers of ϵ:

$$f_\epsilon^{[N+1]} = f_{\epsilon 0}^{[N+1]} + \epsilon f_{\epsilon 1}^{[N+1]} + \epsilon^2 f_{\epsilon 2}^{[N+1]} + \cdots \qquad (12.9.9)$$

The physical distribution function $f^{[N+1]}$ is eventually recovered by relating the quantities $t_0, t_1, t_2, \ldots$ to the physical time t via the rule

$$t_0 = t, \quad t_1 = \epsilon t, \quad t_2 = \epsilon^2 t, \quad \ldots, \quad t_n = \epsilon^n t \qquad (12.9.10)$$

The analogous auxiliary distribution function for the brownian particle, $f_{B\epsilon}(\mathbf{B}; t_0, t_1, t_2, \ldots)$, satisfies the generalisation to multiple time variables of (12.9.7):

$$\left(\frac{\partial}{\partial t_0} + \epsilon \frac{\partial}{\partial t_1} + \epsilon^2 \frac{\partial}{\partial t_2} + \cdots \right) f_{B\epsilon}$$
$$= -\epsilon \frac{\mathbf{P}'}{m} \cdot \frac{\partial}{\partial \mathbf{R}} f_{B\epsilon} - \epsilon \int \mathbf{F}_B \cdot \frac{\partial}{\partial \mathbf{P}'} f_\epsilon^{[N+1]} \, d\mathbf{b}^N \qquad (12.9.11)$$

Term by term integration of (12.9.9) shows that $f_{B\epsilon}$ can be expanded in the form

$$f_{B\epsilon} = f_{B0} + \epsilon f_{B1} + \epsilon^2 f_{B2} + \mathcal{O}(\epsilon^3) \qquad (12.9.12)$$

The crucial difference between (12.9.9) and a conventional perturbation expansion of $f^{[N+1]}$ itself is the fact that the auxiliary function has a physical meaning only along the so-called physical line defined by (12.9.10). We are therefore free to impose whatever boundary conditions are needed to ensure that the expansion is free of secular divergences at successive powers of ϵ; the same is true of the expansion of $f_{B\epsilon}$.

We now restrict the discussion to order ϵ^2, retaining only the three time variables t_0, t_1 and t_2. Then, by substituting (12.9.9) and (12.9.12) in (12.9.11) and equating coefficients of equal powers of ϵ up to $\mathcal{O}(\epsilon^2)$, we arrive at the following results.

To zeroth order in ϵ: Equation (12.9.11) implies that

$$\frac{\partial f_{B0}}{\partial t_0} = 0 \qquad (12.9.13)$$

and hence that $f_{B0} = f_{B0}(\mathbf{R}, \mathbf{P}'; t_1, t_2)$. Similarly, it follows from (12.9.8) that

$$\frac{\partial f_{\epsilon 0}^{[N+1]}}{\partial t_0} = -i \mathcal{L}_b f_{\epsilon 0}^{[N+1]} \qquad (12.9.14)$$

Since the equilibrium, N-particle phase space probability density of the bath particles in the presence of the brownian particle at $\mathbf{R}$ satisfies the relation

$$i \mathcal{L}_b f_0^{[N]}(\mathbf{b}^N | \mathbf{R}) = 0 \qquad (12.9.15)$$

where $f_0^{[N]}$ is given by a minor generalisation of (2.3.1), the solution to (2.9.14) is simply

$$f_{\epsilon 0}^{[N+1]} = f_{B0}(\mathbf{R}, \mathbf{P}'; t_1, t_2) f_0^{[N]}(\mathbf{b}^N | \mathbf{R}) \qquad (12.9.16)$$

We can now exploit the freedom of choice of boundary conditions on the auxiliary function $f_{B\epsilon}$ by imposing the initial condition that

$$f_{B\epsilon}(\mathbf{R}, \mathbf{P}'; t_0 = 0, t_1, t_2) = f_{B0}(\mathbf{R}, \mathbf{P}'; t_1, t_2) \qquad (12.9.17)$$

which in turn implies that

$$f_{Bn}(\mathbf{R}, \mathbf{P}'; t_0 = 0, t_1, t_2) = 0, \qquad n = 1, 2 \qquad (12.9.18)$$

To first order in ϵ: Equations (12.9.8) and (12.9.11) reduce to

$$\frac{\partial f_{\epsilon 1}^{[N+1]}}{\partial t_0} + \frac{\partial f_{\epsilon 0}^{[N+1]}}{\partial t_1} = -i\mathcal{L}_b f_{\epsilon 1}^{[N+1]} - i\mathcal{L}'_B f_{\epsilon 0}^{[N+1]} \qquad (12.9.19)$$

and

$$\frac{\partial f_{B1}}{\partial t_0} + \frac{\partial f_{B0}}{\partial t_1} = \int (-i\mathcal{L}'_B) f_{\epsilon 0}^{[N+1]} \, d\mathbf{b}^N \qquad (12.9.20)$$

Equations (12.9.13) and (12.9.16) show that f_{B0} and $f_{\epsilon 0}^{[N+1]}$ are both independent of t_0. To avoid secular growth of f_{B1} in (12.9.20) it is therefore necessary to impose the condition that

$$\frac{\partial f_{B0}}{\partial t_1} + \int i\mathcal{L}'_B f_{\epsilon 0}^{[N+1]} \, d\mathbf{b}^N = 0 \qquad (12.9.21)$$

which implies that $\partial f_{B1}/\partial t_0 = 0$ for all t_0. Combined with the initial condition (12.9.18) this in turn implies that f_{B1} is identically zero:

$$f_{B1}(\mathbf{R}, \mathbf{P}'; t_0, t_1, t_2) \equiv 0 \qquad (12.9.22)$$

We therefore focus on the time evolution of f_{B0}. Substitution of (12.9.16) in the right-hand side of (12.9.21) and use of the definition (12.9.6) of $\mathcal{L}'_B$ gives

$$\left(\frac{\partial}{\partial t_1} + \frac{\mathbf{P}'}{m} \cdot \frac{\partial}{\partial \mathbf{R}} \right) f_{B0}(\mathbf{R}, \mathbf{P}'; t_1, t_2) = 0 \qquad (12.9.23)$$

Comparison with (7.7.13) shows that on the time scale t_1 the evolution of the distribution function of the brownian particle is the same as that of an ideal gas. Equation (12.9.19) can now be rearranged as

$$\left(\frac{\partial}{\partial t_0} + i\mathcal{L}_b \right) f_{\epsilon 1}^{[N+1]} = -\left(\frac{\partial}{\partial t_1} + i\mathcal{L}'_B \right) f_{\epsilon 0}^{[N+1]}$$

$$= -\mathbf{F}_B \cdot \left(\frac{\beta \mathbf{P}'}{m} + \frac{\partial}{\partial \mathbf{P}'} \right) f_{B0} f_0^{[N]} \qquad (12.9.24)$$

which has the formal solution

$$f_{\epsilon 1}^{[N+1]}(\mathbf{R}, \mathbf{P}', \mathbf{b}^N; t_0, t_1, t_2)$$

$$= -\int_0^{t_0} ds \exp(-i\mathcal{L}_b s) \mathbf{F}_B \cdot \left(\frac{\beta \mathbf{P}'}{m} + \frac{\partial}{\partial \mathbf{P}'} \right)$$

$$\times f_{B0}(\mathbf{R}, \mathbf{P}'; t_1, t_2) f_0^{[N]}(\mathbf{b}^N | \mathbf{R}) \qquad (12.9.25)$$

That this is correct may be checked by direct substitution in (12.9.24).

To second order in ϵ: Equations (12.9.11) and (12.9.22) can be combined to give

$$\frac{\partial f_{B2}}{\partial t_0} + \frac{\partial f_{B0}}{\partial t_2} = -\int i\mathcal{L}'_B f^{[N+1]}_{\epsilon 1} \, db^N \qquad (12.9.26)$$

Since f_{B0} is independent of t_0, secular growth is again suppressed by setting $\partial f_{B2}/\partial t_0 = 0$. On substituting the solution (12.9.25) for $f^{[N+1]}_{\epsilon 1}$ in (12.9.26) we obtain a closed equation for the evolution of $f_{B0}(\mathbf{R}, \mathbf{P}'; t_1, t_2)$:

$$\frac{\partial f_{B0}}{\partial t_2} = \lim_{t_0 \to \infty} \int db^N \, f^{[N]}_0(\mathbf{b}^N | \mathbf{R}) i\mathcal{L}'_B$$
$$\times \int_0^{t_0} ds \exp\left(-i\mathcal{L}_b s\right) \mathbf{F}_B \cdot \left(\frac{\beta \mathbf{P}'}{m} + \frac{\partial}{\partial \mathbf{P}'}\right) f_{B0} \qquad (12.9.27)$$

where the limit $t_0 \to \infty$ can be taken because f_{B0} is independent of t_0. Substitution for $\mathcal{L}'_B$ from (12.9.6) and use of (2.1.14) for the time evolution of the dynamical variable $\mathbf{F}_B$, (12.9.27) shows that

$$\frac{\partial f_{B0}}{\partial t_2} = \lim_{t_0 \to \infty} \int_0^{t_0} ds \, \langle i\mathcal{L}'_B \mathbf{F}_B(-s)\rangle_{\text{bath}} \cdot \left(\frac{\beta \mathbf{P}'}{m} + \frac{\partial}{\partial \mathbf{P}'}\right) f_{B0}$$
$$= \frac{1}{3} \int_0^\infty ds \, \langle \mathbf{F}_B \cdot \mathbf{F}_B(-s)\rangle_{\text{bath}} \frac{\partial}{\partial \mathbf{P}'} \cdot \left(\frac{\beta \mathbf{P}'}{m} + \frac{\partial}{\partial \mathbf{P}'}\right) f_{B0}$$
$$(12.9.28)$$

where $\langle \cdots \rangle_{\text{bath}}$ denotes an equilibrium average over the phase space variables of the bath particles in the external field of the brownian particle.

Bringing together the results to first and second order in ϵ embodied in (12.9.23) and (12.9.28), returning both to physical time via the relation (12.9.10) and to the original momentum variable $\mathbf{P}$, we arrive finally at the Fokker–Planck equation for $f_B(\mathbf{R}, \mathbf{P}; t)$:

$$\frac{\partial f_B(\mathbf{R}, \mathbf{P}; t)}{\partial t} = \left(\epsilon \frac{\partial}{\partial t_1} + \epsilon^2 \frac{\partial}{\partial t_2}\right) f_{B0}(\mathbf{R}, \mathbf{P}; t_1, t_2)\Big|_{t_1 = \epsilon t, t_2 = \epsilon^2 t}$$
$$= \left[-\frac{\mathbf{P}}{M} \cdot \nabla_{\mathbf{R}} + \xi \nabla_{\mathbf{P}} \cdot (\mathbf{P} + M k_B T \nabla_{\mathbf{P}})\right] f_B(\mathbf{R}, \mathbf{P}; t)$$
$$(12.9.29)$$

where the friction coefficient ξ is given by the integral of the autocorrelation function of the force exerted on the brownian particle by the bath:

$$\xi = \frac{\beta}{3M} \int_0^\infty \langle \mathbf{F}_B \cdot \mathbf{F}_B(-s)\rangle_{\text{bath}} \, ds \qquad (12.9.30)$$

The friction coefficient has the same form as that derived from the Langevin equation in Section 7.3 except that the random force in (7.3.8) is replaced by the microscopic force $\mathbf{F}_B$.

The Fokker–Planck equation is commonly written in terms of the velocity $\mathbf{V} = \mathbf{P}/M$ rather than momentum. It is also straightforward to generalise its derivation to the inhomogeneous case in which an external force field $\mathbf{F}_{ext}(\mathbf{R})$ acts on the particle as well as the force due to the bath. With these modifications (12.9.29) takes the from

$$\left(\frac{\partial}{\partial t} + \mathbf{V} \cdot \nabla_{\mathbf{R}} + \frac{\mathbf{F}_{ext}(\mathbf{R})}{M} \cdot \nabla_{\mathbf{V}}\right) f_B(\mathbf{R}, \mathbf{V}; t)$$

$$= \xi \nabla_{\mathbf{V}} \cdot \left(\mathbf{V} + \frac{k_B T}{M} \nabla_{\mathbf{V}}\right) f_B(\mathbf{R}, \mathbf{V}; t) \qquad (12.9.31)$$

The derivation can also be adapted to the technically more difficult case of a large hard sphere in a bath of small spheres, where the instantaneous nature of the collisions means that the Liouville operator no longer has the simple form given by (12.9.4).[119]

Equation (7.3.20) suggests that correlations in the velocity of the brownian particle decay on a time scale $1/\xi$. Hence, in the high-friction limit, the velocity distribution relaxes rapidly and the evolution of $f_B(\mathbf{R}, \mathbf{V}; t)$ at long times reduces to that of the spatial distribution, i.e. the single-particle density given by

$$\rho_B(\mathbf{R}, t) = \int f_B(\mathbf{R}, \mathbf{V}; t) d\mathbf{V} \qquad (12.9.32)$$

An evolution equation[120] for $\rho_B(\mathbf{R}; t)$ can be derived from (12.9.32) by an expansion in powers of the small parameter $(k_B T/M)^{1/2}/\Sigma\xi$ via a multiple time scale analysis[118] similar to that already described; the result is the Smoluchowski equation:

$$\frac{\partial \rho_B(\mathbf{R}, t)}{\partial t} = \nabla_{\mathbf{R}} \cdot D\big(\nabla_{\mathbf{R}} - \beta \mathbf{F}_{ext}(\mathbf{R})\big)\rho_B(\mathbf{R}, t) \qquad (12.9.33)$$

where D, the self-diffusion constant, is related to ξ by (7.3.17). The Smoluchowski equation can be viewed as a generalisation of the diffusion equation (8.2.3) to the case where an external force acts on the brownian particle and is more easily obtained by the route which leads to that result. The current density $\mathbf{j}_B(\mathbf{R}, t)$ of a brownian particle in a dilute solution is given by the constitutive relation (7.7.3), i.e.

$$\mathbf{j}_B(\mathbf{R}, t) = \zeta \rho_B(\mathbf{R}, t)\mathbf{F}_{ext}(\mathbf{R}) - D\nabla\rho_B(\mathbf{R}, t) \qquad (12.9.34)$$

Insertion of this expression in the continuity equation (8.2.1) for $\rho_B(\mathbf{R}, t)$ leads immediately to (12.9.33), since the mobility $\zeta = \beta D$.

To assess the relevance to dilute colloidal dispersions of the evolution equations (12.9.31) and (12.9.33) we must return to the time scale comparison made at the start of this section. Equation (8.4.3) shows that the decay of a transverse current fluctuation of wavenumber k is characterised by a viscous relaxation time τ_η equal to $\rho_m \eta k^2$ where, in the present context, η is the shear viscosity of the suspending fluid and ρ_m is its mass density. For $k \approx 1/\Sigma$ the relaxation time can be interpreted as being roughly the time required for a shear perturbation to propagate over a distance equal to the diameter of the brownian particle. The shortest brownian time scale is $\tau_B = 1/\xi$ which, from Stokes's law (7.3.18), is related to τ_η within a numerical factor by

$$\tau_B \approx \frac{\rho_M}{\rho_m}\tau_\eta \qquad (12.9.35)$$

where $\rho_M = 6M/\pi\Sigma^3$ is the mass density of the brownian particle. Hence a separation of brownian and bath time scales such that $\tau_B \gg \tau_\eta$ would require ρ_M to be much larger than ρ_m, or $M \gg (\Sigma/\sigma)^3$, a condition which is far more difficult to satisfy than the one requiring $(m/M)^{1/2}$ to be much less than unity. To avoid sedimentation of the colloidal particles, experiments are usually carried out on suspensions for which $\rho_M \approx \rho_m$, so that in practice a full separation of time scales cannot be achieved; the dynamics of the bath particles involve slowly decaying shear modes which relax on the same time scale as the velocity of the brownian particle. This means that the Fokker–Planck equation in the local form represented by (12.9.31) is not in general applicable and must be generalised to account for memory effects linked to correlated recollisions between the particles of the bath and the brownian particle.[121] The limitations of the Fokker–Planck equation do not extend to the Smoluchowski equation, since the spatial distribution of the brownian particle relaxes on the much longer, configurational time scale τ_c. It is, however, the Smoluchowski equation which is the more relevant to calculations of experimentally measurable quantities.

The Fokker–Planck equation for the distribution function $f_B^{(n)}(\mathbf{R}^n, \mathbf{V}^n; t)$ of a suspension consisting of n interacting brownian particles in a bath of N much lighter particles can be derived from the Liouville equation for the phase space probability density of the $(N+n)$-particle system by a multiple time scale method or a physically less transparent, projection operator technique.[122–124] The Smoluchowski equation for the n-particle density $\rho^{(n)}(\mathbf{R}^n, t)$ can then be obtained by integration over particle velocities. These manipulations are inevitably more complicated than those leading to the corresponding single-particle expressions. We therefore limit ourselves to sketching what is involved in the case of the Smoluchowski equation. Two modifications of (12.9.33) are required. The first is straightforward. Allowance must be made for the interaction between brownian particles, meaning that the external force in (12.9.33) must be augmented by the sum over ℓ of the force $\mathbf{F}_\ell$ acting on particle ℓ. That force is itself the sum of two terms, since there will be contributions both from direct interactions with other particles and from bath-induced forces, of which the most important is the depletion force.[125] The second modification is of a more subtle nature. As a brownian particle moves through the bath it experiences a frictional force proportional to its velocity, an effect that already appears in the single-particle case. However, the motion of the particle also creates a flow field in the solvent, which influences the motion of other particles, giving rise to an additional, 'hydrodynamic' interaction. On the time scale for which the Smoluchowski equation is valid the time required for the effect of the flow field to reach other brownian particles is so short that the hydrodynamic interaction is essentially instantaneous. The total frictional force on particle ℓ can then be written as

$$\mathbf{F}_\ell^\xi = -M \sum_{m=1}^{n} \boldsymbol{\xi}_{\ell m} \cdot \mathbf{V}_m \qquad (12.9.36)$$

The quantity $\boldsymbol{\xi}_{\ell m}$ is a component of a friction matrix $\boldsymbol{\Xi}$ defined as

$$\boldsymbol{\xi}_{\ell m} = \frac{\beta}{M} \int_0^\infty \langle \delta\mathbf{F}_\ell(0)\delta\mathbf{F}_m(-s) \rangle_{\text{bath}} \, ds \qquad (12.9.37)$$

where $\delta\mathbf{F}_\ell(s)$ is the fluctuating force exerted at time s on particle ℓ by the bath for fixed locations of all brownian particles. The definition provided by (12.9.36) shows that $\boldsymbol{\xi}_{\ell m}$ depends explicitly on the coordinates of particles ℓ and m and implicitly on those of all other brownian particles; the hydrodynamic forces are therefore fundamentally many-body in character. The diagonal element $\boldsymbol{\xi}_{\ell\ell}$ determines the frictional exerted on particle ℓ by virtue of its own velocity, but this is not the same as the friction coefficient of an isolated particle in the same solvent; hydrodynamic interactions around a circuit of particles mean that the motion of particle ℓ can be reflected back on itself, thereby influencing the value of $\boldsymbol{\xi}_{\ell\ell}$. The Smoluchowski equation that emerges when the factors listed have been allowed for, but in the absence of an external force field, is

$$\frac{\partial \rho_{\text{B}}^{(n)}(\mathbf{R}^n, t)}{\partial t} = \sum_\ell \sum_m \boldsymbol{\nabla}_\ell \cdot \mathbf{D}_{\ell m} \cdot \left(\boldsymbol{\nabla}_m - \beta\mathbf{F}_m \right) \rho_{\text{B}}^{(n)}(\mathbf{R}^n, t) \qquad (12.9.38)$$

where $\mathbf{D}_{\ell m}$ is a component of a diffusion tensor matrix $\mathbf{D}$; the matrices $\mathbf{D}$ and $\boldsymbol{\Xi}$ are related by a generalisation of Einstein's relation (7.3.17):

$$\mathbf{D} = \frac{k_{\text{B}}T}{M} \boldsymbol{\Xi}^{-1} \qquad (12.9.39)$$

Calculation of the friction tensors requires the solution of the Navier–Stokes equation for the fluid velocity field, with boundary conditions imposed at the surfaces of the moving particles. This is a highly complex problem that can be solved, even approximately, only in the low velocity, high dilution regime.[123,126] If hydrodynamic interactions are ignored, the diagonal elements of the matrix $\mathbf{D}$ are equal to $D\mathbf{I}$, where $\mathbf{I}$ is the 3×3 identity matrix, and the off-diagonal elements vanish. Equation (12.9.38) then reduces to

$$\frac{\partial \rho_{\text{B}}^{(n)}(\mathbf{R}^n, t)}{\partial t} = \sum_\ell \boldsymbol{\nabla}_\ell \cdot D \left(\boldsymbol{\nabla}_\ell - \beta\mathbf{F}_\ell \right) \rho_{\text{B}}^{(n)}(\mathbf{R}^n, t) \qquad (12.9.40)$$

The Smoluchowski equation can be written more compactly as

$$\frac{\partial \rho_B^{(n)}(\mathbf{R}^n; t)}{\partial t} = -i\mathcal{S}\rho_B^{(n)}(\mathbf{R}^n, t) \qquad (12.9.41)$$

where $\mathcal{S}$ is the Smoluchowski operator. There is an obvious analogy with the Liouville operator $\mathcal{L}$, which has the form appropriate to newtonian mechanics. In the liouvillian description of the dynamics the time evolution of the phase space distribution function is determined by the unitary operator $\exp(-i\mathcal{L}t)$ and that of an arbitrary dynamical variable by the operator $\exp(i\mathcal{L}t)$. Here, however, it can be shown[127] that the evolution of a dynamical variable is controlled not by $\exp(i\mathcal{S}t)$, as analogy would suggest, but by the operator $\exp(-i\tilde{\mathcal{S}}t)$, where $\tilde{\mathcal{S}}$ differs from $\mathcal{S}$ by a change in sign of the $\mathbf{F}_m$ in (12.9.38). In the notation of Section 7.1 the formal expression for the time autocorrelation function of a dynamical variable, A say, is therefore

$$C_{AA}(t) = \langle A(t)A^* \rangle = \left(A, \exp(-i\tilde{\mathcal{S}}t)A \right) \qquad (12.9.42)$$

The difference between newtonian and brownian propagators reflects the irreversibility of brownian dynamics, which has its origin in the frictional forces exerted on the brownian particles by the bath. The prescription provided by (12.9.42), when combined with approximations of the type discussed in Chapters 7–9 provide a basis for the calculation of time correlation functions descriptive, in particular, of concentration fluctuations and collective diffusion in colloidal dispersions.

The Smoluchowski equation was arrived at from what was initially a fully deterministic description of the system by progressive elimination of the degrees of freedom of the bath. The more traditional derivation starts with the stochastic equations of motion of a system of n interacting, brownian particles, corresponding to the high-friction limit of a set of Langevin equations similar to (7.3.1) but expanded to include the effect of interparticle forces. If the friction coefficient in (7.3.1) is sufficiently large, the inertial term on the left-hand side can be neglected. If, in addition, the contribution to the force on particle ℓ arising from interactions with other particles is added to the right-hand side, (7.3.1) can be rearranged to give an a set of coupled equations of motion of the form

$$\dot{\mathbf{R}}_\ell(t) = \zeta\left[-\nabla_\ell V_n(\mathbf{R}^n) + \delta\mathbf{F}_\ell(t)\right] \qquad (12.9.43)$$

where $\zeta = 1/M\xi$ is the mobility defined by (7.7.2) and $\delta\mathbf{F}_\ell$ is the fluctuating force which appears in (12.9.37). The potential energy $V_n(\mathbf{R}^n)$ incorporates both the direct interactions between particles and those induced by the bath, but hydrodynamic interactions have been ignored. These equations can be shown[114] to lead to the Smoluchowski equation in the simplified form given by (12.9.40), but they are important in their own right, not least because their use allows the numerical simulation of brownian motion. The link with (12.9.40) is easily established, given that the n-particle density must satisfy the continuity equation

$$
\frac{\partial \rho_B^{(n)}(\mathbf{R}^n, t)}{\partial t} = -\sum_\ell \nabla_\ell \cdot \dot{\mathbf{R}}_\ell(t) \rho_B^{(n)}(\mathbf{R}^n, t)
$$

$$
= \sum_\ell \nabla_\ell \cdot \zeta \left[\nabla_\ell V_n(\mathbf{R}^n) - \delta \mathbf{F}_\ell \right] \rho_B^{(n)}(\mathbf{R}^n, t) \quad (12.9.44)
$$

As $t \to \infty$, $\rho_B^{(n)}(\mathbf{R}^n, t)$ will approach its equilibrium form, proportional to $\exp\left[-\beta V_n(\mathbf{R}^n)\right]$, the right-hand side of (12.9.44) will vanish, and the two terms within square brackets must therefore cancel each other. It follows that in the long-time limit $\delta \mathbf{F}_\ell = -k_B T \nabla_\ell \ln \rho_B^{(n)}(\mathbf{R}^n)$. Equation (12.9.40) is therefore recovered if it is assumed that the same relationship applies away from equilibrium, i.e. if

$$
\delta \mathbf{F}_\ell(t) = -k_B T \frac{\nabla_\ell \rho_B^{(n)}(\mathbf{R}^n, t)}{\rho_B^{(n)}(\mathbf{R}^n, t)} \quad (12.9.45)
$$

for all t.

12.10 DYNAMICAL DENSITY FUNCTIONAL THEORY

We saw in Chapter 6 that density functional theory can be applied successfully to the calculation of the single particle density $\rho^{(1)}(\mathbf{R})$ and associated, static properties of a fluid under confinement or subject to an external force field, the key requirement being the availability of a good approximation for the free energy functional $\mathcal{F}[\rho^{(1)}]$. It is therefore natural to seek a generalisation of the theory which describes the dynamics of inhomogeneous fluids in cases, for example, where the system is initially out of equilibrium or when it is driven by a time-dependent, external field. That generalisation is provided by *dynamical density functional theory*, or DDFT, which focuses on the time evolution of the time-dependent, single-particle density $\rho(\mathbf{R}, t)$. (As in the previous section, we omit the superscript (1) for sake of notational simplicity, together with the subscript B.) The dynamical theory was originally formulated for systems of interacting brownian particles suspended in a bath of solvent molecules.[128,129] Within the theory, interaction between the brownian particles and the bath is assumed to lead to quasi-instantaneous thermalisation of the particle velocities. This is the situation in which the time dependence of the full, n-particle density is described by the Smoluchowski equation (12.9.38), which provides the starting point for the derivation of the DDFT equation. We shall assume, however, that conditions are such that the many body, hydrodynamic interactions between particles are negligible. The Smoluchowski equation (12.9.38) then takes the simplified form given by (12.9.40).

We start by rewriting (12.9.40) in terms of the mobility:

$$\frac{\partial \rho^{(n)}(\mathbf{R}^n, t)}{\partial t} = \sum_{i=1}^{n} \nabla_i \cdot \zeta \left[k_B T \nabla_i + \nabla_i V_n(\mathbf{R}^n) \right] \rho^{(n)}(\mathbf{R}^n, t) \qquad (12.10.1)$$

where

$$V_n(\mathbf{R}^n) = \sum_i \sum_{j<i} v(\mathbf{R}_i, \mathbf{R}_j) + \sum_i \phi(\mathbf{R}_i) \qquad (12.10.2)$$

is the total potential energy; here the interactions between particles are assumed to be pairwise-additive and ϕ represents any external potential that may be present. The evolution equation for the single-particle density $\rho(\mathbf{R}, t) \equiv \rho^{(1)}(\mathbf{R}; t)$ can be derived from (12.10.1) by integrating both sides over the coordinates of the remaining $(n-1)$ particles. The calculation follows the same lines that lead to the YBG equation (4.2.5), which relates the single-particle and pair densities of a fluid at equilibrium. By proceeding in this way it is found that

$$\frac{\partial \rho(\mathbf{R}_1, t)}{\partial t} = \nabla_1 \cdot \zeta \left[k_B T \nabla_1 \rho(\mathbf{R}_1, t) + \rho(\mathbf{R}_1, t) \nabla_1 \phi(\mathbf{R}_1) \right]$$

$$+ \nabla_1 \cdot \zeta \int \rho^{(2)}(\mathbf{R}_1, \mathbf{R}_2; t) \nabla_1 v(\mathbf{R}_1, \mathbf{R}_2) d\mathbf{R}_2 \quad (12.10.3)$$

The same result can be deduced much more easily from the single-particle Smoluchowski equation (12.9.33) simply by adding the mean force due to interaction with the remaining particles to the external force $\mathbf{F}_{ext} = -\nabla_1 \phi(\mathbf{R}_1)$.

Two approximations are now made. First, the instantaneous pair density $\rho^{(2)}(\mathbf{R}_1, \mathbf{R}_2; t)$ in (12.10.3) is replaced by the pair density $\rho^{(2)}(\mathbf{R}_1, \mathbf{R}_2)$ of a fluid in thermodynamic equilibrium for which the particle density, $\rho(\mathbf{R})$, is equal to $\rho(\mathbf{R}, t)$. This is reasonable assumption for dense fluids, where the pair structure is dominated by excluded volume effects. For a fluid at equilibrium, (3.5.4) and (4.2.5) together imply that

$$\int \rho^{(2)}(\mathbf{R}_1, \mathbf{R}_2) \nabla_1 v(\mathbf{R}_1, \mathbf{R}_2) d\mathbf{R}_2 = -k_B T \rho(\mathbf{R}_1) \nabla c^{(1)}(\mathbf{R}_1) \qquad (12.10.4)$$

where $c^{(1)}(\mathbf{R})$ is the single-particle direct correlation function of an inhomogeneous fluid, defined by (3.5.1). The second approximation is to assume that (3.5.1) carries over to the non-equilibrium situation, with the same, excess free energy functional $\mathcal{F}_{ex}[\rho^{(1)}]$ as in the equilibrium case. Then substitution of (12.10.4) and (3.5.1) in (12.10.3) shows that

$$\frac{\partial \rho(\mathbf{R}, t)}{\partial t} = \nabla \cdot \zeta \left(\rho(\mathbf{R}, t) \nabla \frac{\delta \mathcal{F}[\rho(\mathbf{R}, t)]}{\delta \rho(\mathbf{R}, t)} \right) \qquad (12.10.5)$$

where

$$\mathcal{F} = \mathcal{F}_{id} + \mathcal{F}_{ex} + \int \phi(\mathbf{R}, t) \rho(\mathbf{R}, t) d\mathbf{R} \qquad (12.10.6)$$

is the total free energy functional, in which the ideal contribution is given by (3.1.22) and allowance is made for a possible time dependence of the external potential. Equation (12.10.5) is the DDFT equation. Written in this form it also allows for a spatial dependence of the mobility; in the more commonly occurring situation, ζ appears as a constant prefactor on the right-hand side and the quantity $\xi^{-1} = m\zeta$ serves to define the time scale. If there are no interactions between particles, $\mathcal{F}_{ex} = 0$, and (12.10.5) leads back to (12.9.33). By defining the local chemical potential as $\mu[\rho(\mathbf{R}, t)] = \delta\mathcal{F}[\rho(\mathbf{R}, t)]/\delta\rho(\mathbf{R}, t)$, which is a natural generalisation of the thermodynamic relation (2.3.8), the DDFT equation (for constant ζ) can be written in the intuitively appealing form first suggested by Evans[130]:

$$\zeta^{-1}\frac{\partial}{\partial t}\rho(\mathbf{R}, t) = \nabla\big[\rho(\mathbf{R}, t)\nabla\mu(\mathbf{R}, t)\big] \qquad (12.10.7)$$

The quantity $-\nabla\mu(\mathbf{R}, t)$ plays the role of a driving force that acts on a particle located at $\mathbf{R}$ at time t.

In its applications the DDFT equation must be solved numerically for a given initial condition $\rho(\mathbf{r}, 0)$ and a physically motivated choice of approximate free energy functional; the extension to multi-component systems is straightforward.[131] It also applies[132] to particles obeying newtonian rather than stochastic dynamics in those situations where rapid thermalisation of velocities is ensured by a high collision rate ν, the only difference being that ζ must be replaced by $1/m\nu$. Efforts have been made to go beyond the Smoluchowski regime by shifting the emphasis from the single-particle density to the distribution function $f(\mathbf{R}, \mathbf{V}; t)$, the time evolution of which is described by the Fokker–Planck equation (12.9.31). The local density, local particle current and local stress tensor are all expressible in terms of $f(\mathbf{R}, \mathbf{V}; t)$ and its pair counterpart, $f^{(2)}(\mathbf{R}, \mathbf{V}, \mathbf{R}', \mathbf{V}'; t)$, which allows contact to be made with the flow equations of hydrodynamics.[133]

Equation (12.10.5) and its variants have been applied to a wide range of mostly colloid-related problems. An important, early application concerned the onset of spinodal decomposition in colloidal fluids; the discussion that follows is based on the work of Archer and Evans.[129] Figure 12.18 shows the liquid–gas coexistence curve and spinodal line in the density–temperature plane for a fluid of particles interacting via a Yukawa potential (1.2.2); in the Figure η is the packing fraction of the hard cores and the reduced temperature is expressed in terms of the integrated strength a of the attractive part of the potential:

$$a = -\int_d^\infty v(R)d\mathbf{R} \qquad (12.10.8)$$

The spinodal line, which is also pictured in the density–pressure plane in Figure 5.12, is the locus of points at which the isothermal compressibility diverges or, equivalently, the second derivative of the free energy, $(\partial^2 F/\partial^2 V)_T$

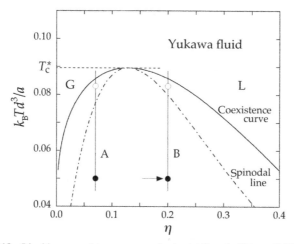

FIGURE 12.18 Liquid–gas coexistence curve and spinodal line of a Yukawa fluid. Open circles mark the regions in which nucleation of liquid droplets may occur, if the fluid is quenched along line A, or of gas bubbles if quenched along line B. Black circles show the region in which spinodal decomposition occurs and the arrow marks the thermodynamic state to which the system is quenched in the calculations described in the text. *Redrawn with permission from Ref. 129 © American Institute of Physics.*

vanishes. In the regions of the density–temperature plane lying between the coexistence curve and the spinodal, which can be reached by quenching the homogeneous liquid or gas along an isochore, the initially homogeneous fluid becomes metastable in the sense that it is stable against small fluctuations in density. Phase separation requires the nucleation and subsequent growth[134] of small liquid droplets, if the quench is made along line A in Figure 12.18, or small gas bubbles, if made along line B. Nucleation is an example of a 'rare event'; the existence of a large free energy barrier means that the metastable state may be long lived. However, if the quench continues beyond the spinodal, the fluid becomes thermodynamically unstable against density fluctuations, however small, and the rapid growth of domains corresponding to one or other of the two emerging phases, liquid in one case and gas in the other, leads to spinodal decomposition. The early and intermediate stages of this process, in which no sharp interfaces have yet formed, can be described within the framework of DDFT. No external field is involved, so the free energy functional is simply $\mathcal{F} = \mathcal{F}_{id} + \mathcal{F}_{ex}$.

We start by supposing that a small, local fluctuation occurs around the initial, bulk density ρ_b:

$$\delta\rho(\mathbf{R}, t) = \rho(\mathbf{R}, t) - \rho_b \tag{12.10.9}$$

Since the fluctuation is small, we can approximate the excess contribution to the free energy function by the quadratic expansion (4.3.11) around the free energy of the homogeneous phase. Then substitution of (3.1.22), (4.3.11) and

(12.10.9) in (12.10.5) leads to an evolution equation for $\delta\rho(\mathbf{R}; t)$:

$$(\zeta k_B T)^{-1} \frac{\partial \delta\rho(\mathbf{R}, t)}{\partial t} = \nabla^2 \delta\rho(\mathbf{R}, t) - \rho_b \nabla^2 \int \delta\rho(\mathbf{R}, t) c^{(2)}(|\mathbf{R} - \mathbf{R}'|; \rho_b)\, d\mathbf{R}'$$
$$-\nabla^2 \left(\delta\rho(\mathbf{R}, t) \int \delta\rho(\mathbf{R}', t) c^{(2)}(|\mathbf{R} - \mathbf{R}'|; \rho_b)\, d\mathbf{R}' \right)$$

$$(12.10.10)$$

At short times it is sufficient to retain only those terms that are linear in $\delta\rho(\mathbf{R}, t)$; the last term on the right-hand side may therefore be discarded. Then, on taking Fourier transforms of the remaining terms, we find that

$$(\zeta k_B T)^{-1} \frac{\partial \rho_k(t)}{\partial t} = -k^2 \rho_k(t)[1 - \rho_b \hat{c}(k)] \qquad (12.10.11)$$

where $\hat{c}(k)$ is the transform of the (pair) direct correlation of the bulk fluid. Equation (12.10.11) has an exponential solution given by

$$\delta\rho_k(t) = \delta\rho_k(0) \exp[-\gamma(k)t] \qquad (12.10.12)$$

with

$$\gamma(k) = \zeta k_B T k^2 [1 - \rho_b \hat{c}(k)] = \frac{\zeta k_B T k^2}{S(k)} \qquad (12.10.13)$$

where $S(k)$ is the structure factor of the bulk fluid, which is related to $\hat{c}(k)$ by (3.6.10). In both the stable and metastable phases $S(k)$ is finite for all k; a small density fluctuation therefore decays on a time scale $1/\gamma(k)$. On approaching the spinodal, $S(k)$ diverges as $k \to 0$, so long-wavelength fluctuations become increasingly long lived. Below the spinodal there is no longer a physically meaningful structure factor, but the damping coefficient $\gamma(k)$ can still be defined by the first equality in (12.10.13) through the definition of the direct correlation function as the second functional derivative (3.5.2) of $\mathcal{F}_{ex}$. An explicit expression can therefore be obtained from an approximate free energy functional appropriate to the system of interest. For a model system, such as the Yukawa fluid, described by a pair potential consisting of a hard-sphere core and an attractive tail, $w(R)$, a suitable choice[129] would be that contained in the grand potential functional (6.6.6), i.e.

$$\mathcal{F}_{ex}[\rho] = \mathcal{F}_{ex,d}[\rho] + \frac{1}{2} \iint \rho(\mathbf{R}) w(\mathbf{R} - \mathbf{R}') \rho(\mathbf{R}) d\mathbf{R}\, d\mathbf{R}' \qquad (12.10.14)$$

where $\mathcal{F}_{ex,d}$ is the excess free energy functional of hard spheres of diameter d; this is well approximated by the fundamental measure theory of Section 6.5. Calculations for the Yukawa potential shows that $\gamma(k)$ is negative for $k < k_c$, where k_c is a critical wavenumber having a value dependent on temperature and bulk density. Density fluctuations in that range of k therefore grow exponentially with time, leading to spinodal instability. An example of

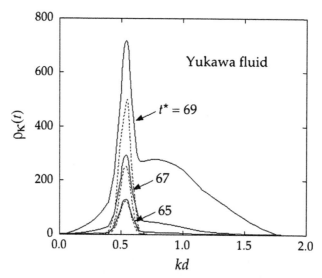

FIGURE 12.19 Predictions of dynamic density functional theory for the growth of Fourier components of the single-particle density of a Yukawa fluid quenched along line B of Figure 12.18. The broken curves show the results of a linear approximation and the full curves include the contribution from a quadratic term; see text for details. The unit of time is the relaxation time τ_c defined by (12.9.1) and $\rho_{\mathbf{k}}(0) = 10^{-8}$. *Redrawn with permission from Ref. 129 © American Institute of Physics.*

the growth of $\rho_{\mathbf{k}}(t)$ when the fluid is quenched along line B in Figure 12.18, is shown in Figure 12.19, where the unit of time is the configurational relaxation time τ_c defined by (12.9.1) and $\rho_{\mathbf{k}}(0)$ is arbitrarily set equal to 10^{-8}. The peak occurs at a wavenumber k_0 corresponding to the minimum in $\gamma(k)$; this provides a measure of the mean size, $\lambda_0 \approx 2\pi/k_0$, of the domains of liquid or, in this case, gas that develop in the early stages of spinodal decomposition. Similar behaviour had been predicted in the pioneering work of Cahn and Hilliard,[135] based on the square-gradient approximation (6.2.12).

At somewhat longer times, use of the linear approximation cannot be justified. The effect of the quadratic correction to the linear theory may be estimated[129] by including the third term on the right-hand side of (12.10.10). The evolution equation for the Fourier components is now

$$(\zeta k_B T)^{-1} \frac{\partial \rho_{\mathbf{k}}(t)}{\partial t} = -k^2 \rho_{\mathbf{k}}(t)[1 - \rho_b \hat{c}(k)]$$
$$+ \frac{1}{(2\pi)^3} \int \mathbf{k} \cdot \mathbf{k}' \rho_{\mathbf{k}'}(t) \hat{c}(k') \rho_{|\mathbf{k}-\mathbf{k}'|}(t) \mathrm{d}\mathbf{k}'$$

$$(12.10.15)$$

The additional, non-linear term acts in a mode coupling sense, since its effect is to couple together density fluctuations of different wavenumber. As Figure 12.19

illustrates, this leads to a marked spread, mostly to larger k, in the spectrum of wavenumbers that contribute to the time evolution of $\rho_{\mathbf{k}}(t)$. A broad component also appears at intermediate times, lying beyond the single peak predicted by the linear approximation. The position of the first peak scarcely changes with time, but the overall width of the non-linear spectrum suggests that there is a wide distribution in size of the domains corresponding to the conjugate phase.

Other problems to which DDFT has been successfully applied include the growth of a colloidal crystal from an initial nucleus[136] and the non-equilibrium sedimentation of interacting colloidal particles under the influence of gravity.[137] The theoretical predictions can in many cases be tested against the results of brownian dynamics simulations.[138] This is a computational technique which in most respects resembles molecular dynamics except that the particle trajectories are computed, not from Newton's equations, but from the coupled, stochastic equations of motion given by (12.9.43). In this way, for example, it has been shown that use of a mean field, excess free energy functional is justified in the case of gaussian-core particles, a system that serves as a model of interacting polymer coils confined to a cavity of variable radius.[139]

REFERENCES

[1] For introductory treatments, see: (a) Barrat, J.L. and Hansen, J.P., 'Basic Concepts for Simple and Complex Fluids'. Cambridge University Press, Cambridge, 2003. (b) Witten, T.A., 'Structured Fluids: Polymers, Colloids, Surfactants'. Oxford University Press, New York, 2004.

[2] Hertlein, C., Helden, L., Gambassi, A., Dietrich, S. and Bechinger, C., *Nature* **45**, 172 (2008).

[3] See, e.g., Shelley, J.C., Shelley, M.Y., Reeder, R.C., Bandyopadhyay, S., Moore, P.B. amd Klein, M.L., *J. Phys. Chem. B* **105**, 9785 (2001).

[4] Noid, W.G., Chu, J.W., Ayton, G.S., Krishna, V., Izekov, S., Voth, G.A., Das, A. and Andersen, H.C., *J. Chem. Phys.* **128**, 244114 (2008).

[5] (a) Doi, M., 'Introduction to Polymer Physics'. Oxford University Press, New York, 1996. (b) Rubinstein, M. and Colby, R.H., 'Polymer Physics'. Oxford University Press, New York, 2003.

[6] (a) Flory, P.J., 'Statistical Mechanics of Chain Molecules'. John Wiley, New York, 1969. (b) de Gennes, P.G., 'Scaling Concepts in Polymer Physics'. Cornell University Press, Ithaca, 1979. (c) Doi, M. and Edwards, S.F., 'The Theory of Polymer Dynamics'. Oxford University Press, New York, 1986. (d) des Cloizeaux, J. and Jannink, G., 'Polymers in Solution'. Oxford University Press, New York, 1989. (e) Grosberg, A.Yu. and Khokhlov, A.R., 'Statistical Physics of Macromolecules'. AIP Press, New York, 1994.

[7] Pelissetto, A. and Hansen, J.P., *J. Chem. Phys.* **122**, 134904 (2005).

[8] Flory, P.J. and Krigbaum, W.R., *J. Chem. Phys.* **18**, 1086 (1950).

[9] Grosberg, A.Y., Khalutur, P.G. and Khoklov, A.R., *Macromol. Chem. Rapid Commun.* **3**, 709 (1982).

[10] Dautenhahn, J. and Hall, C.K., *Macromolecules* **27**, 5399 (1994).

[11] Krakoviack, V., Hansen, J.P. and Louis, A.A., *Phys. Rev. E* **67**, 041801 (2003).

[12] Ruelle, D., 'Statistical Mechanics: Rigorous Results'. W.A. Benjamin, London, 1969.

[13] Grassberger, P. and Hegger, R., J. Chem. Phys. **102**, 6881 (1995).

[14] Stillinger, F.H., J. Chem. Phys. **65**, 3968 (1976).

[15] Lang, A., Likos, C.N., Watzlawek, M. and Löwen, H., J. Phys. Condens. Matter **12**, 5087 (2000).

[16] Louis, A.A., Bolhuis, P.G. and Hansen, J.P., Phys. Rev. E **62**, 7961 (2000).

[17] des Cloizeaux, J., J. Phys. (Paris) **36**, 281 (1975).

[18] Bolhuis, P.G., Louis, A.A., Hansen, J.P. and Meijer, E.J., J. Chem. Phys. **114**, 4296 (2001).

[19] (a) Witten, T.A. and Pincus, P.A., Macromolecules **19**, 2501 (1986). (b) Likos, C.N., Phys. Reports **348**, 267 (2001).

[20] Likos, C.N., Löwen, H., Watzlawek, M., Abbas, B., Jucknische, O., Algaier, J., and Richter, D., Phys. Rev. Lett. **80**, 4450 (1998).

[21] Pierleoni, C., Capone, B. and Hansen, J.P., J. Chem. Phys. **127**, 171102 (2007).

[22] (a) Edwards, S.F., Proc. Phys. Soc. **88**, 265 (1966). (a) Daoud, M., Cotton, J.P., Farnoux, B., Jannink, G., Sarma, G., Benoit, H., Duplessix, C., Picot, C. and de Gennes, P.G., Macromolecules **8**, 804 (1975).

[23] D'Adamo, G., Pelissetto, A. and Pierleoni, C., J. Chem. Phys. **137**, 024901 (2012).

[24] Vettorel, T., Besold, G. and Kremer, K., Soft Matter **6**, 2282 (2010).

[25] (a) Bang, J. and Lodge, T.P., Macromol. Res. **16**, 51 (2005). (b) Capone, B., Coluzza, I. and Hansen, J.P., J. Phys. Condens Matter **23**, 194102 (2011).

[26] Note, however, that some systematic deviations from strictly gaussian statistics have been detected in analysis of data from both neutron scattering experiments and computer simulations. See Beckrich, P., Johner, A., Semenov, A.N., Obukhov, S.P., Benoît, H. and Wittner, J.P., Macromolecules **40**, 3805 (2007).

[27] Ref. 1(a), pp. 103–6.

[28] (a) Schweizer, K.S. and Curro, J.G., Phys. Rev. Lett. **58**, 246 (1987). (b) Curro, J.G., Schweizer, K.S., Grest, G.S and Kremer, K., J. Chem. Phys. **91**, 1357 (1989). (c) Schweizer, K.S. and Curro, J.G., Adv. Chem. Phys. **98**, 1 (1997).

[29] See, e.g., Yethiraj, A., J. Phys. Chem. B **113**, 1539 (2009).

[30] (a) Laria, D., Wu, D. and Chandler, D., J. Chem. Phys. **95**, 4444 (1991). (b) Schweizer, K.S. and Yethiraj, A., J. Chem. Phys. **98**, 9053 (1993).

[31] Krakoviack, V., Hansen, J.P. and Louis, A.A., Europhys. Lett. **58**, 53 (2002).

[32] Clark, A.J. and Guenza, M.G., J. Chem. Phys. **132**, 044902 (2010).

[33] Lekkerkerker, H.N.W. and Tuinier, R., 'Colloids and the Depletion Interaction'. Springer, Heidelberg, 2011.

[34] See, e.g., Hunter, R.J., 'Foundations of Colloid Science', vol. 1, Section 4.4. Clarendon Press, Oxford, 1987.

[35] (a) Alexander, S., J. Physique **38**, 983 (1977). (b) de Gennes, P.G., Macromolecules **13**, 1069 (1980).

[36] Milner, S.T., Witten, T.A. and Cates, M.E., Macromolecules **21**, 2610 (1988).

[37] Coluzza, I. and Hansen, J.P., Phys. Rev. Lett. **100**, 016104 (2008).

[38] de Gennes, P.G. Adv. Colloid Interface Sci. **27**, 189 (1987).

[39] Pusey, P.N. and van Megen, W., Nature **320**, 340 (1986).

[40] Barrat, J.L. and Hansen, J.P., J. Physique **47**, 1547 (1986). See also Pusey, P.N., J. Physique **48**, 709 (1987).

[41] Wilding, N.B. and Sollich, P. J. Chem. Phys. **133**, 224102 (2010).

[42] Biben, T., Hansen, J.P. and Barrat, J.L., J. Chem. Phys. **98**, 7330 (1993).

[43] Buzzaccaro, S., Rusconi, R. and Piazza, R., Phys. Rev. Lett. **99**, 098301 (2007).

[44] Bannerman, M.N., Lue, L. and Woodcock, L.V., *J. Chem. Phys.* **132**, 084507 (2010). The equation of state is of the form proposed by Speedy, R.J., *J. Phys. Condens. Matter* **10**, 4387 (1998) but with revised values of the parameters.

[45] Piazza, R., Buzzaccaro, S., Secchi, E. and Parola, A., *Soft Matter* **8**, 7112 (2012).

[46] (a) Asakura, S. and Oosawa, F., *J. Chem. Phys.* **22**, 1255 (1954). (b) Asakura, S. and Oosawa, F., *J. Polymer Sci.* **33**, 123 (1958). See also Vrij, A., *Pure Appl. Chem.* **48**, 471 (1976).

[47] Lekkerkerker, H.N.W., Poon, W.C.K., Pusey, P.N., Stroobants, A. and Warren, P.B., *Europhys. Lett.* **20**, 559 (1992).

[48] Dijkstra, M., van Rooij, R., Roth, R. and Fortini, A., *Phys. Rev. E* **73**, 0141404 (2006).

[49] Ilett, S.M., Orrock, A., Poon, W.C.K. and Pusey, P.N., *Phys. Rev. E* **51**, 1344 (1995).

[50] Louis, A.A., Bolhuis, P.G., Meijer, E.J. and Hansen, J.P., *J. Chem. Phys.* **117**, 1893 (2002).

[51] Bolhuis, P.G., Louis, A.A. and Hansen, J.P., *Phys. Rev. Lett.* **89**, 128302 (2002).

[52] The discussion here is limited to the case when $q \leq 1$, the 'colloid limit'. For phase behaviour in the extreme 'protein limit', $q \gg 1$ see Bolhuis, P.G., Meijer, E.J. and Louis, A.A., *Phys. Rev. Lett.* **90**, 068304 (2003).

[53] Mao, Y., Cates, M.E. and Lekkerkerker, H.N.W., *Physica A* **222**, 10 (1995).

[54] Dijkstra, M., van Roij, R. and Evans, R., *Phys. Rev. E* **52**, 5744 (1999).

[55] Biben, T. and Hansen, J.P., *J. Phys. Condens. Matter* **3**, F65 (1991).

[56] Vliegenthart, G.A. and Lekkerkerker, H.N.W., *J. Chem. Phys.* **112**, 5364 (2000).

[57] Noro, M.G. and Frenkel, D., *J. Chem. Phys.* **113**, 2941 (2000).

[58] (a) van Roij, R., Dijkstra, M. and Hansen, J.P., *Phys. Rev. E* **59**, 2010 (1999). (b) van Roij, R. and Evans, R., *J. Phys. Condens. Matter* **11**, 1004 (1999).

[59] Löwen, H., Hansen, J.P. and Madden, P.A., *J. Chem. Phys.* **98**, 3275 (1993).

[60] Car, R. and Parrinello, M., *Phys. Rev. Lett.* **55**, 247 (1985).

[61] We ignore any non-coulombic component of the external potential.

[62] Verwey, E.J.W. and Overbeek, J.T.G., 'Theory of Stability of Lyophobic Colloids'. Elsevier, Amsterdam, 1948.

[63] Belloni, L., *Colloids Surf.* **A140**, 227 (1998).

[64] Löwen, H. and Allahyarov, E., *J. Phys. Condens. Matter* **10**, 4147 (1998).

[65] Crocker, J.C. and Grier, D.G., *Phys. Rev. Lett.* **73**, 352 (1994).

[66] Bocquet, L., Trizac, E. and Aubouy, M., *J. Chem. Phys.* **117**, 8138 (2002).

[67] Aubouy, M., Trizac, E. and Bocquet, L., *J. Phys. A: Math. Gen.* **36**, 5835 (2003).

[68] (a) Alexander, A., Chaikin, P.M., Grant, P., Morales, G.J., Pincus, P. and Hone, D.J., *J. Chem. Phys.* **80**, 5776 (1984). (b) Trizac, E., Bocquet, L., Aubouy, M. and von Grünberg, H.H., *Langmuir* **19**, 4027 (2003).

[69] For a review, see Naji, A., Jungblut, S., Moreira, A.G. and Netz, R.R., *Physica A* **352**, 131 (2005).

[70] Rouzina, I. and Bloomfield, V.A., *J. Phys. Chem.* **100**, 9977 (1996).

[71] Samaj, L. and Trizac, E., *Phys. Rev. E* **84** 041401 (2011).

[72] Ruzicka, B., Zaccarelli, E., Zulian, L., Angelini, R., Sztucki, M., Moussaïd, A., Narayanan, T. and Sciortino, F., *Nat. Mater.* **10**, 56 (2011).

[73] Tanaka, H., Meunier, J. and Bonn, D., *Phys. Rev. E* **69**, 031404 (2004).

[74] Leote de Carvalho, R.J.F., Trizac, E. and Hansen, J.P., *Phys. Rev. E* **61**, 1634 (2000).

[75] Chapot, D., Bocquet, L. and Trizac, E., *J. Chem. Phys.* **120**, 3969 (2004).

[76] (a) Chandrasekhar, S., 'Liquid Crystals', 2nd edn. Cambridge University Press, Cambridge, 1992. (b) de Gennes, P.G. and Prost, J., 'The Physics of Liquid Crystals', 2nd edn. Oxford

University Press, New York, 1993. (c) Chaikin, P.M. and Lubensky, T.C., 'Principles of Condensed Matter Physics'. Cambridge University Press, Cambridge, 1995.

[77] (a) Frenkel, D., *In* 'Liquids, Freezing and Glass Transition', Part II (J.P. Hansen, D.Levesque and J. Zinn-Justin, eds). North-Holland, Amsterdam, 1991. (b) Vroege, G.J. and Lekkerkerker, *Rep. Prog. Phys.* **55**, 1241, (1992).

[78] Onsager, L., *Ann. N.Y. Acad. Sci.* **51**, 627 (1949).

[79] Frenkel, D., *J. Phys. Chem.* **92**, 3280 (1988). See also *J. Phys. Chem.* **93**, 5314 (1988) (Erratum).

[80] Kayser, R.F. and Raveché, H.J., *Phys. Rev. A* **17**, 2067 (1978).

[81] (a) Parsons, J.D., *Phys. Rev. A* **19**, 1225 (1979). (b) Lee, S.D., *J. Chem. Phys.* **87**, 4792 (1987).

[82] Colot, J.L., Wu, X.G., Xu, H. and Baus, M., *Phys. Rev. A* **38**, 2022 (1988).

[83] (a) Frenkel, D. and Mulder, B.M., *Mol. Phys.* **55**, 1171 (1985). (b) Frenkel, D., *Mol. Phys.* **60**, 1 (1987).

[84] (a) Somoza, A.M. and Tarazona, P., *J. Chem. Phys.* **91**, 517 (1989). (b) Poniewierski, A. and Sluckin, T.J., *Phys. Rev. A* **43**, 6837 (1991).

[85] (a) Stroobants, A., Lekkerkerker, H.N.W. and Frenkel, D., *Phys. Rev. A* **36**, 2929 (1987). (b) Holyst, R. and Poniewierski, A., *Phys. Rev. A* **39**, 2742 (1989).

[86] Lebowitz, J.L. and Perram, W.J., *Mol. Phys.* **50**, 1207 (1983).

[87] Poniewierski, A. and Holyst, R., *Phys. Rev. A* **41**, 6871 (1990).

[88] Bolhuis, P. and Frenkel, D., *J. Chem. Phys.* **106**, 666 (1997).

[89] Polson, J.M. and Frenkel, D., *Phys. Rev. E* **56**, R6260 (1997).

[90] Eppenga, R. and Frenkel, D., *Mol. Phys.* **52**, 1303 (1984).

[91] Harnau, L., Costa, D. and Hansen, J.P., *Europhys. Lett.* **53**, 729 (2001).

[92] Pickett, G.T. and Schweizer, K.S., *J. Chem. Phys.* **110**, 6597 (1999).

[93] Dijkstra, M., Hansen, J.P. and Madden, P.A., *Phys. Rev. E* **55**, 3044 (1997).

[94] Marechal, M., Cuetos, A., Martinez-Haya, B. and Dijkstra, M., *J. Chem. Phys.* **134**, 094501 (2011).

[95] Veerman, J.A.C. and Frenkel, D., *Phys. Rev. A* **45**, 5632 (1992).

[96] van der Kooij, F.M. and Lekkerkerker, H.N.W., *Phil. Trans. R. Soc. A* **359**, 985 (2001).

[97] Zaccarelli, E., *J. Phys. Condens. Matter* **19**, 323101 (2007).

[98] Sciortino, F. and Zaccarelli, E., *Curr. Opin. Solid State Mater. Sci.* **15**, 246 (2011).

[99] Campbell, A.I., Anderson, V.J., van Duijneveldt, J.S. and Bartlett, P., *Phys. Rev. Lett.* **94**, 208301 (2005).

[100] Stradner, A., Sedgwick, H., Cardinaux, F., Poon, W.C.K., Egelhaaf, S.U. and Schurtenberger, P., *Nature* **432**, 492 (2004).

[101] (a) Sciortino, F., Mossa, S., Zaccarelli, E. and Tartaglia, P., *Phys. Rev. Lett.* **93**, 055701 (2004). (b) Fernandez Toldano, J.C., Sciortino, F. and Zaccarelli, E., *Soft Matter* **5**, 2390 (2009).

[102] Archer, A.J., Pini, D., Evans, R. and Reatto, L., *J. Chem. Phys.* **126**, 014104 (2007).

[103] Sear, R.F. and Gelbart, W.M., *J. Chem. Phys.* **110**, 4582 (1999).

[104] Archer, A.J., Ionescu, C., Pini, D., and Reatto, L., *J. Phys. Condens. Matter* **20**, 415106 (2008).

[105] Bomont, J.M., Bretonnet, J.L. and Costa, D., *J. Chem. Phys.* **132**, 184508 (2010).

[106] Bianchi, E., Blaak, R. and Likos, C.N., *Phys. Chem. Chem. Phys.* **13**, 6397 (2011).

[107] Bianchi, E., Tartaglia, P., Zaccarelli, E. and Sciortino, F., *J. Chem. Phys.* **128**, 144504 (2008).

[108] Bianchi, E., Largo, J., Tartaglia, P., Zaccarelli, E. and Sciortino, F., *Phys. Rev. Lett.* **97**, 168301 (2006).

[109] A point on the percolation line is identified as one for which the bond probability p_B given by (11.10.22) is equal to $1/(f-1)$.

[110] Russo, J., Tartaglia, P. and Sciortino, F., *J. Chem. Phys.* **131**, 014504 (2009).

[111] Weis, J.J. and Levesque, D., *Adv. Polym. Sci.* **185**, 163 (2005).

[112] Miller, M.A., Blaak, R., Lumb, C.N. and Hansen, J.P., *J. Chem. Phys.* **130**, 114507 (2009).

[113] Rovigatti, L. and Sciortino, F., *Phys. Rev. Lett.* **107**, 237801 (2011).

[114] Résibois, P. and De Leener, M., 'Classical Kinetic Theory of Fluids'. Wiley, New York, 1977.

[115] van Kampen, N.G., 'Stochastic Processes in Physics and Chemistry'. North-Holland, Amsterdam, 1981.

[116] For an introduction to multiple time scale methods, see Anderson, J.L., *Am. J. Phys.* **60**, 923 (1992).

[117] Cukier, R.I. and Deutch, J.M., *Phys. Rev.* **177**, 240 (1969).

[118] Bocquet, L., *Am. J. Phys.* **65**, 140 (1997).

[119] Bocquet, L., Piasecki, J. and Hansen, J.P., *J. Stat. Phys.* **76**, 505 and 527 (1994).

[120] Kramers, H.A., *Physica* **7**, 284 (1940).

[121] Bocquet, L. and Piasecki, J., *J. Stat. Phys.* **87**, 1005 (1997).

[122] Mazo, R.M., *J. Stat. Phys.* **1**, 559 (1969).

[123] Deutch, J.M. and Oppenheim, I., *J. Chem. Phys.* **54**, 3547 (1971).

[124] Murphy, T.J. and Aguirre, J.L., *J. Chem. Phys.* **57**, 2098 (1972).

[125] The isotropy of the distribution of bath particles around an isolated B-particle means that there is no depletion force in the case when $n = 1$.

[126] Happel, J. and Brenner, H., 'Low Reynolds Number Hydrodynamics'. Nordhoff, Leiden, 1973.

[127] For a detailed discussion, see Pusey, P.N., *In* 'Liquids, Freezing and the Glass Transition' (J.P. Hansen, D. Levesque and J. Zinn-Justin, eds). North-Holland, Amsterdam, 1991.

[128] Marconi, U.M.B. and Tarazona, P., *J. Chem. Phys.* **110**, 8032 (1999).

[129] Archer, A.J. and Evans, R., *J. Chem. Phys.* **121**, 4246 (2004).

[130] Evans, R., *Adv. Phys.* **28**, 143 (1979).

[131] Archer, A.J., *J. Phys. Condens. Matter* **17**, 1405 (2006).

[132] Archer, A.J., *J. Phys. Condens. Matter* **18**, 5617 (2006).

[133] Marconi, U.M.B. and Melchionna, S., *J. Chem. Phys.* **131**, 04105 (2009).

[134] Onuki, A., 'Phase Transition Dynamics'. Cambridge University Press, Cambridge, 2002.

[135] (a) Cahn, J.W. and Hilliard, J.E., *J. Chem. Phys.* **31**, 688 (1959). (b) Cahn, J.W., *Acta Metall.* **9**, 795 (1961).

[136] van Teefelelen, S., Likos, C.N. and Löwen, H., *Phys. Rev. Lett.* **98**, 188304 (2007).

[137] Royall, C.P., Dzubiella, J., Schmidt, M. and van Blaaderen, A., *Phys. Rev. Lett.* **98**, 188304 (2007).

[138] Ermak, D.L., *J. Chem. Phys.* **62**, 4189 (1975).

[139] Dzubiella, J. and Likos, C.N., *J. Phys. Condens. Matter* **15**, L147 (2003).

Fluctuations

It is shown in Chapter 2 that certain thermodynamic properties are expressible in terms of fluctuations in the microscopic variables of a system. Here we examine the question of fluctuations from a purely thermodynamic point of view.

Consider a subsystem of macroscopic dimensions that forms part of a much larger thermodynamic system. The subsystem is assumed to be in thermal, mechanical and chemical equilibrium with the rest of the system which, being much larger, plays the role of a reservoir. The thermodynamic properties of the subsystem fluctuate around the average values characteristic of the total system and the mean-square deviations from the average values can be derived systematically from the thermodynamic theory of fluctuations.

We assume that the total system is isolated from its surroundings. Then the probability p that a fluctuation will occur is

$$p \propto \exp\left(\Delta S_t / k_B\right)$$

where ΔS_t is the entropy change of the total system due to the fluctuation. Because S_t is a maximum at equilibrium, $\Delta S_t(<0)$ will be a quadratic function of the thermodynamic variables, higher-order terms in the expansion of S_t around its maximum value being negligible for large systems. Let P, T and μ be the average pressure, temperature and chemical potential, respectively, of the reservoir. Then, given that the energy, volume and number of particles of the total system remain constant, the entropy change ΔS_t is

$$\Delta S_t = \Delta S + (-\Delta U - P\Delta V + \mu \Delta N)/T$$

where $\Delta S, \Delta U, \Delta V$ and ΔN are the changes in thermodynamic variables of the subsystem and the second term on the right-hand side represents the entropy change of the reservoir. Since the fluctuations are very small, it is permissible to replace ΔU by an expansion in powers of $\Delta S, \Delta V$ and ΔN truncated at second order, i.e.

$$\Delta U \approx T\Delta S - P\Delta V + \mu \Delta N + \frac{1}{2}(\Delta T \Delta S - \Delta P \Delta V + \Delta \mu \Delta N)$$

Theory of Simple Liquids, Fourth Edition. http://dx.doi.org/10.1016/B978-0-12-387032-2.00021-0

Then

$$p \propto \exp\left[-\frac{1}{2}\beta(\Delta T \Delta S - \Delta P \Delta V + \Delta \mu \Delta N)\right] \qquad (A.1)$$

The subsystem can be defined either by the fraction of volume it occupies in the total system or by the number of particles it contains. In the second case, $\Delta N = 0$, and of the four remaining variables $(P, V, T$ and $S)$ only two are independent. If T and V are chosen as independent variables, and ΔS and ΔP are expressed in terms of ΔT and ΔV, (A.1) becomes

$$p \propto \exp\left(-\frac{\beta C_V}{2T}(\Delta T)^2 + \frac{1}{2}\beta\left(\frac{\partial P}{\partial V}\right)_{N,T}(\Delta V)^2\right) \qquad (A.2)$$

where C_V is the heat capacity at constant volume. The probability that a fluctuation will occur is therefore a gaussian function of the deviations ΔT and ΔV. Equation (A.2) shows that the system is stable against fluctuations in temperature and volume provided $C_V > 0$ and $(\partial P/\partial V)_{N,T} < 0$. The mean-square fluctuations derived from (A.2) are

$$\left\langle(\Delta T)^2\right\rangle = \frac{k_B T^2}{C_V}, \quad \left\langle(\Delta V)^2\right\rangle = -k_B T\left(\frac{\partial V}{\partial P}\right)_{N,T} = V k_B T \chi_T \qquad (A.3)$$

while $\langle \Delta T \Delta V \rangle = 0$. Fluctuations in temperature are therefore independent of those in volume. Alternatively, the choice of S and P as independent variables transforms (A.1) into

$$p \propto \exp\left(-\frac{1}{2k_B C_P}(\Delta S)^2 + \frac{1}{2}\beta\left(\frac{\partial V}{\partial P}\right)_{N,S}(\Delta P)^2\right) \qquad (A.4)$$

where C_P is the heat capacity at constant pressure. The averages calculated from (A.4) are

$$\left\langle(\Delta S)^2\right\rangle = k_B C_P, \quad \left\langle(\Delta P)^2\right\rangle = -k_B T\left(\frac{\partial P}{\partial V}\right)_{N,S} = \frac{k_B T}{V \chi_S}$$

where $\chi_S = -(1/V)(\partial V/\partial P)_{N,S}$ is the adiabatic compressibility, and $\langle \Delta S \Delta P \rangle = 0$. Fluctuations in entropy are therefore independent of those in pressure.

Finally, if the subsystem is defined as occupying a fixed fraction of the total volume, the mean-square fluctuation in the number of particles in the subsystem can be calculated, with the help of (2.4.22), to be

$$\left\langle(\Delta N)^2\right\rangle = k_B T\left(\frac{\partial N}{\partial \mu}\right)_{V,T} = \rho N k_B T \chi_T \qquad (A.5)$$

Equation (A.5) is identical to the statistical mechanical relation (2.4.23), while comparison of (A.5) with (A.3) shows that volume fluctuations at constant N are equivalent to number fluctuations at constant V.

Two Theorems in Density Functional Theory

In this Appendix we prove two of the key results of density functional theory, usually called the Hohenberg-Kohn-Mermin theorems. In doing so we use a simplified notation in which

$$\text{Tr} \cdots \equiv \sum_{N=0}^{\infty} \frac{1}{h^{3N} N!} \iint \cdots \mathrm{d}\mathbf{r}^N \mathrm{d}\mathbf{p}^N$$

This operation is called the 'classical trace', by analogy with the corresponding operation in quantum statistical mechanics. The definition of the grand partition function $\varXi$ and the normalisation of the equilibrium phase space probability density f_0 can then be expressed in the compact form

$$\varXi = \text{Tr} \exp[-\beta(\mathcal{H} - N\mu)], \qquad \text{Tr} f_0 = 1$$

We first prove the following lemma.

Lemma. *Let f be a normalised phase space probability density and let $\Omega[f]$ be the functional defined as*

$$\Omega[f] = \text{Tr} f (\mathcal{H} - N\mu + k_{\mathrm{B}} T \ln f)$$

Then

$$\Omega[f] \geq \Omega[f_0] \qquad (\text{B.1})$$

where f_0 is the equilibrium phase space density.

Proof. From the definition of f_0 in (2.4.5) it follows that

$$\Omega[f_0] = \text{Tr} f_0 (\mathcal{H} - N\mu - k_{\mathrm{B}} T \ln \varXi - \mathcal{H} + N\mu) = -k_{\mathrm{B}} T \ln \varXi \equiv \Omega$$

where Ω is the grand potential. Thus

$$\Omega[f] - \Omega[f_0] = k_{\mathrm{B}} T [\text{Tr}(f \ln f) - \text{Tr}(f \ln f_0)]$$

Theory of Simple Liquids, Fourth Edition. http://dx.doi.org/10.1016/B978-0-12-387032-2.00022-2

The term inside square brackets can be written as

$$\text{Tr}(f \ln f) - \text{Tr}(f \ln f_0) = \text{Tr} f_0[(f/f_0) \ln (f/f_0) - (f/f_0) + 1]$$

The right-hand side is always non-negative, since $x \ln x \geq x - 1$ for any $x > 0$. The inequality (B.1) is thereby verified.

This result is an example of the Gibbs-Bogoliubov inequalities, which are essentially a consequence of the convexity of the exponential function. □

Theorem 1. *For given choices of V_N, T and μ, the intrinsic free energy functional*

$$\mathcal{F}[\rho^{(1)}] = \text{Tr} f_0(K_N + V_N + k_B T \ln f_0) \tag{B.2}$$

is a unique functional of the equilibrium single-particle density $\rho^{(1)}(\mathbf{r})$.

Proof. The equilibrium phase space probability density f_0 is a functional of $\phi(\mathbf{r})$. The same is therefore true of the single-particle density $\rho^{(1)}(\mathbf{r}) = \text{Tr} f_0 \rho(\mathbf{r})$, where $\rho(\mathbf{r})$ is the microscopic density. Let us assume that there exists a different external potential, $\phi'(\mathbf{r}) \neq \phi(\mathbf{r})$, that gives rise to the same $\rho^{(1)}(\mathbf{r})$. With the Hamiltonian $\mathcal{H}' = K_N + V_N + \Phi'_N$ we may associate an equilibrium phase space density f_0' and grand potential Ω'. The inequality (B.1) implies that

$$\Omega' = \text{Tr} f_0'(\mathcal{H}' - N\mu + k_B T \ln f_0') \leq \text{Tr} f_0(\mathcal{H}' - N\mu + k_B T \ln f_0)$$
$$= \Omega + \text{Tr}[f_0(\Phi'_N - \Phi_N)]$$

or

$$\Omega' < \Omega + \int \rho^{(1)}(\mathbf{r})[\phi'(\mathbf{r}) - \phi(\mathbf{r})] \, d\mathbf{r}$$

If the same argument is carried through with primed and unprimed quantities interchanged we find that

$$\Omega < \Omega' + \int \rho^{(1)}(\mathbf{r})[\phi(\mathbf{r}) - \phi'(\mathbf{r})] \, d\mathbf{r}$$

Addition of the two inequalities term by term leads to a contradiction:

$$\Omega + \Omega' < \Omega' + \Omega$$

showing that the assumption concerning $\rho^{(1)}(\mathbf{r})$ must be false. We therefore conclude that there is only one external potential that gives rise to a particular single-particle density. Since f_0 is a functional of $\phi(\mathbf{r})$, it follows that it is also a unique functional of $\rho^{(1)}(\mathbf{r})$. This in turn implies that the intrinsic free energy (B.2) is a unique functional of $\rho^{(1)}(\mathbf{r})$ and that its functional form is the same for all external potentials. □

Theorem 2. *Let $n(\mathbf{r})$ be some average of the microscopic density. Then the functional*

$$\Omega_\phi[n] = \mathcal{F}[n] + \int n(\mathbf{r})\phi(\mathbf{r})d\mathbf{r} - \mu \int n(\mathbf{r})d\mathbf{r}$$

has its minimum value when $n(\mathbf{r})$ coincides with the equilibrium single-particle density $\rho^{(1)}(\mathbf{r})$.

Proof. Let $n(\mathbf{r})$ be the single-particle density associated with a phase space probability density f'. The corresponding grand potential functional is

$$\Omega[f'] = \operatorname{Tr} f'(\mathcal{H} - N\mu + k_{\mathrm{B}}T \ln f')$$

$$= \mathcal{F}[\rho'] + \int n(\mathbf{r})\phi(\mathbf{r})d\mathbf{r} - \mu \int n(\mathbf{r})d\mathbf{r} = \Omega_\phi[n]$$

The inequality (B.1) shows that $\Omega[f_0] \leq \Omega[f']$. It is also clear that $\Omega_\phi[\rho^{(1)}] = \Omega[f_0] = \Omega$. Thus $\Omega_\phi[\rho^{(1)}] \leq \Omega[n]$: the functional $\Omega_\phi[\rho']$ is minimised when $n(\mathbf{r}) = \rho^{(1)}(\mathbf{r})$ and its minimum value is equal to the grand potential. $\qquad\square$

Lemmas on Diagrams

We give here proofs of Lemmas 1, 2 and 4 of Section 3.7; the proofs of Lemmas 3 and 5 are similar to those of 2 and 4, respectively, and are therefore omitted.

Proof of Lemma 1. Let $\{g_1, \ldots, g_N\}$ be the set of diagrams in G (N may be infinite). A typical diagram, Γ, in the set H is the star product of n_1 diagrams g_1, n_2 diagrams $g_2, \ldots$, and n_N diagrams g_N, where some of the numbers n_i may be zero; we express this result symbolically by writing

$$\Gamma = (g_1 * * n_1) * (g_2 * * n_2) * \cdots * (g_N * * n_N)$$

The value of g_i is by definition $[g_i] = I_i/S_i$, where S_i is the symmetry number, I_i is the integral associated with g_i, and we temporarily adopt the notation $[\cdots]$ to denote the value of a diagram. Then the value of Γ is

$$[\Gamma] = I/S = (1/S) \prod_{i=1}^{N} I_i^{n_i} \tag{C.1}$$

where the symmetry number is

$$S = \prod_{i=1}^{N} n_i! \times \prod_{i=1}^{N} S_i^{n_i} \tag{C.2}$$

The factors $n_i!$ take care of the permutations of the n_i identical diagrams g_i; note that (C.2) is true only for diagrams that are star irreducible. Equation (C.1) can be rewritten as

$$[\Gamma] = \prod_{i=1}^{N} I_i^{n_i} S_i^{-n_i} \bigg/ \prod_{i=1}^{N} n_i! = \prod_{i=1}^{N} [g_i]^{n_i} \bigg/ \prod_{i=1}^{N} n_i!$$

We now sum over all diagrams in H and find that

$$\sum_{\Gamma} [\Gamma] = -1 + \sum_{n_1=0}^{\infty} \cdots \sum_{n_N=0}^{\infty} \prod_{i=1}^{N} \frac{[g_i]^{n_i}}{n_i!} = -1 + \prod_{i=1}^{N} \sum_{n_i=0}^{\infty} \frac{[g_i]^{n_i}}{n_i!}$$

$$= \prod_{i=1}^{N} \exp([g_i]) - 1 = \exp\left(\sum_{i=1}^{N} [g_i]\right) - 1$$

Theory of Simple Liquids, Fourth Edition. http://dx.doi.org/10.1016/B978-0-12-387032-2.00023-4

Inclusion of the term -1 in the first equality means that the case when all $n_i = 0$ is omitted.   □

Proof of Lemma 2. If S is the symmetry number and m is the number of black circles of Γ, the number of topologically inequivalent diagrams that are generated by attaching labels $1, \ldots, m$ to the black circles in all possible ways is $v = m!/S$. These diagrams we denote by Γ_i. It follows from the definition of a value of a diagram given by (3.7.3) that

$$\Gamma = \frac{1}{m!} \sum_{i=1}^{v} \Gamma_i \tag{C.3}$$

We now take the functional derivative of Γ with respect to $\gamma(\mathbf{r})$. Since

$$\frac{\delta \gamma(\mathbf{r}_i)}{\delta \gamma(\mathbf{r})} = \delta(\mathbf{r} - \mathbf{r}_i)$$

differentiation corresponds diagrammatically to replacing successively each black γ-circle in (C.3) by a white 1-circle. In this way, vm diagrams are generated, each containing one white circle and $m - 1$ black circles. These we denote by $\Gamma_i^{(j)}$, where j is the label carried by the whitened circle. Thus

$$\frac{\delta \Gamma}{\delta \gamma(\mathbf{r})} = \frac{1}{m!} \sum_{i=1}^{v} \sum_{j=1}^{m} \Gamma_i^{(j)} = \frac{1}{(m-1)!} \sum_{i=1}^{v} \Gamma_i^{(1)}$$

In the second step we have replaced the sum over j by m times the contribution for $j = 1$; this is permissible, since the value of any $\Gamma_i^{(j)}$ is independent of j for given i.

The v diagrams $\Gamma_i^{(1)}$ can now be divided into μ groups, chosen according to the topologically distinct diagrams into which each reduces when the labels of the $m-1$ black circles are removed. If these diagrams are denoted by $\Gamma_1', \ldots, \Gamma_\mu'$, definition (3.7.3) shows that

$$\frac{\delta \Gamma}{\delta \gamma(\mathbf{r})} = \Gamma_1' + \cdots + \Gamma_\mu'$$

which is the required result.   □

Proof of Lemma 4. Let m be the number of black circles in Γ. Any diagram in the set H can be expressed as $h(\Gamma; \{g_i\})$, where $\{g_i\} \equiv \{g_1, \ldots, g_m\}$ is a set of diagrams drawn from G that are attached to the black circles of Γ; some of the g_i may be identical. Two diagrams h obtained from two distinct sets $\{g_i\}$ are not necessarily different. Lemma 4 can then be written in more compact form as

$$\sideset{}{'}\sum_{\{g_i\}} h(\Gamma; \{g_i\}) = \text{[the diagram obtained from } \Gamma \text{ by associating the}$$
$$\text{function } \mathcal{G}(\mathbf{r}) \text{ with each of the black circles]} \tag{C.4}$$

The sum in (C.4) is taken over all sets $\{g_i\}$, with the restriction (denoted by the prime) that the diagrams $h(\Gamma; \{g_i\})$ must be topologically distinct.

Let $S(\Gamma)$ be the symmetry number of Γ, and let $S(g_i)$ and $S(\Gamma; \{g_i\})$ be, respectively, the symmetry numbers of the diagrams in G and H; $S(\Gamma)$ is obviously also the symmetry number of the right-hand side of (C.4). According to the definition (3.7.4):

$$h(\Gamma; \{g_i\}) = \frac{h(\Gamma'; \{g_i'\})}{S(\Gamma; \{g_i\})} \tag{C.5}$$

where $h(\Gamma'; \{g_i'\})$ is a diagram derived from $h(\Gamma; \{g_i\})$ by labelling its black circles in an arbitrary way. Let $h(\Gamma'; \{g_i\})$ be the diagram obtained from $h(\Gamma'; \{g_i'\})$ by removing the labels of the black circles of the g_i', but retaining the labels of the black circles of Γ', and let $S^*(\Gamma; \{g_i\})$ be the number of permutations of the m labels of $h(\Gamma'; \{g_i\})$ that give rise to topologically equivalent diagrams. For each of the S^* permutations there are $\prod_{i=1}^m S(g_i)$ permutations of the black circles of the g_i that yield diagrams equivalent to $h(\Gamma'; \{g_i'\})$. We can therefore write

$$S(\Gamma; \{g_i\}) = S^*(\Gamma; \{g_i\}) \prod_{i=1}^m S(g_i) \tag{C.6}$$

We now require a relation between $S(\Gamma)$ and $S^*(\Gamma; \{g_i\})$. Note that $S(\Gamma) \geq S^*(\Gamma; \{g_i\})$, since the process of decorating the black circles of Γ cannot lead to an increase in symmetry number. Let $n(\Gamma; \{g_i\})$ be the number of labellings that give rise to diagrams $h(\Gamma'; \{g_i\})$ that are topologically inequivalent, but yield diagrams Γ' (i.e. labelled versions of Γ on its own) that are equivalent. Consider now the set of $S(\Gamma)$ diagrams that are obtained from $h(\Gamma'; \{g_i\})$ by making the $S(\Gamma)$ permutations that leave Γ' topologically unaltered. This set can be divided into precisely $n(\Gamma; \{g_i\})$ groups, such that the diagrams in different groups are topologically inequivalent to each other. Each of the $n(\Gamma, \{g_i\})$ groups consists of $S^*(\Gamma; \{g_i\})$ topologically inequivalent diagrams. Thus

$$S(\Gamma) = n(\Gamma; \{g_i\}) S^*(\Gamma; \{g_i\}) \tag{C.7}$$

Illustration

In the example shown, $S(\Gamma) = 6$; $S^*(\Gamma; g_1, g_2, g_3) = 2$, because labels 2 and 3 can be permuted in $h(\Gamma'; g_1, g_2, g_3)$; and $n(\Gamma; g_1, g_2, g_3) = 3$, because permutation of labels 1 and 2 or 1 and 3 in $h(\Gamma'; g_1, g_2, g_3)$ generates diagrams that are topologically inequivalent.

By combining (C.6) and (C.7) we find that

$$S(\Gamma; \{g_i\}) = \frac{S(\Gamma)}{n(\Gamma; \{g_i\})} \prod_{i=1}^{m} S(g_i) \tag{C.8}$$

If use is made of (C.5) and (C.8), the sum on the left-hand side of (C.4) can be rewritten as

$$\sum_{\{g_i\}}' \frac{n(\Gamma; \{g_i\})}{S(\Gamma) \prod_{i=1}^{m} S(g_i)} h(\Gamma'; \{g_i'\})$$

or, from (3.7.4):

$$\sum_{\{g_i\}}' \frac{n(\Gamma; \{g_i\})}{S(\Gamma)} h(\Gamma'; \{g_i\})$$

On recalling the significance of $n(\Gamma; \{g_i\})$ we see that this last result may also be expressed as

$$\sum_{g_i} \cdots \sum_{g_m} \frac{h(\Gamma'; g_1, \ldots, g_m)}{S(\Gamma)} \tag{C.9}$$

where the m summations are now unrestricted. But (C.9) is just a labelled diagram obtained from Γ' by associating the function $\mathcal{G}(\mathbf{r})$ with each black circle and dividing by the symmetry number $S(\Gamma)$. It follows from (3.7.4) that (C.9) is equal to the right-hand side of (C.4).

Solution of the PY Equation for Hard Spheres

The PY closure relation for hard spheres is

$$h(r) = -1, \quad r < d \tag{D.1a}$$
$$c(r) = 0, \quad r > d \tag{D.1b}$$

When substituted in the Ornstein-Zernike relation (3.5.12), this approximation yields an integral equation that can be solved in closed form. We follow here the method of Baxter, which is based on a transformation of the Ornstein-Zernike relation via a so-called Wiener-Hopf factorization of the function $\hat{A}(k)$ defined as

$$\hat{A}(k) = \frac{1}{S(k)} = 1 - \rho\hat{c}(k) = 1 - \frac{4\pi\rho}{k} \int_0^\infty r \sin(kr) c(r) dr \tag{D.2}$$

The three-dimensional Fourier transform of any function f of $r \equiv |\mathbf{r}|$ can be cast in the form

$$\hat{f}(k) = \frac{4\pi}{k} \int_0^\infty r \sin(kr) f(r) dr = 4\pi \int_0^\infty \cos(kr) F(r) dr$$
$$= 2\pi \int_{-\infty}^\infty \exp(ikr) F(r) dr \tag{D.3}$$

where

$$F(r) = \int_r^\infty s f(s) ds = F(-r) \tag{D.4}$$

The second equality in (D.4) follows immediately if the convention that $f(r) = f(-r)$ is followed. Substitution of (D.1b), (D.3) and (D.4) in (D.2) leads to

$$\hat{A}(k) = 1 - 4\pi\rho \int_0^d \cos(kr) S(r) dr = \hat{A}(-k) \tag{D.5}$$

where

$$S(r) = \int_r^d t c(t) dt \tag{D.6}$$

Theory of Simple Liquids, Fourth Edition. http://dx.doi.org/10.1016/B978-0-12-387032-2.00024-6

Similarly:

$$\hat{h}(k) = 2\pi \int_{-\infty}^{\infty} \exp{(ikr)} J(r)\, dr \tag{D.7}$$

with

$$J(r) = \int_{r}^{\infty} sh(s)\, ds \tag{D.8}$$

Consider now the behaviour of the function $\hat{A}(k)$ in the complex k-plane and set $k = x + iy$. Because $\hat{A}(k)$, as given by (D.5), is a Fourier transform over a finite interval, it is regular throughout the complex plane. It also has no zeros on the real axis ($y = 0$), since it is the inverse of the static structure factor; the latter is a finite quantity at all wavenumbers. Moreover, according to (D.5), $\hat{A}(k)$ tends uniformly to unity as $|x| \to \infty$ in any strip $y_1 < y < y_2$. Thus there exists a strip $|y| \le \epsilon$ about the real axis within which $\hat{A}(k)$ has no zeros. The function $\ln \hat{A}(k)$ is therefore regular within that strip and tends uniformly to zero as $|x| \to \infty$. Integrating around the strip and applying Cauchy's theorem, we find that for any $k = x + iy$ such that $|y| < \epsilon$:

$$\ln \hat{A}(k) = \ln \hat{Q}(k) + \ln \hat{P}(k) \tag{D.9}$$

where

$$\ln \hat{Q}(k) = \frac{1}{2\pi i} \int_{-i\epsilon-\infty}^{-i\epsilon+\infty} \frac{\ln \hat{A}(k')}{k' - k} dk' \tag{D.10a}$$

$$\ln \hat{P}(k) = -\frac{1}{2\pi i} \int_{i\epsilon-\infty}^{i\epsilon+\infty} \frac{\ln \hat{A}(k')}{k' - k} dk' \tag{D.10b}$$

Since $\hat{A}(k)$ is an even function of k, (D.10) implies that

$$\ln \hat{P}(k) = \ln \hat{Q}(-k) \tag{D.11}$$

From (D.10a) we see that $\ln \hat{Q}(k)$ is regular in the domain $y > -\epsilon$. It follows from (D.9) and (D.11) that when $|y| < \epsilon$:

$$\hat{A}(k) = \hat{Q}(k)\hat{Q}(-k) \tag{D.12}$$

The function $\hat{Q}(k)$ is regular and has no zeros in the domain $y > -\epsilon$, since it is the exponential of a function that is regular in the same domain. Equation (D.12) is the Wiener-Hopf factorisation of $\hat{A}(k)$.

When $|x| \to \infty$ within the strip $|y| < \epsilon$, it follows from (D.10a) that $\ln \hat{Q}(k) \sim x^{-1}$ and hence that $\hat{Q}(k) \sim 1 - \mathcal{O}(x^{-1})$. The function $1 - \hat{Q}(k)$ is therefore Fourier integrable along the real axis and a function $Q(r)$ can be defined as

$$2\pi \rho Q(r) = \frac{1}{2\pi} \int_{-\infty}^{\infty} \exp{(-ikr)}[1 - \hat{Q}(k)] dk \tag{D.13}$$

Equation (D.10a) shows that that if k is real, the complex conjugate of $\hat{Q}(k)$ is $\hat{Q}(-k)$, and hence that $Q(r)$ is a real function. The same equation also shows that when $y \geq 0$, $\ln \hat{Q}(k) \to 0$, and therefore $\hat{Q}(k) \to 1$, as $k \to \infty$. Thus, if $r < 0$, the integration in (D.13) can be closed around the upper half-plane, where $\hat{Q}(k)$ is regular, to give

$$Q(r) = 0, \quad r < 0 \qquad (D.14)$$

The right-hand side of (D.10a) is a different analytic function of k according to whether $y > -\epsilon$ or $y < -\epsilon$. The analytic continuation of $\hat{Q}(k)$ into the lower half-plane is therefore given, not by (D.10a), but by (D.12), i.e.

$$\hat{Q}(k) = \hat{A}(k)/\hat{Q}(-k) \qquad (D.15)$$

where (D.10a) can be used to evaluate $\hat{Q}(-k)$. Since $\hat{A}(k)$ is regular everywhere, and $\hat{Q}(-k)$ is regular and has no zeros for $y < \epsilon$, we see from (D.15) that $\hat{Q}(k)$ is also regular for $y < \epsilon$. Furthermore, since $\hat{Q}(-k) \to 1$ as $y \to -\infty$, it follows from (D.5) and (D.15) that both $\hat{A}(k)$ and $\hat{Q}(k)$ grow exponentially as $\exp(ikd) = \exp(ixd)\exp(-yd)$ when y becomes large and negative. Thus, when $r > d$, the integration in (D.13) can be closed around the lower half-plane, giving

$$Q(r) = 0, \quad r > d \qquad (D.16)$$

On inversion of the Fourier transform in (D.13), (D.14) and (D.16) together yield

$$\hat{Q}(k) = 1 - 2\pi\rho \int_0^d \exp(ikr)Q(r)dr \qquad (D.17)$$

Substitution in (D.12) of the expressions (D.5) for $\hat{A}(k)$ and (D.17) for $\hat{Q}(k)$, followed by multiplication by $\exp(-ikr)$ and integration with respect to k from $-\infty$ to $+\infty$, shows that

$$S(r) = Q(r) - 2\pi\rho \int_r^d Q(s)Q(s-r)ds, \quad 0 < r < d \qquad (D.18)$$

Equations (3.5.13), (D.2) and (D.12) imply that

$$\hat{Q}(k)[1 + \rho\hat{h}(k)] = 1/\hat{Q}(k) \qquad (D.19)$$

where $\hat{h}(k)$ is given by (D.7). We now multiply both sides of (D.19) by $\exp(-ikr)$ and integrate with respect to k from $-\infty$ to $+\infty$. The contribution from the right-hand side vanishes when $r > 0$, since the integration can then be closed around the lower half-plane, where $\hat{Q}(k)$ is regular, has no zeros and tends to unity at infinity. On substituting (D.7) and (D.17) into the left-hand side of (D.19) and carrying out the integration, we obtain a relation between $Q(r)$ and $J(r)$ for $r > 0$ of the form

$$-Q(r) + J(r) - 2\pi\rho \int_0^d Q(s)J(|r-s|)ds = 0, \quad r > 0 \qquad (D.20)$$

It is clear from (D.6) and (D.18) that $Q(r) \to 0$ as $r \to d$ from below; comparison with (D.16) then shows that $Q(r)$ is continuous at $r = d$.

Equations (D.18) and (D.20) can be expressed in terms of $c(r)$ and $h(r)$, rather than $S(r)$ and $J(r)$, by differentiating them with respect to r. If we use (D.6) and (D.8), and the fact that $Q(d) = 0$, we find after integration by parts that

$$rc(r) = -Q'(r) + 2\pi\rho \int_r^d Q'(s)Q(s-r)ds, \qquad 0 < r < d \qquad (D.21)$$

and

$$rh(r) = -Q'(r) + 2\pi\rho \int_0^d (r-s)h(|r-s|)Q(s)ds, \qquad r > 0 \qquad (D.22)$$

where $Q'(r) \equiv dQ(r)/dr$. Equations (D.21) and (D.22) express $h(r)$ and $c(r)$ in terms of the same function, $Q(r)$, and constitute a reformulation of the Ornstein-Zernike relation that is applicable whenever $c(r)$ vanishes beyond a range d, which is precisely the case with the PY closure. Equation (D.22) is an integral equation for $Q(r)$ that is easy to solve for $0 < r < d$, where $h(r) = -1$ and (D.22) therefore reduces to

$$r = Q'(r) + 2\pi\rho \int_0^d (r-s)Q(s)ds, \qquad 0 < r < d \qquad (D.23)$$

The solution is of the form

$$Q'(r) = ar + b \qquad (D.24)$$

with

$$a = 1 - 2\pi\rho \int_0^d Q(s)ds, \qquad b = 2\pi\rho \int_0^d sQ(s)ds \qquad (D.25)$$

Given the boundary condition $Q(d) = 0$, (D.24) is trivially integrated to yield $Q(r)$. Substitution of the result in (D.25) gives two linear equations, the solutions to which are

$$a = \frac{1 + 2\eta}{(1 - \eta)^2}, \qquad b = \frac{-3d\eta}{2(1 - \eta)^2} \qquad (D.26)$$

where η is the hard-sphere packing fraction. Thus $Q(r)$ is now a known function of r and $c(r)$ can therefore be calculated from (D.21); this leads to the results displayed in (4.4.10) and (4.4.11). The isothermal compressibility is obtained from (3.8.8), (D.2) and (D.15) as

$$\beta/\rho\chi_T = \hat{A}(0) = [\hat{Q}(0)]^2 \qquad (D.27)$$

The function $\hat{Q}(0)$ is easily calculated from (D.17) and the solution for $Q(r)$, leading ultimately to the PY compressibility equation of state (4.4.12).

Scaled Particle Theory

Scaled particle theory is an approximate interpolation scheme that allows the calculation of the work required to create a spherical cavity in a hard-sphere fluid or, equivalently, to insert a solute sphere of the same radius. From this starting point it is possible to derive the equation of state of the fluid. The theory is easily formulated for mixtures but we restrict the discussion here to the one-component case.

Consider a fluid of N hard spheres of diameter $d = 2R$ at a number density ρ. Let $W(R_0)$ be the reversible work required to create a spherical cavity of radius R_0 centred on a point $\mathbf{r}$ within the fluid. According to the basic principles of thermodynamic fluctuation theory, the probability that such a cavity will appear as the result of spontaneous fluctuations within the system is

$$p_0(R_0) = \exp[-\beta W(R_0)] \qquad (E.1)$$

This is the same as the probability that there are no spheres whose centres lie within the spherical region of radius $R_0 + R$ around $\mathbf{r}$. That interpretation can be extended to negative values of R_0 in the range $-R \leq R_0 \leq 0$, in which case the radius of the region of interest is $0 \leq R_0 + R \leq R$. Since overlap of hard spheres is forbidden, there can be at most one particle in such a region, a situation that occurs with probability

$$p_1(R_0) = \frac{4}{3}\pi\rho(R_0 + R)^3 = 1 - p_0(R_0) \qquad (E.2)$$

Combination of (E.1) and (E.2) gives

$$W(R_0) = -k_B T \ln\left[1 - (4\pi\rho/3)(R_0 + R)^3\right], \qquad R_0 \leq 0 \qquad (E.3)$$

In the opposite limit, that of very large cavities, the reversible work required is given by thermodynamics. If P is the pressure of the fluid and $\Delta V_0 = 4\pi R_0^3/3$ is the volume of the cavity, then $W(R_0)$ is the increase in Helmholtz free energy resulting from a reduction equal to ΔV_0 in the volume accessible to the fluid:

$$W(R_0) = P\Delta V_0 = \frac{4}{3}\pi P R_0^3, \qquad R_0 \gg R \qquad (E.4)$$

Theory of Simple Liquids, Fourth Edition. http://dx.doi.org/10.1016/B978-0-12-387032-2.00025-8

The assumption now made is that for $R_0 > 0$, $W(R_0)$ is given by a cubic polynomial in R_0, where the term in R_0^3 (the dominant contribution for large cavities) is given by (E.4), i.e.

$$W(R_0) = w_0 + w_1 R_0 + \frac{1}{2} w_2 R_0^2 + \frac{4}{3} \pi P R_0^3, \qquad R_0 \geq 0 \qquad (E.5)$$

The coefficients w_0, w_1 and w_2 are determined by requiring $W(R_0)$ and its first derivative, as given by (E.3) for $R_0 < 0$ and (E.5) for $R_0 > 0$, to be continuous at $R_0 = 0$. The results obtained in this way are

$$\beta w_0 = -\ln(1 - \eta), \qquad \beta w_1 = \frac{4\pi\rho R^2}{1 - \eta}, \qquad \beta w_2 = \frac{8\pi\rho R}{1 - \eta} + \frac{(4\pi\rho R^2)^2}{(1 - \eta)^2} \qquad (E.6)$$

where η is the hard-sphere packing fraction.

The excess chemical potential of the fluid is the reversible work required to insert a hard sphere of radius $R_0 = R$. Thus, from (E.5) and (E.6):

$$\beta\mu^{ex} = \beta W(R_0) = -\ln(1 - \eta) + \frac{6\eta}{1 - \eta} + \frac{9\eta^2}{2(1 - \eta)^2} + \frac{\beta P\eta}{\rho} \qquad (E.7)$$

Then use of the thermodynamic relation $\partial P/\partial \rho = \rho(\partial \mu/\partial \rho)$ leads to the scaled particle equation of state in the form

$$\frac{\beta P}{\rho} = \frac{1 + \eta + \eta^2}{(1 - \eta)^3} \qquad (E.8)$$

Equation (E.8) is identical to the Percus-Yevick compressibility Eq. (4.4.12). The corresponding expression for the excess free energy is

$$\frac{\beta F^{ex}}{N} = -\ln(1 - \eta) + \frac{3\eta}{1 - \eta} + \frac{3\eta^2}{2(1 - \eta)^2} \qquad (E.9)$$

An Exact Integral Equation for $\rho^{(1)}(\mathbf{r})$

Here we derive an exact integral equation[1] for the equilibrium single-particle density $\rho^{(1)}(\mathbf{r})$; this is a key ingredient in the derivation of a microscopic expression for the surface tension in Section 6.4. The starting point is the relation (3.5.3) between $\rho^{(1)}(\mathbf{r})$ and the single-particle direct correlation function $c^{(1)}(\mathbf{r})$, with $\psi(\mathbf{r}) = \mu - \phi(\mathbf{r})$, where $\phi(\mathbf{r})$ is an external potential. According to the fundamental theorems of density functional theory (Appendix B), $\psi(\mathbf{r})$ uniquely determines $\rho^{(1)}(\mathbf{r})$ for a given Hamiltonian (in the absence of the external potential) and, reciprocally, $\psi(\mathbf{r})$ is a unique functional of $\rho^{(1)}(\mathbf{r})$. Hence, if $\mathbf{s}$ is a translation in space, the functional must satisfy the relation

$$\psi[\rho^{(1)}(\mathbf{r}+\mathbf{s})] = \psi(\mathbf{r}+\mathbf{s})$$

Consider a small displacement $\mathbf{s}$ away from an initial position $\mathbf{r}_1$. Functional expansion to first order in the displacement shows that

$$\psi(\mathbf{r}_1 + \mathbf{s}) - \psi(\mathbf{r}_1) = \int \frac{\partial \psi[\rho^{(1)}(\mathbf{r}_1)]}{\partial \rho^{(1)}(\mathbf{r}_2)} \left(\rho^{(1)}(\mathbf{r}_2 + \mathbf{s}) - \rho^{(1)}(\mathbf{r}_2) \right) d\mathbf{r}_2 \quad \text{(F.1)}$$

On taking the limit $\mathbf{s} \to 0$, (F.1) reduces to

$$\nabla_1 \psi(\mathbf{r}_1) = \int \frac{\partial \psi(\mathbf{r}_1)}{\partial \rho^{(1)}(\mathbf{r}_2)} \nabla_2 \rho^{(1)}(\mathbf{r}_2) d\mathbf{r}_2 \quad \text{(F.2)}$$

From (3.5.9) and the definition of $\psi(\mathbf{r})$ it follows that (F.2) can be rewritten as

$$\nabla_1 \ln \rho^{(1)}(\mathbf{r}_1) + \nabla_1 \beta \phi(\mathbf{r}_1) = \int c^{(2)}(\mathbf{r}_1, \mathbf{r}_2) \nabla_2 \rho^{(1)}(\mathbf{r}_2) d\mathbf{r}_2 \quad \text{(F.3)}$$

or, equivalently, for vanishing external potential ϕ:

$$\int C^{(2)}(\mathbf{r}_1, \mathbf{r}_2) \nabla_2 \rho^{(1)}(\mathbf{r}_2) d\mathbf{r}_2 = 0 \quad \text{(F.4)}$$

Theory of Simple Liquids, Fourth Edition. http://dx.doi.org/10.1016/B978-0-12-387032-2.00026-X

where the function $C^{(2)}$ is defined as

$$C^{(2)}(\mathbf{r}_1, \mathbf{r}_2) = \frac{1}{\rho^{(1)}(\mathbf{r}_1)} \delta(\mathbf{r}_2 - \mathbf{r}_1) - c^{(2)}(\mathbf{r}_1, \mathbf{r}_2)$$

In the special case of a planar, liquid-gas interface in vanishing external field, i.e. $\phi(z) \to 0$, the density profile depends only on the vertical coordinate z, while the function $C^{(2)}$ has the form

$$C^{(2)}(\mathbf{r}_1, \mathbf{r}_2) = C^{(2)}(R \equiv |\mathbf{R}_2 - \mathbf{R}_1|, z_1, z_2)$$

where $\mathbf{R} \equiv (x, y)$ is a two-dimensional vector in the $z = 0$ plane, where translational invariance holds. Under these conditions (F.4) reduces to

$$\int_{-\infty}^{\infty} C_0(z_1, z_2) \frac{d\rho^{(1)}(z_2)}{dz_2} dz_2 = 0$$

where

$$C_0(z_1, z_2) = \int C^{(2)}(R, z_1, z_2) d\mathbf{R}$$

If the direct correlation function is replaced by that of a translationally invariant fluid, $c^{(2)}(\mathbf{r}_1 - \mathbf{r}_2)$, (F.3) can be rewritten as

$$\nabla_1 \left[\ln \rho^{(1)}(\mathbf{r}_1) + \beta\phi(\mathbf{r}_1) \right] = \int c^{(2)}(\mathbf{r}_1 - \mathbf{r}_2) \nabla_2 \rho^{(1)}(\mathbf{r}_2) d\mathbf{r}_2$$

$$= -\int \nabla_2 c^{(2)}(\mathbf{r}_1 - \mathbf{r}_2) \rho^{(1)}(\mathbf{r}_2) d\mathbf{r}_2$$

$$= \nabla_1 \int c^{(2)}(\mathbf{r}_1 - \mathbf{r}_2) \rho^{(1)}(\mathbf{r}_2) d\mathbf{r}_2$$

which in turn can be integrated to give

$$\ln \rho^{(1)}(\mathbf{r}_1) = -\beta\phi(\mathbf{r}_1) + \int c^{(2)}(\mathbf{r}_1 - \mathbf{r}_2) \rho^{(1)}(\mathbf{r}_2) d\mathbf{r}_2 + C \qquad \text{(F.5)}$$

Equation (F.5) has been used to study the stability of a homogeneous liquid relative to a periodic crystal[2] and the asymptotic decay of the density profile of a fluid near a wall or at the liquid-vapour interface.[3]

REFERENCES

[1] Lovett, R.A., Mon, C.Y. and Buff, F.P., *J. Chem. Phys.* **65**, 570 (1976).
[2] Lovett, R.A., *J. Chem. Phys.* **66**, 1225 (1977).
[3] Evans, R., Leote de Carvalho, R.J.F., Henderson, J.R. and Hoyle, D.C., *J. Chem. Phys.* **100**, 591 (1994).

Some Basic Properties of Polymers

Polymers are macromolecules consisting of large numbers of elementary units or *monomers*; chemically they may be either homogeneous (homopolymers) or heterogeneous (heteropolymers). Sections 2 and 3 of Chapter 12 are primarily concerned with flexible, linear polymers, an example of which is shown in Fig. 12.3, but reference is also made to the topologies pictured schematically in Figure G.1: (a) a star polymer, formed from linear branches connected at a common origin; and (b) an AB di-block copolymer, the simplest type of heteropolymer, in which single sequences of monomers A and B are linked together.

The discussion here is limited to dilute solutions of monodisperse, linear polymers, though much of the material is also relevant to concentrated solutions and polymer melts. Consider a single polymer or *coil* consisting of $M \gg 1$ identical monomers linked by $L = M - 1$ bonds of (microscopic) length b. That length may also be regarded as the diameter of the monomers, which are assumed to touch; the *contour length* of the polymer is $\mathcal{L} = Lb$. Some polymers are more flexible than others. Let $\mathbf{u}_\alpha$ and $\mathbf{u}_\beta$ be the unit vectors tangent to the thread-like polymer at the points where monomers α, β are located, and let s be the distance between α and β along the polymer contour. The correlation between the orientations of the two vectors decays exponentially for large s according to the rule

$$\left\langle \mathbf{u}_\alpha \cdot \mathbf{u}_\beta \right\rangle = \langle \cos \theta(\alpha, \beta) \rangle \equiv \langle \cos \theta(s) \rangle = \exp(-s/l)$$

where l is the *persistence length*. If $l \approx b$ the polymer chain is highly flexible and can therefore adopt a large number of conformations. If $l \gg b$ the chain is rigid and locally rod-like; this is the case for DNA, for which $l \approx 50$ nm. For a semi-flexible chain, where l has a value equal to a few bond lengths, it is often convenient to map it onto a chain of effective monomers of length l, called Kuhn segments. This is an example of nanoscale coarse graining, where the spatial resolution is reduced by a factor l/b.

Theory of Simple Liquids, Fourth Edition. http://dx.doi.org/10.1016/B978-0-12-387032-2.00027-1

(a)   **(b)**

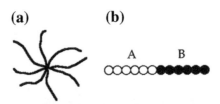

A   B

FIGURE G.1   Examples of polymer topologies. See text for details.

A key length scale of a polymer coil is its radius of gyration R_g, defined as

$$R_g^2 = \frac{1}{2L^2} \left\langle \sum_{\alpha=1}^{M} \sum_{\beta=1}^{M} |\mathbf{r}_\beta - \mathbf{r}_\alpha|^2 \right\rangle = \frac{1}{L} \left\langle \sum_{\alpha=1}^{M} |\mathbf{r}_\alpha - \mathbf{R}_{CM}|^2 \right\rangle \tag{G.1}$$

where $\mathbf{r}_\alpha$ and $\mathbf{R}_{CM} = (1/L) \sum_\alpha \mathbf{r}_\alpha$ are the coordinates, respectively, of monomer α and the centre of mass of the coil; the angular brackets denote a statistical average over polymer conformations. An isolated coil is therefore characterised by three length scales:

$$b \ll R_g \ll Lb$$

An *ideal chain* is one formed from non-interacting monomers. The simplest example is the *freely jointed chain*, for which $\langle \mathbf{r}_\alpha \cdot \mathbf{r}_{\alpha+1} \rangle = 0$ for all α. The end-to-end vector is

$$\mathbf{R} = \sum_{\alpha=1}^{M} \left(\mathbf{r}_{\alpha+1} - \mathbf{r}_\alpha \right) = \mathbf{r}_M - \mathbf{r}_1$$

with $\langle \mathbf{R} \rangle = 0$ and $\langle R^2 \rangle = Lb^2$, results that are characteristic of a random walk. The probability distribution of end-to-end vectors is a gaussian function:

$$P_L(\mathbf{R}) = \left(\frac{3}{2\pi Lb^2} \right)^{3/2} \exp\left(-\frac{3R^2}{2Lb^2} \right) = \exp\left(-\frac{S(R)}{k_B} \right)$$

where $S(R)$ is the conformational entropy of an ideal chain with end-to-end vector $\mathbf{R}$:

$$S(R) = k_B \ln P_L(\mathbf{R}) = \text{constant} - \frac{3k_B R^2}{2Lb^2}$$

The conformational free energy

$$F(R) = -TS(R) = F_0 + \frac{3k_B T R^2}{2Lb^2} \tag{G.2}$$

is the free energy of an elastic spring of end-to-end spring constant $K = 3k_B T/Lb^2$; the elasticity of an ideal chain is of purely entropic origin.

The elastic picture of a polymer may be extended to the microscopic level by introduction of the *gaussian chain* model in which the interaction between adjacent monomers, summed over the length of the chain, is

$$V_L = \frac{1}{2} k \sum_{\alpha=1}^{L} \left(\mathbf{r}_{\alpha+1} - \mathbf{r}_\alpha \right)^2 \tag{G.3}$$

where, by analogy with the global spring constant K, the microscopic spring constant is $k = 3k_B T / b^2$. The normalised Boltzmann distribution

$$P(\mathbf{r}_1, \ldots, \mathbf{r}_M) = \left(\frac{3}{2\pi b^2} \right)^{3/2} \exp \left[-\beta V_L(\mathbf{r}_1, \ldots, \mathbf{r}_M) \right] \tag{G.4}$$

factors into L gaussian distributions, thereby simplifying the calculation of the radius of gyration:

$$
\begin{aligned}
R_g^2 &= \frac{1}{2L^2} \int P(\mathbf{r}_1, \ldots, \mathbf{r}_M) \sum_{\alpha=1}^{M} \sum_{\beta=1}^{M} (\mathbf{r}_\beta - \mathbf{r}_\alpha)^2 d\mathbf{r}_1 \cdots d\mathbf{r}_M \\
&= \frac{1}{6} L b^2 = \frac{1}{6} \left\langle R^2 \right\rangle
\end{aligned} \tag{G.5}
$$

Thus $R_g \sim b L^{1/2}$ for an ideal chain.

By analogy with the structure factor (4.1.3) of an atomic fluid the intramolecular structure factor or *form factor* of a polymer is defined as

$$\hat{\omega}(k) = \frac{1}{M} \left\langle \sum_{\alpha=1}^{M} \sum_{\beta=1}^{M} \exp[-i\mathbf{k} \cdot (\mathbf{r}_\beta - \mathbf{r}_\alpha)] \right\rangle \tag{G.6}$$

or, to quadratic order in k:

$$\hat{\omega}(k) = M \left[1 - \frac{1}{3} R_g^2 k^2 + \mathcal{O}(k^4) \right]$$

For a gaussian chain use of the distribution function (G.4) allows an explicit calculation of $\hat{\omega}(k)$, which takes the Debye form:

$$\hat{\omega}(k) = M f(k R_g) \tag{G.7}$$

where

$$f(x) = \frac{2}{x^4} \left[\exp(-x^2) - 1 + x^2 \right] \approx \frac{1}{1 + \frac{1}{2} x^2} \tag{G.8}$$

Now consider the case of non-ideal polymers where monomers interact. Under 'good solvent' conditions the excluded volume interactions between monomers are dominant. The mutual repulsion leads to a swelling of the coil and the radius of gyration is therefore greater than that predicted for an ideal polymer:

$$R_g \sim b L^\nu, \quad \nu > \frac{1}{2}, \quad L \gg 1$$

A good estimate of the exponent ν is provided by a classic argument due to Flory. The free energy of a non-ideal polymer is the sum $F(R) = F_{\text{elas}}(R) + F_{\text{inter}}(R)$ where $F_{\text{elas}}(R)$ is the elastic component (G.2) and $F_{\text{inter}}(R)$ is the contribution that arises from excluded volume interactions. The interaction term is given within the second virial approximation in (3.9.7) and (3.9.18) by

$$F_{\text{inter}}(R) = M k_B T B_2 \rho_m = M^2 k_B T \frac{v}{2R^3}$$

where ρ_m is the number density of monomers, assumed to be uniformly distributed over a volume R^3, and $v = 4\pi b^3/3$ is the excluded volume of a pair of hard spheres. Minimisation of the total free energy with respect to R for a fixed value of L leads via the relation

$$\left. \frac{\partial F(R)}{\partial R} \right|_{R=R_g} = 0$$

to an estimate of the radius of gyration in the form

$$R_g = \left(\frac{1}{2} v b^2 \right)^{1/5} L^{3/5} \sim b L^{3/5}$$

The 'Flory exponent' is therefore $v = \frac{3}{5}$. This is a universal quantity under good solvent conditions, which generally refer to the high temperature regime, where solvent-induced attractions between monomers can be neglected. The prefactor, on the other hand, is non-universal; it depends on the microscopic nature of the polymer. The Flory argument for arbitrary space dimensionality d leads to a value $v = 3/(d+2)$. In particular, the value for an ideal chain, $v = \frac{1}{2}$, is recovered for $d = 4$. This is reminiscent of the critical exponents for second-order phase transitions, which take their classical, mean field values in four dimensions; see Sections 5.6 and 5.7.

The universality of dimensionless polymer properties means that many theoretical calculations and computer simulations of non-ideal polymers are based on the simplest model, namely the *self-avoiding walk* (SAW) on a cubic lattice, pictured in two-dimensional form in Figure G.2. Within the model each

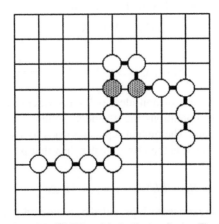

FIGURE G.2 A two-dimensional SAW model of a linear polymer. The shaded circles represent two monomers between which there is an attractive interaction in the augmented version of the model.

site can accommodate at most one monomer. In the augmented version referred to in Section 12.2 the effect of the solvent is accounted for by inclusion of an attractive interaction $-\epsilon$ between non-connected, adjacent monomers. As ϵ increases, a self-avoiding walk will lead to more compact structures; this mimics the transition from good to poor solvent conditions that accompanies a decrease in temperature. Renormalisation group calculations and computer simulations of the SAW model[1] have shown that the radius of gyration behaves as bL^{ν}, where b is the lattice spacing and $\nu \approx 0.5876$, a value very close to that derived by Flory.

REFERENCES

[1] Pelissetto, A. and Vicari, E., *Phys. Rep.* **368**, 549 (2002).

Density Profile of a Polymer Brush

Here we use a simple, density functional argument due to Pincus[1] to show that the monomer density profile $\rho_m(z)$ of a brush of polymers grafted to a planar surface is parabolic, in agreement with the self-consistent field calculation of Milner et al.[2]

The free energy per unit area, f, of the brush is a functional of both $\rho_m(z)$ and the end-monomer profile $\rho_L(z)$ which has the approximate form

$$f = k_B T \int_0^{Lb} \left(\frac{1}{2} v_x [\rho_m(z)]^2 + \frac{z^2}{2Lb^2} \rho_L(z) - \mu \rho_m(z) \right) dz$$

where v_x is the monomer excluded volume and μ is a Lagrange multiplier associated with the constraint that the total number of monomers is equal to L. The first term in the integral is the excluded volume contribution within the second virial approximation; the second term represents the end-to-end elastic free energy, which involves $\rho_L(z)$ rather than $\rho_m(z)$. It is then assumed that $\rho_L(z) = \rho_m(z)/L$; this is reasonable if z is close to the as yet unknown height of the brush. The equilibrium monomer profile is obtained by minimising $f[\rho_m]$ with respect to $\rho_m(z)$ (cf. (3.4.3)), which leads immediately to the parabolic profile:

$$\rho_m(z) = \frac{1}{v_x} \left(\mu - \frac{z^2}{2L^2 b^2} \right), \quad z \le z_0$$
$$= 0, \qquad\qquad\qquad z > z_0$$

where $z_0 = (2L^2 b^2 \mu)^{1/2}$ is the height of the brush, beyond which the monomer density is zero. The Lagrange multiplier is determined by the condition

$$\int_0^{z_0} \rho_m(z) dz = L\sigma$$

where σ is the grafting density and $L\sigma$ is therefore the number of monomers per unit area of the surface. On substituting for $\rho_m(z)$, and setting $v_x = b^3$, we

Theory of Simple Liquids, Fourth Edition. http://dx.doi.org/10.1016/B978-0-12-387032-2.00028-3

find that $\mu = \frac{1}{2}(3\sigma b^2)^{2/3}$. Thus the equilibrium height of the brush is

$$z_0 = Lb(3\sigma b^2)^{1/3} \sim \sigma^{1/3}Lb^{5/3}$$

in agreement with (12.4.3), which was obtained by a scaling argument. The derivation of (12.4.3) was based on the assumption that the profile was rectangular; this explains the difference in prefactors.

REFERENCES

[1] Pincus, P., *Macromolecules* **24**, 2912 (1991).
[2] Milner, S.T., Witten, T.A. and Cates, M.E., *Macromolecules* **21**, 2610 (1988).

Index